Lecture Notes in Computer Science 4646

Jacques Duparc   Thomas A. Henzinger (Eds.)

Computer Science Logic

21st International Workshop, CSL 2007
16th Annual Conference of the EACSL
Lausanne, Switzerland, September 11-15, 2007
Proceedings

Springer

Volume Editors

Jacques Duparc
Université de Lausanne, Ecole des HEC
1015 Lausanne, Switzerland
E-mail: jacques.duparc@unil.ch

Thomas A. Henzinger
EPFL, Station 14
1015 Lausanne, Switzerland
E-mail: tah@epfl.ch

Library of Congress Control Number: 2007934516

CR Subject Classification (1998): F.4.1, F.4, I.2.3-4, F.3

LNCS Sublibrary: SL 1 – Theoretical Computer Science and General Issues

ISSN 0302-9743
ISBN-10 3-540-74914-4 Springer Berlin Heidelberg New York
ISBN-13 978-3-540-74914-1 Springer Berlin Heidelberg New York

Springer is a part of Springer Science+Business Media

springer.com

Printed in Germany

Typesetting: Camera-ready by author, data conversion by Scientific Publishing Services, Chennai, India
Printed on acid-free paper SPIN: 12122257 06/3180 5 4 3 2 1 0

Preface

The Annual Conference of the European Association for Computer Science Logic (EACSL), CSL 2007, was held at the University of Lausanne, September, 11–15 2007. The conference series started as a program of International Workshops on Computer Science Logic, and then in its sixth meeting became the Annual Conference of the EACSL. This conference was the 21st meeting and 16th EACSL conference; it was organized by the Western Swiss Center for Logic, History and Philosophy of Science, and the Information Systems Institute at the Faculty of Business and Economics at the University of Lausanne.

The first day of the conference was a joint event with the European Research Training Network GAMES[1].

The CSL 2007 Program Committee considered 116 submissions from 30 countries during a two-week electronic discussion; each paper was refereed by three to six reviewers. The Program Committee selected 36 papers for presentation at the conference and publication in these proceedings.

The Program Committee invited lectures from Samson Abramsky, Luca de Alfaro, Arnold Beckmann, Anuj Dawar, Orna Kupferman, and Helmut Seidl; the papers provided by the invited speakers appear at the beginning of this volume.

Instituted in 2005, the Ackermann Award is the EACSL Outstanding Dissertation Award for Logic in Computer Science. It was sponsored for the first time by Logitech S.A., headquartered near Lausanne in Romanel-sur-Morges, Switzerland[2]. The award winners for 2007, Dietmar Berwanger, Stéphane Lengrand, and Ting Zhang, were invited to present their work at the conference. Citations for the awards, abstracts of the theses, and biographical sketches of the award winners can be found at the end of the proceedings.

We generously thank the Program Committee and all referees for their work in reviewing the papers, as well as the members of the local organizing team (Christian W. Bach, Jérémie Cabessa, Alessandro Facchini, Elisabeth Fournier, Thomas Strahm, Henri Volken). We also thank the Swiss National Science Foundation, the Swiss Academy of Sciences, and the Western Swiss Center for Logic, History and Philosophy of Science, which provided financial support.

June 2007

Jacques Duparc
Thomas A. Henzinger

---

[1] *Games and Automata for Synthesis and Validation*, coordinated by Erich Grädel. `http://www.games.rwth-aachen.de/`

[2] We would like to thank Daniel Borel, Co-founder and Chairman of the Board of Logitech S.A., for his support of the Ackermann Award for the years 2007–09. For a history of this Swiss company, consult `http://www.logitech.com`.

Organization

Program Committee

Sergei N. Artemov *City University of New York*
Franz Baader *Technische Universität Dresden*
Lev Beklemishev *Steklov Mathematical Institute, Moscow*
Andrei Bulatov *Oxford University*
Michel De Rougemont *Université Paris II, Panthéon-Assas*
Jacques Duparc *Université de Lausanne (Co-chair)*
Erich Grädel *RWTH Aachen*
Thomas A. Henzinger *EPFL, Lausanne (Co-chair)*
Michael Kaminski *Technion-IIT, Haifa*
Stephan Kreutzer *Humboldt-Universität zu Berlin*
Benedikt Löwe *Universiteit van Amsterdam*
Rupak Majumdar *University of California, Los Angeles*
Paul-Andr Melliès *Université Paris Diderot - Paris 7*
Joel Ouaknine *Oxford University*
Jean-Eric Pin *Université Paris Diderot - Paris 7*
Nicole Schweikardt *Humboldt-Universität zu Berlin*
Luc Segoufin *INRIA-Futurs, Orsay*
Thomas Strahm *Universität Bern*
Ashish Tiwari *SRI International, Menlo Park*
Helmut Veith *Technische Universität München*
Igor Walukiewicz *Université Bordeaux-1*

Referees

Parosh Abdulla
Klaus Aehlig
Marco Aiello
Jean-Marc Andreoli
Toshiyasu Arai
Albert Atserias
Philippe Audebaud
Serge Autexier
Jeremy Avigad
Matthias Baaz
Patrick Baillot
Henk Barendregt
Jean-Marie Le Bars
Arnold Beckmann
Emmanuel Beffara
Eli Ben-Sasson
Christoph Benzmüller
Ulrich Berger
Inge Bethke
Lars Birkedal
Alexander Bochman
Manuel Bodirsky
Eduardo Bonelli
Guillaume Bonfante
Richard Bonichon
Ahmed Bouajjani
Patricia Bouyer
Julian Bradfield
James Brotherston
Yegor Bryukhov
Kai Brünnler
Antonio Bucciarelli
Wilfried Buchholz
Olivier Carton
Swarat Chaudhuri
Patrick Chervet
Agata Ciabattoni
Robin Cockett
Simon Colton
S. Barry Cooper
Thierry Coquand
Pierre Corbineau

Bruno Courcelle
Pierre-Louis Curien
Anuj Dawar
Stéphane Demri
Razvan Diaconescu
Lucas Dixon
Dan Dougherty
Gilles Dowek
Roy Dyckhoff
Thomas Ehrhard
Michael Emmi
Wolfgang Faber
Claudia Faggian
Stephan Falke
Christian Fermüller
Maribel Fernandez
Andrzej Filinski
Eldar Fischer
Melvin Fitting
Cormac Flanagan
Nissim Francez
Didier Galmiche
Martin Gebser
Rob van Glabbeek
Guillem Godoy
Valentin Goranko
Bernardo Cuenca Grau
Martin Grohe
Stefano Guerrini
Yuri Gurevich
Arie Gurfinkel
Olivier Hemon
Matt Hague
Russ Harmer
Danny Harnik
Hugo Herbelin
Frédéric Herbreteau
Claudio Hermida
André Hernich
Petr Hlineny
Martin Hofmann
Pieter Hofstra
Markus Holzer
Andreas Holzer
Martin Hyland
Rosalie Iemhoff
Neil Immerman
Benedetto Intrigila
Florent Jacquemard
Radha Jagadeesan
Visar Januzaj
Alan Jeffrey
Emil Jerabek
Thierry Joly
Jean-Pierre Jouannaud
Achim Jung
Ozan Kahramanogullari
Marc Kaplan
Deepak Kapur
Stefan Katzenbeisser
Klaus Keimel
Yechiel Kimchi
Johannes Kinder
Boris Konev
Alexei Kopylov
Vladimir Krupski
Manfred Kufleitner
Oliver Kullmann
Orna Kupferman
Clemens Kupke
Roman Kuznets
Ralf Küsters
Ugo Dal Lago
Jim Laird
Martin Lange
Benoit Larose
Franois Laroussinie
Richard Lassaigne
Olivier Laurent
Hans Leiss
Stéphane Lengrand
Florian Lengyel
Marina Lenisa
Pierre Lescanne
Paul Blain Levy
Jean-Jacques Lévy
Leonid Libkin
Gérard Ligozat
Sébastien Limet
John Longley
Carsten Lutz
Ian Mackie
Frédéric Magniez
Harry Mairson
Zoran Majkic
Johann A. Makowsky
Jose Marcial-Romero
Victor Marek
Edwin Mares
Daniel Marx
Gianfranco Mascari
Ralph Matthes
Damiano Mazza
Guy McCusker
Richard McKinley
François Métayer
Zoltan Miklos
Michael Mislove
David Mitchell
Virgile Mogbil
Eugenio Moggi
Alberto Momigliano
Georg Moser
Andrzej Murawski
Pavel Naumov
Paulin Jacobe de Naurois
Phuong Nguyen
Damian Niwinski
Karim Nour
Luke Ong
Jaap van Oosten
Vincent van Oostrom
Eric Pacuit
Michele Pagani
Erik Palmgren
Jens Palsberg
Francesco Paoli
Luca Paolini
Michel Parigot
Murray Patterson
David Pearce
Marco Pedicini
Mati Pentus
Steven Perron

Sylvain Peyronnet
Brigitte Pientka
Elaine Pimentel
Corin Pitcher
Andrew Pitts
Wojciech Plandowski
Chris Pollett
John Power
Femke van Raamsdonk
Jason Reed
Eike Ritter
Luca Roversi
Eyal Rozenberg
Paul Ruet
Michael Rusinowitch
Andrea Schalk
Christian Schallhart
John Schlipf
Peter Schroeder-Heister
Aleksy Schubert
Phil Scott
Robert Seely
Jonathan Seldin
Peter Selinger
Brian Semmes
Olivier Serre
Natarajan Shankar
Valentin Shehtman
Sharon Shoham
François-Régis Sinot
Evgeny Skvortsov
Sergey Slavnov
Sonja Smets
Colin Stirling
Lutz Straßburger
Thomas Streicher
Grégoire Sutre
Tony Tan
Eugenia Ternovska
Kazushige Terui
Pascal Tesson
Neil Thapen
Hayo Thielecke
Marc Thurley
Michael Tiomkin
Henry Towsner
Yoshihito Toyama
Mathieu Tracol
Nikos Tzevelekos
Michael Ummels
Tarmo Uustalu
Jacqueline Vauzeilles
Adrien Vieilleribiere
Eelco Visser
Heribert Vollmer
Fer-Jan de Vries
Pascal Weil
Benjamin Werner
Claus-Peter Wirth
Ronald de Wolf
Nicolas Wolovick
Stefan Woltran
James Worrell
Tatiana Yavorskaya
Michael Zacharyaschev
Noam Zeilberger
Florian Zuleger

Local Organizing Committee

Christian W. Bach
Jérémie Cabessa
Jacques Duparc, *Chair*
Alessandro Facchini
Thomas Strahm
Henri Volken

Table of Contents

Games and Trees

Logic and Deduction

Lambda Calculus 1

Lambda Calculus 2

Game Semantics

Full Completeness: Interactive and Geometric Characterizations of the Space of Proofs (Abstract)*

Samson Abramsky

Oxford University Computing Laboratory
Wolfson Building, Parks Road, Oxford OX1 3QD, U.K.
samson@comlab.ox.ac.uk

Abstract. We pursue the program of exposing the intrinsic mathematical structure of the "space of a proofs" of a logical system [AJ94b]. We study the case of Multiplicative-Additive Linear Logic (MALL). We use tools from Domain theory to develop a semantic notion of proof net for MALL, and prove a Sequentialization Theorem. We also give an interactive criterion for strategies, formalized in the same Domain-theoretic setting, to come from proofs, and show that a "semantic proof structure" satisfies the geometric correctness criterion for proof-nets if and only if it satisfies the interactive criterion for strategies. We also use the Domain-theoretic setting to give an elegant compositional account of Cut-Elimination. This work is a continuation of previous joint work with Radha Jagadeesan [AJ94b] and Paul-André Melliès [AM99].

1 Introduction

One can distinguish two views on how Logic relates to Structure:

1. **The Descriptive View.** Logic is used to *talk about* structure. This is the view taken in Model Theory, and in most of the uses of Logic (Temporal logics, MSO etc.) in Verification. It is by far the more prevalent and widely-understood view.
2. **The Intrinsic View.** Logic is taken to *embody* structure. This is, implicitly or explicitly, the view taken in the Curry-Howard isomorphism, and more generally in Structural Proof Theory, and in (much of) Categorical Logic. In the Curry-Howard isomorphism, one is not using logic to *talk about* functional programming; rather, logic *is* (in this aspect) functional programming.

If we are to find structure in the proof theory of a logic, we face a challenge. Proof systems are subject to many minor "design decisions", which does not impart confidence that the objects being described — formal proofs — have a robust intrinsic structure. It is perhaps useful to make an analogy with Geometry. A major concern of modern Geometry has been to find *instrinsic*, typically

* This research was partly supported by EPSRC Grant EP/D038987/1.

J. Duparc and T.A. Henzinger (Eds.): CSL 2007, LNCS 4646, pp. 1–2, 2007.

coordinate-free, descriptions of the geometric objects of study. We may view the rôle of *syntax* in Proof Theory as analogous to coordinates in Geometry; invaluable for computation, but an obstacle to finding the underlying invariant structure.

A particularly promising line of progress in finding more intrinsic descriptions of proofs, their geometric structure, and their dynamics under Cut-elimination, has taken place in the study of proof-nets in Linear Logic [Gir87], and the associated study of Geometry of Interaction [Gir89]. On the semantic side, the development of Game Semantics and Full Completeness results [AJ94b] has greatly enriched and deepened the structural perspective.

We build on previous joint work with Radha Jagadeesan [AJ94a] and Paul-André Melliès [AM99]. We study Multiplicative-Additive Linear Logic. We use tools from Domain theory to develop a semantic notion of proof net for MALL, and prove a Sequentialization Theorem for this notion [Abr07]. We also give an interactive criterion for *strategies*, formalized in the same Domain-theoretic setting, to come from proofs, and show that a "semantic proof structure" satisfies the geometric correctness criterion for proof-nets if and only if it satisfies the interactive criterion for strategies. We also use the Domain-theoretic setting to give an elegant compositional account of Cut-Elimination.

References

[AJ94a] Abramsky, S., Jagadeesan, R.: New foundations for the geometry of interaction. Information and Computation 111(1), 53–119 (1994)
[AJ94b] Abramsky, S., Jagadeesan, R.: Games and Full Completeness for Multiplicative Linear Logic. Journal of Symbolic Logic 59(2), 543–574 (1994)
[AM99] Abramsky, S., Melliés, P.-A.: Concurrent games and full completeness. In: Proceedings of the 14th Annual IEEE Symposium on Logic in Computer Science, pp. 431–444. IEEE Computer Society Press, Los Alamitos (1999)
[Abr07] Abramsky, S.: Event Domains, Stable Functions and Proof-Nets. Electronic Notes in Theoretical Computer Science 172, 33–67 (2007)
[Gir87] Girard, J.-Y.: Linear Logic. Theoretical Computer Science, 50, 1–102 (1987)
[Gir88] Girard, J.-Y.: Multiplicatives. Rendiconti del Seminario Matematico (Torino), Universita Pol. Torino (1988)
[Gir89] Girard, J.-Y.: Geometry of Interaction I: Interpretation of System F. In: Ferro, R., et al. (ed.) Logic Colloquium '88, pp. 221–260. North-Holland, Amsterdam (1989)
[Gir95] Girard, J.-Y.: Proof-nets: the parallel syntax for proof theory. In: Logic and Algebra, Marcel Dekker (1995)

The Symbolic Approach to Repeated Games (Abstract)

Luca de Alfaro

Department of Computer Engineering
School of Engineering
1156 High Street MS: SOE3
University of California
Santa Cruz, CA 95064, USA
luca@soe.ucsc.edu

Abstract. We consider zero-sum repeated games with omega-regular goals. Such hgames are played on a finite state space over an infinite number of rounds: at every round, the players select moves, either in turns or simultaneously; the current state and the selected moves determine the successor state. A play of the game thus consists in an infinite path over the state space or, if randomization is present, in a probability distribution over paths. Omega-regular goals generalize the class of regular goals (those expressible by finite automata) to infinite sequences, and include many well-known goals, such as the reachability and safety goals, as well as the Büchi and parity goals.

The algorithms for solving repeated games with omega-regular goals can be broadly divided into *enumerative* and *symbolic/* algorithms. Enumerative algorithms consider each state individually; currently, they achieve the best worst-case complexity among the known algorithms. Symbolic algorithms compute in terms of sets of states, or functions from states to real numbers, rather than single states; such sets or functions can often be represented symbolically (hence the name of the algorithms). Even though symbolic algorithms often cannot match the worst-case complexity of the enumerative algorithms, they are often efficient in practice.

We illustrate how symbolic algorithms provide uniform solutions of many classes of repeated games, from turn-based, non-randomized games where at each state one of the players can deterministically win, to concurrent and randomized games where the ability to win must be characterized in probabilistic fashion. We also show that the symbolic algorithms, and the notation used to express them, are closely related to *game metrics* which provide a notion of distance between game states. Roughly, the distance between two states measures how closely a player can match, from one state, the ability of winning from the other state with respect to any omega-regular goal.

J. Duparc and T.A. Henzinger (Eds.): CSL 2007, LNCS 4646, p. 3, 2007.

Proofs, Programs and Abstract Complexity

Arnold Beckmann

Department of Computer Science
University of Wales Swansea
Singleton Park, Swansea, SA2 8PP, United Kingdom
a.beckmann@swansea.ac.uk

Axiom systems are ubiquitous in mathematical logic, one famous example being first order Peano Arithmetic. Foundational questions asked about axiom systems comprise analysing their provable consequences, describing their class of provable recursive functions (i.e. for which programs can termination be proven from the axioms), and characterising their consistency strength. One branch of *proof theory*, called *Ordinal Analysis*, has been quite successful in giving answers to such questions, often providing a unifying approach to them. The main aim of Ordinal Analysis is to reduce such questions to the computation of so called *proof theoretic ordinals*, which can be viewed as abstract measures of the complexity inherent in axiom systems. Gentzen's famous consistency proof of arithmetic [Gen35, Gen38] using transfinite induction up to (a notation of) Cantor's ordinal ϵ_0, can be viewed as the first computation of the proof theoretic ordinal of Peano Arithmetic.

Bounded Arithmetic, as we will consider it, goes back to Buss [Bus86]. Bounded Arithmetic theories can be viewed as subsystems of Peano Arithmetic which have strong connections to complexity classes like the polynomial time hierarchy of functions. Ever since their introduction, research on Bounded Arithmetic has aimed at obtaining a good understanding of the three questions mentioned above for the Bounded Arithmetic setting, with varying success. While a lot of progress has been obtained in relating definable functions to complexity classes, very little can be said about how the provable consequences are related (this problem is called the separation problem for Bounded Arithmetic), or how the consistency strength of Bounded Arithmetic theories can be characterised.

A natural question to ask is whether proof theoretic ordinals can give answers in the Bounded Arithmetic world. However, results by Sommer [Som93] have shown that this is not the case, proof theoretic ordinals are useless in the setting of Bounded Arithmetic. But there are adaptations of proof theoretic ordinals denoted *dynamic ordinals* which can be viewed as suitable abstract measures of the complexity of Bounded Arithmetic theories [Bec96, Bec03, Bec06]. The simplest definition of a dynamic ordinal is given as follows. Let T be a subsystem of Peano Arithmetic, and $T(X)$ its canonical relativisation using a second order predicate variable X. The dynamic ordinal of T, denoted $\mathcal{DO}(T)$, can be defined as the set of those number theoretic functions $f\colon \mathbb{N} \to \mathbb{N}$ for which the

J. Duparc and T.A. Henzinger (Eds.): CSL 2007, LNCS 4646, pp. 4–5, 2007.

sentence $\forall x\,\mathsf{Min}(f(x), X)$ is a provable consequence of $T(X)$. Here, $\mathsf{Min}(t, X)$ is the formula expressing the following minimisation principle:

$$(\exists y < t)X(y) \rightarrow (\exists y < t)[X(y) \land (\forall z < y)\neg X(z)] \ .$$

In my talk I will draw pictures of this situation, starting from Ordinal Analysis for Peano Arithmetic, via their adaptation to dynamic ordinals, leading to Dynamic Ordinal Analysis for Bounded Arithmetic theories. I will argue that Dynamic Ordinals can equally be viewed as suitable measures of the proof and computation strength of Bounded Arithmetic theories, which can be used to give answers to (some of the) above mentioned questions.

References

[Bec96] Beckmann, A.: Seperating fragments of bounded predicative arithmetic. PhD thesis, Westfälische Wilhelms-Universität, Münster (1996)

[Bec03] Beckmann, A.: Dynamic ordinal analysis. Arch. Math. Logic 42, 303–334 (2003)

[Bec06] Beckmann, A.: Generalised dynamic ordinals-universal measures for implicit computational complexity. In: Logic Colloquium '02, Assoc. Symbol. Logic, La Jolla, CA. Lect. Notes Log, vol. 27, pp. 48–74 (2006)

[Bus86] Buss, S.R.: Bounded arithmetic, Stud. Proof Theory, Lect. Notes. Bibliopolis, Naples, vol. 3 (1986)

[Gen35] Gentzen, G.: Untersuchungen über das logische Schließen i. Math. Z. 39, 176–210 (1935)

[Gen38] Gentzen, G.: Neue Fassung des Widerspruchsfreiheitsbeweises für die reine Zahlentheorie. Forsch. Logik Grundl. exakten Wiss. 4, 19–44 (1938)

[Som93] Sommer, R.: Ordinal arithmetic in $I\Delta_0$. In: Clote, P., Krajíček, J. (eds.) Arithmetic, proof theory, and computational complexity, Oxford Logic Guides, pp. 320–363. Oxford University Press, New York (1993)

Model-Checking First-Order Logic: Automata and Locality

Anuj Dawar

University of Cambridge Computer Laboratory, William Gates Building,
J.J. Thomson Avenue, Cambridge, CB3 0FD, UK.
Anuj.Dawar@cl.cam.ac.uk

The satisfaction problem for first-order logic, namely to decide, given a finite structure $\mathbb{A}$ and a first-order formula φ, whether or not $\mathbb{A} \models \varphi$ is known to be PSpace-complete. In terms of parameterized complexity, where the length of φ is taken as the parameter, the problem is AW[$\star$]-complete and therefore not expected to be fixed-parameter tractable (FPT). Nonetheless, the problem is known to be FPT when we place some structural restrictions on $\mathbb{A}$. For some restrictions, such as when we place a bound on the treewidth of $\mathbb{A}$, the result is obtained as a corollary of the fact that the satisfaction problem for monadic second-order logic (MSO) is FPT in the presence of such restriction [1]. This fact is proved using automata-based methods. In other cases, such as when we bound the degree of $\mathbb{A}$, the result is obtained using methods based on the locality of first-order logic (see [3]) and does not extend to MSO. We survey such fixed-parameter tractability results, including the recent [2] and explore the relationship between methods based on automata, locality and decompositions.

References

1. Courcelle, B.: Graph rewriting: An algebraic and logic approach. In: van Leeuwan, J. (ed.) Handbook of Theoretical Computer Science, Volume B: Formal Models and Sematics (B), pp. 193–242. Elsevier, Amsterdam (1990)
2. Dawar, A., Grohe, M., Kreutzer, S.: Locally excluding a minor. In: Proc. 22nd IEEE Symp. on Logic in Computer Science, IEEE Computer Society Press, Los Alamitos (2007)
3. Seese, D.: Linear time computable problems and first-order descriptions. Math. Struct. in Comp. Science 6, 505–526 (1996)

J. Duparc and T.A. Henzinger (Eds.): CSL 2007, LNCS 4646, p. 6, 2007.

Tightening the Exchange Rates Between Automata

Orna Kupferman

School of Computer Science and Engineering, Hebrew University,
Jerusalem 91904, Israel

Abstract. Automata on infinite objects were the key to the solution of several fundamental decision problems in mathematics and logic. Today, automata on infinite objects are used for formal specification and verification of reactive systems. The practical importance of automata in formal methods has motivated a re-examination of the blow up that translations among different types of automata involve. For most translations, the situation is satisfying, in the sense that even if there is a gap between the upper and the lower bound, it is small. For some highly practical cases, however, the gap between the upper and the lower bound is exponential or even larger. The article surveys several such frustrating cases, studies features that they share, and describes recent efforts (with partial success) to close the gaps.

1 Introduction

Finite *automata on infinite objects* were first introduced in the 60's, and were the key to the solution of several fundamental decision problems in mathematics and logic [3,25,31]. Today, automata on infinite objects are used for *specification* and *verification* of reactive systems. The automata-theoretic approach to verification reduces questions about systems and their specifications to questions about automata [19,41]. Recent industrial-strength property-specification languages such as Sugar, ForSpec, and the recent standard PSL 1.01 [7] include regular expressions and/or automata, making specification and verification tools that are based on automata even more essential and popular.

Early automata-based algorithms aimed at showing decidability. The complexity of the algorithm was not of much interest. For example, the fundamental automata-based algorithms of Büchi and Rabin, for the decidability of S1S and SnS (the monadic second-order theories of infinite words and trees, respectively) [3,31] are of non-elementary complexity (i.e., the complexity can not be bounded by a stack of exponentials of a fixed height [26]). Proving the decidability of a given logic was then done by translating the logic to a monadic second-order theory, ignoring the fact that a direct algorithm could have been more efficient. Things have changed in the early 80's, when decidability of highly expressive logics became of practical importance in areas such as artificial intelligence and formal reasoning about systems. The change was reflected in the development

J. Duparc and T.A. Henzinger (Eds.): CSL 2007, LNCS 4646, pp. 7–22, 2007.

of two research directions: (1) direct and efficient translations of logics to automata [43,37,40], and (2) improved algorithms and constructions for automata on infinite objects [33,10,30].

Both research directions are relevant not only for solving the decidability problem, but also for solving other basic problems in formal methods, such as model checking [5] and synthesis [30]. Moreover, input from the industry continuously brings to the field new problems and challenges, requiring the development of new translations and algorithms.[1] For many problems and constructions, our community was able to come up with satisfactory solutions, in the sense that the upper bound (the complexity of the best algorithm or the blow-up in the best known construction) coincides with the lower bound (the complexity class in which the problem is hard, or the blow-up that is known to be unavoidable). For some problems and constructions, however, the gap between the upper bound and the lower bound is significant. This situation is especially frustrating, as it implies that not only we may be using algorithms that can be significantly improved, but also that something is missing in our understanding of automata on infinite objects.

Before turning to the frustrating cases, let us first describe one "success story" — the complementation construction for nondeterministic Büchi automata on infinite words (NBWs). Translating S1S into NBWs, Büchi had to prove the closure of NBWs under complementation. For that, Büchi suggested in 1962 a doubly-exponential construction. Thus, starting with an NBW with n states, the complementary automaton had $2^{2^{O(n)}}$ states [3]. The lower bound known then for NBW complementation was 2^n, which followed from the complementation of automata on finite words. Motivated by problems in formal methods, Sistla, Vardi, and Wolper developed in 1985 a better complementation construction with only a $2^{O(n^2)}$ blow-up [36]. Only in 1988, Safra introduced a determinization construction for NBWs that enabled a $2^{O(n \log n)}$ complementation construction [33], and Michel proved a matching lower bound [28]. The story, however, was not over. A careful analysis of the lower and upper bounds reveals an exponential gap hiding in the constants of the $O()$ notations. While the upper bound of Safra is n^{2n}, the lower bound of Michel is only $n!$, which is roughly $(n/e)^n$. Only recently, a new complementation construction, which avoids determinization, has led to an improved upper bound of $(0.97n)^n$ [11], and a new concept, of full automata, has led to an improved lower bound of $(0.76n)^n$ [44]. Thus, a gap still exists, but it is an acceptable one, and it probably does not point to a significant gap in our understanding of nondeterministic Büchi automata.

In the article, we survey two representative problems for which the gap between the upper and the lower bound is still exponential. In Section 3, we consider safety properties and the problem of translating safety properties to non-

[1] In fact, the practical importance of automata has lead to a reality in which the complexity of a solution or a construction is only one factor in measuring its quality. Other measures, such as the feasibility of a symbolic implementation or the effectiveness of optimizations and heuristics in the average case are taken into an account too. In this article, however, we only consider worst-case complexity.

deterministic automata on finite words. In Section 4, we consider the problem of translating nondeterministic Büchi word automata to nondeterministic co-Büchi word automata. Both problems have strong practical motivation, and in both progress has been recently achieved. We study the problems, their motivation, and their current status. The study is based on joint work with Moshe Vardi [16], Robby Lampert [14], and Benjamin Aminof [2].

2 Preliminaries

Word automata. An *infinite word* over an alphabet Σ is an infinite sequence $w = \sigma_1 \cdot \sigma_2 \cdots$ of letters in Σ. A *nondeterministic Büchi word automaton* (NBW, for short) is $\mathcal{A} = \langle \Sigma, Q, \delta, Q_0, F \rangle$, where Σ is the input alphabet, Q is a finite set of states, $\delta : Q \times \Sigma \to 2^Q$ is a transition function, $Q_0 \subseteq Q$ is a set of initial states, and $F \subseteq Q$ is a set of accepting states. If $|Q_0| = 1$ and δ is such that for every $q \in Q$ and $\sigma \in \Sigma$, we have that $|\delta(q, \sigma)| = 1$, then $\mathcal{A}$ is a *deterministic* Büchi word automaton (DBW).

Given an input word $w = \sigma_0 \cdot \sigma_1 \cdots$ in Σ^ω, a *run* of $\mathcal{A}$ on w is a sequence $r_0, r_1, \ldots$ of states in Q such that $r_0 \in Q_0$ and for every $i \geq 0$, we have $r_{i+1} \in \delta(r_i, \sigma_i)$; i.e., the run starts in one of the initial states and obeys the transition function. Note that a nondeterministic automaton can have many runs on w. In contrast, a deterministic automaton has a single run on w. For a run r, let $inf(r)$ denote the set of states that r visits infinitely often. That is, $inf(r) = \{q \in Q : r_i = q \text{ for infinitely many } i \geq 0\}$. As Q is finite, it is guaranteed that $inf(r) \neq \emptyset$. The run r is *accepting* iff $inf(r) \cap F \neq \emptyset$. That is, a run r is accepting iff there exists a state in F that r visits infinitely often. A run that is not accepting is *rejecting*. An NBW $\mathcal{A}$ accepts an input word w iff there exists an accepting run of $\mathcal{A}$ on w. The *language* of an NBW $\mathcal{A}$, denoted $\mathcal{L}(\mathcal{A})$, is the set of words that $\mathcal{A}$ accepts. We assume that a given NBW $\mathcal{A}$ has no empty states (that is, at least one word is accepted from each state – otherwise we can remove the state).

Linear Temporal Logic. The logic *LTL* is a linear temporal logic. Formulas of LTL are constructed from a set *AP* of atomic propositions using the usual Boolean operators and the temporal operators G ("always"), F ("eventually"), X ("next time"), and U ("until"). Formulas of LTL describe computations of systems over *AP*. For example, the LTL formula $G(req \to F ack)$ describes computations in which every position in which *req* holds is eventually followed by a position in which *ack* holds. For the detailed syntax and semantics of LTL, see [29]. The *model-checking problem* for LTL is to determine, given an LTL formula ψ and a system M, whether all the computations of M satisfy ψ.

General methods for LTL model checking are based on translation of LTL formulas to nondeterministic Büchi word automata:

Theorem 1. [41] *Given an LTL formula ψ, one can construct an NBW $\mathcal{A}_\psi$ that accepts exactly all the computations that satisfy ψ. The size of $\mathcal{A}_\psi$ is exponential in the length of ψ.*

Given a system M and a property ψ, model checking of M with respect to ψ is reduced to checking the emptiness of the product of M and $\mathcal{A}_{\neg\psi}$ [41]. This check can be performed on-the-fly and symbolically [6,12,38], and the complexity of model checking that follows is PSPACE, with a matching lower bound [35].

3 Translating Safety Properties to Automata

Of special interest are properties asserting that the system always stays within some allowed region, in which nothing "bad" happens. For example, we may want to assert that two processes are never simultaneously in the critical section. Such properties of systems are called *safety properties.* Intuitively, a property ψ is a safety property if every violation of ψ occurs after a finite execution of the system. In our example, if in a computation of the system two processes are in the critical section simultaneously, this occurs after some finite execution of the system.

In this section we study the translation of safety properties to nondeterministic automata on finite words (NFWs). We first define safety and co-safety languages, define bad and good prefixes, and motivate the above construction. We then describe a recent result that describes a construction of an NFW for bad and good prefixes for the case the safety or the co-safety property is given by means on an LTL formula.

3.1 Safety Properties and Their Verification

We refer to computations of a nonterminating system as infinite words over an alphabet Σ. Typically, $\Sigma = 2^{AP}$, where AP is the set of the system's atomic propositions. Consider a language $L \subseteq \Sigma^\omega$ of infinite words over the alphabet Σ. A finite word $x \in \Sigma^*$ is a *bad prefix* for L iff for all $y \in \Sigma^\omega$, we have $x \cdot y \notin L$. Thus, a bad prefix is a finite word that cannot be extended to an infinite word in L. Note that if x is a bad prefix, then all the finite extensions of x are also bad prefixes. A language L is a *safety* language iff every infinite word $w \notin L$ has a finite bad prefix.[2] For a safety language L, we denote by *bad-pref*(L) the set of all bad prefixes for L. For example, if $\Sigma = \{0, 1\}$, then $L = \{0^\omega, 1^\omega\}$ is a safety language. To see this, note that every word not in L contains either the sequence 01 or the sequence 10, and a prefix that ends in one of these sequences cannot be extended to a word in L. Thus, *bad-pref*(L) is the language of the regular expression $(0^* \cdot 1 + 1^* \cdot 0) \cdot (0 + 1)^*$.

For a language $L \subseteq \Sigma^\omega$ (Σ^*), we use *comp*(L) to denote the complement of L; i.e., *comp*$(L) = \Sigma^\omega \setminus L$ $(\Sigma^* \setminus L$, respectively). We say that a language $L \subseteq \Sigma^\omega$ is a *co-safety* language iff *comp*(L) is a safety language. (The term used in [23] is *guarantee* language.) Equivalently, L is co-safety iff every infinite word $w \in L$ has a *good prefix* $x \in \Sigma^*$: for all $y \in \Sigma^\omega$, we have $x \cdot y \in L$. For a co-safety

[2] The definition of safety we consider here is given in [1], it coincides with the definition of limit closure defined in [9], and is different from the definition in [20], which also refers to the property being closed under stuttering.

language L, we denote by *good-pref*(L) the set of good prefixes for L. Note that for a safety language L, we have that *good-pref*$(comp(L))$ = *bad-pref*(L). Thus, in order to construct the set of bad prefixes for a safety property, one can construct the set of good prefixes for its complementary language.

We say that an NBW $\mathcal{A}$ is a safety (co-safety) automaton iff $\mathcal{L}(\mathcal{A})$ is a safety (co-safety) language. We use *bad-pref*$(\mathcal{A})$, *good-pref*$(\mathcal{A})$, and *comp*$(\mathcal{A})$ to abbreviate *bad-pref*$(\mathcal{L}(\mathcal{A}))$, *good-pref*$(\mathcal{L}(\mathcal{A}))$, and *comp*$(\mathcal{L}(\mathcal{A}))$, respectively.

In addition to proof-based methods for the verification of safety properties [23,24], there is extensive work on model checking of safety properties. Recall that general methods for model checking of linear properties are based on a construction of an NBW $\mathcal{A}_{\neg\psi}$ that accepts exactly all the infinite computations that violate the property ψ and is of size exponential in ψ [22,41]. Verification of a system M with respect to ψ is then reduced to checking the emptiness of the product of M and $\mathcal{A}_{\neg\psi}$ [39].

When ψ is a safety property, the NBW $\mathcal{A}_{\neg\psi}$ can be replaced by *bad-pref*$(\mathcal{A}_\psi)$ – an NFW that accepts exactly all the bad prefixes of ψ [16]. This has several advantages, as reasoning about finite words is simpler than reasoning about infinite words: symbolic reasoning (in particular, bounded model checking procedures) need not look for loops and can, instead, apply backward or forward reachability analysis [4]. In fact, the construction of *bad-pref*$(\mathcal{A}_\psi)$ reduces the model-checking problem to the problem of invariance checking [23], which is amenable to both model-checking techniques and deductive verification techniques. In addition, using *bad-pref*$(\mathcal{A}_\psi)$, we can return to the user a finite error trace, which is a bad prefix, and which is often more helpful than an infinite error trace.

Consider a safety NBW $\mathcal{A}$. The construction of *bad-pref*$(\mathcal{A})$ was studied in [16]. If $\mathcal{A}$ is deterministic, we can construct a *deterministic* automaton on finite words (DFW) for *bad-pref*$(\mathcal{A})$ by defining the set of accepting states to be the set of states s for which $\mathcal{A}$ with initial state s is empty. Likewise, if $\mathcal{A}$ is a co-safety automaton, we can construct a DFW for *good-pref*$(\mathcal{A})$ by defining the set of accepting states to be the set of states s for which $\mathcal{A}$ with initial state s is universal.

When $\mathcal{A}$ is nondeterministic, the story is more complicated. Even if we are after a nondeterministic, rather than a deterministic, automaton for the bad or good prefixes, the transition from infinite words to finite words involves an exponential blow-up. Formally, we have the following.

Theorem 2. [16] *Consider an NBW $\mathcal{A}$ of size n.*

1. *If $\mathcal{A}$ is a safety automaton, the size of an NFW for bad-pref$(\mathcal{A})$ is $2^{\Theta(n)}$.*
2. *If $\mathcal{A}$ is a co-safety automaton, the size of an NFW for good-pref$(\mathcal{A})$ is $2^{\Theta(n)}$.*

The lower bound in Theorem 2 for the case $\mathcal{A}$ is a safety automaton is not surprising. Essentially, it follows from the fact that *bad-pref*$(\mathcal{A})$ refers to words that are not accepted by $\mathcal{A}$. Hence, it has the flavor of complementation, and complementation of nondeterministic automata involves an exponential blow-up [27]. The second blow up, however, in going from a co-safety automaton to a nondeterministic automaton for its good prefixes is surprising. Its proof in [16] highlights the motivation behind the definition of a *fine automaton* for safety properties, and we describe it below.

For $n \geq 1$, let $\Sigma_n = \{1, \ldots, n, \&\}$. We define L_n as the language of all words $w \in \Sigma_n^\omega$ such that w contains at least one & and the letter after the first & is either & or it has already appeared somewhere before the first &. The language L_n is a co-safety language. Indeed, each word in L_n has a good prefix (e.g., the one that contains the first & and its successor). We can recognize L_n with an NBW with $O(n)$ states (the NBW guesses the letter that appears after the first &). Obvious good prefixes for L_n are 12&&, 123&2, etc. That is, prefixes that end one letter after the first &, and their last letter is either & or has already appeared somewhere before the &. We can recognize these prefixes with an NFW with $O(n)$ states. But L_n also has some less obvious good prefixes, like $1234 \cdots n\&$ (a permutation of $1 \ldots n$ followed by &). These prefixes are indeed good, as every suffix we concatenate to them would start with & or with a letter in $\{1, \ldots, n\}$, which has appeared before the &. To recognize these prefixes, an NFW needs to keep track of subsets of $\{1, \ldots, n\}$, for which it needs 2^n states. Consequently, an NFW for *good-pref*(L_n) must have at least 2^n states.

It is also shown in [16] that the language L_n can be encoded by an LTL formula of length quadratic in n. This implies that the translation of safety and co-safety LTL formulas to NFWs for their bad and good prefixes is doubly exponential. Formally, we have the following.

Theorem 3. [16] *Given a safety LTL formula ψ of size n, the size of an NFW for bad-pref(ψ) is $2^{2^{\Omega(\sqrt{n})}}$.*

3.2 Fine Automata and Their Construction

As described in the proof of Theorem 2, some good prefixes for L_n (the "obvious prefixes") can be recognized by a small NFW. What if we give up the non-obvious prefixes and construct an NFW $\mathcal{A}'$ that accepts only the "obvious subset" of L_n? It is not hard to see that each word in L_n has an obvious prefix. Thus, while $\mathcal{A}'$ does not accept all the good prefixes, it accepts at least one prefix of every word in L. This useful property of $\mathcal{A}'$ is formalized below.

Consider a safety language L. We say that a set $X \subseteq$ *bad-pref*(L) is a *trap* for L iff every word $w \notin L$ has at least one bad prefix in X. Thus, while X need not contain all the bad prefixes for L, it must contain sufficiently many prefixes to "trap" all the words not in L. Dually, a trap for a co-safety language L is a set $X \subseteq$ *good-pref*(L) such that every word $w \in L$ has at least one good prefix in X. We denote the set of all the traps, for an either safety or co-safety language L, by *trap*(L).

An NFW $\mathcal{A}$ is *fine* for a safety or a co-safety language L iff $\mathcal{A}$ accepts a trap for L. For example, an NFW that accepts $0^* \cdot 1 \cdot (0+1)$ does not accept all the bad prefixes of the safety language $\{0^\omega\}$; in particular, it does not accept the minimal bad prefixes in $0^* \cdot 1$. Yet, such an NFW is fine for $\{0^\omega\}$. Indeed, every infinite word that is different from 0^ω has a prefix in $0^* \cdot 1 \cdot (0+1)$. Likewise, the NFW is fine for the co-safety language $0^* \cdot 1 \cdot (0+1)^\omega$. In practice, almost all the benefit that one obtains from an NFW that accepts all the bad/good prefixes can also be obtained from a fine automaton. It is shown in [16] that for natural safety formulas ψ, the construction of an NFW fine for ψ is as easy as the construction of $\mathcal{A}_{\neg\psi}$. In more details, if we regard $\mathcal{A}_{\neg\psi}$ as an NFW, with an appropriate definition of the set of accepting states, we get an automaton fine for ψ. For general safety formulas, the problem of constructing small fine automata was left open in [16] and its solution in [14] has led to new mysteries in the context of safety properties. Let us first describe the result in [14].

Recall that the transition from a safety NBW to an NFW for its bad prefixes is exponential, and that the exponential blow up follows from the fact that a complementing NBW can be construction from a tight NFW. When we consider fine automata, things are more complicated, as the fine NFW need not accept all bad prefixes. As we show below, however, a construction of fine automata still has the flavor of complementation, and must involve an exponential blow up.

Theorem 4. [14] *Given a safety NBW $\mathcal{A}$ of size n, the size of an NFW fine for $\mathcal{A}$ is $2^{\Theta(n)}$.*

We now move on to consider co-safety NBWs. Recall that, as with safety properties and bad prefixes, the transition from a co-safety NBW to an NFW for its good prefixes is exponential. We show that a fine NFW for a co-safety property can be constructed from the NBWs for the property and its negation. The idea is that it is possible to bound the number of times that a run of $\mathcal{A}$ visits the set of accepting states when it runs on a word not in $\mathcal{L}(\mathcal{A})$. Formally, we have the following:

Lemma 1. [14] *Consider a co-safety NBW $\mathcal{A}$. Let F be the set of accepting states of $\mathcal{A}$ and let $\overline{\mathcal{A}}$ be an NBW with $\overline{n}$ states such that $\mathcal{L}(\overline{\mathcal{A}}) = comp(\mathcal{L}(\mathcal{A}))$. If a run of $\mathcal{A}$ on a finite word $h \in \Sigma^*$ visits F more than $|F| \cdot \overline{n}$ times, then h is a good prefix for $\mathcal{L}(\mathcal{A})$.*

Consider a co-safety NBW $\mathcal{A}$ with n states, m of them accepting. Let $\overline{\mathcal{A}}$ be an NBW with $\overline{n}$ states such that $\mathcal{L}(\overline{\mathcal{A}}) = comp(\mathcal{L}(\mathcal{A}))$. Following Lemma 1, we can construct an NFW fine for $\mathcal{A}$ by taking $(m \cdot \overline{n}) + 1$ copies of $\mathcal{A}$, and defining the transition function such that when a run of $\mathcal{A}'$ visits F in the j-th copy of $\mathcal{A}$, it moves to the $(j+1)$-th copy. The accepting states of $\mathcal{A}'$ are the states of F in the $(m \cdot \overline{n} + 1)$-th copy. This implies the following theorem.

Theorem 5. [14] *Consider a co-safety NBW $\mathcal{A}$ with n states, m of them accepting. Let $\overline{\mathcal{A}}$ be an NBW with $\overline{n}$ states such that $\mathcal{L}(\overline{\mathcal{A}}) = comp(\mathcal{L}(\mathcal{A}))$. There exists an NFW $\mathcal{A}'$ with $n \cdot (m \cdot \overline{n} + 1)$ states such that $\mathcal{A}'$ is fine for $\mathcal{L}(\mathcal{A})$.*

Given a safety NBW, its complement NBW is co-safety. Thus, dualizing Theorem 5, we get the following.

Theorem 6. [14] *Consider a safety NBW with n states. Let $\overline{\mathcal{A}}$ be an NBW with $\overline{n}$ states, $\bar{m}$ of them accepting, such that $\mathcal{L}(\overline{\mathcal{A}}) = comp(\mathcal{L}(\mathcal{A}))$. There exists an NFW $\mathcal{A}'$ with $\bar{n} \cdot (\bar{m} \cdot n + 1)$ states such that $\mathcal{A}'$ is fine for $\mathcal{L}(\mathcal{A})$.*

By Theorem 1, given an LTL formula ψ, we can construct NBWs $\mathcal{A}_\psi$ and $\mathcal{A}_{\neg\psi}$ for ψ and $\neg\psi$, respectively. The number of states in each of the NBWs is at most $2^{O(|\psi|)}$. Hence, by Theorem 5, we can conclude:

Theorem 7. [14] *Consider a safety LTL formula φ of length n. There exists an NFW fine for φ with at most $2^{O(n)}$ states.*

It follows from Theorem 7 that the transition from a tight NFW (one that accepts exactly all bad or good prefixes) to a fine NFW is significant, as it circumvents the doubly exponential blow-up in Theorem 3.

3.3 Discussion

The work in [14] has answered positively the question about the existence of exponential fine automata for general safety LTL formulas, improving the doubly-exponential construction in [16]. Essentially, the construction adds a counter on top of the NBW for the formula. The counter is increased whenever the NBW visits an accepting state, and a computation is accepted after the counter reaches a bound that depends on the size of the formula. For a discussion on the application of the result in the context of bounded model checking and run-time verification see [14]. Here, we discuss the theoretical aspects of the result.

While [14] has solved the problem of constructing exponential fine automata for LTL formulas, the problem of constructing polynomial fine automata for co-safety NBW is still open. The challenge here is similar to other challenges in automata-theoretic constructions in which one needs both the NBW and its complementing NBW — something that is easy to have in the context of LTL, but difficult in the context of NBW. More problems in this status are reported in [18]. For example, the problem of deciding whether an LTL formula ψ can be translated to a DBW can be solved by reasoning about the NBWs for ψ and $\neg\psi$. This involves an exponential blow up in the length of ψ, but, as in our case, no additional blow-up for complementation. The problem of deciding whether an NBW can be translated to a DBW cannot be solved using the same lines, as here complementation does involve an exponential blow up. From a practical point of view, however, the problem of going from a co-safety automaton to a fine NFW is of less interest, as users that use automata as their specification formalism are likely to start with an automaton for the bad or the good prefixes anyway. Thus, the problem about the size of fine automata is interesting mainly for the specification formalism of LTL, which [14] did solve.

4 From Büchi to co-Büchi Automata

The second open problem we describe is the problem of translating, when possible, a nondeterministic Büchi word automaton to an equivalent nondeterministic co-Büchi word automaton (NCW). The *co-Büchi* acceptance condition is dual to the Büchi acceptance condition. Thus, $F \subseteq Q$ and a run r is accepting if it visits F only finitely many times. Formally, $inf(r) \cap F = \emptyset$. NCWs are less expressive than NBWs. For example, the language $\{w : w$ has only finitely many 0s$\} \subseteq \{0,1\}^\omega$ cannot be recognized by an NCW. In fact, NCWs are as expressive as deterministic co-Büchi automata (DCWs). Hence, as DBWs are dual to DCWs, a language can be recognized by an NCW iff its complement can be recognized by a DBW.

The best translation of NBW to NCW (when possible) that is currently known actually results in a deterministic co-Büchi automaton (DCW), and it goes as follows. Consider an NBW $\mathcal{A}$ that has an equivalent NCW. First, co-determinize $\mathcal{A}$ and obtain a deterministic Rabin automaton (DRW) $\tilde{\mathcal{A}}$ for the complement language. By [13], DRWs are *Büchi type*. That is, if a DRW has an equivalent DBW, then the DRW has an equivalent DBW on the same structure. Let $\tilde{\mathcal{B}}$ be the DBW equivalent to $\tilde{\mathcal{A}}$ (recall that since $\mathcal{A}$ can be recognized by an NCW, its complement can be recognized by a DBW). By dualizing $\tilde{\mathcal{B}}$ one gets a DCW equivalent to $\mathcal{A}$. The co-determinization step involves an exponential blowup in the number of states [33]. Hence, starting with an NBW with n states, we end up with an NCW with $2^{O(n \log n)}$ states. This is particularly annoying as even a lower bound showing that an NCW needs one more state is not known. As we discuss below, the translation of NBW to an equivalent NCW is of practical importance because of its relation to the problem of translating LTL formulas to equivalent alternation-free μ-calculus (AFMC) formulas (when possible).

It is shown in [17] that given an LTL formula ψ, there is an AFMC formula equivalent to $\forall\psi$ iff ψ can be recognized by a DBW. Evaluating specifications in the alternation-free fragment of μ-calculus can be done with linearly many symbolic steps. In contrast, direct LTL model checking reduces to a search for bad-cycles and its symbolic implementation involves nested fixed-points, and is typically quadratic [32]. Hence, identifying LTL formulas that can be translated to AFMC, and coming up with an optimal translation, is a problem of great practical importance. The best known translations of LTL to AFMC first translates the LTL formula ψ to a DBW, which is then linearly translated to an AFMC formula for $\forall\psi$. The translation of LTL to DBW, however, is doubly exponential, thus the overall translation is doubly-exponential, with only an exponential matching lower bound [17].

The reason that current translations go through an intermediate deterministic automaton is the need to run this automaton on all the computations of the system in a way that computations with the same prefix follow the same run. A similar situation exists when we expand a word automaton to a tree automaton [8] — the word automaton cannot be nondeterministic, as different branches of the tree that have the same prefix u may be accepted by runs of the word automaton that do not agree on the way they proceed on u. A promising direction for coping with this situation was suggested in [17]: Instead of translating the LTL formula ψ to a DBW, one can translate $\neg\psi$ to an NCW. This can be done either directly, or by translating the NBW for $\neg\psi$ to an equivalent NCW. Then, the NCW can be linearly translated to an AFMC formula for $\exists\neg\psi$, whose negation is equivalent to $\forall\psi$. The fact that the translation can go through a nondeterministic rather than a deterministic automaton is very promising, as nondeterministic automata are typically exponentially more succinct than deterministic ones.[3] Nevertheless, the problem of translating LTL formulas to NCWs of exponential size is still open.[4] The best translation that is known today involves a doubly-exponential blow up, and it actually results in a DCW, giving up the idea that the translation of LTL to AFMC can be exponentially more efficient by using intermediate nondeterministic automata. Note that a polynomial translation of NBW to NCW will imply a singly-exponential translation of LTL to AFMC, as the only exponential step in the procedure will be the translation of LTL to NBW.[5]

Recall that while the best upper bound for an NBW to NCW translation is $2^{O(n \log n)}$, we do not even have a single example to a language whose NBW is smaller than its NCW. In fact, it was only recently shown that NBWs are not *co-Büchi-type*. That is, there is an NBW $\mathcal{A}$ such that $L(\mathcal{A})$ can be recognized by an NCW, but an NCW for $L(\mathcal{A})$ must have a different structure than $\mathcal{A}$. We describe such an NBW in the proof below (the original proof, in [15], has a different example).

Lemma 2. [15] *NBWs are not co-Büchi-type.*

Proof: Consider the NBW described in Figure 1. Note that the NBW has two initial states. The NBW recognizes the language L of all words with at least one a and at least one b. This language can be recognized by an NCW, yet it is easy to see that there is no way to define F on top of $\mathcal{A}$ such that the result is an NCW that recognizes L. □

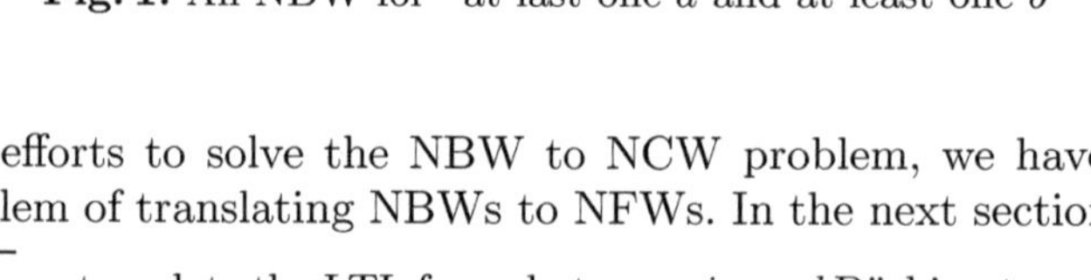

Fig. 1. An NBW for "at last one a and at least one b"

During our efforts to solve the NBW to NCW problem, we have studied the related problem of translating NBWs to NFWs. In the next section we describe

[3] Dually, we can translate the LTL formula to a *universal* Büchi automaton and translate this automaton to an AFMC formula. The universal Büchi automaton for ψ is dual to the nondeterministic co-Büchi automaton for $\neg\psi$.

[4] As mentioned above, not all LTL formulas can be translated to NCWs. When we talk about the blow up in a translation, we refer to formulas for which a translation exists.

[5] Wilke [42] proved an exponential lower-bound for the translation of an NBW for an LTL formula ψ to and AFMC formula equivalent to $\forall\psi$. This lower-bound does not preclude a polynomial upper-bound for the translation of an NBW for $\neg\psi$ to an AFMC formula equivalent to $\exists\neg\psi$, which is our goal.

the problem, its relation to the NBW to NCW problem, and our partial success in this front.

4.1 From Büchi to Limit Finite Automata

Recall that DBWs are less expressive than NBWs. Landweber characterizes languages $L \subseteq \Sigma^\omega$ that can be recognized by a DBW as those for which there is a regular language $R \subseteq \Sigma^*$ such that L is the *limit* of R. Formally, w is in the limit of R iff w has infinitely many prefixes in R [21]. It is not hard to see that a DBW for L, when viewed as a DFW, recognizes a language whose limit is L, and vice versa – a DFW for R, when viewed as a DBW, recognizes the language that is the limit of R. What about the case R and L are given by nondeterministic automata? It is not hard to see that the simple transformation between the two formalisms no longer holds. For example, the NBW $\mathcal{A}$ in Figure 2 recognizes the language L of all words with infinitely many bs, yet when viewed as an NFW, it recognizes $(a+b)^+$, whose limit is $(a+b)^\omega$. As another example, the language of the NBW $\mathcal{A}'$ is empty, yet when viewed as an NFW, it recognizes the language $(a+b)^* \cdot b$, whose limit is L. As demonstrated by the examples, the difficulty of the nondeterministic case originates from the fact that different prefixes of the infinite word may follow different accepting runs of the NFW, and there is no guarantee that these runs can be merged into a single run of the NBW. Accordingly, the best translation that was known until recently for going from an NFW to an NBW accepting its limit, or from an NBW to a limit NFW, is to first determinize the given automaton. This involves a $2^{O(n \log n)}$ blow up and gives up the potential succinctness of the nondeterministic model. On the other hand, no lower bound above $\Omega(n \log n)$ is known.

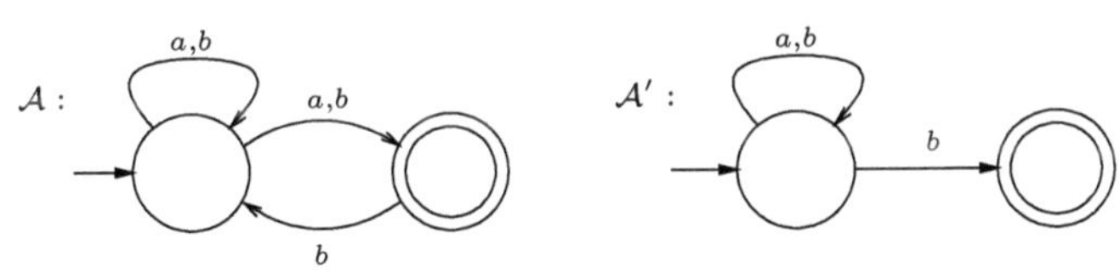

Fig. 2. Relating NBWs and limit NFWs

In [2] we study this exponential gap and tried to close it. In addition to the limit operator introduced by Landweber, we introduce and study two additional ways to induce a language of infinite words from a language of finite words: the *co-limit* of R is the set of all infinite words that have only finitely many prefixes in R. Thus, co-limit is dual to Landweber's limit. Also, the *persistent limit* of R is the set of all infinite words that have only finitely many prefixes not in R. Thus, eventually all the prefixes are in R. Formally, we have the following.

Definition 1. *Consider a language $R \subseteq \Sigma^*$. We define three languages of infinite words induced by R.*

1. **[limit]** *$lim(R) \subseteq \Sigma^\omega$ is the set of all words that have infinitely many prefixes in R. I.e., $lim(R) = \{w \mid w[1,i] \in R$ for infinitely many i's$\}$* [21].
2. **[co-limit]** *$co\text{-}lim(R) \subseteq \Sigma^\omega$ is the set of all words that have only finitely many prefixes in R. I.e., $co\text{-}lim(R) = \{w \mid w[1,i] \in R$ for finitely many i's$\}$.*
3. **[persistent limit]** *$plim(R) \subseteq \Sigma^\omega$ is the set of all words that have only finitely many prefixes not in R. I.e., $plim(R) = \{w \mid w[1,i] \in R$for almost alli's$\}$.*

For example, for $R = (a+b)^*b$, the language $lim(R)$ consists of all words that have infinitely many b's, $co\text{-}lim(R)$ is the language of words that have finitely many b's, and $plim(R)$ is the language of words that have finitely many a's. For an NFW $\mathcal{A}$, we use $lim(\mathcal{A})$, $co\text{-}lim(\mathcal{A})$, and $plim(\mathcal{A})$, to denote $lim(L(\mathcal{A}))$, $co\text{-}lim(L(\mathcal{A}))$, and $plim(L(\mathcal{A}))$, respectively. The three limit operators are dual in the sense that for all $R \subseteq \Sigma^*$, we have $comp(lim(R)) = co\text{-}lim(R) = plim(comp(R))$.

Below we describe the main results of [2], which studies the relative succinctness of NBWs, NCWs, and NFWs whose limit, co-limit, and persistent limit correspond to the NBW and NCW.

We first need some notations. Consider an NFW $\mathcal{A} = \langle \Sigma, Q, \delta, Q_0, F \rangle$. For two sets of states $P, S \subseteq Q$, we denote by $L_{P,S}$ the language of $\mathcal{A}$ with initial set P and accepting set S. Theorem 8 is a key theorem in beating the "determinize first" approach. It implies that the transition from an NFW $\mathcal{A}$ to an NBW for $lim(\mathcal{A})$ need not involve a determinization of $\mathcal{A}$. Indeed, we can specify $lim(\mathcal{A})$ as the union of languages that are generated by automata with a structure similar to the structure of $\mathcal{A}$. Formally, we have the following.

Theorem 8. [2] *For every NFW $\mathcal{A} = \langle \Sigma, Q, \delta, Q_0, F \rangle$,*

$$lim(\mathcal{A}) = \bigcup_{p \in Q} L_{Q_0,\{p\}} \cdot (L_{\{p\},\{p\}} \cap L_{\{p\},F})^\omega.$$

Given $\mathcal{A}$ with n states, Theorem 8 implies that an NBW accepting $lim(\mathcal{A})$ can be constructed by intersection, application of ω, concatenation, and union, starting with NFWs with n states. Exploiting the the similarity in the structure of the involved NFWs, the resulting NBW has $O(n^2)$ states.

Corollary 1. *Given an NFW $\mathcal{A}$ with n states, there is an NBW $\mathcal{A}'$ with $O(n^2)$ states such that $L(\mathcal{A}') = lim(L(\mathcal{A}))$.*

Corollary 1 implies that going from an NFW to an NBW for its limit, it is possible to do better than determinize the NFW. On the other hand, it is shown in [2] that going from an NFW to an NCW for its co-limit or persistent limit, an exponential blow-up cannot be avoided, and determinization is optimal.

Further results of [2] study succinctness among NFWs to which different limit operators are applied. For example, in Theorem 9 below we prove that going from a persistent limit NFW to a limit NFW involves an exponential blow up. In other words, given an NFW $\mathcal{A}$ whose persistent limit is L, translating $\mathcal{A}$ to an NFW whose limit is L may involve an exponential blow up. Note that persistent limit and limit are very similar – both require the infinite word to have infinitely many prefixes in $L(\mathcal{A})$, only that the persistent limit requires, in addition, that only finitely many prefixes are not in $L(\mathcal{A})$. This difference, which is similar to the difference between NBW and NCW, makes persistent limit exponentially more succinct. Technically, it follows from the fact that persistent limit NFWs inherit the power of alternating automata. In a similar, though less surprising way, co-limit NFWs inherit the power of complementation, and are also exponentially more succinct.

Theorem 9. [2] *For every $n \geq 1$, there is a language $L_n \subseteq \Sigma^\omega$ such that there are NFWs $\mathcal{A}$ with $O(n)$ states, and $\mathcal{A}'$ with $O(n^2)$ states, such that co-lim($\mathcal{A}$) = plim($\mathcal{A}'$) = L_n but an NFW $\mathcal{A}''$ such that lim($\mathcal{A}''$) = L_n must have at least 2^n states.*

Proof: Consider the language $L_n \subseteq \{0,1\}^\omega$ of all words w such that $w = uuz$, with $|u| = n$. We prove that an NFW $\mathcal{A}''$ such that $lim(\mathcal{A}'') = L_n$ must remember subsets of size n, and thus must have at least 2^n states. In order to construct small NFW for the co-limit and persistent limit operators, we observe that a word w is in L_n iff $\bigwedge_{i=1}^{n}(w[i] = w[n+i])$. In the case of co-limit, we can check that only finitely many (in fact, 0) prefixes h of an input word are such that $h[i] \neq h[i+n]$ for some $1 \leq i \leq n$. The case of persistent limit is much harder, as we cannot use the implicit complementation used in the co-limit case. Instead, we use the universal nature of persistence. We define the NFW $\mathcal{A}'$ as a union of n NFWs $\mathcal{A}'_1, \ldots, \mathcal{A}'_n$. The NFW $\mathcal{A}'_i$ is responsible for checking that $w[i] = w[n+i]$. In order to make sure that the conjunction on all $1 \leq i \leq n$ is satisfied, we further limit $\mathcal{A}'_i$ to accept only words of length i mod n. Hence, $\mathcal{A}'_i$ accepts a word $u \in \Sigma^*$ iff $u[i] = u[n+i] \wedge |u| = i \bmod n$. Thus, $plim(\mathcal{A}') = L_n$. □

4.2 Discussion

The exponential gap between the known upper and lower bounds in the translation of NBW to NCW is particularly annoying: the upper bound is $2^{O(n \log n)}$ and for the lower bound we do not even have an example of a language whose NCW needs one more state than the NBW. The example in the proof of Lemma 2 shows an advantage of the Büchi condition. In a recent work with Benjamin Aminof and Omer Lev, we hope to turn this advantage into a lower bound. The idea is as follows. NCWs cannot recognize the language of all words that have infinitely many occurrences of some letter. Indeed, DBWs cannot recognize the complement language [21]. Thus, a possible way to obtain a lower bound for the translation of NBW to NCW is to construct a language that is recognizable by an NCW, but for which an NCW needs more states than an NBW due to its inability to recognize infinitely many occurrences of a letter. One such candidate is the family of languages $L_1, L_2, \ldots$ over the alphabet $\{a, b\}$, where L_k contains exactly all words that have at least k occurrences of the letter a and at least k

occurrences of the letter b. An NBW can follow the idea of the NBW in Figure 1: since every infinite word has infinitely many a's or infinitely many b's, the NBW for L can guess which of the two letters occurs infinitely often, and count k occurrences of the second letter. Thus, the NBW is the union of two components, one looking for k occurrences of a followed by infinitely many b's and the other looking for k occurrences of b followed by infinitely many a's. This can be done with $2k + 1$ states. We conjecture that an NCW needs more than two counters. The reason is that an NCW with less than k states accepting all words with infinitely many a's, inevitably also accepts a word with less than k a's.

References

1. Alpern, B., Schneider, F.B.: Defining liveness. IPL 21, 181–185 (1985)
2. Aminof, B., Kupferman, O.: On the succinctness of nondeterminizm. In: Graf, S., Zhang, W. (eds.) ATVA 2006. LNCS, vol. 4218, pp. 125–140. Springer, Heidelberg (2006)
3. Büchi, J.R.: On a decision method in restricted second order arithmetic. In: Proc. Int. Congress on Logic, Method, and Philosophy of Science. 1960, pp. 1–12. Stanford University Press (1962)
4. Burch, J.R., Clarke, E.M., McMillan, K.L., Dill, D.L., Hwang, L.J.: Symbolic model checking: 10^{20} states and beyond. I&C 98(2), 142–170 (1992)
5. Clarke, E.M., Grumberg, O., Peled, D.: Model Checking. MIT Press, Cambridge (1999)
6. Courcoubetis, C., Vardi, M.Y., Wolper, P., Yannakakis, M.: Memory efficient algorithms for the verification of temporal properties. FMSD 1, 275–288 (1992)
7. Eisner, C., Fisman, D.: A Practical Introduction to PSL. Springer, Heidelberg (2006)
8. Emerson, A.E., Sistla, A.P.: Deciding full branching time logics. I&C 61(3), 175–201 (1984)
9. Emerson, E.A.: Alternative semantics for temporal logics. TCS 26, 121–130 (1983)
10. Emerson, E.A., Jutla, C.: The complexity of tree automata and logics of programs. In: Proc. 29th FOCS, pp. 328–337 (1988)
11. Friedgut, E., Kupferman, O., Vardi, M.Y.: Büchi complementation made tighter. In: Wang, F. (ed.) ATVA 2004. LNCS, vol. 3299, pp. 64–78. Springer, Heidelberg (2004)
12. Gerth, R., Peled, D., Vardi, M.Y., Wolper, P.: Simple on-the-fly automatic verification of linear temporal logic. In: Protocol Specification, Testing, and Verification, pp. 3–18. Chapman & Hall, Sydney, Australia (1995)
13. Krishnan, S.C., Puri, A., Brayton, R.K.: Deterministic ω-automata vis-a-vis deterministic Büchi automata. In: Du, D.-Z., Zhang, X.-S. (eds.) ISAAC 1994. LNCS, vol. 834, pp. 378–386. Springer, Heidelberg (1994)
14. Kupferman, O., Lampert, R.: On the construction of fine automata for safety properties. In: Graf, S., Zhang, W. (eds.) ATVA 2006. LNCS, vol. 4218, pp. 110–124. Springer, Heidelberg (2006)
15. Kupferman, O., Morgenstern, G., Murano, A.: Typeness for ω-regular automata. In: Wang, F. (ed.) ATVA 2004. LNCS, vol. 3299, pp. 324–338. Springer, Heidelberg (2004)
16. Kupferman, O., Vardi, M.Y.: Model checking of safety properties. FMSD 19(3), 291–314 (2001)

17. Kupferman, O., Vardi, M.Y.: From linear time to branching time. ACM TOCL 6(2), 273–294 (2005)
18. Kupferman, O., Vardi, M.Y.: Safraless decision procedures. In: Proc. 46th FOCS, pp. 531–540 (2005)
19. Kurshan, R.P.: Computer Aided Verification of Coordinating Processes. Princeton Univ. Press, Princeton (1994)
20. Lamport, L.: Logical foundation. In: Alford, M.W., Hommel, G., Schneider, F.B., Ansart, J.P., Lamport, L., Mullery, G.P., Liskov, B. (eds.) Distributed Systems. LNCS, vol. 190, Springer, Heidelberg (1985)
21. Landweber, L.H.: Decision problems for ω–automata. Mathematical Systems Theory 3, 376–384 (1969)
22. Lichtenstein, O., Pnueli, A.: Checking that finite state concurrent programs satisfy their linear specification. In: Proc. 12th POPL, pp. 97–107 (1985)
23. Manna, Z., Pnueli, A.: The Temporal Logic of Reactive and Concurrent Systems: Specification. Springer, Heidelberg (1992)
24. Manna, Z., Pnueli, A.: The Temporal Logic of Reactive and Concurrent Systems: Safety. Springer, Heidelberg (1995)
25. McNaughton, R.: Testing and generating infinite sequences by a finite automaton. I&C 9, 521–530 (1966)
26. Meyer, A.R.: Weak monadic second order theory of successor is not elementary recursive. In: Proc. Logic Colloquium. LNM 453, pp. 132–154. Springer, Heidelberg (1975)
27. Meyer, A.R., Fischer, M.J.: Economy of description by automata, grammars, and formal systems. In: Proc. 12th SWAT, pp. 188–191 (1971)
28. Michel, M.: Complementation is more difficult with automata on infinite words. CNET, Paris (1988)
29. Pnueli, A.: The temporal semantics of concurrent programs. TCS 13, 45–60 (1981)
30. Pnueli, A., Rosner, R.: On the synthesis of a reactive module. In: Proc. 16th POPL, pp. 179–190 (1989)
31. Rabin, M.O.: Decidability of second order theories and automata on infinite trees. Transaction of the AMS 141, 1–35 (1969)
32. Ravi, K., Bloem, R., Somenzi, F.: A comparative study of symbolic algorithms for the computation of fair cycles. In: Johnson, S.D., Hunt Jr., W.A. (eds.) FMCAD 2000. LNCS, vol. 1954, pp. 143–160. Springer, Heidelberg (2000)
33. Safra, S.: On the complexity of ω-automata. In: Proc. 29th FOCS, pp. 319–327 (1988)
34. Sistla, A.P.: Safety, liveness and fairness in temporal logic. Formal Aspects of Computing 6, 495–511 (1994)
35. Sistla, A.P., Clarke, E.M.: The complexity of propositional linear temporal logic. JACM 32, 733–749 (1985)
36. Sistla, A.P., Vardi, M.Y., Wolper, P.: The complementation problem for Büchi automata with applications to temporal logic. In: Brauer, W. (ed.) Automata, Languages and Programming. LNCS, vol. 194, pp. 465–474. Springer, Heidelberg (1985)
37. Street, R.S., Emerson, E.A.: An elementary decision procedure for the μ-calculus. In: Paredaens, J. (ed.) Automata, Languages, and Programming. LNCS, vol. 172, pp. 465–472. Springer, Heidelberg (1984)
38. Touati, H.J., Brayton, R.K., Kurshan, R.: Testing language containment for ω-automata using BDD's. I&C 118(1), 101–109 (1995)
39. Vardi, M.Y., Wolper, P.: An automata-theoretic approach to automatic program verification. In: Proc. 1st LICS, pp. 332–344 (1986)

40. Vardi, M.Y., Wolper, P.: Automata-theoretic techniques for modal logics of programs. JCSS 32(2), 182–221 (1986)
41. Vardi, M.Y., Wolper, P.: Reasoning about infinite computations. I&C 115(1), 1–37 (1994)
42. Wilke, T.: CTL^+ is exponentially more succinct than CTL. In: Pandu Rangan, C., Raman, V., Ramanujam, R. (eds.) Foundations of Software Technology and Theoretical Computer Science. LNCS, vol. 1738, pp. 110–121. Springer, Heidelberg (1999)
43. Wolper, P., Vardi, M.Y., Sistla, A.P.: Reasoning about infinite computation paths. In: Proc. 24th FOCS, pp. 185–194 (1983)
44. Yan, Q.: Lower bounds for complementation of ω-automata via the full automata technique. In: Bugliesi, M., Preneel, B., Sassone, V., Wegener, I. (eds.) ICALP 2006. LNCS, vol. 4052, pp. 589–600. Springer, Heidelberg (2006)

Precise Relational Invariants Through Strategy Iteration

Thomas Gawlitza and Helmut Seidl

TU München, Institut für Informatik, I2
85748 München, Germany
{gawlitza, seidl}@in.tum.de

Abstract. We present a practical algorithm for computing exact least solutions of systems of equations over the rationals with addition, multiplication with positive constants, minimum and maximum. The algorithm is based on strategy improvement combined with solving linear programming problems for each selected strategy. We apply our technique to compute the abstract least fixpoint semantics of affine programs over the relational template constraint matrix domain [20]. In particular, we thus obtain practical algorithms for computing the abstract least fixpoint semantics over the zone and octagon abstract domain.

1 Introduction

Abstract interpretation aims at inferring run-time invariants of programs [5]. Such an invariant may state, e.g., that a variable $\mathbf{x}$ is always contained in the interval $[2, 99]$ whenever a specific program point is reached. In order to compute such invariants, often an abstract semantics is considered which for each program point over-approximates the collecting semantics at this program point. Technically, the abstract semantics is given as the least fixpoint of an appropriate system of in-equations over a complete lattice. *Any* solution of this system provides safe information while only the *precision* of the information returned by the analysis depends on computing as small solutions as possible. In the example of interval analysis, clearly, the smaller the interval which the analysis returns for a given variable at a program point, the better is the information.

Thus, any ambitious program analyzer aims at computing *least* solutions of the systems of in-equations in question. Since ordinary fixpoint iteration does not provide a terminating algorithm in general, *widening* combined with *narrowing* has been proposed to accelerate fixpoint computation and thus guarantee termination of the analysis algorithm at a moderate loss of precision [7,8]. Finding useful widening and narrowing operators, however, is a kind of a black art and it is not a priori clear whether the chosen heuristics will be sufficient for a given program. As an alternative to the general technique of widening and narrowing, we are interested in methods which allow to compute least solutions of in-equations *precisely* – at least for certain interesting cases.

Here, we are interested in computing precise abstract least fixpoint semantics of affine programs over a certain *relational* domain which enables us to describe (certain) relations between the values of program variables. Our key techniques refer to the *template constraint matrix (TCMs) abstract domain* introduced by Sankaranarayanan et al.

J. Duparc and T.A. Henzinger (Eds.): CSL 2007, LNCS 4646, pp. 23–40, 2007.

[20]. Polyhedra of a predefined fixed shape can be represented through elements of this domain. As a particular case, we obtain practical precise algorithms also for intervals, the zone abstract domain and octogons [16,15].

The key idea for our precise algorithm for equations over rationals is *strategy iteration*. Recently, strategy iteration (called policy iteration in [4,10]) has been suggested by Costan et al. as an alternative method for the widening and narrowing approach of Cousot and Cousot [7,8] for computing (hopefully small) solutions of systems of in-equations. Originally, strategy iteration has been introduced by Howard for solving stochastic control problems [13,19] and is also applied to zero-sum two player games [12,18,22] or fixpoints of min-max-plus systems [3]. In general, though, naive strategy iteration will only find some fixpoint — not necessarily the least one [4].

In [4] Costan et al. consider systems of equations over integer intervals. The authors then generalize their idea in [10] to the *zone- and octagon-domain* [16,15] as well as to the *TCM domain* [20]. Their strategy iteration scheme can be applied to monotone self maps F satisfying a *selection property*. This selection property states that the self map F can be considered as the infimum of a set of simpler self maps. Then the selection property enables to compute a fixpoint of F by successively computing least fixpoints of the simpler maps. In certain cases, e.g., for non-expansive self maps on $\overline{\mathbb{R}}^n$, this approach returns the *least* fixpoint. In many practical cases, however, this cannot be guaranteed. In [11], we provide a practical algorithm for computing least solutions of (in-)equations over integer intervals. This algorithm crucially exploits the fact that the interval bounds are integers. Interestingly, it is not applicable to (in-)equations of intervals with rational bounds or multiplication with fractions such as 0.5.

In contrast to [4,10] and similar to [11] we do not apply strategy iteration directly to systems of equations over the interval, the zone/octagon or the TCM domain. Instead, we design just one strategy improvement algorithm for computing least solutions of systems of rational equations. Technically, our algorithm in [11] relies on an *instrumentation* of the underlying lattice [11]. This instrumentation is no longer possible for rationals. Our main technical contribution therefore is to construct a precise strategy iteration *without* extra instrumentation. For solving the subsystems selected by a strategy, we use *linear programming* [14,21]. Using a similar reduction as in [11] for integer intervals, systems of rational equations can be used for interval analysis with rational bounds. Because of lack of space, we do not present this reduction here. Instead, by additionally allowing a (monotone) *linear programming operator* in right-hand sides of equations, we use our techniques for computing abstract least fixpoint semantics of affine programs over the TCM domain. We emphasize that our methods return *precise* answers and do not rely on widening or narrowing. Using the simplex algorithm for solving the occurring linear programs, our algorithm is even uniform, i.e., the number of arithmetic operations does not depend on the sizes of occurring numbers.

The paper is organized as follows. Section 2 introduces systems of rational equations and basic notations. Section 3 presents our strategy improvement algorithm for systems of rational equations. Affine programs are discussed in section 4. There we show how to compute the abstract semantics over the TCM domain using systems of rational equations extended with linear programming operators. Solving these systems is discussed in section 5. Finally, we conclude with section 6.

2 Systems of Rational Equations

We are interested in computing least solutions of systems of equations over the rationals. Since the least upper bound of a bounded set of rationals need not be rational any longer, we consider the complete lattice $\overline{\mathbb{R}} = \mathbb{R} \cup \{-\infty, \infty\}$ of real numbers equipped with the natural ordering $\leq$ and extended with $-\infty$ as least and ∞ as greatest element. On $\overline{\mathbb{R}}$ we consider the operations addition, multiplication with positive constants, minimum "$\wedge$" and maximum "$\vee$" which are extended to operands "$-\infty$" and "∞" as follows. We set $x + (-\infty) = y \cdot (-\infty) = -\infty$ for $x \in \overline{\mathbb{R}}, y \geq 0$; we set $x + \infty = y \cdot \infty = \infty$ for $x \in \overline{\mathbb{R}}, y > 0$; and we set $0 \cdot x = 0$ for $x > -\infty$. For $c > 0$, the operations $+$ and $c\cdot$ distribute over $\vee$ and $\wedge$. Moreover $+$ distributes over $c\cdot$. A system of rational equations is a sequence $\mathbf{x}_1 = e_1, \ldots, \mathbf{x}_n = e_n$ of *rational equations* where $\mathbf{x}_1, \ldots, \mathbf{x}_n$ are pairwise distinct variables, and the right-hand sides are expressions e' built up from constants and variables by means of addition, multiplication with positive constants, minimum "$\wedge$" and maximum "$\vee$". Thus, an expression is defined by the grammar

$$e' ::= a \mid \mathbf{x}_i \mid e'_1 + e'_2 \mid b \cdot e' \mid e'_1 \vee e'_2 \mid e'_1 \wedge e'_2$$

where $a \in \overline{\mathbb{Q}}$, $b \in \mathbb{Q}^{>0}$, $\mathbf{x}_i$ is a variable and e', e'_1, e'_2 are expressions. Note that all occurring constants are rationals. We call a system $\mathcal{E}$ of rational equations *conjunctive* (resp. *disjunctive*) iff no right-hand side of $\mathcal{E}$ contains the maximum-operator "$\vee$" (resp. minimum-operator "$\wedge$"). A system without occurrences of minimum and maximum operators is called *basic*. As usual, every expression e evaluates to a value $[\![e]\!]\mu \in \overline{\mathbb{R}}$ under a *variable assignment* $\mu : \mathbf{X} \to \overline{\mathbb{R}}$. Thus, e.g., $[\![e'_1 + e'_2]\!]\mu = [\![e'_1]\!]\mu + [\![e'_2]\!]\mu$ where e'_1, e'_2 are expressions. Assume that $\mathcal{E}$ denotes the system $\mathbf{x}_1 = e_1, \ldots, \mathbf{x}_n = e_n$ of rational equations. A variable assignment μ which satisfies all equations of $\mathcal{E}$, i.e., $\mu(\mathbf{x}_i) = [\![e_i]\!]\mu$ for $i = 1, \ldots, n$, is called a *solution* of $\mathcal{E}$. Accordingly, we call a variable assignment μ a *pre-solution* of $\mathcal{E}$ iff $\mu(\mathbf{x}_i) \leq [\![e_i]\!]\mu$ for $i = 1, \ldots, n$ and a *post-solution* of $\mathcal{E}$ iff $\mu(\mathbf{x}_i) \geq [\![e_i]\!]\mu$. A solution of $\mathcal{E}$ is a fixpoint of the function given through the right-hand sides of $\mathcal{E}$. Since every right-hand side e_i induces a monotonic function $[\![e_i]\!] : (\mathbf{X} \to \overline{\mathbb{R}}) \to \overline{\mathbb{R}}$, every system $\mathcal{E}$ of rational equations has a least solution. We write $\mu \ll \mu'$ iff $\mu(\mathbf{x}) < \mu'(\mathbf{x})$ for all variables $\mathbf{x}$. Moreover, we write $\underline{-\infty}$ (resp. $\underline{\infty}$) for the variable assignment which maps every variable to $-\infty$ (resp. ∞).

We remark, that least solutions of systems of rational equations cannot effectively be computed by performing ordinary Kleene fixpoint iteration. Even if the least solution is finite, infinitely many iterations may be necessary. A simple example is the equation $\mathbf{x} = 0.5 \cdot \mathbf{x} + 1 \vee 0$, whose least solution maps $\mathbf{x}$ to 2.

As a start, we consider disjunctive systems of rational equations. We recall from [10] that computing the least solution for such a system can be reduced to solving linear programs (LPs). For a set S and a matrix $A \in S^{m \times n}$, we write $A_{i\cdot}$ for the i-th row of A and $A_{\cdot j}$ for the j-th column of A. Accordingly $A_{i \cdot j}$ denotes the element in row i and column j. As usual we identify $S^{m \times 1}$ with S^m. We denote the transposed of A by A^T. For $A \in \mathbb{R}^{m \times n}$ and $c \in \mathbb{R}^n$ we define the operator $LP_{A,c} : \overline{\mathbb{R}}^m \to \overline{\mathbb{R}}$ by

$$LP_{A,c}(b) = \bigvee \{c^T x \mid x \in \mathbb{R}^n, Ax \leq b\}$$

for $b \in \mathbb{R}^m$. This operator is monotone and represents a linear program. If the program is *infeasible*, i.e., $Ax \leq b$ for no x, $LP_{A,c}(b)$ returns $-\infty$. If the program is unbounded, i.e., for all $r \in \mathbb{R}$, $c^T x > r$ for some x satisfying $Ax \leq b$, $LP_{A,c}(b)$ returns ∞.

Our goal is to compute the least solution of a system $\mathcal{E}$ of disjunctive rational equations. For simplicity, we assume that all maximum operators in right-hand sides of $\mathcal{E}$ occur on top-level such as in:

$$\mathbf{x}_1 = \tfrac{1}{3}\mathbf{x}_2 + 3 \vee 1 \qquad \mathbf{x}_2 = 2\mathbf{x}_1 - 6 \vee 5\mathbf{x}_2 - 1$$

Assume that $\mathcal{E}$ has n variables and the least solution is given by μ^*. In the first step, we compute the set of variables $\mathbf{x}_i$ with $\mu^*(\mathbf{x}_i) = -\infty$. This can be done in time $\mathcal{O}(n \cdot |\mathcal{E}|)$ by performing n rounds of fixpoint iteration which results in a variable assignment μ with $\mu(\mathbf{x}) = -\infty$ iff $\mu^*(\mathbf{x}) = -\infty$ for all variables $\mathbf{x}$. Accordingly, the least solution of the example system returns values exceeding $-\infty$ for both $\mathbf{x}_1$ and $\mathbf{x}_2$.

Having determined the set of variables $\mathbf{x}_i$ with $\mu^*(\mathbf{x}_i) = -\infty$, we can remove these variables from our system of equations. Therefore, we now w.l.o.g. may assume that $\mu^* \gg \underline{-\infty}$. Also, we may assume that the constant $-\infty$ does not occur in $\mathcal{E}$. For a moment assume furthermore that $\mu^* \ll \underline{\infty}$. From the set of equations we can extract a set of constraints (here in-equations) which are satisfied exactly by all post-solutions of $\mathcal{E}$. In our example these are given by:

$$\mathbf{x}_1 \geq \tfrac{1}{3}\mathbf{x}_2 + 3 \qquad \mathbf{x}_1 \geq 1 \qquad \mathbf{x}_2 \geq 2\mathbf{x}_1 - 6 \qquad \mathbf{x}_2 \geq 5\mathbf{x}_2 - 1$$

Since $\underline{-\infty} \ll \mu^* \ll \underline{\infty}$, the least solution μ^* can be characterized as the (unique) vector $x = (x_{1\cdot}, \ldots, x_{n\cdot}) \in \mathbb{R}^n$ that represents a solution of the above constraints and for which $-(x_{1\cdot} + \cdots + x_{n\cdot})$ is maximal. Thus, x can be determined by solving the appropriate LP. In the example, this results in the vector $x = (3, 0)$.

In general, it might not be the case that $\mu^* \ll \underline{\infty}$. If this is not the case, the LP corresponding to the system $\mathcal{E}$ is not feasible. In order to deal with this case as well, we consider the variable dependency graph $G = (V, \rightarrow)$ of $\mathcal{E}$ where the set of vertices V is the set of variables and the set of edges $\rightarrow \subseteq V^2$ is the smallest set s.t. $\mathbf{x}_j \rightarrow \mathbf{x}_i$ iff $\mathbf{x}_i = e_i \in \mathcal{E}$ and $\mathbf{x}_j$ occurs in e_i. Since $\mu^* \gg \underline{-\infty}$ and $-\infty$ does not occur as a constant in $\mathcal{E}$, $[\![e]\!]\mu^* > -\infty$ for every subexpression occurring in $\mathcal{E}$. Thus, $\mu^*(\mathbf{x}_j) = \infty$ and $\mathbf{x}_j \rightarrow^* \mathbf{x}_i$ implies $\mu^*(\mathbf{x}_i) = \infty$. In particular if $\mu^*(\mathbf{x}_i) = \infty$ for some variable $\mathbf{x}_i$ of a strongly connected component (SCC), then $\mu^*(\mathbf{x}_j) = \infty$ for every variable $\mathbf{x}_j$ of the same SCC. Therefore, we proceed by processing one maximal SCC after the other. Thereby we start with a maximal SCC $G' = (V', \rightarrow')$ without in-going edges. The least solution of the subsystem of $\mathcal{E}$ described by G' can be computed using linear programming as sketched above. If the corresponding LP is infeasible, then $\mu^*(\mathbf{x}_i) = \infty$ for all variables $\mathbf{x}_i$ of the SCC and in fact for all variables $\mathbf{x}_i$ reachable from this SCC. The corresponding LP cannot be unbounded, since this would be a contradiction to $\mu^* \gg \underline{-\infty}$.

Having computed the values of all variables in the first maximal SCC, we replace all occurrences of these variables in the remaining equations by their values and proceed with another maximal SCC without in-going edges. In essence, this is the algorithm of [10] simplified for systems of rational constraints. Summarizing, we have:

Theorem 1 (Costan et al. 2007). *The least solution of a disjunctive system $\mathcal{E}$ of rational equations can be computed by solving linearly many LPs of polynomial sizes.* □

This theorem results in a polynomial algorithm if we apply interior point methods for solving the occurring LPs [14,21,1]. Note, however, that the run-time then crucially depends on the sizes of occurring numbers. At the danger of an exponential run-time in contrived cases, we can also rely on the simplex algorithm instead: the advantage of the latter algorithm is that its run-time is *uniform*, i.e., independent of the sizes of occurring numbers (given that arithmetic operations, comparison, storage and retrieval for numbers are counted for $\mathcal{O}(1)$).

3 Least Solutions of Systems of Rational Equations

In this section we provide techniques for computing least solutions of systems of rational equations. Our techniques are based on (max-) strategy improvement. Let $M_\vee(\mathcal{E})$ denote the set of all maximum subexpressions occurring in $\mathcal{E}$. A *(max-)strategy* π is a function mapping every expression $e_1 \vee e_2$ in $M_\vee(\mathcal{E})$ to one of the subexpressions e_1, e_2. Given a max-strategy π together with an expression e, we write $e\,\pi$ for the expression obtained by recursively replacing every maximum expression in $\mathcal{E}$ by the respective subexpression selected by π. Assuming that $\mathcal{E}$ is the system $\mathbf{x}_i = e_i, i = 1, \ldots, n$, we write $\mathcal{E}(\pi)$ for the system $\mathbf{x}_i = e_i\,\pi, i = 1, \ldots, n$. Thus $\mathcal{E}(\pi)$ is extracted from $\mathcal{E}$ via the strategy π. Note that $\mathcal{E}(\pi)$ is conjunctive.

Example 1. Consider the system $\mathcal{E}$ of rational equations given by the equation $\mathbf{x} = (2 \cdot \mathbf{x} - 2 \wedge 10) \vee 4$. Consider the max-strategy π which maps the top-level expression $(2 \cdot \mathbf{x} - 2 \wedge 10) \vee 4$ to the expression 4. Then the system $\mathcal{E}(\pi)$ of conjunctive equations is given by the equation $\mathbf{x} = 4$. □

Assume that μ^* denotes the least solution of the system $\mathcal{E}$ of rational equations. Our goal is to construct a strategy improvement algorithm for computing μ^*. The algorithm maintains a current max-strategy π and a current variable assignment μ. The current variable assignment μ is a pre-solution of $\mathcal{E}$ which is less than or equal to μ^*. For a current max-strategy π and a current variable assignment μ, the algorithm performs an accelerated least fixpoint computation on the system $\mathcal{E}(\pi)$ which starts with μ. This fixpoint computation results in a variable assignment μ' which is a a solution of $\mathcal{E}(\pi)$ and a pre-solution of $\mathcal{E}$ and moreover is still less than or equal to μ^*. If μ' is not a solution of $\mathcal{E}$, a new improved max-strategy π' is determined and the algorithm re-starts with π' as current max-strategy and μ' as current variable assignment. These steps are repeated until the least fixpoint of $\mathcal{E}$ is reached.

Given a current max-strategy π and a solution μ of $\mathcal{E}(\pi)$, we pursue the policy to improve π at all expressions $e'_1 \vee e'_2$ where $[\![e'_1 \vee e'_2]\!]\mu > [\![(e'_1 \vee e'_2)\,\pi]\!]\mu$ simultaneously. Formally, we introduce an improvement operator $P_\vee$ by:

$$P_\vee(\pi, \mu)(e_1 \vee e_2) = \begin{cases} e_1 & \text{if } [\![e_1]\!]\mu > [\![e_2]\!]\mu \\ e_2 & \text{if } [\![e_1]\!]\mu < [\![e_2]\!]\mu \\ \pi(e_1 \vee e_2) & \text{if } [\![e_1]\!]\mu = [\![e_2]\!]\mu \end{cases}$$

Note that the strategy $P_\vee(\pi, \mu)$ differs from π only if μ is not a solution of $\mathcal{E}$.

Example 2. Consider the system $\mathcal{E}$ and the max-strategy π from example 1. Let μ denote the unique solution of $\mathcal{E}(\pi)$, i.e., $\mu(\mathbf{x}) = 4$. The variable assignment μ is less than

Algorithm 1. Least Solution of The System $\mathcal{E}$ of Rational Equations

$\pi \leftarrow \pi_{-\infty}$; $\mu \leftarrow \underline{-\infty}$;
while (μ is not a solution of $\mathcal{E}$) {
  $\pi \leftarrow P_{\vee}(\pi, \mu)$; $\mu \leftarrow$ least solution of $\mathcal{E}(\pi)$ that is greater than or equal to μ;
}
return μ

or equal to the least solution of $\mathcal{E}$ and the max-strategy $\pi' := P_{\vee}(\pi, \mu) \neq \pi$ leads to the system $\mathcal{E}(\pi')$ given by the equation $\mathbf{x} = (2 \cdot \mathbf{x} - 2 \wedge 10)$. □

In order to formulate our strategy improvement algorithm, we do not consider the original system $\mathcal{E}$. Instead, we replace every equation $\mathbf{x}_i = e_i$ of $\mathcal{E}$ by $\mathbf{x}_i = e_i \vee -\infty$. For simplicity, we denote the resulting system again by $\mathcal{E}$. Our algorithm starts with the max-strategy that maps every top-level expression to $-\infty$. We denote this max-strategy by $\pi_{-\infty}$. Then, our strategy improvement algorithm is given as algorithm 1.

Clearly, if algorithm 1 terminates, it returns a solution of $\mathcal{E}$. It returns the *least* one, since for every strategy π the least solution μ' of $\mathcal{E}(\pi)$ with $\mu' \geq \mu$ is less than or equal to the least solution μ'' of $\mathcal{E}$ with $\mu'' \geq \mu$. Therefore the value of the program variable μ is always less than or equal to μ^*.

Two things remain to be explained. First, we need an algorithm for computing the least solution μ' of a conjunctive system such as $\mathcal{E}(\pi)$ with $\mu' \geq \mu$ for a given variable assignment μ. Here, we will exploit that every μ to be considered is not arbitrary but a *consistent pre-solution* (see below) of $\mathcal{E}(\pi)$. Secondly, we must prove that every strategy π occurs only finitely often during the strategy iteration. Before going further, we illustrate algorithm 1 by an example.

Example 3. Consider the system $\mathcal{E}$ of rational equations shown on the right. Algorithm 1 computes the least solution μ^* using 4 max-strategies $\pi_1, \ldots, \pi_4$. The strategies π_i lead to the systems $\mathcal{E}(\pi_i)$ shown on the right. Let us consider the system $\mathcal{E}(\pi_3)$. The only solution maps every variable to 2. Thus, the improvement step leads to the system $\mathcal{E}(\pi_4)$ for which we must compute the least solution which maps every variable to values greater than or equal to 2. This solution maps $\mathbf{x}_1$ to 500 and $\mathbf{x}_2$ to 100 and is also the least solution of $\mathcal{E}$. □

$\mathcal{E}$	$\equiv$	$\mathbf{x}_1 = 0.8 \cdot \mathbf{x}_1 + \mathbf{x}_2 \vee 2 \vee -\infty$	$\mathbf{x}_2 = (\mathbf{x}_2 + 1 \wedge 100) \vee \mathbf{x}_1 \vee -\infty$
$\mathcal{E}(\pi_1)$	$\equiv$	$\mathbf{x}_1 = -\infty$	$\mathbf{x}_2 = -\infty$
$\mathcal{E}(\pi_2)$	$\equiv$	$\mathbf{x}_1 = 2$	$\mathbf{x}_2 = -\infty$
$\mathcal{E}(\pi_3)$	$\equiv$	$\mathbf{x}_1 = 2$	$\mathbf{x}_2 = \mathbf{x}_1$
$\mathcal{E}(\pi_4)$	$\equiv$	$\mathbf{x}_1 = 0.8 \cdot \mathbf{x}_1 + \mathbf{x}_2$	$\mathbf{x}_2 = \mathbf{x}_2 + 1 \wedge 100$

Assume that $\mathcal{E}$ denotes the conjunctive system $\mathbf{x}_i = e_i$, $i = 1, \ldots, n$ and that μ is a pre-fixpoint of $\mathcal{E}$. We define the set $\mathcal{D}_\mu(\mathcal{E})$ of *derived constraints* as the smallest set of constraints of the form $\mathbf{x} \leq e$ such that

- $\mathbf{x}_i \leq e' \in \mathcal{D}_\mu(\mathcal{E})$ whenever $\mathbf{x}_i = e_i$ with $\mu(\mathbf{x}_i) < \infty$ can be rewritten (using distributivity) into $\mathbf{x}_i = e' \wedge e''$ where e' does not contain $\wedge$-operators;
- $\mathbf{x}_i \leq \frac{1}{1-c} \cdot e \in \mathcal{D}_\mu(\mathcal{E})$ whenever $\mathbf{x}_i \leq c \cdot \mathbf{x}_i + e \in \mathcal{D}_\mu(\mathcal{E})$ where $0 < c < 1$; and
- $\mathbf{x}_i \leq c \cdot e' + e \in \mathcal{D}_\mu(\mathcal{E})$ whenever $\mathbf{x}_i \leq c \cdot \mathbf{x}_j + e \in \mathcal{D}_\mu(\mathcal{E})$ and $\mathbf{x}_j \leq e' \in \mathcal{D}_\mu(\mathcal{E})$.

Lemma 1. *Assume that μ is a pre-solution of the conjunctive system $\mathcal{E}$. Then $\mu(\mathbf{x}) \leq [\![e]\!]\mu$ for every $\mathbf{x} \leq e \in \mathcal{D}_\mu(\mathcal{E})$.* □

We call the pre-solution μ of $\mathcal{E}$ *($\mathcal{E}$-)consistent* iff

- $[\![e]\!]\mu = -\infty$ implies $e = -\infty$ for every expression e occurring in $\mathcal{E}$.
- $\mathcal{D}_\mu(\mathcal{E})$ does not contain a derived constraint $\mathbf{x}_i \leq c \cdot \mathbf{x}_i + e \in \mathcal{D}_\mu(\mathcal{E})$ with $c \geq 1$ and $\mu(\mathbf{x}_i) = [\![c \cdot \mathbf{x}_i + e]\!]\mu$. (We call such a constraint μ-critical).

Example 4. The pre-solution $\mu = \{\mathbf{x}_1 \mapsto 2, \mathbf{x}_2 \mapsto 2\}$ is *not* $\mathcal{E}$-consistent for the conjunctive system $\mathcal{E}$ given by the equations $\mathbf{x}_1 = 0.75{\cdot}\mathbf{x}_1 + 0.25{\cdot}\mathbf{x}_2$ and $\mathbf{x}_2 = 4{\cdot}\mathbf{x}_1 - 6$, because $\mathbf{x}_1 \leq 1.75 \cdot \mathbf{x}_1 - 1.5 \in \mathcal{D}_\mu(\mathcal{E})$ and $\mu(\mathbf{x}_1) = 2 = [\![1.75 \cdot \mathbf{x}_1 - 1.5]\!]\mu$. However, the pre-solution $\mu' = \{\mathbf{x}_1 \mapsto 3, \mathbf{x}_2 \mapsto 4\}$ is $\mathcal{E}$-consistent. □

We claim that algorithm 1 computes least solutions μ' of $\mathcal{E}(\pi)$ with $\mu' \geq \mu$ for variable assignments μ which are consistent pre-solutions of $\mathcal{E}$, only. Since $\underline{-\infty}$ is a consistent pre-solution of $\mathcal{E}(\pi_{-\infty})$, this follows inductively using the following two lemmas.

Lemma 2. *Let $\mathcal{E}$ be a conjunctive system and μ be a consistent pre-solution of $\mathcal{E}$. Every pre-solution $\mu' \geq \mu$ of $\mathcal{E}$ is consistent.* □

Lemma 3. *Assume that $\mathcal{E}$ is a system, π a max-strategy, μ a consistent pre-solution of $\mathcal{E}(\pi)$ and $\pi' = P_\vee(\pi, \mu)$. Then μ is a consistent pre-solution of $\mathcal{E}(\pi')$.* □

It remains to provide a method for computing the least solution μ' with $\mu' \geq \mu$ of a conjunctive system $\mathcal{E}$ for a *consistent* pre-solution μ of $\mathcal{E}$.

3.1 Systems of Conjunctive Equations

In this subsection we consider conjunctive systems $\mathcal{E}$ of rational equations. Of a particular interest are *feasible* systems. We call $\mathcal{E}$ *feasible* iff there exists a consistent pre-solution $\mu \ll \underline{\infty}$ of $\mathcal{E}$. It turns out that feasible systems enjoy the property to have a *least consistent* solution. The main challenge and the goal of this section therefore is to derive a method for computing the least consistent solution of feasible systems. This method then will be used to compute the least solution μ' of $\mathcal{E}$ with $\mu' \geq \mu$ provided that μ is a consistent pre-solution of $\mathcal{E}$ with $\mu \ll \underline{\infty}$.

The restriction to consistent pre-solutions with $\mu \ll \underline{\infty}$ can be lifted as follows. Assume that μ denotes an arbitrary consistent pre-solution of $\mathcal{E}$. Let $\mathbf{X}^\infty$ be the set of variables $\mathbf{x}$ with $\mu(\mathbf{x}) = \infty$. Let $\mathcal{E}'$ denote the system obtained from $\mathcal{E}$ by (1) removing every equation $\mathbf{x} = e$ with $\mathbf{x} \in \mathbf{X}^\infty$ and (2) replacing every variable $\mathbf{x} \in \mathbf{X}^\infty$ by the constant ∞. Then $\mu|_{\mathbf{X}\setminus\mathbf{X}^\infty}$ is a consistent pre-solution of $\mathcal{E}'$ with $\mu|_{\mathbf{X}\setminus\mathbf{X}^\infty} \ll \underline{\infty}$ and thus the least solution of $\mathcal{E}'$ with $\mu' \geq \mu|_{\mathbf{X}\setminus\mathbf{X}^\infty}$ is the least consistent solution of $\mathcal{E}'$. Finally, the least solution μ^* of $\mathcal{E}$ with $\mu^* \geq \mu$ is then given by $\mu^*(\mathbf{x}) = \infty$ for $\mathbf{x} \in \mathbf{X}^\infty$ and $\mu^*(\mathbf{x}) = \mu'(\mathbf{x})$ for $\mathbf{x} \notin \mathbf{X}^\infty$. In the following, we only consider *feasible* systems of conjunctive rational equations. Furthermore, we assume that the constant $-\infty$ does not occur in the systems under consideration. In a first step we consider systems of *basic* equations, i.e., systems in which neither $\vee$ nor $\wedge$ occur. The following lemma implies that every feasible system of *basic* equations has a least consistent solution.

Lemma 4. *Assume that $\mathcal{E}$ is a feasible system of basic equations. Assume that $\mu \ll \underline{\infty}$ is a pre-solution of $\mathcal{E}$ and μ' a consistent solution of $\mathcal{E}$. Then $\mu \leq \mu'$.*

Proof. Assume that $\mathcal{E}$ denotes the system $\mathbf{x}_i = e_i, i = 1, \ldots, n$. We proceed by induction on the number of variables occurring in right-hand sides of $\mathcal{E}$. If no variable occurs in a right-hand side of $\mathcal{E}$, then μ' is the only solution of $\mathcal{E}$. Thus $\mu \leq \mu'$, since otherwise μ would not be a pre-solution of $\mathcal{E}$. For the induction step, consider an equation $\mathbf{x}_i = e_i$ of $\mathcal{E}$ where $\mathbf{x}_i$ occurs in a right-hand side e_j of $\mathcal{E}$.

Case 1: e_i does not contain $\mathbf{x}_i$

We obtain a system $\mathcal{E}'$ from $\mathcal{E}$ by replacing all occurrences of $\mathbf{x}_i$ in right-hand sides with e_i. Since $\mathcal{D}_{\mu'}(\mathcal{E}') \subseteq \mathcal{D}_{\mu'}(\mathcal{E})$, μ' is a consistent solution of $\mathcal{E}'$. Since μ is also a pre-solution of $\mathcal{E}'$ and the system $\mathcal{E}'$ contains one variable less in right-hand sides we get $\mu \leq \mu'$ by induction hypothesis.

Case 2: e_i contains $\mathbf{x}_i$

Using distributivity, we rewrite the equation $\mathbf{x}_i = e_i$ equivalently into

$$\mathbf{x}_i = c \cdot \mathbf{x}_i + e$$

where $c \in \mathbb{R}^{>0}$ and e does not contain $\mathbf{x}_i$. Then we obtain the systems $\mathcal{E}_1$ and $\mathcal{E}_2$ from $\mathcal{E}$ by replacing the equation $\mathbf{x}_i = c \cdot \mathbf{x}_i + e$ by $\mathbf{x}_i = \infty$ and $\mathbf{x}_i = \frac{1}{1-c} \cdot e$, respectively. Then we obtain systems $\mathcal{E}_1'$ and $\mathcal{E}_2'$ from $\mathcal{E}_1$ and $\mathcal{E}_2$ by replacing all occurrences of the variable $\mathbf{x}_i$ in right-hand sides with ∞ and $\frac{1}{1-c} \cdot e$, respectively.

First consider the case $c < 1$. Since μ' is consistent we get that $\mu'(\mathbf{x}_i) > -\infty$. Thus, $\mu'(\mathbf{x}_i) \in \{[\![\frac{1}{1-c} \cdot e]\!]\mu', \infty\}$. If $\mu'(\mathbf{x}_i) = \infty$, we conclude that, since $\mathcal{D}_{\mu'}(\mathcal{E}_1') \subseteq \mathcal{D}_{\mu'}(\mathcal{E}_1) \subseteq \mathcal{D}_{\mu'}(\mathcal{E})$, μ' is a consistent solution of $\mathcal{E}_1'$. Since μ is a pre-solution of $\mathcal{E}_1'$ and $\mathcal{E}_1'$ has at least one variable less in right-hand sides than $\mathcal{E}$, we get $\mu \leq \mu'$ by induction hypothesis. If $\mu'(\mathbf{x}_i) = [\![\frac{1}{1-c} \cdot e]\!]\mu'$, we conclude that since $\mathcal{D}_{\mu'}(\mathcal{E}_2') \subseteq \mathcal{D}_{\mu'}(\mathcal{E}_2) \subseteq \mathcal{D}_{\mu'}(\mathcal{E})$, μ' is a consistent solution of $\mathcal{E}_2'$. Since μ is a pre-solution of $\mathcal{E}_2'$ and $\mathcal{E}_2'$ has at least one variable less in right-hand sides than $\mathcal{E}$, we get $\mu \leq \mu'$ by induction hypothesis.

Now consider the case $c \geq 1$. Again, $\mu'(\mathbf{x}_i) > -\infty$, since μ' is consistent. It follows $\mu'(\mathbf{x}_i) = \infty$. Otherwise μ' would not be consistent, since then $\mu'(\mathbf{x}_i) = [\![c \cdot \mathbf{x}_i + e]\!]\mu'$ and thus $\mathbf{x}_i \leq c \cdot \mathbf{x}_i + e \in \mathcal{D}_{\mu'}(\mathcal{E})$ would be μ'-critical. Note that, since $\mathcal{D}_{\mu'}(\mathcal{E}_1') \subseteq \mathcal{D}_{\mu'}(\mathcal{E}_1) \subseteq \mathcal{D}_{\mu'}(\mathcal{E})$, μ' is a consistent solution of $\mathcal{E}_1'$. Since μ is a pre-solution of $\mathcal{E}_1'$ and $\mathcal{E}_1'$ has at least one variable less in right-hand sides than $\mathcal{E}$, we get $\mu \leq \mu'$ by induction hypothesis. □

We now extend this result to systems of conjunctive equations.

Lemma 5. *Assume that $\mathcal{E}$ is a feasible system of conjunctive equations. Assume that $\mu \ll \underline{\infty}$ is a pre-solution of $\mathcal{E}$ and μ' is a consistent solution of $\mathcal{E}$. Then $\mu \leq \mu'$. Moreover, there exists at most one consistent solution μ' with $\mu' \ll \underline{\infty}$.*

Proof. There exists a min-strategy (min-strategies are defined analog to max-strategies) π s.t. μ' is a consistent solution of the system $\mathcal{E}(\pi)$ of basic equations. Then $\mu \ll \underline{\infty}$ is a pre-solution of $\mathcal{E}(\pi)$ by monotonicity. Thus, $\mu \leq \mu'$ by lemma 4. In order to show the second statement, assume that $\mu' \ll \underline{\infty}$ and let $\mu'' \ll \underline{\infty}$ denote a consistent solution of $\mathcal{E}$. Then $\mu' \leq \mu''$ and $\mu'' \leq \mu'$ implying $\mu' = \mu''$. □

Using lemma 5 we conclude that every feasible conjunctive system has a least consistent solution. The following theorem states this fact and moreover observes that the least consistent solution is given by the least solution which is bounded below by a consistent pre-solution μ with $\mu \ll \underline{\infty}$.

Theorem 2. *Assume that $\mathcal{E}$ is a feasible conjunctive system, and $\mu \ll \underline{\infty}$ is a consistent pre-solution of $\mathcal{E}$. Then there exists a least consistent solution μ^* of $\mathcal{E}$ which equals the least solution μ' of $\mathcal{E}$ with $\mu' \geq \mu$.* □

In order to simplify complexity estimations, we state the following corollary explicitly.

Corollary 1. *Assume that $\mathcal{E}$ denotes a conjunctive system with n variables. Let $(\mu_i)_{i \in \mathbb{N}}$ denote an increasing sequence of consistent pre-solutions of $\mathcal{E}$. Let μ_i' denote the least solution of $\mathcal{E}$ with $\mu_i' \geq \mu_i$ for $i \in \mathbb{N}$. Then $|\{\mu_i' \mid i \in \mathbb{N}\}| \leq n$.* □

We now use the results above in order to compute the least consistent solution μ^* of the feasible conjunctive system $\mathcal{E}$. We first restrict our consideration to the case $\mu^* \ll \underline{\infty}$. Since, by lemma 5, μ^* is the *only* solution of $\mathcal{E}$ with $\mu \leq \mu^* \ll \underline{\infty}$, μ^* is in particular the *greatest* solution of $\mathcal{E}$ with $\mu^* \ll \underline{\infty}$. We compute μ^* by solving a linear program which maximizes the sum of the values of the variables occurring in $\mathcal{E}$. Assume w.l.o.g. that $\mathcal{E}$ is given by $\mathbf{x}_i = e_i^{(1)} \wedge \cdots \wedge e_i^{(k_i)}$ for $i = 1, \ldots, n$ where $e_i^{(j)}$ do not contain $\wedge$-operators, i.e., $\mathcal{E}$ is in normal form. (This form can be achieved from a general form in linear time by introducing at most $m_\wedge$ auxiliary variables and equations, where $m_\wedge$ denotes the number of $\wedge$-subexpressions.) We define $\mathcal{C}_\mathcal{E}$ as the following system of rational *constraints*:

$$\mathbf{x}_i \leq e_i^{(j)} \qquad \text{for } i = 1, \ldots, n,\ j = 1, \ldots, k_i.$$

Then we must maximize $\sum_{\mathbf{x} \in \mathbf{X}} \mu(\mathbf{x})$ under the restriction that μ is a solution of $\mathcal{C}_\mathcal{E}$.

Lemma 6. *Assume that $\mathcal{E}$ denotes a feasible conjunctive system and that $\mu^* \ll \underline{\infty}$ denotes the least consistent solution of $\mathcal{E}$. Then there exists a solution μ' of $\mathcal{C}_\mathcal{E}$ with $\mu' \ll \underline{\infty}$ which maximizes the sum $\sum_{\mathbf{x} \in \mathbf{X}} \mu'(\mathbf{x})$. Furthermore, $\mu' = \mu^*$. Thus, μ^* can be computed by solving a single LP which can be extracted from $\mathcal{E}$ in linear time.* □

Example 5. Consider the system $\mathcal{E}(\pi_4)$ from example 3. Our goal is to compute the least solution μ' with $\mu' \geq \mu = \{\mathbf{x}_1 \mapsto 2, \mathbf{x}_2 \mapsto 2\}$. Theorem 2 implies that μ' is given as the *least consistent solution*. Assuming that $\mu' \ll \underline{\infty}$, i.e., μ' maps all variables to finite values, lemma 6 implies that μ' is given as the *unique* solution of the LP

$$\bigvee \{\mathbf{x}_1 + \mathbf{x}_2 \mid \mathbf{x}_1 \leq 0.8 \cdot \mathbf{x}_1 + \mathbf{x}_2,\ \mathbf{x}_2 \leq \mathbf{x}_2 + 1,\ \mathbf{x}_2 \leq 100\}.$$

Thus, μ' maps $\mathbf{x}_1$ to 500 and $\mathbf{x}_2$ to 100. □

Until now, we can only deal with feasible systems $\mathcal{E}$ whose least consistent solution μ^* does not map any variable to ∞. In order to lift this restriction, we first have to determine the set $\mathbf{X}^{*\infty} := \{\mathbf{x} \in \mathbf{X} \mid \mu^*(\mathbf{x}) = \infty\}$. Given $\mathbf{X}^{*\infty}$ we can remove each equation $\mathbf{x}_i = e_i$ with $\mathbf{x}_i \in \mathbf{X}^{*\infty}$ and thus obtain a system whose least consistent solution $\mu^{*\prime}$ does not map any variable to ∞. Moreover $\mu * |_{\mathbf{X} \setminus \mathbf{X}^{*\infty}} = \mu^{*\prime}$.

We reduce the problem of determining $\mathbf{X}^{*\infty}$ to the problem of computing the greatest solution of an *abstracted* system of rational equations for which we know that the greatest solution does not map any variable to ∞ or $-\infty$. Therefore, this *abstracted* system can be solved again by linear programming. We first define a transformation $[\cdot]^\infty$ which maps the constant ∞ to 1 and every finite constant to 0 while preserving all multiplicative factors and variable occurrences (recall that $-\infty$ does not occur in the expressions under consideration):

$$\begin{array}{llllll} [\mathbf{x}]^\infty &= \mathbf{x} & [a]^\infty &= 0 & [\infty]^\infty &= 1 \\ [c \cdot e]^\infty &= c \cdot [e]^\infty & [e_1 + e_2]^\infty &= [e_1]^\infty + [e_2]^\infty & [e_1 \wedge e_2]^\infty &= [e_1]^\infty \wedge [e_2]^\infty \end{array}$$

where $a < \infty$, $0 < c < \infty$, $\mathbf{x}$ is a variable and e, e_1, e_2 are expressions. Assuming that $\mathcal{E}$ denotes the system $\mathbf{x}_1 = e_1, \ldots, \mathbf{x}_n = e_n$ we write $[\mathcal{E}]^\infty$ for the system

$$\mathbf{x}_1 = [e_1]^\infty \wedge 1, \ldots, \mathbf{x}_n = [e_n]^\infty \wedge 1.$$

The next lemma states that the set $\mathbf{X}^{*\infty}$ can be read off the greatest solution μ^∞ of $[\mathcal{E}]^\infty$. Thus our problem reduces to computing μ^∞. Since by construction $0 \leq \mu^\infty(\mathbf{x}) \leq 1$ for every variable $\mathbf{x}$, this can be done using linear programming, i.e., we have to compute a solution μ^∞ of $\mathcal{C}_{\mathcal{E}^\infty}$ which maximizes the sum $\sum_{\mathbf{x} \in \mathbf{X}} \mu^\infty(\mathbf{x})$. There exists only one such solution and this solution is the greatest solution of $\mathcal{E}^\infty$. We have:

Lemma 7. *Assume that μ^* denotes the least consistent solution of the feasible conjunctive system $\mathcal{E}$. Let μ^∞ denote the greatest solution of $[\mathcal{E}]^\infty$. Then $\mu^*(\mathbf{x}) = \infty$ iff $\mu^\infty(\mathbf{x}) > 0$ for all variables $\mathbf{x}$. Furthermore, μ^∞ and thus $\{\mathbf{x} \in \mathbf{X} \mid \mu^*(\mathbf{x}) = \infty\}$ can be computed by solving a single LP which can be extracted from $\mathcal{E}$ in linear time.* □

Example 6. Consider again the system $\mathcal{E}(\pi_4)$ from example 3. As we already know, the system is feasible and we are interested in computing the least consistent solution μ^*. In order to compute the set of variables which μ^* maps to ∞, we construct the abstracted system $x_1 = 0.8 \cdot x_1 + x_2 \wedge 1, \quad x_2 = x_2 \wedge 0 \wedge 1$ for which we must compute the greatest solution μ^∞. Then μ^∞ can be computed using linear programming. More exactly, μ^∞ is given as the *unique determined* solution which maximizes the sum $\sum_{\mathbf{x} \in \mathbf{X}} \mu^\infty(\mathbf{x})$. Here, obviously, μ^∞ maps every variable to 0. Thus, according to lemma 7, μ^* maps all variables to finite values — implying that the finiteness assumption in example 5 is justified. □

In conclusion, our method for computing the least consistent solution μ^* of a feasible conjunctive system $\mathcal{E}$ works as follows. Using lemma 7, we first determine the set $\mathbf{X}^{*\infty}$ of variables $\mathbf{x}$ with $\mu^*(\mathbf{x}) = \infty$. After that we obtain a system $\mathcal{E}'$ of conjunctive equations from $\mathcal{E}$ by (1) removing all equations $\mathbf{x} = e$ with $\mu^*(\mathbf{x}) = \infty$ and (2) replacing all expressions e with $[\![e]\!]\mu$ by ∞. Then $\mu^*|_{\mathbf{X} \setminus \mathbf{X}^{*\infty}}$ is the least consistent solution of $\mathcal{E}'$ and moreover $\mu^*|_{\mathbf{X} \setminus \mathbf{X}^{*\infty}} \ll \underline{\infty}$. By lemma 6, $\mu^*|_{\mathbf{X} \setminus \mathbf{X}^{*\infty}}$ and thus μ^* can be determined by solving an appropriate LP. We arrive at our result for feasible conjunctive systems:

Theorem 3. *The least consistent solution of a feasible conjunctive system $\mathcal{E}$ can be computed by solving two LPs each of which can be extracted from $\mathcal{E}$ in linear time.* □

3.2 The Result

Consider again algorithm 1. Assume w.l.o.g. that $\mathcal{E}$ denotes the system $\mathbf{x}_i = e_i \vee -\infty$, $i = 1, \ldots, n$ with least solution μ^* and $m_\vee = m + n$ $\vee$-expressions. In order to give a precise characterization of the run-time, let $\Pi(m_\vee)$ denote the maximal number of updates of strategies necessary for systems with $m_\vee$ maximum expressions.

Let π_i denote the max-strategy π after the execution of the first statement in the i-th iteration. Accordingly, let μ_i denote the variable assignment μ at this point and let μ'_i denote the variable assignment μ after the i-th iteration. It remains to show that algorithm 1 always terminates. Lemmas 2 and 3 imply that μ_i is a consistent pre-solution of $\mathcal{E}(\pi_i)$ with $\mu_i \leq \mu^*$ for every i. By theorem 3 μ'_i can be computed by solving two appropriate LP problems extracted from $\mathcal{E}$. The sequence (μ_i) is strictly increasing until the least solution is reached. Moreover, every strategy π is contained at most n times in the sequence (π_i). Otherwise, there would be more than n least solutions of the conjunctive system $\mathcal{E}(\pi)$ exceeding some consistent pre-solution contained in (μ_i). This would be a contradiction to corollary 1. Therefore, the number of iterations of the loop executed by algorithm 1 is bounded by $n \cdot \Pi(m+n)$. Summarizing, we have:

Theorem 4. *The least solution of a system $\mathcal{E}$ of rational equations with n variables and m maximum expressions can be computed by solving $2n \cdot \Pi(m+n)$ LPs each of which can be extracted from $\mathcal{E}$ in linear time.* □

All practical experiments with strategy iteration we know of seem to indicate that the number of strategy improvements $\Pi(m+n)$ (at least practically) grows quite slowly in the number of maximums m and the number of variables n. Interestingly, though, it is still open whether (or: under which circumstances) the trivial upper bound of 2^{m+n} for $\Pi(m+n)$ can be significantly improved [22,2]. For a small improvement, we notice that for expressions $e_1 \vee e_2$ in which e_2 is an expression without variables, all strategies considered by algorithm 1 after e_1 evaluates to a greater value than e_2 will always select e_1. This in particular holds for the n $\vee$-expressions $e \vee -\infty$ at the top-level introduced in order to deal with $-\infty$. Thus, $\Pi(m_\vee + n)$ in our complexity estimation can be replaced with $n \cdot 2^{m_\vee}$.

4 Analyzing Affine Programs

In this section we discuss affine programs, their collecting semantics as well as their abstract semantics over the template constraint matrix domain [20] which subsumes the interval as well as the zone- and octagon domains [16,15]. We use similar notations as in [17]. Let $\mathbf{X}_G = \{\mathbf{x}_1, \ldots, \mathbf{x}_n\}$ be the set of variables the program operates on and let $\mathbf{x} = (\mathbf{x}_1, \ldots, \mathbf{x}_n)$ denote the vector of variables. We assume that the variables take values in $\mathbb{R}$. Then in a concrete semantics a *state* assigning values to the variables is conveniently modeled by a vector $x = (x_1, \ldots, x_n) \in \mathbb{R}^n$; x_i is the value assigned to variable $\mathbf{x}_i$. Note that we distinguish variables and their values by using a different font. Statements in affine programs are of the following forms:

$$(1) \quad \mathbf{x} := A\mathbf{x} + b \qquad (2) \quad \mathbf{x}_j := ? \qquad (3) \quad A\mathbf{x} + b \geq 0$$

where $A \in \mathbb{R}^{n\times n}$, $b \in \mathbb{R}^n$. Statements of the form (1), (2) and (3) are called *affine assignments*, *non-deterministic assignments* and *guards*, respectively. Non-deterministic assignments are necessary to model input routines returning unknown values or variable assignments whose right-hand sides are not affine expressions. Such a statement may update $\mathbf{x}_i$ in the current state with any possible value. We denote the set of all statements by Stmt.

As common in flow analysis, we use the program's collecting semantics which associates a set of vectors $x = (x_1, \ldots, x_n) \in \mathbb{R}^n$ to each program point. Each statement $s \in \mathsf{Stmt}$ induces a transformation $[\![s]\!] : 2^{\mathbb{R}^n} \to 2^{\mathbb{R}^n}$, given by

$$\begin{array}{ll} [\![\mathbf{x} := A\mathbf{x} + b]\!]X = \{Ax{+}b \mid x \in X\} & [\![A\mathbf{x} + b \geq 0]\!]X = \{x \in X \mid Ax{+}b \geq 0\} \\ [\![\mathbf{x}_k :=?]\!]X \quad = \{x{+}\delta\mathbf{1}_k \mid x \in X, \delta \in \mathbb{R}\} & \end{array}$$

for $X \subseteq \mathbb{R}^n$ where $\mathbf{1}_k$ denotes the vector whose components are zero beside the k-th component which is 1. The branching of an *affine program* is non-deterministic. Formally, an *affine program* is given by a *control flow graph* $G = (N, E, \mathsf{st})$ that consists of a set N of *program points*, a set $E \subseteq N \times \mathsf{Stmt} \times N$ of *(control flow) edges* and a special *start point* $\mathsf{st} \in N$. Then, the collecting semantics V is characterized as the least solution of the constraint system

$$\mathbf{V}[\mathsf{st}] \supseteq \mathbb{R}^n \qquad\qquad \mathbf{V}[v] \supseteq [\![s]\!](\mathbf{V}[u]) \quad \text{for each } (u, s, v) \in E$$

where the variables $\mathbf{V}[v]$, $v \in N$ take values in $2^{\mathbb{R}^n}$. We denote the components of the collecting semantics V by $V[v]$ for $v \in N$.

Example 7. Let $G = (N, E, \mathsf{st})$ denote the affine program shown on the right and let V denotes the collecting semantics of G. For simplicity, we do not use matrices in the control-flow graph. However, all statements can be considered as affine assignments and guards, respectively. The statement $(\mathbf{x}_2, \mathbf{x}_3) := (1, 2 \cdot \mathbf{x}_1)$, for instance, represents the affine assignment

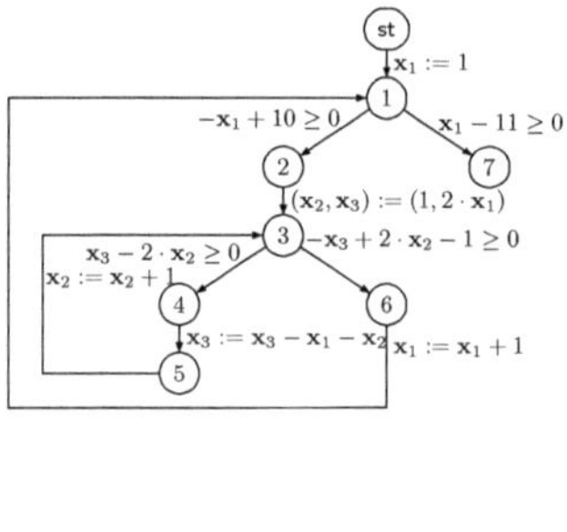

$$\mathbf{x} := \begin{pmatrix} 1\,0\,0 \\ 0\,0\,0 \\ 2\,0\,0 \end{pmatrix} \mathbf{x} + \begin{pmatrix} 0 \\ 1 \\ 0 \end{pmatrix}$$

A program analysis could, for instance, aim to answer the question whether at program point 5 the program variable $\mathbf{x}_3$ takes values within the interval $[0, 9]$, only. Formally, this is the question whether $V[5] \subseteq \{(x_1, x_2, x_3) \mid 0 \leq x_3 \leq 9,\ x_1, x_2 \in \mathbb{R}\}$ — which is the case here. □

We now consider an abstract semantics which is an over-approximation of the *collecting semantics*. We assume that we are given a complete lattice $\mathbb{D}$ of abstract values (with partial ordering $\sqsubseteq$). Assume that we are given a function $\alpha_{\mathbb{D}} : 2^{\mathbb{R}^n} \to \mathbb{D}$ (the abstraction) and a function $\gamma_{\mathbb{D}} : \mathbb{D} \to 2^{\mathbb{R}^n}$ (the concretization) which form a Galois-connection. The elements in $\alpha_{\mathbb{D}}(2^{\mathbb{R}^n})$ are called *open* (see e.g. [9]). The best abstract transformer $[\![s]\!]^\sharp_{\mathbb{D}} : \mathbb{D} \to \mathbb{D}$ for a statement s (see, e.g., [6]) is given by

$$[\![s]\!]^\sharp_{\mathbb{D}} = \alpha_{\mathbb{D}} \circ [\![s]\!] \circ \gamma_{\mathbb{D}}.$$

In particular, $[\![s]\!]^\sharp_{\mathbb{D}}$ always returns open elements. We emphasize that we are concerned with *best abstract transformers* only. The *abstract semantics* $V^\sharp_{\mathbb{D}}$ of the affine program $G = (N, E, \mathsf{st})$ over $\mathbb{D}$ is given as the least solution of the system of constraints

$$\mathbf{V}^\sharp_{\mathbb{D}}[\mathsf{st}] \sqsupseteq \top_{\mathbb{D}} \qquad\qquad \mathbf{V}^\sharp_{\mathbb{D}}[v] \sqsupseteq [\![s]\!]^\sharp(\mathbf{V}^\sharp_{\mathbb{D}}[u]) \quad \text{for each } (u, s, v) \in E$$

where the variables $\mathbf{V}^\sharp_{\mathbb{D}}[v]$, $v \in N$ take values in $\mathbb{D}$ and $\top_{\mathbb{D}}$ denotes the greatest element of $\mathbb{D}$. We denote the components of the abstract semantics $V^\sharp_{\mathbb{D}}$ by $V^\sharp_{\mathbb{D}}[v]$ for $v \in N$. $V^\sharp_{\mathbb{D}}$ represents an over-approximation of the collecting semantics V [7], i.e., $V^\sharp_{\mathbb{D}}[v] \sqsupseteq \alpha_{\mathbb{D}}(V[v])$ and $\gamma_{\mathbb{D}}(V^\sharp_{\mathbb{D}}[v]) \supseteq V[v]$ for every $v \in N$. Since every transformer $[\![s]\!]^\sharp_{\mathbb{D}}$ always returns open elements, we deduce from the theory of Galois-connections (see e.g. [9]) that $V^\sharp_{\mathbb{D}}[v], v \in N$ are open.

In this paper we consider the complete lattice introduced in [20]. For that, we consider a fixed *template constraints matrix* $T \in \mathbb{R}^{m \times n}$. Each row in this matrix represents a linear combination of variables of interest. Special cases of this domain are intervals, zones and octagons [16,15,20]. All these domains represent subclasses of convex polyhedra in the vector space $\mathbb{R}^n$ (n the number of variables). Let us w.l.o.g. assume that T does not contain rows consisting of zeros only. The set $\mathcal{T}_T := \overline{\mathbb{R}}^m$ together with the component-wise partial ordering $\leq$ forms a complete lattice. The *concretization* $\gamma_{\mathcal{T}_T} : \mathcal{T}_T \to 2^{\mathbb{R}^n}$ and the *abstraction* $\alpha_{\mathcal{T}_T} : 2^{\mathbb{R}^n} \to \mathcal{T}_T$ are defined by

$$\gamma_{\mathcal{T}_T}(c) = \{x \in \mathbb{R}^n \mid Tx \leq c\} \qquad \alpha_{\mathcal{T}_T}(X) = \bigwedge\{c \in \overline{\mathbb{R}}^m \mid \gamma_{\mathcal{T}_T}(c) \supseteq X\}$$

for $c \in \overline{\mathbb{R}}^m$, $X \subseteq \mathbb{R}^n$. As shown in [20], $\alpha_{\mathcal{T}_T}$ and $\gamma_{\mathcal{T}_T}$ form a Galois-connection. Thus, the abstract semantics $V^\sharp_{\mathcal{T}_T}$ of an affine program $G = (N, E, \mathsf{st})$ is well-defined.

In [20] the author allows one template constraint matrix for each program point. For simplicity and similar to [10], we consider one global template constraint matrix only. Note also that open elements of $\mathcal{T}_T$ are called *canonical* in [20].

We now show how to compute the abstract semantics $V^\sharp_{\mathcal{T}_T}$ of the affine program $G = (N, E, \mathsf{st})$ which uses variables $\mathbf{X}_G = \{\mathbf{x}_1, \ldots, \mathbf{x}_n\}$. First of all we have to describe the abstract effect $[\![s]\!]^\sharp_{\mathcal{T}_T}$ for each statement s by a linear program. We have:

Lemma 8. *Let $c \in \mathcal{T}_T$, $A \in \mathbb{R}^{n \times n}$, $b \in \mathbb{R}^n$, $\mathbf{x} = (\mathbf{x}_1, \ldots, \mathbf{x}_n)^T$ and $i = 1, \ldots, m$. Then:*

1. $([\![\mathbf{x} := A\mathbf{x} + b]\!]^\sharp_{\mathcal{T}_T}\, c)_{i\cdot} = T_{i\cdot}b + LP_{T,(T_{i\cdot}A)^T}(c)$
2. $([\![A\mathbf{x} + b \geq 0]\!]^\sharp_{\mathcal{T}_T}\, c)_{i\cdot} = LP_{A',T^T_{i\cdot}}(c')$ *where* $A' := \begin{pmatrix} T \\ -A \end{pmatrix}$ *and* $c' := \begin{pmatrix} c \\ b \end{pmatrix}$.
3. $[\![\mathbf{x}_k :=?]\!]^\sharp_{\mathcal{T}_T}\, c \leq \mathit{forget}_{T,k} + c$. *Moreover* $[\![\mathbf{x}_k :=?]\!]^\sharp_{\mathcal{T}_T}\, c = \mathit{forget}_{T,k} + c$ *whenever c is open. Thereby the vector $\mathit{forget}_{T,k} \in \mathcal{T}_T$ is defined by*

$$(\mathit{forget}_{T,k})_{i\cdot} = \begin{cases} \infty & \text{if } T_{i\cdot k} \neq 0 \\ 0 & \text{if } T_{i\cdot k} = 0. \end{cases}$$

□

Note that the post operator in [20] combines an affine assignment and a guard. In order to compute the abstract semantics $V^\sharp_{\mathcal{T}_T}$ of G over $\mathcal{T}_T$, we rely on our methods for systems of rational equations presented in section 3. We additionally allow the LP operator to occur in right-hand sides, i.e., we additionally allow subexpressions of the form: $LP_{A,b}(e_1,\dots,e_m)$ where $A \in \mathbb{R}^{m\times n}$, $b \in \mathbb{R}^n$ and e_i are expressions. We call such expressions and equations *with LP*. We define:

$$[\![LP_{A,b}(e_1,\dots,e_m)]\!]\mu = LP_{A,b}\left(([\![e_1]\!]\mu,\dots,[\![e_m]\!]\mu)^T\right)$$

Since again all operators in expressions with LP are monotone, every system of rational equations with LP has a least solution. For the computation of $V^\sharp_{\mathcal{T}_T}$, we construct a system $\mathcal{C}_G$ of rational constraints with LP which uses variables $\mathbf{X} = \{\mathbf{x}_{v,i} \mid v \in N, i = 1,\dots,m\}$ (m is the number of rows of T) as follows. For the start point st of the affine program we introduce the constraints $\mathbf{x}_{\mathsf{st},i} \geq \infty$ for $i = 1,\dots,m$. According to lemma 8 we introduce a constraint for every control flow edge $(u,s,v) \in E$ and every $i = 1,\dots,m$ as shown in the following table.

control flow edge	constraint
$(u,\ \mathbf{x} := A\mathbf{x}+b,\ v)$	$\mathbf{x}_{v,i} \geq T_{i\cdot}b + LP_{T,(T_{i\cdot}A)^T}(\mathbf{x}_{u,1},\dots,\mathbf{x}_{u,m})$
$(u,\ A\mathbf{x}+b \geq 0,\ v)$	$\mathbf{x}_{v,i} \geq LP_{\begin{pmatrix}T\\-A\end{pmatrix},T_{i\cdot}^T}(\mathbf{x}_{u,1},\dots,\mathbf{x}_{u,m},b_{1\cdot},\dots,b_{n\cdot})$
$(u,\ \mathbf{x}_k :=?,\ v)$	$\mathbf{x}_{v,i} \geq (forget_{T,k})_{i\cdot} + \mathbf{x}_{u,i}$

The correctness follows from lemma 8 and the fact that $V^\sharp_{\mathcal{T}_T}[v], v \in N$ are open.

Theorem 5. *Let $V^\sharp_{\mathcal{T}_T}$ be the abstract semantics of the affine program $G = (N,E,\mathsf{st})$ over $\mathcal{T}_T$ and let μ^* be the least solution of the corresponding system $\mathcal{C}_G$ of rational constraints with LP. Then $(V^\sharp_{\mathcal{T}_T}[v])_{i\cdot} = \mu^*(\mathbf{x}_{v,i})$ for $v \in N, i = 1,\dots,m$.* □

Example 8.

Let V denote the collecting semantics of the affine program $G = (N,E,\mathsf{st})$ of example 7. For our analysis we choose the set of constraints shown on the right which lead to the template constraint matrix T. Our goal is to determine for every program point v a vector $(c_1,\dots,c_8)$ which is as small as possible and for which every vector $(x_1,x_2,x_3) \in V[v]$ fulfills the constraints. Let us consider the edge $(1, -\mathbf{x}_1+10 \geq 0, 2) \in E$ which is an abbreviation for $(1, A\mathbf{x}+b \geq \mathbf{0}, 2)$ (using the matrices above). This edge leads amongst others to the constraint

set of constraints:

$$\begin{aligned} x_1 &\leq c_1 \\ -x_1 &\leq c_2 \\ 2x_2 &\leq x_3 + c_3 \\ -x_2 &\leq c_4 \\ x_3 &\leq 2x_1 + c_5 \\ -x_3 &\leq -2x_1 + c_6 \\ x_3 &\leq c_7 \\ -x_3 &\leq c_8 \end{aligned} \qquad T = \begin{pmatrix} 1 & 0 & 0 \\ -1 & 0 & 0 \\ 0 & 2 & -1 \\ 0 & -1 & 0 \\ -2 & 0 & 1 \\ 2 & 0 & -1 \\ 0 & 0 & 1 \\ 0 & 0 & -1 \end{pmatrix} \qquad A = \begin{pmatrix} -1 & 0 & 0 \\ 0 & 1 & 0 \\ 0 & 0 & 1 \end{pmatrix} \qquad b = \begin{pmatrix} 10 \\ 0 \\ 0 \end{pmatrix}$$

$$\mathbf{x}_{2,1} \geq LP_{\begin{pmatrix}T\\-A\end{pmatrix},\begin{pmatrix}1\\0\\0\end{pmatrix}}(\mathbf{x}_{1,1},\mathbf{x}_{1,2},\dots,\mathbf{x}_{1,8},10,0,0)$$

Here, for instance, evaluating the right-hand side of the constraint above under the variable assignment $\underline{\infty}$ results in the value 10. Finally, the whole system $\mathcal{C}_G$ describes the abstract semantics $V^\sharp_{\mathcal{T}_T}$. Here, in particular, $(V^\sharp_{\mathcal{T}_T}[5])_{7\cdot} = 9$ and $(V^\sharp_{\mathcal{T}_T}[5])_{8\cdot} = 0$ which means that the value of the program variable $\mathbf{x}_3$ is between 0 and 9 at program point 5. This result is optimal and could not be established using interval analysis. □

By theorem 5, our problem reduces to computing least solutions of systems of rational equations *with LP*. Such systems will be discussed in the next section.

5 Systems of Rational Equations with LP

Our goal is to apply algorithm 1 also for computing the least solution μ^* of a system $\mathcal{E}$ of rational equations *with LP*. In order to use the results from section 3, we state the following lemma which can be shown using the duality theorem for linear programming.

Lemma 9. *Let* $A \in \mathbb{R}^{m \times n}$ *with* $A_{i\cdot} \neq (0, \dots, 0)$ *for* $i = 1, \dots, m$ *and* $b \in \mathbb{R}^n$. *There exists a finite (possibly empty) set* $mult(LP_{A,b}) = \{y_1, \dots, y_k\} \subseteq \mathbb{R}^m$ *with* $y_1, \dots, y_k \geq 0$ *and* $A^T y_1 = \dots = A^T y_k = b$ *such that for every* $c \in \overline{\mathbb{R}}^m$ *with* $LP_{A,b}(c) > -\infty$ *it holds that* $LP_{A,b}(c) = \bigwedge_{y \in mult(LP_{A,b})} c^T y$. □

We also extent the definition of $\mathcal{E}(\pi)$ for a strategy π and the definition of the improvement operator $P_\vee$ in the natural way. Moreover, for a *conjunctive* system $\mathcal{E}$, we extend the notion of *consistency* to a notion of *LP-consistency*.

In order to define *LP-consistency* let us, for every $LP_{A,b}$-operator, fix a finite set $mult(LP_{A,b})$ of vectors which satisfies the claims of lemma 9. Let $\mathcal{E}$ denote a conjunctive system of rational equations with LP. We first define the transformation $[\cdot]$ by:

$$[a] = a \quad [\mathbf{x}] = \mathbf{x} \quad [e_1 + e_2] = [e_1] + [e_2] \quad [c \cdot e] = c \cdot [e] \quad [e_1 \wedge e_2] = [e_1] \wedge [e_2]$$
$$[LP_{A,b}(e_1, \dots, e_m)] = \bigwedge_{y \in mult(LP_{A,b})} ([e_1], \dots, [e_m])y$$

where $a \in \overline{\mathbb{R}}$, $c \in \mathbb{R}^{>0}$, $\mathbf{x}$ is a variable and e_i are expressions. Thereby $([e_1], \dots, [e_m])y$ denotes the expression $y_{1\cdot} \cdot [e_1] + \dots + y_{m\cdot} \cdot [e_m]$ and we assume that an expression $0 \cdot e_i$ is simplified to 0 (This is correct, since e_i does not evaluate to $-\infty$ in the cases which have to be considered). Assuming that $\mathcal{E}$ denotes the system $\mathbf{x}_i = e_i, i = 1, \dots, n$, we write $[\mathcal{E}]$ for the system $\mathbf{x}_i = [e_i], i = 1, \dots, n$. Then, we call a pre-solution μ of $\mathcal{E}$ *LP-consistent* iff $[\![LP_{A,b}(e_1, \dots, e_m)]\!]\mu > -\infty$ for every subexpression $LP_{A,b}(e_1, \dots, e_m)$ and μ is a consistent pre-solution of $[\mathcal{E}]$.

We have to ensure that μ will be a LP-consistent pre-solution of $\mathcal{E}$ whenever algorithm 1 computes the least solution μ' of $[\mathcal{E}]$ with $\mu' \geq \mu$. This is fulfilled, since lemmas 2 and 3 can be formulated literally identical for LP-consistency instead of consistency.

Assume that μ is a LP-consistent pre-solution of $\mathcal{E}$. It remains to compute the least solution μ' of $\mathcal{E}$ with $\mu' \geq \mu$. Since $\mathcal{E}$ is LP-consistent and thus in particular $[\![LP_{A,b}(e_1, \dots, e_m)]\!]\mu > -\infty$ for every subexpression $LP_{A,b}(e_1, \dots, e_m)$, lemma 9 implies that μ' is the least solution of $[\mathcal{E}]$ with $\mu' \geq \mu$. Since $[\mathcal{E}]$ denotes a conjunctive system *without LP*, we can compute it and moreover corollary 1 implies that every conjunctive system $\mathcal{E}$ is considered at most n times in algorithm 1. We find:

Lemma 10. *Assume that* $\mathcal{E}$ *denotes a conjunctive system with LP which uses* n *variables and* m $\vee$*-expressions. Algorithm 1 computes at most* $n \cdot \Pi(m+n)$ *times the least solution* μ' *of* $\mathcal{E}(\pi)$ *with* $\mu' \geq \mu$ *for some* π *and some LP-consistent pre-solution* μ *of* $\mathcal{E}(\pi)$. *After that, it returns the least solution of* $\mathcal{E}$. □

We want to compute the least solution μ' of $\mathcal{E}$ with $\mu' \geq \mu$ which is also the least solution of $[\mathcal{E}]$ with $\mu' \geq \mu$. Recall from section 3 that for this purpose we essentially have to compute least consistent solutions of feasible systems of conjunctive equations. Writing down the system $[\mathcal{E}]$ explicitly and solving it after this would be too inefficient.

Therefore we aim at computing the least consistent solution of the feasible system $[\mathcal{E}]$ without explicit representation. For that purpose, assume w.l.o.g. that $\mathcal{E}$ is given by

$$\begin{array}{ll} \mathbf{x}_i = e_i^{(1)} \wedge \cdots \wedge e_i^{(k_i)} & \text{for } i = 1, \ldots, n' \\ \mathbf{x}_i = LP_{A_i,b_i}(\mathbf{x}'_1, \ldots, \mathbf{x}'_{m_i}) & \text{for } i = n'+1, \ldots, n \end{array}$$

where the $e_i^{(j)}$ do neither contain $\wedge$- nor $LP_{A,b}$-operators. This form can be achieved by introducing variables. Recall from section 3 that the system $\mathcal{C}_{[\mathcal{E}]}$ is given by

$$\begin{array}{ll} \mathbf{x}_i \leq [e_i^{(j)}] & \text{for } i = 1, \ldots, n',\ j = 1, \ldots, k_i \\ \mathbf{x}_i \leq \bigwedge_{y \in mult(LP_{A_i,b_i})}(\mathbf{x}'_1, \ldots, \mathbf{x}'_{m_i})y & \text{for } i = n'+1, \ldots, n \end{array}$$

We define the system $\mathcal{C}_{\mathcal{E}}^{LP}$ of rational constraints as the system:

$$\begin{array}{ll} \mathbf{x}_i \leq e_i^{(j)} & \text{for } i = 1, \ldots, n',\ j = 1, \ldots, k_i \\ \mathbf{x}_i \leq LP_{A_i,b_i}(\mathbf{x}'_1, \ldots, \mathbf{x}'_{m_i}) & \text{for } i = n'+1, \ldots, n \end{array}$$

Using lemma 9 we conclude that the sets of solutions μ' with $\mu' \geq \mu$ of $\mathcal{C}_{\mathcal{E}}^{LP}$ and of $\mathcal{C}_{[\mathcal{E}]}$ are equal. In particular, the sets of solutions μ' with $\mu' \geq \mu$ which maximize the sum $\sum_{\mathbf{x} \in \mathbf{X}} \mu'(\mathbf{x})$ are equal.

As in section 3 we first assume that the least consistent solution μ^* of $\mathcal{E}$ does not map any variable to ∞. In this situation, the above considerations and lemma 6 implies, that μ^* is the uniquely determined solution of $\mathcal{C}_{\mathcal{E}}^{LP}$ which maximizes the sum $\sum_{\mathbf{x} \in \mathbf{X}} \mu^*(\mathbf{x})$. In order to compute it using linear programming, we have to eliminate all occurrences of $LP_{A,b}$-operators. Therefore, consider a constraint

$$\mathbf{x} \leq LP_{A,b}(\mathbf{x}_1, \ldots, \mathbf{x}_m)$$

occurring in $\mathcal{C}_{\mathcal{E}}^{LP}$. Using the definition of the $LP_{A,b}$-operator we can conceptually replace the right-hand side with $\bigvee\{b^T y \mid y \in \mathbb{R}^n, Ay \leq (\mathbf{x}_1, \ldots, \mathbf{x}_m)^T\}$. Since we are interested in maximizing the value of the variable $\mathbf{x}$ anyway we replace the above constraint with the constraints

$$\mathbf{x} \leq b_{1\cdot} \cdot \mathbf{y}_1 + \cdots + b_{n\cdot} \cdot \mathbf{y}_n, \qquad A_{i\cdot 1} \cdot \mathbf{y}_1 + \cdots + A_{i\cdot n} \cdot \mathbf{y}_n \leq \mathbf{x}_i \quad \text{for } i = 1, \ldots, m$$

where $\mathbf{y}_1, \ldots, \mathbf{y}_n$ are fresh variables. This replacement step preserves the solution μ^* which maximizes the sum $\sum_{\mathbf{x} \in \mathbf{X}} \mu^*$. Doing this for every $LP_{A,b}$-expression in $\mathcal{E}$ we obtain a system of constraints without $LP_{A,b}$-expressions. Thus, we can compute μ^* by linear programming. We have:

Lemma 11. *Assume that $\mu \ll \underline{\infty}$ is a LP-consistent pre-solution of the conjunctive system $\mathcal{E}$ with LP. Assume that $\mu^* \ll \underline{\infty}$ is the least consistent solution of $[\mathcal{E}]$. Then μ^* can be computed by solving one LP which can be extracted from $\mathcal{E}$ in linear time.* □

Until now we have assumed that the least consistent solution μ^* of $[\mathcal{E}]$ maps every variable to values strictly smaller then ∞. As in section 3, we have to identify the variables $\mathbf{x}$ with $\mu^*(\mathbf{x}) = \infty$ in order to lift this restriction. For that, by lemma 7, we must compute the greatest solution μ^∞ of the system $[\ [\mathcal{E}]\ ]^\infty$. For this purpose we extend the abstraction $[\cdot]^\infty$ by setting $[LP_{A,b}(e_1,\ldots,e_m)]^\infty = LP_{A,b}([e_1]^\infty,\ldots,[e_m]^\infty)$. It turns out that μ^∞ is also the greatest solution of $[\mathcal{E}]^\infty$. Since μ^∞ maps every variable to a finite value, μ^∞ is the only solution finite solution of $\mathcal{C}_{[\mathcal{E}]^\infty}$ which maximizes the sum $\sum_{\mathbf{x}\in\mathbf{X}} \mu^\infty(\mathbf{x})$. Thus, μ^∞ can again be computed using linear programming. Since we can identify the set $\{\mathbf{x} \in \mathbf{X} \mid \mu^*(\mathbf{x}) = \infty\}$ in this way, we can lift the restriction to systems with finite least consistent solutions in lemma 11. We have:

Lemma 12. *Assume that $\mu \ll \underline{\infty}$ denotes a LP-consistent pre-solution of the conjunctive system $\mathcal{E}$ with LP. Let μ^* denote the least consistent solution of $[\mathcal{E}]$. Then μ^* can be computed by solving two LP problems which can be extracted from $\mathcal{E}$ in linear time.* □

In conclusion, we obtain our main result for systems of rational equations with LP:

Theorem 6. *The least solution of a system $\mathcal{E}$ of rational equations with LP which uses n variables and m maximum expressions can be computed by solving $3n \cdot \Pi(m+n)$ LPs each of which can be extracted from $\mathcal{E}$ in linear time.* □

Finally, combining theorem 5 and 6, we derive our main result for the analysis of affine programs:

Theorem 7. *Assume that $G = (N, E, \mathsf{st})$ denotes an affine program. Let $T \in \mathbb{R}^{m\times n}$, $indeg(v) := \{(u,s,v') \in E \mid v' = v\}$ and $m_\vee := m \cdot \sum_{v\in N} max(indeg(v) - 1, 0)$. The abstract fixpoint semantics of G over $\mathcal{T}_T$ can be computed by solving at most $3m|N| \cdot \Pi(m_\vee + m|N|)$ LPs.* □

It remains to emphasize that all least solutions (resp. abstract semantics) computed by our methods are rational whenever all numbers occurring in the input are rational.

6 Conclusion

We presented a practical strategy improvement algorithm for computing exact least solutions of systems of equations over the rationals with addition, multiplication with positive constants, maximum and minimum. The algorithm is based on strategy improvement combined with LP solving for each selected strategy where each strategy can be selected only linearly often. We extended the method in order to deal a special LP-operator in right-hand sides of equations. We applied our techniques to compute the abstract least fixpoint semantics of affine programs over the template constraint matrix domain. In particular, we thus obtain practical algorithms for dealing with zones and octagons. It remains for future work to experiment with practical implementations of the proposed approaches.

References

1. GNU Linear Programming Kit, `http://www.gnu.org/software/glpk`
2. Bjorklund, H., Sandberg, S., Vorobyov, S.: Complexity of Model Checking by Iterative Improvement: the Pseudo-Boolean Framework. In: Broy, M., Zamulin, A.V. (eds.) PSI 2003. LNCS, vol. 2890, pp. 381–394. Springer, Heidelberg (2004)

3. Cochet-Terrasson, J., Gaubert, S., Gunawardena, J.: A Constructive Fixed Point Theorem for Min-Max Functions. Dynamics and Stability of Systems 14(4), 407–433 (1999)
4. Costan, A., Gaubert, S., Goubault, E., Martel, M., Putot, S.: A Policy Iteration Algorithm for Computing Fixed Points in Static Analysis of Programs. In: Etessami, K., Rajamani, S.K. (eds.) CAV 2005. LNCS, vol. 3576, pp. 462–475. Springer, Heidelberg (2005)
5. Cousot, P., Cousot, R.: Static Determination of Dynamic Properties of Recursive Procedures. In: Neuhold, E.J. (ed.) IFIP Conf. on Formal Description of Programming Concepts, pp. 237–277. North-Holland, Amsterdam (1977)
6. Cousot, P., Cousot, R.: Systematic Design of Program Analysis Frameworks. In: 6th ACM Symp. on Principles of Programming Languages (POPL), pp. 238–352 (1979)
7. Cousot, P., Cousot, R.: Static Determination of Dynamic Properties of Programs. In: Second Int. Symp. on Programming, Dunod, Paris, France, pp. 106–130 (1976)
8. Cousot, P., Cousot, R.: Comparison of the Galois Connection and Widening/Narrowing Approaches to Abstract Interpretation. In: JTASPEFL '91, Bordeaux. BIGRE, vol. 74, pp. 107–110 (1991)
9. Erné, M., Koslowski, J., Melton, A., Strecker, G.E.: A Primer On Galois Connections (1992)
10. Gaubert, S., Goubault, E., Taly, A., Zennou, S.: Static Analysis by Policy Iteration on Relational Domains. In: De Nicola, R. (ed.) ESOP 2007. LNCS, vol. 4421, pp. 237–252. Springer, Heidelberg (2007)
11. Gawlitza, T., Seidl, H.: Precise Fixpoint Computation Through Strategy Iteration. In: De Nicola, R. (ed.) ESOP 2007. LNCS, vol. 4421, pp. 300–315. Springer, Heidelberg (2007)
12. Hoffman, A.J., Karp, R.M.: On Nonterminating Stochastic Games. Management Sci. 12, 359–370 (1966)
13. Howard, R.: Dynamic Programming and Markov Processes. Wiley, New York (1960)
14. Megiddo, N.: On the Complexity of Linear Programming. In: Bewley, T. (ed.). Advances in Economic Theory: 5th World Congress, pp. 225–268. Cambridge University Press, Cambridge (1987)
15. Miné, A.: The Octagon Abstract Domain in Analysis, Slicing and Transformation. In: IEEE Working Conf. on Reverse Engineering, pp. 310–319. IEEE Computer Society Press, Los Alamitos (2001)
16. Miné, A.: A new numerical abstract domain based on difference-bound matrices. In: Danvy, O., Filinski, A. (eds.) PADO 2001. LNCS, vol. 2053, pp. 155–172. Springer, Heidelberg (2001)
17. Müller-Olm, M., Seidl, H.: Precise Interprocedural Analysis through Linear Algebra. In: 31st ACM Symp. on Principles of Programming Languages (POPL), pp. 330–341 (2004)
18. Puri, A.: Theory of Hybrid and Discrete Systems. PhD thesis, University of California, Berkeley (1995)
19. Puterman, M.L.: Markov Decision Processes: Discrete Stochastic Dynamic Programming. Wiley, New York (1994)
20. Sankaranarayanan, S., Sipma, H.B., Manna, Z.: Scalable analysis of linear systems using mathematical programming. In: Cousot, R. (ed.) VMCAI 2005. LNCS, vol. 3385, pp. 25–41. Springer, Heidelberg (2005)
21. Schrijver, A.: Theory of linear and integer programming. John Wiley & Sons, New York, NY, USA (1986)
22. Vöge, J., Jurdzinski, M.: A Discrete Strategy Improvement Algorithm for Solving Parity Games. In: Emerson, E.A., Sistla, A.P. (eds.) CAV 2000. LNCS, vol. 1855, pp. 202–215. Springer, Heidelberg (2000)

Omega-Regular Half-Positional Winning Conditions

Eryk Kopczyński*

Institute of Informatics, Warsaw University
erykk@mimuw.edu.pl

Abstract. We study infinite games where one of the players always has a positional (memory-less) winning strategy, while the other player may use a history-dependent strategy. We investigate winning conditions which guarantee such a property for all arenas, or all finite arenas. Our main result is that this property is decidable in single exponential time for a given prefix independent ω-regular winning condition. We also exhibit a big class of winning conditions (XPS) which has this property.

Keywords: automata, infinite games, omega-regular languages, positional strategies, winning condtions.

1 Introduction

The theory of infinite games is relevant for computer science because of its potential application to verification of interactive systems. In this approach, the system and environment are modeled as players in an infinite game played on a graph (called *arena*) whose vertices represent possible system states. The players (conventionally called Eve and Adam) decide which edge (state transition, or *move*) to choose; each edge has a specific *color*. The desired system's behavior is expressed as a winning condition of the game — the winner depends on the sequence of colors which appear during an infinite play. If a winning strategy exists in this game, the system which implements it will behave as expected. Positional strategies, also called *memoryless* — ones that depend only on the current position, not on the history of play — are of special interest here, because of their good algorithmic properties which can lead to an efficient implementation.

Infinite games are strongly linked to automata theory. An accepting run of an alternating automaton (on a given tree) can be presented as a winning strategy in a certain game between two players. Parity games are related to automata on infinite structures with parity acceptance condition. For example, positional determinacy of parity games ([Mos91], [EJ91], [McN93]) is used in modern proofs of Rabin's complementation theorem for finite automata on infinite trees with Büchi or parity acceptance condition. See [GTW02] for more links between infinite games, automata, and logic.

* Supported by KBN grant N206 002 32/0472 and the EC RTN GAMES.

J. Duparc and T.A. Henzinger (Eds.): CSL 2007, LNCS 4646, pp. 41–53, 2007.

An interesting question is, what properties of the winning condition are sufficient for the winning condition to be positionally determined, i.e. guarantee positional winning strategies independently on the arena on which the game is played. Note that although the parity condition gives such strategies for both players, often we need it only for one player (say, Eve). This is true both for interactive systems (we need a simple strategy for ourselves only, we do not care about the environment) and for automata theoretic applications mentioned above (like Rabin's complementation theorem). We call winning conditions with such property *half-positional*, to make it clear that we require a positional strategy for only one of the players.

Recently some interesting characterizations of *positional* winning conditions have been found ([CN06], [GZ04], [GZ05], [GW06]; see also [Gra04] for a survey of results on positional determinacy). Our work attempts to obtain similar characterizations and find interesting properties (e.g. closure and decidability properties) of *half-positional* winning conditions. In this paper we concentrate on winning conditions which are ω-regular.

We show that finite half-positional determinacy of such winning conditions is decidable in singly exponential time. This is done by showing that if a winning condition is not half-positional then there exists a simple arena witnessing this. Then it is possible to check all of such simple arenas. We also show some constructions which lead to half-positional ω-regular winning conditions, such as extended positional/suspendable conditions, concave conditions, and monotonic automata.

This work is a continuation of [Kop06]. Also see the draft [Kop07], which is a superset of both this paper and [Kop06] and contains full proofs which had to be omitted from here due to space limitations.

2 Preliminaries

We consider perfect information antagonistic infinite games played by two players, called conventionally Adam and Eve. Let C be a set of **colors** (possibly infinite).

An **arena** over C is a tuple $G = (\mathrm{Pos}_A, \mathrm{Pos}_E, \mathrm{Mov})$, where:

- Elements of $\mathrm{Pos} = \mathrm{Pos}_E \cup \mathrm{Pos}_A$ are called **positions**; Pos_A and Pos_E are disjoint sets of Adam's positions and Eve's positions, respectively.
- Elements of $\mathrm{Mov} \subseteq \mathrm{Pos} \times \mathrm{Pos} \times (C \cup \{\epsilon\})$ are called **moves**; (v_1, v_2, c) is a move from v_1 to v_2 colored by c. We denote $\mathrm{source}(v_1, v_2, c) = v_1$, $\mathrm{target}(v_1, v_2, c) = v_2$, $\mathrm{rank}(v_1, v_2, c) = c$.
- ϵ denotes an empty word, i.e. a colorless move. There are no infinite paths of such colorless moves.

A game is a pair (G, W), where G is an arena, and W is a winning condition. A **winning condition** W over C is a subset of C^ω which is *prefix independent*, i.e., $u \in W \iff cu \in W$ for each $c \in C, u \in C^\omega$. We name specific winning conditions WA, WB, ... For example, the **parity condition** of rank n is the winning condition over $C = \{0, 1, \ldots, n\}$ defined with

$$WP_n = \{w \in C^\omega : \limsup_{i \to \infty} w_i \text{ is even}\}. \tag{1}$$

The game (G, W) carries on in the following way. The play starts in some position v_1. The owner of v_1 (e.g. Eve if $v_1 \in \mathrm{Pos}_E$) chooses one of the moves leaving v_1, say (v_1, v_2, c_1). If the player cannot choose because there are no moves leaving v_1, he or she loses. The next move is chosen by the owner of v_2; denote it by (v_2, v_3, c_2). And so on: in the n-th move the owner of v_n chooses a move (v_n, v_{n+1}, c_n). If $c_1 c_2 c_3 \ldots \in W$, Eve wins the infinite play; otherwise Adam wins.

A **play** in the arena G is any sequence of moves π such that $\mathrm{source}(\pi_{n+1}) = \mathrm{target}(\pi_n)$. By $\mathrm{source}(\pi)$ and $\mathrm{target}(\pi)$ we denote the initial and final position of the play, respectively. The play can be finite ($\pi \in \mathrm{Pos} \cup \mathrm{Mov}^+$, where by $\pi \in \mathrm{Pos}$ we represent the play which has just started in the position π) or infinite ($\pi \in \mathrm{Mov}^\omega$; infinite plays have no target).

A **strategy for player** X (i.e. $X \in \{\mathrm{Eve}, \mathrm{Adam}\}$) is a partial function $s : \mathrm{Pos} \cup \mathrm{Mov}^+ \to \mathrm{Mov}$. Intuitively, $s(\pi)$ for π endining in Pos_X says what X should do next. We say that a play π is **consistent** with strategy s for X if for each prefix π' of π such that $\mathrm{target}(\pi') \in \mathrm{Pos}_X$ the next move is given by $s(\pi')$.

A strategy s is **winning** (for X) from the position v if $s(\pi)$ is defined for each finite play π starting in v, consistent with s, and ending in Pos_X, and each infinite play starting in v consistent with s is winning for X.

A strategy s is **positional** if it depends only on $\mathrm{target}(\pi)$, i.e., for each finite play π we have $s(\pi) = s(\mathrm{target}(\pi))$.

A game is **determined** if for each starting position one of the players has a winning strategy. This player may depend on the starting position in the given play. Thus if the game is determined, the set Pos can be split into two sets Win^E and Win^A and there exist strategies s_E and s_A such that each play π with $\mathrm{source}(\pi) \in \mathrm{Win}^X$ and consistent with s_X is winning for X. All games with a Borel winning condition are determined [Mar75], but there exist (exotic) games which are not determined. A winning condition W is **determined** if for each arena G the game (G, W) is determined.

We are interested in games and winning conditions for which one or both of the players have positional winning strategies. A **determinacy type** is given by three parameters: admissible strategies for Eve (positional or arbitrary), admissible strategies for Adam (positional or arbitrary), and admissible arenas (finite or infinite). We say that a winning condition W is (α, β, γ)**-determined** if for every γ-arena G the game (G, W) is (α, β)**-determined**, i.e. for every starting position either Eve has a winning α-strategy, or Adam has a winning β-strategy. Clearly, there are 8 determinacy types in total. For short, we call (positional, positional, infinite)-determined winning conditions **positionally determined** or just **positional**, (positional, arbitrary, infinite)-determined winning conditions **half-positional**, (arbitrary, positional, infinite)-determined winning conditions **co-half-positional**. If we restrict ourselves to finite arenas, we add **finitely**, e.g. (positional, arbitrary, finite)-determined conditions are called **finitely half-positional**. For a determinacy type $D = (\alpha, \beta, \gamma)$, D-arenas mean γ-arenas, and D-strategies mean α-strategies (if they are strategies for Eve) or β-strategies (for Adam).

It can be easily proven that a determined game (G, W) is half-positionally determined (i.e. (positional, arbitrary)-determined) iff Eve has a single positional strategy which is winning from each starting position $v \in \mathrm{Win}^E(G, W)$ (see e.g. Lemma 5 in [CN06]).

Note that if a game (G, W) is (α, β)-determined, then its dual game obtained by using the complement winning condition and switching the roles of players is (β, α)-determined. Thus, W is (α, β, γ)-determined iff its complement is (β, α, γ)-determined.

We also need following definitions from automata theory.

Definition 1. *A* **deterministic finite automaton on infinite words with parity acceptance condition** *is a tuple* $A = (Q, q_I, \delta, \mathrm{rank})$, *where* Q *is a finite set of states,* $q_I \in Q$ *the initial state,* $\mathrm{rank} : Q \to \{0, \dots, d\}$, *and* $\delta : Q \times C \to Q$. *We extend the definition of* δ *to* $\delta : Q \times C^* \to Q$ *by* $\delta(q, \epsilon) = q, \delta(q, wu) = \delta(\delta(q, w), u)$ *for* $w \in C^*, u \in C$. *For* $w \in C^\omega$, *let* $q_0(w) = q_I$ *and* $q_{n+1}(w) = \delta(q_n, w_{n+1}) = \delta(q_I, w_0 \dots w_{n+1})$. *We say that the word* $w \in C^\omega$ *is* **accepted** *by* A *iff* $\limsup_{n\to\infty} \mathrm{rank}(q_n(w))$ *is even. The set of all words accepted by* A *is called* **language accepted by** A *and denoted* L_A. *We say that a language* $L \subseteq C^\omega$ *is* ω-**regular** *if it is accepted by some automaton.*

3 Types of Arenas

In the games defined above the moves are colored, and it is allowed to have moves without colors. In the literature, several types of arenas are studied.

- ϵ-arenas (C), like the ones described above.
- *Move-colored arenas* (B). In this setting each move needs to be assigned a color. Moves labelled with ϵ are not allowed.
- *Position-colored arenas* (A). In this setting, colors are assigned to positions instead of colors. Instead of $\mathrm{Mov} \subseteq \mathrm{Pos} \times \mathrm{Pos} \times C$ we have $\mathrm{Mov} \subseteq \mathrm{Pos} \times \mathrm{Pos}$ and a function $\mathrm{rank} : \mathrm{Pos} \to C$. Again, each positions needs to be assigned a color. The winner of a play in such games is defined similarly.

If we take a position-colored arena and color each move m with the color rank (source(m)), we obtain an equivalent move-colored arena. Therefore position-colored arenas are a subclass of move-colored arenas. Obviously, move-colored arenas are also a subclass of ϵ-colored arenas. When speaking about a determinacy type where we restrict arena to position-colored or move-colored arenas, or we want to emphasize that we allow ϵ-arenas, we add the letter A, B or C (e.g. A-half-positional conditions when we restrict to position-colored arenas).

Hence C-half-positional conditions are a subclass of B-half-positional conditions, and B-half-positional conditions are a subclass of A-half-positional conditions. In the first case the inclusion is proper: there is no way to transform a move-colored arena into a position-colored one such that nothing changes with respect to positional strategies (we can split a position into several new positions according to colors of moves which come into them, but then we obtain new positional strategies which were not positional previously). We know examples of winning conditions which are A-positional but not B-positional. One of them is $C^*(01)^*$; for position-colored arenas we know from our current position to which color we should move next, but not for edge-colored arenas. Another example is min-parity [GW06]. B-positional determinacy has been characterized in [CN06]; this result can be easily generalized to ϵ-arenas. ϵ-arenas have been studied in [Zie98].

In the second case the question whether the inclusion is proper remains open. Note that when we allow ϵ labels there is no difference whether we label positions or moves: we can replace each move $v_1 \to v_2$ colored with c in an ϵ-arena with $v_1 \to v \to v_2$, color v with c, and leave all the original positions colorless.

In this paper we concentrate on ϵ-arenas since we think that this class gives the least restriction on arenas. As shown by the example above, $C^*(01)^*$, positional strategies for move-colored games are „more memoryless" than for position-colored games since they do not even remember the last color used, although winning conditions for position-colored games (like min-parity) may also be interesting. As we will see, it is natural to allow having colorless moves (or equivalently positions).

4 Simplifying the Witness Arena

To show that finite half-positional determinacy of winning conditions which are prefix independent ω-regular languages is decidable, we will first need to show that if W is not finitely half-positional, then it is witnessed by a simple arena.

Theorem 1. *Let W be a winning condition accepted by a deterministic finite automaton with parity acceptance condition $A = (Q, q_I, \delta, \mathrm{rank} : Q \to \{0 \dots d\})$ (see Definition 1). If W is not finitely half-positional then there is a witness arena (i.e. such that Eve has a winning strategy, but no positional winning strategy) where there is only one Eve's position, and only two moves from this position. (There is no restriction on Adam's moves and positions.)*

Proof. Let G be any finite witness arena. First, we will show how to reduce the number of Eve's positions to just one. Then, we will show how to remove unnecessary moves.

Let $G^0 = (\mathrm{Pos}_A \times Q, \mathrm{Pos}_E \times Q, \mathrm{Mov}^0)$ and $G^1 = (\mathrm{Pos}_A \times Q, \mathrm{Pos}_E \times Q, \mathrm{Mov}^1)$ where for each move (v_1, v_2, c) in G and each state q we have corresponding moves $((v_1, q), (v_2, \delta(q, c), c))$ in Mov^0 and $((v_1, q), (v_2, \delta(q, c), \mathrm{rank}(q)))$ in Mov^1. One can easily see that the three games (G, W), (G^0, W) and (G^1, WP_d) are equivalent: each play in one of them can be interpreted as a play in each another, and the winner does not change for infinite plays. The games G^0 and G are equivalent because in G^0 we just replace each position with a set, and G^0 and

G^1 are equivalent because A's acceptance parity condition in (G^0, W) uses the same ranks as the parity condition in (G^1, WP_d).

Since Eve has a winning strategy in (G, W), she also has a winning strategy in (G^1, WP_d). This game is positionally determined, so she also has a positional strategy here. She can use it in (G^0, W) too.

Let s be Eve's positional winning strategy in G^0. Let

$$N(s) = \{v : \exists q_1 \exists q_2\ \pi_1(\text{target}(s(v, q_1))) \neq \pi_1(\text{target}(s(v, q_2)))\},$$

i.e. the set of positions where s is not positional as a strategy in G. Since the arena is finite, we can assume without loss of generality that there is no positional winning strategy s' in G^0 such that $N(s') \subsetneq N(s)$.

Obviously, the set $N(s)$ is non-empty. Let $v_0 \in N(s)$. We construct a new arena G^2 from G^0 in two steps.

First, merge $\{v_0\} \times Q$ into a single position v_0. Eve can transform s into a winning strategy s_1 in this new game — the only difference is that in v_0 she needs to remember in what state q she is currently, and move according to $s(v_0, q)$.

Then, remove all Eve's moves which are not used by s from all Eve's positions except v_0, and transfer these positions to Adam. Eve still wins, because s_1 remains a winning strategy after this operation. Thus, we obtained an arena G^2 with only one Eve's position v_0 where she has a winning strategy from v_0.

Eve has no positional strategy in G^2. Otherwise this strategy could be simulated without changing the winner (in the natural way) by a strategy s_1 in G which is positional in all positions except $N(s) - \{v_0\}$. This means that there is a positional strategy s_2 in G^0 for which $N(s_2) \subseteq N(s) - \{v_0\}$. (We obtain s_2 by removing from G moves which are not used by s_1 (s_1 remains a winning strategy) and repeating the construction of G^0 on the result arena (obtaining a positional strategy on the new arena), and bringing the removed moves back (so our strategy is on G^0).) This contradicts our assumption that $N(s)$ is minimal.

Hence, we found a witness arena where $|Pos_E| = 1$. To see that we can assume that Eve has at most $|Q|$ moves here, it is enough to see that Eve's finite memory strategy cannot use more than $|Q|$ different moves from this position, hence we can remove the ones which she does not use.

Now, suppose that G is a witness arena with only one Eve's position. We will construct a new arena with only two possible moves for Eve. The construction goes as follows:

- We start with $G^3 = G^0$. Let s be Eve's winning strategy in G^3.
- For each of Eve's $|Q|$ positions, we remove all moves except the one which is used by s.
- (*) Let v_1 and v_2 be two Eve's positions in G^3.
- We merge Eve's positions v_1 and v_2 into one, v_0.
- Eve still has a winning strategy everywhere in this new game (by reasoning similar to one we used for G^2). We check if Eve has a positional winning strategy.
- If yes, we remove the move which is not used in v_0, and go back to (*). (Two distinct Eve's positions in G^3 must still exist — if we were able to

merge all Eve's positions into one, it would mean that G^3 was positionally determined.)

- Otherwise we transfer all Eve's positions other than v_0 to Adam. There was only one move from each of these positions, hence G^3 still is a witness arena.
- In G^3 we have now only one Eve's position (v_0) and only two Eve's moves — one inherited from v_1 and one inherited from v_2.

□

5 Decidability

Theorem 2. *Let W be a (prefix independent) ω-regular winning condition recognized by a DFA with parity acceptance condition $A = (Q, q_I, \delta, \mathrm{rank} : Q \to \{0 \dots d\})$ with n states. Then finite half-positional determinacy of W is decidable in time $O(n^{O(n^2)})$.*

Proof. It is enough to check all possible witness arenas which agree with the hypothesis of Theorem 1. Such arena consists of (the only) Eve's position E from which she can move to A_1 by move m_1 or to A_2 by move m_2. (In general it is possible that A_1 or A_2 is equal to E. However, this is a simple case and we will concentrate on the one where A_1 and A_2 are Adam's positions.) Adam has a choice of word w by which he will return to E from A_i. (In general it is possible that Adam can choose to never return to E. However, if such infinite path would be winning for Eve he would not choose it, and if it would be winning for Adam Eve would never hope to win by choosing to move to A_i.) Let L_i be the set of all possible Adam's return words from A_i to E.

Let $T(w) : Q \to \{0, \dots, d\} \times Q$ be the function defined as follows: $T(w)(q) = (r, q')$ iff $\delta(q, w) = q'$ and the greatest rank visited during these transitions is r. The function $T(w)$ contains all the information about $w \in L_i$ which is important for our game: if $T(w_1) = T(w_2)$ then it does not matter whether Adam chooses to return by w_1 or w_2 (the winner does not change). Thus, instead of Adam choosing a word w from L_i, we can assume that Adam chooses a function t from $T(L_i) \subseteq T(C^*) \subseteq P((Q \times Q)^{\{0,\dots,d\}})$.

For non-empty $R \subseteq \{0, \dots, d\}$, let $\mathrm{best}^{\mathrm{A}}(\mathrm{R})$ be the priority which is the best for Adam, i.e. the greatest odd element of R, or the smallest even one if there are no odd priorities in R. We also put $\mathrm{best}^{\mathrm{A}}(\emptyset) = \perp$.

For $T \subseteq P((Q \times Q)^{\{0,\dots,d\}})$, let

$$U(T) = \mathrm{best}^{\mathrm{A}}(\{\mathrm{d} : \exists \mathrm{t} \in \mathrm{T}\ \mathrm{t}(\mathrm{q}_1) = (\mathrm{d}, \mathrm{q}_2)\}).$$

Again, the function $U_i = U(T(L_i)) : Q \times Q \to \{\perp, 0, \dots, d\}$ contains all the information about L_i which is important for our game — if Adam can go from q_1 to q_2 by one of two words w_1 and w_2 having the highest priorities d_1 or d_2, respectively, he will never want to choose the one which is worse to him.

Our algorithm will check all possible functions U_i. For this, we need to know whether a particular function $U : Q \times Q \to \{\perp, 0, \dots, d\}$ is of form $U(T(L_i))$ for some L_i. This can be done in the following way. We start with $V(q, q) = \perp$.

Generate all elements of $T(L_i)$. This can be done by doing a search (e.g. breadth first search) on the graph whose vertices are $T(w)$ and edges are $T(w) \to T(wc)$ ($T(wc)$ obviously depends only on $T(w)$). For each of these elements, we check if it does not give Adam a better option than U is supposed to give — i.e. for some q_1 we have $T(wc)(q_1) = (q_2, d)$ and $d = \mathrm{best^A(d, U(q_1, q_2))}$. If it does not, we add $T(w)$ to our set T and update V: for each q_1, $T(wc)(q_1) = (q_2, d)$, we set $V(q_1, q_2) := \mathrm{best^A(d, V(q_1, q_2))}$. If after checking all elements of $T(L_i)$ we get $V = U$, then $U = U(T)$. Otherwise, there is no L such that $U = U(T(L))$.

The general algorithm is as follows:

- Generate all possible functions U of form $U(T(L))$.
- For each possible function U_1 describing Adam's possible moves after Eve's move m_1 such that Eve cannot win by always moving with m_1:
- For each U_2 (likewise):
- Check if Eve can win by using a non-positional strategy. (This is done easily by constructing an equivalent parity game which has $3|Q|$ vertices: $\{E, A_1, A_2\} \times Q$.) If yes, then we found a witness arena.

Time complexity of the first step is $O(d^{O(|Q|^2)}(d|Q|)^{|Q|}|C|)$ (for each of $d^{O(|Q|^2)}$ functions, we have to do a BFS on a graph of size $(d|Q|)^{|Q|}$). The parity game in the fourth step can be solved with one of the known algorithm for solving parity games, e.g. with the classical one in time $O(O(|Q|)^{d/2})$. This is done $O(d^{O(|Q|^2)})$ times. Thus, the whole algorithm runs in time $O(d^{O(|Q|^2)}|Q|^{|Q|}|C|)$. □

In the proof above the witness arena we find is an ϵ-arena: we did not assign any colors to moves m_1 and m_2. If we want to check whether the given condition is A-half-positional or B-half-positional, similar constructions work. For B-half-positional determinacy, we need to not only choose the sets U_1 and U_2, but also assign specific colors c_1 and c_2 to both moves m_1 and m_2 in the algorithm above. For A-half-positional determinacy, we need to assign specific colors for targets of these two moves, and also a color for Eve's position E.

6 Examples of ω-Regular Half-Positional Winning Conditions

In this section we give some examples of ω-regular half-positional winning conditions. First, we show a class of half-positional winning conditions which generalizes the well known Rabin conditions (see [Kla92, Gra04]) and positional/ suspendable conditions ([Kop06]), and also has some nice closure properties. Note that most of these examples are not restricted to finite arenas. The only exception are the concave conditions, which are only finitely half-positional.

Definition 2. *For $S \subseteq C$, WB_S is the set of infinite words where elements of S occur infinitely often, i.e. $(C^*S)^\omega$. Winning conditions of this form are called* Büchi conditions. *Complements of Büchi conditons, $WB'_S = C^*(C-S)^\omega$ are called* co-Büchi conditions.

Theorem 3 ([Kop06]). *Let D be a determinacy type. Let $W \subseteq C^\omega$ be a winning condition, and $S \subseteq C$. If W is D-determined, then so is $W \cup WB_S$.*

Note that, by duality, Thm 3 implies that if W is D-determined, then so is $W \cap WB'_S$. This yields an easy proof of positional determinacy of parity conditions. It is enough to start with an empty winning condition (which is positionally determined) and apply Thm 3 and its dual n times.

Definition 3 ([Kop06]). *A* **suspendable winning strategy** *for player X is a pair (s, Σ), where $s : \mathrm{Pos} \cup \mathrm{Mov}^+ \to \mathrm{Mov}$ is a strategy, and $\Sigma \subseteq \mathrm{Mov}^*$, such that:*

- *s is defined for every finite play π such that $\mathrm{target}(\pi) \in \mathrm{Pos}_X$.*
- *every infinite play π that is consistent with s from some point t* [1] *has a prefix longer than t which is in Σ;*
- *Every infinite play π that has infinitely many prefixes in Σ is winning for X.*

We say that X has a **suspendable winning strategy in** Win_X *when he has a suspendable winning strategy in the arena $(\mathrm{Pos}_A \cap \mathrm{Win}_X, \mathrm{Pos}_E \cap \mathrm{Win}_X, \mathrm{Mov} \cap \mathrm{Win}_X \times \mathrm{Win}_X \times C)$.*

A winning condition W is **positional/suspendable** *if for each arena G in the game (G, W) Eve has a positional winning strategy in Win_E and Adam has a suspendable winning strategy in Win_A.*

Intuitively, if at some moment X decides to play consistently with s, the play will eventually reach Σ; Σ is the set of moments when X can temporarily suspend using the strategy s and return to it later without a risk of ruining his or her victory.

A suspendable winning strategy is a winning strategy, because the conditions above imply that each play which is always consistent with s has infinitely many prefixes in Σ, and thus is winning for X.

The parity condition WP_2 is positional, but not positional/suspendable, because a suspendable strategy cannot be winning for Adam — it is possible that the play enters state 2 infinitely many times while it is suspended. However, we will now extend the class of positional/suspendable conditions to include also parity (and Rabin) conditions.

Definition 4. *The class of* **extended positional/suspendable (XPS)** *conditions over C is the smallest set of winning conditions that contains all Büchi and positional/ suspendable conditions, is closed under intersection with co-Büchi conditions, and is closed under finite union.*

Theorem 4. *All XPS conditions are half-positional.*

---

[1] That is, for each prefix u of π which is longer than t and such that $\mathrm{target}(u) \in \mathrm{Pos}_X$, the next move is given by $s(u)$.

The proof is a modification and generalization of proof of half-positional determinacy of Rabin conditions from [Gra04].

Proof. We use the following lemma:

Lemma 1 ([Kop06]). *Let D be a determinacy type. Let $W \subseteq C^\omega$ be a winning condition. Suppose that, for each non-empty D-arena G over C, there exists a non-empty subset $M \subseteq G$ such that in game (G, W) one of the players has a D-strategy winning from M. Then W is D-determined.*

Let W be an XPS condition. The proof is by induction over construction of W.

We know that Büchi conditions and positional/suspendable conditions are half-positional.

If W is a finite union of simpler XPS conditions, and one of them is a Büchi condition WB_S, then $W = W' \cup WB_S$. Then W' is half-positional since it is a simpler XPS condition, and from Theorem 3 we get that W is also half-positional.

Otherwise, $W = W' \cup \bigcup_{k=1}^{n}(W_k \cap WB'_{S_k})$, where W' is a positional/ suspendable condition, W_k is a simpler XPS condition, and WB'_{S_k} is a co-Büchi condition. (It is also possible that there is no W', but it is enough to consider this case since it is more general. A union of several positional/suspendable conditions is also positional/suspendable [Kop06].) To apply Lemma 1 we need to show that either Eve has a positional winning strategy from some position in the arena, or Adam has a winning strategy everywhere.

For $m = 1, \ldots, n$ let $W^{(m)} = W' \cup W_m \cup \bigcup_{k \neq m}(W_k \cap WB'_{S_k})$. We know that $W^{(m)}$ is half-positional since it is a simpler XPS condition.

Let G be an arena. Let H_m be the greatest subgraph of G which has no moves colored with any of the colors from S_m, Adam's positions where he can make a move not in H_m, and moves which lead to positions which are not in H_m (i.e. it is the subgraph where Adam is unable to force doing a move colored with a color from S_m). If Eve has a positional winning strategy from some starting position v in $(H_m, W^{(m)})$, then she can use the same strategy in (G, W) and win (Eve has more options in G hence she can use the same strategy, and this strategy forces moves colored with S_m to never appear).

Assume that Eve has no positional strategy for any starting position and m. Then Adam has a following winning strategy in (G, W):

- Adam uses his suspendable strategy (s, Σ) for the game (G, W'), until the play reaches Σ.
- For $m = 1, \ldots, n$:
  - Let v be the current position.
  - If $v \in H_m$ then Adam uses his winning strategy s'_m in $(H_m, W^{(m)})$. (Adam forgets what has happened so far in order to use s'_m.) If Eve never makes a move which does not belong to H_m then Adam wins. Otherwise, he stops using s'_m in some position v.
  - If $v \notin H_m$ then Adam performs a sequence of moves which finally lead to a move colored with S_m. (He does not need to do that if Eve made a move which is not in H_m, since it is already colored with S_m.)
- Repeat.

If finally the game remains in some H_m, then Adam wins since he is using a winning strategy in $(H_m, W^{(m)})$. Otherwise, Adam wins W' and all the co-Büchi conditions WB'_{S_k} for $k = 1, \ldots, n$, hence he also wins $W \subseteq W' \cup \bigcup_{k=1}^{n} WB'_{S_k}$. □

Among positional/suspendable (and thus XPS) conditions we have monotonic conditions and geometrical conditions, introduced in [Kop06]. For convenience, we have included the definition of monotonic conditions since they are ω-regular. Monotonic conditions are closed under finite union and include co-Büchi conditions. Another example of a monotonic condition is the complement of the set of words containing a^n (alternatively ba^n or $a^n b$) infinitely many times.

Definition 5. *A* **monotonic automaton** $A = (n, \delta)$ *over an alphabet* C *is a deterministic finite automaton where:*

- *the set of states is* $Q = \{0, \ldots, n\}$*;*
- *the initial state is 0, and the accepting state is n;*
- *the transition function* $\delta : Q \times C \to Q$ *is monotonic in the first component, i.e.,* $q \leq q'$ *implies* $\delta(q, c) \leq \delta(q', c)$*.*

The language accepted by A (L_A) *is the set of words* $w \in C^*$ *such that* $\delta(0, w) = n$*. A* **monotonic condition** *is a winning condition of form* $WM_A = C^\omega - L_A^\omega$ *for some monotonic automaton A.*

Another interesting class of finitely half-positional winning conditions are the concave conditions, also introduced in [Kop06]. This class is closed under union and includes parity conditions, and ω-regular concave conditions have nice algorithmic properties. Note that concave conditions do not need to be *infinitely* half-positional (although counterexamples known to us are not ω-regular).

Definition 6. *A word* $w \in \Sigma^* \cup \Sigma^\omega$ *is a (proper)* **combination** *of words* w_1 *and* w_2*, iff for some sequence of words* (u_n), $u_n \in \Sigma^*$, $w = \prod_{k \in \mathbb{N}} u_k = u_0 u_1 u_2 u_3 u_4 u_5 u_6 u_7 u_8 \ldots$, $w_1 = \prod_{k \in \mathbb{N}} u_{2k+1} = u_1 u_3 u_5 u_7 \ldots$, $w_2 = \prod_{k \in \mathbb{N}} u_{2k} = u_0 u_2 u_4 u_6 \ldots$*. A winning condition W is* **convex** *if as a subset of* C^ω *it is closed under combinations, and* **concave** *if its complement is convex.*

Proposition 6.1 ([Kop06]) *Suppose that a winning condition W is given by a deterministic parity automaton on infinite words using s states and d ranks. Then there exists a polynomial algorithm determining whether W is concave (or convex). If W is concave and G is an arena with n positions, then the winning sets and Eve's positional strategy can be found in time* $O(n(ns)^{d/2} \log s)$.

7 Conclusion and Future Work

The main result of this paper is decidability of finite half-positional determinacy of ω-regular conditions (Theorem 2). We also have introduced a large class of half-positionally determined conditions (XPS). This is a generalization of Rabin conditions which includes positional/suspendable ones, such as monotonic conditions introduced in [Kop06].

There are still many questions open. Is our algorithm optimal? Showing lower complexity bounds by reducing other known hard problems (e.g. solving parity games) seems difficult in this case. We would have to construct a relevant winning condition (do not forget about prefix independence!), but *not* the arena — our winning condition is required to have the desired properties for *all* arenas. It is still possible that simpler characterizations of ω-regular half-positional winning conditions exist.

We have only shown that *finite* half-positional determinacy is decidable. What about infinite arenas? We conjecture that the classes of finitely half-positional and half-positional conditions coincide for ω-regular languages, since all examples of finitely half-positional conditions which are not half-positional known to us are not ω-regular. However, a proof is required.

There was also an interesting question raised in section 3. In this paper we assume that not all moves need to be colored. Does this assumption in fact reduce the class of half-positional conditions? The algorithm given in Theorem 2 is more natural when we allow colorless moves, which motivates our opinion that C-half-positional determinacy is a natural notion.

XPS is a very big class, since it generalizes both Rabin conditions and positional/suspendable conditions (including monotonic conditions). This class is closed under finite union; however, we would also like to know whether the whole class of (finitely) half-positional conditions is closed under finite and countable union. In [Kop06] we have shown some more cases where union is also half-positional, and also that it is not closed under uncountable union.

References

[CN06] Colcombet, T., Niwiński, D.: On the positional determinacy of edge-labeled games. Theor. Comput. Sci. 352, 190–196 (2006)

[EJ91] Emerson, E.A., Jutla, C.S.: Tree automata, mu-calculus and determinacy. In: Proceedings 32th Annual IEEE Symp. on Foundations of Comput. Sci., pp. 368–377. IEEE Computer Society Press, Los Alamitos (1991)

[Gra04] Grädel, E.: Positional Determinacy of Infinite Games. In: Diekert, V., Habib, M. (eds.) STACS 2004. LNCS, vol. 2996, pp. 4–18. Springer, Heidelberg (2004)

[GTW02] Grädel, E., Thomas, W., Wilke, T. (eds.): Automata, Logics, and Infinite Games. LNCS, vol. 2500. Springer, Heidelberg (2002)

[GW06] Grädel, E., Walukiewicz, I.: Positional determinacy of games with infinitely many priorities. Logical Methods in Computer Science 2(4:6), 1–22 (2006)

[GZ04] Gimbert, H., Zielonka, W.: When can you play positionally? In: Fiala, J., Koubek, V., Kratochvíl, J. (eds.) MFCS 2004. LNCS, vol. 3153, pp. 686–697. Springer, Heidelberg (2004)

[GZ05] Gimbert, H., Zielonka, W.: Games Where You Can Play Optimally Without Any Memory. In: Abadi, M., de Alfaro, L. (eds.) CONCUR 2005. LNCS, vol. 3653, pp. 428–442. Springer, Heidelberg (2005)

[Kla92] Klarlund, N.: Progress measures, immediate determinacy, and a subset construction for tree automata. In: Proc. 7th IEEE Symp. on Logic in Computer Science (1992)

[Kop06] Kopczyński, E.: Half-Positional Determinacy of Infinite Games. In: Bugliesi, M., Preneel, B., Sassone, V., Wegener, I. (eds.) ICALP 2006. LNCS, vol. 4052, pp. 336–347. Springer, Heidelberg (2006)

[Kop07] Kopczyński, E.: Half-positional determinacy of infinite games. Draft. `http://www.mimuw.edu.pl/~erykk/papers/hpwc.ps`

[Mar75] Martin, D.A.: Borel determinacy. Ann. Math. 102, 363–371 (1975)

[McN93] McNaughton, R.: Infinite games played on finite graphs. Annals of Pure and Applied Logic 65, 149–184 (1993)

[Mos91] Mostowski, A.W.: Games with forbidden positions. Technical Report 78, Uniwersytet Gdański, Instytut Matematyki (1991)

[Zie98] Zielonka, W.: Infinite Games on Finitely Coloured Graphs with Applications to Automata on Infinite Trees. Theor. Comp. Sci. 200(1-2), 135–183 (1998)

Clique-Width and Parity Games

Jan Obdržálek*

Faculty of Informatics, Masaryk University, Brno, Czech Republic
obdrzalek@fi.muni.cz

Abstract. The question of the exact complexity of solving parity games is one of the major open problems in system verification, as it is equivalent to the problem of model-checking the modal μ-calculus. The known upper bound is NP∩co-NP, but no polynomial algorithm is known. It was shown that on tree-like graphs (of bounded tree-width and DAG-width) a polynomial-time algorithm does exist. Here we present a polynomial-time algorithm for parity games on graphs of bounded clique-width (class of graphs containing e.g. complete bipartite graphs and cliques), thus completing the picture. This also extends the tree-width result, as graphs of bounded tree-width are a subclass of graphs of bounded clique-width. The algorithm works in a different way to the tree-width case and relies heavily on an interesting structural property of parity games.

1 Introduction

Parity games are infinite two-player games of perfect information played on directed graphs where the vertices are labelled by priorities. The players take turns in pushing token along the edges of the graph. The winner is determined by the parity of the highest priority occurring infinitely often in this infinite play. The significance of parity games comes from the area of model-checking and system verification. The modal μ-calculus, a fixed-point logic of programs, was introduced by Kozen in [13] and subsumes most of the used modal and temporal logics. The problem of determining, given a system $\mathcal{A}$ and a formula φ of the modal μ-calculus, whether or not $\mathcal{A}$ satisfies φ can be translated to the problem of solving a parity game – see e.g. [9,16] (the converse is also true).

The exact complexity of solving parity games is a long-standing open problem which received a large amount of attention. The best known upper bound is NP∩co-NP (more precisely UP∩co-UP [10]), but no polynomial-time algorithm is known. Up till recently the best known deterministic algorithm was the small progress measures algorithm [11] with time complexity $\mathcal{O}(dm \cdot (2n/d)^{d/2})$, where n is the number of vertices, m the number of edges and d the number of priorities. The first truly sub-exponential algorithm was the randomised algorithm by Björklund et al. [3], with the complexity bounded by $2^{\mathcal{O}(\sqrt{n \log n})}$. Recently Jurdziński et al. [12] came with a very simple deterministic sub-exponential

* Supported by the Institute of Theoretical Computer Science, project 1M0545.

J. Duparc and T.A. Henzinger (Eds.): CSL 2007, LNCS 4646, pp. 54–68, 2007.

algorithm of complexity roughly matching the upper bound of the randomised algorithm. Finally there is the strategy improvement algorithm of Vöge and Jurdziński [19] (and some others - e.g. [16]) which behaves extremely well in practice, but for which we do not have better than exponential upper bound.

Since so far we have not been able to find a polynomial-time algorithm for solving parity games on general graphs, the question is whether we can do better for some restricted classes of graphs. In [15] the author showed that there is a polynomial-time algorithm for solving parity games on graphs of bounded tree-width. (Tree-width, introduced by Robertson and Seymour in [18], is a well known and phenomenally successful connectivity measure for undirected graphs. Check [16] for a more elegant proof.) The result of [15] does not follow from the general result of Courcelle [4] which states that for a fixed formula φ of MSO and a graph of bounded tree-width the model-checking problem can be solved in linear time in the size of the graph. The important difference is that [4] considers the formula to be fixed, whereas in parity games the size of the formula describing the winning condition depends on the number of priorities, which can be as high as the number of vertices. (The constant factor of the algorithm presented in [4] depends on both n and k, and it is not even elementary [8]).

The way the concept of tree-width is applied to directed graphs is that we forget the orientation of the edges. That unfortunately means that tree-width can be high even for a very simple directed graphs (e.g. directed cliques). Therefore a new connectivity measure called DAG-width has been developed [17,1] to better describe the connectivity of directed graphs. In the latter paper it was shown that on graphs of bounded DAG-width parity games can be solved in polynomial time. Another class of directed graphs for which we have a polynomial-time algorithm are graphs of bounded entanglement [2].

Both tree-width and DAG-width measure how close a graph is to a relatively simple/sparse graph (e.g. tree, DAG). Clique-width, defined in [6], stands on the opposite end of the spectrum – it measures how close is a graph to a complete bipartite graph (e.g. complete bipartite graphs and cliques have clique-width 2). In [6] it was also shown that bounded tree-width implies bounded clique-width, and so clique-width can be seen both as a generalisation and as a "dual notion" to tree-with. In this paper we present a polynomial-time algorithm for solving parity games on graphs of bounded clique-width. As argued above this both extends and complements our understanding of the complexity of solving parity games. As is the case for tree-width (see above), the result is not a consequence of a general result saying that MSO logic is decidable in linear time on graphs of bounded clique-width [5]. The reasons are the same as for tree-width.

The work is organised as follows: In Section 2 we present parity games, and prove a general property of parity games which is in the core our algorithm (but can be likely of use elsewhere). In Section 3 we present the definition of clique-width and some extra notation. Section 4 then contains the main result – the algorithm for solving parity games on graphs of bounded clique-width.

2 Parity Games

A *parity game* $\mathcal{G} = (V, E, \lambda)$ consists of a finite directed graph $G = (V, E)$, where V is a disjoint union of V_0 and V_1 (we assume that this partition is implicit), and a parity function $\lambda : V \to \mathbb{N}$ (we assume $0 \notin \mathbb{N}$). As it is usually clear from the context, we sometimes talk about a parity game G – i.e. we identify the game with its game graph. For technical reasons we also assume that the edge relation $E : V \times V$ is total: that is, for all $u \in V$ there is $v \in V$ such that $(u, v) \in E$. The game $\mathcal{G}$ is played by two players P_0 and P_1 (also called EVEN and ODD), who move a single token along edges of the graph G. The game starts in an initial vertex and players play indefinitely as follows: if the token is on a vertex $v \in V_0$ ($v \in V_1$), then P_0 (P_1) moves it along some edge $(v, w) \in E$ to w. As a result, a play of $\mathcal{G}$ is an infinite path $\pi = \pi_1\pi_2\ldots$, where $\forall i > 0.(\pi_i, \pi_{i+1}) \in E$.

Let $Inf(\pi) = \{v \in V \mid v$ appears infinitely often in $\pi\}$. Player P_0 *wins the play* π if $\max\{\lambda(v) \mid v \in Inf(\pi)\}$ is even, and otherwise player P_1 wins. A *(total) strategy* σ (τ) for P_0 (P_1) is a function $\sigma : V^*V_0 \to V$ $(\tau : V^*V_1 \to V)$ which assigns to each play $\pi.v \in V^*V_0$ $(\in V^*V_1)$ a vertex w such that $(v, w) \in E$. A player *uses a strategy* σ in the play $\pi = \pi_1\pi_1\ldots\pi_k\ldots$, if $\pi_{k+1} = \sigma(\pi_1\ldots\pi_k)$ for each vertex $\pi_k \in V_i$. A strategy σ is *winning* for a player and a vertex $v \in V$ if she wins every play that starts from v using σ.

If we fix an initial vertex v, then we say player P_i *wins the game* $\mathcal{G}(v)$ if he has a strategy σ such that using σ he wins every play starting in v. Finally we say that player *wins the game* $\mathcal{G}$ if he has a strategy σ such that using σ he wins the game $\mathcal{G}(v)$ for each $v \in V$. By *solving* the game $\mathcal{G}$ we mean finding the winner of $\mathcal{G}(v)$ for each vertex $v \in V$.

A *memoryless strategy* σ (τ) for P_i $(i \in \{0, 1\})$ is a function $\sigma : V_0 \to V$ $(\tau : V_1 \to V)$ which assigns each $v \in V_i$ a vertex w such that $(v, w) \in E$. I.e. memoryless strategies do not consider the history of the play so far, but only the vertex the play is currently in. We use Σ_i to denote the set of memoryless strategies of player P_i, and $\overline{\Sigma_i}$ the sets of all strategies. For a strategy $\sigma \in \Sigma_0$ and $(v, w) \in V_0 \times V$ we define a strategy $\sigma[v \to w]$ which agrees with σ on all vertices except for v by the prescription

$$\sigma[v \to w](x) = \begin{cases} w & \text{if } x = v \\ \sigma(x) & \text{otherwise} \end{cases}$$

Parity games are memorylessly determined. By that we mean the following theorem.

Theorem 1 ([7,14]). *For each parity game $\mathcal{G} = (V, E, \lambda)$ we can partition the set V into two sets W_0 and W_1 such that the player P_i has a memoryless winning strategy for all vertices in W_i.*

Example 1. Fig. 1 shows a parity game of six vertices. Circles denote the vertices of player P_0 and boxes the vertices of player P_1. Priorities are written inside vertices. In this game player P_0 can win from the shaded vertices by forcing a play to the vertex with priority four. Player P_1 has no choice in that vertex and

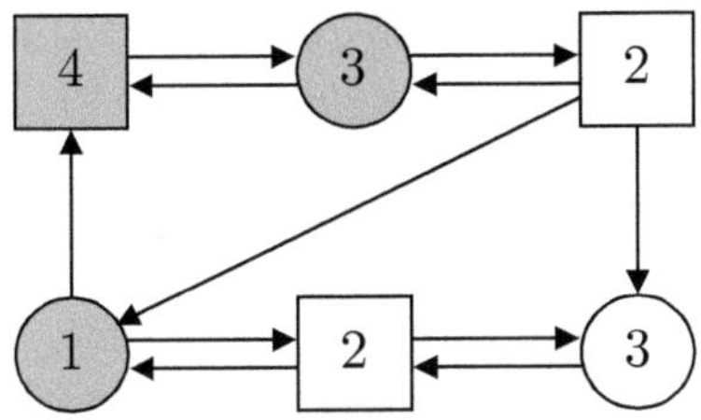

Fig. 1. A parity game

must play to the vertex with priority three. The play will stay in the cycle with the highest priority four and therefore P_0 wins.

Definition 1 (G_σ)**.** *For game $\mathcal{G} = (V, E, \lambda)$ and a strategy $\sigma \in \Sigma_0$ we define G_σ to be the subgraph of G induced by σ - i.e. $G_\sigma = (V, E_\sigma)$, where $E_\sigma = E \setminus \{(v, w) \in E \mid v \in V_0 \text{ and } \sigma(v) \neq w\}$.*

An immediate observation is that $\sigma \in \Sigma_0$ is winning for P_0 from v iff no cycle such that the highest priority on this cycle is odd is reachable from v in G_σ.

In addition to the standard ordering of priorities (by the relation "$<$"), it is often useful to have priorities ordered from the point of their "attractiveness" for one of the players. I.e. for player P_0 a high even priority is more attractive than a low even one, which is still more attractive than any odd priority. We define the order $\sqsubseteq$ in the following way:

Definition 2 ($\sqsubseteq$)**.** *For two priorities $p, q \in \mathbb{N}$ we write $p \sqsubset q$ if p is odd and v is even, or $p > q$ and p, q are odd, or $p < q$ and p, q are even. We write $p \sqsubseteq q$ if $p \sqsubset q$ or $p = q$.*

We need a way of describing partial results for plays. More specifically we need to know the result for all paths between two specified vertices within a subgraph G' of G when a strategy σ of P_0 is fixed. Let $\Pi_\sigma(v, w, G')$ be the set of all paths in G'_σ from v to w. If $\Pi_\sigma(v, w, G') \neq \emptyset$ we define

$$result_\sigma(v, w, G') = \min_{\sqsubseteq}\{\max\{\lambda(\pi_i) \mid 0 \leq i \leq j\} \mid v = \pi_0, \pi_1, \ldots, \pi_j = w \in \Pi_\sigma(v, w, G')\}$$

If $G' = G$ we write just $result_\sigma(v, w)$. We also write $x \to_\sigma y$ if $(x, y) \in E(G_\sigma)$, and use $\to_\sigma^*$ to denote the reflexive and transitive closure of $\to_\sigma$.

2.1 Joint Choice Property

The following theorem is in the core of the proof of the main result of this paper. It basically says that if we have a set $U \subseteq V_0$ of vertices of P_0 such that they have the same sets of successors and a memoryless strategy σ winning for at least one of these vertices, then there is a memoryless strategy σ' which assigns to each $u \in U$ the same successor and is winning for all the vertices of U and all the vertices for which σ is a winning strategy for P_0.

Theorem 2. *Let $\mathcal{G} = (V, E, \lambda)$ be a parity game, $U \subseteq V_0$ a set of vertices of P_0 and $\sigma \in \Sigma_0$ a strategy of P_0. Also denote $\sigma(U) = \bigcup_{u \in U} \sigma(u)$ and require that $\forall u \in U\ \forall v \in \sigma(U).(u,v) \in E$. If σ is a winning strategy for some vertex $w \in U$ and P_0, then there exists a vertex $\overline{u}$ in $\sigma(U)$ such that the strategy σ' defined as*

$$\sigma'(v) = \begin{cases} \overline{u} & \text{if } v \in U \\ \sigma(v) & \text{otherwise} \end{cases}$$

is a winning strategy for P_0 from all vertices of U. Moreover for all $v \in V(G)$ if σ is winning for v and P_0, so is σ'.

Proof. First note that we can assume that σ is winning for all vertices of U. If it is not, we can simply switch the strategy into any winning successor. Let $\sigma \in \Sigma_0$ be fixed and let us take a set $U \subseteq V_0$ satisfying the requirements above. We will proceed by induction on the size of U. If U is a set with a single vertex u, then we just put $\sigma' = \sigma$ as σ must be a winning strategy for P_0 and u.

Therefore assume $|U| = k + 1$ for some $k \in \mathbb{N}$. Let $U = X \cup \{x'\}$, where $X \subset U$ has k elements. By induction hypothesis there exists $y \in \sigma(X)$ such that the strategy $\overline{\sigma}$ defined as $\overline{\sigma}(x) = y$ for $x \in X$ and identical with σ on all other vertices is winning for P_0 from all vertices of X.

If $\sigma(x') = \overline{\sigma}(x') = y$ we put $\sigma' = \overline{\sigma}$ and we are done, since $\overline{\sigma}$ is also winning from x' and satisfies the requirements. Otherwise $y' = \overline{\sigma}(x') \neq y$ and there are three cases to be considered (see Fig. 2):

1. If $y \not\to^*_{\overline{\sigma}} x'$ we can safely set $\sigma' = \overline{\sigma}[x' \to y]$, as this change of strategy cannot introduce a new cycle to $G_{\overline{\sigma}}$. Obviously σ' is winning for x' as $\overline{\sigma}$ is winning for y.
2. If $y \to^*_{\overline{\sigma}} x'$ but $y' \not\to^*_{\overline{\sigma}} x$ for all $x \in X$ we can similarly put $\sigma' = \overline{\sigma}_{x \in X}[x \to y']$.
3. The last case is there are both a path from y to x' and a set of paths from y' to X in the graph $G_{\overline{\sigma}}$. As in the previous case $\overline{\sigma}$ is winning for both y and y', and therefore no odd cycle is reachable from y and y' in $G_{\overline{\sigma}}$. Let us put $p_1 = \mathit{result}_{\overline{\sigma}}(y, x')$ and $p_2 = \min_{\sqsubseteq} \bigcup_{x \in X} \mathit{result}_{\sigma}(y', x)$. Note that $p = \max(p_1, p_2)$ must be even, otherwise there would be an odd cycle through x' in $G_{\overline{\sigma}}$, which is not possible since $\overline{\sigma}$ is winning for all $x \in X$. Now if $p = p_1$ put $\sigma' = \overline{\sigma}[x' \to y]$, otherwise $\sigma' = \overline{\sigma}_{x \in X}[x \to y']$.
   Let us assume the first case ($p = p1$). Then all newly created cycles must use the edge (x, y'), and for the highest priority p on such a cycle it must be the case that $p_1 \sqsubseteq p$. By definition of p_1 all the newly created cycles are winning for P_0. The case $p = p_2$ is similar.

3 Clique Width

The notion of clique-width was first introduced by Courcelle and Olariu in [6]. There are actually two definitions of clique-width in that paper, one for directed

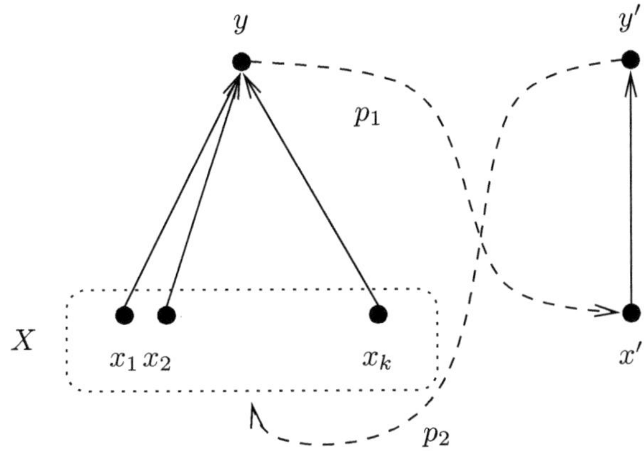

Fig. 2. Illustration of the proof of Theorem 2

graphs and one for undirected graphs. We will present the former definition, as it is more suitable for parity games, graphs of which are directed.

Let k be a positive integer. We call (G, γ) a *k-graph* if G is a graph and $\gamma : V(G) \to \{1, 2, \ldots, k\}$ is a mapping. (As γ is usually fixed, we often write just G for the k-graph (G, γ).) We call $\gamma(v)$ for $v \in V(G)$ the *label* of a vertex v (also the *colour* of v). We also define the following operators:

1. For $i \in \{1, 2, \ldots, k\}$ let $[i]$ denote an isolated vertex labelled by i.
2. For $i, j \in \{1, 2, \ldots, k\}$ with $i \neq j$ we define a unary operator $\alpha_{i,j}$ which adds edges between all pairs of vertices with label i and j. Formally
$$\alpha_{i,j}(G, \gamma) = (G', \gamma)$$
where $V(G') = V(G)$ and $E(G') = E(G) \cup \{(v, w) \mid v, w \in V, \gamma(v) = i \text{ and } \gamma(w) = j\}$.
3. For $i, j \in \{1, 2, \ldots, k\}$ let $\rho_{i \to j}$ be the unary operator which relabels every vertex labelled by i to j. Formally
$$\rho_{i \to j}(G, \gamma) = (G, \gamma')$$
where
$$\gamma'(v) = \begin{cases} j & \text{if } \gamma(v) = i \\ \gamma(v) & \text{otherwise} \end{cases}$$
4. Finally $\oplus$ is a binary operation which makes a disjoint union of two k-graphs. Formally
$$(G_1, \gamma_1) \oplus (G_2, \gamma_2) = (G, \gamma)$$
where $V(G) = V(G_1) \uplus V(G_2)$, $E(G) = E(G_1) \uplus E(G_2)$ and
$$\gamma(v) = \begin{cases} \gamma_1(v) & \text{if } v \in V(G_1) \\ \gamma_2(v) & \text{if } v \in V(G_2) \end{cases}$$

A *k-expression* is a well formed expression t written using the operators defined above. The k-graph produced by performing these operations in order therefore has as a vertex set the set of all occurrences of the constant symbols in t, and this k-graph (or any k-graph isomorphic to it) is called the *value* $val(t)$ of t. If a k-expression t has value (G, γ) we say that t is a *k-expression corresponding to G*. The *clique-width* of G, written as $cwd(G)$, is the minimum k such that there is a k-expression corresponding to G.

Example 2. $\alpha_{1,2}(\rho_{1\to 2}(\alpha_{1,2}(\rho_{1\to 2}(\alpha_{1,2}([1] \oplus [2])) \oplus [1])) \oplus [1])$ is a 2-expression corresponding to a directed clique of size 4. See Fig. 3.

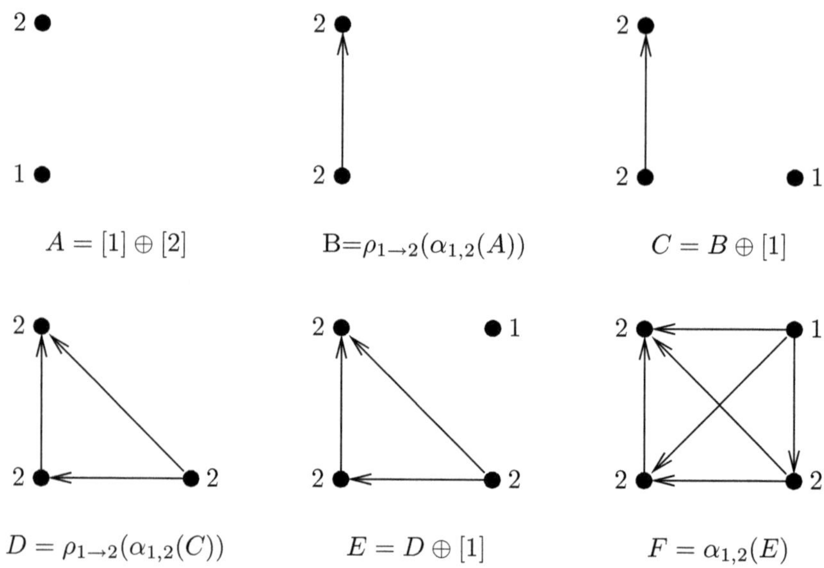

Fig. 3. Construction of the directed clique of size 4

Note aside: For undirected graphs the clique-width is defined in just the same way, except for the operator $\alpha_{i,j}$. In undirected case it is replaced by an operator $\eta_{i,j}$ which creates undirected edges between all vertices with colours i and j.

It is quite natural to view a k-expression t_G corresponding to G as a tree T with nodes labelled by subterms of t_G (t_G is the root), together with a bijection between the leaves of the tree and vertices of G. In this setting the *type* of each node $t \in V(T)$ is the top-level operator of t, so we have four different node types.

In the rest of this paper we will always have G and a k-term t_G (and therefore also its tree T) corresponding to G fixed. We will also use the following notation: For a node $t \in V(T)$ let $G[t]$ be the subgraph of G given by t, and $V[t]$ ($E[t]$) the set of its vertices (edges). Slightly abusing the notation we use $\gamma(v, t)$ to denote the colour of v in the node t, $\gamma(i, t)$ to denote the set of vertices of G which have the colour i at the node t, and $\gamma(t)$ to denote the set of all colours whose associated vertex sets for t contain at least one vertex.

3.1 k-Expressions and t-Strategies

For the purposes of our algorithm we have to consider a special kind of memoryless strategies, called *t-strategies*. Such strategies assign to all free (see below) even vertices of the same colour the same successor. Let $\mathcal{G} = (V, E, \lambda)$ be a parity game, where $G = (V, E)$ is of clique-width k and t is the corresponding k-expression (and T its associated tree). A vertex $v \in \gamma(i, t)$ is *free* for $\sigma \in \Sigma_0$ if $(v, \sigma(v)) \notin E[t]$. A strategy $\sigma \in \Sigma_0$ is called a *t-strategy* if for each $t' \in V(T)$ of type $\alpha_{i,j}(t'')$ we have that for all free $v, w \in \gamma(i, t'') \cap V_0$ it is true that $\sigma(v) = \sigma(w)$. We use Σ_0^t to denote the set of all t-strategies.

Theorem 3. *Let $\mathcal{G} = (V, E, \lambda)$ be a parity game, and t the corresponding k-expression to $G = (V, E)$. Let $\sigma \in \Sigma_0$ be a strategy winning for P_0 from $W \subseteq V$. Then there is a t-strategy $\sigma' \in \Sigma_0^t$ such that σ' is winning from W.*

Proof (sketch). By repeated application of Theorem 2 at nodes of type $\alpha_{i,j}(t')$ following the structure of t from the leaves upwards. Let $X \subseteq \gamma(i, t')$ be the set of all free vertices for σ. Then from now on they can only be connected to the same common successors. Note that edges from X to $\sigma(X)$ must be added either in this node or somewhere higher up the tree T. If σ is winning for some $x \in X$, then we can apply the Theorem 2 to find the common winning successor $\overline{u}$, and change the strategy σ. If this $\overline{u}$ is in $\gamma(j, t')$ we are done for this node. Otherwise we deal with X higher up in T. If σ is not winning for any vertex of X, we select a common successor at random.

4 Algorithm

In this section we present the main result – that parity games can be solved in polynomial time on graphs of bounded clique-width. The algorithm described here is in spirit similar to the one for graphs of bounded tree-width [15] and bounded DAG-width [1,16]. However there are many conceptual differences, as there is no small set of vertices forming an *interface* between an already processed subgraph and the rest of the graph. To the contrary, in one step we can connect an unbounded number of edges of G. The problem is we cannot keep the results for all the individual vertices of one colour. This is the main obstacle we have to deal with.

For the rest of this section let us fix a parity game $\mathcal{G} = (V, E, \lambda)$, where $G = (V, E)$ is of clique-width k and t_G is the corresponding k-expression (and T its associated tree). In addition we will assume that $\mathcal{G}$ is bipartite - i.e. that $E \subseteq (V_0 \times V_1) \cup (V_1 \times V_0)$. This requirement implies that for each $t \in V(T)$ and each colour i we have that either $\gamma(i, t) \subseteq V_0$ or $\gamma(i, t) \subseteq V_1$. (If vertices of both players have the same colour, then there cannot be any edge added to or from this colour.) We therefore talk about *even colours* – containing the vertices from V_0, and similarly of *odd colours*. This will allow us to simplify our construction. At the end of this section we discuss how we can modify the construction so it does not require this assumption.

In the previous section we have seen the definition of $result_\sigma$ for the set of all paths between two vertices in a subgraph G' of G. We need to extend this definition to colours and subgraphs $G[t]$ for $t \in V(T)$.

For a vertex $v \in V[t]$ and strategy $\sigma \in \overline{\Sigma_0}$ we want to find the best result the player P_1 can achieve against this strategy for a play starting from v within $G[t]$. If there is a winning cycle of P_1 (i.e. the highest priority on this cycle is odd) reachable from v in the game $G[t]$ when P_0 uses σ, then we know P_1 can win against the strategy σ if starting in v in the game G. Falling short of this objective the player P_1 can force a play to some vertex $w \in \gamma(i,t)$, with the hope of extending this play to a winning play as more edges and vertices are added. The 'value' of such play is the highest priority of a vertex on this path. However note that there can be more paths starting in v which lead to a vertex of a given colour i. In that case it is in the player P_1's interest to choose the one with the lowest score w.r.t. the '$\sqsubseteq$' ordering. For each colour in $\gamma(t)$ we remember the best such result. But not all vertices w can be used to extend a play. As σ is fixed, only vertices from V_1 and those vertices of V_0 at the end of a maximal finite path (i.e. free vertices) qualify.

To formalise the description above, we need to extend the '$\sqsubseteq$' ordering to pairs (i,p). For two such pairs (i,p),(j,q) we put $(i,p) \sqsubseteq (j,q)$ iff $i = j$ and $p \sqsubseteq q$. Specifically if $i \neq j$ then the two pairs are incomparable. We extend the ordering $\sqsubseteq$ by adding the minimal element $\bot$. For a set $X \subseteq (\{1,\dots,k\} \times \mathbb{N}) \cup \{\bot\}$ we denote $\min_\sqsubseteq X$ to be the set of $\sqsubseteq$-minimal elements of X. Note that $\min_\sqsubseteq X = \{\bot\}$ iff $\bot \in X$. Moreover for $Z = \min_\sqsubseteq X$ if $Z \neq \{\bot\}$ then Z contains at most k pairs (i,p), one for each colour i.

We consider the maximal plays first. Starting from a vertex $v \in V[t]$ and given a strategy $\sigma \in \overline{\Sigma_0}$ let Π be the set of maximal paths in $G[t]$ using σ. For a finite path $\pi = \pi_0,\dots,\pi_j$ we define the operator $r(\pi) = (\gamma(\pi_j), \max\{\lambda(\pi_i) \mid 0 \leq i \leq j\})$. Then we put

$$result_\sigma(v, G[t]) = \begin{cases} \bot & \text{if } \exists \pi \in \Pi \text{ s.t. } \pi \text{ is infinite and winning for } P_1 \\ \min_\sqsubseteq \bigcup_{\pi \in \Pi} r(\pi) & \text{otherwise} \end{cases}$$

To the set defined above we add also the results of all finite plays in $G[t]$ starting in v and ending in a vertex in V_1 (the set X). Finally we take only the minimal elements.

$$\begin{aligned} result_\sigma(v,t) &= \min_\sqsubseteq(X \cup Y) \\ X &= \{(\gamma(w), result_\sigma(v,w,G[t])) \mid w \in V_1 \cap V[t]\} \\ Y &= result_\sigma(v, G[t]) \end{aligned}$$

Note: The set $result_\sigma(v,t)$ will never be empty for $v \in V[t]$. Even when all the reachable cycles are won by P_0, there must be vertices in $V_1 \cap V[t]$ reachable from v since the game is bipartite.

Finally we want to say what happens when the play is restricted to $G[t]$ and starts in any vertex of colour i. The definition of $result_\sigma(i,t)$ depends on whether

i is odd or even colour. The reason for this is the following: if i is an odd colour, then all edges to vertices of colour i are from even vertices. Using a memoryless strategy the player P_0 chooses *one* vertex of colour i as a successor. On the other hand if i is an even colour, then the player P_1, playing against σ, can choose *any* vertex of colour i in the next step. This is formalised as follows:

$$\mathit{result}_\sigma(i,t) = \begin{cases} \{\mathit{result}_\sigma(v,t) \mid v \in \gamma(i,t)\} & \text{if } i \text{ is odd colour in } t \\ \min_{\sqsubseteq} \bigcup_{v\in\gamma(i,t)} \mathit{result}_\sigma(v,t) & \text{if } i \text{ is even colour in } t \end{cases}$$

For $t \in V(T)$ and $i \in \gamma(t)$ we define the set $\mathit{Result}(i,t)$ as the set of all possible results when starting in a (vertex of) colour i and restricted to $G[t]$.

$$\mathit{Result}(i,t) = \{\mathit{result}_\sigma(i,t) \mid \sigma \in \overline{\Sigma_0}\}$$

Note that in the definition above we do not require σ to be memoryless.

In our algorithm we use dynamic programming on T to compute sets $R(i,t)$ for each node t of T from the bottom up. We distinguish four cases (as we have four types of nodes). For each node we have to prove (by induction from the leaves) that the following two properties hold:

P1. $\forall i \in \gamma(t).R(i,t) \subseteq \mathit{Result}(i,t)$
P2. $\{\mathit{result}_\sigma(i,t) \mid \sigma \in \Sigma_0^t\} \subseteq R(i,t)$

The property P1 states that each member of $R(i,t)$ is a result for some strategy $\sigma \in \overline{\Sigma_0}$. The property P2 then means that the results for all t-strategies are included.

The following lemma states that if the sets $R(i,t)$ satisfy properties P1 and P2, then we can effectively solve the game $\mathcal{G}(v)$. Here we assume that v is the only vertex of colour i, it is created by the $[i]$ operator, and keeps its colour all the time till the end of the construction. It is not difficult to modify t (by adding an extra colour) so it satisfies this requirement.

Lemma 1. *Let $\mathcal{G} = (V,E,\lambda)$ be a parity game where $G = (V,E)$ is of clique-width k and t is the associated k-expression corresponding to G. Let $R(i,t)$ be a set satisfying P1 and P2, and $\gamma(i,t) = \{v\}$. Then P_0 wins the game starting in v iff $R(i,t) \neq \{\bot\}$.*

Proof. Let $\sigma \in \Sigma_0$ be a winning strategy of P_0 from v. By Theorem 3 there must be $\sigma' \in \Sigma_0^t$ such that σ' is also winning from v. By P2 $\mathit{result}_{\sigma'}(i,t) \in R(i,t)$ and by definition of $\mathit{result}_{\sigma'}(i,t)$ we must have $\mathit{result}_{\sigma'}(i,t) \neq \{\bot\}$.

On the other hand if $R(i,t) \neq \{\bot\}$, then by P1 there must be a strategy σ s.t. $\mathit{result}_\sigma(i,t) = \mathit{result}_\sigma(v,t) \neq \{\bot\}$. By definition of $\mathit{result}_\sigma(v,t)$ there is no odd cycle reachable from v if P_0 is using σ and therefore P_0 wins the game G for v.

4.1 Algorithms for Operators

Here we present the algorithms for all four types of nodes. Once the algorithm is given the proof is just a simple, if tedious, work and follows from the definition of result_σ. To keep the presentation clear and tidy we only sketch some of the proofs.

t is of type $[i]$. Let v be the corresponding vertex of G. Then we put $R(i,t) = \{(i, \lambda(v))\}$. In this case obviously $R(i,t) = \mathit{Result}(i,t)$ and therefore both P1 and P2 hold.

t is of type $\rho_{i\to j}(t')$. We have to deal with two separate issues: choose the "better" result (for P_1) for paths ending in i and j and "join" $R(i,t')$ and $R(j,t')$. To address the first issue we define the operation *mod* which takes $\alpha \in R(l,t)$ and works as follows: If α contains both (i,p_i) and (j,p_j), it replaces these two pairs by a single pair $(j, \min_{\sqsubseteq}(p_i,p_j))$ (and if α contains only a pair (i,p), it is replaced by (j,p)).

$$mod(\alpha) = \begin{cases} \min_{\sqsubseteq}\left((\alpha \cup \{(j,p)\}) \setminus \{(i,p)\}\right) & \text{if } (i,p) \in \alpha \text{ for some } p \\ \alpha & \text{otherwise} \end{cases}$$

The sets $R(l,t)$ for $l \neq i,j$ are then defined in a straightforward way:

$$R(l,t) = \{mod(\alpha) \mid \alpha \in R(l,t')\}$$

That $R(l,t)$ satisfies P1 and P2 follows from the fact that $R(l,t')$ does and the definition of $\mathit{result}_\sigma(l,t)$.

Now it remains to show how to compute the set $R(j,t)$. Let $S_i = \{mod(\alpha) \mid \alpha \in R(i,t')\}$ and $S_j = \{mod(\beta) \mid \beta \in R(j,t')\}$ the sets modified as above. To produce the set $R(j,t)$ we have to merge S_i and S_j. The way we do this depends on whether i,j are even or odd colours (by definition they must be either both odd or both even).

$$R(j,t) = \begin{cases} \{\min_{\sqsubseteq}(\alpha \cup \beta) \mid \alpha \in S_i, \beta \in S_j\} & \text{if } i,j \text{ are even} \\ S_i \cup S_j & \text{if } i,j \text{ are odd} \end{cases}$$

Proposition 1. *$R(j,t)$ satisfies P1 and P2.*

Proof. For odd colours i,j P1 follows from the definition of result_σ. For P2 let $\sigma \in \Sigma_0^t$. By P2 for t' we have $\mathit{result}_\sigma(i,t') \in R(i,t')$ and $\mathit{result}_\sigma(j,t') \in R(j,t')$. The rest follows from the fact that $\gamma(j,t) = \gamma(i,t') \cup \gamma(j,t')$.

For even i,j take $\alpha \in S_i$ and $\beta \in S_j$. Then by induction hypothesis $\alpha = mod(\mathit{result}_{\sigma_1}(i,t'))$ for some σ_1 and $\beta = mod(\mathit{result}_{\sigma_2}(j,t'))$ for some σ_2. Let σ be a strategy which behaves like σ_1 if we start in a vertex of $\gamma(i,t')$ and behaves like σ_2 if we start in a vertex of $\gamma(j,t')$. Then $\mathit{result}_\sigma(j,t) = \min_{\sqsubseteq}(\alpha \cup \beta)$. This proves P1.

For P2 and a strategy $\sigma \in \Sigma_0^t$ take $\alpha = \mathit{result}_\sigma(i,t')$ and $\beta = \mathit{result}_\sigma(j,t')$. By P2 for t' we have $\alpha \in R(i,t')$ and $\beta \in R(j,t')$. The rest follows from the definition of result_σ.

t is of type $\alpha_{i,j}(t')$. We are adding edges between colours i and j in this step. We start with computing the set $R(j,t)$ first. For each $\alpha \in R(j,t')$ we construct

α' by taking $\alpha' = \alpha$ and looking for any new winning cycles for P_1: If there is a pair $(i,p) \in \alpha$ we put $\bot$ into α' if p is odd. We then chose the minimal pairs in α' – the set $\min_{\sqsubseteq}(\alpha')$ – and call the result $mod_1(\alpha)$. This works well for odd i. However for even i if we prolong the path through i we have in addition to remove the pair (i,p) (as there is not now any free vertex of colour i) – and we call such set $mod_2(\alpha)$. The last case is we do not add the edges between the vertices of colour i, j. Altogether:

$$R(j,t) = \begin{cases} \{mod_1(\alpha) \mid \alpha \in R(j,t')\} & \text{if } i \text{ is odd} \\ \{mod_2(\alpha) \mid \alpha \in R(j,t')\} \cup R(j,t') & \text{if } i \text{ is even} \end{cases}$$

Note: We do not add the set $R(j,t')$ if there is no $w \in V \setminus V[t]$ such that $(u,w) \in E$ for $u \in \gamma(i,t)$. The reason is that we cannot postpone the choice of strategy for free vertices in $\gamma(i,t)$ as there is no choice left.

Proposition 2. *P1 and P2 hold for $R(j,t)$*

Proof. For i odd (implies j is even) and P1 take $\alpha \in R(j,t')$. By P1 for t' $\alpha = result_\sigma(j,t')$ for some σ. But then $mod_1(\alpha) = result_\sigma(j,t)$ from the definition of $result_\sigma$.

So let i be even (this implies j is odd) and $\alpha \in R(j,t')$. By P1 for t' $\alpha = result_\sigma(v,t')$ for some σ and $v \in \gamma(i,t')$. For the first part of the union take σ' to be the strategy which behaves like σ until the play reaches a vertex in i, then goes to v and then continues again as σ. Then $result_{\sigma'}(i,t) = mod_2(\alpha)$. Finally for the second part consider σ' which behaves like σ and to free $v \in \gamma(i,t)$ it assigns a successor not in $V[t]$. Then $result_{\sigma'}(i,t) = \alpha$.

For P2 the proofs are very similar, the difference being that for each $\sigma \in \Sigma_0^t$ we take $\alpha = result_\sigma(i,t')$ to start with.

Once the set $R(j,t)$ is computed, the sets $R(l,t)$ for $l \in \gamma(t) \setminus \{j\}$ are computed in the following way. Take an element $\alpha \in R(l,t')$. If α contains a pair (i,p) for some p, then consider all possibilities of extending a path from l to i by an element of $R(j,t)$. The way of combining $R(l,t')$ and $R(j,t)$ is defined using the operator *weld*:

Definition 3 (*weld*). *Let $\alpha \in R(l,t')$, $\beta \in R(j,t)$ and $l \neq j$. Then*

$$weld_{i,j}(\alpha,\beta) = \begin{cases} \min_{\sqsubseteq}((\alpha \cup \beta_p)) & \text{if } (i,p) \in \alpha \text{ for some } p \\ \alpha & \text{otherwise} \end{cases}$$

where $\beta_p = \{(k, \max\{p,q\}) \mid (k,q) \in \beta\}$.

Let $weld'_{i,j}$ be defined in exactly the same way, except for it removes the pair (i,p) from α. Finally we put

$$R(l,t) = \begin{cases} \{weld_{i,j}(\alpha,\beta) \mid \alpha \in R(l,t'), \beta \in R(j,t)\} & \text{if } i \text{ is odd} \\ \{weld'_{i,j}(\alpha,\beta) \mid \alpha \in R(l,t'), \beta \in R(j,t)\} \cup R(l,t') & \text{if } i \text{ is even} \end{cases}$$

Note: We do not add the set $R(l,t')$ if i is an even colour and there is no $w \in V \setminus V[t]$ such that $(u,w) \in E$ for $u \in \gamma(i,t)$. The reason is that we cannot postpone the choice of strategy for free vertices in $\gamma(i,t)$ as there is no choice left.

Proposition 3. *P1 and P2 hold for* $R(l,t)$

Proof (sketch). Similar to the proof of Proposition 2. For P1 let $\alpha \in R(l,t')$ and $\beta \in R(j,t)$. By P1 and the construction for $R(j,t)$ above we have $\alpha = \mathit{result}_{\sigma_1}(l,t')$ and $\beta = \mathit{result}_{\sigma_2}(j,t)$ for some σ_1, σ_2. For the proof we take σ to be the strategy that behaves like σ_1 until it reaches i and like σ_2 afterwards.

For P2 and each $\sigma \in \Sigma_0^t$ we take $\alpha = \mathit{result}_\sigma(l,t')$ and $\beta = \mathit{result}_\sigma(j,t)$, which must be in $R(l,t')$ and $R(j,t)$ by induction hypothesis and construction. The rest follows from the definition of result_σ.

***t* is of type $t_1 \oplus t_2$.** This node type is pretty straightforward. What we do depends on the colour – for a colour i we put

$$R(i,t) = \begin{cases} R(i,t_1) \cup R(i,t_2) & \text{if } i \text{ is an odd colour} \\ \{\min_{\sqsubseteq}(\alpha_1 \cup \alpha_2) \mid \alpha_1 \in R(i,t_1), \alpha_2 \in R(i,t_2)\} & \text{if } i \text{ is an even colour} \end{cases}$$

The proof that both P1 and P2 hold is follows immediately from the definition of $\mathit{result}_\sigma(i,t)$.

4.2 Complexity

Let us have a look at the size of the sets $R(i,t)$. First note that the set $\min_{\sqsubseteq} X$ for $X \subseteq (\{1,\ldots,k\} \times \mathbb{N})$ can have at most k elements, each chosen from the set of available priorities, size of which can be bounded by $n = |V|$. Therefore the set $R(i,t)$ can contain at most $(n+1)^k + 1$ different elements. As we have k different colours, the size of the data computed in each step is bounded by $k \cdot (n+1)^k + 1$.

The number of steps is equal to the size (the number of operators) of the k-expression t corresponding to G (without loss of generality we can assume that $|t|$ is linear in $|V|$). Finally we see that the running time of the algorithms for different operators is at most the square of $|R(i,t)|$, as in some cases we have to take all pairs of members of $R(i,t)$. This, together with Lemma 1, proves the main result:

Theorem 4. *Let $\mathcal{G} = (V,E,\lambda)$ be a parity game where $G = (V,E)$ is of clique-width k and t is the associated k-expression corresponding to G. Then there is an algorithm which solves the parity game $\mathcal{G}$ in time polynomial in n.*

4.3 Dropping the Restriction on Colours

In the algorithm above we restricted ourselves to bipartite parity games, where players alternate their moves. Here we briefly discuss how to drop this restriction.

The first step is to split each colour i into two colours i_0 and i_1. The colour i_0 will will be used only vertices from V_0, and the colour i_1 similarly for vertices from V_1. This at most doubles the number of colours. The second difference is that for each colour i_x we will keep two sets $R_0(i_x, t)$ and $R_1(i_x, t)$. One will be computed like it was a set for an even colour, and the other one as a set for odd colour. This again at most doubles the amount of information we need to keep. Altogether we store four times more information that in the basic algorithm.

The algorithms for the different node types are modified as follows. A vertex v created by the operator $[i]$ is added to the set it belongs to. For the $\oplus$ operator we join separately the sets for even and odd versions of the same colour. For renaming $\rho_{i \to j}$ we again rename separately. Finally we have to deal with the $\alpha_{i,j}$ operator. The odd and even version of the i colour are treated separately. For i_0 we connect first to $R_1(j_0, t)$ and then to $R_1(j_1, t)$. Similarly for i_1 we connect first to $R_0(j_0, t)$ and then to $R_0(j_1, t)$.

Altogether the number of steps and the size of the sets we have to remember is within a constant factor of the original algorithm, so the running time remains polynomial.

References

1. Berwanger, D., Dawar, A., Hunter, P., Kreutzer, S.: DAG-width and parity games. In: Durand, B., Thomas, W. (eds.) STACS 2006. LNCS, vol. 3884, pp. 524–536. Springer, Heidelberg (2006)
2. Berwanger, D., Grädel, E.: Entanglement – a measure for the complexity of directed graphs with applications to logic and games. In: Baader, F., Voronkov, A. (eds.) LPAR 2004. LNCS (LNAI), vol. 3452, pp. 209–223. Springer, Heidelberg (2005)
3. Björklund, H., Sandberg, S., Vorobyov, S.: A discrete subexponential algorithm for parity games. In: Alt, H., Habib, M. (eds.) STACS 2003. LNCS, vol. 2607, pp. 663–674. Springer, Heidelberg (2003)
4. Courcelle, B.: Graph rewriting: An algebraic and logic approach. In: van Leeuwen, J. (ed.) Handbook of Theoretical Computer Science: Volume B: Formal Models and Semantics, pp. 193–242. Elsevier, Amsterdam (1990)
5. Courcelle, B., Makowsky, J.A., Rotics, U.: Linear time solvable optimization problems on graphs of bounded clique-width. Theory Comput. Syst. 33(2), 125–150 (2000)
6. Courcelle, B., Olariu, S.: Upper bounds to the clique width of graphs. Discrete Appl. Math. 101(1-3), 77–114 (2000)
7. Emerson, E.A., Jutla, C.S.: Tree automata, mu-calculus and determinacy. In: Proc. 5th IEEE Foundations of Computer Science, pp. 368–377. IEEE Computer Society Press, Los Alamitos (1991)
8. Frick, M., Grohe, M.: The complexity of first-order and monadic second-order logic revisited. In: LICS 2002, pp. 215–224. IEEE Computer Society Press, Los Alamitos (2002)
9. Grädel, E., Thomas, W., Wilke, T. (eds.): Automata, Logics, and Infinite Games. LNCS, vol. 2500. Springer, Heidelberg (2002)
10. Jurdziński, M.: Deciding the winner in parity games is in UP ∩ co-UP. Information Processing Letters 68(3), 119–124 (1998)

11. Jurdziński, M.: Small progress measures for solving parity games. In: Reichel, H., Tison, S. (eds.) STACS 2000. LNCS, vol. 1770, pp. 290–301. Springer, Heidelberg (2000)
12. Jurdziński, M., Paterson, M., Zwick, U.: A deterministic subexponential algorithm for solving parity games. In: Proceedings of ACM-SIAM Symposium on Discrete Algorithms, pp. 117–123. ACM Press, New York (2006)
13. Kozen, D.: Results on the propositional μ-calculus. Theoretical Computer Science 27, 333–354 (1983)
14. Mostowski, A.W.: Games with forbidden positions. Technical Report 78, Instytut Matematyki, Uniwersytet Gdański, Poland (1991)
15. Obdržálek, J.: Fast mu-calculus model checking when tree-width is bounded. In: Hunt Jr., W.A., Somenzi, F. (eds.) CAV 2003. LNCS, vol. 2725, pp. 80–92. Springer, Heidelberg (2003)
16. Obdržálek, J.: Algorithmic Analysis of Parity Games. PhD thesis, University of Edinburgh (2006)
17. Obdržálek, J.: DAG-width – connectivity measure for directed graphs. In: Proceedings of ACM-SIAM Symposium on Discrete Algorithms, pp. 814–821. ACM Press, New York (2006)
18. Robertson, N., Seymour, P.D.: Graph Minors. III. Planar tree-width. Journal of Combinatorial Theory, Series B 36, 49–63 (1984)
19. Vöge, J., Jurdziński, M.: A discrete strategy improvement algorithm for solving parity games. In: Emerson, E.A., Sistla, A.P. (eds.) CAV 2000. LNCS, vol. 1855, pp. 202–215. Springer, Heidelberg (2000)

Logical Refinements of Church's Problem*

Alexander Rabinovich[1] and Wolfgang Thomas[2]

[1] Tel Aviv University, Department of Computer Science
rabino@math.tau.ac.il
[2] RWTH Aachen, Lehrstuhl Informatik 7, 52056 Aachen, Germany
thomas@informatik.rwth-aachen.de

Abstract. Church's Problem (1962) asks for the construction of a procedure which, given a logical specification φ on sequence pairs, realizes for any input sequence X an output sequence Y such that (X, Y) satisfies φ. Büchi and Landweber (1969) gave a solution for MSO specifications in terms of finite-state automata. We address the problem in a more general logical setting where not only the specification but also the solution is presented in a logical system. Extending the result of Büchi and Landweber, we present several logics $\mathcal{L}$ such that Church's Problem with respect to $\mathcal{L}$ has also a solution in $\mathcal{L}$, and we discuss some perspectives of this approach.

1 Introduction

An influential paper in automata theory is the address of A. Church to the Congress of Mathematicians in Stockholm (1962) [3]. Church discusses systems of restricted arithmetic, used for conditions on pairs (X, Y) of infinite sequences over two finite alphabets. As "solutions" of such a condition φ he considers "restricted recursions" (of which the most natural example is provided by finite automata with output), which realize letter-by-letter transformations of input sequences X to output sequences Y, such that φ is satisfied by (X, Y). By "Church's Problem" one understands today the question whether a condition in MSO (the monadic second-order theory of the structure $(\mathbb{N}, <)$) can be realized in this sense by a finite automaton with output, and in this case to synthesize it. Büchi and Landweber solved this problem, and in recent years many authors took up the question in various applications of the algorithmic theory of infinite games (see e.g. [6]).

In the original problem, a precise relation between a specification and its solution was not addressed, maybe due to the fact that specifications and solutions were considered to be in different domains. However, by a well-known correspondence between finite automata and monadic second-order logic (established by Büchi, Elgot, and Trakhtenbrot), the Büchi-Landweber solution for MSO-specifications can again be presented as MSO-formulas. In this paper we analyze

* This paper was written during a visit of the first author in Aachen in February 2007, funded by the European Science Foundation ESF in the Research Networking Programme AutoMathA (Automata: From Mathematics to Applications).

J. Duparc and T.A. Henzinger (Eds.): CSL 2007, LNCS 4646, pp. 69–83, 2007.

this view and generalize the result to a number of sublogics of MSO. We exhibit natural logics $\mathcal{L}$ such that Church's Problem for conditions in $\mathcal{L}$ is solvable again in $\mathcal{L}$. Moreover, a slightly sharpened statement holds: If a condition is not solvable, then again a procedure definable in $\mathcal{L}$ exists to prohibit satisfaction of the condition. In shorter words one says that infinite games defined in $\mathcal{L}$ are determined with $\mathcal{L}$-definable winning strategies.

We shall show the result for the following logics:

1. MSO, monadic second-order logic over $(\mathbb{N}, <)$.
2. FO($<$), first-order logic over $(\mathbb{N}, <)$ (with free set variables, similarly for the logics below).
3. FO($<$)+MOD, the extension of FO($<$) by modular counting quantifers,
4. FO(S), first-order logic over $(\mathbb{N}, S)$ with successor relation S,
5. strictly bounded logic, which is quantifier-free logic over $(\mathbb{N}, 0, +1)$.

The first three are treated in one proof, the last two handled separately.

We also exhibit examples of logics where the statement of the theorem fails, among them the extension of FO(S) by the quantifier $\exists^\omega$ ("there exist infinitely many") and Presburger arithmetic.

The study of the Büchi-Landweber Theorem for subsystems of MSO was started in a recent paper by Selivanov [15]. He showed that specifications given by aperiodic regular ω-languages can be solved with aperiodic transducers, which is a setting semantically equivalent to FO($<$) but relying on different techniques. The present paper uses a different concept of definability and a general proof method based on Ehrenfeucht-Fraissé type equivalences of the logics under consideration. The essential proof ingredients for the logics 1. - 4. above are the following:

1. A normal form of $\mathcal{L}$-formulas in the form of Boolean combinations of formulas $\exists^\omega x \varphi(x)$, respectively $\exists x \varphi(x)$, where $\varphi(x)$ is an $\mathcal{L}$-formula bounded in x,
2. a refinement of these normal forms corresponding to parity automata, respectively weak parity automata, whose states are $\mathcal{L}$-definable equivalence types of finite words (the equivalence typically based on the undistinguishability by $\mathcal{L}$-formulas of a given quantifier rank),
3. the construction of (weak) parity games over game graphs whose vertices are essentially the mentioned equivalence classes, and the application of the positional determinacy of (weak) parity games.

These results motivate a closer study of refined uniformization problems that already have a tradition in recursion theory and descriptive set theory (however, there in the context of degrees of unsolvability). We mention some perspectives in the Conclusion.

In the subsequent Section 2 we introduce the terminology and state the main result, adding a discussion of the possible concepts of definability for solutions of Church's Problem. In Section 3 we recall the prerequisites on (weak) parity games, and in Section 4 we develop the above-mentioned normal forms. The proof of the main result follows in Section 5.

2 Preliminaries and Main Result

2.1 Church's Problem, Games, and Strategies

Church's Problem deals with sequence pairs over alphabets of the form $\{0,1\}^n$. An instance is (the formal definition of) a set S of pairs (α, β) where say $\alpha \in (\{0,1\}^{m_1})^\omega$ and $\beta \in (\{0,1\}^{m_2})^\omega$. We identify this set with the ω-language over $\{0,1\}^{m_1+m_2}$ which contains for each pair (α, β) the ω-word $\alpha(0)\hat{}\beta(0)$, $\alpha(1)\hat{}\beta(1)\ldots$ (where $u\hat{}v$ denotes the concatenation of the vectors u and v).

In the original form of Church's Problem, the set S is defined in a restricted system of arithmetic. In this context, it is convenient to use the standard correspondence between a sequence $\alpha \in (\{0,1\}^n)^\omega$ and an m-tuple $(P_1,\ldots,P_m)$ of predicates P_i over the natural numbers, with $i \in P_j$ iff the j-th component of $\alpha(i)$, short $(\alpha(i))_j$, is equal to 1. The underlying mathematical model is $(\mathbb{N},<)$. We assume that the reader is acquainted with MSO logic over this structure (see [6]). We indicate first-order variables by $x, y\ldots$ and monadic second-order variables by $X, Y, \ldots$ Then an instance of Church's Problem is a formula $\varphi(\overline{X}, \overline{Y})$ with tuples of free set variables $\overline{X} = (X_1, \ldots, X_{m_1})$ and $\overline{Y} = (Y_1, \ldots, Y_{m_2})$, interpreted by tuples $\overline{P}$ and $\overline{Q}$ of predicates. For simplicity we write the ω-word associated with $\overline{P}$ as $\overline{P}(0)\overline{P}(1)\ldots$, each letter being a bit vector of length m_1, similarly for $\overline{Q}$.

A solution of Church's problem for the formula $\varphi(\overline{X}, \overline{Y})$ is an operator F mapping an m_1-tuple $\overline{P}$ of predicates to an m_2-tuple $\overline{Q}$, subject to the important restriction that F should be *causal*, i.e. $\overline{Q}(n)$ should only depend on the segment $\overline{P}(0), \ldots, \overline{P}(n)$ of $\overline{P}$. [1] This corresponds to the view that $\varphi(\overline{X}, \overline{Y})$ defines an infinite two-person game in which a play is built up as a sequence $\overline{P}(0), \overline{Q}(0), \overline{P}(1), \overline{Q}(1)$ etc., where the players 1 and 2 supply their choices $\overline{P}(i)$, respectively $\overline{Q}(i)$, in alternation. A strategy for Player 2 is given by a causal operator F, and a strategy for Player 1 by a *strongly causal* operator $G : \overline{Q} \mapsto \overline{P}$ (where $\overline{P}(n)$ only depends on the prefix $\overline{Q}(0), \ldots, \overline{Q}(n-1)$) of $\overline{Q}$. The causal operator F is a winning strategy for Player 2 in the game defined by φ (in Church's words: a solution to the condition φ) if $\forall \overline{X} \varphi(\overline{X}, F(\overline{X}))$ holds, similarly the strongly causal operator G is a winning strategy for Player 1 if we have $\forall \overline{Y} \neg\varphi(G(\overline{Y}), \overline{Y})$. holds. All games considered in this paper are determined, i.e. either Player 1 or Player 2 has a winning strategy.

We define a causal operator $F : \overline{P} \mapsto \overline{Q}$ in terms of a word function f that assigns to any finite sequence

$$\binom{\overline{P}(0)}{\overline{Q}(0)} \cdots \binom{\overline{P}(n-1)}{\overline{Q}(n-1)} \binom{\overline{P}(n)}{*}$$

a vector from $\{0,1\}^{m_2}$ (which is then taken as $\overline{Q}(n)$). By $F(\overline{P})$ we denote the sequence $\overline{Q}$ that is generated by applying f successively to $\binom{\overline{P}(0)}{*}$, to $\binom{\overline{P}(0)}{\overline{Q}(0)}\binom{\overline{P}(1)}{*}$ etc. Similarly, a strongly causal operator G is presented as a word function g

[1] So F is a special kind of continuous operator in the Cantor topology over infinite sequences.

which assigns to a sequence as above, with the last column letter missing, a vector from from $\{0,1\}^{m_1}$ (which is then taken as $\overline{P}(n)$).

For the definability of an operator F in a logical system $\mathcal{L}$, we assume that $\mathcal{L}$-formulas can be interpreted in finite word models of the form displayed above. Formally, we consider structures $([0,n], <, (\overline{P} \cap [0,n]), (\overline{Q} \cap [0,n-1]), n)$ for causal operators $F : \overline{P} \mapsto \overline{Q}$, and $([0,n], <, (\overline{P} \cap [0,n-1]), (\overline{Q} \cap [0,n-1]), n)$ for strongly causal operators $G : \overline{Q} \mapsto \overline{P}$. (For the latter case we have to provide means to cover the case $n = 0$, in order to fix $\overline{P}(0)$. This is done by a Boolean combination of formulas $X_i(x)$ and the statement that x has no predecessor.) We say that a function f over finite words is $\mathcal{L}$-definable if there are $\mathcal{L}$-formulas $\psi_1(\overline{X}, \overline{Y}, x), \ldots, \psi_{m_2}(\overline{X}, \overline{Y}, x)$ such that

$$([0,n], <, (\overline{P} \cap [0,n]), (\overline{Q} \cap [0,n-1]), n) \models \psi_j(\overline{X}, \overline{Y}, z)$$

iff the j-component of f applied to $\binom{\overline{P}(0)}{\overline{Q}(0)} \cdots \binom{\overline{P}(n-1)}{\overline{Q}(n-1)} \binom{\overline{P}(n)}{*}$ is equal to 1, and we call then also the associated causal operator F $\mathcal{L}$-definable. Similarly the definability of strongly causal operators is defined.

We say that an $\mathcal{L}$-defined game is *determined with $\mathcal{L}'$-definable winning strategies* if for each $\mathcal{L}$-formula $\varphi(\overline{X}, \overline{Y})$, there is either an $\mathcal{L}'$-definable causal operator as winning strategy of Player 2, or an $\mathcal{L}'$-definable strongly causal operator as winning strategy for Player 1.

2.2 Fragments of MSO and Main Result

The systems MSO, FO($<$), and FO(S) are all well-known from the literature. In the first two cases, the underlying model is $(\mathbb{N}, <)$, in the third case $(\mathbb{N}, S)$ with the successor relation S. We use free predicate variables $X, Y, \ldots$ (for monadic predicates) in all cases; in MSO we allow also quantification over them. We write atomic formulas in the form $x = y$ (equality is included), $x < y, S(x,y)$, and $X(y)$, and use the standard connectives $\neg, \vee, \wedge, \rightarrow, \leftrightarrow$ and quantifiers $\exists, \forall$. MSO is strictly more expressive than FO($<$) (as exemplified by the formula saying "the minimum of X is even"), and FO($<$) is strictly more expressive than FO(S) (as exemplified by "there is a Y element between two X-elements").

The logic FO($<$)+MOD is obtained from FO($<$) by adjoining the quantifiers $\exists^{r,q}x$, for $q > 1$ and $0 \leq r < q$, meaning "there is a finite number n of elements x with $n \equiv r (\mathrm{mod}\ q)$". FO($<$)+MOD is a system located in expressive power strictly between FO($<$) and MSO (see [17]). Note that over $\mathbb{N}$, the quantifier allows to express $\exists^\omega x$ ("there exist infinitely many x") by negating $\exists^{r,q}$ for all $r = 0, \ldots, q-1$. Denote by FO(S)+$\exists^\omega$ the extension of FO(S) by $\exists^\omega x$.

By *strictly bounded logic* mean the quantifier-free fragment of FO(0, +1). This logic characterizes the properties of sequences that are determined by their prefixes of fixed given length (or, in other words, are both open and closed in the Cantor topology). Corresponding formulas are called "bounded specifications" in [9].

Theorem 1. (Main Theorem)
Let $\mathcal{L}$ be any of the logics MSO, FO(<), FO(<)+MOD, FO(S), or strictly bounded logic. Then each $\mathcal{L}$-definable game is determined with $\mathcal{L}$-definable winning strategies.

Theorem 2. *If $\mathcal{L}$ is FO(S)+$\exists^\omega$ or FO(S)+MOD, then there are $\mathcal{L}$-definable games that are determined, however not with $\mathcal{L}$-definable winning strategies.*

2.3 On Definability of Causal Operators

It is useful to discuss the (in general non-equivalent) options for defining causal and strictly causal operators.

An operator $F : \Sigma_1^\omega \to \Sigma_2^\omega$ is *implicitly defined* by a formula $\psi(\overline{X}, \overline{Y})$ over the structure $(\mathbb{N}, \ldots)$ if for any $\overline{P}, \overline{Q}$ we have

$$F(\overline{P}) = \overline{Q} \quad \text{iff} \quad (\mathbb{N}, \ldots) \models \psi[\overline{P}, \overline{Q}]$$

and F is said *implicitly $\mathcal{L}$ definable* iff it is defined by a formula in the logic $\mathcal{L}$. An operator F is *explicitly defined* by the formulas $\varphi_1(X_1, \ldots, X_{m_1}, x), \ldots, \varphi_{m_2}(X_1, \ldots X_{m_1}, x)$ over the structure $(\mathbb{N}, \ldots)$ if for every $\bar{P} = (P_1, \ldots, P_{m_1}) \in \mathbb{P}(\mathbb{N})^{m_1}$ and $\bar{Q} = (Q_1, \ldots, Q_{m_2}) \in \mathbb{P}(\mathbb{N})^{m_2}$ the following holds:

$$\overline{Q} = F(\overline{P}) \text{ iff } (\mathbb{N}, \ldots) \models \textstyle\bigwedge_i \forall x (Q_i(x) \leftrightarrow \varphi_i(\overline{P}, x)).$$

Note that F is implicitly MSO-definable iff F is explicitly MSO-definable. However, for the fragments of MSO considered here, implicit $\mathcal{L}$-definability is (in general: strictly) more general than $\mathcal{L}$-explicit definability. As an example consider a constant operator $G : \{0,1\}^\omega \to \{0,1\}^\omega$ defined as $G(a_0, \ldots a_i \ldots) = b_0 \ldots b_i \ldots$, where $b_i = 1$ iff i is even. It is definable in $FO[<]$ implicitly but not explicitly.

Both notions are not adequate for our purpose since non causal and even non-continous functions can be definable (e.g. $F : P \mapsto Q$ with $Q = \mathbb{N}$ if P is infinite and else $Q = \emptyset$).

An operator F is *explicitly definable by bounded formulas in $\mathcal{L}$* if there are $\mathcal{L}$-formulas $\psi_i(\overline{X}, x)$ where atomic formulas $X(t)$ are only allowed for $t \leq x$ such that

$$F(\overline{P}) = \overline{Q} \quad \text{iff} \quad (\mathbb{N}, \ldots) \models \bigwedge_i \forall x (Q_i(x) \leftrightarrow \psi_i(\overline{P}, x))$$

Let $\mathcal{L}$ be one of the following logics MSO, $FO[<]$ and for $FO[<]$+MOD. Clearly, if an operator is $\mathcal{L}$-definable by bounded formulas, then it is causal.

However, this notion is insufficient for weak logics such as FO(<), due to a lack of reference to previous values; note that in MSO a sequence of values $Q(y)$ for $y < x$ is accessible by an auxiliary second-order quantification. An option to implement this reference is to use the full play prefix $P(0)Q(0)P(1)Q(1)\ldots P(n-1)Q(n-1)P(n)$ when defining $Q(n)$. But this does not allow (in FO(<) and FO(S)) to determine the origin of, say, a single bit 1 (assuming 0 elsewhere), namely whether this bit belongs to the $P(i)$ or to the $Q(i)$. Thus we use the vector representation for the pairs $(\overline{P}(i), \overline{Q}(i))$.

Finally, we remark that the notion of bounded formula is meaningful only when the order relation $<$ (or $\leq$) is in the signature. In order to cover also the successor logic FO(S), we pass to finite prefixes of infinite sequences as the models of formulas that define strategies, shifting boundedness from the syntactic to the semantic level. We thus arrive at the concept of definability for causal and strongly clausal operators as used above in Theorem 1.

3 Preliminaries on Games and Automata

3.1 Muller and Parity Games and Their Weak Versions

A directed bipartite graph $G = (V_1, V_2, E)$ is called a *game arena* if the outdegree of every vertex is at least one. We assume in this paper that $V := V_1 \cup V_2$ is finite. If G is an arena, a game on G is defined by an initial node $v_{init} \in V_1$ and a winning condition (by convention for Player 2). A play over G is an infinite sequence $\rho \in V^\omega$ starting in v_{init}, built up by two agents called player 1 and 2 that choose edges in alternation (player i from vertices in V_i). By $\mathrm{Inf}(\rho)$, respectively $\mathrm{Occ}(\rho)$ we denote the set of vertices from V which occur infinitely often, respectively just occur, in ρ. A winning condition decides who wins a play; we declare Player 2 to be the winner iff it is satisfied. (Here we follow the convention above that the "good" player is the second, having to react move by move to Player 1.) First we consider two winning conditions, called Muller condition, respectively weak Muller condition (the latter is also called Staiger-Wagner-condition in the literature). In both cases, the condition is specified by a family $\mathcal{F}$ of subsets of V. A play ρ satisfies the Muller (resp. weak Muller) condition $\mathcal{F}$ if $\mathrm{Inf}(\rho) \in \mathcal{F}$ (resp. if $\mathrm{Occ}(\rho) \in \mathcal{F}$).

As usual, a strategy f for Player 1 (Player 2) is a function which assigns to every path of odd (positive even) length a node adjacent to the last node of the path. (We assume that the inital vertex is given, and that player 1 starts from there.) A play $v_{init}v_1v_2\ldots$ is played according to a strategy f_1 of Player 1 (strategy f_2 of Player 2) if for every prefix $\pi = v_{init}v_1 \ldots v_n$ of odd (even) length we have $v_{n+1} = f_1(\pi)$ (respectively, $v_{n+1} = f_2(\pi)$). A strategy is winning for Player 2 (respectively, for Player 1) if all the plays played according to this strategy satisfy the winning condition under consideration. A strategy is *memoryless* if it depends only on the respective last node in the path.

In a *parity game* (resp. *weak parity game*), the winning condition refers to a coloring $c : V_1 \cup V_2 \to \{0, 1, \ldots m\}$ of the game graph. For a play $\rho = v_0v_1\ldots$ let $C_\omega(\rho)$, resp. $C(\rho)$, be the maximal color occurring infinitely often, resp. occurring at all, in the sequence $c(v_0)c(v_1)\ldots$. A play ρ is won according to the parity condition, resp. weak parity condition, if $C_\omega(\rho)$, resp. $C(\rho)$, is even.

The following theorem, due to Emerson/Jutla and Mostowski, is fundamental (see [6,11] and e.g. [21] for the weak case):

Theorem 3 (Memoryless Determinacy for (Weak) Parity Games)
In a parity game, one of the two players has a winning strategy which moreover is memoryless. The same statement holds for weak parity games.

3.2 From (Weak) Muller to (Weak) Parity Conditions

There are well-known reductions of Muller games and weak Muller games to parity games, respectively weak parity games. We recall these constructions here as a background to the next section. The idea is to transform an infinite sequence (e.g., an infinite play) ρ over the set V into a new sequence ρ' over a larger alphabet, which results from V by adding a "memory component". A coloring is associated with each of the new letters, making use of the state-set collection $\mathcal{F}$ that defines the (weak) Muller condition under consideration. The aim is to obtain the following equivalence: The sequence ρ satisfies the (weak) Muller condition iff ρ' satisfies the (weak) parity condition.

First we treat the case of weak Muller conditions, say for a collection $\mathcal{F} \subseteq 2^V$. We associate with each sequence $\rho = v_0 v_1 \dots$ over V a sequence over $V \times 2^V$, where the first components give a copy of ρ and a second component $P \subseteq V$ indicates the set of previously visited V-elements. We thus speak of the data structure *appearence record*, short AR. So the sequence ρ' starts with $(v_0, \emptyset)$, and for a step from v to v' in ρ we pass in ρ' from (v, P) to $(v', P \cup \{v\})$. The sequence ρ' will have, from some point onwards, a constant second component, which equals $\mathrm{Occ}(\rho)$. If we attach to (v, P) the color $2|P|$ when $P \in \mathcal{F}$ and $2|P| - 1$ when $p \notin \mathcal{F}$, then we have that $\mathrm{Occ}(\rho) \subset \mathcal{F}$ iff $C(\rho)$ is even (which means that the weak parity condition is satisfied).

A similar reduction is known for the Muller condition (see [19]). Consider again $\mathcal{F} \subseteq 2^V$, where $V = \{v_1, \dots, v_n\}$. Now not only the visited V-elements but also the order of their visits is recorded (*latest appearance record*, short LAR [7]). Let $\rho = v_0 v_1 \dots v_j \dots$ be a sequence over V. The associated latest appearance record at time point j is a sequence $(v_j, v_{i_1}, \dots, v_{i_k})$ where $S = \{v_{i_1}, v_{i_2}, \dots, v_{i_k}\}$ is the appearance record at j, i.e., the set of states visited before j. We list them in the order of latest appearance before v_j (most recently visited vertices listed first): Every state from S appears exactly once in the sequence $v_{i_1}, v_{i_2}, \dots, v_{i_k}$, and we have a sequence $j > j_1 > j_2 > \dots > j_k$ such that: (1) v_{i_m} appears at j_m in ρ $(m = 1, \dots k)$; (2) There is no occurrence of v_{i_m} between positions j_m and j in ρ $(m = 1, \dots k)$.

If in the LAR $(v, v_{i_1}, \dots v_{i_m})$, v occurs as entry v_{i_h}, we call h the "index" of the LAR (otherwise let $h = 0$). Using the index we equip a LAR with a color as follows. Assign to $(v, v_{i_1}, \dots v_{i_m})$ of index h the color $2h$ if $\{v_{i_1}, \dots, v_{i_h}\} \in \mathcal{F}$ and otherwise $2h - 1$. Then the sequence ρ satisfies the Muller condition w.r.t. $\mathcal{F}$ iff the induced sequence ρ' satisfies the parity condition with respect to the defined coloring.

Two observations are of interest in the sequel:

1. The description of the LAR structure requires the order relation of the underlying model, while this is irrelevant for the AR structure.
2. Together with the positional determinacy of (weak) parity games, the reductions yield finite-state automata that execute winning strategies in (weak) Muller games; the state set (consisting of LAR's, respectively AR's) and the transition function of a strategy automaton depend only on the game graph, whereas for the output function the winning condition is also used.

4 Normal Forms

As a tool to analyze logical formulas we recall well-known equivalences between models based on indistinguishability by formulas up to some quantifier depth k. We call these equivalence classes "k-types". The models under consideration are expansions of finite orderings $(A, <)$ or of $(\mathbb{N}, <)$ by tuples $\overline{P}$ of unary predicates. For m-tuples $\overline{P}$ we speak of m-chains.

The logics $\mathcal{L}$ to be considered below are MSO, FO(<), FO(<)+MOD, and FO(S).

4.1 Types

Two m-chains M, M' are k-equivalent for $\mathcal{L}$ (written: $M \equiv_k^{\mathcal{L}} M'$) if $M \models \varphi \Leftrightarrow M' \models \varphi$ for every $\mathcal{L}$-formula $\varphi(\overline{X})$ of quantifier depth k. This is an equivalence relation between m-chains; its equivalence classes are called k-types for $\mathcal{L}$ (and for the considered signature and m unary predicates). We denote the (finite) set of k-types by H_k, usually suppressing an index for the logic under consideration. (In the case of FO[<]+MOD, we assume that also a maximal divisor q is fixed in advance.) If a formula φ is true in every model of type t we say that "t implies φ".

Let us list some fundamental and well-known properties of k-types for any of the mentioned logics $\mathcal{L}$ above; we suppress the reference to $\mathcal{L}$ for simplicity of notation.

Proposition 4. *1. For every m and k there are only finitely many k-types of m-chains.*
2. For each k-type t there is a "characteristic formula" which defines t (i.e., is satisfied by an m-chain iff it belongs to t). For given k and m, a finite list of characteristic formulas for all the possible k-types can be computed.
3. Each formula $\varphi(\overline{X})$ is equivalent to a (finite) disjunction of characteristic formulas; moreover, this disjunction can be computed from φ.

The proofs of these facts can be found in several sources, we mention [8,16,20] for MSO and FO(<), [17] for FO(S) and FO(<)+MOD.

Given m-labelled chains M_0, M_1 we write $M_0 + M_1$ for their concatenation (ordered sum). In our context, M_0 will always be finite and M_1 finite or of order type ω. Similarly, if for $i \geq 0$ the chain M_i is finite, the model $\Sigma_{i \in \mathbb{N}} M_i$ is obtained by the concatenation of all M_i in the order given by the index structure $(\mathbb{N}, <)$.

We need the following composition theorem on ordered sums ([16]):

Proposition 5. (Composition Theorem)
Let $\mathcal{L}$ be any of the logics considered above.

(a) *The k-types of m-chains M_0, M_1 for $\mathcal{L}$ determine the k-type of the ordered sum $M_0 + M_1$ for $\mathcal{L}$, which moreover can be computed from the k-types of M_0, M_1.*
(b) *If the m-chains $M_0, M_1, \ldots$ all have the same k-type for $\mathcal{L}$, then this k-type determines the k-type of $\Sigma_{i \in \mathbb{N}} M_i$, which moreover can be computed from the k-type of M_0.*

Part (a) of the theorem justifies the notation $s+t$ for the k-type of an m-chain which is the sum of two m-chains of k-types s and t, respectively.

4.2 Strong Normal Forms

In this subsection we deal with the logics MSO, FO(<)+MOD, and FO(<), and we denote here by $\mathcal{L}$ any of these three logics.

A formula $\psi(x)$ with at most one free individual variable x is (syntactically) *bounded* if all its first-order quantifiers are of the form $\exists^{\leq x} y \ldots$ (short for $\exists y(y \leq x \wedge \ldots)$ and $\forall^{\leq x} y \ldots$ (short for $\forall y(y \leq x \rightarrow \ldots)$).

We use a normal form proved in [18] for MSO and FO(<) which easily extends to FO(<)+MOD. For all logics that satisfy the Composition Theorem and extends FO(<), it provides a first-order description of the fact (for given k) that the infinite m-chain under consideration can be cut into a sequence $w_0, w_1, w_2, \ldots$ such that all w_i for $i>0$ share the same k-type t. (The fact that such a decomposition exists is clear from Ramsey's Theorem; the obvious formalization, however, uses a second-order quantifier.)

Proposition 6 ([18]). *Every $\mathcal{L}$-formula $\varphi(\overline{Z})$ is equivalent (over the class of m-chains with domain $\mathbb{N}$) to a formula in* bounded normal form, *more specifically of the form*

$$\bigvee_{i=1}^{n} (\exists^{\omega} z \psi_i(\overline{Z}, z) \ \wedge \neg \exists^{\omega} z \psi_i'(\overline{Z}, z))$$

where the ψ_i, ψ_i' are bounded.

In a next step we sharpen this normal form to a "parity normal form".

Formulas in parity normal form involve a coloring of a certain finite set of bounded formulas $\Delta = \{\varphi_1(\overline{Z}, x), \ldots, \varphi_{n+1}(\overline{Z}, x)\}$ such that an m-chain can satisfy at most one of them. We denote the color of φ_i by $col(\varphi_i)$ and call Δ_j the formulas from Δ of color $\leq j$.

Lemma 7. (Parity Normal Form)
Every $\mathcal{L}$-formula $\varphi(\overline{Z})$ is equivalent (over the class of m-chains with domain $\mathbb{N}$) to a formula in parity normal form:

$$\bigvee_{i=0}^{n/2} (\bigvee_{\varphi \in \Delta_{2i}} \exists^{\omega} x \ \varphi(\overline{Z}, x) \ \wedge \bigwedge_{\varphi \in \Delta_{2i+1}} \neg \exists^{\omega} x \ \varphi(\overline{Z}, x)))$$

where Δ is a finite set of bounded formulas, such that an ω-chain satisfies at most one of them, and $\Delta_0 \subseteq \ldots \subseteq \Delta_{n+1} = \Delta$. (This formula will be denoted $Parity_{\varphi}(\Delta, col)$.)

Proof. The proof follows the mentioned transformation from Muller automata to parity automata. The only important observation is that the latest appearance record LAR can be formalized in FO(<).

We let $\varphi(\overline{Z})$ be a formula and let ψ_i, ψ'_i be formulas as in Proposition 6. Let k be the maximal quantifier depth of the formulas $\psi_i(\overline{Z}, x), \psi'_i(\overline{Z}, x)$ of the above bounded normal form of $\varphi(\overline{Z})$. Again, let H_k be the set of k-types over $\overline{Z}$.

Let us define the following Muller acceptance condition $\mathcal{F} \subseteq \mathbb{P}(H_k)$: We include a subset R of H_k in the set $\mathcal{F}$ iff for some j, R contains some type implying ψ_j but contains no type implying ψ'_j. Note that a structure $M = (\mathbb{N}, \ldots)$ satisfies φ iff the set of k-types that are satisfiable cofinally often on the initial substructures of M is in $\mathcal{F}$.

For every latest appearance record $lar = (t, t_{i_1}, \ldots, t_{i_m})$ over H_k, we can write a formula $\varphi_{lar}(\overline{Z}, x)$ such that $(\mathbb{N}, <) \models \varphi_{lar}(\overline{P}, j)$ iff lar is the latest appearance record of the sequence $t_1 \ldots t_j$, where t_i is the k-type of of the submodel of $(\mathbb{N}, <, \overline{P})$ over the interval $[0, i]$. For every k-type t one can write in these logics a formula expressing that the k-type of an interval $[0, x]$ is t (We denote this formula by $T^k(x) = t$, suppressing $\overline{Z}$). Then φ_{lar} can be easily expressed as explained in Subsection 3.2. For example, if $lar = (t, t_{i_1}, t, t_{i_2})$, then φ_{lar} is the conjunction of the following formulas:

$$T^k(x) = t \wedge \forall y < x(T^k(y) = t \vee T^k(y) = t_{i_1} \vee T^k(y) = t_{i_2})$$

$$\begin{aligned}\exists y_1 y_2 y(y_2 < y < y_1 < x) \wedge (T^k(y_1) = t_{i_1} \wedge T^k(y_2) = t_{i_2} \wedge T^k(y) = t = T^k(x)\\ \wedge \forall z\big((x > z > y) \rightarrow \neg T^k(z) = t\big)\\ \wedge \forall z\big((x > z > y_1) \rightarrow \neg T^k(z) = t_{i_1}\big)\\ \wedge \forall z\big((x > z > y_2) \rightarrow \neg T^k(z) = t_{i_2}\big)\end{aligned}$$

Now let $\mathcal{F}$ be the acceptance Muller conditions over H_k which corresponds to presentation of φ in bounded normal form. We transform Muller conditions into a coloring of LAR over H_k exactly like explained in Section 3.2, and for $lar \in LAR$ will assign to φ_{lar} the color of lar. Let Δ be the set of formulas $\{\varphi_{lar} : lar$ is a latest appearance record over $H_k\}$. It is easy to verify that φ is equivalent to the formula $Parity_{\varphi}(\Delta, col)$.

4.3 Weak Normal Forms

For the logic FO(S), we consider models $M = (\mathbb{N}, S, \overline{P})$ and $M = ([0, n], S, \overline{P})$ with an m-tuple $\overline{P}$. We first recall the model theoretic analysis of FO(S) which relies on the first-order model theory of finite graphs (due to Hanf, see e.g. [5]).

By the r-sphere around the element x we mean the submodel, pointed at x, consisting of x with its next k neighbours to the left and to the right (as far as these neighbors exist). We call a sphere right-complete (left-complete) if these k elements exist to the right (left) of x, and complete if both properties apply. By σ we denote an isomorphism type of an r-sphere; the set of all possible r-sphere isomorphism types is denoted S_r.

A (r, K)-type of a model M is given, for each $\sigma \in S_r$, by the numbers n_σ of occurrences of σ counted up to threshold K. So it is a vector $(n_\sigma)_{\sigma \in S_r}$ of values in $[0, K+1]$ and defined by a conjunction of FO(S)-formulas "there are precisely k elements x with: "r-sphere type of $x = \sigma$" where $k < K$, and "there are $K+1$ elements x with "r-sphere type of $x = \sigma$".

We need the following known facts which give a rather direct description of k-types for FO(S):

Lemma 8. (Hanf, see [5])
For each k, there are r, K such that each (r, K)-type implies a fixed k-type (so (r, K)-types refine k-ypes), or in other words: Truth of a FO(S)-formula of quantifier depth k over a model M as considered here is determined by the (r, K)-type of M.

Now we consider models with domain $\mathbb{N}$. Then only right-complete spheres are relevant. We use this fact to introduce a restricted version of type for the finite prefixes (ignoring the right-incomplete spheres of prefixes), which induces a monotonicity property when we let the prefixes increase. For the initial segment M_s of a model $M = (\mathbb{N}, S, \overline{P})$ up to number s we denote by $\pi(s) = (n_\sigma)_\sigma$ the vector of natural numbers in $[0, K+1]$ which lists, in some fixed order of the r-sphere types σ that are right-complete, the numbers of their occurrences in M_s counted up to threshold value K (again representing any number $> K$ by $K+1$). Call $\pi(s)$ the "(r, K)-profile" of M_s, and π the (r, K)-profile of the whole structure M. ¿From the lemma it follows (under the given conventions for the parameters k, r, K) that the (r, K)-profile π determines truth of formulas $\varphi(\overline{Z})$ in M up to quantifier depth k.

When s increases, the profiles $\pi(s)$ can only increase as well (componentwise). At some point s_0, the value $\pi(s_0) = (n_\sigma^{s_0})_\sigma$ is equal to the M-profile π and stays constant. Let us write $\pi' > \pi$ if for all components the π'-value is $\geq$ the corresponding π-value, and for some component we have a strict inequality $>$. We can write down FO(S)-formulas φ_τ, $\varphi_{>\tau}$ expressing in a segment model M_s that its (r, K)-profile is τ, respectively $> \tau$.

We obtain the following "normal form" for FO(S)-formulas; note that we write it down on the semantical level since bounded formulas are not available in FO(S).

Lemma 9. ("Weak Normal Form")
Let $\varphi(\overline{Z})$ be a FO(S)-formula. Then for each model $M = (\mathbb{N}, S, \overline{P})$, we have $M \models \varphi(\overline{Z})$ iff

$$\bigvee_{\tau \in \Pi(\varphi)} (\exists s\ M_s \models \varphi_\tau \ \wedge\ \neg \exists s\ M_s \models \varphi_{>\tau})$$

where $\Pi(\varphi)$ is the set of (r, K)-profiles that imply φ.

A small further step gives us a parity normal form. For this we consider an extension of the partial order of (r, K)-profiles to a linear order, giving each profile an index h. We associate now colors with the formulas $\varphi_\tau, \varphi_{>\tau}$ above, by giving a profile of index h the color $2h$ if it belongs to $\Pi(\varphi)$, otherwise $2h-1$. Assume the colors range from 0 to $n+1$. Then we obtain the following, using again the notation of Δ_j for the formulas of color $\leq j$:

Lemma 10. ("Weak Parity Normal Form")
Let $\varphi(\overline{Z})$ be a FO(S)-formula. For each model $M = (\mathbb{N}, S, \overline{P})$, $M \models \varphi(\overline{Z})$ iff

$$\bigvee_{i=0}^{n/2} \Big(\bigvee_{\varphi_\tau \in \Delta_{2i}} \exists s\ M_s \models \varphi_\tau \ \wedge \bigwedge_{\varphi_\sigma \in \Delta_{2i+1}} \neg \exists s\ M_s \models \varphi_\sigma \Big)$$

5 Proof of Theorems 1 and 2

5.1 Logics with Strong or Weak Normal Form

Let $\mathcal{L}$ be any of the logics MSO, FO(<), FO(<)+MOD. In the previous section we have shown that each $\mathcal{L}$-formula $\varphi(\overline{Z})$ can be transformed into an equivalent parity automaton $\mathcal{A}_\varphi$ whose states are k-types for the logic $\mathcal{L}$, for suitable k. (In order to include the empty model as initial state, the state space of $\mathcal{A}_\varphi$ is $H_k \cup \{\epsilon\}$.) After scanning an initial segment $\overline{P}(0) \ldots \overline{P}(n)$, the automaton assumes just the k-type (for $\mathcal{L}$) of the model $([0,n], <, (\overline{P} \cap [0,n]))$. By construction (and by the properties of k-types), each state is $\mathcal{L}$-definable.

This transformation of a specification (game definition) $\varphi(\overline{Z})$ into an automaton is independent of the game theoretical context. Now we emphasize this aspect again. The m-tuple $\overline{Z}$ is split into two blocks $\overline{X}, \overline{Y}$ of length m_1, m_2, respectively, the specification reads $\varphi(\overline{X}, \overline{Y})$, and predicates $\overline{P}, \overline{Q}$ used for the interpretation of $\overline{X}, \overline{Y}$ are built up step by step in alternation.

Following this splitted construction, we introduce a game graph, where the vertices from $H_k \cup \{\epsilon\}$ are accompanied by extra vertices in $(H_k \cup \{\epsilon\}) \times \{0,1\}^{m_1}$, which serve to represent the "intermediate steps" reached by Player 1 after his choice of a m_1-tuple $\overline{P}(n)$.

Formally, we define the game graph $G_\varphi = (V_1, V_2, E)$ by

- $V_1 = H_k \cup \{\epsilon\}$, $V_2 = V_1 \times \{0,1\}^{m_1}$
- the edge set E with an edge from $t \in V_1$ to $(t, \overline{a})$ for each $t \in V_1$ and each $\overline{a} \in \{0,1\}^{m_1}$, and an edge from each $(t, \overline{a})$ to the k-type $t + (\overline{a}, \overline{b})$ for each $\overline{b} \in \{0,1\}^{m_2}$. (Recall that $t + (\overline{a}, \overline{b})$ is the k-type of a model which results from a model of type t by concatenating the m-tuple $(\overline{a}, \overline{b})$.)

In order to obtain a parity game, we have to define a coloring c. For this, we associate to both t and each $(t, \overline{a})$ the same color as given for t in the automaton $\mathcal{A}_\varphi$. Then it is obvious that a sequence $(\overline{P}, \overline{Q})$ satisfies φ iff Player 2 wins the parity game over G_φ with the coloring c. Assume that Player 2 wins. Then we can fix a winning strategy by choosing one m_2-tuple $\overline{b}$ for each vertex $(t, \overline{a})$ in V_2. Denote the i-th component of this vector $\overline{b}$ by $b_i(t, \overline{a})$. We show that this strategy is $\mathcal{L}$-definable. For this, recall that we can express by an $\mathcal{L}$-formula $\psi_t(\overline{X}, \overline{Y}, x)$ that "the k-type of $([0, x-1], (\overline{X} \cap [0, x-1]), (\overline{Y} \cap [0, x-1], x)$ is t". We define the winning strategy by the following $\mathcal{L}$-formulas $\psi_i(\overline{X}, \overline{Y}, x)$:

$$\bigvee_{(t,\overline{a}) \in V_2} (\psi_t(\overline{X}, \overline{Y}, x) \wedge \overline{X}(x) = \overline{a} \wedge \text{"}b_i(t, \overline{a}) = 1\text{"})$$

Here $\overline{X}(x) = \overline{a}$ stands for $\bigwedge_j [\neg] X_j(x)$ with negations inserted for those j where $a_j = 0$, and $b_i(t, \overline{a}) = 1$ for "true" or "false" depending on the value of b_i. The proof for definability of a winning strategy for Player 1 works similarly.

For the logic FO(S), we proceed in exact analogy, invoking the weak parity normal form, constructing a weak parity game consisting of FO(S)-definable states, and using positional determinacy of weak parity games. Since compositionality of types is needed in the definition of the game graph, one uses (r, K)-types as vertices, but the induced (r, K)-profiles for the winning condition.

5.2 Strictly Bounded Logic

For a specification $\varphi(\overline{X}, \overline{Y})$ of strictly bounded logic (the quantifier-free fragment of FO(0, +1)) with an m_1-tuple $\overline{X}$ and a m_2-tuple $\overline{Y}$, let k be the maximal nesting of the function symbol $+1$ in φ. It is easy to show that satisfaction of φ in $(\mathbb{N}, 0, +1, \overline{P}, \overline{Q})$ only depends on the prefix of $(\overline{P}, \overline{Q})$ up to position k. Collect the finite set L_0 of these prefixes (of length $k+1$) such that all their extensions to ω-sequences satisfy φ. We consider the game graph with vertices $w \in \{0,1\}^{m_1+m_2}$ of length $\leq k+1$ (for Player 1) and $(w, \overline{a})$ for these w and $\overline{a} \in \{0,1\}^{m_1}$ (for Player 2). The standard attractor construction (see [6]) yields a positional winning strategy for either of the two players which is clearly definable in strictly bounded logic.

5.3 Proof of Theorem 2

We consider a game due to Dziembowski, Jurdziński, and Walukiewicz [4]. For better readability we use the alphabet $\{a, b, c\}$ for X and $\{0, 1, 2\}$ for Y. Let $\varphi(X, Y)$ be the following condition: "If a, b occur infinitely often in X, then 2 occurs infinitely often in Y; if only one of a, b occurs infinitely often, then 1, but not 2, occurs infinitely often in Y; otherwise Y is ultimately 0". This condition is expressible in FO(S)+$\exists^\omega$ (even without S).

The game is solvable by means of the LAR over $\{a, b, c\}$. The output is 0 if c is the current X-value, it is 1 if a is the current value with b occurring after a in the current LAR-list, and otherwise 2 (dually for letter b).

Assume that a winning strategy is definable in FO(S)+$\exists^\omega$. Since the underlying models are finite words, it is easy to transform the definition into an equivalent FO(S)-definition. Choose (r, K) as in the weak normal form theorem and consider the word $u = (c^{2r} a c^{2r} b)^K c^{2r}$ over $\{a, b, c\}$. Now we apply a case distinction concerning the output values of the strategy for words in uac^* after u. If the maximum is 0 or 2, we obtain a contradiction to the assumption that the strategy wins, by considering $P = u(ac^{2r})^\omega$; namely, the strategy will yield 0, respectively 2, as the maximal output repeated infinitely often but should do this with value 1. Similarly, one argues for the case of words in ubc^*. It remains to consider the situation that for both cases the maximum output is 1. Then we obtain a contradiction for $u(ac^{2r}bc^{2r})^\omega$; the strategy yields 1 as maximal output repeated infinitely often but should produce value 2.

The proof for FO(S)+MOD works similarly (since again the quantifier $\exists^\omega$ is implicitly present), however with a more involved case distinction which is omitted here.

6 Discussion and Perspectives

Based on a natural concept of logical definability of winning strategies in infinite games, we exhibited several fragments $\mathcal{L}$ of MSO logic such that the $\mathcal{L}$-definable games are determined with $\mathcal{L}$-definable winning strategies.

Let us add remarks on possible extensions of these results.

Our formulation of Church's Problem can be viewed as a task to transform ω-languages (specifications) to tuples of standard languages (defining the output functions, essentially by descriptions of mutually disjoint languages of play prefixes for the different output letters). This study can be pursued in language theory, and further types of properties can be considered, like "locally testable" or "piecewise testable" (see [12]). These properties are not captured by logics but can be analyzed in a very similar way by approriate equivalence relations. We leave a more detailed treatment (which involves some extension of the method of the present work) to a future paper.

In [14] the Church problem for MSO over expansions of ω by monadic predicates (i.e., over structures $(\mathbb{N}, <, \mathbb{P}_1, \ldots, \mathbb{P}_n)$ with fixed subsets $\mathbb{P}_i$ as "parameters") was investigated. It was shown that for every MSO formula $\psi(X, Y, P_1, \ldots, P_n)$ and structure $M = (\mathbb{N}, <, \mathbb{P}_1, \ldots, \mathbb{P}_n)$ there is either an MSO-definable (in M) causal operator, as winning strategy of Player 2, or an MSO-definable (in M) strongly causal operator, as winning strategy for Player 1. There, similarly to our paper, for an instance of Church's problem a parity game graph is constructed. However, unlike the case considered here, this graph is infinite. The finiteness of the game graph is crucial to our proof of definability of the winning strategy (see section 5.1). So it remains an open question whether the results of our paper can be extended to the expansions of $(\mathbb{N}, <)$ by monadic predicates.

Let us consider Church's Problem in the framework of another decidable theory: Preburger arithmetic, the first-order theory of addition over $\mathbb{N}$. Here Theorem 1 fails:[2] It is easy to write down in Presburger arithmetic a formula $\varphi(Y_0)$ which says that Y_0 is the set Squ of squares (use the fact that the distances of successive squares increase by 2). The strategy to generate Squ is also Presburger definable. Invoking the fact that multiplication is FO-definable in $(\mathbb{N}, +, \mathrm{Squ})$ ([13]), the FO(+)-specifications are the arithmetical relations in $(\overline{X}, Y_0, \overline{Y})$ where $Y_0 = \mathrm{Squ}$. On the other hand, it is known ([10]) that there are specifications $\exists^\omega x R(\overline{X}, \overline{Y}, x)$ even with recursive R which (are determined but) do not allow an arithmetical winning strategy. This leads us to the question: Are

[2] The short abstract [2] of Büchi, Elgot, and Wright (without a corresponding paper) claims that specifications in Presburger arithmetic do not have, in general, MSO-definable solutions; the reference to [13] given in [2] seems to point to a similar argument as sketched here.

there natural logics $\mathcal{L}$ which are not covered by the proof method of this paper and still satisfy Theorem 1?

References

1. Büchi, J.R., Landweber, L.H.: Solving sequential conditions by finite-state strategies. Trans. Amer. Math. Soc. 138, 367–378 (1969)
2. Büchi, J.R., Elgot, C.C., Wright, J.B.: The nonexistence of certain algorithms of finite automata theory. Notices of the AMS 5, 98 (1958)
3. Church, A.: Logic, arithmetic, and automata. In: Proc. Int. Congr. Math. 1962, Inst. Mittag-Lefler, Djursholm, Sweden, pp. 23–35 (1963)
4. Dziembowski, S., Jurdziński, M., Walukiewicz, I.: How much memory is needed to win infinite games? In: Proc. 12th LICS, pp. 99–110 (1997)
5. Ebbinghaus, H.D., Flum, J.: Finite Model Theory. Springer, Heidelberg (1995)
6. Grädel, E., Thomas, W., Wilke, T. (eds.): Automata, Logics, and Infinite Games. LNCS, vol. 2500. Springer, Heidelberg (2002)
7. Gurevich, Y., Harrington, L.: Trees, automata, and games. In: Proc. 14th STOC, pp. 60–65. ACM Press, New York (1982)
8. Gurevich, Y.: Monadic second order theories. In: Barwise, J., Feferman, S. (eds.) Model Theoretic Logics, pp. 479–506. Springer, Heidelberg (1986)
9. Kupferman, O., Vardi, M.Y.: On Bounded Specifications. In: Nieuwenhuis, R., Voronkov, A. (eds.) LPAR 2001. LNCS (LNAI), vol. 2250, pp. 24–38. Springer, Heidelberg (2001)
10. Moschovakis, Y.: Descriptive Set Theory. North-Holland, Amsterdam (1980)
11. Perrin, D., Pin, J.E.: Infinite Words. Elsevier, Amsterdam (2004)
12. Pin, J.E.: Varieties of Formal Languages. North Oxford and Pergamon, Oxford (1986)
13. Putnam, H.: Decidability and essential undecidability. JSL 22, 39–54 (1957)
14. Rabinovich, A.: Church Synthesis Problem with Parameters. In: Ésik, Z. (ed.) CSL 2006. LNCS, vol. 4207, pp. 546–561. Springer, Heidelberg (2006)
15. Selivanov, V.: Fine hierarchy of regular aperiodic ú- languages. In: Harju, T., et al. (eds.) DLT 2007. LNCS, vol. 4588, Springer, Heidelberg (2007)
16. Shelah, S.: The monadic theory of order. Ann. of Math. 102, 349–419 (1975)
17. Straubing, H.: Finite Automata, Formal Logic, and Circuit Complexity, Birkhäuser, Boston (1994)
18. Thomas, W.: A combinatorial approach to the theory of ω-automata. Inf. Contr. 48, 261–283 (1981)
19. Thomas, W.: On the synthesis of strategies in infinite games. In: Mayr, E.W., Puech, C. (eds.) STACS 95. LNCS, vol. 900, pp. 1–13. Springer, Heidelberg (1995)
20. Thomas, W.: Languages, automata, and logic. In: Rozenberg, G., Salomaa, A. (eds.) Handbook of Formal Languages, vol. 3, pp. 389–455. Springer, Heidelberg (1997)
21. Thomas, W.: Complementation of Büchi automata revisited. In: Contributions on Theoretical Computer Science in Honor of Arto Salomaa, pp. 109–120. Springer, Heidelberg (1999)

The Power of Counting Logics on Restricted Classes of Finite Structures*

Anuj Dawar[1] and David Richerby[2,**]

[1] University of Cambridge Computer Laboratory, William Gates Building, J.J. Thomson Avenue, Cambridge, CB3 0FD, UK.
Anuj.Dawar@cl.cam.ac.uk

[2] Department of Mathematics, University of Athens, Panepistimioupolis, GR157-84 Athens, Greece
davidr@math.uoa.gr

Abstract. Although Cai, Fürer and Immerman have shown that fixed-point logic with counting (IFP + C) does not express all polynomial-time properties of finite structures, there have been a number of results demonstrating that the logic does capture **P** on specific classes of structures. Grohe and Mariño showed that IFP + C captures **P** on classes of structures of bounded treewidth, and Grohe showed that IFP + C captures **P** on planar graphs. We show that the first of these results is optimal in two senses. We show that on the class of graphs defined by a non-constant bound on the tree-width of the graph, IFP + C fails to capture **P**. We also show that on the class of graphs whose local tree-width is bounded by a non-constant function, IFP + C fails to capture **P**. Both these results are obtained by an analysis of the Cai–Fürer–Immerman (CFI) construction in terms of the treewidth of graphs, and cops and robber games; we present some other implications of this analysis. We then demonstrate the limits of this method by showing that the CFI construction cannot be used to show that IFP + C fails to capture **P** on proper minor-closed classes.

1 Introduction

The central open problem in descriptive complexity theory is whether there exists a logic that can express exactly the polynomial-time decidable properties of unordered structures. For some time it was conjectured that the extension of fixed-point logic with counting (IFP + C) would be such a logic, but this was shown not to be the case by a construction due to Cai, Fürer and Immerman [4], which we refer to below as the CFI construction. Nonetheless, IFP + C provides a natural level of expressiveness within the complexity class **P** which has

* Research partly carried during the programme 'Logic and Algorithms' at the Isaac Newton Institute for Mathematical Sciences, University of Cambridge, 20 Clarkson Road, Cambridge, CB3 0EH, UK.

** Funded by the European Social Fund and Greek National Resources — (EPEAEK II) PYTHAGORAS II.

J. Duparc and T.A. Henzinger (Eds.): CSL 2007, LNCS 4646, pp. 84–98, 2007.

been explored in its own right [20]. It has also been shown that, on certain restricted classes of structures, IFP + C is indeed powerful enough to express all polynomial-time properties. In particular, Immerman and Lander have shown that IFP + C defines exactly the polynomial-time properties of trees [16] and Grohe and Mariño [14] extended this to show that on any class of structures of bounded tree-width, IFP + C captures **P**. Grohe also showed that IFP + C captures **P** on the class of planar graphs [12] and, more generally, on classes of embeddable graphs [13]. In particular, these results imply that the CFI construction cannot be carried out when restricting ourselves to such classes of structures.

There is a growing body of work studying the finite model theory of restricted classes of structures, where the restrictions are essentially borrowed from graph structure theory. Such graph-theoretic restrictions, such as bounding the tree-width of a graph or restricting to planar graphs (or, more generally, proper minor-closed classes of graphs), often yield classes with good algorithmic properties and there has been an effort to explore whether these also correspond to interesting model-theoretic properties which may be tied to the good algorithmic behaviour. In many cases, the logical or model-theoretic view provides a clean general "explanation" of the algorithmic properties of a class. Examples of such meta-theorems are Courcelle's theorem [5], which shows that any property definable in monadic second-order logic is decidable in linear time on classes of bounded tree-width, and the result of Dawar *et al.* [7] that first-order definable optimization problems admit polynomial-time approximation schemes on proper minor-closed classes. At the same time, ever more expansive (i.e., less restricted) classes of structures have been studied such as classes of bounded local tree-width [11], classes that locally exclude a minor [6] and classes of bounded expansion [18]. Our aim in this paper is to explore the boundary of the classes where IFP + C captures **P**. In particular, we wish to determine on which of these various classes the CFI construction can be carried out.

The CFI construction relies on the fact that every formula of IFP + C is equivalent to one of $\mathcal{C}^{\omega}_{\infty\omega}$, the infinitary logic with counting. A separator of a graph $G = (V, E)$ is a set $S \subseteq V$ of vertices whose deletion from the graph leaves no connected component with more than $|V|/2$ vertices. Cai *et al.* show that for each graph G, we can construct two graphs $X(G)$ and $\widetilde{X}(G)$ such that, if G has no separator of size k, then $X(G)$ and $\widetilde{X}(G)$ cannot be distinguished by any formula of $\mathcal{C}^{k}_{\infty\omega}$, the k-variable fragment of $\mathcal{C}^{\omega}_{\infty\omega}$. Since $X(G)$ and $\widetilde{X}(G)$ are distinguished by a polynomial-time algorithm, it follows that IFP + C does not capture **P** on any class of graphs that includes both $X(G)$ and $\widetilde{X}(G)$ for graphs G with arbitrarily large minimal size separators. As Cai *et al.* already noted, this includes the class of graphs with degree bounded by 3. We show (in Section 3) that the assumption that G has no separator of size k can be replaced by the weaker requirement that the tree-width of G is at least k. This is established by a game construction that combines the cops-and-robber game of Seymour and Thomas [24] with the bijection game of Hella [15] (see [1] for another application of the same idea).

An immediate consequence is that IFP + C does not capture **P** on any class of graphs that includes $X(G)$ and $\widetilde{X}(G)$ for graphs of unbounded tree-width. As a corollary, we show that the result of Grohe and Mariño is, in a sense, optimal. For a function $f : \mathbb{N} \to \mathbb{N}$, let TW_f be the class of all graphs G such that the tree-width of G is at most $f(|G|)$. Grohe and Mariño show that, if f is bounded above by a constant, then IFP + C captures **P** on TW_f. On the other hand, we show that, no matter how slowly f grows, if it is unbounded, then IFP + C does not capture **P** on TW_f. Note that this does not show that IFP + C fails to capture **P** on *any* class of graphs of unbounded tree-width. Indeed, planar graphs have unbounded tree-width but IFP + C capture **P** on this class. However, if the class contains all graphs of tree-width bounded by f, we show that the CFI construction applies.

Instead of restricting tree-width as a function of the order of the graph, we can consider graphs where tree-width grows as a function of the diameter. Recall that the r-neighbourhood of a vertex v in a graph is the subgraph induced by the vertices within distance r of v. For a non-decreasing function $f : \mathbb{N} \to \mathbb{N}$, let LTW_f be the class of graphs G such that, for all $r \geqslant 1$, the r-neighbourhood of every vertex in G has tree-width at most $f(r)$. (Eppstein introduces these classes as graphs with the 'diameter-treewidth property' [10] and the restriction is termed bounded local tree-width in [11]). For any such graph, we have $\mathrm{tw}(G) \leqslant f(|G|)$ so $\mathrm{LTW}_f \subseteq \mathrm{TW}_f$. We show, in Section 4 that, analogous to the case for global tree-width, IFP + C captures **P** on LTW_f if, and only if, $f = \mathcal{O}(1)$. Thus, the result of Grohe and Mariño is optimal in a stronger sense.

Grohe [13] has conjectured that IFP + C captures **P** on any proper minor-closed class of finite graphs. We show, in Section 5 that the CFI construction cannot be used to refute this conjecture. That is, we show that for any graph G and any graph H of sufficient tree-width G is a minor of both $X(H)$ and $\widetilde{X}(H)$. Thus, if a class of graphs forbids G as a minor, it excludes $X(H)$ and $\widetilde{X}(H)$ for all graphs H except those of some fixed tree-width.

There are several generalizations of the concept of tree-width to directed graphs including that of directed tree-width [17], DAG-width [2,19] and entanglement [3]. In each of these measures, the class of directed acyclic graphs (DAGs) has width 1. Since the CFI construction works in the class of DAGs (see Section 3) it follows that our results do not extend to these measures.

2 Background

The notion of a relational structure $\mathfrak{A} = (A, R_1^{\mathfrak{A}}, \ldots, R_s^{\mathfrak{A}})$ over vocabulary $\sigma = (R_1^{r_1}, \ldots, R_s^{r_s})$ is standard. All structures and graphs in this paper are finite; we treat graphs as structures with a single binary relation symbol E, interpreted by an irreflexive relation that, in the case of undirected graphs, is also symmetric. All graphs mentioned in this paper are undirected unless specifically stated to be directed.

2.1 Counting Logics

IFP + C is the extension of first-order logic with inflationary fixed-points and a mechanism for counting. For formal definitions, which we will not need in this paper, we refer the reader to [9]. It is known that every class of structures definable in IFP + C is decidable in polynomial time.

The formulas of the logic $C_{\infty\omega}$ are obtained from the atomic formulas using negation, infinitary conjunction and disjunction, and counting quantifiers ($\exists^i x\, \varphi$ for every integer $i \geqslant 0$). The fragment $\mathcal{C}^k_{\infty\omega}$ consists of those formulas of $C_{\infty\omega}$ in which only k distinct variables appear and $\mathcal{C}^\omega_{\infty\omega} = \bigcup_{k\in\omega} \mathcal{C}^k_{\infty\omega}$. The significance of $\mathcal{C}^\omega_{\infty\omega}$ for our purposes is every formula of IFP + C is equivalent to one of $\mathcal{C}^\omega_{\infty\omega}$.

Hella shows that definability in $\mathcal{C}^k_{\infty\omega}$ is characterized by the *k-pebble bijection game* [15]. The game is played on structures $\mathfrak{A}$ and $\mathfrak{B}$ by two players, the spoiler and the duplicator, using pebbles $a_1, \ldots, a_k$ on $\mathfrak{A}$ and $b_1, \ldots, b_k$ on $\mathfrak{B}$. If $|A| \neq |B|$, the spoiler wins immediately; otherwise, each move is made as follows:

- the spoiler chooses a pair of pebbles a_i and b_i;
- the duplicator chooses a bijection $h : A \to B$ such that for pebbles a_j and b_j $(j \neq i)$, $h(a_j) = b_j$; and
- the spoiler chooses $a \in A$ and places a_i on a and b_i on $h(a)$.

If, after this move, the map $a_1 \ldots a_k \mapsto b_1 \ldots b_k$ is not a partial isomorphism $\mathfrak{A} \to \mathfrak{B}$, the game is over and the spoiler wins; the duplicator wins all infinite plays. Hella shows that the duplicator has a winning strategy in the k-pebble bijection game on $\mathfrak{A}$ and $\mathfrak{B}$ if, and only if, the two structures agree on every formula of $\mathcal{C}^k_{\infty\omega}$, in which case, we write $\mathfrak{A} \equiv^{\mathcal{C}^k_{\infty\omega}} \mathfrak{B}$. In order to show that a class Q of structures is not definable in $\mathcal{C}^\omega_{\infty\omega}$ (and, hence, not definable in IFP + C), it suffices to demonstrate, for each $k \geqslant 1$, structures $\mathfrak{A}_k \in Q$ and $\mathfrak{B}_k \notin Q$ on which the duplicator has a winning strategy in the k-pebble bijection game.

By a result of Otto [20, Theorem 4.22] we have the following:

Theorem 1 (Otto). *If* IFP + C *captures* **P** *on a class $\mathcal{C}$ of structures that is closed under disjoint unions, then there is a k such that $\equiv^{\mathcal{C}^k_{\infty\omega}}$ coincides with isomorphism on $\mathcal{C}$.*

Thus, to show that IFP + C does not capture **P** on a class of structures $\mathcal{C}$, it suffices to show that for every k $\mathcal{C}$ contains a pair of non-isomorpic structures H and H', such that $H \equiv^{\mathcal{C}^k_{\infty\omega}} H'$.

2.2 Tree-Width

Tree-width was introduced by Robertson and Seymour [21] as a key component of the proof of the Graph Minor Theorem. A *tree decomposition* of a graph $G = (V, E)$ is a pair $(T, \{\, B_t : t \in T \,\})$, where T is a tree, $B_t \subseteq V$ and

- $\bigcup_{t\in T} B_t = V$;
- if there is an edge $uv \in E$ then $\{\, u, v \,\} \subseteq B_t$ for some t; and
- for each $v \in V$, the set $\{\, t : v \in B_t \,\}$ is connected in T.

The *width* of a tree decomposition is $\max\{\, |B_t| : t \in T \,\} - 1$ and the *tree-width* of G is $\mathrm{tw}(G)$, the least k for which G has a tree decomposition of width k.

We will use the following game-theoretic characterization of tree width, due to Seymour and Thomas [24]. The k cops and robber game is played by two players, the cops and the robber, on a graph $G = (V, E)$. At each move, the cops player either removes a cop from the graph or takes a cop not currently on the graph and places him on some vertex v. The robber may then move along any cop-free path in the graph. If the cops' move was to place a cop on v, that vertex counts as cop-free for this turn. If a cop moves to the vertex occupied by the robber and the robber has no non-trivial legal move, the cops win; the robber wins if he can stay on the run indefinitely. Seymour and Thomas show that the cops have a winning strategy in the k cops and robber game on G if, and only if, $\mathrm{tw}(G) \leqslant k - 1$.

For a positive integer k, we write TW_k for the class of graphs of tree width at most k. For a function $f : \mathbb{N} \to \mathbb{N}$, TW_f denotes the class of all graphs G such that the tree-width of G is at most $f(|G|)$.

Given a graph G and $r \geqslant 0$, for each vertex $v \in G$, let $N_G^r(x)$ be the subgraph of G induced by the vertices at distance at most r from v. The *local tree-width* of a graph [11] is the function

$$\mathrm{ltw}_G(r) = \max\{\, \mathrm{tw}(N_G^r(v)) : v \in G \,\}.$$

A class $\mathcal{G}$ of graphs is said to have *bounded local tree-width* if there is a (non-decreasing) function $f : \mathbb{N} \to \mathbb{N}$ such that, for all $G \in \mathcal{G}$ and all $r \geqslant 0$, $\mathrm{ltw}_G(r) \leqslant f(r)$. Classes of graphs of bounded local tree-width are introduced by Eppstein [10], who refers to such classes as having the 'diameter-treewidth property'.

2.3 Graph Minors

We say that a graph G is a *minor* of H, and write $G \preccurlyeq H$, if there is a map that associates with each vertex v of G a non-empty, connected subgraph H_v of H such that H_u and H_v are disjoint for $u \neq v$ and if there is an edge between u and v in G then there is an edge in H between some vertex in H_u and some vertex in H_v. We refer to the sets H_v as the branch sets witnessing that G is a minor of H. An equivalent characterization (see [8]) states that G is a minor of H if G can be obtained from a subgraph of H by contracting edges. The contraction of an edge consists of identifying its two endpoints into a single vertex and removing the resulting loop.

We collect here a few facts about graph minors that we will need. All of these can be found in [8]. Note that if $G \preccurlyeq H$ then $\mathrm{tw}(G) \leqslant \mathrm{tw}(H)$. By the well-known Kuratowski–Wagner theorem, a graph G is planar if, and only if, neither K_5 nor $K_{3,3}$ is a minor of G. Robertson and Seymour [23] showed that any class of graphs that is closed under taking minors and is not the class of all graphs is characterized by a finite set of forbidden minors. We call such a class of graphs a *proper minor-closed class*.

For any $n > 1$, let G_n be the $n \times n$ grid graph, i.e., the graph with vertex set $\{1, \ldots, n\}^2$ and all edges of the form $\{\, (i, j), (i + 1, j) \,\}$ and $\{\, (i, j), (i, j + 1) \,\}$.

$\Delta(G_n) = 4$ (where $\Delta(G)$ denotes the maximum degree of any vertex in G) and it is easy to see from the cops and robber game that $\mathrm{tw}(G_n) = n$. Also, it can be shown that for any planar graph G, there is an n such that $G \preccurlyeq G_n$.

2.4 CFI Graphs

The graphs we describe in this section are a minor variation on the graphs used by Cai *et al.* to separate IFP + C from **P** [4] and proofs of all the results we quote here can be found there. The difference is that Cai *et al.* do not have the '*c*' and '*d*' vertices but, instead, colour the vertices on the understanding that the colours can be replaced by appropriate gadgets. The gadgets are simple and we will use some of their properties later on so we prefer to make them explicit.

Let $G = (V, E)$ be a graph in which every vertex has degree at least two. In the following discussion, we will assume that G is connected but there are easy component-wise extensions in the case where G is not connected. For each $v \in V$ let $\Gamma(v) = \{ u : uv \in E \}$ and let $\widehat{v}$ be the set of new vertices,

$$\widehat{v} = \{ a_{vw}, b_{vw}, c_{vw}, d_{vw} : w \in \Gamma(v) \} \\ \cup \{ v^X : X \subseteq \Gamma(v) \text{ and } |X| \equiv 0 \pmod 2 \} .$$

Call the v^X *inner vertices* and the other members of $\widehat{v}$ *outer vertices.* Let $X_\emptyset(G)$ be the graph with vertices $\bigcup_{v \in V} \widehat{v}$ and edges as follows:

- edges $a_{vw}c_{vw}$, $b_{vw}c_{vw}$ and $c_{vw}d_{vw}$ for each edge $vw \in E$;
- an edge $a_{vw}v^X$ whenever $w \in X$;
- an edge $b_{vw}v^X$ whenever $w \in \Gamma(v) \setminus X$; and
- edges $a_{vw}a_{wv}$ and $b_{vw}b_{wv}$ for each edge $vw \in E$.

The subgraph of $X_\emptyset(G)$ induced by $\widehat{v}$ for a vertex v of G with three neighbours w_1, w_2, w_3 is illustrated in Fig. 1, where the dashed lines indicate edges connecting this subgraph to the rest of $X_\emptyset(G)$.

For any $S \subseteq E$, let $X_S(G)$ be $X_\emptyset(G)$ with the edges $a_{vw}a_{wv}$ and $b_{vw}b_{wv}$ deleted and edges $a_{vw}b_{wv}$ and $b_{vw}a_{wv}$ added, for every edge $vw \in S$. We say that the edges in S have been *twisted.* Cai *et al.* show that $X_S(G) \cong X_T(G)$ if, and only if, $|S| \equiv |T| \pmod 2$. This being the case, we write $X(G)$ for the graph $X_\emptyset(G)$ and write $\widetilde{X}(G)$ for $X_{\{e\}}(G)$ for any edge e and call these, respectively, the *untwisted* and *twisted CFI graphs of G.*

For distinct edges e and f of G, we can obtain an isomorphism between $X_{\{e\}}(G)$ and $X_{\{f\}}(G)$ as follows. Note that, for each $v \in V$ and $N \subseteq \Gamma(v)$ with $|N| \equiv 0 \pmod 2$, there is an automorphism $\eta_{v,N}$ of the subgraph of $X_S(G)$ induced by $\widehat{v}$ that exchanges a_{vw} and b_{vw}, for each $w \in N$ (and there is no such automorphism if $|N| \equiv 1 \pmod 2$). Let e be the edge uv and f be the edge xy. If $v = x$, then the required isomorphism is just the map $\eta_{v,\{u,y\}}$. Otherwise, if the four vertices are distinct then, by the assumption that G is connected, there is a simple path from one endpoint of e to an endpoint of f that does not pass through the other endpoints. Without loss of generality let $v_1 \dots v_\ell$ be a simple

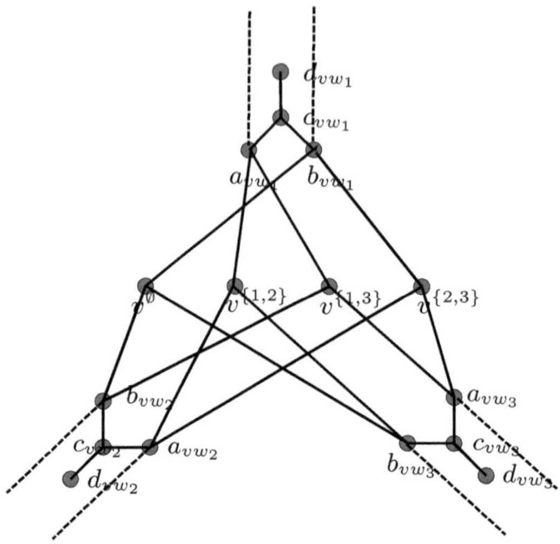

Fig. 1. The graph on the vertices $\widehat{v}$ in $X_\emptyset(G)$

path with $v = v_1$ and $x = v_\ell$ such that neither u nor y occurs on the path. Then, the required isomorphism from $X_{\{e\}}(G)$ to $X_{\{f\}}(G)$ is

$$\eta = \eta_{v_1,\{u,v_2\}} \circ \eta_{v_2,\{v_1,v_3\}} \circ \cdots \circ \eta_{v_{\ell-1},\{v_{\ell-2},v_k\}} \circ \eta_{v_\ell,\{v_{\ell-1},y\}}\,.$$

3 Tree-Width

A *separator* of a graph $G = (V, E)$ is a set $S \subseteq V$ such that every connected component of $G - S$ has at most $|V|/2$ vertices. Cai, Fürer and Immerman prove that, if G is connected, has minimal degree $\delta(G) \geqslant 2$ and has no separator of size k, then $X(G) \equiv^{\mathcal{C}^k_{\infty\omega}} \widetilde{X}(G)$. It follows that IFP + C does not define all polynomial-time queries on graphs, for instance by Theorem 1.

The following lemma on the relationship between separators and tree-width is a special case of [21, Theorem 2.5].

Lemma 2. *Every graph G of tree-width k has a separator of size at most $k+1$.*

So, any graph that has no separator of size k must have tree-width at least k. On the other hand, for all k there are connected graphs of tree-width k that have separators of size one (and, of course, disconnected graphs of tree-width k with $\emptyset$ as a separator): take any order-n graph G of tree-width k, choose a vertex $v \in G$ and add n new vertices and an edge from each new vertex to v. The resulting graph still has tree-width k but $\{v\}$ is a separator. Therefore, requiring that G have tree-width at least $k - 1$ is weaker than requiring it to have no separator of size k.

Theorem 3. *Let G be any connected graph with $\delta(G) \geqslant 2$ and* $\mathrm{tw}(G) \geqslant k$. $X(G) \equiv^{\mathcal{C}^{k}_{\infty\omega}} \widetilde{X}(G)$.

Proof. We exhibit a winning strategy for the duplicator in the k-pebble bijection game on $X(G)$ and $\widetilde{X}(G)$.

Given $u, v \in V(G)$, let $\sigma[u, v]$ be the permutation of $V(G)$ that exchanges u and v and fixes every other vertex. For a vertex u of G, we say that a bijection $h : X(G) \to \widetilde{X}(G)$ is *good except at* u if it satisfies the following conditions:

- for every vertex v of G, $h\widehat{v} = \widehat{v}$;
- h maps inner vertices to inner vertices and outer vertices to outer vertices;
- h is an isomorphism between the graphs $X(G) \setminus u_{\mathrm{I}}$ and $\widetilde{X}(G) \setminus u_{\mathrm{I}}$, where u_{I} is the set of inner vertices in $\widehat{u}$; and
- for every pair a_{uv}, b_{uv} in $\widehat{u}$, $h \circ \sigma[a_{uv}, b_{uv}]$ is an isomorphism from $X(G)[\widehat{u}]$ to $\widetilde{X}(G)[\widehat{u}]$, where $X(G)[\widehat{u}]$ is the subgraph of $X(G)$ induced by $\widehat{u}$.

For concreteness, say $\widetilde{X}(G)$ is the graph $X_{\{uv\}}(G)$. Then $\sigma[a_{uv}, b_{uv}]$ is a bijection that is good except at u; similarly, $\sigma[a_{vu}, b_{vu}]$ is good except at v. Note that if η is an automorphism of $\widetilde{X}(G)$ that fixes $\widehat{v}$ (set-wise) for every $v \in G$ and h is a bijection that is good except at u, then $h \circ \eta$ is also good except at u. We claim that, if h is a bijection that is good except at u and there is a simple path P from u to v, then there is a bijection h' that is good except at v such that for all vertices w not in P and all $x \in \widehat{w}$, $h'(x) = h(x)$.

To prove the claim, let the path P be $v_1 \ldots v_\ell$ with $u = v_1$ and $v = v_\ell$. Let η_P be the permutation

$$\eta_P = \sigma[a_{uv_1}, b_{uv_1}] \circ \eta_{v_2,\{v_1,v_3\}} \circ \cdots \circ \eta_{v_{l-1},\{v_{l-2},v_l\}} \circ \sigma[a_{v,v_{l-1}}, b_{v,v_{l-1}}].$$

The properties of the graphs $X(G)$ and $\widetilde{X}(G)$ then ensure that taking $h' = h \circ \eta_P$ satisfies the claim.

The duplicator's strategy can now be described as always playing a bijection that is good except at u for some u. The vertex u is given by the position of the robber in the cops and robber game played on G where the positions of the cops are $v_1 \ldots v_k$ when the pebbles in the bijection game are in the sets $\widehat{v}_1 \ldots \widehat{v}_k$.

Initially, the duplicator plays a bijection that is good except at u for one of the endpoints u of the twisted edge in $\widetilde{X}(G)$. At the same time, she initiates a cops and robber game on G with the robber initially at u. At each subsequent move, when the spoiler moves one of the pebbles, the duplicator moves the corresponding cop in the cops and robber game. This yields a path P for the robber to move along to a vertex v. By the claim, this yields a bijection h' that the duplicator can play. Since P, by definition, does not go through any of the cop positions, this means that h' agrees with h on all currently pebbled positions in the bijection game as required. Also, since h is, at all times, an isomorphism everywhere except at the inner vertices of $\widehat{v}$, for the current robber position v, it follows that it must be a partial isomorphism on the pebbled vertices, as required. □

The following corollary is immediate from Theorem 3 and Theorem 1.

Corollary 4. IFP + C *does not capture* **P** *on any class of graphs containing* $X(G)$ *and* $\widetilde{X}(G)$ *for graphs* G *of unbounded tree width.*

Let $\Delta(G)$ be the maximal degree of any vertex in a given graph G. The following lemma shows that $\mathrm{tw}(X(G))$ can be bounded in terms of tree-width and maximal degree of G. The same bounds apply to $\mathrm{tw}(\widetilde{X}(G))$.

Lemma 5. $\mathrm{tw}(G) \leqslant \mathrm{tw}(X(G)) \leqslant (4\Delta(G) + 2^{\Delta(G)-1})\mathrm{tw}(G)$.

Proof. The first inequality holds because $G \preccurlyeq X(G)$: contracting along every edge in each $\widehat{v}$ in $X(G)$ gives G. The second holds because if $(T, \{ B_t : t \in T \})$ is a tree decomposition of G then $(T, \{ (\bigcup_{v \in B_t} \widehat{v}) : t \in T \})$ is a tree decomposition of $X(G)$. □

For any function $f : \mathbb{N} \to \mathbb{N}$, let $\mathrm{TW}_f = \{ G : \mathrm{tw}(G) \leqslant f(|G|) \}$. Grohe and Mariño show that, if f is any constant function, IFP + C captures **P** on TW_f [14]. We show that this result is, in a sense, optimal: if f is unbounded then IFP + C does not capture **P** on TW_f.

Recall that G_n is the $n \times n$ grid graph and that $\mathrm{tw}(G_n) = n$. For $n \geqslant 3$, $\Delta(G_n) = 4$.

Theorem 6. IFP + C *captures* **P** *on* TW_f *if, and only if,* $f = \mathcal{O}(1)$.

Proof. The 'if' direction is the Grohe–Mariño theorem. Conversely, if $f \neq \mathcal{O}(1)$ then, for any n, there is some $k > |X(G_n)|$ such that $f(k) \geqslant 24n$. Since $\mathrm{tw}(X(G_n)) \leqslant 24n$ (Lemma 5), it follows that TW_f contains $X(G)$ and $\widetilde{X}(G)$ (possibly padded with some number of isolated vertices) for graphs of arbitrary tree width and so, by Corollary 4, IFP + C does not capture **P** on TW_f. □

Notice that we do not claim that IFP + C fails to capture **P** on any class of graphs containing graphs of unbounded tree-width. For example, the complete graph on n vertices has tree-width $n - 1$ so the class of all complete graphs contains graphs of arbitrarily high tree-width but IFP + C does capture **P** on this class. Similarly, the class of planar graphs contains graphs of unbounded tree-width (it contains G_n for all n), but Grohe has shown that IFP + C captures **P** on this class [12]. However, if $\mathcal{C}$ contains all graphs whose tree-width is bounded by the function f, then the CFI construction applies.

Lemma 7. *If* G *is bipartite,* $X(G)$ *and* $\widetilde{X}(G)$ *are bipartite.*

Proof. Let $G = (V, E)$ with bipartition V_0, V_1. Then $X(G)$ and $\widetilde{X}(G)$ have bipartition W_0, W_1, where W_i consists of all inner and 'c' vertices corresponding to elements of V_i and all 'a', 'b' and 'd' vertices from V_{1-i}. □

Corollary 8. IFP + C *does not capture* **P** *on the class of bipartite graphs.*

Proof. For any n, G_n is bipartite so, by Lemmata 5 and 7, the class of bipartite graphs contains $X(G)$ and $\widetilde{X}(G)$ for graphs of unbounded tree width. □

Moreover, IFP + C does not capture **P** on the class of graphs of chromatic number k, for any fixed $k \geqslant 2$, as the disjoint union of $X(G_n)$ and K_k has chromatic number k but no formula of IFP + C can distinguish it from $\widetilde{X}(G_n) \cup K_k$ if n is large enough.

One might hope that Theorem 6 could be extended to measures of graph connectivity on directed graphs such as directed tree-width [17], DAG-width [2] or entanglement [3] but this is not the case. All directed acyclic graphs (DAGs) have low width in all of these measures (directed tree-width zero, DAG-width one and entanglement zero) but there are polynomial-time queries on DAGs not definable in IFP + C.

Theorem 9. IFP + C *does not capture* **P** *on the class of DAGs.*

Proof. Let D be a directed graph. Define $X'(D)$ and $\widetilde{X}'(D)$ in the same way as for undirected graphs but with the following directions on the edges:

- edges from inner vertices to outer vertices are directed that way;
- edges between outer vertices in the same $\widehat{v}$ are directed ac, bc and dc;
- any edges between a_{uv} or b_{uv}, and a_{vu} or b_{vu} have the same direction as the corresponding edge in D between u and v.

Note that, if D contains edges uv and vu then $\widehat{u}$ will contain two sets of outer vertices associated with v: one for each edge. Observe that $X'(D)$ and $\widetilde{X}'(D)$ are DAGs. Clearly, there is a polynomial-time algorithm that distinguishes $X'(D)$ from $\widetilde{X}'(D)$ — just forget the orientation of the edges and use the algorithm that distinguishes $X(G)$ from $\widetilde{X}(G)$. Suppose the query $\{ X'(D) : D \text{ is a DAG} \}$ is defined by some sentence $\varphi \in \text{IFP} + \text{C}$ for DAGs.

Fix any vertex $v \in G_n$. There are no edges in G_n between vertices at the same distance from v so the orientation $D(G_n, v)$ of G_n that orients every edge from its end further from v to the end nearer v is a DAG. There is an IFP formula $\psi(xy, v)$ that, given some vertex $v \in X(G_n)$ (respectively, $\widetilde{X}(G_n)$) as a parameter, defines the edge relation of $X'(D(G_n, v))$ (respectively, $\widetilde{X}'(D(G_n, v))$). Let $\chi \equiv \exists v\, \varphi[\psi(xy, v)/E(xy)]$, where $\varphi[\cdots]$ is the result of replacing every subformula $E(xy)$ with $\psi(xy, v)$. Then χ distinguishes $X(G_n)$ from $\widetilde{X}(G_n)$ for all n, contradicting Theorem 3. □

Corollary 8 and Theorem 9 are to be expected. The relation $\equiv^{\mathcal{C}^k_{\infty\omega}}$ can be tested in polynomial time by means of a colour-refinement algorithm (see, e.g., [20]). Therefore, by Theorem 1, it follows that if, IFP + C captures **P** on a class of structures $\mathcal{C}$ (closed under disjoint unions), then $\mathcal{C}$ admits a polynomial-time isomorphism test. It is not difficult to see that bipartite graphs and DAGs admit such a test if, and only if, all graphs do. Indeed, given an undirected graph $G = (V, E)$, let G' be the directed graph whose vertex set is $V \cup E$ and with a directed edge from $v \in V$ to $e \in E$ exactly when v is one of the ends of e in G. Clearly, G is acyclic and $G \cong H$ if, and only if, $G' \cong H'$. The undirected version of G' (known as the *incidence graph* of G) is bipartite.

4 Bounded Local Tree-Width

Given a non-decreasing function $f : \mathbb{N} \to \mathbb{N}$, let LTW_f be the class of graphs whose local tree-width is bounded by f. In this section, we extend the results of Section 3 to show that IFP + C captures **P** on LTW_f if, and only if, $f = \mathcal{O}(1)$.

For a graph $G = (V, E)$ and a positive integer r, we define the graph $r(G)$ to have vertex set $V \cup \{(u, v, i) : u, v \in V, 1 \leqslant i \leqslant r\}$ and edges:

- $\{ u, (u, v, 1) \}$ for all $u, v \in V$;
- $\{ (u, v, i), (u, v, i + 1) \}$ for all $u, v \in V$ and $1 \leqslant i < r$; and
- $\{ (u, v, r), (v, u, r) \}$ for all edges $uv \in E$.

Recall that a *subdivision* of a graph G is any graph H formed by replacing each edge $uv \in G$ with a u–v path, such that the paths in H corresponding to distinct edges in G are internally disjoint. We can think of $r(G)$ as a graph that is obtained from a subdivision of G (where each edge is replaced by a path of length $2r$) by further adding, for each pair u, v that is not an edge in G, two simple paths of length r — one originating at u and one at v — that do not meet.

The properties of the graphs $r(G)$ that we need are established in the following lemmata.

Lemma 10. $\mathrm{tw}(r(G)) = \mathrm{tw}(G)$.

Proof. $\mathrm{tw}(G) \leqslant \mathrm{tw}(r(G))$ because $G \preccurlyeq r(G)$. For the converse, if $\mathrm{tw}(G) = 1$, then G is a forest and $r(G)$ is a forest as well, so $\mathrm{tw}(H) = 1$, as required. Now, suppose $\mathrm{tw}(G) = k > 1$. Let $(T, \{ B_t : t \in T \})$ be a width-k tree decomposition of G. We construct a width-k tree decomposition of H as follows. For each edge $uv \in G$ there must, by definition, be some $t \in T$ such that $\{ u, v \} \subseteq B_t$. Add to T a path $tt_1 \ldots t_{2r}$ and set $B_{t_1} = \{ u, v, (u, v, 1) \}$; for $2 \leqslant i \leqslant r$ set $B_{t_i} = \{ v, (u, v, i - 1), (u, v, i) \}$; $B_{t_{r+1}} = \{ v, (u, v, r), (v, u, r) \}$; and for $r + 2 \leqslant i \leqslant 2r$ set $B_{t_i} = \{ v, (v, u, 2r - i + 2), (v, u, 2r - i + 1) \}$. Finally, if uv is not an edge in G, choose any $t \in T$ such that $u \in B_t$ and add a path $tt_1 \ldots t_{r-1}$ to T with $B_{t_i} = \{ u, (u, v, i), (u, v, i + 1) \}$. □

For any vertex w in $r(G)$, we write $\pi w = w$, if $w \in V(G)$, and $\pi(u, v, i) = u$.

Lemma 11. *If* $G \equiv^{C^k_{\infty\omega}} H$*, then* $r(G) \equiv^{C^k_{\infty\omega}} r(H)$.

Proof. By the assumption $G \equiv^{C^k_{\infty\omega}} H$, the duplicator has a winning strategy in the k-pebble bijection game played on these two graphs. This winning strategy is easily adapted to a winning strategy on the pair of graphs $r(G)$ and $r(H)$. For any bijection h between the vertices of G and the vertices of H, consider the extension h_r of h to the vertices of of $r(G)$ given by

$$h_r(w) = \begin{cases} h(w) & \text{if } w \in V^G \\ (h(u), h(v), i) & \text{if } w = (u, v, i). \end{cases}$$

The duplicator's strategy is to maintain the condition that, at any point in the game, if the pebbles are on the vertices $s_1, \ldots, s_k$ in $r(G)$ and $t_1, \ldots, t_k$ in $r(H)$, then $\pi s_1, \ldots, \pi s_k$ and $\pi t_1, \ldots, \pi t_k$ is a winning position in the game played on G and H. It is now easily verified that, if the duplicator's strategy called for playing the bijection h in the latter game, then the bijection h_r will maintain this condition in the game on $r(G)$ and $r(H)$. □

We are now ready to prove the strengthening of Theorem 6.

Theorem 12. *Let $f : \mathbb{N} \to \mathbb{N}$ be any non-decreasing function.* IFP + C *captures* **P** *on* LTW_f *if, and only if,* $f = \mathcal{O}(1)$.

Proof. $\mathrm{LTW}_f \subseteq \mathrm{TW}_f$ so, if $f = \mathcal{O}(1)$, then IFP + C defines all polynomial-time properties over LTW_f by the Grohe–Mariño theorem.

Suppose $f \neq \mathcal{O}(1)$. For any k, let G by a graph with $tw(G) \geqslant k$ and $\delta(G) \geqslant 2$. Now, there is some r such that $f(2r) \geqslant \mathrm{tw}(X(G)) = \mathrm{tw}(\widetilde{X}(G))$. Let $H = r(X(G))$ and $H' = r(\widetilde{X}(G))$. By Lemma 10, $\mathrm{tw}(H) = \mathrm{tw}(H') = \mathrm{tw}(X(G))$. Notice that ltw_H and $\mathrm{ltw}_{H'}$ are bounded by the function

$$h(x) = \begin{cases} 1 & \text{if } x < 2r \\ \mathrm{tw}(X(G)) & \text{otherwise,} \end{cases}$$

and that this function is, in turn, bounded by f. Therefore, $H, H' \in \mathrm{LTW}_f$. Since, $\mathrm{tw}(G) \geqslant k$ we have (by Theorem 3) that $H \equiv^{C^k_{\infty\omega}} H'$. The result now follows from Theorem 1, □

5 Graph Minors

Grohe has conjectured that IFP + C captures **P** on any proper minor-closed class of graphs [13], i.e., any minor-closed class except the class of all graphs. In this section, we show that this conjecture cannot be refuted by the CFI construction. Specifically, we show that any class $\mathcal{C}$ of graphs containing at least one of $X(G)$ and $\widetilde{X}(G)$ for graphs G of unbounded tree-width has no forbidden minors. Since any proper minor-closed class must have at least one forbidden minor, it follows that $\mathcal{C}$ is either the class of all graphs or is not minor-closed. Note that the requirement that $\mathcal{C}$ contain CFI graphs derived from graphs of unbounded tree-width is crucial here: it does not suffice to require merely that $\mathcal{C}$ contain graphs of unbounded tree-width. For example, the class of planar graphs does not have bounded tree-width.

We wish to show that, for any graph G, if H is a graph with $\mathrm{tw}(H)$ large enough, relative to G, then $X(H)$ and $\widetilde{X}(H)$ contain G as a minor. To do this, we will first produce a planar graph G' such that $G' \preccurlyeq H$. The graph G' is obtained from a plane drawing of G by inserting new vertices at crossing points of edges. The assumption on the tree-width of H will ensure that any such planar graph is a minor of H. The paths in G' corresponding to distinct edges in G will be edge-disjoint but not necessarily independent: two of the paths may cross at some vertex. To show that $X(H)$ and $\widetilde{X}(H)$ contain G as a minor, we need to show that, even if edge-disjoint paths P_1 and P_2 meet at a vertex u, $X(H)$ and $\widetilde{X}(H)$ contain corresponding independent paths.

Lemma 13. *Let $u \in G$ be a vertex of degree 4, with neighbours w, x, y, z. For each $v \in \{w, x, y, z\}$, choose $v' \in \{a_{uv}, b_{uv}\}$. $\widehat{u}$ contains vertices v_1 and v_2 such that $X(G)$ and $\widetilde{X}(G)$ contain disjoint paths $w'v_1x'$ and $y'v_2z'$.*

Proof. Note that $X(G)$ and $\widetilde{X}(G)$ have the same edges within each $\widehat{u}$. The values of v_1 and v_2 are given in the following table.

w'	x'	y'	z'	v_1	v_2
a_{uw}	a_{ux}	a_{uy}	a_{uz}	$v^{\{w,x\}}$	$v^{\{y,z\}}$
a_{uw}	a_{ux}	a_{uy}	b_{uz}	$v^{\{w,x,y,z\}}$	$v^{\{x,y\}}$
a_{uw}	b_{ux}	a_{uy}	b_{uz}	$v^{\{w,z\}}$	$v^{\{x,y\}}$

The other cases are symmetric, either by permutations of $\{w, x, y, z\}$ or the automorphisms of $\widehat{u}$ that exchange the 'a' and 'b' vertices for even-cardinality subsets of $\{w, x, y, z\}$. □

We first restrict attention to minors of CFI graphs of grids. Recall that G_r is the $r \times r$ grid graph.

Theorem 14. *Let G be any graph. For sufficiently large grids G_r, $G \preccurlyeq X(G_r)$ and $G \preccurlyeq \widetilde{X}(G_r)$.*

Proof. Let $V(G) = \{v_1, \ldots, v_n\}$ and $E(G) = \{e_1, \ldots, e_m\}$. We first produce a drawing G^* of G. Choose a set $V^* = \{v_1^*, \ldots, v_n^*\}$ of distinct points in $\mathbb{R}^2$ to represent the vertices of G and a set $E^* = \{e_1^*, \ldots, e_m^*\}$ of distinct, simple, piecewise-linear curves to represent the edges, such that:

- if $e_i = v_j v_k$ then the endpoints of e_i^* are v_j^* and v_k^*;
- no e_i^* contains any v_j^* except its endpoints;
- for $i \neq j$, $e_i^* \cap e_j^*$ is finite; and
- no point in $\mathbb{R}^2 \setminus V^*$ appears in more than two of the e_i^*.

We can now produce a planar graph G' whose vertices are the points of intersection of the e_i^* (i.e., $V(G') = \bigcup_{1 \leqslant i < j \leqslant m}(e_i^* \cap e_j^*)$, including V) and whose edges are precisely those pairs $\{x, y\}$ such that G^* contains an x–y curve that passes through no other points in $V(G')$.

Since G' is planar, we have $G' \preccurlyeq G_r$ for large enough r. Unless G is, itself, planar, we cannot have $G \preccurlyeq G_r$; however, we claim that $G \preccurlyeq X(G_r)$ and $G \preccurlyeq \widetilde{X}(G_r)$. To this end, let H be $X(G_r)$ or $\widetilde{X}(G_r)$.

Let $\{V_x : x \in G'\}$ be the branch sets witnessing that G' is a minor of G_r. For each $x \in G'$, let $\widehat{V}_x = \bigcup_{y \in V_x} \widehat{y} \subseteq V(H)$. To show that $G \preccurlyeq H$, we proceed as follows. First, for each $x \in G$, contract all the edges in the subgraph induced by $\widehat{V}_x$, calling the resulting vertex v_x. We now show that there is a system of independent paths P_{xy}, from v_x to v_y, for each edge $xy \in G$.

For each $xy \in G$, let Q_{xy} be the x–y path in G' corresponding to the edge $xy \in G^*$. These paths are not necessarily independent: in particular, they share the vertices of $V(G') \setminus V(G)$. Lemma 13 shows that, even if Q_{wx} and Q_{yz} meet at vertex u, H contains paths $\widehat{w}$–$\widehat{x}$ and $\widehat{y}$–$\widehat{z}$ that are independent. This completes the proof. □

The significance of grids is given by the following theorem of Robertson and Seymour. (Note that the tree-width required grows rapidly with r: Diestel shows that $r^{4r^4(r+2)}$ suffices [8].)

Theorem 15 ([22]). *For every $r > 1$, every graph of sufficiently high tree-width contains G_r as a minor.*

Theorem 16. *The only minor-closed class of graphs that contains $X(G)$ or $\widetilde{X}(G)$ for graphs G of unbounded tree-width is the class of all graphs.*

Proof. Let $\mathcal{C}$ be a minor-closed class of graphs containing $X(G)$ or $\widetilde{X}(G)$ for graphs G of unbounded tree-width and let H be any graph. By Theorem 14, $X(G_r)$ and $\widetilde{X}(G_r)$ contain H as a minor, for large enough grids G_r. By Theorem 15, any graph of large enough tree-width contains G_r as a minor. Therefore, $\mathcal{C}$ contains a graph $X \in \{X(G), \widetilde{X}(G)\}$ for some graph G containing G_r as a minor, so $H \preccurlyeq X$. But $\mathcal{C}$ is minor-closed so $H \in \mathcal{C}$. □

A consequence of this theorem is that any attempt to refute Grohe's conjecture that IFP + C captures **P** on all non-trivial minor-closed classes of graphs cannot rely on Cai–Fürer–Immerman graphs. For, to use the CFI construction (in the form in Theorem 3), we need precisely to find for each k, a graph G of tree-width at least k such that $X(G)$ and $\widetilde{X}(G)$ are both in the class.

Acknowledgment. We would like to thank Stephan Kreutzer for fruitful discussions, especially in connection with the construction in Section 5.

References

1. Atserias, A., Bulatov, A., Dawar, A.: Affine systems of equations and counting infinitary logic. In: Proc. 34th International Colloquium on Automata, Languages and Programming. LNCS, vol. 4596, pp. 558–570. Springer, Heidelberg (to appear, 2007)
2. Berwanger, D., Dawar, A., Hunter, P., Kreutzer, S.: DAG-width and parity games. In: Durand, B., Thomas, W. (eds.) STACS 2006. LNCS, vol. 3884, pp. 524–536. Springer, Heidelberg (2006)
3. Berwanger, D., Grädel, E.: Entanglement: A measure for the complexity of directed graphs with applications to logic and games. In: Baader, F., Voronkov, A. (eds.) LPAR 2004. LNCS (LNAI), vol. 3452, pp. 209–223. Springer, Heidelberg (2005)
4. Cai, J.-Y., Fürer, M., Immerman, N.: An optimal lower bound on the number of variables for graph identification. Combinatorica 12(4), 389–410 (1992)

5. Courcelle, B.: Graph rewriting: An algebraic and logic approach. In: van Leeuwan, J. (ed.) Handbook of Theoretical Computer Science, Volume B: Formal Models and Sematics (B), pp. 193–242. Elsevier, Amsterdam (1990)
6. Dawar, A., Grohe, M., Kreutzer, S.: Locally excluding a minor. In: Proc. 22nd IEEE Annual Symposium on Logic in Computer Science, IEEE Computer Society, Los Alamitos (to appear, 2007)
7. Dawar, A., Grohe, M., Kreutzer, S., Schweikardt, N.: Approximation schemes for first-order definable optimisation problems. In: Proc. 21st IEEE Annual Symposium on Logic in Computer Science, pp. 411–420. IEEE Computer Society, Los Alamitos (2006)
8. Diestel, R.: Graph Theory, 3rd edn. Springer, Heidelberg (2005)
9. Ebbinghaus, H.-D., Flum, J.: Finite Model Theory, 2nd edn. Springer, Heidelberg (1999)
10. Eppstein, D.: Diameter and treewidth in minor-closed graph families. Algorithmica 27, 275–291 (2000)
11. Frick, M., Grohe, M.: Deciding first-order properties of locally tree-decomposable structures. Journal of the ACM 48, 1184–1206 (2001)
12. Grohe, M.: Fixed-point logics on planar graphs. In: Proc. 13th IEEE Annual Symposium on Logic in Computer Science, pp. 6–15. IEEE Computer Society, Los Alamitos (1998)
13. Grohe, M.: Isomorphism testing for embeddable graphs through definability. In: Proc. 32nd ACM Symposium on Theory of Computing, pp. 63–72. ACM, New York (2000)
14. Grohe, M., Mariño, J.: Definability and descriptive complexity on databases of bounded tree-width. In: Beeri, C., Bruneman, P. (eds.) ICDT 1999. LNCS, vol. 1540, pp. 70–82. Springer, Heidelberg (1998)
15. Hella, L.: Logical hierarchies in PTIME. Information and Computation 129(1), 1–19 (1996)
16. Immerman, N., Lander, E.S.: Describing graphs: A first-order approach to graph canonization. In: Selman, A. (ed.) Complexity Theory Retrospective, pp. 59–81. Springer, Heidelberg (1990)
17. Johnson, T., Robertson, N., Seymour, P.D., Thomas, R.: Directed tree-width. Journal of Combinatorial Theory Series B 82, 138–155 (2001)
18. Nešetřil, J., de Ossona Mendez, P.: The grad of a graph and classes with bounded expansion. In: Raspaud, A., Delmas, O. (eds.) International Colloquium on Graph Theory. Electronic Notes in Discrete Mathematics, vol. 22, pp. 101–106. Elsevier, Amsterdam (2005)
19. Obdržálek, J.: DAG-width: Connectivity measure for directed graphs. In: Proc. 17th Annual ACM-SIAM Symposium on Discrete Algorithms, pp. 814–821 (2006)
20. Otto, M.: Bounded Variable Logics and Counting — A Study in Finite Models. Lecture Notes in Logic, vol. 9. Springer, Heidelberg (1997)
21. Robertson, N., Seymour, P.D.: Graph minors II: Algorithmic aspects of tree-width. Journal of Algorithms 7(3), 307–322 (1986)
22. Robertson, N., Seymour, P.D.: Graph minors V: Excluding a planar graph. Journal of Combinatorial Theory Series B 41(1), 91–114 (1986)
23. Robertson, N., Seymour, P.D.: Graph minors XX: Wagner's conjecture. Journal of Combinatorial Theory Series B 92(2), 325–357 (2004)
24. Seymour, P.D., Thomas, R.: Graph searching and a min-max theorem for tree-width. Journal of Combinatorial Theory Series B 58, 22–33 (1993)

Comparing the Expressive Power of Well-Structured Transition Systems

Parosh Aziz Abdulla[1], Giorgio Delzanno[2], and Laurent Van Begin[3,⋆]

[1] Uppsala University, Sweden
parosh@it.uu.se
[2] Università di Genova, Italy
giorgio@disi.unige.it
[3] Université Libre de Bruxelles, Belgium
lvbegin@ulb.ac.be

Abstract. We compare the expressive power of a class of well-structured transition systems that includes relational automata, Petri nets, lossy channel systems, and constrained multiset rewriting systems. For each one of these models we study the class of languages generated by labelled transition systems describing their semantics. We consider here two types of accepting conditions: coverability and reachability of a given configuration. In both cases we obtain a strict hierarchy in which constrained multiset rewriting systems is the the most expressive model.

Keywords: Expressiveness, well-structured systems, language theory.

1 Introduction

The theory of well-structured transition systems [1,13] is a powerful tool for studying the decidability of verification problems of infinite-state systems. A system is well-structured when its transition relation is monotonic with respect to a well-quasi ordering defined over configurations. A well-known example of well-structured system is that of *Petri nets* [19] equipped with marking inclusion [1,13]. For a well-structured transition system, the *coverability problem* can be decided by the symbolic backward reachability algorithm scheme proposed in [1]. Since checking safety properties can be translated into instances of the coverability problem, an algorithm for coverability like that proposed in [1] can be used for automatic verification of an infinite-state system. This connection has been exploited in order to develop automatic verification procedures for Petri nets and their extensions [10,11], for abstract models of imperative programs called *relational automata* [9], for abstract models of unreliable communication systems called *lossy (FIFO) channel systems* [5,8], and for *constrained multiset rewriting systems* [2]. The latter model is an extension of Petri nets in which tokens are colored with natural numbers and in which transitions have numerical conditions defined over variables representing colors. The resulting model can

⋆ Research fellow supported by the FNRS.

J. Duparc and T.A. Henzinger (Eds.): CSL 2007, LNCS 4646, pp. 99–114, 2007.

be applied to model parameterized systems in which individual processes have local data that range over an infinite domain.

Although several efforts have been spent in studying the expressive power of variations of Petri nets (see e.g. [10,12,14]), a comparison of the relative expressiveness of the class of well-structured transition systems is still missing. Such a comparison is a challenging research problem with a possible practical impact. Indeed, it can be useful to extend the applicability of a verification method (e.g. a particular instance of the scheme of [1]) to an entire class of models.

In this paper we apply tools of language theory to formally compare the expressive power of a large class of well-structured infinite-state systems that includes constrained multiset rewriting systems, lossy channel systems, (extensions of) Petri nets, and relational automata. To achieve this goal, for each one of these models we study the class of languages generated by labeled transition systems describing their semantics. We consider here two types of accepting conditions: coverability (with respect to a fixed ordering) and reachability of a given configuration. Two models are considered to be equivalent if they generate the same class of languages.

For coverability accepting conditions, we obtain the following classification. We first prove that lossy channel systems are equivalent to a syntactic fragment of constrained multiset rewriting, we named Γ_0. The fragment Γ_0 is obtained by restricting conditions of a rule in such a way that equalities cannot be used as guards. Furthermore, we prove that lossy channel systems are strictly less expressive than the full model of constrained multiset rewriting systems. We then show that Petri nets are equivalent to a syntactic fragment of constrained multiset rewriting systems, we named Γ_1, obtained by considering nullary predicates only. We also prove that Petri nets are strictly less expressive than lossy channel systems. We then prove that relational automata are equivalent to a syntactic fragment of constrained multiset rewriting, we named Γ_2, obtained by imposing an upper bound on the size (number of predicates) of reachable configurations. Finally, we prove that Γ_2 generates the class of regular languages. This implies that relational automata are strictly less expressive than Petri nets. In the paper we also extend the comparison to extensions of Petri nets like *transfer/reset nets* and *broadcast protocols* [10,11] and to *lossy vector addition systems* [18]. Specifically, we prove that all these models are strictly less expressive than constrained multiset rewriting systems.

For reachability accepting conditions, we obtain a slightly different classification. First, we prove that Γ_0 is equivalent to constrained multiset rewriting systems and *two counter machines*. Thus, with reachability acceptance, Γ_0 and constrained multiset rewriting systems turn out to be strictly more expressive than lossy channel systems. On the contrary, Γ_1 is still equivalent to Petri nets and strictly less expressive than Γ_0 and Γ_2 is still equivalent to relational automata and to finite automata. Finally, we show that lossy channel systems and Petri nets define incomparable classes of languages.

Concerning related work, the relative expressiveness of well-structured systems has been investigated for a limited number of extensions of Petri nets with reset, transfer, and non-blocking arcs in [12,14]. Classical results on finite and infinite languages generated by Petri nets can be found, e.g., in [15]. A classification of infinite-state systems in terms of structural properties and decidable verification problems is presented in [16]. The classification is extended to well-structured systems in [7]. A classification of the complexity of the decision procedures for coverability is studied in [17]. In contrast with the aforementioned work, we provide here a strict classification of the expressive power of several well-structured transition systems built with the help of tools of language theory. An extended version of the present paper is available as technical report [3].

Outline. In Section 2, we give some preliminary notions on well-structured transition systems. In Section 3, we give some first results on the class of languages accepted by constrained multiset rewriting systems. In Section 4, 5, and 6, we compare the class of languages recognized by constrained multiset rewriting systems and, respectively, lossy channel systems, Petri nets, and relational automata. Finally, in Section 7 we discuss some final remarks.

2 Preliminaries on Well-Structured Transition Systems

In this section we recall some definitions taken from [1]. A *transition system* is a tuple $T = (S, R)$ where S is a (possibly infinite) set of configurations, R is a finite set of transitions where each $\xrightarrow{\sigma} \in R$ is a binary relation over S, i.e. $\xrightarrow{\sigma} \subseteq S \times S$. We use $\gamma \xrightarrow{\sigma} \gamma'$ to denote $(\gamma, \gamma') \in \xrightarrow{\sigma}$, and $\gamma \xrightarrow{\rho_1 \cdots \rho_k} \gamma'$ to denote that there exist $\gamma_1, \ldots, \gamma_{k-1}$ such that $\gamma \xrightarrow{\rho_1} \gamma_1 \ldots \xrightarrow{\rho_{k-1}} \gamma_{k-1} \xrightarrow{\rho_k} \gamma'$. A quasi-ordering $(S, \preceq)$ is a *well-quasi ordering* if for any infinite sequence $s_1 s_2 \ldots s_i \ldots$ there exist indexes $i < j$ such that $s_i \preceq s_j$. A transition system $T = (S, R)$ is *well-structured* with respect to a quasi-order $\preceq$ on S iff: (i) $\preceq$ is a well-quasi ordering; (ii) for any $\xrightarrow{\sigma} \in R$ and $\gamma_1, \gamma_1', \gamma_2$ s.t. $\gamma_1 \preceq \gamma_1'$ and $\gamma_1 \xrightarrow{\sigma} \gamma_2$, there exists γ_2' s.t. $\gamma_1' \xrightarrow{\sigma} \gamma_2'$ and $\gamma_2 \preceq \gamma_2'$, i.e., T is *monotonic*. We use $T = (S, R, \preceq)$ to indicate a *well-structured transition system* (wsts for short).

To formalize the comparison between models, a wsts $T = (S, R, \preceq)$ can be viewed as a language acceptor. For this purpose, we assume a finite alphabet Σ and a labelling function $\lambda : R \mapsto \Sigma$ that associates to each transition of R a symbol of $\Sigma \cup \{\epsilon\}$, where ϵ denotes the empty sequence ($w \cdot \epsilon = \epsilon \cdot w = w$ for any $w \in \Sigma^*$). In the following, we use $\gamma_1 \xrightarrow{w} \gamma_2$ with $w \in \Sigma^*$ to denote that $\gamma_1 \xrightarrow{\rho_1 \cdots \rho_k} \gamma_2$ and $\lambda(\xrightarrow{\rho_1}) \cdots \lambda(\xrightarrow{\rho_k}) = w$. Furthermore, we associate to T an *initial* configuration $\gamma_{init} \in S$ and a *final* configuration $\gamma_{acc} \in S$ and assume an accepting relation $\bowtie: S \times S$. For a fixed accepting relation $\bowtie$, we define the language accepted (generated) by $T = (S, R, \preceq, \gamma_{init}, \gamma_{acc})$ as:

$$L(T) = \{w \in \Sigma^* \mid \gamma_{init} \xrightarrow{w} \gamma \text{ and } \gamma_{acc} \bowtie \gamma\}$$

In this paper we consider two types of accepting relations:

- *Coverability*: the accepting relation $\bowtie$ is defined as $\gamma_{acc} \preceq \gamma$.
- *Reachability*: the accepting relation $\bowtie$ is defined as $\gamma_{acc} = \gamma$.

Let $\mathcal{M}$ be a wsts model (e.g. Petri nets) and let T be one of its instances (i.e. a particular net). We define $L_c(T)$, resp $L_r(T)$, as the language accepted by T with accepting relation $\bowtie_c$, resp. $\bowtie_r$. We say that L is a c-language, resp. r-language, of $\mathcal{M}$ if there is an instance T of $\mathcal{M}$ such that $L = L_c(T)$, resp. $L = L_r(T)$. We use $L_c(\mathcal{M})$, resp. $L_r(\mathcal{M})$, to denote the class of c-languages, resp. r-languages, of $\mathcal{M}$. Finally, we use $\mathcal{L}_1 \not\sim \mathcal{L}_2$ to denote that $\mathcal{L}_1$ and $\mathcal{L}_2$ are incomparable classes of languages.

3 Constrained Multiset Rewriting Systems

In this section we recall the main definitions and prove the first results for *constrained multiset rewriting systems* [2]. Let us first give some preliminary definitions. We use $\mathbb{N}$ to denote the set of natural numbers and $\overline{n}$ to denote the interval $[0, \ldots, n]$ for any $n \in \mathbb{N}$. We assume a set $\mathbb{V}$ of variables which range over $\mathbb{N}$, and a set $\mathbb{P}$ of unary predicate symbols. For a set A, we use A^* and $A^{\otimes}$ to denote the sets of (finite) words and (finite) multisets over A respectively. Sometimes, we write multisets as lists, so $[1, 5, 5, 1, 1]$ represents a multiset with three occurrences of 1 and two occurrences of 5; $[\,]$ represents the empty multiset. We use the usual relations and operations such as $\leq$ (inclusion), $+$ (union), and $-$ (difference) on multisets. For a set $V \subseteq \mathbb{V}$, a *valuation* *Val* of V is a mapping from V to $\mathbb{N}$. A *condition* is a finite conjunction of *gap order* formulas of the forms: $x <_c y$, $x \leq y$, $x = y$, $x < c$, $x > c$, $x = c$, where $x, y \in \mathbb{V}$ and $c \in \mathbb{N}$. Here $x <_c y$ stands for $x + c < y$. We often use $x < y$ instead of $x <_0 y$. Sometimes, we treat a condition ψ as a set, and write e.g. $(x <_c y) \in \psi$ to indicate that $x <_c y$ is one of the conjuncts in ψ. We use *true* to indicate an empty set of conditions. A *term* is of the form $p(x)$ where $p \in \mathbb{P}$ and $x \in \mathbb{V}$. A *ground term* is of the form $p(c)$ where $p \in \mathbb{P}$ and $c \in \mathbb{N}$. We sometimes say that a predicate symbol is *nullary* to mean that its parameter is not relevant (hence may be omitted).

A *constrained multiset rewriting system (CMRS)* $\mathcal{S}$ consists of a finite set of *rules* each of the form $L \rightsquigarrow R : \psi$, where L and R are multisets of terms, and ψ is a condition. We assume that ψ is consistent (otherwise, the rule is never enabled). For a valuation *Val*, we use $Val(\psi)$ to denote the result of substituting each variable x in ψ by $Val(x)$. We use $Val \models \psi$ to denote that $Val(\psi)$ evaluates to *true*. For a multiset T of terms we define $Val(T)$ as the multiset of ground terms obtained from T by replacing each variable x by $Val(x)$. A *configuration* is a multiset of ground terms. Each rule $\rho = L \rightsquigarrow R : \psi \in \mathcal{S}$ defines a relation between configurations. More precisely, $\gamma \xrightarrow{\rho} \gamma'$ if and only if there is a valuation *Val* s.t. the following conditions are satisfied: (*i*) $Val \models \psi$, (*ii*) $\gamma \geq Val(L)$, and (*iii*) $\gamma' = \gamma - Val(L) + Val(R)$. As an example, consider the rule:

$$\rho = [p(x)\ ,\ q(y)] \rightsquigarrow [q(z)\ ,\ r(x)\ ,\ r(w)]\ :\ \{x <_2 y\ ,\ x <_4 z\ ,\ z <_0 w\}$$

A valuation which satisfies the condition is $Val(x) = 1$, $Val(y) = 4$, $Val(z) = 8$, $Val(w) = 10$, Therefore, we have that $[p(1), p(3), q(4)] \xrightarrow{\rho} [p(3), q(8), r(1), r(10)]$.

A run σ is a sequence of transitions $\gamma_0 \xrightarrow{\rho_1} \gamma_1 \xrightarrow{\rho_2} \ldots \xrightarrow{\rho_n} \gamma_n$; where $\lambda(\rho_1) \cdot \ldots \lambda(\rho_n)$ is the word associated to σ for some labelling function λ.

Let us fix a CMRS $\mathcal{S}$ operating on a set of predicate symbols $\mathbb{P}$. Let $cmax$ be the maximal constant which appears in the rules of $\mathcal{S}$; $cmax$ is equal to 0 if there are no constant in $\mathcal{S}$. We now define an ordering $\preceq_c$ on configurations extracted from the ordering defined in [2] to solve the coverability problem.

Definition 1. *Given a configuration γ, we define the* index *of γ, $index(\gamma)$, to be a word of the form $D_0 \cdots D_{cmax}\ d_0\ B_0\ d_1\ B_1\ d_2 \cdots d_n\ B_n$ where*

- *$D_0, \ldots, D_{cmax}, B_0, \ldots, B_n \in \mathbb{P}^{\otimes}$ and $d_0, d_1, \ldots, d_n \in \mathbb{N} \setminus \{0\}$;*
- *B_i must not be empty for $0 \leq i \leq n$;*
- *for each $p \in \mathbb{P}$, D_i contains k occurrences of predicate p iff $p(i)$ occurs k times in γ for $0 \leq i \leq cmax$;*
- *given $v_0 = cmax + d_0$, for each $p \in \mathbb{P}$, B_0 contains k occurrences of predicate p iff $p(v_0)$ occurs k times in γ;*
- *given $v_{i+1} = v_i + d_{i+1}$, for each $p \in \mathbb{P}$, B_{i+1} contains k occurrences of predicate p iff $p(v_{i+1})$ occurs k times in γ for all $0 \leq i < n$;*
- *for all $p(v) \in \gamma$ with $v > cmax$, there exists $i : 0 \leq i \leq n$ such that $v = cmax + d_0 + d_1 + \ldots + d_i$.*

The ordering $\preceq_c$ is defined as follows.

Definition 2. *Let $D_0\ D_1 \cdots D_{cmax}\ d_0\ B_0\ d_1\ B_1\ d_2 \cdots d_n\ B_n$ be the index of a configuration γ_1 and $D'_0\ D'_1 \cdots D'_{cmax}\ d'_0\ B'_0\ d'_1\ B'_1\ d'_2 \cdots d'_m\ B'_m$ be the index of a configuration γ_2. Then, $\gamma_1 \preceq_c \gamma_2$ iff $D_i \leq D'_i$ for $0 \leq i \leq cmax$ and there exists a strictly monotone injection $h : \overline{n} \mapsto \overline{m}$ such that $B_0 \leq B'_{h(0)}$, $B_i \leq B'_{h(i)}$, $d_0 \leq \sum_{k=0}^{h(0)} d'_k$, and $d_i \leq \sum_{k=h(i-1)+1}^{h(i)} d'_k$ for $1 \leq i \leq n$.*

In the rest of the paper we assume that the values appearing in the initial configuration γ_{init} and in the accepting configuration γ_{acc} are smaller or equal than $cmax$. The ordering $\preceq_c$ is obtained by composing string embedding and multiset inclusion. From standard properties of orderings, it follows that $\preceq_c$ is a well-quasi ordering. Furthermore, a CMRS is monotonic with respect to corresponding ordering $\preceq_c$. The following property then holds.

Proposition 1. *A CMRS $\mathcal{S}$ equipped with $\preceq_c$ is well-structured.*

We now define a restriction $\ll$ of the relation $\preceq_c$ in which we require that the distribution of predicates in two configurations has the same structure but larger gaps. Formally, under the assumptions of Def. 2, $\gamma_1 \ll \gamma_2$ iff $n = m$, $D_i = D'_i$ for $0 \leq i \leq cmax$, $B_j = B'_j$ for $0 \leq j \leq n$, and $d_k \leq d'_k$ for $0 \leq k \leq n$. A CMRS $\mathcal{S}$ satisfies then the following property (the proof is given in [3]).

Proposition 2. *Let $\gamma_0 \xrightarrow{\rho_0} \gamma_1 \xrightarrow{\rho_1} \ldots \xrightarrow{\rho_{k-1}} \gamma_k \xrightarrow{\rho_k} \gamma$ be a run of $\mathcal{S}$. For any γ' s.t. $\gamma \ll \gamma'$ there exist $\gamma'_1, \ldots, \gamma'_k$ such that $\gamma_i \ll \gamma'_i$ for $i : 1 \leq i \leq k$ and $\gamma_0 \xrightarrow{\rho_1} \gamma'_1 \xrightarrow{\rho_1} \ldots \xrightarrow{\rho_{k-1}} \gamma'_k \xrightarrow{\rho_k} \gamma'$ is still a run of $\mathcal{S}$.*

In other words, with coverability accepting conditions a CMRS $\mathcal{S}$ can recognize a word w passing through configurations where gaps between parameters strictly greater than $cmax$ can be arbitrarily large. As discussed in the rest of the paper this property is very important to compare c-languages accepted by (fragments of) CMRS with those accepted by other wsts.

We are ready now to give a first characterization for the expressive power of CMRS. In [14, Prop. 4], the authors show that there exists a recursively enumerable (RE) language that cannot be recognized by any wsts with coverability acceptance. Hence, the following proposition holds.

Theorem 1. *$L_c(CMRS) \subset RE$.*

With reachability as accepting condition, CMRS recognize instead the class of recursively enumerable languages (RE).

Theorem 2. *$L_r(CMRS) = RE$.*

Proof. We prove that CMRS can weakly simulate 2-counter machines. A *2-counter machine (CM)* operates on two counters and on a finite set Q of control states. A transition updates the control state and executes either an increment, a decrement, or a zero-test of one of the two counters. Operations and tests on counters have their usual semantics, assuming that the values of counters are natural values. In the initial configuration the counters are set to zero. A 2-counter machine accepts an execution if it ends into the control state q_f. Assume a CM $\mathcal{M}$. The CMRS $\mathcal{S}$ that weakly simulates $\mathcal{M}$ operates in a sequence of phases indexed by natural numbers. Counters are represented as a multiset of terms of the form $cnt_1(c)$ and $cnt_2(c)$ where c denotes the current phase. During each phase, $\mathcal{S}$ simulates increment and decrement transitions of $\mathcal{M}$. As soon as $\mathcal{M}$ performs a zero-test of a counter, $\mathcal{S}$ enters an intermediate stage. After conclusion of the intermediate stage, a new phase is started and the index phase is increased. Transitions from q_1 to q_2 that update the current value of a counter are encoded by Γ_0 rules of the following form (they have the same labels as the corresponding CM transitions):

$$(q_1, cnt_i{:=}cnt_i{+}1, q_2) \Rightarrow \quad [q_1, phase(x)] \rightsquigarrow [q_2, phase(x), cnt_i(x)] : true$$
$$(q_1, cnt_i{:=}cnt_i{-}1, q_2) \Rightarrow [q_1, phase(x), cnt_i(x)] \rightsquigarrow [q_2, phase(x)] \quad : true$$

In these rules we update the value of the i-th counter by adding or deleting one occurrence of the term $cnt_i(c)$. Notice that the parameter c must be equal to the current phase index. A transition $(q_1, cnt_1 = 0?, q_2)$ labeled with a is encoded by the following Γ_0 rules (the two first labeled with ϵ, the last one with a):

$$[q_1, phase(x), phase'(x)] \rightsquigarrow [q_2', phase(y), phase'(x)] \quad : \{x < y\}$$
$$[q_2', cnt_2(x), phase(y), phase'(x)] \rightsquigarrow [q_2', cnt_2(y), phase(y), phase'(x)] : true$$
$$[q_2', phase(y), phase'(x)] \rightsquigarrow [q_2, phase(y), phase'(y)] \quad : true$$

(The Γ_0 rules encoding the test on cnt_2 are obtained from the previous ones by replacing predicate cnt_2 with cnt_1.) In the first rule we store the current index using $phase'$, and generate a new index which is strictly larger than the current one. This resets counter cnt_1 since all ground terms in its encoding will now have too small arguments for other rules in $\mathcal{S}$ to modify them. With the second rule, we change the arguments of (some of) the ground terms encoding cnt_2 to the new index. The third rule terminates the simulation of the zero-test.

Finally, we add to $\mathcal{S}$ the following rules (all labeled by ϵ) for $i \in \{1, 2\}$:

$$\begin{array}{rcll} [q_{fin}] & \rightsquigarrow & [q''_{fin}] & : \ true \\ [q''_{fin}\ ,\ phase(x)\ ,\ cnt_i(x)] & \rightsquigarrow & [q''_{fin}\ ,\ phase(x)] & : \ true \\ [q''_{fin}\ ,\ phase(x)\ ,\ phase'(y)] & \rightsquigarrow & [q''_{fin}] & : \ true \end{array}$$

By means of these additional rules, when we reach state q_{fin} we can move to q''_{fin} and erase the ground terms corresponding to the counters. The key observation here is that ground terms with parameters strictly less than the current phase are not removed during the simulation procedure described above. This implies that there exists an execution where $\mathcal{S}$ recognizes a word w that reaches $[q_f]$ iff there exists an execution where CM recognizes the word w that reaches q_f. Finally, the class of languages accepted by 2-counter machines with reachability accepting condition is RE. □

4 Lossy FIFO Channel Systems

In this section we study the relationship between a fragment of CMRS, we named Γ_0, and *lossy (FIFO) channel systems* (LCS) [5].

In the fragment Γ_0 of CMRS every rule $L \rightsquigarrow R : \psi$ satisfies the following conditions: every variable x occurs at most once in L and at most once in R, and ψ does not contain equality constraints. As an example, $[p(x), r(y)] \rightsquigarrow [q(x), r(z)] : x < y, y < z$ is a rule in Γ_0, whereas $[p(x), q(x)] \rightsquigarrow [q(y)] : true$ and $[p(x)] \rightsquigarrow [q(y), r(y)] : true$ are not in Γ_0.

A *Lossy FIFO Channel System* (LCS) consists of an asynchronous parallel composition of finite-state machines that communicate through sending and receiving messages via a finite set of unbounded lossy FIFO channels (in the sense that they can non-deterministically lose messages). Formally, an LCS $\mathcal{F}$ is a tuple (Q, C, N, δ) where Q is a finite set of control states (the Cartesian product of those of each finite-state machine), C is a finite set of channels, M is a finite set of messages, δ is a finite set of transitions, each of which is of the form (q_1, Op, q_2) where $q_1, q_2 \in Q$, and Op is a mapping from channels to channel operations. For any $c \in C$ and $a \in M$, an operation $Op(c)$ is either a *send* operation $!a$, a *receive* operation $?a$, the *empty* test $\epsilon?$, or the *null* operation *nop*. A configuration γ is a pair (q, w) where $q \in Q$, and w is a mapping from C to M^* giving the content of each channel. The initial configuration γ_{init} of $\mathcal{F}$ is the pair (q_0, ε) where $q_0 \in Q$, and ε denotes the mapping that assigns the empty sequence ϵ to each channel. The (strong) transition relation (that defines the semantics of machines with *perfect* FIFO channels) is defined as follows: $(q_1, w_1) \xrightarrow{\sigma} (q_2, w_2)$ if and only if $\sigma = (q_1, Op, q_2) \in \delta$ such that if $Op(c) = !a$, then $w_2(c) = w_1(c) \cdot a$; if $Op(c) = ?a$, then $w_1(c) = a \cdot w_2(c)$; if $Op(c) = \epsilon?$ then $w_1(c) = \epsilon$ and $w_2(c) = \epsilon$; if $Op(c) = nop$, then $w_2(c) = w_1(c)$. Now let $\preceq_l$ be the quasi ordering on LCS configurations such that $(q_1, w_1) \preceq_l (q_2, w_2)$ iff $q_1 = q_2$ and $\forall c \in C : w_1(c) \preceq_w w_2(c)$ where $\preceq_w$ indicates the subword relation. By Higman's theorem, we know that $\preceq_l$ is a well-quasi ordering. We introduce then the weak transition relation $\stackrel{\sigma}{\Longrightarrow}$ that defines the semantics of LCS: we have $\gamma_1 \stackrel{\sigma}{\Longrightarrow} \gamma_2$ iff there exists γ_1' and γ_2' s.t. $\gamma_1' \preceq_l \gamma_1$, $\gamma_1' \stackrel{\sigma}{\longrightarrow} \gamma_2'$, and $\gamma_2 \preceq_l \gamma_2'$. Thus, $\gamma_1 \stackrel{\sigma}{\Longrightarrow} \gamma_2$ means that γ_2 is reachable from γ_1 by first losing messages from the channels and reaching γ_1', then performing a transition, and, thereafter losing again messages from channels. As shown in [5], an LCS is well-structured with respect to $\preceq_l$. Furthermore, as shown in [6], in presence of transitions labeled with ϵ, we can restrict our attention to systems with only one channel. As a last remark, notice that for any model with lossy semantics like LCS, e.g. lossy vector addition systems [18], the class of c-languages coincide with the class of r-languages, i.e., $L_r(LCS) = L_c(LCS)$.

Our first result is that Γ_0 and LCS define the same class of c-languages.

Theorem 3. $L_c(\Gamma_0) = L_c(LCS)$.

Proof. The proof is based on encodings of LCS into Γ_0 and of Γ_0 into LCS. We next sketch the main ideas behind the two encodings (the complete proof is in [3]). In the encoding of an LCS in Γ_0, we represent the content $a_1 \ldots a_n$ of a channel c as a multiset $M_c = [h_c(x), a_1(x_1), \ldots, a_n(x_n), t_c(y)]$ where $x < x_1 < \ldots < x_n < y$. The predicates h_c (head) and t_c (tail) are used as sentinels to mark the two ends of the queue. The operation $!a$ on channel c is implemented by adding a new ground term with predicate a to M_c and by moving the tail to the right. The operation $?a$ on channel c is implemented by consuming an element with predicate a chosen non-deterministically from the multiset M_c and moving the head to the right. This operation simulates a lossy channel in the sense that when we update h_c we forget all elements to the left of the deleted element. Finally, the empty test is simulated by a reset of the channel. Formally, we encode LCS transitions operating on channel c into the following Γ_0 rules with the same labels:

$$
\begin{array}{lcrl}
(q_1, !a, q_2) & \Rightarrow & [q_1, t_c(x)] \rightsquigarrow [q_2, a(x), t_c(y)] & : \ \{x < y\} \\
(q_1, ?a, q_2) & \Rightarrow & [q_1, h_c(x), a(y)] \rightsquigarrow [q_2, h_c(y)] & : \ \{x < y\} \\
(q_1, \epsilon?, q_2) & \Rightarrow & [q_1, h_c(x), tail_c(y)] \rightsquigarrow [q_2, h_c(x'), tail(y')] & : \ \{y < x', x' < y'\}
\end{array}
$$

It is easy to verify that the resulting Γ_0 model accepts the same language as the original LCS.

The encoding of Γ_0 into LCS is more complicate and exploits special properties of Γ_0. We first exploits Prop. 2 to observe that for any CMRS $\mathcal{S}$ with initial and accepting configuration γ_{init} and γ_{acc} if we replace each gap order formula $x <_c y$ in $\mathcal{S}$ by $x < y$ we obtain a CMRS $\mathcal{S}'$ such that $L_c(\mathcal{S}) = L_c(\mathcal{S}')$. Hence, we assume w.l.o.g. that there is no gap order formula $x <_c y$ with $c > 0$ in $\mathcal{S}$. Secondly, when considering c-languages of a Γ_0 model, we can always restrict our attention to configurations in which ground terms (with parameter greater than *cmax*) are totally ordered with respect to $<$, i.e. in which there cannot be two ground terms (with parameter greater than *cmax*) with the same parameter. The proof of this property requires some attention. A Γ_0-rule may produce indeed a configuration containing two or more ground terms with the same parameter (e.g. when the right-hand side contains unconstrained variables as in $[\,] \rightsquigarrow [p(x), p(y)] : true$). We notice however that Γ_0-rules cannot use equality as guard in a condition. Thus, we can always choose a different evaluation for the variables in a condition such that ground terms assume distinct values and such that the word accepted by the corresponding execution remains the same. As a consequence, a Γ_0 configuration can be represented as a word of predicate symbols (We recall that CMRS configurations are words of multisets of predicate symbols as shown by the definition of $index(\cdot)$). Thus, a Γ_0-rule operates on configurations as a transformation of words. With these properties in mind, it comes natural to build an LCS that uses a lossy channel to encode a configuration and operations on (auxiliary) channels to simulate the transformations on words defined by a Γ_0-rule with lossy semantics. The thesis follows by noticing that, as for any other wsts, a version of Γ_0 with lossy semantics recognizes the same *c*-languages as those accepted by Γ_0. □

We show next that CMRS are strictly more expressive than LCS and Γ_0.

Theorem 4. $L_c(LCS) \subset L_c(CMRS)$.

Proof. We define a language L_{ent} which is accepted by a CMRS and that cannot be accepted by any LCS. Assume a finite alphabet Σ such that $\{\$, \#\} \not\subseteq \Sigma$. For each $w = a_1 \cdots a_k \in \Sigma^*$, we interpret w in the following as the multiset $[a_1, \ldots, a_k]$. Hence, we do not distinguish words in Σ^* from the multiset they represent, and vice versa. In particular, we will use the notation $a_1 \cdots a_k \leq a'_1 \cdots a'_l$ to denote that $[a_1, \ldots, a_k] \leq [a'_1, \ldots, a'_l]$. Define V to be the set of words of the form $w_1 \# w_2 \# \cdots \# w_n$ where $w_i \in \Sigma^*$ for each $i : 1 \leq i \leq n$. Consider $v = w_1 \# w_2 \# \cdots \# w_m \in V$ and $v' = w'_1 \# w'_2 \# \cdots \# w'_n \in V$. We write $v \sqsubseteq v'$ to denote that there is an injection $h: \{1, \ldots, m\} \mapsto \{1, \ldots, n\}$ such that

1. $1 \leq i < j \leq m$ implies $h(i) < h(j)$ (h is monotonic) and
2. $w_i \leq w'_{h(i)}$ ($\leq$ is multiset inclusion) for each $i : 1 \leq i \leq m$.

We now define the language $L_{ent} = \{v\$v' \mid v' \sqsubseteq v\} \subseteq (\Sigma \cup \{\#, \$\})^*$. As an example, given $\Sigma = \{a, b\}$, we have that $[a, b, b]\#[a, b, b]\#[a, a]\$[b, a]\#[a, a]$ is in L_{ent}, whereas $[a, b, b]\#[b, a, b]\#[a, a]\$[a, a]\#[a, b]$ is not in L_{ent}.

We now exhibit a CMRS $\mathcal{S}$ with $L_c(\mathcal{S}) = L_{ent}$. The set of predicate symbols which appear in $\mathcal{S}$ consists of (i) a predicate symbol a for each $a \in \Sigma$, and (ii) the symbols *guess*, *check*, $sep_\#$ and *ok*. The initial configuration γ_{init} is defined as $[guess(0)]$. Furthermore, we have the following rules:
(1) For each $a \in \Sigma$, we have a rule labelled with a and which is of the form

$$[guess(x)] \rightsquigarrow [guess(x)\ ,\ a(x)]\ :\ true$$

Rules of this form are used to guess the letters in w_i in the first part of a word in L_{ent}. We keep track of the symbols inside w_i through their argument. These arguments are all the same by definition of the rule.
(2) A rule labelled with # of the form:

$$[guess(x)] \rightsquigarrow [sep_{\#}(x)\ ,\ guess(y)] \ :\ \{x < y\}$$

This rule is used to switch from the guessing of the part w_i to the guessing of the next part w_{i+1}. $sep_{\#}(x)$ remembers the parameter on which the switch has been executed.
(3) A rule labelled with \$ of the form:

$$[guess(x)\ ] \rightsquigarrow [check(y)\ ,\ sep_{\#}(x)] \ :\ \{y = 0\}$$

This rule is used to switch from the guessing of the part $w_1\#\ldots\#w_n$ to the selection of the second part of the word. The parameter of *check* is equal to the initial value of *guess*, i.e. to 0. This way, we can scan the word stored in the first phase from left-to-right, i.e., working on the argument order we define a monotonic injective mapping h.
(4) For each $a \in \Sigma$, we have a rule labelled with a which is of the form

$$[check(y)\ ,\ a(y)] \rightsquigarrow [check(y)] \ :\ true$$

This rule is used to read a word (multiset) u_i contained in $w_{h(i)}$.
(5) A rule labelled with # of the form:

$$[check(x)\ ,\ sep_{\#}(x)\ ,\ sep_{\#}(y)\ ] \rightsquigarrow [check(y)\ ,\ sep_{\#}(y)] \ :\ \{x < y\}$$

This rule is used to pass from u_i to u_{i+1} for $i \geq 1$.
(6) A rule labelled with ϵ of the form:

$$[check(x)\ ] \rightsquigarrow [ok(y)] \ :\ \{y = 0\}$$

This rule is used to non-deterministically terminate the checking phase. The accepting configuration γ_{acc} is defined as $[ok(0)]$.

Assuming that $\Sigma = \{a, b\}$, we now show that L_{ent} is not an LCS language. Suppose that $L_c(\mathcal{F}) = L_{ent}$ for some LCS $\mathcal{F} = (Q, \{c\}, M, \delta)$. We show that this leads to a contradiction. Let γ_{init} be the initial global state in $\mathcal{F}$ and γ_{acc} be the accepting global state. We use a binary encoding $enc : Q \cup M \mapsto \Sigma^*$ such that $enc(m) \not\leq enc(m')$ if $m \neq m'$. We will also use a special word $v_{init} \in \Sigma^*$ such that $v_{init} \not\leq enc(m)$ for each $m \in Q \cup M$. It is clear that such enc function and v_{init} exist. As an example, if $|Q \cup M| = n$ then we define enc as an injective map from $Q \cup M$ to multisets of $n+1$ elements with $i+1$ occurrences of a and $n-i$ occurrences of b for $0 \leq i \leq n$, and we use the multiset with $n+1$ occurrences of b for v_{init}. For instance, for $n = 2$ we use $[a, a, a], [a, a, b], [a, b, b]$ for control states and messages and $[b, b, b]$ for v_{init}. We extend enc to global states such that if $\gamma = (q, m_1 m_2 \cdots m_n)$ then

$$enc(\gamma) = enc(q)\#enc(m_1)\#enc(m_2)\#\cdots\#enc(m_n)$$

Observe that (i) $enc(\gamma) \in V$; (ii) for global states γ_1 and γ_2, it is the case that $\gamma_1 \preceq_l \gamma_2$ iff $enc(\gamma_1) \sqsubseteq enc(\gamma_2)$; and (iii) $v_{init} \not\sqsubseteq enc(\gamma)$ for each global state γ.

Since $L_{ent} = L_c(\mathcal{F})$ and $v\$v \in L_{ent}$ for each $v \in V$, it follows that for each $v \in V$, there is a global state γ such that $\gamma_{init} \xrightarrow{v} \gamma \xrightarrow{\$v} \gamma'$ with $\gamma_{acc} \preceq_l \gamma'$. We use $reach(v)$ to denote γ. We define two sequences $\gamma_0, \gamma_1, \gamma_2, \ldots$ of global states, and $v_0, v_1, v_2, \ldots$ of words in V such that $v_0 = v_{init}$, $\gamma_i = reach(v_i)$, and $v_{i+1} = enc(\gamma_i)$ for each $i \geq 0$. By Higman's theorem we know that there is a j such that $\gamma_i \preceq_l \gamma_j$ for some $i < j$. Let j be the smallest natural number satisfying this property. First, we show that $v_i \not\sqsubseteq v_j$. There are two cases: if $i = 0$ then $v_i \not\sqsubseteq v_j$ by *(iii)*; if $i > 0$ then we know that $\gamma_{i-1} \not\preceq_l \gamma_{j-1}$ and hence, following *(ii)*, $v_i = enc(\gamma_{i-1}) \not\sqsubseteq enc(\gamma_{j-1}) = v_j$. Since $\gamma_j = reach(v_j)$, we know that $\gamma_{init} \xrightarrow{v_j} \gamma_j$. By monotonicity, $\gamma_i \xrightarrow{\$v_i} \gamma_i', \gamma_{acc} \preceq_l \gamma_i', \gamma_i \preceq_l \gamma_j$ implies $\gamma_j \xrightarrow{\$v_i} \gamma_j'$ with $\gamma_{acc} \preceq_l \gamma_i' \preceq_l \gamma_j'$. We conclude that $\gamma_{init} \xrightarrow{v_j} \gamma_j \xrightarrow{\$v_i} \gamma_j'$ with $\gamma_{acc} \preceq_l \gamma_j'$. Hence, $v_j\$v_i \in L_c(\mathcal{F}) = L_{ent}$ which is a contradiction since $v_i \not\sqsubseteq v_j$. □

Let us now consider r-languages. As mentioned at the beginning of the section, the expressive power of LCS remains the same as for coverability accepting conditions, However, this property does not hold anymore for Γ_0.

Proposition 3. $L_c(\Gamma_0) \subset L_r(\Gamma_0) = L_r(CMRS) = RE$.

Proof. It is well known that *perfect* FIFO channel systems with reachability accepting condition recognize the class RE. We prove that perfect channel systems accept the same languages as Γ_0 with reachability accepting condition. Given an LCS $\mathcal{F}$, let $\mathcal{S}$ be the Γ_0 used to encode an LCS in the proof of Theorem 3. In each step of a run σ in $\mathcal{S}$ the head and tail delimiters are moved to the right of their current positions. Thus, a "lost" ground term to left of the head delimiter, i.e. with parameter smaller than that of h_c, can never be removed in successive steps of σ. This implies that an accepting configuration in which all ground terms have parameters strictly greater than the parameter of the head delimiter characterize reachable configurations of a perfect FIFO channel system. □

Hence, we have the following property.

Corollary 1. $L_r(LCS) \subset L_r(CMRS)$.

5 Petri Nets and Their Extensions

Petri nets (PN), a well-known model of concurrent computation [19], can naturally be reformulated in a multiset rewriting system operating on nullary predicates only (i.e. predicates with no parameters). Let us call Γ_1 this fragment of CMRS. It is easy to see that, if we associate a predicate symbol to each place of a net, configurations and rules of a Γ_1 model are just alternative representations of markings and transitions of a Petri net. As an immediate consequence of this connection, we have that $L_c(\Gamma_1) = L_c(PN)$ and $L_r(\Gamma_1) = L_r(PN)$. To formally compare Γ_1 with the other models, we use some known results on languages accepted by extensions of Petri nets. A *lossy Petri net with inhibitor arcs* (LN) is a Petri net in which it is possible to test if a place has no tokens and in which tokens may get lost before and after executing a transition. A *transfer net* [10] (TN) is a Petri net extended with transfer arcs. A transfer arc specifies an atomic transfer of all tokens in a given set of source places to a given target place. Finally, a *reset net* [10] is a Petri net in which it is possible to atomically remove all tokens from a given place. LN, TN, and RN are well-structured with respect to the inclusion ordering of markings (see, e.g., [10,11]). For these models, it is simple to verify that $L_c(LN) = L_c(RN) = L_c(TN)$, $L_c(LN) \subseteq L_c(LCS)$, and, as for LCS, $L_r(LN) = L_c(LN)$ (see for [3] for formal proofs). Furthermore, in [14] the authors proved that $L_c(PN) \subset L_c(TN)$. From all these properties, we obtain the following result.

Theorem 5. $L_c(\Gamma_1) \subset L_c(\Gamma_0)$.

For r-languages, the classification changes as follows.

Theorem 6. $L_r(\Gamma_1) \not\sim L_r(LCS)$*,* $L_r(\Gamma_1) \not\sim L_r(LN)$*, and* $L_r(\Gamma_1) \subset L_r(\Gamma_0)$*.*

Proof. We first prove that $L_r(\Gamma_1) = L_r(PN) \not\subseteq L_c(LCS) = L_r(LCS)$, hence $L_r(\Gamma_1) \not\subseteq L_c(LN) = L_r(LN)$ since $L_c(LN) \subseteq L_c(LCS) = L_r(LCS)$. Consider the language $L = \{a^n b^n \mid n \geq 0\}$. It is easy to verify that there exists a Petri net $\mathcal{N}$ such that $L_r(\mathcal{N}) = L$. We now prove that $L \notin L_r(LCS)$. Per absurdum, suppose there exists an LCS $\mathcal{F}$ such that $L_c(\mathcal{F}) = L$. For any $k \geq 1$, let γ_k and γ'_k be two global states s.t. γ_{init} leads to γ_k by accepting the word a^k, γ_k leads to γ'_k by accepting the word b^k, and $\gamma_{acc} \preceq_l \gamma'_k$. Since $\preceq_l$ is a well-quasi ordering, there exists $i < j$ such that $\gamma_i \preceq_l \gamma_j$. By monotonicity of $\mathcal{F}$, we have γ_j leads to γ'' by accepting the word b^i and $\gamma_{acc} \preceq_l \gamma'_i \preceq_l \gamma''$. We conclude that $a^j b^i \in L_c(\mathcal{F})$ with $i < j$, which gives us a contradiction.

We now prove that $L_c(LN) \not\subseteq L_r(\Gamma_1)$, hence $L_c(LCS) \not\subseteq L_r(\Gamma_1)$. Let $\Sigma = \{a, b\}$ and let L_{par} be the language over the alphabet $\Sigma \cup \{\#\}$ that contains all the words $w_1\#\ldots\#w_n$ with $n \geq 0$ such that $w_i \in \Sigma^*$ and there is no prefix of w_i that contains more occurrences of symbol b than those of symbol a, for $i : 1 \leq i \leq n$. Notice that the number of occurrences of symbols a and b in w_i may be different. The language can be accepted by a LN defined as follows. When we accept the symbol a we add one token in a special place p_a. To accept the symbol b, we remove one token from p_a. To pass from w_i to w_{i+1}, we accept symbol # whenever p_a is empty (in LN the empty test is just a reset).

We now show that L_{par} cannot be recognized by a Petri net. Suppose that there exists a Petri net $\mathcal{N}$ such that $L_r(\mathcal{N}) = L_{par}$. Starting from $\mathcal{N}$, we build a net $\mathcal{N}_1$ by adding a new place d that keeps track of the difference between the number of occurrences of symbols a and b in the prefix of the word that is being processed in $\mathcal{N}$. Furthermore, we add the condition that d is empty to the accepting marking of $\mathcal{N}$. It is easy to verify that $\mathcal{N}_1$ accepts the language L_{bal} consisting of words of the form $w = w_1\#\cdots\#w_n$ where w_i belongs the the language of *balanced parentheses* on the alphabet Σ for $i : 1 \leq i \leq n$. We exploit now [15, Lemma 9.8] that states that L_{bal} cannot be recognized by a Petri net with reachability accepting condition, which gives us a contradiction.

Finally, the property $L_r(\Gamma_1) = L_r(PN) \subset L_r(\Gamma_0)$ follows from [15, Lemma 9.8] and Prop. 3, Indeed, we have that $L_{bal} \in L_r(\Gamma_0) = RE$ and $L_{bal} \notin L_r(\Gamma_1)$. □

Finally, we observe that we can use an argument similar to that used in the proof of Theorem 6 to show that $L_r(PN) \not\sim L_c(CMRS)$.

6 (Integral) Relational Automata

In this section we compare the class of languages accepted by a fragment of CMRS, called Γ_2, with those accepted by *relational automata* [9].

The fragment Γ_2 is defined as follows. Let us first use $|B|$ to denote the cardinality of a multiset B. Γ_2 is the fragment of CMRS in which a rule $L \rightsquigarrow R : \psi$ satisfies the condition $|R| \leq |L|$. In other words, in Γ_2 the cardinality of a reachable configuration is always bounded by the cardinality of the initial configuration.

An *(integral) relational automaton* (RA) operates on a finite set X of positive integer variables, and is of the form (Q, δ) where Q and δ are finite sets of control states and transitions respectively. A transition is a triple (q_1, op, q_2) where $q_1, q_2 \in Q$ and op is of one of the following three operations: (i) *reading*: $read(x)$ reads a new value of variable x (i.e. assigns a non-deterministically chosen value to x), (ii) *assignment*: $x := y$ assigns the value of variable y to x; (ii) *testing*: $x < y$, $x = y$, $x < c$, $x = c$, and $x > c$ are guards which compare the values of variables x, y and the natural constant c. Assume a RA $\mathcal{A} = (Q, \delta)$. A *valuation* v is a mapping form X to $\mathbb{N}$. A *configuration* is of the form (q, v), where $q \in Q$ and v is a valuation. We define γ_{init} to be (q_{init}, v_{init}) where $q_{init} \in Q$ and $v_{init}(x) = 0$ for all $x \in X$. For a transition $\rho \in \delta$ of the form (q_1, op, q_2), we let $\gamma_1 \xrightarrow{\rho} \gamma_2$ if and only if $\gamma_1 = (q_1, v_1)$, $\gamma_2 = (q_2, v_2)$, and one of the following holds: $op = read(x)$ and $v_2(y) = v_1(y)$ for each $y \in X - \{x\}$; $op = (y := x)$, $v_2(z) = v_1(z)$ for each $z \in X - \{y\}$, and $v_2(y) = v_1(x)$; $op = (x < y)$, $v_2 = v_1$, and $v_1(x) < v_1(y)$. Other testing operations are defined in a similar manner. In [9] Cerans has shown that RA equipped with the *sparser-than* order of tuples of natural numbers are well-structured. In the case of RA with that order, the coverability accepting condition is equivalent to the control state acceptance, i.e., a word is accepted if it is recognized by an execution ending in a particular control state $q_{acc} \in Q$.

As stated in the following propositions, RA and Γ_2 define the same class of c- and r-languages.

Proposition 4. $L_c(\Gamma_2) = L_c(RA)$.

Proof. Given an RA $\mathcal{A} = (Q, \delta)$ over the set of variables X, we can build the Γ_2 $\mathcal{S}$ defined below. The set of predicate symbols in $\mathcal{S}$ consists of the following: (i) for each $q \in Q$, there is a predicate symbol q in $\mathcal{S}$; and (ii) for each variable x

in X, there is a predicate symbol q_x in $\mathcal{S}$. Transitions in δ are encoded via the following CMRS rules (with the same labels)

$$\begin{array}{lcl}
(q_1, read(x), q_2) & \Rightarrow & [q_1, p_x(z)] \rightsquigarrow [q_2, p_x(w)] : true \\
(q_1, x := y, q_2) & \Rightarrow & [q_1, p_x(z), p_y(w)] \rightsquigarrow [q_2, p_x(w), p_y(w)] : true \\
(q_1, x < y, q_2) & \Rightarrow & [q_1, p_x(z), p_y(w)] \rightsquigarrow [q_2, p_x(z), p_y(w)] : \{z < w\}
\end{array}$$

For $X = \{x_1, \ldots, x_n\}$, the initial configuration is $\gamma_{init} = [q_0, p_{x_1}(0), \ldots, p_{x_n}(0)]$. The accepting configuration γ_{acc} is the multiset $[q_{acc}]$.

For the other inclusion, by using Prop. 2, we assume w.l.o.g. that there is no gap order formula $x <_c y$ with $c > 0$ in $\mathcal{S}$ and that $\gamma_{acc} = [ok]$. To justify the second assumption, notice that we can always introduce new predicate symbols *ko* and *ok* and a new rule that can be executed only on a configuration γ with $\gamma_{acc} \preceq_c \gamma$ and that replace *ko* with *ok*. All other rules are modified to be enabled only at configurations containing *ko*. Finally, we also observe that we can assume that all configurations of $\mathcal{S}$ have the same size (the size of the initial configuration of the Γ_2 model). Thus, we associate a variable of X to each ground term of the initial CMRS configuration and compose the predicate symbols in a CMRS configuration to form a single control state. CRMS rules can then be simulated in several steps by operations on variables and updates of control states. To each control state containing *ok*, we add a transition labeled with ϵ to the accepting control state q_{acc}. □

Theorem 7. $L_c(\Gamma_2) =$ *Regular Languages.*

To prove this claim, we define a finite state automaton where states are abstractions of configurations in which we only keep the order on parameters and not their exact values (when parameters are greater than *cmax*). The relation transition of the symbolic graph mimics the transition relation of $\mathcal{S}$. Then, we show (by using Prop. 2) that the symbolic graph contains exactly the information we need to characterize the language recognized by $\mathcal{S}$. Finiteness of the graph allows us to conclude that Γ_2 corresponds to the class of regular languages. The complete construction is given in [3].

We are ready now to compare Γ_2 (hence RA) with the other models studied in this paper. For this purpose, we first observe that Petri nets can accept regular languages (finite automata can be encoded as Petri nets). Furthermore, it is straightforward to build a Petri net that accepts a non-regular language like $L = \{a^n \# b^m \mid n \geq m\}$. As a consequence of this observation and of Theorem 7, we have the following result.

Corollary 2. $L_c(\Gamma_2) \subset L_c(\Gamma_1)$.

Let us now consider the reachability accepting condition. We first notice that $L_c(\Gamma_2) = L_r(\Gamma_2) = L_c(RA) = L_r(RA)$. Indeed, in both cases of Γ_2 and RA we can encode the reachability acceptance into the coverability acceptance by adding transitions (labelled with ϵ) that can be fired only from the accepting configuration and leads to a configuration with control state q_{acc} in the case of

RA and a configuration containing a special accepting predicate symbol in the case of Γ_2. Thus, we have the following property.

Theorem 8. $L_r(\Gamma_2) \subset L_r(\Gamma_1)$.

7 Conclusions

In this paper we have compared the class of languages recognized with coverability and reachability as accepting conditions by *relational automata* (RA), *Petri nets* (PN), *lossy channel systems* (LCS), and *constrained multiset rewriting systems* (CMRS). With both accepting conditions, CMRS turns out to be the most expressive model among the different well-structured systems considered in the paper. Indeed, with coverability as accepting condition we have that $FA = RA < PN < LCS < CMRS < CM$, whereas with reachability we have that $FA = RA < PN, LCS < CMRS = CM$ and PN and LCS are incomparable models. Here FA and CM denote resp. *finite automata* (they recognize regular languages) and *counter automata* (they recognize recursively enumerable languages), and $<$ means "strictly less expressive than". We also prove that *transfer nets*, *reset nets*, *broadcast protocols* and *lossy vector addition systems* are strictly less expressive than CMRS. CMRS can thus be viewed as a unified model for analysis and verification of a large class of infinite-state models.

References

1. Abdulla, P.A., Čerāns, K., Jonsson, B., Yih-Kuen, T.: General decidability theorems for infinite-state systems. In: LICS, pp. 313–321 (1996)
2. Abdulla, P.A., Delzanno, G.: On the coverability problem for constrained multiset rewriting. In: AVIS 2006, an ETAPS Workshop (2006)
3. Abdulla, P.A., Delzanno, G., Van Begin, L.: Comparing the expressive power of well-structured transition systems. Technical Report, DISI (June 2007)
4. Abdulla, P.A., Jonsson, B.: Model checking of systems with many identical timed processes. TCS 290(1), 241–264 (2003)
5. Abdulla, P.A., Jonsson, B.: Verifying programs with unreliable channels. Inf. Comput. 127(2), 91–101 (1996)
6. Abdulla, P.A., Jonsson, B.: Undecidable verification problems for programs with unreliable channels. Inf. Comput. 130(1), 71–90 (1996)
7. Bertrand, N., Schnoebelen, Ph.: A short visit to the STS hierarchy. ENTCS 154(3), 59–69 (2006)
8. Cécé, G., Finkel, A., Iyer, S.P.: Unreliable channels are easier to verify than perfect channels. Inf. Comput. 124(1), 20–31 (1996)
9. Čerāns, K.: Deciding properties of integral relational automata. In: Shamir, E., Abiteboul, S. (eds.) ICALP 1994. LNCS, vol. 820, pp. 35–46. Springer, Heidelberg (1994)
10. Dufourd, C., Finkel, A., Schnoebelen, P.: Reset nets between decidability and undecidability. In: Larsen, K.G., Skyum, S., Winskel, G. (eds.) ICALP 1998. LNCS, vol. 1443, pp. 103–115. Springer, Heidelberg (1998)

11. Esparza, J., Finkel, A., Mayr, R.: On the verification of broadcast protocols. In: LICS 1999, pp. 352–359 (1999)
12. Finkel, A., Geeraerts, G., Raskin, J.-F., Van Begin, L.: On the ω-language expressive power of extended petri nets. TCS 356(3), 374–386 (2006)
13. Finkel, A., Schnoebelen, Ph.: Well-structured transition systems everywhere! TCS 256(1-2), 63–92 (2001)
14. Geeraerts, G., Raskin, J.-F., Van Begin, L.: Well-structured languages. Technical report 542, ULB (2005)
15. Hack, M.: Petri net languages. Technical report 159, MIT, Cambridge (1976)
16. Henzinger, T.A., Majumdar, R., Raskin, J.-F.: A classification of symbolic transition systems. ACM Trans. Comput. Log. 6(1), 1–32 (2005)
17. Lazič, R., Newcomb, T., Ouaknine, J., Roscoe, A.W., Worell, J.: Nets with tokens which carry data. In: ATPN 2007 (to appear, 2007)
18. Mayr, R.: Undecidable problems in unreliable computations. In: Gonnet, G.H., Viola, A. (eds.) LATIN 2000. LNCS, vol. 1776, pp. 377–386. Springer, Heidelberg (2000)
19. Petri, C.A.: Kommunikation mit Automaten. PhD Thesis. Univ. of Bonn (1962)

There Exist Some ω-Powers of Any Borel Rank

Olivier Finkel[1] and Dominique Lecomte[2]

[1] Equipe Modèles de Calcul et Complexité
Laboratoire de l'Informatique du Parallélisme*
CNRS et Ecole Normale Supérieure de Lyon
46, Allée d'Italie 69364 Lyon Cedex 07, France
Olivier.Finkel@ens-lyon.fr
[2] Equipe d'Analyse Fonctionnelle
Université Paris 6
4, place Jussieu, 75 252 Paris Cedex 05, France
lecomte@moka.ccr.jussieu.fr

Abstract. The operation $V \to V^\omega$ is a fundamental operation over finitary languages leading to ω-languages. Since the set Σ^ω of infinite words over a finite alphabet Σ can be equipped with the usual Cantor topology, the question of the topological complexity of ω-powers of finitary languages naturally arises and has been posed by Niwinski [Niw90], Simonnet [Sim92] and Staiger [Sta97a]. It has been recently proved that for each integer $n \geq 1$, there exist some ω-powers of context free languages which are $\mathbf{\Pi}^0_n$-complete Borel sets, [Fin01], that there exists a context free language L such that L^ω is analytic but not Borel, [Fin03], and that there exists a finitary language V such that V^ω is a Borel set of infinite rank, [Fin04]. But it was still unknown which could be the possible infinite Borel ranks of ω-powers.

We fill this gap here, proving the following very surprising result which shows that ω-powers exhibit a great topological complexity: for each non-null countable ordinal ξ, there exist some $\mathbf{\Sigma}^0_\xi$-complete ω-powers, and some $\mathbf{\Pi}^0_\xi$-complete ω-powers.

Keywords: Infinite words; ω-languages; ω-powers; Cantor topology; topological complexity; Borel sets; Borel ranks; complete sets.

1 Introduction

The operation $V \to V^\omega$ is a fundamental operation over finitary languages leading to ω-languages. It produces ω-powers, i.e. ω-languages in the form V^ω, where V is a finitary language. This operation appears in the characterization of the class REG_ω of ω-regular languages (respectively, of the class CF_ω of context free ω-languages) as the ω-Kleene closure of the family REG of regular finitary languages (respectively, of the family CF of context free finitary languages) [Sta97a].

Since the set Σ^ω of infinite words over a finite alphabet Σ can be equipped with the usual Cantor topology, the question of the topological complexity of ω-powers of finitary languages naturally arises and has been posed by Niwinski [Niw90], Simonnet

* UMR 5668 - CNRS - ENS Lyon - UCB Lyon - INRIA.

J. Duparc and T.A. Henzinger (Eds.): CSL 2007, LNCS 4646, pp. 115–129, 2007.

[Sim92], and Staiger [Sta97a]. A first task is to study the position of ω-powers with regard to the Borel hierarchy (and beyond to the projective hierarchy) [Sta97a, PP04].

It is easy to see that the ω-power of a finitary language is always an analytic set because it is either the continuous image of a compact set $\{0, 1, \ldots, n\}^\omega$ for $n \geq 0$ or of the Baire space ω^ω.

It has been recently proved, that for each integer $n \geq 1$, there exist some ω-powers of context free languages which are $\mathbf{\Pi}^0_n$-complete Borel sets, [Fin01], and that there exists a context free language L such that L^ω is analytic but not Borel, [Fin03]. Notice that amazingly the language L is very simple to describe and it is accepted by a simple 1-counter automaton.

The first author proved in [Fin04] that there exists a finitary language V such that V^ω is a Borel set of infinite rank. However the only known fact on their complexity is that there is a context free language W such that W^ω is Borel above $\mathbf{\Delta}^0_\omega$, [DF06].

We fill this gap here, proving the following very surprising result which shows that ω-powers exhibit a great topological complexity: for each non-null countable ordinal ξ, there exist some $\mathbf{\Sigma}^0_\xi$-complete ω-powers, and some $\mathbf{\Pi}^0_\xi$-complete ω-powers. For that purpose we use a theorem of Kuratowski which is a level by level version of a theorem of Lusin and Souslin stating that every Borel set $B \subseteq 2^\omega$ is the image of a closed subset of the Baire space ω^ω by a continuous bijection. This theorem of Lusin and Souslin had already been used by Arnold in [Arn83] to prove that every Borel subset of Σ^ω, for a finite alphabet Σ, is accepted by a non-ambiguous finitely branching transition system with Büchi acceptance condition and our first idea was to code the behaviour of such a transition system. This way, in the general case, we can manage to construct an ω-power of the same complexity as B.

The paper is organized as follows. In Section 2 we recall basic notions of topology and in particular definitions and properties of Borel sets. We proved our main result in Section 3.

2 Topology

We first give some notations for finite or infinite words we shall use in the sequel, assuming the reader to be familiar with the theory of formal languages and of ω-languages, see [Tho90, Sta97a, PP04]. Let Σ be a finite or countable alphabet whose elements are called letters. A non-empty finite word over Σ is a finite sequence of letters: $x = a_0.a_1.a_2 \ldots a_n$ where $\forall i \in [0; n]\ a_i \in \Sigma$. We shall denote $x(i) = a_i$ the $(i+1)^{th}$ letter of x and $x\lceil(i+1) = x(0) \ldots x(i)$ for $i \leq n$, is the beginning of length $i+1$ of x. The length of x is $|x| = n+1$. The empty word will be denoted by $\emptyset$ and has 0 letters. Its length is 0. The set of finite words over Σ is denoted $\Sigma^{<\omega}$. A (finitary) language L over Σ is a subset of $\Sigma^{<\omega}$. The usual concatenation product of u and v will be denoted by $u^\frown v$ or just uv. If $l \in \omega$ and $(a_i)_{i<l} \in (\Sigma^{<\omega})^l$, then ${}^\frown_{i<l}\, a_i$ is the concatenation $a_0 \ldots a_{l-1}$.

The first infinite ordinal is ω. An ω-word over Σ is an ω -sequence $a_0 a_1 \ldots a_n \ldots$, where for all integers $i \geq 0 \quad a_i \in \Sigma$. When σ is an ω-word over Σ, we write $\sigma = \sigma(0)\sigma(1)\ldots\sigma(n)\ldots$ and $\sigma\lceil(n+1) = \sigma(0)\sigma(1)\ldots\sigma(n)$ the finite word of length $n+1$, prefix of σ. The set of ω-words over the alphabet Σ is denoted by Σ^ω. An

ω-language over an alphabet Σ is a subset of Σ^ω. If $\forall i \in \omega \quad a_i \in \Sigma^{<\omega}$, then $\frown_{i \in \omega} a_i$ is the concatenation $a_0 a_1 \ldots$. The concatenation product is also extended to the product of a finite word u and an ω-word v: the infinite word $u.v$ or $u^\frown v$ is then the ω-word such that: $(uv)(k) = u(k)$ if $k < |u|$, and $(u.v)(k) = v(k - |u|)$ if $k \geq |u|$.

The prefix relation is denoted $\prec$: the finite word u is a prefix of the finite word v (respectively, the infinite word v), denoted $u \prec v$, if and only if there exists a finite word w (respectively, an infinite word w), such that $v = u^\frown w$.

If $s \prec \alpha = \alpha(0)\alpha(1)...$, then $\alpha - s$ is the sequence $\alpha(|s|)\alpha(|s|+1)...$

For a finitary language $V \subseteq \Sigma^{<\omega}$, the ω-power of V is the ω-language

$$V^\omega = \{u_1 \ldots u_n \ldots \in \Sigma^\omega \mid \forall i \geq 1 \;\; u_i \in V\}$$

We recall now some notions of topology, assuming the reader to be familiar with basic notions which may be found in [Kur66, Mos80, Kec95, LT94, Sta97a, PP04].

There is a natural metric on the set Σ^ω of infinite words over a countable alphabet Σ which is called the prefix metric and defined as follows. For $u, v \in \Sigma^\omega$ and $u \neq v$ let $d(u, v) = 2^{-l_{pref(u,v)}}$ where $l_{pref(u,v)}$ is the first integer n such that the $(n+1)^{th}$ letter of u is different from the $(n+1)^{th}$ letter of v. The topology induced on Σ^ω by this metric is just the product topology of the discrete topology on Σ. For $s \in \Sigma^{<\omega}$, the set $N_s := \{\alpha \in \Sigma^\omega \mid s \prec \alpha\}$ is a basic clopen (i.e., closed and open) set of Σ^ω. More generally open sets of Σ^ω are in the form $W^\frown \Sigma^\omega$, where $W \subseteq \Sigma^{<\omega}$.

The topological spaces in which we will work in this paper will be subspaces of Σ^ω where Σ is either finite having at least two elements or countably infinite.

When Σ is a finite alphabet, the prefix metric induces on Σ^ω the usual Cantor topology and Σ^ω is compact.

The Baire space ω^ω is equipped with the product topology of the discrete topology on ω. It is homeomorphic to $P_\infty := \{\alpha \in 2^\omega \mid \forall i \in \omega \; \exists j \geq i \;\; \alpha(j) = 1\} \subseteq 2^\omega$, via the map defined on ω^ω by $H(\beta) := 0^{\beta(0)} 1 0^{\beta(1)} 1 \ldots$

We define now the **Borel Hierarchy** on a topological space X:

Definition 1. *The classes* $\mathbf{\Sigma}^0_n(X)$ *and* $\mathbf{\Pi}^0_n(X)$ *of the Borel Hierarchy on the topological space* X *are defined as follows:*

$\mathbf{\Sigma}^0_1(X)$ *is the class of open subsets of* X.
$\mathbf{\Pi}^0_1(X)$ *is the class of closed subsets of* X.
And for any integer $n \geq 1$*:*
$\mathbf{\Sigma}^0_{n+1}(X)$ *is the class of countable unions of* $\mathbf{\Pi}^0_n$*-subsets of* X.
$\mathbf{\Pi}^0_{n+1}(X)$ *is the class of countable intersections of* $\mathbf{\Sigma}^0_n$*-subsets of* X.
The Borel Hierarchy is also defined for transfinite levels. The classes $\mathbf{\Sigma}^0_\xi(X)$ *and* $\mathbf{\Pi}^0_\xi(X)$*, for a non-null countable ordinal* ξ*, are defined in the following way:*
$\mathbf{\Sigma}^0_\xi(X)$ *is the class of countable unions of subsets of* X *in* $\cup_{\gamma<\xi}\mathbf{\Pi}^0_\gamma$.
$\mathbf{\Pi}^0_\xi(X)$ *is the class of countable intersections of subsets of* X *in* $\cup_{\gamma<\xi}\mathbf{\Sigma}^0_\gamma$.

Suppose now that $X \subseteq Y$; then $\mathbf{\Sigma}^0_\xi(X) = \{A \cap X \mid A \in \mathbf{\Sigma}^0_\xi(Y)\}$, and similarly for $\mathbf{\Pi}^0_\xi$, see [Kec95, Section 22.A]. Notice that we have defined the Borel classes $\mathbf{\Sigma}^0_\xi(X)$ and $\mathbf{\Pi}^0_\xi(X)$ mentioning the space X. However when the context is clear we will sometimes omit X and denote $\mathbf{\Sigma}^0_\xi(X)$ by $\mathbf{\Sigma}^0_\xi$ and similarly for the dual class.

The Borel classes are closed under finite intersections and unions, and continuous preimages. Moreover, $\boldsymbol{\Sigma}^0_\xi$ is closed under countable unions, and $\boldsymbol{\Pi}^0_\xi$ under countable intersections. As usual the ambiguous class $\boldsymbol{\Delta}^0_\xi$ is the class $\boldsymbol{\Sigma}^0_\xi \cap \boldsymbol{\Pi}^0_\xi$.

The class of **Borel sets** is $\boldsymbol{\Delta}^1_1 := \bigcup_{\xi<\omega_1} \boldsymbol{\Sigma}^0_\xi = \bigcup_{\xi<\omega_1} \boldsymbol{\Pi}^0_\xi$, where ω_1 is the first uncountable ordinal.

The **Borel hierarchy** is as follows:

$$\begin{array}{ccccccccc} & \boldsymbol{\Sigma}^0_1\!=\!\text{open} & & \boldsymbol{\Sigma}^0_2 & \ldots & & \boldsymbol{\Sigma}^0_\omega & \ldots & \\ \boldsymbol{\Delta}^0_1\!=\!\text{clopen} & & \boldsymbol{\Delta}^0_2 & & & \boldsymbol{\Delta}^0_\omega & & & \boldsymbol{\Delta}^1_1 \\ & \boldsymbol{\Pi}^0_1\!=\!\text{closed} & & \boldsymbol{\Pi}^0_2 & \ldots & & \boldsymbol{\Pi}^0_\omega & \ldots & \end{array}$$

This picture means that any class is contained in every class to the right of it, and the inclusion is strict in any of the spaces Σ^ω.

For a countable ordinal α, a subset of Σ^ω is a Borel set of *rank* α iff it is in $\boldsymbol{\Sigma}^0_\alpha \cup \boldsymbol{\Pi}^0_\alpha$ but not in $\bigcup_{\gamma<\alpha}(\boldsymbol{\Sigma}^0_\gamma \cup \boldsymbol{\Pi}^0_\gamma)$.

We now define completeness with regard to reduction by continuous functions. For a countable ordinal $\alpha \geq 1$, a set $F \subseteq \Sigma^\omega$ is said to be a $\boldsymbol{\Sigma}^0_\alpha$ (respectively, $\boldsymbol{\Pi}^0_\alpha$)-*complete set* iff for any set $E \subseteq Y^\omega$ (with Y a finite alphabet): $E \in \boldsymbol{\Sigma}^0_\alpha$ (respectively, $E \in \boldsymbol{\Pi}^0_\alpha$) iff there exists a continuous function $f : Y^\omega \to \Sigma^\omega$ such that $E = f^{-1}(F)$. $\boldsymbol{\Sigma}^0_n$ (respectively, $\boldsymbol{\Pi}^0_n$)-complete sets, with n an integer ≥ 1, are thoroughly characterized in [Sta86].

Recall that a set $X \subseteq \Sigma^\omega$ is a $\boldsymbol{\Sigma}^0_\alpha$ (respectively $\boldsymbol{\Pi}^0_\alpha$)-complete subset of Σ^ω iff it is in $\boldsymbol{\Sigma}^0_\alpha$ but not in $\boldsymbol{\Pi}^0_\alpha$ (respectively in $\boldsymbol{\Pi}^0_\alpha$ but not in $\boldsymbol{\Sigma}^0_\alpha$), [Kec95].

For example, the singletons of 2^ω are $\boldsymbol{\Pi}^0_1$-complete subsets of 2^ω. The set P_∞ is a well known example of a $\boldsymbol{\Pi}^0_2$-complete subset of 2^ω.

If $\boldsymbol{\Gamma}$ is a class of sets, then $\check{\boldsymbol{\Gamma}} := \{\neg A \mid A \in \boldsymbol{\Gamma}\}$ is the class of complements of sets in $\boldsymbol{\Gamma}$. In particular, for every non-null countable ordinal α, $\check{\boldsymbol{\Sigma}}^0_\alpha = \boldsymbol{\Pi}^0_\alpha$ and $\check{\boldsymbol{\Pi}}^0_\alpha = \boldsymbol{\Sigma}^0_\alpha$.

There are some subsets of the topological space Σ^ω which are not Borel sets. In particular, there exists another hierarchy beyond the Borel hierarchy, called the projective hierarchy. The first class of the projective hierarchy is the class $\boldsymbol{\Sigma}^1_1$ of **analytic** sets. A set $A \subseteq \Sigma^\omega$ is analytic iff there exists a Borel set $B \subseteq (\Sigma \times Y)^\omega$, with Y a finite alphabet, such that $x \in A \leftrightarrow \exists y \in Y^\omega$ such that $(x, y) \in B$, where $(x, y) \in (\Sigma \times Y)^\omega$ is defined by: $(x, y)(i) = (x(i), y(i))$ for all integers $i \geq 0$.

A subset of Σ^ω is analytic if it is empty, or the image of the Baire space by a continuous map. The class of analytic sets contains the class of Borel sets in any of the spaces Σ^ω. Notice that $\boldsymbol{\Delta}^1_1 = \boldsymbol{\Sigma}^1_1 \cap \boldsymbol{\Pi}^1_1$, where $\boldsymbol{\Pi}^1_1$ is the class of co-analytic sets, i.e. of complements of analytic sets.

The ω-power of a finitary language V is always an analytic set because if V is finite and has n elements then V^ω is the continuous image of a compact set $\{0, 1, \ldots, n-1\}^\omega$ and if V is infinite then there is a bijection between V and ω and V^ω is the continuous image of the Baire space ω^ω, [Sim92].

3 Main Result

We now state our main result, showing that ω-powers exhibit a very surprising topological complexity.

Theorem 2. *Let ξ be a non-null countable ordinal.*
(a) There is $A \subseteq 2^{<\omega}$ such that A^ω is $\mathbf{\Sigma}^0_\xi$-complete.
(b) There is $A \subseteq 2^{<\omega}$ such that A^ω is $\mathbf{\Pi}^0_\xi$-complete.

To prove Theorem 2, we shall use a level by level version of a theorem of Lusin and Souslin stating that every Borel set $B \subseteq 2^\omega$ is the image of a closed subset of the Baire space ω^ω by a continuous bijection, see [Kec95, p.83]. It is the following theorem, proved by Kuratowski in [Kur66, Corollary 33.II.1]:

Theorem 3. *Let ξ be a non-null countable ordinal, and $B \in \mathbf{\Pi}^0_{\xi+1}(2^\omega)$. Then there is $C \in \mathbf{\Pi}^0_1(\omega^\omega)$ and a continuous bijection $f : C \to B$ such that f^{-1} is $\mathbf{\Sigma}^0_\xi$-measurable (i.e., $f[U]$ is $\mathbf{\Sigma}^0_\xi(B)$ for each open subset U of C).*

The existence of the continuous bijection $f : C \to B$ given by this theorem (without the fact that f^{-1} is $\mathbf{\Sigma}^0_\xi$-measurable) has been used by Arnold in [Arn83] to prove that every Borel subset of Σ^ω, for a finite alphabet Σ, is accepted by a non-ambiguous finitely branching transition system with Büchi acceptance condition. Notice that the sets of states of these transition systems are countable.

Our first idea was to code the behaviour of such a transition system. In fact this can be done on a part of ω-words of a special compact set $K_{0,0}$. However we shall have also to consider more general sets $K_{N,j}$ and then we shall need the hypothesis of the $\mathbf{\Sigma}^0_\xi$-measurability of the function f.

We now come to the proof of Theorem 2.

Let $\mathbf{\Gamma}$ be the class $\mathbf{\Sigma}^0_\xi$, or $\mathbf{\Pi}^0_\xi$. We assume first that $\xi \geq 3$.

Let $B \subseteq 2^\omega$ be a $\mathbf{\Gamma}$-complete set. Then B is in $\mathbf{\Gamma}(2^\omega)$ but not in $\check{\mathbf{\Gamma}}(2^\omega)$. As $B \in \mathbf{\Pi}^0_{\xi+1}$, Theorem 3 gives $C \in \mathbf{\Pi}^0_1(P_\infty)$ and f. By Proposition 11 in [Lec05], it is enough to find $A \subseteq 4^{<\omega}$. The language A will be made of two pieces: we will have $A = \mu \cup \pi$. The set π will code f, and π^ω will look like B on some nice compact sets $K_{N,j}$. Outside this countable family of compact sets we will hide f, so that A^ω will be the simple set μ^ω.

• We set $Q := \{(s,t) \in 2^{<\omega} \times 2^{<\omega} \mid |s| = |t|\}$. We enumerate Q as follows. We start with $q_0 := (\emptyset, \emptyset)$. Then we put the sequences of length 1 of elements of 2×2, in the lexicographical ordering: $q_1 := (0,0)$, $q_2 := (0,1)$, $q_3 := (1,0)$, $q_4 := (1,1)$. Then we put the 16 sequences of length 2: $q_5 := (0^2, 0^2)$, $q_6 := (0^2, 01)$, ... And so on. We will sometimes use the coordinates of $q_N := (q^0_N, q^1_N)$. We put $M_j := \Sigma_{i<j}\, 4^{i+1}$. Note that the sequence $(M_j)_{j\in\omega}$ is strictly increasing, and that q_{M_j} is the last sequence of length j of elements of 2×2.

• Now we define the "nice compact sets". We will sometimes view 2 as an alphabet, and sometimes view it as a letter. To make this distinction clear, we will use the boldface notation **2** for the letter, and the lightface notation 2 otherwise. We will have the same distinction with 3 instead of 2, so we have $2 = \{0,1\}, 3 = \{0,1,\mathbf{2}\}, 4 = \{0,1,\mathbf{2},\mathbf{3}\}$. Let N, j be non-negative integers with $N \leq M_j$. We set

$$K_{N,j} := \{\, \gamma = \mathbf{2}^N \frown [\, \frown_{i\in\omega}\ m_i\ \mathbf{2}^{M_j+i+1}\ \mathbf{3}\ \mathbf{2}^{M_j+i+1}\,] \in 4^\omega \mid \forall i \in \omega\ \ m_i \in 2 = \{0,1\}\}.$$

As the map $\varphi_{N,j}: K_{N,j} \rightarrow 2^\omega$ defined by $\varphi_{N,j}(\gamma) := \frown_{i\in\omega} m_i$ is a homeomorphism, $K_{N,j}$ is compact.

• Now we will define the sets that "look like B".

- Let $l \in \omega$. We define a function $c_l : B \rightarrow Q$ by $c_l(\alpha) := [f^{-1}(\alpha), \alpha]\lceil l$. Note that Q is countable, so that we equip it with the discrete topology. In these conditions, we prove that c_l is $\mathbf{\Sigma}^0_\xi$-measurable.

If $l \neq |q^0| = |q^1|$ then $c_l^{-1}(q)$ is the empty set. And for any $q \in Q$, and $l = |q^0| = |q^1|$, it holds that $c_l^{-1}(q) = \{\alpha \in B \mid [f^{-1}(\alpha), \alpha]\lceil l = q\} = \{\alpha \in B \mid \alpha\lceil l = q^1 \text{ and } f^{-1}(\alpha)\lceil l = q^0\}$. But $\alpha\lceil l = q^1$ means that α belongs to the basic open set N_{q^1} and $f^{-1}(\alpha)\lceil l = q^0$ means that $f^{-1}(\alpha)$ belongs to the basic open set N_{q^0} or equivalently that $\alpha = f(f^{-1}(\alpha))$ belongs to $f(N_{q^0})$ which is a $\mathbf{\Sigma}^0_\xi$-subset of B. So $c_l^{-1}(q) = N_{q^1} \cap f(N_{q^0})$ is a $\mathbf{\Sigma}^0_\xi$-subset of B and c_l is $\mathbf{\Sigma}^0_\xi$-measurable.

- Let N be an integer. We put

$$E_N := \{\ \alpha \in 2^\omega \mid q^1_N\alpha \in B \ \text{ and } \ c_{|q^1_N|}(q^1_N\alpha) = q_N\ \}.$$

Notice that $E_0 = \{\ \alpha \in 2^\omega \mid \alpha \in B \text{ and } c_0(\alpha) = \emptyset\} = B$.

As $c_{|q^1_N|}$ is $\mathbf{\Sigma}^0_\xi$-measurable and $\{q_N\} \in \mathbf{\Delta}^0_1(Q)$, we get $c^{-1}_{|q^1_N|}(\{q_N\}) \in \mathbf{\Delta}^0_\xi(B) \subseteq \mathbf{\Gamma}(B)$. Therefore there is $G \in \mathbf{\Gamma}(2^\omega)$ with $c^{-1}_{|q^1_N|}(\{q_N\}) = G \cap B$. Thus $c^{-1}_{|q^1_N|}(\{q_N\}) \in \mathbf{\Gamma}(2^\omega)$ since $\mathbf{\Gamma}$ is closed under finite intersections. Note that the map S associating $q^1_N\alpha$ with α is continuous, so that $E_N = S^{-1}[c^{-1}_{|q^1_N|}(\{q_N\})]$ is in $\mathbf{\Gamma}(2^\omega)$.

• Now we define the transition system obtained from f.

- If $m \in 2$ and $n, p \in \omega$, then we write $n \overset{m}{\rightarrow} p$ if $q^0_n \prec q^0_p$ and $q^1_p = q^1_n m$.

- As f is continuous on C, the graph $\text{Gr}(f)$ of f is a closed subset of $C \times 2^\omega$. As C is $\mathbf{\Pi}^0_1(P_\infty)$, $\text{Gr}(f)$ is also a closed subset of $P_\infty \times 2^\omega$. So there is a closed subset F of $2^\omega \times 2^\omega$ such that $\text{Gr}(f) = F \cap (P_\infty \times 2^\omega)$. We identify $2^\omega \times 2^\omega$ with $(2 \times 2)^\omega$, i.e., we view (β, α) as $[\beta(0), \alpha(0)], [\beta(1), \alpha(1)], \ldots$ By [Kec95, Proposition 2.4], there is $R \subseteq (2 \times 2)^{<\omega}$, closed under initial segments, such that $F = \{(\beta, \alpha) \in 2^\omega \times 2^\omega \mid \forall k \in \omega\ \ (\beta, \alpha)\lceil k \in R\}$; notice that R is a tree whose infinite branches form the set F. In particular, we get

$$(\beta, \alpha) \in \text{Gr}(f) \ \Leftrightarrow\ \beta \in P_\infty \ \text{ and } \ \forall k \in \omega\ \ (\beta, \alpha)\lceil k \in R.$$

- Set $Q_f := \{(t, s) \in R \mid t \neq \emptyset \ \text{ and } \ t(|t| - 1) = 1\}$. Notice that Q_f is simply the set of pairs $(t, s) \in R$ such that the last letter of t is a 1.

We have in fact already defined the transition system $\mathcal{T}$ obtained from f. This transition system has a countably infinite set Q of states and a set Q_f of accepting states. The initial state is $q_0 := (\emptyset, \emptyset)$. The input alphabet is $2 = \{0, 1\}$ and the transition relation

$\delta \subseteq Q \times 2 \times Q$ is given by: if $m \in 2$ and $n, p \in \omega$ then $(q_n, m, q_p) \in \delta$ iff $n \overset{m}{\to} p$. Recall that a run of $\mathcal{T}$ is said to be Büchi accepting if final states occur infinitely often during this run. Then the set of ω-words over the alphabet 2 which are accepted by the transition system $\mathcal{T}$ from the initial state q_0 with Büchi acceptance condition is exactly the Borel set B.

• Now we define the finitary language π.

- We set

$$\pi := \left\{ \begin{array}{l} s \in 4^{<\omega} \mid \exists j, l \in \omega \ \exists (m_i)_{i \leq l} \in 2^{l+1} \ \exists (n_i)_{i \leq l}, (p_i)_{i \leq l}, (r_i)_{i \leq l} \in \omega^{l+1} \\ \qquad\qquad n_0 \leq M_j \\ \qquad\qquad \text{and} \\ \qquad\qquad \forall i \leq l \ n_i \overset{m_i}{\to} p_i \text{ and } p_i + r_i = M_{j+i+1} \\ \qquad\qquad \text{and} \\ \qquad\qquad \forall i < l \ p_i = n_{i+1} \\ \qquad\qquad \text{and} \\ \qquad\qquad q_{p_l} \in Q_f \\ \qquad\qquad \text{and} \\ \qquad\qquad s = {}^\frown_{i \leq l} \ \mathbf{2}^{n_i} \ m_i \ \mathbf{2}^{p_i} \ \mathbf{2}^{r_i} \ \mathbf{3} \ \mathbf{2}^{r_i} \end{array} \right\}.$$

• Let us show that $\varphi_{N,j}[\pi^\omega \cap K_{N,j}] = E_N$ if $N \leq M_j$.

Let $\gamma \in \pi^\omega \cap K_{N,j}$, and $\alpha := \varphi_{N,j}(\gamma)$. We can write

$$\gamma = {}^\frown_{k \in \omega} \left[{}^\frown_{i \leq l_k} \ \mathbf{2}^{n_i^k} \ m_i^k \ \mathbf{2}^{p_i^k} \ \mathbf{2}^{r_i^k} \ \mathbf{3} \ \mathbf{2}^{r_i^k} \ \right].$$

As this decomposition of γ is in π, we have $n_i^k \overset{m_i^k}{\to} p_i^k$ if $i \leq l_k$, $p_i^k = n_{i+1}^k$ if $i < l_k$, and $q_{p_{l_k}^k} \in Q_f$, for each $k \in \omega$. Moreover, $p_{l_k}^k = n_0^{k+1}$, for each $k \in \omega$, since $\gamma \in K_{N,j}$ implies that $p_{l_k}^k + r_{l_k}^k = r_{l_k}^k + n_0^{k+1} = M_{j+1+m}$ for some integer m. So we get

$$N \overset{\alpha(0)}{\to} p_0^0 \overset{\alpha(1)}{\to} \ldots \overset{\alpha(l_0)}{\to} p_{l_0}^0 \overset{\alpha(l_0+1)}{\to} p_0^1 \overset{\alpha(l_0+2)}{\to} \ldots \overset{\alpha(l_0+l_1+1)}{\to} p_{l_1}^1 \cdots$$

In particular we have

$$q_N^0 \prec q_{p_0^0}^0 \prec \ldots \prec q_{p_{l_0}^0}^0 \prec q_{p_0^1}^0 \prec \ldots \prec q_{p_{l_1}^1}^0 \cdots$$

because $n \overset{m}{\to} p$ implies that $q_n^0 \prec q_p^0$. Note that $|q_{p_{l_k}^k}^1| = |q_N^1| + \Sigma_{j \leq k} \ (l_j + 1)$ because $n \overset{m}{\to} p$ implies that $|q_p^1| = |q_n^1| + 1$, so that the sequence $(|q_{p_{l_k}^k}^0|)_{k \in \omega}$ is strictly increasing since $|q_n^0| = |q_n^1|$ for each integer n. This implies the existence of $\beta \in P_\infty$ such that $q_{p_{l_k}^k}^0 \prec \beta$ for each $k \in \omega$. Note that $\beta \in P_\infty$ because, for each integer k, $q_{p_{l_k}^k} \in Q_f$. Note also that $(\beta, q_N^1 \alpha)\lceil k \in R$ for infinitely many k's. As R is closed under initial segments, $(\beta, q_N^1 \alpha)\lceil k \in R$ for every $k \in \omega$, so that $q_N^1 \alpha = f(\beta) \in B$. Moreover,

$$c_{|q_N^1|}(q_N^1\alpha)=(\beta\lceil|q_N^1|,q_N^1)=(q_N^0,q_N^1)=q_N,$$

and $\alpha\in E_N$.

Conversely, let $\alpha\in E_N$. We have to see that $\gamma:=\varphi_{N,j}^{-1}(\alpha)\in\pi^\omega$. As $\gamma\in K_{N,j}$, we are allowed to write $\gamma = \mathbf{2}^N \frown [\frown_{i\in\omega}\ \alpha(i)\ \mathbf{2}^{M_{j+i+1}}\ \mathbf{3}\ ^{M_{j+i+1}}]$. Set $\beta:=f^{-1}(q_N^1\alpha)$. There is a sequence of integers $(k_l)_{l\in\omega}$ such that $q_{k_l}=(\beta,q_N^1\alpha)\lceil l$. Note that $N \overset{\alpha(0)}{\rightarrow} k_{|q_N^1|+1} \overset{\alpha(1)}{\rightarrow} k_{|q_N^1|+2}\ldots$ As $N\leq M_j$ we get $k_{|q_N^1|+i+1}\leq M_{j+i+1}$. So we can define $n_0:=N$, $p_0:=k_{|q_N^1|+1}$, $r_0:=M_{j+1}-p_0$, $n_1:=p_0$. Similarly, we can define $p_1:=k_{|q_N^1|+2}$, $r_1:=M_{j+2}-p_1$. We go on like this until we find some q_{p_i} in Q_f. This clearly defines a word in π. And we can go on like this, so that $\gamma\in\pi^\omega$.

Thus $\pi^\omega\cap K_{N,j}$ is in $\mathbf{\Gamma}(K_{N,j})\subseteq\mathbf{\Gamma}(4^\omega)$. Notice that we proved, among other things, the equality $\varphi_{0,0}[\pi^\omega\cap K_{0,0}]=B$. In particular, $\pi^\omega\cap K_{0,0}$ is not in $\check{\mathbf{\Gamma}}(4^\omega)$.

Notice that π^ω codes on $K_{0,0}$ the behaviour of the transition system accepting B. In a similar way π^ω codes on $K_{N,j}$ the behaviour of the same transition system but starting this time from the state q_N instead of the initial state q_0. But some ω-words in π^ω are not in $K_{0,0}$ and even not in any $K_{N,j}$ and we do not know what is exactly the complexity of this set of ω-words. However we remark that all words in π have the same form $\mathbf{2}^N \frown [\frown_{i\leq l}\ m_i\ \mathbf{2}^{P_i}\ \mathbf{3}\ \mathbf{2}^{R_i}]$.

• We are ready to define μ. The idea is that an infinite sequence containing a word in μ cannot be in the union of the $K_{N,j}$'s. We set

$$\mu^0=\left\{\begin{array}{r} s\in 4^{<\omega}\mid \exists l\in\omega\ \exists(m_i)_{i\leq l+1}\in 2^{l+2}\ \exists N\in\omega\ \exists(P_i)_{i\leq l+1},(R_i)_{i\leq l+1}\in\omega^{l+2}\\ \forall i\leq l+1\ \exists j\in\omega\ P_i=M_j\\ \text{and}\\ P_l\neq R_l\\ \text{and}\\ s=\mathbf{2}^N\frown[\frown_{i\leq l+1}\ m_i\ \mathbf{2}^{P_i}\ \mathbf{3}\ \mathbf{2}^{R_i}]\end{array}\right\},$$

$$\mu^1=\left\{\begin{array}{r} s\in 4^{<\omega}\mid \exists l\in\omega\ \exists(m_i)_{i\leq l+1}\in 2^{l+2}\ \exists N\in\omega\ \exists(P_i)_{i\leq l+1},(R_i)_{i\leq l+1}\in\omega^{l+2}\\ \forall i\leq l+1\ \exists j\in\omega\ P_i=M_j\\ \text{and}\\ \exists j\in\omega\ (P_l=M_j\ \text{and}\ P_{l+1}\neq M_{j+1})\\ \text{and}\\ s=\mathbf{2}^N\frown[\frown_{i\leq l+1}\ m_i\ \mathbf{2}^{P_i}\ \mathbf{3}\ \mathbf{2}^{R_i}]\end{array}\right\},$$

$\mu:=\mu^0\cup\mu^1$.

All the words in A will have the same form $\mathbf{2}^N \frown [\frown_{i\leq l}\ m_i\ \mathbf{2}^{P_i}\ \mathbf{3}\ \mathbf{2}^{R_i}]$. Note that any finite concatenation of words of this form still has this form. Moreover, such a concatenation is in μ^i if its last word is in μ^i.

• Now we show that μ^ω is "simple". The previous remarks show that

$$\mu^\omega=\{\,\gamma\in 4^\omega \mid \exists i\in 2\ \forall j\in\omega\ \exists k,n\in\omega\ \exists t_0,t_1,\ldots,t_n\in\mu^i\ n\geq j \text{ and } \gamma\lceil k={}^\frown{}_{l\leq n}\, t_l\,\}.$$

This shows that $\mu^\omega\in\mathbf{\Pi}^0_2(4^\omega)$.

Notice again that all words in A have the same form $\mathbf{2}^N\frown[{}^\frown{}_{i\leq l}\ m_i\ \mathbf{2}^{P_i}\ \mathbf{3}\ \mathbf{2}^{R_i}\ ]$. We set

$$P:=\{\mathbf{2}^N\frown[{}^\frown{}_{i\in\omega}\ m_i\ \mathbf{2}^{P_i}\ \mathbf{3}\ \mathbf{2}^{R_i}\ ]\in 4^\omega \mid N\in\omega \text{ and } \forall i\in\omega\ m_i\in 2,\ \ P_i,R_i\in\omega \text{ and } \forall i\in\omega\ \exists j\in\omega\ P_i=M_j\}.$$

We define a map $F:P\backslash\mu^\omega\rightarrow(\{\emptyset\}\cup\mu)\times\omega^2$ as follows.
Let $\gamma:=\mathbf{2}^N\frown[{}^\frown{}_{i\in\omega}\ m_i\ \mathbf{2}^{P_i}\ \mathbf{3}\ \mathbf{2}^{R_i}\ ]\in P\backslash\mu^\omega$, and $j_0\in\omega$ with $P_0=M_{j_0}$. If $\gamma\in K_{N,j_0-1}$, then we put $F(\gamma):=(\emptyset,N,j_0)$. If $\gamma\notin K_{N,j_0-1}$, then there is an integer l maximal for which $P_l\neq R_l$ or there is $j\in\omega$ with $P_l=M_j$ and $P_{l+1}\neq M_{j+1}$. Let $j_1\in\omega$ with $P_{l+2}=M_{j_1}$. We put

$$F(\gamma):=(\mathbf{2}^N\frown[{}^\frown{}_{i\leq l}\ m_i\ \mathbf{2}^{P_i}\ \mathbf{3}\ \mathbf{2}^{R_i}\ ]\frown m_{l+1}\ \mathbf{2}^{P_{l+1}}\ \mathbf{3},R_{l+1},j_1).$$

• Fix $\gamma\in A^\omega$. If $\gamma\notin\mu^\omega$, then $\gamma\in P\backslash\mu^\omega$, $F(\gamma):=(t,S,j)$ is defined. Note that $t\ \mathbf{2}^S\prec\gamma$, and that $j>0$. Moreover, $\gamma-t\ \mathbf{2}^S\in K_{0,j-1}$. Note also that $S\leq M_{j-1}$ if $t=\emptyset$, and that $t\ \mathbf{2}^S\ \gamma(|t|+S)\ \mathbf{2}^{M_j}\ \mathbf{3}\notin\mu$. Moreover, there is an integer $N\leq\min(M_{j-1},S)$ ($N=S$ if $t=\emptyset$) such that $\gamma-t\ \mathbf{2}^{S-N}\in\pi^\omega\cap K_{N,j-1}$, since the last word in μ in the decomposition of γ (if it exists) ends before $t\ \mathbf{2}^S$.

• In the sequel we will say that $(t,S,j)\in(\{\emptyset\}\cup\mu)\times\omega^2$ is <u>*suitable*</u> if $S\leq M_j$ if $t=\emptyset$, $t(|t|-1)=\mathbf{3}$ if $t\in\mu$, and $t\ \mathbf{2}^S\ m\ \mathbf{2}^{M_{j+1}}\ \mathbf{3}\notin\mu$ if $m\in 2$. We set, for (t,S,j) suitable,

$$P_{t,S,j}:=\{\,\gamma\in 4^\omega \mid t\ \mathbf{2}^S\prec\gamma \text{ and } \gamma-t\ \mathbf{2}^S\in K_{0,j}\,\}.$$

Note that $P_{t,S,j}$ is a compact subset of $P\backslash\mu^\omega$, and that $F(\gamma)=(t,S,j+1)$ if $\gamma\in P_{t,S,j}$. This shows that the $P_{t,S,j}$'s, for (t,S,j) suitable, are pairwise disjoint. Note also that μ^ω is disjoint from $\bigcup_{(t,S,j)\text{ suitable}} P_{t,S,j}$.

• We set, for (t,S,j) suitable and $N\leq\min(M_j,S)$ ($N=S$ if $t=\emptyset$),

$$A_{t,S,j,N}:=\{\,\gamma\in P_{t,S,j} \mid \gamma-t\ \mathbf{2}^{S-N}\in\pi^\omega\cap K_{N,j}\,\}.$$

Note that $A_{t,S,j,N}\in\mathbf{\Gamma}(4^\omega)$ since $N\leq M_j$.

• The previous discussion shows that

$$A^\omega=\mu^\omega\cup\bigcup_{(t,S,j)\text{ suitable}}\ \bigcup_{\substack{N\leq\min(M_j,S)\\ N=S \text{ if } t=\emptyset}} A_{t,S,j,N}.$$

As $\boldsymbol{\Gamma}$ is closed under finite unions, the set

$$A_{t,S,j} := \bigcup_{\substack{N \leq \min(M_j, S) \\ N = S \text{ if } t = \emptyset}} A_{t,S,j,N}$$

is in $\boldsymbol{\Gamma}(4^\omega)$. On the other hand we have proved that $\mu^\omega \in \boldsymbol{\Pi}^0_2(4^\omega) \subseteq \boldsymbol{\Gamma}(4^\omega)$, thus we get $A^\omega \in \boldsymbol{\Gamma}(4^\omega)$ if $\boldsymbol{\Gamma} = \boldsymbol{\Sigma}^0_\xi$.

Consider now the case $\boldsymbol{\Gamma} = \boldsymbol{\Pi}^0_\xi$. We can write

$$A^\omega = \mu^\omega \setminus \left(\bigcup_{(t,S,j) \text{ suitable}} P_{t,S,j} \right) \cup \bigcup_{(t,S,j) \text{ suitable}} A_{t,S,j} \cap P_{t,S,j}.$$

Thus

$$\neg A^\omega = \neg \left[\mu^\omega \cup \left(\bigcup_{(t,S,j) \text{ suitable}} P_{t,S,j} \right) \right] \cup \bigcup_{(t,S,j) \text{ suitable}} P_{t,S,j} \setminus A_{t,S,j}.$$

Here $\neg \left[\mu^\omega \cup \left(\bigcup_{(t,S,j) \text{ suitable}} P_{t,S,j} \right) \right] \in \boldsymbol{\Delta}^0_3(4^\omega) \subseteq \check{\boldsymbol{\Gamma}}(4^\omega)$ because μ^ω is a $\boldsymbol{\Pi}^0_2$-subset of 4^ω and $(\bigcup_{(t,S,j) \text{ suitable}} P_{t,S,j})$ is a $\boldsymbol{\Sigma}^0_2$-subset of 4^ω as it is a countable union of compact hence closed sets. On the other hand $P_{t,S,j} \setminus A_{t,S,j} \in \check{\boldsymbol{\Gamma}}(4^\omega)$, thus $\neg A^\omega$ is in $\check{\boldsymbol{\Gamma}}(4^\omega)$ and $A^\omega \in \boldsymbol{\Gamma}(4^\omega)$. Moreover, the set $A^\omega \cap P_{\emptyset,0,0} = \pi^\omega \cap P_{\emptyset,0,0} = \pi^\omega \cap K_{0,0}$ is not in $\check{\boldsymbol{\Gamma}}$. This shows that A^ω is not in $\check{\boldsymbol{\Gamma}}$. Thus A^ω is in $\boldsymbol{\Gamma}(4^\omega) \setminus \check{\boldsymbol{\Gamma}}$.

We can now end the proof of Theorem 2.

(a) If $\xi = 1$, then we can take $A := \{s \in 2^{<\omega} \mid 0 \prec s \text{ or } \exists k \in \omega \ \ 10^k 1 \prec s\}$ and $A^\omega = 2^\omega \setminus \{10^\omega\}$ is $\boldsymbol{\Sigma}^0_1 \setminus \boldsymbol{\Pi}^0_1$.

• If $\xi = 2$, then we will see in Theorem 4 the existence of $A \subseteq 2^{<\omega}$ such that A^ω is $\boldsymbol{\Sigma}^0_2 \setminus \boldsymbol{\Pi}^0_2$.

• So we may assume that $\xi \geq 3$, and we are done.

(b) If $\xi = 1$, then we can take $A := \{0\}$ and $A^\omega = \{0^\omega\}$ is $\boldsymbol{\Pi}^0_1 \setminus \boldsymbol{\Sigma}^0_1$.

• If $\xi = 2$, then we can take $A := \{0^k 1 \mid k \in \omega\}$ and $A^\omega = P_\infty$ is $\boldsymbol{\Pi}^0_2 \setminus \boldsymbol{\Sigma}^0_2$.

• So we may assume that $\xi \geq 3$, and we are done. □

As we have said above it remains a Borel class for which we have not yet got a complete ω-power: the class $\boldsymbol{\Sigma}^0_2$. Notice that it is easy to see that the classical example of $\boldsymbol{\Sigma}^0_2$-complete set, the set $2^\omega \setminus P_\infty$, is not an ω-power. However we are going to prove the following result.

Theorem 4. *There is a context-free language $A\subseteq 2^{<\omega}$ such that $A^\omega \in \mathbf{\Sigma}^0_2\setminus\mathbf{\Pi}^0_2$.*

Proof. By Proposition 11 in [Lec05], it is enough to find $A\subseteq 3^{<\omega}$. We set, for $j<3$ and $s\in 3^{<\omega}$,

$$n_j(s) := \text{Card}\{i<|s| \mid s(i)=j\},$$

$$T := \{\alpha\in 3^{\leq\omega} \mid \forall l<1+|\alpha| \;\; n_2(\alpha\lceil l)\leq n_1(\alpha\lceil l)\}.$$

• We inductively define, for $s\in T\cap 3^{<\omega}$, $s^{\hookleftarrow}\in 2^{<\omega}$ as follows:

$$s^{\hookleftarrow} := \begin{cases} \emptyset \text{ if } s=\emptyset, \\ t^{\hookleftarrow}\varepsilon \text{ if } s=t\varepsilon \text{ and } \varepsilon<2, \\ t^{\hookleftarrow}\text{, except that its last 1 is replaced with 0, if } s=t\mathbf{2}. \end{cases}$$

• We will extend this definition to infinite sequences. To do this, we introduce a notion of limit. Fix $(s_n)_{n\in\omega}$ a sequence of elements in $2^{<\omega}$. We define $\lim_{n\to\infty} s_n \in 2^{\leq\omega}$ as follows. For each $t\in 2^{<\omega}$,

$$t\prec \lim_{n\to\infty} s_n \Leftrightarrow \exists n_0\in\omega \;\; \forall n\geq n_0 \;\; t\prec s_n.$$

• If $\alpha\in T\cap 3^\omega$, then we set $\alpha^{\hookleftarrow} := \lim_{n\to\infty} (\alpha\lceil n)^{\hookleftarrow}$. We define $e : T\cap 3^\omega \to 2^\omega$ by $e(\alpha):=\alpha^{\hookleftarrow}$. Note that $T\cap 3^\omega\in\mathbf{\Pi}^0_1(3^\omega)$, and e is a $\mathbf{\Sigma}^0_2$-measurable partial function on $T\cap 3^\omega$, since for $t\in 2^{<\omega}$ we have

$$t\prec e(\alpha) \Leftrightarrow \exists n_0\in\omega \;\; \forall n\geq n_0 \;\; t\prec(\alpha\lceil n)^{\hookleftarrow}.$$

• We set $E:=\{s\in T\cap 3^{<\omega} \mid n_2(s)=n_1(s) \text{ and } s\neq\emptyset \text{ and } 1\prec[s\lceil(|s|-1)]^{\hookleftarrow}\}$. Note that $\emptyset\neq s^{\hookleftarrow}\prec 0^\omega$, and that $s(|s|-1)=\mathbf{2}$ changes $s(0)=[s\lceil(|s|-1)]^{\hookleftarrow}(0)=1$ into 0 if $s\in E$.

• If $S\subseteq 3^{<\omega}$, then $S^*:=\{{}^\frown_{i<l}\, s_i\in 3^{<\omega} \mid l\in\omega \text{ and } \forall i<l \;\; s_i\in S\}$. We put

$$A:=\{0\}\cup E\cup\{{}^\frown_{j\leq k}\,(c_j 1)\in 3^{<\omega} \mid [\forall j\leq k \;\; c_j\in(\{0\}\cup E)^*] \text{ and } [k>0 \text{ or } (k=0 \text{ and } c_0\neq\emptyset)]\}.$$

• In the proof of Theorem 2.(b) we met the set $\{s\in 2^{<\omega} \mid 0\prec s \text{ or } \exists k\in\omega \;\; 10^k 1\prec s\}$. We shall denoted it by B in the sequel. We have seen that $B^\omega=2^\omega\setminus\{10^\omega\}$ is $\mathbf{\Sigma}^0_1\setminus\mathbf{\Pi}^0_1$. Let us show that $A^\omega=e^{-1}(B^\omega)$.

- By induction on $|t|$, we get $(st)^{\hookleftarrow} = s^{\hookleftarrow}t^{\hookleftarrow}$ if $s,t\in T\cap 3^{<\omega}$. Let us show that $(s\beta)^{\hookleftarrow}=s^{\hookleftarrow}\beta^{\hookleftarrow}$ if moreover $\beta\in T\cap 3^\omega$.

Assume that $t\prec(s\beta)^{\hookleftarrow}$. Then there is $m_0\geq|s|$ such that, for $m\geq m_0$,

$$t\prec[(s\beta)\lceil m]^{\hookleftarrow}=[s\beta\lceil(m-|s|)]^{\hookleftarrow}=s^{\hookleftarrow}[\beta\lceil(m-|s|)]^{\hookleftarrow}.$$

This implies that $t \prec s^{\hookleftarrow}\beta^{\hookleftarrow}$ if $|t|<|s^{\hookleftarrow}|$. If $|t|\geq|s^{\hookleftarrow}|$, then there is $m_1\in\omega$ such that, for $m\geq m_1$, $\beta^{\hookleftarrow}\lceil(|t|-|s^{\hookleftarrow}|)\prec[\beta\lceil(m-|s|)]^{\hookleftarrow}$. Here again, we get $t\prec s^{\hookleftarrow}\beta^{\hookleftarrow}$. Thus $(s\beta)^{\hookleftarrow}=s^{\hookleftarrow}\beta^{\hookleftarrow}$.

Let $(s_i)_{i\in\omega}$ be a sequence such that for each integer $i\in\omega$, $s_i\in T\cap 3^{<\omega}$. Then $\frown_{i\in\omega}\ s_i\in T$, and $(\frown_{i\in\omega}\ s_i)^{\hookleftarrow}=\frown_{i\in\omega}\ s_i^{\hookleftarrow}$, by the previous facts.

- Let $(a_i)_{i\in\omega}$ be a sequence such that for each integer $i\in\omega$, $a_i\in A\setminus\{\emptyset\}$ and $\alpha:=\frown_{i\in\omega}\ a_i$. As $A\subseteq T$, $e(\alpha)=(\frown_{i\in\omega}\ a_i)^{\hookleftarrow}=\frown_{i\in\omega}\ a_i^{\hookleftarrow}$.
If $a_0\in\{0\}\cup E$, then $\emptyset\neq a_0^{\hookleftarrow}\prec 0^\omega$, thus $e(\alpha)\in N_0\subseteq 2^\omega\setminus\{10^\omega\}=B^\omega$.
If $a_0\notin\{0\}\cup E$, then $a_0=\frown_{j\leq k}\ (c_j1)$, thus $a_0^{\hookleftarrow}=\frown_{j\leq k}\ (c_j^{\hookleftarrow}1)$.
If $c_0\neq\emptyset$, then $e(\alpha)\in B^\omega$ as before.
If $c_0=\emptyset$, then $k>0$, so that $e(\alpha)\neq 10^\omega$ since $e(\alpha)$ has at least two coordinates equal to 1.
We proved that $A^\omega\subseteq e^{-1}(B^\omega)$.

- Assume now that $e(\alpha)\in B^\omega$. We have to find $(a_i)_{i\in\omega}\subseteq A\setminus\{\emptyset\}$ with $\alpha=\frown_{i\in\omega}\ a_i$. We split into cases:

1. $e(\alpha)=0^\omega$.
1.1. $\alpha(0)=0$.
In this case $\alpha-0\in T$ and $e(\alpha-0)=0^\omega$. Moreover, $0\in A$. We put $a_0:=0$.

1.2. $\alpha(0)=1$.
In this case there is a coordinate j_0 of α equal to **2** ensuring that $\alpha(0)$ is replaced with a 0 in $e(\alpha)$. We put $a_0:=\alpha\lceil(j_0+1)$, so that $a_0\in E\subseteq A$, $\alpha-a_0\in T$ and $e(\alpha-a_0)=0^\omega$.

Now the iteration of the cases 1.1 and 1.2 shows that $\alpha\in A^\omega$.

2. $e(\alpha)=0^{k+1}10^\omega$ for some $k\in\omega$.

As in case 1, there is $c_0\in(\{0\}\cup E)^*$ such that $c_0\prec\alpha$, $c_0^{\hookleftarrow}=0^{k+1}$, $\alpha-c_0\in T$ and $e(\alpha-c_0)=10^\omega$. Note that $\alpha(|c_0|)=1$, $\alpha-(c_01)\in T$ and $e[\alpha-(c_01)]=0^\omega$. We put $a_0:=c_01$, and argue as in case 1.

3. $e(\alpha)=(\frown_{j\leq l+1}\ 0^{k_j}1)0^\omega$ for some $l\in\omega$.

The previous cases show the existence of $(c_j)_{j\leq l+1}$, where for each $j\leq l+1$ $c_j\in(\{0\}\cup E)^*$ such that :
$a_0:=\frown_{j\leq l+1}\ c_j1\prec\alpha$, $\alpha-a_0\in T$ and $e(\alpha-a_0)=0^\omega$. We are done since $a_0\in A$.

4. $e(\alpha)=\frown_{j\in\omega}\ 0^{k_j}1$.

An iteration of the discussion of case 3 shows that we can take a_i of the form $\frown_{j\leq l+1}\ c_j1$.

• The previous discussion shows that $A^\omega = e^{-1}(B^\omega)$. As B^ω is an open subset of 2^ω and e is $\mathbf{\Sigma}^0_2$-measurable, the ω-power $A^\omega = e^{-1}(B^\omega)$ is in $\mathbf{\Sigma}^0_2(3^\omega)$.

It remains to see that $A^\omega = e^{-1}(B^\omega) \notin \mathbf{\Pi}^0_2$. We argue by contradiction.

Assume on the contrary that $e^{-1}(B^\omega) \in \mathbf{\Pi}^0_2(3^\omega)$. We know that $B^\omega = 2^\omega \setminus \{10^\omega\}$ so $e^{-1}(\{10^\omega\}) = (T \cap 3^\omega) \setminus e^{-1}(B^\omega)$ would be a $\mathbf{\Sigma}^0_2$-subset of 3^ω since $T \cap 3^\omega$ is closed in 3^ω. Thus $e^{-1}(\{10^\omega\})$ would be a countable union of compact subsets of 3^ω.

Consider now the **cartesian product** $(\{0\} \cup E)^{\mathbb{N}}$ of countably many copies of $(\{0\} \cup E)$. The set $(\{0\} \cup E)$ is countable and it can be equipped with the discrete topology. Then the product $(\{0\} \cup E)^{\mathbb{N}}$ is equipped with the product topology of the discrete topology on $(\{0\} \cup E)$. The topological space $(\{0\} \cup E)^{\mathbb{N}}$ is homeomorphic to the Baire space ω^ω.

Consider now the map $h : (\{0\} \cup E)^{\mathbb{N}} \to e^{-1}(\{10^\omega\})$ defined by $h(\gamma) := 1[\frown_{i\in\omega} \gamma_i]$ for each $\gamma = (\gamma_0, \gamma_1, \ldots, \gamma_i, \ldots) \in (\{0\} \cup E)^{\mathbb{N}}$. The map h is a homeomorphism by the previous discussion. As $(\{0\} \cup E)^{\mathbb{N}}$ is homeomorphic to the Baire space ω^ω, the Baire space ω^ω is also homeomorphic to the space $e^{-1}(\{10^\omega\})$, so it would be also a countable union of compact sets. But this is absurd by [Kec95, Theorem 7.10].

It remains to see that A is context-free. It is easy to see that the language E is in fact accepted by a 1-counter automaton: it is the set of words $s \in 3^{<\omega}$ such that :

$$\forall l \in [1; |s|[\;\; n_2(s\lceil l) < n_1(s\lceil l) \text{ and } n_2(s) = n_1(s) \text{ and } s(0) = 1 \text{ and } s(|s|-1) = \mathbf{2}.$$

This implies that A is also accepted by a 1-counter automaton because the class of 1-counter languages is closed under concatenation and star operation. In particular A is a context-free language because the class of languages accepted by 1-counter automata form a strict subclass of the class of context-free languages, [ABB96]. □

Remark 5. *The operation $\alpha \to \alpha^{\hookleftarrow}$ we have defined is very close to the erasing operation defined by Duparc in his study of the Wadge hierarchy, [Dup01]. However we have modified this operation in such a way that $\alpha^{\hookleftarrow}$ is always infinite when α is infinite, and that it has the good property with regard to ω-powers and topological complexity.*

4 Concluding Remarks and Further Work

It is natural to wonder whether the ω-powers obtained in this paper are effective. For instance could they be obtained as ω-powers of recursive languages ?

In the long version of this paper we prove effective versions of the results presented here. Using tools of effective descriptive set theory, we first prove an effective version of Kuratowski's Theorem 3. Then we use it to prove the following effective version of Theorem 2, where Σ^0_ξ and Π^0_ξ denote classes of the hyperarithmetical hierarchy and ω_1^{CK} is the first non-recursive ordinal, usually called the Church-kleene ordinal.

Theorem 6. *Let ξ be a non-null ordinal smaller than ω_1^{CK}.*
(a) There is a recursive language $A \subseteq 2^{<\omega}$ such that $A^\omega \in \Sigma^0_\xi \setminus \mathbf{\Pi}^0_\xi$.
(b) There is a recursive language $A \subseteq 2^{<\omega}$ such that $A^\omega \in \Pi^0_\xi \setminus \mathbf{\Sigma}^0_\xi$.

The question, left open in [Fin04], also naturally arises to know what are all the possible infinite Borel ranks of ω-powers of finitary languages belonging to some natural class like the class of context free languages (respectively, languages accepted by stack automata, recursive languages, recursively enumerable languages, ...).
We know from [Fin06] that there are ω-languages accepted by Büchi 1-counter automata of every Borel rank (and even of every Wadge degree) of an effective analytic set. Every ω-language accepted by a Büchi 1-counter automaton can be written as a finite union $L = \bigcup_{1 \leq i \leq n} U_i^\frown V_i^\omega$, where for each integer i, U_i and V_i are finitary languages accepted by 1-counter automata. And the supremum of the set of Borel ranks of effective analytic sets is the ordinal γ_2^1. This ordinal is defined by A.S. Kechris, D. Marker, and R.L. Sami in [KMS89] and it is proved to be strictly greater than the ordinal δ_2^1 which is the first non Δ_2^1 ordinal. Thus the ordinal γ_2^1 is also strictly greater than the first non-recursive ordinal ω_1^{CK}. From these results it seems plausible that there exist some ω-powers of languages accepted by 1-counter automata which have Borel ranks up to the ordinal γ_2^1, although these languages are located at the very low level in the complexity hierarchy of finitary languages.

Another question concerns the Wadge hierarchy which is a great refinement of the Borel hierarchy. It would be interesting to determine the Wadge hierarchy of ω-powers. In the full version of this paper we give many Wadge degrees of ω-powers and this confirms the great complexity of these ω-languages.

References

[Arn83] Arnold, A.: Topological Characterizations of Infinite Behaviours of Transition Systems, Automata, Languages and Programming, J. In: Díaz, J. (ed.) Automata, Languages and Programming. LNCS, vol. 154, pp. 28–38. Springer, Heidelberg (1983)

[ABB96] Autebert, J.-M., Berstel, J., Boasson, L.: Context Free Languages and Pushdown Automata. In: Handbook of Formal Languages, vol. 1, Springer, Heidelberg (1996)

[Dup01] Duparc, J.: Wadge Hierarchy and Veblen Hierarchy: Part 1: Borel Sets of Finite Rank. Journal of Symbolic Logic 66(1), 56–86 (2001)

[DF06] Duparc, J., Finkel, O.: An ω-Power of a Context-Free Language Which Is Borel Above $\Delta^0{}_\omega$. In: submitted to the Proceedings of the International Conference Foundations of the Formal Sciences V: Infinite Games, Bonn, Germany (November 26-29, 2004)

[Fin01] Finkel, O.: Topological Properties of Omega Context Free Languages. Theoretical Computer Science 262(1-2), 669–697 (2001)

[Fin03] Finkel, O.: Borel Hierarchy and Omega Context Free Languages. Theoretical Computer Science 290(3), 1385–1405 (2003)

[Fin04] Finkel, O.: An omega-Power of a Finitary Language Which is a Borel Set of Infinite Rank. Fundamenta Informaticae 62(3-4), 333–342 (2004)

[Fin06] Finkel, O.: Borel Ranks and Wadge Degrees of Omega Context Free Languages. Mathematical Structures in Computer Science 16(5), 813–840 (2006)

[HU69] Hopcroft, J.E., Ullman, J.D.: Formal Languages and their Relation to Automata. Addison-Wesley Publishing Company, Reading, Massachussetts (1969)

[Kec95] Kechris, A.S.: Classical Descriptive Set Theory. Springer, Heidelberg (1995)
[KMS89] Kechris, A.S., Marker, D., Sami, R.L.: $\Pi_1{}^1$ Borel Sets. The Journal of Symbolic Logic 54(3), 915–920 (1989)
[Kur66] Kuratowski, K.: Topology, vol. 1. Academic Press, New York (1966)
[Lec01] Lecomte, D.: Sur les Ensembles de Phrases Infinies Constructibles a Partir d'un Dictionnaire sur un Alphabet Fini, Séminaire d'Initiation a l'Analyse, vol. 1, année (2001-2002)
[Lec05] Lecomte, D.: Omega-Powers and Descriptive Set Theory. Journal of Symbolic Logic 70(4), 1210–1232 (2005)
[LT94] Lescow, H., Thomas, W.: Logical Specifications of Infinite Computations. In: de Bakker, J.W., de Roever, W.-P., Rozenberg, G. (eds.) A Decade of Concurrency. LNCS, vol. 803, pp. 583–621. Springer, Heidelberg (1994)
[Mos80] Moschovakis, Y.N.: Descriptive Set Theory. North-Holland, Amsterdam (1980)
[Niw90] Niwinski, D.: Problem on ω-Powers posed in the Proceedings of the 1990 Workshop Logics and Recognizable Sets (Univ. Kiel)
[PP04] Perrin, D., Pin, J.-E.: Infinite Words, Automata, Semigroups, Logic and Games. Pure and Applied Mathematics 141 (2004)
[Sim92] Simonnet, P.: Automates et Théorie Descriptive, Ph. D. Thesis, Université Paris 7(March 1992)
[Sta86] Staiger, L.: Hierarchies of Recursive ω-Languages. Jour. Inform. Process. Cybernetics EIK 22(5/6), 219–241 (1986)
[Sta97a] Staiger, L.: ω-Languages, Chapter of the Handbook of Formal Languages. In: Rozenberg, G., Salomaa, A. (eds.), vol. 3, Springer, Heidelberg (1997)
[Sta97b] Staiger, L.: On ω-Power Languages, in New Trends in Formal Languages, Control, Coperation, and Combinatorics. In: Păun, G., Salomaa, A. (eds.) New Trends in Formal Languages. LNCS, vol. 1218, pp. 377–393. Springer, Heidelberg (1997)
[Tho90] Thomas, W.: Automata on Infinite Objects. In: Van Leeuwen, J. (ed.) Handbook of Theoretical Computer Science, vol. B, pp. 133–191. Elsevier, Amsterdam (1990)

Satisfiability of a Spatial Logic with Tree Variables

Emmanuel Filiot[1], Jean-Marc Talbot[2], and Sophie Tison[1]

[1] INRIA Futurs, Mostrare Project, University of Lille 1 (LIFL, UMR 8022 of CNRS)
[2] University of Provence (LIF, UMR 6166 of CNRS), Marseille

Abstract. We investigate in this paper the spatial logic TQL for querying semi-structured data, represented as unranked ordered trees over an infinite alphabet. This logic consists of usual Boolean connectives, spatial connectives (derived from the constructors of a tree algebra), tree variables and a fixpoint operator for recursion. Motivated by XML-oriented tasks, we investigate the guarded TQL fragment. We prove that for closed formulas this fragment is MSO-complete. In presence of tree variables, this fragment is strictly more expressive than MSO as it allows for tree (dis)equality tests, i.e. testing whether two subtrees are isomorphic or not. We devise a new class of tree automata, called TAGED, which extends tree automata with global equality and disequality constraints. We show that the satisfiability problem for guarded TQL formulas reduces to emptiness of TAGED. Then, we focus on bounded TQL formulas: intuitively, a formula is bounded if for any tree, the number of its positions where a subtree is captured by a variable is bounded. We prove this fragment to correspond with a subclass of TAGED, called bounded TAGED, for which we prove emptiness to be decidable. This implies the decidability of the bounded guarded TQL fragment. Finally, we compare bounded TAGED to a fragment of MSO extended with subtree isomorphism tests.

1 Introduction

In this paper, we consider the spatial logic TQL [7]. Spatial logics have been used to express properties of structures such as trees [7], graphs [6,12] and heaps [19]. The main ingredients of spatial logics are spatial connectives: roughly speaking, these connectives are derived from operators that can be used to generate the domain of interpretation.

The logic we consider here is interpreted over hedges (*i.e.* unranked ordered trees) labelled over an infinite alphabet. The logic integrates Boolean connectives, spatial connectives (derived from the constructors of an unranked ordered tree algebra), tree variables and a fixpoint operator for recursion.

We focus on the satisfiability problem of TQL. It is quite simple to prove that the full TQL logic over unranked ordered trees even without tree variables is undecidable. We then focus on the guarded fragment which ensures that recursive variables have to be guarded by tree extension. We show that guarded TQL logic without tree variables is equivalent to the monadic second order logic (MSO).

However, when tree variables are considered, things are getting more complicated. Indeed, we can express that two non-empty paths starting from a node of a tree lead to two isomorphic subtrees, which goes beyond regularity over unranked trees.

J. Duparc and T.A. Henzinger (Eds.): CSL 2007, LNCS 4646, pp. 130–145, 2007.

Still about expressiveness of this logic, an infinite alphabet and the ability to test for tree equality allow us to consider some data values. We can write formulas whose models are hedges which violate some key constraints or some functional dependencies.

We focus on bounded TQL formulas: intuitively, a formula is bounded if for any tree, the number of equalities and disequalities that have to be tested – to check non-deterministically that the tree is a model of the formula – is bounded.

We introduce a new class of tree automata, called *tree automata with global equalities and disequalities* (TAGED for short), which extends unranked tree automata A with an equality relation $=_A$ and a disequality relation $\neq_A$ on states. Subtrees of some tree t which evaluates by A to states which are in relation by $=_A$ (resp. by $\neq_A$) have to be isomorphic (resp. non isomorphic). Naturally, $=_A$ induces an equivalence relation on a subset of nodes of t, but the number of classes of this relation is bounded. E.g., TAGED can express that all subtrees of height n, for some fixed natural n, are equal, but not that for each node of the tree, all the subtrees rooted at its sons are equal. Although it is a natural extension of tree automata, this extension has never been considered.

We show that satisfiability of guarded TQL formulas reduces to emptiness of TAGED. We define a subclass of TAGED, called *bounded TAGED*, for which we can decide emptiness. Intuitively, boundedness ensures that the cardinality of every equivalence class is bounded, which may not be the case for full TAGED. We show emptiness decidability of bounded TAGED.

We complete our proof by constructing a TAGED from a guarded and bounded TQL formula. This construction extends the one from [4] with tree variables. This extension is non-trivial as the automata we have to consider are non-deterministic.

Finally, we define an extension of MSO with a binary relation $\sim$ between nodes; two nodes are in relation if they are roots of two isomorphic subtrees. We consider MSO formulas extended with the predicate $\sim$. It is easy to see that this extension renders MSO undecidable. However, we prove that if the relation $\sim$ concerns only variables belonging to a prefix of existentially quantified first-order variables, then this extension is decidable. The proof works by reduction to emptiness of bounded TAGED.

Automata dealing with data values have been studied in [18,3]. However, our motivations are different and we obtain the capability to manage data values as a side-effect. In [3], the authors study two-variables FO logic extended with an equality relation on data values. The expressiveness of this formalism and the one presented here are not comparable: we can test tree isomorphisms while they can test data value equality only, but restricting our logic to data-value equality is strictly less expressive, as we do not have quantifiers.

The paper is organised as follows: in Section 2 we recall definitions for hedges, hedge automata and monadic second order logic. Section 3 describes the TQL logic and results we obtain concerning its satisfiability. Section 4 is dedicated to the tool we use to solve the satisfiability problem, namely tree automata with global equalities and disequalities (TAGED). In Section 5 we relate satisfiability of TQL formulas and emptiness of TAGED. Finally, in Section 6 we propose an extension of MSO with isomorphism tests whose satisfiability problem is decidable.

2 Preliminaries

We consider an infinite set of labels Λ.

Hedges - Trees. Let Σ be the signature $\{0, |\} \cup \{a \mid a \in \Lambda\}$, where 0 is a constant, $|$ a binary symbol and as unary symbols. We call *hedge* an element of the Σ-algebra Hedge obtained by quotienting the free Σ-algebra by the following three axioms:

$$0|h = h \qquad h|0 = h \qquad (h|h'')|h''' = h|(h''|h''')$$

0 will be the *empty hedge*. We call respectively *trees* and *leaves* hedges of the form $a(h)$ and $a(0)$. We may omit $|$ and write $a(h)b(h')c(h'')$ instead of $a(h)|b(h')|c(h'')$. We define roots(h) as the word from Λ^* defined recursively as : (i) roots$(0) = \epsilon$, (ii) roots$(a(h')) = a$ and, (iii) roots$(h_1|h_2) =$ roots(h_1)roots(h_2).

We will also adopt the graph point of view and consider hedges as a set of vertices V, two disjoint sets of directed edges E_c, E_s and a mapping λ from V to Λ. In a hedge h, one associates a vertex with each occurrence of elements of Λ. There is an edge from E_c (resp. from E_s) from an occurrence a_1 to an occurrence a_2 if the hedge contains a pattern of the form $a_1(h_1|a_2(h)|h_2)$ for some hedges h_1, h_2 (resp. $a_1(h_1)|a_2(h_2)$ for some hedges h_1, h_2).

For every hedge $h = (V, E_c, E_s, \lambda)$, we denote by $\mathtt{nodes}(h)$ the set V and by $\mathtt{lab}_h(u)$ the label $\lambda(u)$, for $u \in V$. We denote $h|_u$ the subtree of h rooted at u, and by $\leq$ the reflexive-transitive closure of E_c. E.g., the root is minimal for $\leq$ in a tree.

For a set of labels L, we denote $\mathbb{H}_L$ (resp. $\mathbb{T}_L$) the set of hedges (resp. trees) with nodes labelled by elements in L.

Hedge automata [17]. A *hedge automaton* A is a 4-tuple (Λ, Q, F, Δ) where Δ is a finite set of rules $\alpha(L) \to q$ where α is a finite or cofinite set of labels, $L \subseteq Q^*$ is a regular language over states from Q, and $F \subseteq Q^*$ is an accepting regular language.

Definition 1 (runs). *Let h be a hedge and A be a hedge automaton. The set of runs $R_A(h) \subseteq \mathbb{H}_Q$ of A on h is the set of hedges over Q inductively defined by:*

$$\begin{array}{rcl} R_A(h_1|h_2) &=& \{r_1|r_2 \mid r_1 \in R_A(h_1), r_2 \in R_A(h_2)\} \qquad R_A(0) = \{0\} \\ R_A(a(h)) &=& \{q(r) \mid \exists \alpha(L) \to q \in \Delta, a \in \alpha, r \in R_A(h), \mathit{roots}(r) \in L\} \end{array}$$

Let $\overline{q}$ be a word of states, we denote by $R_{A,\overline{q}}(h) \subseteq R_A(h)$ the set of runs r such that roots$(r) = \overline{q}$, and often say that h evaluates to $\overline{q}$ by A. The set of accepting runs of A on h, denoted by $R_A^{acc}(h)$, is defined by $\{r \mid \exists \overline{q} \in F, r \in R_{A,\overline{q}}(h)\}$.

The language accepted by A, denoted $L(A)$, is defined by $\{h \mid R_A^{acc}(h) \neq \varnothing\}$.

Testing emptiness of the language accepted by a hedge automaton is decidable [5].

MSO. The logic MSO (Monadic Second Order logic) is the extension of the first-order logic FO with the possibility to quantify over unary relations, *i.e.* over sets.

Let σ be the signature $\{\mathsf{lab}_a \mid a \in \Lambda\}$ where lab_as are unary predicates. We associate with a hedge $h = (V, E_c, E_s, \lambda)$ a finite σ-structure $\mathcal{S}^h = \langle V, \{\mathsf{ch}, \mathsf{ns}\} \cup \{\mathsf{lab}_a^h \mid a \in \Lambda\}\rangle$, such that $\mathsf{lab}_a^h(v)$ (resp. $\mathsf{ch}(v, v')$, $\mathsf{ns}(v, v')$) holds in $\mathcal{S}^h$ if $\lambda(v) = a$ (resp. $(v, v') \in E_c$, $(v, v') \in E_s$).

We assume a countable set of first-order variables ranging over by $x, y, z, \ldots$ and a countable set of second-order variables ranging over by $X, Y, Z, \ldots$.

(a) Syntax

$$
\begin{array}{lll}
\phi ::= & 0 & \text{empty hedge} \\
& \top & \text{truth} \\
& \alpha[\phi] & \text{extension} \\
& \phi|\phi & \text{composition} \\
& \neg\phi & \text{negation} \\
& \phi \vee \phi & \text{disjunction} \\
& X & \text{tree variable} \\
& \xi & \text{recursion variables} \\
& \mu\xi.\phi & \text{least fixpoint} \\
& \phi^* & \text{iteration}
\end{array}
$$

(b) Semantics

$$
\begin{array}{ll}
[\![0]\!]_{\rho,\delta} & = \{0\} \\
[\![\top]\!]_{\rho,\delta} & = \mathbb{H}_\Lambda \\
[\![\alpha[\phi]]\!]_{\rho,\delta} & = \{a(h) \mid h \in [\![\phi]\!]_{\rho,\delta},\ a \in \alpha\} \\
[\![\phi|\phi']\!]_{\rho,\delta} & = \{h|h' \mid h \in [\![\phi]\!]_{\rho,\delta}, h' \in [\![\phi']\!]_{\rho,\delta}\} \\
[\![\neg\phi]\!]_{\rho,\delta} & = \mathbb{H}_\Lambda \backslash [\![\phi]\!]_{\rho,\delta} \\
[\![\phi \vee \phi']\!]_{\rho,\delta} & = [\![\phi]\!]_{\rho,\delta} \cup [\![\phi']\!]_{\rho,\delta} \\
[\![X]\!]_{\rho,\delta} & = \{\rho(X)\} \\
[\![\xi]\!]_{\rho,\delta} & = \delta(\xi) \\
[\![\mu\xi.\phi]\!]_{\rho,\delta} & = \bigcap\{S \subseteq \mathbb{H}_\Lambda \mid [\![\phi]\!]_{\rho,\delta[\xi\mapsto S]} \subseteq S\} \\
[\![\phi^*]\!]_{\rho,\delta} & = 0 \cup \bigcup_{i>0} \underbrace{[\![\phi]\!]_{\rho,\delta}|\cdots|[\![\phi]\!]_{\rho,\delta}}_{i \text{ times}}
\end{array}
$$

Fig. 1. TQL logic

MSO formulas are given by the following syntax:

$$\psi ::= \mathsf{lab}_a(x) \mid \mathsf{ch}(x,y) \mid \mathsf{ns}(x,y) \mid x \in X \mid \psi \vee \psi \mid \neg\psi \mid \exists x.\psi \mid \exists X.\psi$$

Let $\mathcal{S}$ be a σ-structure with domain V. Let ρ be a valuation mapping first-order variables to elements from V and second-order variables to subsets of V. We write $\mathcal{S} \models_\rho \psi$ when the structure $\mathcal{S}$ is a model of the formula ψ under the valuation ρ; this is defined in the usual Tarskian manner and we have in particular, (i) $\psi \models_\rho \mathsf{lab}_a(x)$ if $\mathsf{lab}_a(\rho(x))$ holds in $\mathcal{S}$, (ii) $\psi \models_\rho \mathsf{ch}(x,y)$ if $\mathsf{ch}(\rho(x),\rho(y))$ holds in $\mathcal{S}$, (iii) $\psi \models_\rho \mathsf{ns}(x,y)$ if $\mathsf{ns}(\rho(x),\rho(y))$ holds in $\mathcal{S}$.

A set of hedges S is *MSO -definable* if there is an MSO sentence ψ such that $S = \{h \mid h \models \psi\}$. It is well-known that a language is accepted by some hedge automata iff it is MSO-definable.

3 The Tree Query Logic

We consider here a fragment of the TQL logic defined in [7] and adapt it to unranked ordered trees.

Syntax. We assume a countable set $\mathcal{T}$ of tree variables ranging over by X, Y, and a countable set $\mathcal{R}$ of recursion variables ranging over by ξ. Let α be a finite or co-finite set of labels from Λ. Formulas ϕ from TQL are given by the syntax on Figure 1(a). We allow cofinite sets in extensions, otherwise we could not express formula $\Lambda[0]$.

We assume that μ is the binder for recursion variables and the notions of bound and free variables are defined as usual. To ensure the existence of fixpoint, we will assume that in formulas $\mu\xi.\phi$, the recursion variable ξ occurs under an even number of negations. A formula is said to be *recursion-closed* if all the occurrences of its recursion variables are bound. A TQL *sentence* is a recursion-closed formula that does not contain tree variables. A TQL formula ϕ is *guarded* if for any of its subformula $\mu\xi.\phi'$, the variable ξ occurs in the scope of some extension operator $\alpha[\,]$ in ϕ'.

We assume from now on that recursion variables are bound only once in formulas and denote recvar(ϕ) (resp. var(ϕ)) the set of recursion variables (resp. tree variables) occurring in ϕ. We may write $a[\phi]$ instead of $\{a\}[\phi]$.

Semantics. Interpretation maps a TQL formula to a set of hedges. Let ρ be an assignment of tree variables into $\mathbb{T}_\Lambda$ and δ be an assignment of the recursion variables into sets of hedges. The interpretation of the formula ϕ under ρ and δ, denoted by $[\![\phi]\!]_{\rho,\delta}$ is inductively defined and given on Figure 1(b).

Examples. Let us consider the following formulas:

$$\phi = a[\top]|\top \quad (1)$$
$$\phi_s = \mu\xi.(a[\top]|\xi \vee 0) \quad (2)$$
$$\phi_{dtd} = (employee[name[\Lambda[0]] \mid dpt[\Lambda[0]] \mid manager[\Lambda[0]]])^* \quad (3)$$
$$\phi_{dd} = \phi_{dtd} \wedge \top \mid employee[X] \mid \top \mid employee[X] \mid \top \quad (4)$$
$$\phi_{path(a),0} = \mu\xi.((\top|a[\xi]|\top) \vee 0) \quad (5)$$

The above formula ϕ is interpreted as the set of hedges of length at least 1, such that the root of the first tree is labelled a. The formula ϕ_s is interpreted as the set of hedges $\{a[h_1]|\dots|a[h_n] \mid h_i \in \mathbb{H}_\Lambda, n \geq 0\}$. The formula ϕ_{dtd} is interpreted as the set of hedges defining employees by their name, the department they work in, and their manager whereas the formula ϕ_{dd} expresses that an employee occurs twice in the database. Finally, the models of the formula $\phi_{path(a),0}$ are hedges with a path labelled by as from one of the roots to some leaf (*i.e.* the empty hedge 0).

The formula ϕ_{odd} is interpreted as the set of hedges having an odd number of nodes:

$$\phi_{odd} = \mu\xi_o.(\Lambda[0] \vee \Lambda[\phi_{even}(\xi_o)]|\phi_{even}(\xi_o) \vee \Lambda[\xi_o]|\xi_o$$
$$\text{where } \phi_{even}(\xi_o) = \mu\xi_e.(0 \vee (\Lambda[\xi_o]|\xi_e \vee \Lambda[\xi_e]|\xi_o))$$

Let us denote $\phi_{path(L),\psi} = \mu\xi.((\top|L[\xi]|\top) \vee \psi)$ the formula whose models are hedges containing a path labelled by elements from L from one of the roots to a hedge satisfying ψ. The models of the following formula are the hedges having two non-empty paths labelled respectively by as and by bs from two of the top-level nodes, those two paths leading to some identical subtrees

$$\top \mid (a[\phi_{path(a),X}]|\top|b[\phi_{path(b),X}] \vee b[\phi_{path(b),X}]|\top|a[\phi_{path(a),X}])| \top$$

The formula $\phi_{id\_not\_key}$ is interpreted as the set of hedges for which two nodes labelled id have identical subtrees (roughly speaking "the (data) value of the element `id` can not be used as a key in this XML document")

$$\phi_{id\_not\_key} = \top|\Lambda[\phi_{path(\Lambda),id[X]}]|\top|\Lambda[\phi_{path(\Lambda),id[X]}]|\top$$

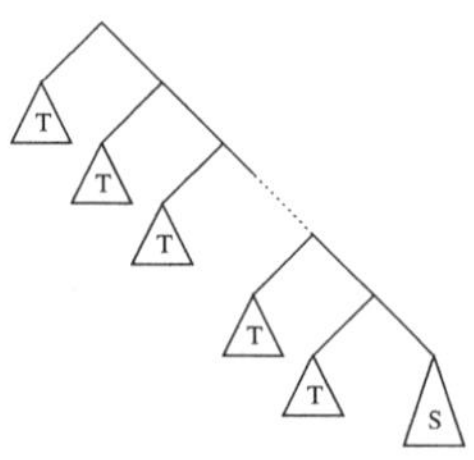

Fig. 2. A tree with $T \neq S$

The following formula states that two trees $employee$ have identical subtrees rooted by dpt but different subtrees rooted by $manager$

$$\phi_{dtd} \wedge \top \mid employee[\top \mid dpt[X]|manager[Y]] \mid \top$$
$$\mid employee[\top \mid dpt[X]|manager[\neg Y]] \mid \top$$

A hedge satisfying this formula may be considered as ill-formed assuming the existence of some functional dependency stating that *department* has only one *manager*.

Models of the formula ϕ_{branch} are the trees whose shapes are described on Figure 2.

$$\phi_{branch} = a[X|\mu\xi.(a[X \mid \xi] \vee \neg X \wedge \Lambda[\top])]$$

ϕ_s and ϕ_{odd} are the only two formulas that are not guarded; however, ϕ_s is equivalent to $a[\top]^*$, which is guarded and ϕ_{odd} to the following guarded formula

$$\phi_{odd} = \mu\xi_o.\Lambda[\phi_{even}(\xi_o) \vee (\xi_o|\phi_{even}(\xi_o)^*|\xi_o)^*]$$
$$\phi_{even}(\xi_o) = \mu\xi_e.\Lambda[\xi_e^*\xi_o|(\phi_{even}(\xi_o) \vee (\xi_o|\phi_{even}(\xi_o)^*|\xi_o)^*) \vee 0$$

But the formula $\mu\xi.(a[0]|\xi|b[0] \vee 0)$ is neither guarded nor equivalent to any guarded formula.

Definition 2 (satisfiability). *A recursion-closed TQL formula ϕ is satisfiable if there exists a hedge h and an assignment ρ such that $h \in [\![\phi]\!]_\rho$.*

TQL sentences. It is easy to prove that TQL sentences can describe sets of hedges that are not MSO-definable; for instance, the TQL sentence $\mu\xi.(a[0]|\xi|b[0] \vee 0)$ describes a "flat" hedge of the form $(a[0]^n b[0]^n)$ for $n \in \mathbb{N}$.

Theorem 1. *For any set of hedges S, there exists a guarded TQL sentence ϕ such that $[\![\phi]\!] = S$ iff S is MSO-definable.*

As a consequence of Theorem 4, satisfiability is decidable for guarded TQL sentences. This restriction is crucial, since, by reduction of emptiness problem for the intersection of two context-free grammars, one can prove that:

Theorem 2. *Satisfiability for TQL sentences is undecidable.*

Adding quantification As in [7], one could also consider quantification over tree variables $\exists X$ and $\forall X$ with the following semantics:

$$[\![\exists X.\phi]\!]_{\rho,\delta} = \bigcup_{t\in\mathbb{T}} [\![\phi]\!]_{\rho[X\to t],\delta} \qquad [\![\forall X.\phi]\!]_{\rho,\delta} = \bigcap_{t\in\mathbb{T}} [\![\phi]\!]_{\rho[X\to t],\delta}$$

where $\rho[X \to t]$ is the assignment identical to ρ except that it associates t with X.

Hence, the satisfiability problem from Definition 2 is equivalent to the one of closed formulas of the form $\exists X_1 \ldots \exists X_n \phi$ where ϕ is a recursion-closed TQL formula, *i.e.* the existence of a tree satisfying this formula.

For more complicated alternation of quantifiers, one can easily adapt the proof from [8] about the undecidability of the fragment of TQL without iteration, recursion and tree variable but with quantification over labels to prove that

Theorem 3. *Satisfiability for recursion-closed TQL formulas with quantification is undecidable (this holds even for recursion-free formulas).*

Bounded TQL formulas. In bounded formulas, variables can occur in recursion only in a restricted manner: intuitively a formula is bounded if there exists a bound on the number of equality test performed to (non-deterministically) verify that a hedge is a model of the formula. Boundedness appears naturally in unification problems and in pattern languages (where variables appears a bounded number of times in terms or patterns). But defining boundedness in the presence of recursion is a bit more technical.

In the examples we have given so far, the only formula that is not bounded is ϕ_{branch}. The following formula is also not bounded as it asks each subtree of the hedge to be different from X: $\neg(\mu\xi.\top|(X \vee \Lambda[\xi])|\top)$

We let β be a recursion-closed formula s.t. no recursion variable is bound twice, and denote by ϕ_ξ the formula s.t. $\mu\xi.\phi_\xi$ is a subformula of β. To define *boundedness* formally, we introduce, for every subformula ϕ of β, the *generalized free variables* of ϕ, denoted $\text{var}^*_\beta(\phi)$, as the least solution of the following (recursive) equations:

$$\text{var}^*_\beta(0) = \text{var}^*_\beta(\top) = \varnothing \quad \text{var}^*_\beta(a[\phi]) = \text{var}^*_\beta(\neg\phi) = \text{var}^*_\beta(\mu\xi.\phi) = \text{var}^*_\beta(\phi^*) = \text{var}^*_\beta(\phi)$$
$$\text{var}^*_\beta(\xi) = \text{var}^*_\beta(\phi_\xi) \quad \text{var}^*_\beta(X) = \{X\} \quad \text{var}^*_\beta(\phi \vee \phi) = \text{var}^*_\beta(\phi|\phi') = \text{var}^*_\beta(\phi) \cup \text{var}^*_\beta(\phi')$$

Note that the least solution of these equations is computable. An operator occurs positively (resp. negatively) in a formula if it occurs under an even (resp. odd) number of negations.

A formula β is *bounded* if it satisfies the following properties:

1. for any subformula ϕ^*, $\text{var}^*_\beta(\phi) = \varnothing$;
2. for any subformula $\phi|\phi'$ where $|$ occurs negatively, $\text{var}^*_\beta(\phi) = \text{var}^*_\beta(\phi') = \varnothing$.
3. each formula $\phi|\phi'$ where $|$ occurs positively and each formula $\phi \vee \phi'$ where $\vee$ occurs negatively satisfy $\forall\xi \in \text{recvar}(\phi)$, $\text{var}^*_\beta(\xi) \cap \text{var}^*_\beta(\phi') = \varnothing$, and $\forall\xi' \in \text{recvar}(\phi')$, $\text{var}^*_\beta(\xi') \cap \text{var}^*_\beta(\phi) = \varnothing$.

As a consequence of Theorem 1, this fragment is strictly more expressive than guarded TQL sentences, as it allows for tree isomorphism tests. Combining Theorems 6 and 5 of the next two sections proves our main result:

Theorem 4. *Satisfiability is decidable for bounded guarded TQL formulas.*

Remarks Our logic can be seen as an extension of the (recursive) pattern-language of XDuce [13]. The main difference here is that we allow Boolean operators and drop the *linear* condition for variables of XDuce. The pattern-matching mechanism of CDuce [1] extends the one from XDuce with Boolean operations and weaker conditions on variables. However, no equality tests between terms can be performed making our logic more powerful. Since we consider an infinite alphabet and we allow equality tests between trees, we can, as a side effect, simulate data values as done in some of the examples we gave.

4 Tree Automata with Global Equalities and Disequalities

In this section, we present a new extension of hedge automata (called TAGED) with the ability to test tree equalities or disequalities globally on the run. We prove decidability of emptiness for a particular class of TAGED, called bounded TAGED, which we use to decide satisfiability of bounded TQL formulas.

4.1 Definitions

Definition 3 (TAGED). *A tree automaton with global equalities and disequalities (TAGED) is a 6-tuple* $A = (\Lambda, Q, F, \Delta, =_A, \neq_A)$ *such that:*

- (Λ, Q, F, Δ) *is a hedge automaton;*
- $=_A$ *is an equivalence relation on a subset of* Q*;*
- $\neq_A$ *is a non-reflexive symmetric binary relation on* Q*;*

A TAGED is *positive* if $\neq_A = \varnothing$, and is simply denoted by $A = (\Lambda, Q, F, \Delta, =_A)$. The set $\{q \mid \exists q' \in Q,\ q =_A q'\}$ is denoted by E, and for all states $q \in E$, we denote by $[q]$ its equivalence class. The set $\{q \mid \exists q' \in Q,\ q \neq_A q'\}$ is denoted by D. The notion of run differs from hedge automata as we add equality and disequality constraints.

Definition 4 (runs). *Let* $A = (\Lambda, Q, F, \Delta, =_A, \neq_A)$ *be a TAGED. A run* r *of the hedge automaton* (Λ, Q, F, Δ) *on a hedge* h *satisfies the equality constraints if the following holds:* $\forall i \in \{1, \ldots, n\}$, $\forall u, v \in \mathtt{nodes}(h)$, $\mathtt{lab}_r(u) =_A \mathtt{lab}_r(v) \implies h|_u = h|_v$.

Similarly, the run r *on* h *satisfies the disequality constraints if the following holds:* $\forall i \in \{1, \ldots, n\}$, $\forall u, v \in \mathtt{nodes}(h)$, $\mathtt{lab}_r(u) \neq_A \mathtt{lab}_r(v) \implies h|_u \neq h|_v$.

The set of accepting runs of A *on* h*, denoted* $R_A^{acc}(h)$*, is the set of accepting runs* r *of* (Λ, Q, F, Δ) *which satisfy the equality and disequality constraints. The language accepted by* A*, denoted* $L(A)$*, is the set of hedges* h *such that* $R_A^{acc}(h) \neq \varnothing$.

Remark that $L(A)$ is not necessarily regular, as illustrated by Example 1.

Example 1. Let Λ be an infinite alphabet. Let $Q = \{q, q_X, q_f\}$, $F = \{q_f\}$, and let Δ be defined as the set of following rules: $\Lambda(q^*) \to q \quad \Lambda(q^*) \to q_X \quad a(q_X q_X) \to q_f$ Let A_1 be the positive TAGED $(\Lambda, Q, F, \Delta, \{q_X =_{A_1} q_X\})$. Then $L(A_1)$ is the set $\{a(t|t) \mid a \in \Lambda, t \in \mathbb{T}_\Lambda\}$, which is known to be non regular [10].

Example 2. Let $Q = \{q, q_X, q_{\overline{X}}, q_f\}$, $F = \{q_f\}$, and let Δ defined as the set of following rules:

$$\Lambda(q^*) \to q_{\overline{X}} \qquad \Lambda(q^*) \to q \qquad \Lambda(q^*) \to q_X \qquad a(q_X(q_{\overline{X}} + q_f)) \to q_f$$

Let A_2 be the TAGED $(\Lambda, Q, F, \Delta, \{q_X =_{A_2} q_X\}, \{q_X \neq_{A_2} q_{\overline{X}}\})$. Then $L(A_2)$ is the set of hedges whose shapes are described on Figure 2.

Remarks. Extensions of tree automata which allow for syntactic equality and disequality tests between subterms have already been defined by adding constraints to automaton rules. E.g., adding the constraint $1.2 = 2$ to a rule means that one can apply the rule at position π only if the subterm at position $\pi.1.2$ is equal to the subterm at position $\pi.2$. Testing emptiness of the recognized language is undecidable in general [16] but two classes with a decidable emptiness problem have been emphasised. In the first class, automata are deterministic and the number of equality tests along a path is bounded [11] whereas the second restricts tests to sibling subterms [2]. This latter class has recently been extended to unranked trees [15], the former one has been extended to equality modulo equational theories [14]. But, contrarily to TAGED, in all these classes, tests are performed locally, typically between sibling or cousin subterms. Finally, automata with local and global equality tests, using one memory, have been considered in [9]. Their emptiness problem is decidable, and they can simulate positive TAGEDs which use at most one state per runs to test equalities, but not general positive TAGEDs.

Definition 5 (bounded TAGED). *A bounded TAGED is a 7-tuple $A = (\Lambda, Q, F, \Delta, =_A, \neq_A, k)$ where $A' = (\Lambda, Q, F, \Delta, =_A, \neq_A)$ is a TAGED and $k \in \mathbb{N}$ is a natural. An accepting run r of A on a tree t is an accepting run of A' on t such that the following is true: $|\{u \mid \mathtt{lab}_r(u) \in E \cup D\}| \leq k$.*

We say that A and its accepting runs are k-bounded and may write (A', k) instead of A. We say that a TAGED B is equivalent to a bounded TAGED A if $L(A) = L(B)$.

The TAGED of Example 1 is equivalent to the 2-bounded TAGED $(A_1, 2)$, whereas one can prove that the one of Example 2 is not equivalent to any bounded TAGED.

Theorem 5 (emptiness of bounded TAGED). *Let A be a bounded TAGED. It is decidable to know whether $L(A) = \varnothing$ holds or not.*

The rest of this subsection is dedicated to the proof of this theorem, first for positive bounded TAGED, then for full bounded TAGED.

4.2 Configurations

We define a tool called *configurations* used to decide emptiness of positive bounded TAGED. In this subsection, the 6-tuple $A = (\Lambda, Q, F, \Delta, =_A, k)$ always denotes a positive bounded TAGED. We suppose that A accepts trees only, i.e. $F \subseteq Q$. It is not difficult to adapt the decidability result to hedge acceptors. Moreover, we suppose that for any run r -even non accepting- on a tree, the cardinal of the set of nodes labelled by states of E is at most k. Indeed, it is easy to transform A to ensure this property by enriching states with a counter up to k. We show how to decompose a positive TAGED into an equivalent and computable finite set of configurations. Since testing emptiness of configurations is easily decidable, we get the decidability result for positive TAGED. Informally, configurations are (non-regular) tree acceptors which make explicit parent or ancestor relations between nodes for which equality tests are performed by A. These are DAG-like structures labelled by sets of states of A. A tree t is accepted by some configuration c if the unfolding of c can be embedded into a run r of A on t, such that labels of r belong to labels of c. By putting suitable rules on how sets of states occur as labels of c, we can enforce r to respect equality constraints.

Definition 6 (configurations). *A configuration c of A is a rooted directed acyclic graph such that: (i) every node carries a symbol from 2^Q, (ii) outgoing edges of a node are ordered, and (iii) for every equivalence class $[q]$ of $=_A$, there is at most one set $P \subseteq Q$ such that $[q] \cap P \neq \varnothing$ and P is a label of c.*

Nodes of c are denoted by $\mathtt{nodes}(c)$. For every node $u \in \mathtt{nodes}(c)$, we denote by $c|_u$ the subgraph of c induced by the nodes reachable from u in c, and by $u_1, \ldots, u_n$ the n successor nodes of u given in order. Note that it might be the case that $u_i = u_j$ for some $i, j \in \{1, \ldots, n\}$. Finally, we always denote by $\mathtt{lab}_c(u) \subseteq Q$ the label of node u in c. Fig. 3 illustrates a configuration whose nodes are labelled either by set of states $\{q, q'\}$, $\{s\}$, $\{p\}$ or $\{r\}$. Note that the second successor of the root is the node labelled $\{r\}$, while its first and third successor is the node labelled $\{q, q'\}$.

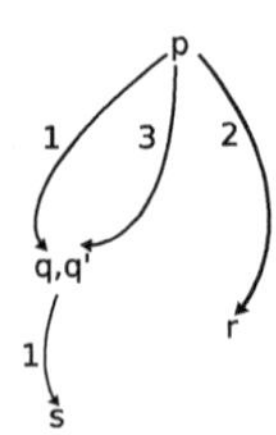

Fig. 3. A configuration (naturals represent the order on edges)

In order to define semantics of configurations, we first introduce some useful notions about contexts. For $n \geq 0$, we define n-ary contexts Cs as usual, and the hole substitution with trees $t_1, \ldots, t_n$ is denoted by $C[t_1, \ldots, t_n]$. Note that 0-ary contexts are just trees. See [10] for a formal definition. Given n states $p_1, \ldots, p_n \in Q$ and an n-ary context C, we denote by $C[p_1, \ldots, p_n]$ the tree over $\Lambda \cup Q$, where each p_i is viewed as a constant symbol. We let A^* be the TAGED over alphabet $\Lambda \cup Q$ which is just A with additional rules $q(\epsilon) \to q$, for every $q \in Q$. We say that $C[p_1, \ldots, p_n]$ evaluates to p, denoted $C[p_1, \ldots, p_n] \to_A p$ if there is a run r of A^* on $C[p_1, \ldots, p_n]$ such that $\text{roots}(r) = p$. For any set $S \subseteq Q$, we write $C[p_1, \ldots, p_n] \to_{Q \setminus S} q$ if states from S does not occur in r, except at the root and at the leaves labelled $p_1, \ldots, p_n$.

We now view configurations as a way to combine contexts to form trees t, with additional requirements which enforce existence of a run r of A on t. Condition (iii) of Definition 6 ensures r satisfies the equality constraints. This motivates the semantics of configurations. More formally, let c be a configuration of A. A mapping λ from $\texttt{nodes}(c)$ into contexts over Λ is an *interpretation* of c if for every node $u \in \texttt{nodes}(c)$, if u has exactly n successor nodes $u_1, \ldots, u_n$ in c, then $\lambda(u)$ is an n-ary context. Moreover, λ must satisfy the following: for every node $u \in \texttt{nodes}(c)$ and every nodes $u_1, \ldots, u_n \in \texttt{nodes}(c)$, if $u_1, \ldots, u_n$ are the successor nodes of u then the following holds (called condition (P)):

$$\forall p \in \texttt{lab}_c(u),\ \exists p_1 \in \texttt{lab}_c(u_1) \ldots \exists p_n \in \texttt{lab}_c(u_n),\ \lambda(u)[p_1, \ldots, p_n] \to_{Q \setminus (E \cup D)} p$$

As A is positive, the set D is empty, but we keep this definition for uniformity reasons when dealing with disequalities. Every node $u \in \texttt{nodes}(c)$, together with the mapping λ, define a tree $t(u, \lambda)$ over Λ as follows: $t(u, \lambda) = \lambda(u)[t(u_1, \lambda), \ldots, t(u_n, \lambda)]$, where $n \in \mathbb{N}$ and $u_1, \ldots, u_n$ are the successor nodes of u in c. Note that this definition is well-founded since c is a DAG. As we will see, condition (P) ensures the existence of a run of A on $t(u_0, \lambda)$, where u_0 is the root of c.

Definition 7 (tree language recognized by a configuration). *Let c be a configuration of A. The tree language recognized by c, denoted $L(A, c)$, is defined by the set of trees $t(u_0, \lambda)$, where u_0 is the root of c, and λ is an interpretation of c.*

Trees accepted by configuration of Fig. 3 are necessarily of form $C[C'[t], t', C'[t]]$, for some contexts C, C', C'' and trees t, t'. As already said, the constraints on the contexts and the configuration ensure the existence of a run on the trees of $L(A, c)$ which satisfies the equality constraints. In particular, we can prove the following:

Proposition 1. *Let c be a configuration of A such that $L(A, c)$ is nonempty. Let t be a tree of $L(A, c)$, and $u_0 \in \texttt{nodes}(c)$ be the root of c. For every $p \in \texttt{lab}_c(u_0)$, there is a run $r \in R_{A,p}(t)$ which respects the equality constraints.*

The converse holds too, and we can bound the size of configurations:

Proposition 2. *Let $t \in \mathbb{T}_\Lambda$ be a tree such that $t \in L(A)$. Then there is a configuration c of size at most $|Q|.k^{|Q|}$ such that $t \in L(A, c)$.*

Sketch of proof We start from an accepting run r of A on t and define an equivalence relation on a subset of $\texttt{nodes}(t)$. Informally, two nodes u, v are equivalent if an equality test between $t|_u$ and $t|_v$ is performed in r. This is the case for instance when $\mathtt{lab}_r(u) =_A \mathtt{lab}_r(v)$. Each equivalence class will represent a node of c, to enforce equalities. □

Hence, we can finitely represent the language recognized by A as a computable set of configurations of A, as stated below:

Corollary 1. *Let A be a k-bounded positive TAGED. We let $\mathcal{D}(\mathcal{A})$ be the set of configurations of A whose sizes are bounded by $|Q|.k^{|Q|}$. The following holds:*

$$L(A) = \bigcup_{c \in \mathcal{D}(\mathcal{A})} L(A, c)$$

Moreover, we can decide emptiness of the language recognized by any configuration.

Lemma 1. *Given a configuration c, it is decidable to know whether $L(A, c) = \varnothing$ holds.*

Proof (Sketch). For all nodes $u, u_1, \dots, u_n$ s.t. $u_1, \dots, u_n$ are the successors of u, it suffices to test whether there is an n-ary context C s.t. for all state $p \in \mathtt{lab}_c(u)$, there are $p_1 \in \mathtt{lab}_c(u_1)$,...,$p_n \in \mathtt{lab}_c(u_n)$ s.t. $C[p_1, \dots, p_n] \rightarrow_{Q \backslash E} p$. We can represent the set of contexts C such that $C[p_1, \dots, p_n] \rightarrow_{Q \backslash E} p$ by a tree automaton $A_{(p_i)_i, p}$. Then, it suffices to test emptiness of $\bigcap_{p \in P} \bigcup_{(p_i)_i \in \prod_i \mathtt{lab}_c(u_i)} L(A_{(p_i)_i, p})$, which is decidable, since regular languages are closed by Boolean operations.

As a corollary of Lemma 1 and Corollary 1, we get the following:

Proposition 3. *Emptiness of positive bounded TAGED is decidable.*

4.3 Adding Disequalities to Positive Bounded TAGED

In the previous section, we have shown that emptiness of positive bounded TAGED is decidable. In this section, we extend this result to full bounded TAGED. A always denotes a k-bounded TAGED. The definition of configurations of A remains unchanged, and the set $\mathcal{D}(\mathcal{A})$ still denotes the set of configurations of A whose size is bounded by $|Q|.k^{|Q|}$. We have the following inclusion: $L(A) \subseteq \bigcup_{c \in \mathcal{D}(\mathcal{A})} L(A, c)$, but the other one does not hold in general, since configurations do not require disequality constraints to be satisfied. We show how an additional test on configurations c allows us to decide whether $L(A) \cap L(A, c)$ is empty, which will be sufficient to decide whether $L(A)$ is empty. Informally, let c be a configuration and λ be an interpretation of c. We say that λ satisfies the disequalities of c if for all nodes $u, v \in \mathtt{nodes}(c)$, if there are $p \in \mathtt{lab}_c(u)$ and $q \in \mathtt{lab}_c(v)$ such that $p \neq_A q$, then $t(u, \lambda) \neq t(v, \lambda)$.

We now relate the problem of finding such an interpretation to context disunification. For all nodes $u \in \mathtt{nodes}(c)$, we let $\mathrm{cxt}_c(u)$ be the set of contexts satisfying condition (P) of the definition of interpretation. We define the notion of *partial interpretation* β of c, as a mapping from $\mathtt{nodes}(c)$ into contexts, such that it maps every node u such that $\mathrm{cxt}_c(u)$ is finite into a context of $\mathrm{cxt}_c(u)$, and every other node v to a context $@_v(\bullet, \dots, \bullet)$ with n holes (if v has n successors), where $@_v$ is a fresh symbol such that $@_v \notin \Lambda$. Note that trees $t(u, \beta)$ are trees over alphabet $\Lambda \cup \{@_v \mid v \in \mathtt{nodes}(c)\}$. We can show the following, by using context disunification (symbols $@_v$ are viewed as context variables):

Lemma 2. *Let A be a bounded TAGED. We have $L(A) \neq \varnothing$ iff there exist a configuration $c \in \mathcal{D}(\mathcal{A})$ and a partial interpretation β of c such that β satisfies the disequality constraints of c. Moreover, it is decidable to know whether such an interpretation β exists.*

As a corollary, by combining Lemma 2 and Lemma 1, we get the proof of Theorem 5.

5 From TQL to Automata

In this section, ϕ denotes a recursion-closed and guarded TQL formula over tree variables $X_1, \ldots, X_n$. We sketch the construction of a TAGED A_ϕ such that for all hedges $h \in \mathbb{H}_\Lambda$, we have $h \in L(A_\phi)$ iff there exists a valuation $\rho : \text{var}(\phi) \to \mathbb{T}_\Lambda$ such that $h \in [\![\phi]\!]_\rho$. Moreover, we prove A_ϕ to be equivalent to a computable bounded TAGED whenever ϕ is bounded.

In a first step, we transform ϕ into an equivalent system of fixpoint equations, and then sketch the construction of A_ϕ starting from this system. This construction extends the construction of [4]. This extension is non-trivial, since it manages tree variables, which induce non-determinism in the produced tree automaton. Moreover, even for sentences, this construction is different, since trees are ordered.

System of equations. We define dual connectives for parallel composition and Kleene star. We let $\phi_1||\phi_2$ as a shortcut for $\neg(\neg\phi_1|\neg\phi_2)$, $\phi_1^\diamond$ for $\neg(\neg\phi_1)^*$ and $\overline{X}$ for $\neg X \wedge \Lambda[\top]$. A *system of fixpoint equations* Σ is a sequence of equations $\xi_1 = \mathsf{rhs}_1, \ldots, \xi_n = \mathsf{rhs}_n$ where every rhs_i has one of the following form:

$$\mathsf{0} \quad \overline{\mathsf{0}} \quad \xi \vee \xi' \quad \xi \wedge \xi' \quad \alpha[\xi] \quad \xi|\xi' \quad \xi||\xi' \quad X \quad \overline{X} \quad \xi^* \quad \xi^\diamond$$

The last fixpoint variable, ξ_n, is denoted by last(Σ). The set of tree variables occurring in Σ is denoted by var(Σ). Systems of equations are interpreted over the complete lattice $(2^{\mathbb{H}_\Lambda}, \bigcup, \bigcap)$, modulo an assignment $\rho : \text{var}(\Sigma) \to \mathbb{T}_\Lambda$. We consider the following monotonic operations over $2^{\mathbb{H}_\Lambda}$, modulo ρ: $\mathsf{0}$ is interpreted as $\{\mathsf{0}^{\mathbb{H}_\Lambda}\}$, $\overline{\mathsf{0}}$ as $\overline{\{\mathsf{0}^{\mathbb{H}_\Lambda}\}}$ (the overline denotes the complement in $\mathbb{H}_\Lambda$), $\vee$ by $\cup$, $\wedge$ by $\cap$, $\alpha[.]$ as the unary operator which maps any set of hedges $H \subseteq \mathbb{H}_\Lambda$ into $\{a[h] \mid a \in \alpha, h \in H\}$, $.|.$ as the binary operator which maps two sets of hedges H, H' into $H|H' = \{h|h' \mid h \in H, h' \in H'\}$, its dual $.||.$ maps H, H' into $H||H' = \overline{\overline{H}|\overline{H'}}$. The Kleene star $.^*$ and its dual $.^\diamond$ are interpreted similarly. Finally, X is interpreted by $\rho(X)$ and $\overline{X}$ by $\mathbb{T}_\Lambda \backslash \{\rho(X)\}$.

The solution of Σ over $(2^{\mathbb{H}_\Lambda}, \bigcup, \bigcap)$ modulo ρ is a mapping from fixpoint variables of Σ into $2^{\mathbb{H}_\Lambda}$, and is denoted by $\text{Sol}_{\mathbb{H}}(\Sigma, \rho)$. We can push down the negations to the leaves of ϕ, using the dual connectives, and introduce fixpoint variables for every position in ϕ, which allow to construct a system of equations S_ϕ such that var$(S_\phi) =$ var(ϕ) and the following holds:

Lemma 3. *For all valuations $\rho : \text{var}(S_\phi) \to \mathbb{T}_\Lambda$, we have* $[\![\phi]\!]_\rho = \text{Sol}_{\mathbb{H}}(S_\phi, \rho)(\text{last}(S_\phi))$

E.g., the equation system associated with the formula $\mu\xi.(a[\xi] \vee (\mu\xi'.(b[\xi'] \vee X)))$ is $\{\xi' = \xi_2 \vee \xi_3;\ \ \xi_2 = b[\xi'];\ \ \xi_1 = a[\xi];\ \ \xi_3 = X;\ \ \xi = \xi_1 \vee \xi'\}$.

Ideas of automaton construction for sentences. In this paragraph, we assume that ϕ is a sentence. Checking whether a hedge h is a solution of S_ϕ is similar to a run of some hedge automaton on h. Let us consider the system $S = \{\xi = \xi_1 \vee \xi_2; \xi_1 = a[\xi]; \xi_2 = 0\}$, where the last variable is ξ. Solutions of S are chains labelled by as. To check whether hedge $a(0)$ is a solution of ξ, first verify that $a(0)$ is a solution of ξ_1 or a solution of ξ_2. One can easily see that $a(0)$ is not a solution of ξ_2. It remains to verify whether $a(0)$ is a solution of $\xi_1 = a[\xi]$, by verifying that 0 is a solution of ξ, etc... This can be done by an automaton with transition rules $a(\epsilon + q_{a[\xi]}) \to q_{a[\xi]}$, where $q_{a[\xi]}$ is a final state. We define the set of (final) states by $Q = F = \{q_{a[\xi]}\}$. Let us interpret S over 2^{Q^*}, where Q^* is the set of words over Q. We interpret $\vee$ by $\cup$, 0 by $\{\epsilon\}$, and $a[\xi]$ by $\{q_{a[\xi]}\}$. Solutions of ξ are denoted by $s(\xi)$ (and similarly for other variables). Hence we get $s(\xi_2) = \{\epsilon\}$, $s(\xi_1) = \{q_{a[\xi]}\}$, and $s(\xi) = \{\epsilon, q_{a[\xi]}\}$. Which trees evaluate to $q_{a[\xi]}$? Trees of the form $a(h)$ where h evaluates to some word of states from $s(\xi)$. Hence, we can define transition $a(s(\xi)) \to q_{a[\xi]}$.

Things get more complicated when adding intersection. For instance, consider the system $S' = \{\xi_1 = a[\xi]; \xi_2 = a[\xi']; \xi' = 0; \xi = \xi_1 \wedge \xi_2\}$. If we interpret this system as before, with states $q_{a[\xi]}$ and $q_{a[\xi']}$, we would get $s(\xi) = \varnothing$. Hence, states should carry enough information to know if the current tree is a solution of $a[\xi]$, $a[\xi']$, or both. In the construction we provide, every state is a tuple of atoms of the form $\alpha[\xi]$, $\overline{\alpha}[\top]$, or $\alpha[\neg\xi]$, for every right-hand side of the form $\alpha[\xi]$ occuring in the system. If some tree t evaluates to a tuple which has a component equal to $\alpha[\xi]$, it means that t is of the form $a(h)$, where $a \in \alpha$ and h is solution of ξ. If the component is $\overline{\alpha}[\top]$ or $\alpha[\neg\xi]$, it means that $a(h)$ is not a solution of $\alpha[\xi]$, because, in the first case, $a \notin \alpha$, and in the second one, $a \in \alpha$, but h is not a solution of ξ. Knowing this complete information, i.e. which right-hand sides of the form $\alpha[\xi]$ are satisfied or not by the current tree, we are able to construct exactly one rule per state, by solving the system on sets of words of states, with suitable interpretations. As the formula is guarded, solutions of the system are regular state word languages. We then get a deterministic hedge automaton whose accepted trees are the solutions of the system.

Adding tree variables. When adding variables, we cannot keep the automaton deterministic, since subtrees will be captured non-deterministically. For instance, consider the system $S = \{\xi'' = X; \xi' = \xi''|\xi''; \xi_X = a[\xi']; \xi_2 = 0; \xi_3 = \xi_\top^*; \xi_1 = \Lambda[\xi_3]; \xi_\top = \xi_1 \vee \xi_2; \xi = \xi_\top \wedge \xi_X\}$, where the last variable is ξ. Given a valuation $\rho : \text{var}(S) \to \mathbb{T}_\Lambda$, the system S has a unique solution, modulo ρ, which is $a(\rho(X)|\rho(X))$. A TAGED accepting the solutions of S is the TAGED of Example 1 of Section 4. It is non-deterministic, since it has to choose to go in a state q_X which will enforce the TAGED to test whether the two sons of the root are equal.

As tree variables induce a kind of non-determinism, we emphasize two kinds of recursion variables: deterministic recursion variables, for which the problem of checking whether a given hedge is a solution does not involve tree variables, such as $\xi_\top$, ξ_1, ξ_2 and ξ_3 in our example; and non-deterministic one: ξ,ξ_X,ξ' and ξ''.

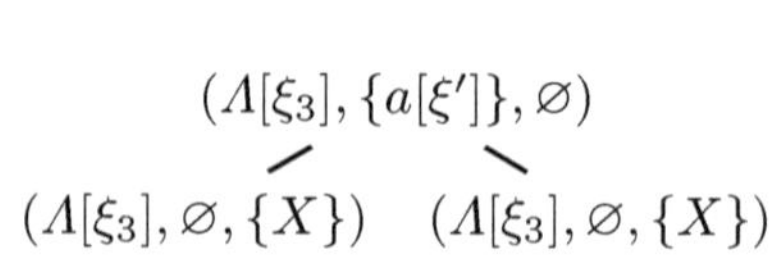

Fig. 4. Run of A_ϕ on $a(a(0)a(0))$

States of the automaton we construct have three components. The first component is induced by deterministic recursion variables and simulate a classical hedge automaton, as for the case of sentences. The second component is induced by non-deterministic recursion variables, and collects atoms of the form $\alpha[\xi]$, where ξ is non-deterministic, for which the current tree is the solution. In other words, it guesses the positions in the tree under which capture variables occur. Third components are sets of variables X or $\overline{X}$, meaning that equality or disequality tests are performed on the current tree.

Transition rules are obtained by suitable interpretation of the system over words of states. Finally, two states are equivalent for the automaton if their third component shares a tree variable. Disequalities are defined similarly. For instance, an accepting run of the automaton for S, on the tree $a(a(0)a(0))$, is represented in Fig. 4. The state $(\Lambda[\xi_3], \varnothing, \{X\})$ is equivalent to itself.

Theorem 6. *Let ϕ be a guarded TQL formula. There is a computable TAGED A_ϕ such that for all hedges h, we have $h \in L(A_\phi)$ iff there exists a valuation $\rho : \mathrm{var}(\phi) \to \mathbb{T}_\Lambda$ such that $h \in [\![\phi]\!]_\rho$.*

Moreover, if ϕ is bounded, the TAGED A_ϕ is equivalent to the bounded TAGED (A_ϕ, B), for some computable bound $B \in \mathbb{N}$.

To compute the bound B, we interpret the system S_ϕ on naturals, with suitable interpretations (for instance, X is interpreted by 1, $\wedge$ by $+$, and $\vee$ by max).

6 Extending MSO with Tree Isomorphism Tests

In this section, we propose an extension of MSO for unranked trees with isomorphism tests between trees.

Let $\sim$ be a binary predicate s.t. for a structure $\mathcal{S}^h$ associated with a hedge h and a mapping ρ from $\{x, y\}$ to nodes of h, $\mathcal{S}^h \models_\rho x \sim y$ holds if the two subtrees rooted at respectively $\rho(x)$ and $\rho(y)$ in h are isomorphic. We consider sentences of the form

$$Q_1 x_1 Q_2 x_2 \dots Q_n x_n \psi$$

where $Q_i \in \{\exists, \forall\}$ and ψ is an MSO formula extended with atoms $x_i \sim x_j$ $(1 \le i, j \le n)$. We call $MSO_\sim$ this logic. We will also consider the fragment $MSO_\sim^\exists$ for which formulas satisfy $Q_1 = Q_2 = \dots = Q_n = \exists$. Remark that ψ can again contain quantifiers. Hence, since tree isomorphism cannot be expressed in MSO [10], $MSO_\sim$ and $MSO_\sim^\exists$ are stricly more expressive than MSO. By reduction of the Post correspondence problem, we can prove that:

Theorem 7. *Satisfiability for $MSO_\sim$ is undecidable.*

However, $MSO_\sim^\exists$ and bounded TAGED are equally expressive: for any formula φ in $MSO_\sim^\exists$, one can compute a bounded TAGED, whose size is non-elementary in the size of φ, accepting the models of φ. The converse holds too. As a consequence of decidability of emptiness for bounded TAGED, we have:

Theorem 8. *Satisfiability is decidable for $MSO_\sim^\exists$.*

7 Conclusion

In this paper, we have considered the spatial logic TQL with tree variables. We have proved that for the guarded fragment when variables appear in a bounded way then the satisfiability problem is decidable. To do so, we have introduced a new class of tree automata, called TAGED, permitting to test global equalities and disequalities on the accepted trees. Finally, we have used TAGED to prove decidability for an extension of MSO with isomorphism tests interpreted over unranked trees.

We speculate that the boundedness condition is not required for the decidability of emptiness of TAGED, as pumping technics dealing with constraints may be applicable. However, it is non-trivial, since TAGED are not determinizable in general. This would imply that the full guarded TQL with trees variables is decidable. Another extension would be to consider hedge variables. This problem seems to be non trivial as the satisfiability problem for such formulas could encode word equations.

We emphasized a correspondence between $MSO^{\exists}_{\sim}$ and bounded TAGED. It would be interesting to exhibit a fragment of $MSO_{\sim}$ equivalent to full TAGED.

The TQL system also includes a transformation language; we aim at using TAGED automata to type these transformations and more generally, tree transducers.

References

1. Benzaken, V., Castagna, G., Frisch, A.: CDuce: an XML-centric general-purpose language. In: 8th ACM International Conf. on Functional Programming, pp. 51–63 (2003)
2. Bogaert, B., Tison, S.: Equality and disequality constraints on direct subterms in tree automata. In: Finkel, A., Jantzen, M. (eds.) STACS 92. LNCS, vol. 577, pp. 161–171. Springer, Heidelberg (1992)
3. Bojanczyk, M., David, C., Muscholl, A., Schwentick, T., Segoufin, L.: Two-variable logic on data trees and xml reasoning. In: ACM 25th Symp. on Principles of database systems, pp. 10–19 (2006)
4. Boneva, I., Talbot, J.M., Tison, S.: Expressiveness of a spatial logic for trees. In: 20th IEEE Symposium on Logic in Computer Science, pp. 280–289 (2005)
5. Brüggemann-Klein, A., Murata, M., Wood, D.: Regular tree languages over non-ranked alphabets. unpublished manuscript (1998)
6. Cardelli, L., Gardner, P., Ghelli, G.: A spatial logic for querying graphs. In: Widmayer, P., Triguero, F., Morales, R., Hennessy, M., Eidenbenz, S., Conejo, R. (eds.) ICALP 2002. LNCS, vol. 2380, pp. 597–610. Springer, Heidelberg (2002)
7. Cardelli, L., Ghelli, G.: TQL: A Query Language for Semistructured Data Based on the Ambient Logic. Mathematical Structures in Computer Science 14, 285–327 (2004)
8. Charatonik, W., Talbot, J.: The Decidability of Model Checking Mobile Ambients. In: Fribourg, L. (ed.) CSL 2001 and EACSL 2001. LNCS, vol. 2142, pp. 339–354. Springer, Heidelberg (2001)
9. Comon, H., Cortier, V.: Tree automata with one memory, set constraints and cryptographic protocols. TCS 331(1), 143–214 (2005)
10. Comon, H., Dauchet, M., Gilleron, R., Jacquemard, F., Lugiez, D., Tison, S., Tommasi, M.: Tree automata techniques and applications, 1997 (October 1st, 2002)
11. Dauchet, M., Caron, A.-C., Coquidé, J.-L.: Reduction properties and automata with constraints. 20, 215–233 (1995)

12. Dawar, A., Gardner, P., Ghelli, G.: Expressiveness and complexity of graph logic. In: Information and Computation, vol. 205, pp. 263–310 (2007)
13. Hosoya, H., Pierce, B.C.: XDuce: A statically typed xml processing language. ACM Trans. Internet Techn. 3(2), 117–148 (2003)
14. Jacquemard, F., Rusinowitch, M., Vigneron, L.: Tree automata with equality constraints modulo equational theories. Research Report LSV-06-07, LSV, ENS Cachan, France (2006)
15. Karianto, W., Löding, C.: Unranked tree automata with sibling equalities and disequalities. In: 34th International Colloquium on Automata, Languages and Programming (2007)
16. Mongy, J.: Transformation de noyaux reconnaissables d'arbres. Forêts RATEG. PhD thesis, Université de Lille (1981)
17. Murata, M.: Hedge automata: A formal model for xml schemata. Technical report, Fuji Xerox Information Systems (1999)
18. Neven, F., Schwentick, T., Vianu, V.: Towards regular languages over infinite alphabets. In: Sgall, J., Pultr, A., Kolman, P. (eds.) MFCS 2001. LNCS, vol. 2136, pp. 560–572. Springer, Heidelberg (2001)
19. Reynolds, J.C.: Separation logic: A logic for shared mutable data structures. In: 17th IEEE Symp. on Logic in Computer Science, pp. 55–74. IEEE, Los Alamitos (2002)

Forest Expressions

Mikołaj Bojańczyk*

Warsaw University

Abstract. We define regular expressions for unranked trees (actually, ordered sequences of unranked trees, called forests). These are compared to existing regular expressions for trees. On the negative side, our expressions have complementation, and do not define all regular languages. On the positive side, our expressions do not use variables, and have a syntax very similar to that of regular expressions for word languages.

We examine the expressive power of these expressions, especially from a logical point of view. The class of languages defined corresponds to a form of chain logic [5,6]. Furthermore, the star-free expressions coincide with first-order logic. Finally, we show that a concatenation hierarchy inside the expressions corresponds to the quantifier prefix hierarchy for first-order logic, generalizing a result of Thomas.

1 Introduction

We define a new type of regular expressions for forests, i.e. ordered sequences of unranked trees. Like word regular expressions and unlike the known tree regular expressions, our expressions do not use variables. Similar to CTL*, we have two sorts of expressions: one describing forests, and one describing contexts (forests with a hole, which can be used for substitution). The two sorts are defined by mutual recursion, and each allows concatenation, star, and boolean operations (including complementation and intersection). Forest expressions do not capture all regular languages of unranked forests (or even trees), but they do have a characterization in terms of logic:

Theorem 1. *Forest expressions have the same expressive power as extended chain logic. Star-free forest expressions have the same expressive power as first-order logic.*

In the above theorem, the models for the logic are forests. The signature has label tests, and two partial orders: vertical and horizontal. The vertical order corresponds to descendants, and the horizontal order is used to say when one sibling is to the left of another. Extended chain logic is a fragment of MSO, where set quantification is restricted to two types of sets: chains (sets linearly ordered by the vertical order) and sibling sets (sets linearly ordered by the horizontal order). Over binary trees, the second quantification is superfluous, and the logic collapses to chain logic as defined by Thomas in [5]. The second statement in

* Supported by Polish goverment grant no. N206 008 32/0810.

J. Duparc and T.A. Henzinger (Eds.): CSL 2007, LNCS 4646, pp. 146–160, 2007.

the above theorem can be seen as a tree extension of [3], where it is shown that star-free word expressions have the same expressive power as first-order logic.

Furthermore, we show that the correspondence between the quantifier alternation hierarchy, and the concatenation hierarchy, which has been shown for words by Thomas in [4], also works with our expressions:

Theorem 2. *A forest language is on level $n + 1/2$ if and only if it is defined by a sentence in Σ_n.*

This result is the second main contribution of this paper.

2 The Regular Expressions

In the formal syntax of expressions, there will be two sorts of expressions, one for forests and one fo contexts. Before we present this syntax in Definition 1, we will gradually introduce the operations allowed in the expressions.

We use $+$ to denote concatenation of forests (sequences of trees); while $\vee$ is used in the regular expressions for language union (likewise $\wedge, \neg$ for intersection and complement). For instance, both the expressions

$$a(a + a) \vee a(a) \qquad a((a + a) \vee a)$$

denote the same language containing two trees:

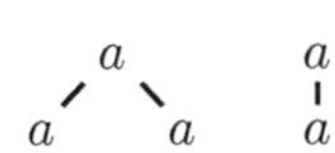

Concatenation of trees can be iterated using the Kleene star; for instance the expression

$$a((a \vee b)^*)$$

denotes all trees of depth at most two that have a in the root and a, b in the leaves. If we want to define languages of unbounded tree depth, we must also have some form of vertical iteration. Here, we use iteration of contexts, i.e. forests with a single hole. The hole is denoted by $\square$, and is used for substitution. For instance, the set of a-labeled trees that have a single leaf is defined by the expression:

$$(a\square)^*a \ .$$

Context composition is written multiplicatively, for instance $(aa\square)^*a$ denotes trees with a single path of odd length.

The kind of a subexpression can be determined by the way it is used in the larger expression: if it is an argument of $+$ then it must be a forest expression, if it is an argument of the multiplicative context concatenation, then it must be a context expression. The only possible ambiguity is when the expression does not contain forest concatenation or context composition, eg. $a \vee (\neg b)$; by convention we assume the expression describes forests and not contexts.

Even with the two Kleene stars (for forests and contexts); our expressions cannot define trees with a large amount of branching. In particular, the set of all trees cannot be defined. Therefore, as in star-free word languages, the key to success is judicious use of complementation. As usual with complementation, it is important that an alphabet A is specified beforehand; let us fix $A = \{a, b\}$ for the next several examples. For instance, $a(\neg\emptyset)$ defines the set of all trees over A that have a in the root (by convention, $\emptyset$ is the empty forest – and not context – language).

We have already discussed concatenation and Kleene star for forests and contexts, and the boolean operations. We have also implicitly attached forests to contexts, for instance in $(a\Box)^*a$ we have attached the forest a to the context $(a\Box)^*$. The last remaining operation is one that embeds a forest within a context; we use $+$ to add a context to a forest, with the result being a context. For instance,

$$\neg\emptyset + \Box + \neg\emptyset$$

denotes all contexts where the hole is at the root. In particular,

$$\big((a\Box \ \vee \ b\Box \ \vee \ (\neg\emptyset + \Box + \neg\emptyset)\big)^*$$

denotes all contexts. An equivalent expression would be $(\Box \wedge \ \neg\Box)$. Using this expression and complementation, we can define all contexts where only the letter a occurs.

Below is the formal definition of the expressions:

Definition 1. *The syntax of forest and context expressions over an alphabet A is defined by the following grammar:*

$$\mathcal{F} \to \emptyset \mid \epsilon \mid a \mid \mathcal{F}+\mathcal{F} \mid \mathcal{F}^* \mid \mathcal{C}\mathcal{F} \mid \mathcal{F}\vee\mathcal{F} \mid \mathcal{F}\wedge\mathcal{F} \mid \neg\mathcal{F}$$
$$\mathcal{C} \to \emptyset \mid \Box \mid a\Box \mid \mathcal{C}\mathcal{C} \mid \mathcal{C}^* \mid \mathcal{F}+\mathcal{C} \mid \mathcal{C}+\mathcal{F} \mid \mathcal{C}\vee\mathcal{C} \mid \mathcal{C}\wedge\mathcal{C} \mid \neg\mathcal{C}$$

In the above, a stands for any letter in the alphabet A. An expression is called star-free *if it does not use the productions $\mathcal{F}^*$ or $\mathcal{C}^*$.*

The semantics have already been defined in the discussion leading to the above definition. We only omitted ϵ, which stands for the empty forest. To avoid excess parentheses, we assume the following binding precedence order: context composition, $+$, $\neg$, $\wedge$ and finally $\vee$.

Note that Definition 1 contains some redundant rules. For instance, $F \wedge G$ is equivalent to $((F+\Box) \wedge (G+\Box))\epsilon$. Using similar methods, we can actually eliminate all rules on the right-hand side of $\mathcal{F} \to$, except $\mathcal{F} \to \epsilon$ and $\mathcal{F} \to \mathcal{C}\mathcal{F}$. However, we choose to keep all the rules to have a more convenient notation.

Before continuing with our discussion of these expressions, we would like to comment on complexity issues. Emptiness of a forest expression is EXPTIME hard, since emptiness of an alternating polynomial space Turing machine can be reduced to emptiness of a forest expression. A matching EXPTIME upper bound can be found by compiling an expression into a bottom-up automaton with an exponential state space. Membership, on the other hand, is polynomial: one can check in polynomial time if a given forest belongs to a given forest expression by using a dynamic algorithm.

2.1 Comparison with Existing Expressions

The regular expressions usually considered for trees, see eg. Chapter 2 of [2], are different from our expressions in that:

1. They use arbitrarily many types of hole $\square_1, \ldots, \square_n$ (instead of one $\square$).
2. The contexts can have holes of different types, and with multiple copies.
3. There is no negation or intersection.

Another difference is that we consider unranked trees instead of ranked trees, but the definitions from [2] can be adapted to the unranked setting, with similar results. The differences 1,2 can be seen as advantages of our expressions, while the difference 3 is a disadvantage. Another disadvantage is that our expressions cannot define all regular languages, as opposed to the standard expressions. For instance, the language "positive boolean formulas that evaluate to true" is not definable via our forest expressions (otherwise, this language would be definable in chain logic, which it is not), while it can be done using even with one type of hole, but with multiple copies of this hole. We present this expression below, to give the reader a taste of the expressions with many holes. We use the alphabet $\{OR, AND, 0, 1\}$ here to avoid confusion with $\vee, \wedge$ found in the syntax of expressions. The expression

$$E \quad = \quad (OR\square \;\vee\; AND\square \;\vee\; 0 \vee 1 \;\vee\; \square + \square)\ .$$

describes forests where a non-leaf node has label OR or AND, while a leaf has label $0, 1$ or possibly $\square$. Note that the hole $\square$ may appear in many leaves, or possibly no leaf at all. The following expression denotes contexts that take 1 to 1 (i.e. forests with many holes that will give a true formula if all holes are substituted by true boolean formula):

$$F^* \qquad \text{where } F = AND(\square^*) \;\vee\; OR((E \vee \square)^* + \square + (E \vee \square)^*)\ .$$

By induction on formula size, one can show that every true formula can be decomposed as F^*1.

It should be remarked here that our forest expressions are not meant to replace the standard regular expressions with many holes. We only claim that they have a convenient syntax and, most importantly, disallowing multiple holes gives a close correspondence to logic, as witnessed by Theorems 1 and 2.

3 Forest Automata

We denote forests by letters s, t, with forest concatenation denoted by $s + t$. We denote contexts by letters p, q, r, with context composition written multiplicatively as pq. The result of placing a forest s in the hole of a context p is denoted by ps. We use the standard relationships of nodes in a forest/context: descendant, ancestor, child, parent, sibling, leftmost sibling, rightmost sibling, left neighbor (closest sibling to the left), right neighbor.

Recall that a monoid is a set H along with an associative operation, which we denote here by $+$ to underline the connection with forest concatenation. Furthermore, each monoid has a unit element $\epsilon \in H$, which satisfies $\epsilon + h = h + \epsilon = h$ for all $h \in H$. We will denote monoids by F, G, H and their elements by f, g, h (there are two justifications here: first, we want to avoid confusion with trees s, t, u; second we want to be consistent with notation in forest algebra, see [1]). A *monoid forest automaton* is a deterministic word automaton where the states have a monoid structure. In other words, the automaton has two components: a finite monoid H and a function $\delta : A \times H \to H$, where A is the label alphabet. The automaton assigns to each forest t a unique element $\mathcal{A}(t)$ of H; called the *value* of the forest. The empty forest is assigned the unit of the monoid, while for larger forests we have:

$$\mathcal{A}(t + s) = \mathcal{A}(t) + \mathcal{A}(t) \qquad \mathcal{A}(a(t)) = \delta(a, \mathcal{A}(t)) \ .$$

In the above, the left $+$ is forest concatenation, while the right $+$ is the monoid operation. Thanks to associativity, the above definition is nonambiguous. The automaton recognizes a forest language L if membership in L only depends on the value of a forest. From now on, we will only be using monoid forest automata; and we will simply call them forest automata.

Let $\mathcal{A} = (H, \delta)$ be a forest automaton, and let $g, h \in H$. We say a context p takes g to h if there is some forest s with value g such that ps has value h. Note that in the above we could have equivalently written "for any forest with value h".

Definition 2. *We say a forest automaton $\mathcal{A} = (H, \delta)$ is* represented by expressions *if for all possible values $g, h \in H$*

- *There is an forest expression L_h describing forests with value h.*
- *There is a context expression L_g^h describing contexts p that take h to g.*

A forest language L is represented by expressions if some forest automaton recognizing L is. Clearly, a language represented by expressions can be defined by a forest expression: it suffices to take the union of L_m for all values m of forests in the language. Furthermore, the proof of Theorem 1 also gives the converse: if a language is defined by an expression, then its also represented by expressions.

4 Equivalence with First-Order Logic

When speaking of forest languages defined in first-order logic, we refer to the usual interpretation: a forest is considered as a model for logic, whose universe consists of nodes in the forest. A given formula can be true in such a model or not, and therefore each formula naturally induces a forest language. The signature contains a unary relation for each possible label, and two orders: the descendant order, and the sibling-to-the-left order. As far as first-order logic and extensions are concerned, the latter order could be replaced by the lexicographic order, without changing expressive power. Later on, in Section 6, we will be considering fragments where more care is needed when chosing the signature.

It is fairly easy to show the left to right inclusions from Theorem 1: extended chain logic and first-order logic can capture forest expressions and star-free expressions, respectively. We only give the proof of the more difficult right to left inclusions here.

Proposition 1. *Forest languages definable in first-order logic can be represented by expressions.*

The proof of this proposition is rather standard and proceeds by induction on the formula of first-order logic. Later on in the paper, we present a stronger result, however we include the below proof since it is significantly shorter.

To go through the induction step, we need to prove the statement also for formulas with free variables. A formula

$$\varphi(x_1, \ldots, x_n)$$

with free variables $x_1, \ldots, x_n$ can be seen as a forest language $L(\varphi)$ over an extended alphabet $A \times \{0,1\}^n$:

$$t \in L(\varphi) \qquad \text{iff} \qquad t, \nu \models \varphi \ ,$$

where ν is the valuation that assigns to x_i the unique node with 1 on the i-th $\{0,1\}$ coordinate. (If the forest has 0 or at least 2 nodes with 1 on a given coordinate, the valuation ν is undefined and therefore $t, \nu \models \varphi$ cannot hold.)

The proof of Proposition 1 is a rather standard induction on the size of the formula φ. The only nontrivial step is when passing from φ to $\exists x.\varphi$ (we eliminate $\forall$ using negation). Let then $\mathcal{A} = (H, \delta)$ be an automaton recognizing a forest language L over an alphabet $A \times \{0,1\}$. We need to show that if $\mathcal{A}$ is represented by star-free expressions, then so is an automaton recognizing the language

$$K = \{t : t[x] \in L \text{ for some node } x \text{ of } t\} \ .$$

In the above, $t[x]$ is the forest over over $A \times \{0,1\}$ obtained from t by adding label 1 on the second coordinate of the node x, and label 0 for the remaining nodes. We first define a forest automaton $\mathcal{B}$ that recognizes the language K; then we will show star-free expressions that represent this automaton. The value under $\mathcal{B}$ of a forest in t is a pair (h, G). The first coordinate is the value (in $\mathcal{A}$) of $t[\emptyset]$, while the second coordinate is the set of values (in $\mathcal{A}$) of forests $t[x]$ for all possible nodes x. The reader will easily fill in the monoid operation on states of $\mathcal{B}$, and its transition function.

We need to show that the automaton $\mathcal{B}$ is represented by star-free expressions. First, we will define some auxiliary star-free expressions. We begin by describing, for every $g, h \in H$, the forests and contexts where the node x is not present:

- K_h is a star-free forest expression describing forests t over A where $t[\emptyset]$ has value h.
- K_g^h is a star-free context expression describing contexts p over A where $p[\emptyset]$ takes g to h.

These star-free expressions are easily obtained from the star-free expressions representing the automaton $\mathcal{A}$, by writing $a \in A$ instead of $(a, 0), (a, 1)$ and then intersecting with a language that forbids labels of the form $(a, 1)$. Next, we write for each $g, h \in H$ star-free expressions that describe forests and context where the node x is present. These are defined by distinguishing the node x:

- L_h is star-free forest expression describing forests t over A where $t[x]$ has value h, for some node x of t. This is the union of star-free expressions

$$K_g^h a K_f$$

over $a \in A$ and $f, g \in H$ that satisfy $\delta((a,1), f) = g$
- L_g^h is a star-free context expression describing contexts p over A where $p[x](t)$ takes g to h. This is the union of star-free expressions

$$K_f^h a K_g^{f'}$$

over $a \in A$ and $f, f' \in H$ that satisfy $\delta((a,1), f') = f$.

Once the auxiliary star-free expressions have been found, the star-free expressions defining the automaton $\mathcal{B}$ can be easily obtained by using boolean combinations. The forests that have value (h, G) are described by the star-free expression:

$$L_h \quad \cap \quad \bigcap_{g \in G} K_g \quad \setminus \quad \bigcup_{g \notin G} K_g \ ,$$

while the contexts that take the automaton from value (f, F) to value (h, G) are described by the star-free expression:

$$L_f^h \quad \cap \quad \bigcap_{g \in G} (K_f^g \cup \bigcup_{g' \in F} L_{g'}^g) \quad \setminus \quad \bigcup_{g \notin G} (K_f^g \cup \bigcup_{g' \in F} L_{g'}^g) \ .$$

5 Equivalence with Chain Logic

Proposition 2. *Forest languages definable in extended chain logic can be represented by expressions.*

The proof is similar to the one in the previous section. We only comment on chain quantification, denoted here by $\exists^c X\ \varphi$; quantification over sibling sets is done in a similar manner.

Similar to the notation in Section 4, if t is a forest over A, and X is a set of nodes in X, then $t[X]$ is the forest over over $A \times \{0, 1\}$ obtained from t by adding label 1 on the second coordinate of nodes in X, and label 0 for the remaining nodes. If L is a forest language over $A \times \{0,1\}$, the *chain projection* is defined to be the set of forests t such that $t[X]$ belongs to L for some chain X. The language defined by $\exists^c X\varphi$ is the chain projection of the language defined by φ.

The inductive step in Proposition 2 that goes from φ to $\exists^c\varphi$ will follow once we show that languages represented by expressions are closed under chain projection. The proof of this closure is similar to the one in Section 4. We only sketch a key step, leaving the remaining details to the reader. In the following, $\mathcal{A} = (H, \delta)$ is a forest automaton over the alphabet $A \times \{0,1\}$ that is represented by expressions.

A set of nodes X in a context p is called a p-chain if all of its elements are ancestors of the hole in p. We say a context over A *chains* $g \in H$ *to* $h \in H$ if there is some p-chain X such that the context $p[X]$ takes g to h. The following statement is the key step in showing closure under chain projection, the rest of the proof is left to the reader:

Lemma 1. *For every $g, h \in H$, the set of contexts over A that chains g to h is described by context expression.*

thmheadfont Proof

The proof proceeds in two steps.

In the first step, we write a forest expression for "small" contexts that chain g to h. We will denote this language by K_g^h. A "small" context is of the form:

$$s + a\Box + t \ ,$$

where $a \in A$ is a label and s, t are forests over A. This context chains g to h if and only if for some $i = 0, 1$ we have

$$f + \delta((a,i), g) + f' = h \ , \tag{1}$$

where $f, f' \in H$ are the values under $\mathcal{A}$ of the trees $s[\emptyset], t[\emptyset]$. Recall that $\delta : (A \times \{0,1\}) \times H \to H$ is the transition function in the automaton A; therefore the parameter i specifies whether the chain contains the a node or not. Therefore, the set of "small" contexts that chain g to h is the (finite) union of context expressions

$$L_f + a\Box + L_{f'}$$

such that for some $i = 0, 1$, condition (1) is satisfied. Here L_f (likewise for $L_{f'}$) is the language of forests s such that $s[\emptyset]$ has value f in $\mathcal{A}$; this language has a forest expression by assumption on $\mathcal{A}$ being fully defined by forest expressions.

We now proceed to the second step. We will only describe an expression for contexts where the hole has no siblings; the more general case can be easily obtained using techniques as above. Contexts that chain g to h and have no siblings of the hole can be described by composing small contexts:

$$\bigcup\{K_{f_{k-1}}^{f_k} K_{f_{k-2}}^{f_{k-1}} \cdots K_{f_2}^{f_3} K_{f_1}^{f_2} : g = f_1, \ldots, h = f_k\} \ .$$

Although the above is an infinite expression, it can be easily rendered finite using the Kleene star. □

6 Concatenation Hierarchy

In this section, we show that our expressions admit a hierarchy similar to the concatenation (Straubing) hierarchy for word languages. Also similar to the word case, this hierarchy coincides with the quantifier hierarchy of first-order logic with order.

In the context of forest and context expressions, the name concatenation hierarchy would be confusing, since we have two types of concatenation: forest concatenation and context composition. To make things even more confusing, the role of language concatenation in the word concatenation hierarchy is played here by context composition, and not forest concatenation. Therefore, below we refrain from using the name concatenation hierarchy.

The definition below is for a hierarchy of context languages. The hierarchy is extended to forest languages by substituting the empty forest ϵ for the hole: a forest language L is said to be on level n (or $n+1/2$) if there is a context language K on level n (or $n+1/2$) such that $L = \{p\epsilon : p \in K\}$.

We fix an alphabet A.

- Level 0 of the hierarchy contains two context languages: the set of all contexts over A, and the empty set.
- Level $n+1/2$ of the hierarchy is defined as follows.
  1. Every context language on level n is on level $n+1/2$.
  2. Context languages on level $n+1/2$ are closed under finite union.
  3. If $a \in A$, K is a context language on level $n+1/2$ and L is a context language on level n, then LaK is a context language on level $n+1/2$.
  4. If L is a forest language of level $n+1/2$, and K is a context language of level $n+1/2$, then $K+L$ and $L+K$ are context languages of level $n+1/2$.
- Level $n+1$ is the boolean closure of level $n+1/2$.

The above definition is the same as for word languages, except for clause 4, which introduces branching into the expressions. Clearly each language in the hierarchy—both for forests and contexts–can be described by a star-free expression. The hierarchy for words can be recovered by taking context languages, and looking at the word that labels the path from root to hole.

Recall that Σ_n is the class of existential first-order formulas—possibly with free variables—whose quantifier prefix has $n-1$ alternations. For instance, Σ_2 are the formulas with quantifier prefix $\exists^*\forall^*$. Over words, a result of Thomas [4] shows that Σ_n defines the same languages as level $n+1/2$ (the signature contains the linear order $<$ on word positions, but not the successor relation $x = y+1$).

We want to have the same result. However, we need to be careful about the picking the correct signature. We cannot have only the descendant order on tree nodes, since this would give commutative languages; while the forest concatenation $K+L$ gives non-commutative languages already on level $1/2$. Furthermore, the hierarchy is sensitive to slight changes in the signature. For instance, the language "the root has label a" will be on level Σ_1 if the signature contains the "root" unary relation; however if the "root" relation is dropped this language moves up to level Σ_2.

We chose two relations: the lexicographic order $\sqsubseteq$ and the greatest common ancestor $gca(x,y) = z$. The descendant order $x \leq y$ can be defined in terms of gca without using quantifiers, since y is a descendant of x if and only if $gca(x,y) = x$. However, to define gca in terms of the descendant, we need a universal quantifier, to enforce minimality. Therefore, the quantifier alternation hierarchies are different for the signatures $\{gca, \sqsubseteq\}$ and $\{\leq, \sqsubseteq\}$. From now on, we will be using the signature $\{gca, \sqsubseteq\}$ when talking about the quantifier hierarchy.

We repeat the statement of our result relating the hierarchies in expressions and logic:

Theorem 2
A forest language is on level $n + 1/2$ if and only if it is defined by a sentence in Σ_n.

We present the two implications in the next two sections. A consequence of the above theorem is that expressions on level $n + 1/2$ are closed under intersection, which is not immediately apparent from the syntax. Another consequence is Proposition 1, since every first-order formula can be found on some level Σ_n.

It will be convenient to have a notion of formulas describing contexts. Here a formula for contexts has the same syntax as a formula for forests; except that there is a free variable x, which corresponds to the hole. The formula describes those contexts where it holds under the valuation that maps x to the hole. For instance, the formula $\forall y\ x = y$ is true only in the empty context, which maps every forest to itself. The hierarchy Σ_n is defined for context formulas in the same way as for forest formulas. The correspondence in Theorem 2 also extends to contexts.

Level $n + 1/2$ is definable in Σ_{n+1}. The proof is by induction, first on n and then on the size of the expression. The induction base of $n = 0$ is fairly simple, and we omit the details.

The key step is to show how to construct formulas for the steps LaK and $L + K$ in the definition of level $n + 1/2$.

The step for LaK is a simple relativization technique, and works the same way as for words. Let then $a \in A$, and let L, K be context languages of levels n and $n + 1/2$ respectively. We will write a formula of Σ_{n+1} that defines the context language LaK. By induction assumption, there are formulas $\varphi(x) \in \Sigma_n, \psi(x) \in \Sigma_{n+1}$ for the languages L, K. The language LaK is defined by the formula

$$\exists y \quad a(y) \wedge \varphi'(y) \wedge \psi'(x) \qquad \in \Sigma_{n+1} .$$

Here φ' is obtained from φ by relativizing every quantifier $\exists z$ to $\exists z \not\geq y$; similarly for ψ' and $z > y$. Note that the descendant order $\leq$ is definable in terms of gca without quantifiers, so this relativization does not change the position in the hierarchy.

The case of $L + K$ requires a little more attention. Let then L be a forest language defined by Σ_{n+1} formula φ, and let K be a context language defined by a Σ_{n+1} formula ψ. We will write a formula for $L + K$ (the case of $K + L$ is done the same way), and we also want to use relativization. However, the node that divides L from K must be on the root: we use the leftmost root node x of K here (this node is lexicographically the first one in K). The formula is:

$$\exists x \qquad (\forall y\ gca(x,y) = y \Rightarrow x = y)\ \wedge\ \varphi'\ \wedge\ \psi' \qquad \in \Sigma_{n+1}\ .$$

The first conjunct says that x is one of the roots; the universal quantifier in this conjunct does not increase the hierarchy level of the larger formula, since we are working at or above level Σ_2 (recall that the base case of $n = 0$ was done separately). The relativized formulas are defined as in the previous case, only this time we use the lexicographic order.

Formulas in Σ_{n+1} are captured by level $n + 1/2$. As a warm-up, we do the case of $n = 0$. Let then

$$\varphi = \exists x_1, \ldots, x_k \psi$$

be a formula, with φ quantifier-free. By choice of signature, φ can only say what is the lexicographic order on the nodes $x_1, \ldots, x_k$, and which quantified nodes are greatest common ancestors of other quantified nodes. The basic idea is that in a forest of the form $pa(s + t)$, with p a context, a a label, and s, t forests, the greatest common ancestor of a node from s and a node from t must be the node a. The lexicographic order can also be expressed, since the expressions are ordered. We only present the construction on an example:

$$\exists x, y, z \quad a(x) \wedge b(y) \wedge c(z) \wedge gca(x,y) = z$$

Since the labels a, b, c are different, the nodes x, y, z are different. The appropriate expression is:

$$Kc((KaK\epsilon) + (KbK\epsilon)) \qquad \cup \qquad Kc((KbK\epsilon) + (KaK\epsilon))\ .$$

In the above, K is the level 0 expression describing all contexts.

We would like to continue along these lines for the classes Σ_2, Σ_3, etc. Unfortunately we run across a technical problem, which does not occur for words, and makes the proof rather tedious. If we have nodes $x_1, \ldots, x_k$ in a word w (say, ordered from left to right), then these partition the word into words of the form $w[0..x_1], w[x_1 + 1..x_2], \ldots, w[x_{k-1} + 1..x_k]$. Unfortunately, this is no longer the case for forests.

A first obstacle is how to use two nodes x, y to cut out a piece from a larger forest. If we want to cut out a forest, it would be reasonable to expect the two nodes x, y to be siblings; but the sibling relationship is not quantifier-free in our signature. A second problem is that a given set of nodes may not partition a forest: for this the set must satisfy certain closure properties (such as containing greatest common ancestors).

We write $x \preceq y$ to denote that x is a sibling to the right of y; this formula needs a universal quantifier (to say that all ancestors of x are also ancestors of y) and therefore is a negation of a Σ_1 formula. Note that this is *not* the lexicographic order. A formula $x \preceq y$ is called a *forest link*. The ideas that the nodes x, y can be used to *induce* a subforest. If $x = y$ then the induced subforest contains proper descendants z of x; if $x \prec y$ then the induced subforest contains nodes z that lexicographically between x, y, but are not descendants of x. A formula $x < y$ is called a *context link*. Similarly to forest links, the nodes x, y *induce* a context: the nodes of this context are nodes z with $x < z$ and $y \not\leq z$, and the hole is in y. We will use letters α, β to refer to links, be they context or forest. In either case, the property that z belongs to to the induced forest/context of a link α is expressed by a quantifier-free formula in x, y, z.

Types
Let X be set of variables. A *pre-type over X* is a conjunction, which for every variables $x, y, z \in X$ specifies the label of this variable, and which of the following hold, and which do not hold: x is a leftmost sibling, x is a rightmost sibling, $x \leq y$, $x \preceq y$, x is a child of y, x is the left neighbor of y, z is the greatest common ancestor of x, y. Since a pre-type specifies all this information, it makes sense to write "a node $x \in \tau$ has a child in τ"—likewise for the other relations described in a pre-type—with the meaning being that $x \in X$ and for some node $y \in X$, the pre-type τ specifies that y is the child of x. Note that a pre-type is a boolean combination of Σ_1 formulas, since universal quantification is necessary to talk about left/right siblings, etc.

Take a link α—either a forest link $x \preceq y$ or a context link $x < y$. A *type local to α* is a type $\tau \ni x, y$ where every variable $z \in \tau$ distinct from x, y is in the (forest/context) induced by α. Furthermore, we require the saturation properties listed below; these ensure that the nodes induced by α can be partitioned into links (both context and forest) given by variables from τ.

- If $z_1, z_2 \in \tau$ are distinct siblings, but not siblings of x and y (possible only when α is a forest link), then the parent of z_1, z_2 belongs to τ.
- If $z \in \tau$ has its parent in τ, then its leftmost, rightmost siblings are in τ.
- If $z_1, z_2 \in \tau$ are not related by the descendant relation, then there are siblings $z'_1, z'_2 \in \tau$ that are ancestors of z_1, z_2, respectively.

A consequence of the properties above is that nodes described by a type τ are closed under greatest common ancestor, as long as this greatest common ancestor stays within the link α.

Below we define the *joints* of type τ. The idea is that these are context/forest links that are closest to each other and contain some nodes in their induced forest/context.

- Let $x, y \in \tau$ be such that y is a descendant of x, and minimal for this property. If y is not a child of x, then $x < y$ is called a *(context) joint of τ*.
- There are two kinds of *forest joints*. If $x, y \in \tau$ are distinct siblings, with no sibling from τ strictly between them, then $x \prec y$ is a forest joint of τ. Also, if $x \in \tau$ has no proper descendants in τ, then $x \preceq x$ is a joint of τ.

The saturation properties in the definition of a type are chosen so that the joints partition a type, i.e.:

Lemma 2. *Let τ be a type local to a link α. The type τ entails that any node in the (context/forest) induced by α is either one of the nodes from τ, or in exactly one (context/forest) induced by a joint of τ.*

thmheadfont Proof
Using the saturation properties. We consider here the case when $\alpha = x \preceq y$. Let then z be a node not in the forest induced by the link $x \preceq y$. Assume furthermore that z is not equal to one of the nodes from τ. Consider two cases:

- There are siblings of z in τ. We claim that there are nodes $z_1, z_2 \in \tau$ with $z_1 \prec z \prec z_2$. If $x \prec z \prec y$ then we are done. Otherwise, since z has siblings in τ, then the parent of z belongs to τ thanks to the saturation properties. This entails that also the leftmost and rightmost siblings of z belong to τ. Let then $z_1, z_2 \in \tau$ be such that $z_1 \prec z \prec z_2$, and chosen closest to z for this property. Therefore, $z_1 \preceq z_2$ is a joint of τ, and its induced forest contains z.
- There are no siblings of z in τ. First we claim that z has an ancestor in τ. If x is an ancestor of z, we are done. Otherwise, x and z are incomparable by the descendant relation, and therefore by the saturation properties, τ must contain siblings that are ancestors of x, z, which concludes the claim. Let z' be the ancestor in τ that is closest to z. If z has a descendant z'' in τ, then z belongs to the context induced by $z' < z''$, if z'' is taken closest to z. Otherwise, we show that z' has no descendants in τ, and hence z belongs to the forest induced by the link $z' \preceq z'$. Assume for the sake of contradiction that τ contains a descendant z'' of z' that is not a descendant of z. By definition of z', z'' must be incomparable with z. Using the last saturation property, we get a node in τ that is strictly between z' and z, a contradiction. □

Let X be a set of variables, and let α be a link (using variables from X). A type *local to α generated by X* is defined to be any type τ local to α over variables $Y \supseteq X$ that is minimal for this property: i.e. there is no type σ local to α over variables Z, with $X \subseteq Z \subsetneq Y$ such that τ and σ are consistent. If τ is generated, then $Y \setminus X$ is called the set of *auxiliary* variables of τ, and is denoted by Y_τ. The idea is that the variables in Y_τ are used to add the saturation properties required in a type. Since the saturation properties require adding at most a linear number of additional nodes, there are finitely many nonequivalent types generated by a given set X and a link α; furthermore any two nonequivalent ones are inconsistent.

Normal Form
A formula is said to be *local* to a link α if its quantified and free variables are relativized to nodes induced by the link α. Let α be a link. We define two types of normal form: $(n-1/2, \alpha)$-normal form, which corresponds to Σ_n formulas local to α, and (n, α)-normal form, which corresponds to boolean combinations of the former. We define $(0, \alpha)$-normal form to be all quantifier-free formulas local to α. For $n \geq 1$, a formula is said to be in $(n - 1/2, \alpha)$ *normal form* if it is a positive boolean combination of formulas $\exists x_1, \ldots, x_k \psi$, with ψ in $(n-1, \alpha)$ normal form. A formula—with free variables X—is in (n, α)-*normal form* if it is of the form $\exists y_1, \ldots, y_k\ \tau \wedge \psi$, where

- τ is a type local to α generated by X with auxiliary variables $y_1, \ldots, y_k$. To abbreviate the notation, we write $\exists y_1, \ldots, y_k$ as $\exists Y_\tau$.
- For some joint β of τ, the formula ψ is either in $(n - 1/2, \beta)$-normal form, or a negation thereof.

Lemma 3. *For $n \geq 1$, every formula in Σ_n local to α is equivalent to a formula in $(n - 1/2, \alpha)$ normal form.*

thmheadfont Proof
The proof is a by induction on n. The only nontrivial part is showing that positive boolean combinations of formulas in (n, α)-normal form are closed under negation. By De Morgan laws, we only need to do the negation of a single formula:

$$\neg(\exists Y_\tau\ \tau \wedge \psi) \qquad \Leftrightarrow \qquad (\exists Y_\tau\ \tau \wedge \neg\psi)\ \vee\ \bigvee_\sigma (\exists Y_\sigma\ \sigma \wedge \mathit{true})\ ,$$

where the disjunction above ranges over the—finitely many—types σ generated by X that are inconsistent with τ.

The case of existential quantification, which corresponds to going from (n, α) normal form to $(n + 1/2, \alpha)$-normal form, is a straightforward consequence of Lemma 2. □

From Normal Form to Expressions
We will now rewrite a formula in $(n + 1/2, \alpha)$ normal form into an expression on level $n + 1/2$. The expression will be a context or forest expression, depending on the kind of link α. It will be convenient to use an intermediate form, where logical formulas and regular expressions can be mixed. For this purpose we extend first-order logic with the following type of formulas as follows. Let $\alpha = x < y$ be a context link. If K is a context expression, then $K_\alpha(x, y)$ is a formula. This formula holds if K contains the context induced by α. The definition for a forest link $x \preceq y$ and a forest expression L is analogous. Together with Lemma 3, the following result concludes the "if" part of Theorem 2.

Lemma 4. *Let $n \geq 0$. Let $\alpha = x < y$ be a context link, and let φ be a formula in $(n + 1/2, \alpha)$ normal form. There is a context expression K of level $n + 1/2$ such that φ and $K_\alpha(x, y)$ are equivalent. Likewise for forests.*

thmheadfont Proof (Sketch)
Induction on n, with plenty case analysis. The base case $n = 0$ is done as in the beginning of this section. We only do the induction step for a context link $x < y$, leaving the proof for forests to the reader. Let then $n \geq 1$. Since the expressions are closed under union, and types are either equal or inconsistent, we only need consider a formula φ be of the form:

$$\varphi = \exists x_1, \ldots, x_k\ \tau\ \wedge\ \psi\ ,$$

with τ a type over variables $x_1, \ldots, x_k, x, y$ and ψ a conjunction of formulas in $(n - 1/2, \beta)$ normal form (or their negations), with β ranging over joints of τ.

Using the induction assumption, we may replace ψ be a conjunction $\bigwedge_\beta K_\beta$, where β ranges over joints of τ, and each K_β is a forest or context expression of level n, depending on the kind of β. (Recall that expressions of level n are closed under boolean combinations, so we need not bother with the negations in conjuncts in ψ.)

The proof then proceeds via a second induction, this time on the number of variables k. Recall the link $\alpha = x < y$ to which the formula φ is local. The proof is a case analysis. We only consider here the case when τ contains the parent of y, but not the grandparent of y. Let then $x_1 \in \tau$ be the parent of y, and let $x_2 < x_1$ be the closest ancestor of x_1 that is in τ. We leave to the reader the special cases when $x_1 = x$, and therefore x_2 is undefined, and also when y is either a leftmost or rightmost sibling. Let then

$$z_1 \prec \cdots \prec z_i \prec y \prec z'_1 \prec \cdots \prec z'_j \tag{2}$$

be all the distinct siblings of y given in the type τ. Using the induction assumption, we may assume that $i, j = 1$, and that the links $z_1 \preceq z_1$, $z_1 \prec y$, $y \prec z'_1$, and $z'_1 \preceq z'_1$ are described by expressions L, M, L', M' respectively. Finally, let K be the context expression describing the joint $x_2 < x_1$, and let a be the label of the node x_1, which can be read from τ. We can now lose all the variables from (2), by describing the context link $x_2 < y$ via the expression

$$Ka(L + M + \Box + L' + M') \ .$$

The above is a level $n + 1/2$ expression, since $n \geq 1$ and $\Box$—an expression that describes only the empty context $\Box$—is a level $1 + 1/2$ context expression. $\Box$

References

1. Bojańczyk, M., Walukiewicz, I.: Forest algebras (unpublished manuscript), `http://hal.archives-ouvertes.fr/ccsd-00105796`
2. Comon, H., Dauchet, M., Gilleron, R., Jacquemard, F., Lugiez, D., Tison, S., Tommasi, M.: Tree automata techniques and applications, 2002 (October 1st 2002), Available on: `http://www.grappa.univ-lille3.fr/tata`
3. McNaughton, R., Papert, S.: Counter-Free Automata. MIT Press, Cambridge (1971)
4. Thomas, W.: Classifying regular events in symbolic logic. Journal of Computer and System Sciences 25, 360–375 (1982)
5. Thomas, W.: Logical aspects in the study of tree languages. In: Colloquium on Trees and Algebra in Programming, pp. 31–50 (1984)
6. Thomas, W.: On chain logic, path logic, and first-order logic over infinite trees. In: Logic in Computer Science, pp. 245–256 (1987)

MSO on the Infinite Binary Tree: Choice and Order

Arnaud Carayol and Christof Löding

RWTH Aachen, Informatik 7, 52056 Aachen, Germany
{carayol,loeding}@i7.informatik.rwth-aachen.de

Abstract. We give a new proof showing that it is not possible to define in monadic second-order logic (MSO) a choice function on the infinite binary tree. This result was first obtained by Gurevich and Shelah using set theoretical arguments. Our proof is much simpler and only uses basic tools from automata theory. We discuss some applications of the result concerning unambiguous tree automata and definability of winning strategies in infinite games. In a second part we strengthen the result of the non-existence of an MSO-definable well-founded order on the infinite binary tree by showing that every infinite binary tree with a well-founded order has an undecidable MSO-theory.

1 Introduction

Our main purpose is to present a simple proof for the fact (first shown by Gurevich and Shelah in [GS83]) that on the infinite binary tree there is no choice function that can be defined in monadic second-order logic (MSO), i.e., in the extension of first-order logic by quantification over sets of elements. A choice function on the infinite binary tree is a mapping assigning to each nonempty set of nodes one element from this set, i.e., the function chooses for each set one of its elements. Such a function is MSO-definable if there is an MSO-formula with one free set variable X and one free element variable x such that for each nonempty set U of nodes there is exactly one element $u \in U$ such that the formula is satisfied if X is interpreted as U and x is interpreted as u.

The question of the existence of an MSO-definable choice function over the infinite binary tree can be seen as a special instance of the more general uniformization problem, which asks, given a relation that is defined by a formula with free variables, whether it is possible to define by another formula a function that is compatible with this relation. More precisely, given a formula $\phi(\bar{X}, \bar{Y})$ with vectors $\bar{Y}, \bar{X}$ of free variables, such that $\forall \bar{X} \exists \bar{Y} \phi(\bar{X}, \bar{Y})$ is satisfiable, uniformization asks for a formula $\phi^*(\bar{X}, \bar{Y})$ such that

1. ϕ^* implies ϕ (each interpretation of $\bar{Y}, \bar{X}$ making ϕ^* true also makes ϕ true),
2. and ϕ^* defines a function in the sense that for each interpretation of $\bar{X}$ there is exactly one interpretation of $\bar{Y}$ making ϕ^* true.

J. Duparc and T.A. Henzinger (Eds.): CSL 2007, LNCS 4646, pp. 161–176, 2007.

The question of the existence of a choice function is the uniformization problem for the formula $\phi(X, y) = X \neq \emptyset \to y \in X$.

The infinite line, i.e., the structure $(\omega, succ)$ of the naturals with the successor function is known to have the uniformization property for MSO [Sie75]. On the infinite binary tree MSO is known to be decidable [Rab69] but it does not have the uniformization property. This was conjectured in [Sie75] and proved in [GS83] where it is shown that there is no MSO-definable choice function on the infinite binary tree.

The reason for the present paper is that the proof in [GS83] uses complex set theoretical arguments, whereas it appears that the result can be obtained by much more basic techniques. We show that this is indeed true and present a proof that only relies on the equivalence of MSO and automata over infinite trees and otherwise only uses basic techniques from automata theory. Besides its simplicity, another advantage of the proof is that we provide a concrete family of sets (parameterized by natural numbers) such that each formula will fail to make a choice for those sets with the parameters chosen big enough. We use this fact when we discuss two applications of the result concerning unambiguous tree automata (as presented in [NW]) and the definability of strategies in infinite games.

The subject of MSO-definability of choice functions on trees has been studied in more depth in [LS96], where the authors consider more general trees not only the infinite binary tree. They show the following dichotomy: for a tree it is either not possible to define a choice function in MSO, or it is possible to define a well-ordering on the domain of the tree.

We strengthen this result by showing that extending the infinite binary tree by any well-ordering leads to a structure with undecidable MSO-theory. As a consequence we obtain that each structure in which we can MSO-define a well-ordering and MSO-interpret the infinite binary tree must have an undecidable MSO-theory.

The article is structured as follows. In the next section we introduce some notations and basic terminology. In Section 3 we give the proof that there is no MSO-definable choice function on the infinite binary tree and discuss applications of the result. In Section 4 the undecidability of the MSO-theory of the infinite binary tree augmented by any well-ordering is shown.

2 Preliminaries

Words. For a finite alphabet Σ, we write Σ^* for the set of all words over Σ. The length of a word $u \in \Sigma^*$ is denoted by $|u|$ and ε is the empty word. For all words $u, v \in \Sigma^*$, u is a *prefix* of v (written $u \sqsubseteq v$) if there exists $w \in \Sigma^*$ such that $v = uw$. If $w \in \Sigma^+$ then u is a *strict prefix* of v (written $u \sqsubset v$). The *greatest common prefix* of two words u and v (written $u \wedge v$) is the longest word which is a prefix of u and v.

Relational structures. A *signature* is a ranked set of symbol $\mathcal{S}$, where for all $R \in \mathcal{S}$, $|R|$ denotes the arity (which is ≥ 1) of the symbol R. A *relational structure* $\mathcal{R}$ over the *signature* $\mathcal{S}$ is given by a tuple $(U, (R^{\mathcal{R}})_{R\in\mathcal{S}})$ where U is the *universe* of $\mathcal{R}$ and where for all $R \in \mathcal{S}$, $R^{\mathcal{R}}$ (which is also called the interpretation of R in $\mathcal{R}$) is a subset of $U^{|R|}$. When $\mathcal{R}$ is clear from the context, we will simply write R instead of $R^{\mathcal{R}}$.

Monadic second-order logic. We adopt the definition of *monadic second-order logic (MSO)* over relational structures with the standard syntax and semantics (see e.g. [EF95] for a detailed presentation). We write $\varphi(X_1, \dots, X_n, y_1, \dots, y_m)$ to denote that the free variables of the formula φ are among $X_1, \dots, X_n$ (monadic second-order) and $y_1, \dots, y_m$ (first-order) respectively. A formula without free variables is called a *sentence.*

For a relational structure $\mathcal{R}$ and a sentence φ, we write $\mathcal{R} \models \varphi$ if $\mathcal{R}$ *satisfies* the formula φ. The *MSO-theory* of $\mathcal{R}$ is the set of sentences satisfied by $\mathcal{R}$. For every formula $\varphi(X_1, \dots, X_n, y_1, \dots, y_m)$, all subsets $U_1, \dots, U_n$ of the universe of $\mathcal{R}$ and all elements $v_1, \dots, v_m$ of the universe of $\mathcal{R}$, we write $\mathcal{R} \models \varphi[U_1, \dots, U_n, v_1, \dots, v_m]$ to express that φ holds in $\mathcal{R}$ when X_i is interpreted as U_i for all $i \in [1, n]$ and y_j is interpreted as v_j for all $j \in [1, m]$.

Infinite binary labelled trees. An (infinite binary) tree labeled by a finite alphabet Σ is a mapping $t : \{0, 1\}^* \to \Sigma$. We denote by T_Σ the set of all trees labeled by Σ. For a set $U \subseteq \{0, 1\}^*$, we write $t(U) \subset T_{\{0,1\}}$ for the characteristic tree of U, i.e., the tree which labels all nodes in U with 1 and all the other nodes with 0. This notation is extended to the case of several sets. The characteristic tree of $U_1, \dots, U_n \subseteq \{0, 1\}^*$ is the tree labeled by $\{0, 1\}^n$ written $t(U_1, \dots, U_n)$ and defined for all $u \in \{0, 1\}^*$ by $t(U_1, \dots, U_n)(u) := (b_1, \dots, b_n)$ where for all $i \in [1, n]$, $b_i = 1$ if $u \in U_i$ and $b_i = 0$ otherwise.

To every tree t labeled by $\Sigma = \{a_1, \dots, a_n\}$, we associate a canonical structure over the signature $\{E_0, E_1, P_{a_1}, \dots, P_{a_n}\}$ where E_0 and E_1 are binary symbols and the P_{a_i} are predicates. The universe of this structure is $\{0, 1\}^*$. The symbols E_0 and E_1 are respectively interpreted as $\{(w, w0) \mid w \in \{0, 1\}^*\}$ and $\{(w, w1) \mid w \in \{0, 1\}^*\}$. Finally for all $i \in [1, n]$, P_{a_i} is interpreted as $\{u \in \Sigma^* \mid t(u) = a_i\}$. In the following, we will not distinguish between a tree and its canonical relational structure.

In particular, for a formula $\varphi(X_1, \dots, X_n)$ and sets $U_1, \dots, U_n \subseteq \{0, 1\}^*$, we write $t(U_1, \dots, U_n) \models \varphi$ to indicate that the infinite binary tree satisfies φ when X_i is interpreted by U_i.

3 Choice

As described in the introduction, an MSO-definable choice function is given by an MSO-formula $\phi(X, x)$ such that

$$\forall X \exists x.\ X \neq \emptyset \to (x \in X \land \phi(X, x) \land \forall y.\ \phi(X, y) \to x = y)$$

is true over the infinite binary tree. This section is mainly devoted to the proof of the following theorem of Gurevich and Shelah.

Theorem 1 ([GS83]). *There is no MSO-definable choice function on the infinite binary tree.*

The technical formulation of the result we prove is given in Theorem 3, where we concretely provide counter examples for which a given formula cannot choose a unique element. As a machinery for the proof we use tree automata, which are easier to manipulate (at least for our purpose) than formulas. In the following we give some basic definitions. More details on automata for infinite trees can be found in [Tho97].

In this section we take the view of labeled trees as mappings from $\{0,1\}^*$ to the label alphabet (which usually is $\{0,1\}^n$ for some number n).

A parity automaton on Σ-labeled trees is a tuple $\mathcal{A} = (Q, \Sigma, q_0, \Delta, \Omega)$ with a finite set Q of states, initial state $q_0 \in Q$, transition relation $\Delta \subseteq Q \times \Sigma \times Q \times Q$, and a priority function $\Omega : Q \to \mathbb{N}$. A run of $\mathcal{A}$ on a tree $t \in T_\Sigma$ from a state $q \in Q$ is a tree $\rho \in T_Q$ such that $\rho(\varepsilon) = q$, and for each $u \in \{0,1\}^*$ we have $(\rho(u), t(u), \rho(u0), \rho(u1)) \in \Delta$. We say that ρ is accepting if on each path the minimal priority appearing infinitely often is even. If we only speak of a run of $\mathcal{A}$ without specifying the state at the root, we implicitly refer to a run from q_0.

We extend this model to automata that do not only accept or reject trees but also mark some of the nodes with special marking states. A marking parity automaton (MPA) is a tuple $\mathcal{A} = (Q, \Sigma, q_0, \Delta, \Omega, P)$ with an additional set of marking states $P \subseteq Q$. A run ρ of such an automaton defines a set $U_\rho \subseteq \{0,1\}^*$ as the set of those nodes that are labeled by a marking state, i.e., $U_\rho = \rho^{-1}(P)$.

Given a set $U \subseteq \{0,1\}^*$ and an MPA $\mathcal{A}$ on $\{0,1\}$-labeled trees, we define

$$T(\mathcal{A}, U) = \{U_\rho \mid \rho \text{ is an accepting run of } \mathcal{A} \text{ on } t(U)\}.$$

That is, the set of all sets that are marked by $\mathcal{A}$ in an accepting run on $t(U)$.

Theorem 2 ([Rab69]). *For each MSO-formula $\phi(X, Y)$ there is an MPA $\mathcal{A}_\phi$ such that $T(\mathcal{A}_\phi, U) = \{U' \subseteq \{0,1\}^* \mid t(U, U') \models \phi\}$ for each $U \subseteq \{0,1\}^*$.*

For a set $U \subseteq \{0,1\}^*$ we say that $\mathcal{A}$ marks an element u of U if $u \in U$ and there is an accepting run ρ of $\mathcal{A}$ on $t(U)$ such that $U_\rho = \{u\}$. Note that with other runs $\mathcal{A}$ might mark other elements or sets of elements.

For two trees t, t' we say that they are $\mathcal{A}$-equivalent, written as $t \equiv_\mathcal{A} t'$, if for each state q of $\mathcal{A}$ there is an accepting run from q on t iff there is an accepting run from q on t'. Intuitively, this means that $\mathcal{A}$ cannot distinguish the two trees.

3.1 Undefinability of Choice Functions

We now define a family $(U_{M,N})_{M,N}$ of sets such that for each MPA we can find M and N such that this MPA cannot mark a unique element of $U_{M,N}$. To achieve this, we "hide" the elements from the set very deep in the tree such that MPAs up to a certain size are not able to uniquely choose an element that they can mark.

For $M, N \in \mathbb{N}$ the set $U_{M,N} \subseteq \{0,1\}^*$ is defined by the following regular expression $U_{M,N} = \{0,1\}^*(0^N 0^* 1)^M \{0,1\}^*$. Let $t_{M,N} = t(U_{M,N})$. This tree can be obtained by unfolding the finite graph $G_{M,N}$ depicted in Figure 1 from x_M. In this picture, the dashed arrows represent 1-labeled edges leading back to the node x_M. The chains of 0-edges between x_{k+1} and x_k have length N. All nodes in this graph are labeled 0 except x_0, which is labeled by 1.

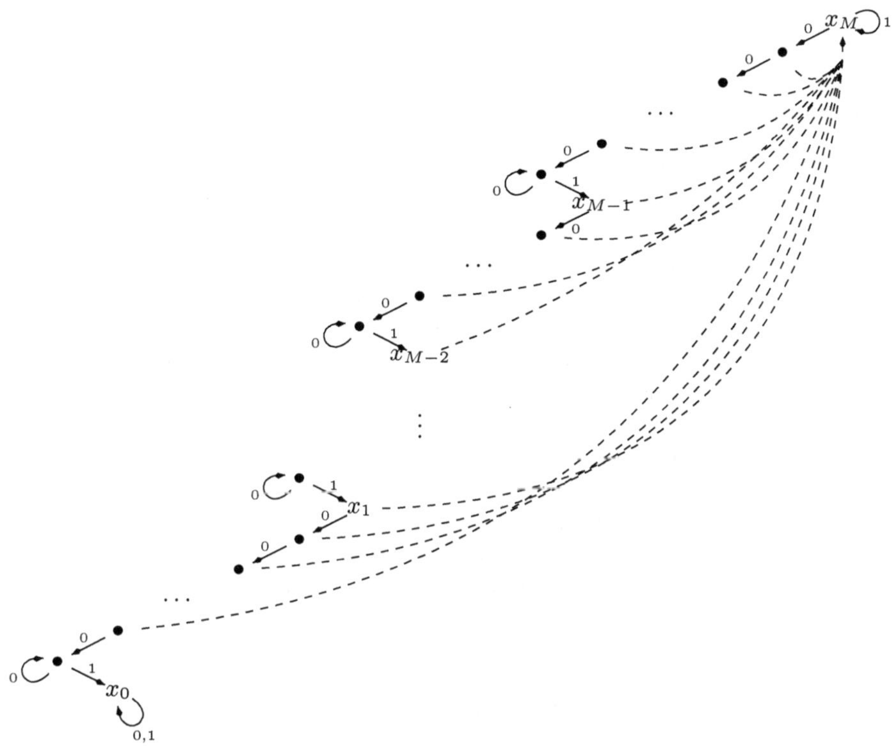

Fig. 1. A representation of the regular tree $t_{M,N}$ by the graph $G_{M,N}$

It is easy to verify that x_0 is reachable from x_M by exactly those paths whose sequence of edge labels is in the set $U_{M,N}$. So $G_{M,N}$ can be viewed as the minimal DFA accepting the language $U_{M,N}$ where x_0 is the only final state. Let $t_{k,M,N}$ denote the tree that we obtain by unfolding the graph $G_{M,N}$ from the node x_k.

We now fix an MPA $\mathcal{A} = (Q, \{0,1\}, q_0, \Delta, \Omega, P)$ on $\{0,1\}$-labeled trees and take $M = 2^{|Q|} + 1$ and $N = |Q| + 1$. For these fixed parameters we simplify the notation by letting $t_k = t_{k,M,N}$. In particular, $t_M = t_{M,M,N} = t_{M,N}$.

We say that a subtree (of some tree t) that is isomorphic to t_k for some k is of type t_k.

Our aim is to trick the automaton $\mathcal{A}$ to show that it cannot choose a unique element from the set $U_{M,N}$. This is done by modifying a run that marks an element u of $U_{M,N}$ such that we obtain another run marking something different.

To understand the general idea, consider the path between x_{k+1} and x_k for some k. If we take a 1-edge before having reached the end of the 0-chain, i.e., if we take a dashed edge in the picture, then we reach a subtree of type t_M. But if we walk to the end of the 0-chain and then move to the right using a 1-edge, then we arrive at a subtree of type t_k. If we show that there is $\ell < M$ such that t_M and t_ℓ are $\mathcal{A}$-equivalent, then this means that $\mathcal{A}$ has no means to identify when it enters the part where taking a 1-edge leads to subtree of type t_ℓ. We then exploit this fact by pumping the run on this part of the tree such that we obtain another run marking something different.

Lemma 1. *There exists $\ell < M$ such that $t_M \equiv_{\mathcal{A}} t_\ell$.*

Proof. We consider for each tree $t \in T_{\{0,1\}}$ the function $f_t : Q \to \{a, r\}$ with $f_t(q) = a$ if there is an accepting run from q on t, and $f_t(q) = r$ otherwise. By definition, two trees t, t' are $\mathcal{A}$-equivalent if $f_t = f_{t'}$. There are at most $2^{|Q|}$ different such functions. By the choice of M there are $1 \leq k_1 < k_2 \leq M$ such that $t_{k_1} \equiv_{\mathcal{A}} t_{k_2}$. Let $k = k_2 - k_1$ and $\ell = M - k$.

We show that we can obtain t_ℓ from t_M by substituting some of the subtrees of type t_{k_2} in t_M by subtrees of type t_{k_1}. As we have seen that t_{k_1} and t_{k_2} are $\mathcal{A}$-equivalent, this suffices to show that t_M and t_ℓ are also $\mathcal{A}$-equivalent.

We know that t_M is obtained by unraveling the graph $G_{M,N}$ (Figure 1) from x_M. One way to obtain t_ℓ is the following. We take a second copy of $G_{M,N}$ and denote in this copy the vertices corresponding to $x_0, \ldots, x_M$ by $x'_0, \ldots, x'_M$. Now we redirect the edge leading to x_{k_2} in the first copy to point to x'_{k_1} in the second copy. It is easy to verify that unravelling this new graph from x_M (in the first copy) yields the tree t_ℓ. And furthermore, this shows that t_ℓ can be obtained by replacing some subtrees of type t_{k_2} in t_M by subtrees of type t_{k_1}. Hence, $t_M \equiv_{\mathcal{A}} t_\ell$. □

The following lemma states that it is impossible for $\mathcal{A}$ to distinguish a unique element of $U_{M,N}$, i.e., it is not possible that $T(\mathcal{A}, U_{M,N}) = \{\{u\}\}$ for some $u \in U_{M,N}$.

Lemma 2. *If $\mathcal{A}$ marks an element of $U_{M,N}$, then $|T(\mathcal{A}, U_{M,N})| > 1$*

Proof. Assume that that there is an accepting run ρ of $\mathcal{A}$ on $t_M = t_{M,N}$ such that $U_\rho = \{u\}$ with $u \in U_{M,N}$. From ρ we construct another accepting run marking a different set of nodes.

For $0 \leq k \leq M$ let u_k denote the maximal prefix of u such that the subtree at u_k is of type t_k. Let ℓ be as in Lemma 1. For $i \geq 0$ we let $v_i = u_{\ell+1}0^i$ and $v'_i = v_i 1$. Note that $v_0 = u_{\ell+1}$ and that for $0 \leq i < N$ the subtrees at v'_i are of type t_M, and for $i \geq N$ the subtrees at v'_i are of type t_ℓ.

From Lemma 1 we know that $t_\ell \equiv_{\mathcal{A}} t_M$. Hence, for each accepting run ρ_q of $\mathcal{A}$ from q on t_M we can pick an accepting run ρ'_q of $\mathcal{A}$ from q on t_ℓ.

By the choice of N there are $0 \leq j < j' < N$ such that $\rho(v_j) = \rho(v_{j'})$. For the moment, consider only the transitions taken in ρ on the sequence $v_0, v_1, \ldots$, i.e., on the infinite branch to the left starting from v_0. We now simply repeat the part of the run between v_j and $v_{j'}$ once. The effect is that some of the states that were at a node v'_i for $i < N$ are pushed to nodes v'_i for $i \geq N$, i.e., the states are moved from subtrees of type t_M to subtrees of type t_ℓ. But for those states q we can simply plug the runs ρ'_q that we have chosen above.

More formally, we define the new run ρ' of $\mathcal{A}$ on t_M as follows. On the part that is not in the subtree below v_0 the run ρ' corresponds to ρ. In the subtree at v_0 we make the following definitions, where $h = j' - j$.

- For $i < j'$ let $\rho'(v_i) = \rho(v_i)$ and $\rho'(v'_i) = \rho(v'_i)$.
- For $i \geq j'$ let $\rho'(v_i) = \rho(v_{i-h})$ and $\rho'(v'_i) = \rho(v'_{i-h})$.
- For the subtrees at v'_i for $i < j'$ we take the subrun of ρ at v'_i.
- For the subtrees at v'_i for $j' \leq i < N$ or $i \geq N + h$ we take the subrun of ρ at v'_{i-h}. This is justified because in these cases $\rho'(v'_i) = \rho(v'_{i-h})$ and the subtrees at v'_i and v'_{i-h} are of the same type (both of type t_M or both of type t_ℓ).
- For the subtrees at v'_i for $N \leq i < N+h$ we take the runs ρ'_{q_i} for $q_i = \rho'(v'_i)$. This is justified as follows. From $q_i = \rho'(v'_i)$ and the definition of ρ' we know that $\rho(v'_{i-h}) = q_i$. Hence, there is an accepting run of $\mathcal{A}$ from q_i on t_M. Thus, ρ'_{q_i} as chosen above is an accepting run of $\mathcal{A}$ from q_i on t_ℓ.

This run ρ' is accepting. Furthermore, the state marking u in the run ρ has been moved to another subtree: There are $n \geq N$ and $w \in \{0,1\}^*$ such that $u = u_{\ell+1}0^n w$. In ρ' the state marking u is at $u' = u_{\ell+1}0^{n+h}w$. Hence, we have constructed an accepting run marking a set different from $\{u\}$. □

Of course, the statement is also true if we increase the value of M or N, e.g., if we let $N = M = 2^{|Q|+1}$. Thus, combining Theorem 2 and Lemma 2 we obtain the following.

Theorem 3. *Let $\phi^*(X, x)$ by an MSO-formula. There exists $n \in \mathbb{N}$ such that for each $u \in U_{n,n}$ with $t(U_{n,n}, u) \models \phi^*$ there is $u' \neq u$ with $t(U_{n,n}, u') \models \phi^*$.*

A direct consequence is the theorem of Gurevich and Shelah. The advantage of our proof is that we obtain a rather simple family of counter examples (the sets $U_{M,N}$).

An easy reduction allows us to extend the non-existence of an MSO-definable choice function to the case where we allow a finite number of fixed predicates as parameters. This result has already been shown in [LS98] in an even more general context, but again relying on the methods employed in [GS83].

Corollary 1. *Let $P_1, \ldots, P_n \subseteq \{0,1\}^*$ be arbitrary predicates. There is no MSO-formula $\phi^*(X_1, \ldots, X_n, X, x)$ such that for each nonempty set U there is exactly one $u \in U$ with $t(P_1, \ldots, P_n, U, u) \models \phi^*$.*

Proof. Assume that there are $P_1, \ldots, P_n \subseteq \{0,1\}^*$ and $\phi^*(X_1, \ldots, X_n, X, x)$ such that for each set U there is exactly one $u \in U$ with $t(P_1, \ldots, P_n, U, u) \models \phi^*$. Then the formula

$$\exists X_1, \ldots, X_n \forall X \exists x\ \phi^*(X_1, \ldots, X_n, X, x) \wedge \forall y\ \phi^*(X_1, \ldots, X_n, X, y) \rightarrow x = y$$

is satisfiable. Hence, by the Rabin Basis Theorem (cf. [Tho97]), there are regular predicates $P_1, \ldots, P_n$ such that

$$t(P_1, \ldots, P_n) \models \forall X \exists x \ \phi^*(X_1, \ldots, X_n, X, x) \wedge \forall y \ \phi^*(X_1, \ldots, X_n, X, y) \to x = y.$$

Regular predicates are MSO-definable, so let $\psi_1(X_1), \ldots, \psi_n(X_n)$ be formulas defining $P_1, \ldots, P_n$, respectively. Then the formula $\phi'(X, x)$ defined as

$$\exists X_1, \ldots, X_n \ \phi^*(X_1, \ldots, X_n, X, y) \wedge \bigwedge_{i=1}^{n} \psi_i(X_i)$$

describes a choice function, contradicting Theorem 3. □

We point out here that this method only relies on the fact that the property of being a choice function is MSO-definable. So this way of reducing the case with parameters to the parameter free case can be applied whenever the properties of the object under consideration are MSO-definable (at the end of the last section we briefly mention another application of this technique).

3.2 Applications of the Result and Its Proof

We now discuss a few applications of the results presented so far in this section. One immediate application concerns the non-definability of well-founded orders over the infinite binary tree. We skip this subject here because it is treated in detail in the next section.

The first application is about unambiguous tree automata. It is well known that parity automata on infinite trees cannot be determinized. A weaker requirement than determinism is unambiguity. An automaton is called unambiguous if for each object that it accepts there is exactly one accepting run. For example, it is known that all regular languages of infinite words can be accepted by an unambiguous Büchi automaton [CM03] (and deterministic Büchi automata do not suffice to accept all regular ω-languages).

In an unpublished note [NW] Niwiński and Walukiewicz have shown that not every parity tree automaton is equivalent to an unambiguous one.

Theorem 4 ([NW]). *There is no unambiguous parity automaton accepting exactly those $\{0, 1\}$-labeled trees in which at least one node is labeled* 1.

The underlying idea is that the set of 1-labeled nodes represents the set from which an element has to be chosen. Now we assume that $\mathcal{A}$ is a parity automaton accepting the language used in Theorem 4. Using a game that is similar to the emptiness game for tree automata (cf. [Tho97]) one can show that each accepting run on a tree allows us to pick a unique 1-labeled node from the accepted tree in an MSO-definable way. If $\mathcal{A}$ is unambiguous, then this means that for each accepted tree there is a unique run and hence we can build an MSO-formula picking exactly one 1-labeled node for each accepted tree. This yields an MSO-definable choice function.

In fact, the proof in [NW] yields a more general result: Assume that $\mathcal{A}$ is a parity tree automaton (over the alphabet $\{0,1\}$) that does not accept the tree that is completely labeled by 0, i.e., the tree $t(\emptyset)$. Then there is a formula $\phi_{\mathcal{A}}(X,x)$ such that for each $U \subseteq \{0,1\}^*$ for which there is a unique accepting run of $\mathcal{A}$ on $t(U)$, there is a unique element $u \in U$ such that $t(U,u) \models \phi_{\mathcal{A}}$.

We can use this fact to show the following result.

Theorem 5. *There is a regular language $T \subseteq T_{\{0,1\}}$ and a tree $t \in T$ such that there is no parity automaton accepting T that has a unique accepting run for t.*

Proof. Consider the language T of trees with the property that each subtree rooted at a node of the form 1^*0 contains a node labeled 1. The tree t is defined to have all nodes of the form 1^* labeled 0, and as the subtree rooted at the nodes 1^n0 we plug the trees $t_{n,n}$. As each $t_{n,n}$ contains a node labeled 1, we have $t \in T$.

Assume that there is parity automaton $\mathcal{A}$ accepting T that has a unique run on t. Let q be a state of $\mathcal{A}$ that occurs at infinitely many nodes of the form 1^*0 in this run. Let $\mathcal{A}'$ be the automaton $\mathcal{A}$ with initial state q. As $\mathcal{A}$ accepts T it is clear that $\mathcal{A}'$ does not accept the tree $t(\emptyset)$. Furthermore, as the run of $\mathcal{A}$ on t is unique, there are infinitely many n such that $\mathcal{A}'$ has a unique run on $t_{n,n}$. In combination with the result from [NW] that we discussed above, this gives a contradiction to Theorem 3. □

Another application of Theorem 1 concerns the definability of winning strategies in infinite games. In the following we show that there exist game trees that do not admit the definition of winning strategies in MSO.

A game tree is a tuple $\Gamma = (U_1, U_2, \Omega, W)$ where $U_1, U_2 \subseteq \{0,1\}^*$ form a partition of $\{0,1\}^*$, $\Omega : \{0,1\}^* \to \{0,\ldots,n\}$ maps the nodes to a finite set of natural numbers, and $W \subseteq \{0,\ldots,n\}^\omega$ is the winning condition. A play of Γ starts in ε. If the play is currently in $u \in \{0,1\}^*$, then Player 1 or Player 2, depending on whether $u \in U_1$ or $u \in U_2$, chooses $b \in \{0,1\}$ and the next game position is ub. In the limit, such a play forms an infinite word in $\{0,1\}^\omega$. This infinite word corresponds to an infinite sequence over $\{0,\ldots,n\}$ by applying Ω to each prefix. If this sequence is in W, then Player 1 wins, and otherwise Player 2 wins. We identify a play with the corresponding infinite word in $\{0,1\}^\omega$.

A strategy for Player i is a function $f_i : U_i \to \{0,1\}$ and a play γ is played according to f_i if for each prefix $u \in U_i$ of γ, Player 1 uses the strategy to determine the next move, i.e., if $uf_i(u)$ is a prefix of γ. A strategy f_i is winning for Player i if each play γ that is played according to f_i is winning for Player i.

If $W \subseteq \{0,\ldots,n\}^\omega$ is a regular ω-language (i.e. MSO-definable), then for each game tree $\Gamma = (U_1, U_2, \Omega, W)$ one of the players has a winning strategy ([BL69]). If we represent the mapping Ω by sets $\Omega_0, \ldots, \Omega_n \subseteq \{0,1\}^*$, each Ω_i corresponding to the set of nodes mapped to i by Ω, the set of all trees $t(U_1, U_2, \Omega_0, \ldots, \Omega_n)$ such that Player 1 has a winning strategy in $\Gamma = (U_1, U_2, \Omega, W)$ is MSO-definable.

This raises the question whether it is also possible to define winning strategies in MSO. Note that we can represent a strategy for Player 1 by a subset of nodes that contains ε, for each $u \in U_1$ one successor $u0$ or $u1$, and for each $u \in U_2$ both successors $u0$ and $u1$. The plays according to this strategy are exactly the infinite paths contained in this set. If the winning condition is MSO-definable, then the set of all winning strategies for Player 1 is also MSO-definable, i.e., there is a formula $\phi(X_1, X_2, Y_0, \ldots, Y_n, X)$ such that $t(U_1, U_2, \Omega_0, \ldots, \Omega_n, U) \models \phi$ iff U is a winning strategy for Player 1 (for the fixed winning condition W).

We now look at the case where we want to select a single strategy, i.e., we are interested in a formula that defines exactly one strategy. We show in the following that there is even a fixed game tree on which no formula can define a single winning strategy (this tree is similar to the one used in the proof of Theorem 5).

Theorem 6. *There is a game tree $\Gamma = (U_1, U_2, \Omega, W)$ with an MSO-definable winning condition W such that Player 1 has a winning strategy for Γ but there is no MSO-definable winning strategy for Player 1.*

Proof. Consider the following $\{0,1,2\}$-labeled tree t such that for each $n \in \mathbb{N}$ the subtree at the node 1^n0 is isomorphic to $t_{n,n}$, and all nodes of the form 1^n are labeled by 2. The labeling of t defines the mapping Ω, i.e., $\Omega(u) = t(u)$. We let $U_2 = \{1^n \mid n \in \mathbb{N}\}$ and $U_1 = \{0,1\}^* \setminus U_2$. The winning condition W contains all infinite words over $\{0,1,2\}$ that do not contain 0 or that contain a 1.

Intuitively, Player 2 can move along the right branch of the game tree. If he continues like this forever, then he loses because only nodes labeled 2 are visited during the play. Otherwise, he moves to the left at some position, that is, to the root of a subtree $t_{n,n}$. Now Player 1 chooses all the following moves and wins if a node labeled 1 is reached eventually. As each subtree $t_{n,n}$ contains a node labeled 1 it is obvious that Player 1 has a winning strategy.

Assume that there is a winning strategy on t that is MSO-definable by a formula $\phi(X)$ (as the game tree is fixed we omit the other free variables). This formula is equivalent to a parity automaton $\mathcal{A}$ that accepts exactly one strategy labeling of t, corresponding to the winning strategy defined by ϕ.

Using $\mathcal{A}$ we can construct a formula $\phi^*(X, x)$ that chooses exactly one $u \in U_{n,n}$ for each $n \in \mathbb{N}$, contradicting Theorem 3. For this we fix an arbitrary order on the states of $\mathcal{A}$. For each n there is at least one state q of $\mathcal{A}$ such that $\mathcal{A}$ accepts exactly one winning strategy on the subtree $t_{n,n}$ of t (namely the state assumed at the root of the subtree $t_{n,n}$ in an accepting run for the unique winning strategy on Γ that is accepted by $\mathcal{A}$). The formula ϕ^* picks the smallest state q with this property and then chooses the element of $U_{n,n}$ that is described by the unique winning strategy accepted by $\mathcal{A}$ from q on $t_{n,n}$. It is not difficult to verify that this is indeed possible in MSO. □

One should note here that the tree constructed in the proof of Theorem 6 (and also the one from Theorem 5) is not too complicated: it belongs to the Caucal hierarchy[1] ([Cau02]). This means that it can be obtained from a regular tree by a finite number of applications of MSO-interpretations and unfoldings, or

[1] In particular, as all graphs in the Caucal hierarchy have a decidable MSO-theory, the tree constructed in the proof of Theorem 6 also has a decidable MSO-theory.

equivalently, it is the transition graph of a higher-order pushdown automaton ([CW03]).

4 Order

A direct consequence of Theorem 1 is that there exists no MSO-definable well-founded order on the nodes of the infinite binary tree. In fact from an MSO-formula $\varphi_{\leq}(x,y)$ defining a well-founded order on the nodes of the infinite binary tree, a choice function $\varphi_{\text{choice}}(x,X) := x \in X \wedge \forall y, y \in X \rightarrow x \leq y$ is easily defined by taking the smallest element of the set. In this section, we prove the following stronger result.

Theorem 7. *The MSO-theory of the full-binary tree together with any well-founded order is undecidable.*

As the infinite binary tree has a decidable MSO-theory [Rab69], the existence of an MSO-definable well-order would contradict Theorem 7. In the particular case of t_{llex}, the infinite binary tree with length-lexicographic order (formally defined below), this result is well-known [BG00]. We show that t_{llex} can be MSO-interpreted in the infinite binary tree with any well-founded order.

Theorem 8. *There exists an MSO-interpretation $\mathcal{I}$ such that for every well-ordered infinite binary tree t, $\mathcal{I}(t)$ is isomorphic to t_{llex}*

As MSO-interpretations preserve the decidability of MSO, Theorem 7 follows from the undecidability of the MSO-theory of t_{llex}. The rest of this section is dedicated to the proof of Theorem 8.

4.1 Well-Ordered Trees

We consider structures over the binary signature $\mathcal{S} = \{E_0, E_1, \leq\}$. We say that an $\mathcal{S}$-structure t is a well-ordered (infinite binary) tree if it is isomorphic to a well-ordered tree with universe $\{0,1\}^*$ where E_0 and E_1 are respectively interpreted as $\{(u, u0) \mid u \in \{0,1\}^*\}$ and $\{(u, u1) \mid u \in \{0,1\}^*\}$, and $\leq$ is interpreted as a well-founded order on $\{0,1\}^*$. Such an $\mathcal{S}$-structure will be referred to as a canonical well-ordered tree. Up to isomorphism, a well-ordered tree is entirely characterized by the well-founded order on the set of words over $\{0,1\}$.

For example, consider the length-lexicographic order $\leq_{\text{llex}}$ defined by: $u \leq_{\text{llex}} v \Leftrightarrow |u| < |v|$ or ($|u| = |v|$ and $u \leq_{\text{lex}} v$) where $\leq_{\text{lex}}$ refers to the standard lexicographic order. This order is well-founded and we write t_{llex} the canonical well-ordered tree associated to $\leq_{\text{llex}}$. The key property of t_{llex} is that it is MSO-definable (up to isomorphism) in the class of well-ordered trees[2].

Proposition 1. *There exists an MSO-formula φ_{llex} such that for every well-ordered tree t, $t \models \varphi_{llex} \Leftrightarrow t \cong t_{llex}$.*

[2] The class of well-ordered trees is itself MSO-definable in the class of $\mathcal{S}$-structures.

Proof. Consider the MSO-formula φ_{llex} over $\mathcal{S}$ expressing that: the root ε is the smallest element for $\leq$, for all nodes $u \in \{0,1\}^*$ and $v \in 1^*$, the successor of the node $u0v$ is the smallest element for the prefix order of the set $u10^* \setminus \{\text{Succ}(u0v') \mid v' \sqsubset v\}$ where $\text{Succ}(u)$ the successor of u for the order $\leq$ and for all nodes $u \in 1^*$, the successor of u is the smallest element for the prefix order of the set $0^+ \setminus \{\text{Succ}(u') \mid u' \sqsubset u\}$.

It is easy to see that t_{llex} satisfies the formula φ_{llex}. It remains to show that for every well-ordered tree t, if t satisfies φ_{llex} then t is isomorphic to t_{llex}.

Let t be a well-founded tree satisfying φ_{llex}. We can assume w.l.o.g that t is a canonical well-ordered tree. It is therefore enough to show that $t = t_{\text{llex}}$.

For all nodes $u \in \{0,1\}$, we write $\text{Succ}_{\text{llex}}(u)$ for the successor of u for the order $\leq_{\text{llex}}$. As by condition 1 the root ε of t is the minimal element of $\leq$, it is enough to establish that for all $u \in \{0,1\}^*$, $\text{Succ}(u) = \text{Succ}_{\text{llex}}(u)$.

Assume by contradiction that this property is not satisfied. Let u_0 be the smallest node u for the order $\leq_{\text{llex}}$ such that $\text{Succ}(u) \neq \text{Succ}_{\text{llex}}(u)$ We distinguish two cases depending whether u_0 contains an occurrence of 0 or not.

If u_0 contains an occurrence of 0 then u_0 can be uniquely written as $u0v$ with $u \in \{0,1\}^*$ and $v \in 1^*$. The successor of u_0 in the order $\leq_{\text{llex}}$ is $\text{Succ}_{\text{llex}}(u_0) = u10^{|v|}$. By condition 2 of the definition of φ_{llex}, $\text{Succ}(u_0)$ is the smallest element for the prefix order of the set $u10^* \setminus \{\text{Succ}(u0v') \mid v' \sqsubset v\}$.

By minimality (for the order $\leq_{\text{llex}}$) of u_0, we have for all $v' \sqsubset v$ that $\text{Succ}(u0v') = \text{Succ}_{\text{llex}}(u0v') = u10^{|v'|}$. Therefore, $\text{Succ}(u_0)$ is the minimal element for the prefix order of the set $u10^* \setminus \{u10^{|v'|} \mid v' \sqsubset v\}$. This implies that $\text{Succ}(u_0) = u10^{|v|} = \text{Succ}_{\text{llex}}(u_0)$ which contradicts the definition of u_0.

If u_0 does not contain an occurrence of 0 then $u_0 \in 1^*$. The successor of u_0 for the order $\leq_{\text{llex}}$ is $0^{|u_0|+1}$. By condition 3 of the definition of φ_{llex}, $\text{Succ}(u_0)$ is the smallest element for the prefix order of the set $0^+ \setminus \{\text{Succ}(u) \mid u \sqsubset u_0\}$.

By minimality (for the order $\leq_{\text{llex}}$) of u_0, we have for all $u \sqsubset u_0$ that $\text{Succ}(u) = \text{Succ}_{\text{llex}}(u) = 0^{|u|+1}$. Therefore, $\text{Succ}(u_0)$ is the minimal element for the prefix order of the set $0^+ \setminus \{0^{|u|+1} \mid u \sqsubset u_0\}$. This implies that $\text{Succ}(u_0) = 0^{|u_0|+1} = \text{Succ}_{\text{llex}}(u_0)$ which contradicts the choice of u_0. □

4.2 Interpreting t_{llex}

We now define the notion of induced well-ordered tree. Consider a canonical well-ordered tree t and a set $U \subseteq \{0,1\}^*$ of nodes which is closed under greatest common prefix (i.e. $u \in U \wedge v \in V \rightarrow u \wedge v \in U$) and such that for all $u \in U$, $u0\{0,1\}^* \cap U \neq \emptyset$ and $u1\{0,1\}^* \cap U \neq \emptyset$. The well-ordered tree $t|_U$ induced by U in t has universe U and its signature is interpreted as $E_i^{t|_U} = \{(u,v) \in U^2 \mid v \text{ is the smallest element for } \sqsubseteq \text{ of } ui\{0,1\}^* \cap U\}$ for $i \in \{0,1\}$ and $\leq^{t|_U} = \{(u,v) \in U^2 \mid u \leq^t v\}$. It is easy to check that $t|_U$ is a well-ordered tree.

Lemma 3. *For every MSO-formula φ over $\mathcal{S}$, there exists a formula $\varphi^*(X)$ such that for every canonical well-ordered tree t and set U, $t \models \varphi^*[U]$ if and only if the set U induces a well-ordered tree $t|_U$ on t and $t|_U \models \varphi$.*

Proof. Consider an MSO-formula φ over $\mathcal{S}$. Let $\varphi_{\text{ind}}(X)$ be an $\mathcal{S}$-formula expressing that X satisfies the conditions to induce a full binary tree and let $\varphi'(X)$ be the formula obtained from φ by relativizing the quantifications to X and by replacing $E_i(x, y)$ with $y \in xi\{0,1\}^* \cap X \wedge \forall z, z \in xi\{0,1\}^* \cap X \rightarrow y \sqsubseteq z$ for $i \in \{0,1\}$. It is easy to check that the formula $\varphi^*(X) := \varphi_{\text{ind}}(X) \wedge \varphi'(X)$ satisfies the property stated in the lemma. □

We first show that for every canonical well-ordered tree t, there exists a subset $U \subseteq \{0,1\}^*$ such that $t|_U$ is isomorphic to t_{llex}. To construct such a set we need the following technical definition. A node $u \in \{0,1\}^*$ of a canonical well-ordered tree t is *mixed* if for all $v \sqsupseteq u, v' \sqsupseteq u \in \{0,1\}^*$ there exists a $w \in \{0,1\}^*$ such that $v < v'w$.

Lemma 4. *For every canonical well-ordered tree t, there exists a mixed node.*

Proof. Let t be a canonical well-ordered tree. Assume by contradiction that t does not have any mixed nodes. We construct by induction two sequences of nodes $(u_i)_{i \in \mathbb{N}}$ and $(v_i)_{i \in \mathbb{N}}$ such that for all $i \geq 0$, $u_i > u_{i+1}$ and $u_i \geq v_i w$ for all $w \in \{0,1\}^*$.

As ε is not mixed there exist two nodes u and v such that $u \geq vw$ for all $w \in \{0,1\}^*$. We take $u_0 = u$ and to ensure that $u_0 > v_0 w$ for all $w \in \{0,1\}^*$, we pick as v_0 an element of $v\{0,1\}^* \setminus \{u' \mid u' \sqsubseteq u_0\}$.

Assume that both sequences are constructed up to rank $i \geq 0$, we define u_{i+1} and v_{i+1}. As v_i is not mixed, there exists two nodes $u \sqsupseteq v_i$ and $v \sqsupseteq v_i$ such that $u \geq vw$ for all $w \in \{0,1\}^*$. We take u_{i+1} equal to u and v_{i+1} an element of $v\{0,1\}^* \setminus \{u' \mid u' \sqsubseteq u_{i+1}\}$ thus ensuring that for all $w \in \{0,1\}^*$, $u_{i+1} > v_{i+1}w$. By induction hypothesis, we have that $u_i > u_{i+1}$.

The sequence $(u_i)_{i \in \mathbb{N}}$ is an infinite strictly decreasing sequence which contradicts the fact that $\leq$ is a well-founded order. □

Proposition 2. *For every canonical well-ordered tree t, there exists a set of nodes U inducing a well-ordered tree $t|_U$ isomorphic to t_{llex}.*

Proof. Let t be a canonical well-ordered tree. We construct a sequence of nodes $(u_w)_{w \in \{0,1\}^*}$ indexed by the set of words over $\{0,1\}^*$ such that: for all $w, w' \in \{0,1\}^*$, $w \leq_{\text{llex}} w'$ implies $u_w \leq u_{w'}$, and for all $w \in \{0,1\}^*$ and $i \in \{0,1\}$, $u_{wi} \in u_w i\{0,1\}^*$.

If we assume that this sequence has been constructed and we take $U := \{u_w \mid w \in \{0,1\}^*\}$, it is easy to check that U is closed by greatest common prefix and hence U induces a full binary tree on t. Furthermore the mapping from $\{0,1\}^*$ to U associating w to u_w is an isomorphism from t_{llex} to $t|_U$.

We now construct the sequence $(u_w)_{w \in \{0,1\}^*}$ by induction on the length-lexicographic order $\leq_{\text{llex}}$. By Lemma 4, the tree t has a mixed node. We take u_ε to be a mixed node of t. Assume that the sequence has been constructed up to $w_0 \in \{0,1\}^*$, we construct the element u_{w_1} where w_1 is the successor of w_0 for the length-lexicographic order. From the definition of $\leq_{\text{llex}}$, it follows that the element w_1 is equal to $w_2 i$ for some $i \in \{0,1\}$ and $w_2 \leq_{\text{llex}} w_0$. As u_ε is mixed, there exists a v such that $u_{w_0} < u_{w_2}iv$. We take $u_{w_1} = u_{w_2}iv$. By induction hypothesis, for all $w \leq_{\text{llex}} w_1$, we have $u_w \leq u_{w_1}$. □

Note that Theorem 7 can already be derived from the above proposition. For every formula φ over $\mathcal{S}$, consider the formula $\varphi^*(X)$ obtained from φ by Lemma 3 and $\varphi^*_{\text{llex}}(X)$ obtained from the formula φ_{llex} of Proposition 1. By Proposition 2, for every well-ordered tree t, $t \models \exists X,\ \varphi^*_{\text{llex}}(X) \wedge \varphi^*(X)$ if and only if $t_{\text{llex}} \models \varphi$. As the formula $\varphi^*(X)$ can be effectively constructed from the formula φ, it follows that the MSO-theory of t_{llex} is recursive in the MSO-theory of any well-ordered tree t.

We now strengthen the result of Proposition 2 by showing that in every well-ordered tree t there exists an MSO-definable set of nodes inducing a well-ordered tree isomorphic to t_{llex}.

Proposition 3. *For every canonical well-ordered tree, there exists an MSO-definable set of nodes U_0 inducing a well-ordered tree isomorphic t_{llex}.*

Proof. Consider the following MSO-formula $\psi(X)$ defined by:

$$\varphi^*_{\text{llex}}(X) \ \wedge\ \forall x \in X, \forall Z, \quad (X_{<x} < Z \wedge \varphi^*_{\text{llex}}(X_{<x} \cup Z)) \ \rightarrow\ x \leq \min Z$$

where $X_{<x} = \{x' \in X \mid x' < x\}$, $X < Y$ stands for $\forall x \in X, \forall y \in Y, x < y$ and $\min Z$ designates the smallest element of the set Z for the order $\leq$.

Let t be a well-ordered tree. We claim that $t \models \exists^{=1} X, \psi(X)$ (where $\exists^{=1}$ stands for there exists a unique), which establishes the MSO-definability of a set of nodes U_0 inducing on t a well-ordered tree isomorphic to t_{llex}.

The first step is to show that $t \models \exists X, \psi(X)$. For this, we define a sequence of nodes $(u_w)_{w \in \{0,1\}^*}$ of t by induction on the order $\leq_{\text{llex}}$. The node u_ε is the smallest element of $\{\min Z \mid Z \subseteq \{0,1\}^* \wedge t \models \varphi^*_{\text{llex}}[Z]\}$. Proposition 2 guarantees that this set is not empty. Assume that the sequence has been constructed up to the element u_{w_0}. We define u_{w_1} where w_1 is the successor of w_0 for the order $\leq_{\text{llex}}$. We take u_{w_1} as the smallest element for $\leq$ of the set $\{\min Z \mid w_0 < Z \wedge \varphi^*_{\text{llex}}(\{u_w \mid w \leq w_0\} \cup Z\})\}$. This set is not empty by definition of u_{w_0}.

Consider the set $U_0 := \{u_w \mid w \in \{0,1\}^*\}$. It is straightforward to show that U_0 induces a well-ordered tree on t isomorphic to t_{llex}. It follows that $t \models \psi[U_0]$. Hence $t \models \exists X, \psi(X)$.

We now show that $t \models \exists^{=1} X, \psi(X)$. It is enough to show that for all $U_1 \subseteq \{0,1\}^*$, $t \models \psi[U_1]$ implies $U_1 = U_0$. For all $u \in U_0$ (resp. $u \in U_1$), we write $\text{Succ}_0(u)$ (resp. $\text{Succ}_1(u)$) the smallest element of U_0 (resp. of U_1) strictly greater than u.

By induction on w for the order $\leq_{\text{llex}}$, it is easy to show that $u_w \in U_1$ for all $w \in \{0,1\}$ and hence $U_0 \subseteq U_1$. Assume by contradiction that $U_0 \subsetneq U_1$. Let u be the smallest element of $U_1 \setminus U_0$. It is easy to see that both sets have the same minimal element and therefore $u \neq \min U_1$. As U_1 induces a well-ordered tree isomorphic to t_{llex} and since $u \neq \min U_1$, there exists $v \in U_1$ such that $u = \text{Succ}_1(v)$. By minimality of u, v belongs to U_0 and for all $v' \leq v$, $v' \in U_1$ implies $v' \in U_0$. Furthermore $\text{Succ}_0(v)$ which belongs to U_0 is different from

$\mathrm{Succ}_1(v)$ which belongs to U_1. Remarking that U_1 is equal to $(U_1)_{<u} \cup (U_1)_{\geq u} = (U_0)_{\leq v} \cup (U_1)_{\geq u}$and using the fact that $t \models \psi[U_0]$, we obtain that $\mathrm{Succ}_0(v) \leq u$. Similarly U_0 is equal to $(U_0)_{<\mathrm{Succ}_0(v)} \cup (U_0)_{\geq \mathrm{Succ}_0(v)} = (U_1)_{\leq v} \cup (U_0)_{\geq \mathrm{Succ}_0(v)}$ and using the fact that $t \models \psi[U_1]$, we obtain that $u = \mathrm{Succ}_1(v) \leq \mathrm{Succ}_0(v)$. It follows that $u = \mathrm{Succ}_0(v)$ which brings the contradiction. □

Theorem 8 directly follows from Proposition 3. An immediate consequence of this result is that the infinite binary tree cannot be MSO-interpreted in $(\omega, succ)$, i.e., the natural numbers with successor. As a well-founded order can be defined in MSO on $(\omega, succ)$, one could interpret the full binary tree with a well-founded order. From Theorem 7, this structure has an undecidable MSO-theory which would contradict the fact that the MSO-theory of $(\omega, succ)$ is decidable [Büc62]. More generally, we obtain that the infinite binary tree cannot be MSO-interpreted in any structure having both a decidable MSO-theory and an MSO-definable well-founded order. Combining the result that the infinite binary tree cannot be interpreted in $(\omega, succ)$ with the same technique as in the proof of Corollary 1, we can also show that the infinite binary tree cannot be interpreted in $(\omega, succ, P_1, \ldots, P_n)$ for arbitrary fixed predicates P_i.

References

[BG00] Blumensath, A., Grädel, E.: Automatic structures. In: Proc. of LICS '00, pp. 51–62. IEEE Computer Society Press, Los Alamitos (2000)

[BL69] Büchi, J.R., Landweber, L.H.: Solving sequential conditions by finite-state strategies. Transactions of the AMS 138, 295–311 (1969)

[Büc62] Büchi, J.R.: On a decision method in restricted second order arithmetic. In: International Congress on Logic, Methodology and Philosophy of Science, pp. 1–11. Stanford University Press (1962)

[Cau02] Caucal, D.: On infinite terms having a decidable monadic theory. In: Diks, K., Rytter, W. (eds.) MFCS 2002. LNCS, vol. 2420, pp. 165–176. Springer, Heidelberg (2002)

[CM03] Carton, O., Michel, M.: Unambiguous büchi automata. Theor. Comput. Sci. 297(1–3), 37–81 (2003)

[CW03] Carayol, A., Wöhrle, S.: The Caucal hierarchy of infinite graphs. In: Pandya, P.K., Radhakrishnan, J. (eds.) FST TCS 2003: Foundations of Software Technology and Theoretical Computer Science. LNCS, vol. 2914, pp. 112–123. Springer, Heidelberg (2003)

[EF95] Ebbinghaus, H.D., Flum, J.: Finite Model Theory. Springer, Heidelberg (1995)

[GS83] Gurevich, Y., Shelah, S.: Rabin's uniformization problem. J. Symb. Log. 48(4), 1105–1119 (1983)

[LS96] Lifsches, S., Shelah, S.: Uniformization, choice functions and well orders in the class of trees. J. Symb. Log. 61(4), 1206–1227 (1996)

[LS98] Lifsches, S., Shelah, S.: Uniformization and skolem functions in the class of trees. J. Symb. Log. 63(1), 103–127 (1998)

[NW] Niwiński, D., Walukiewicz, I.: Ambiguity problem for automata on infinite trees.(Unpublished note)

[Rab69] Rabin, M.O.: Decidability of second-order theories and automata on infinite trees. Transactions of the American Mathematical Society 141, 1–35 (1969)
[Sie75] Siefkes, D.: The recursive sets in certain monadic second order fragments of arithmetic. Arch. für mat. Logik und Grundlagenforschung 17, 71–80 (1975)
[Tho97] Thomas, W.: Languages, automata, and logic. In: Handbook of Formal Language Theory, vol. III, pp. 389–455 (1997)

Classical and Intuitionistic Logic Are Asymptotically Identical

Hervé Fournier[1], Danièle Gardy[1], Antoine Genitrini[1], and Marek Zaionc[2]

[1] Laboratoire PRiSM
Université de Versailles St-Quentin en Yvelines, 45 av. des États-Unis, 78035 Versailles, France
{herve.fournier, daniele.gardy, antoine.genitrini}@prism.uvsq.fr
[2] Theoretical Computer Science
Jagiellonian University, Gronostajowa 3, 30-387 Kraków, Poland
zaionc@tcs.uj.edu.pl

Abstract. This paper considers logical formulas built on the single binary connector of implication and a finite number of variables. When the number of variables becomes large, we prove the following quantitative results: *asymptotically, all classical tautologies are simple tautologies.* It follows that *asymptotically, all classical tautologies are intuitionistic.*

Keywords: Implicational formulas; Tautologies; Intuitionistic logic; Analytic combinatorics.

1 Introduction

We investigate the proportion between the number of formulas of size n that are tautologies against the number of all formulas of size n for propositional formulas built on implication and k variables. Our interest lays in proving the existence and computing the limit of that fraction when n grows to infinity. This limit can be called the density of truth for the logic with k variables. After isolating the special class of formulas called simple tautologies, of density $1/k + O(1/k^2)$, we exhibit some families of non-tautologies whose cumulated density is $1 - 1/k + O(1/k^2)$. It follows that the fraction of tautologies, for large k, is very close to the lower bound determined by simple tautologies. A consequence is that classical and intuitionistic logics are close to each other when the number of propositional variables is large.

This work is a part of the research in which the likelihood of truth is estimated for the propositional logic with a restricted number of variables. We refer to Gardy [4] for a survey on probability distribution on Boolean functions induced by random Boolean expressions. For the purely implicational logic of one variable, and at the same time simple type systems, the exact value of the density of truth was computed in the paper of Moczurad, Tyszkiewicz and Zaionc [9]. The classical logic of one variable and the two connectors implication and negation was studied in Zaionc [12]. Over the same language, the exact proportion between intuitionistic and classical logics has been determined in Kostrzycka and

J. Duparc and T.A. Henzinger (Eds.): CSL 2007, LNCS 4646, pp. 177–193, 2007.

Zaionc [6]. Some variants involving formulas with other logical connectives have also been considered. The case of and/or connectors received much attention – see Lefmann and Savický [7], Chauvin, Flajolet, Gardy and Gittenberger [1] and Gardy and Woods [5]. Matecki [8] considered the case of the equivalence connector.

We next give a couple of definitions. Section 2 briefly presents the use of enumeration via generating functions and analytic combinatorics, which constitutes the main tool we shall use. The different classes of formulas we consider are described in Section 3, while Section 4 is devoted to the enumeration of these classes and the computation of their densities.

Definition 1. *Let $\{x_1, x_2, \ldots, x_k\}$ a set of Boolean propositional variables. We define $\mathcal{F}_k$ to be the set of all Boolean expressions (or formulas) over these variables and the implication connector $\rightarrow$. Boolean expressions are defined recursively from Boolean variables and the implication connector by the following grammar: $F := x_1 \mid \ldots \mid x_k \mid (F \rightarrow F)$.*

Obviously the expressions can be represented by binary planar trees, suitably labelled: their internal nodes are labelled by the connector $\rightarrow$ and their leaves by some Boolean variables. By $\|\phi\|$ we mean the *size* of expression ϕ which we define as the total number of occurrences of propositional variables in the expression (or leaves in the tree representation of the expression). Parentheses which are sometimes necessary and the implication sign itself are not included in the size of expression. Formally,

$$\|x_i\| = 1 \text{ and } \|\phi \rightarrow \psi\| = \|\phi\| + \|\psi\| .$$

We denote by $\mathcal{F}_k^n$ the set of expressions of $\mathcal{F}_k$ of size n.

We can now define the *canonical form of an expression.* Let T be an expression. It can be decomposed with respect to its right branch – see Figure 1. Hence it is of the form

$$A_1 \rightarrow (A_2 \rightarrow (\ldots \rightarrow (A_p \rightarrow r(T))\ldots));$$

we shall write it

$$T = A_1, \ldots, A_p \rightarrow r(T).$$

The formulas A_i are called the *premises* of T and $r(T)$, the rightmost leaf of the tree, is called the *goal* of T. Of course the expression $T = A_1 \rightarrow (A_2 \rightarrow (\ldots \rightarrow (A_p \rightarrow r(T))\ldots))$ is logically equivalent with $\overline{A_1} \vee \overline{A_2} \vee \ldots \vee \overline{A_p} \vee r(T)$, where $\overline{A_i}$ stands for negation of A_i.

For a subset $\mathcal{A} \subseteq \mathcal{F}_k$ we define the *density* $\mu(\mathcal{A})$ as:

$$\mu(\mathcal{A}) = \lim_{n\rightarrow\infty} \frac{|\{t \in \mathcal{A} : \|t\| = n\}|}{|\{t \in \mathcal{F}_k : \|t\| = n\}|}$$

if the limit exists. The number $\mu(\mathcal{A})$ if it exists is an asymptotic probability (with respect to uniform distribution) of finding a formula from the class $\mathcal{A}$ among all

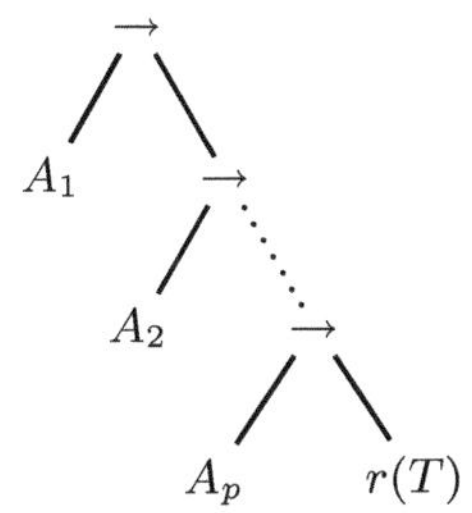

Fig. 1. The canonical decomposition of a tree

formulas from $\mathcal{F}_k$; it can be interpreted as the asymptotic density of the set $\mathcal{A}$ in the set $\mathcal{F}_k$. It can be seen immediately that the density μ is finitely additive so if $\mathcal{A}$ and $\mathcal{B}$ are disjoint classes of formulas such that $\mu(\mathcal{A})$ and $\mu(\mathcal{B})$ exist then $\mu(\mathcal{A} \cup \mathcal{B})$ also exists and $\mu(\mathcal{A} \cup \mathcal{B}) = \mu(\mathcal{A}) + \mu(\mathcal{B})$.

2 Generating Functions

In this paper we investigate the proportion between the number of formulas of size n that are tautologies against the number of all formulas of size n for propositional formulas of the language $\mathcal{F}_k$. Our interest lays in finding the limit of that fraction when n grows to infinity. For this purpose analytic combinatorics has developed an extremely powerful tool, in the form of generating series and generating functions. A nice exposition of the method can be found in Wilf [11], or in Flajolet, Sedgewick [2,3]; see also Gardy [4, 5.2] for a systematic application of these techniques to densities for Boolean functions. As the reader may now expect, while working with propositional logic we will be often concerned with complex analysis, analytic functions and their singularities.

Let $A = (A_0, A_1, A_2, \ldots)$ be a sequence of real numbers. The *ordinary generating series* for A is the formal power series $\sum_{n=0}^{\infty} A_n z^n$. And, of course, formal power series are in one-to-one correspondence to sequences. However, considering z as a complex variable, this series, as known from the theory of analytic functions, converges uniformly to a function $f_A(z)$ in some open disc $\{z \in \mathcal{C} : |z| < R\}$ of maximal diameter, and $R \geqslant 0$ is called its radius of convergence. So with the sequence A we can associate a complex function $f_A(z)$, called the *ordinary generating function* for A, defined in a neighbourhood of 0. This correspondence is one-to-one again (unless $R = 0$), since, as it is well known from the theory of analytic functions, the expansion of a complex function $f(z)$, analytic in a neighbourhood of z_0, into a power series $\sum_{n=0}^{\infty} A_n(z - z_0)^n$ is unique. For F a function in z analytic in a neighbourhood of 0, we shall denote by $[z^n]F$ the coefficient of z^n in the series expansion of F.

Many questions concerning the asymptotic behavior of A can be efficiently resolved by analyzing the behavior of f_A at the complex circle $|z| = R$. This is the approach we take to determine the asymptotic fraction of tautologies and many other classes of formulas among all formulas of a given size.

Each set of expressions is defined recursively from simpler sets; we build the generating functions enumerating the elements of these sets by size (number of leaves), using univariate functions with the variable z marking the leaves, and obtain a generating function $\phi(z)$ for the set under consideration. We then extract the coefficient $[z^n]\phi(z)$ and obtain the density of the set under study as $\lim_{n\to\infty}[z^n]\phi(z)/[z^n]f_k(z)$, $f_k(z)$ being the generating function for the set of all expressions of $\mathcal{F}_k$.

We now recall three constructions on classes of combinatorial objects, and how they translate into ordinary generating functions. Let A and B be two classes of combinatorial objects, with generating functions $f_A(z)$ and $f_B(z)$. The first construction, called *combinatorial sum*, captures the union of disjoint sets. The generating function of the combinatorial sum of A and B is $f_A(z) + f_B(z)$. The second construction called *cartesian product* forms all possible ordered pairs of objects from A and B – the size of (a, b) being the sum of the size of a and b. The generating function enumerating this class is $f_A(z)f_B(z)$. Finally the *sequence construction* builds all sequences of objects from A. Again the size of a sequence of objects is the sum of their size. The generating function enumerating this class is $1/(1 - f_A(z))$.

The Catalan number C_n is defined as the number of full binary trees (i.e. every vertex has either two children or no children) with n internal nodes and $n + 1$ leaves. Basic results about Catalan numbers and its generating function are summarized below.

Proposition 1. *Let $C(z)$ be the generating function enumerating full binary trees with respect to the number of leaves; it satisfies:*

$$C(z) = z + C(z)^2,$$

and is equal to:

$$C(z) = \frac{1 - \sqrt{1 - 4z}}{2}.$$

Its coefficients are

$$[z^{n+1}]C(z) = C_n = \frac{1}{n+1}\binom{2n}{n}.$$

It follows that the number of Boolean expressions of size n over k variables is $k^n C_{n-1}$, since such an expression is obtained by labelling the n leaves with any of the variables $x_1, \ldots, x_k$.

As an example, in the rest of this section we show how we can obtain the generating function $f_k(z)$ for the set of all the expressions built on k variables and the implication connector, before defining several subsets of expressions in Section 3 and computing their generating functions in Section 4.

Proposition 2. *The generating function enumerating the set $\mathcal{F}_k$ of all Boolean expressions over k variables is*

$$f_k(z) = kz\,C(kz) = \frac{1 - \sqrt{1 - 4kz}}{2}.$$

Proof. Using the canonical form of an expression, we know that a tree is a (possibly empty) sequence of trees, followed by a leaf – see Figure 1. The function $f_k(z)$ thus satisfies

$$f_k(z) = \frac{kz}{1 - f_k(z)}, \ ie \ f_k(z) = kz + f_k(z)^2.$$

Solving the equation and choosing between the two possibilities ($f_k(0) = 0$) gives the solution. □

Proposition 2 gives another way to obtain the number of expressions of size n by extracting the coefficients from $f_k(z)$. In the rest of the paper, f_k is abbreviated to f.

Finally, the following basic computations will be used intensively in the rest of the paper. First notice that for all j,

$$\lim_{i \to \infty} \frac{C_i}{C_{i+j}} = \frac{1}{4^j}.$$

Furthermore,

$$[z^n]\sqrt{1 - 4kz} = (4k)^n [z^n]\sqrt{1 - z} = -2k^n C_{n-1}.$$

3 Tautologies and Non-tautologies

Let us now define several classes of expressions, all of them being special kinds of either tautologies or non-tautologies.

Definition 2. *We define the following subsets of $\mathcal{F}_k$:*

- Cl_k *is the set of all* classical tautologies *i.e. formulas which are true under any valuation.*
- Int_k *is the set of all* intuitionistic tautologies *i.e. formulas for which there are closed lambda terms (constructive proofs) of type identical with the formula.*
- $Pierce_k$ *is the set of all* Pierce expressions *i.e. classical tautologies which are not intuitionistic ones.*
- SN_k *is the set of* simple expressions which are not classical tautologies, *defined as*

  $$T = A_1, \dots, A_p \to r(T),$$

  such that for all i, $r(A_i) \neq r(T)$.

- G_k *is the set of* simple tautologies *i.e. expressions that can be written as*

  $$T = A_1, \dots, A_p \to r(T),$$

  where there exists i *such that* A_i *is a variable equal to* $r(T)$.

– LN_k *is the set of* less simple expressions that are not classical tautologies, *defined as the set of trees of the form*

$$T = B_1, \ldots, B_{i-1}, C, B_i, \ldots, B_p \rightarrow r(T),$$

such that

$$C = C_1, C_2, \ldots, C_q \rightarrow r(C),$$

where $r(C) = r(T)$, $q \geqslant 1$, *and*

$$C_1 = D_1, D_2, \ldots, D_s \rightarrow r(D),$$

where $r(D) \neq r(T)$, $s \geqslant 0$, *and the following holds: for all* j, $r(B_j) \notin \{r(T), r(D)\}$ *and* $r(D_j) \notin \{r(T), r(D)\}$.

Adding a superscript n to the sets we have just defined means that we consider only expressions of size exactly n (the tree that represents the expression has n leaves).

Note that simple tautologies are instuitionistic ones since one of the premises is equal to the goal. The obvious relations between classes above are the following.

$$\begin{aligned} SN_k \cup LN_k &\subset \mathcal{F}_k \setminus Cl_k \\ SN_k \cap LN_k &= \emptyset \\ G_k \subsetneq Int_k \subsetneq\ Cl_k &\subsetneq\ \mathcal{F}_k \setminus (SN_k \cup LN_k) \\ Pierce_k &= Cl_k \setminus Int_k \end{aligned}$$

Our aim in the rest of this paper will be to compute the densities of these sets. Results are summed up in Figure 2; proofs are given in the following section. As a consequence, we obtain the following result, giving a positive answer to the conjecture of [9, page 593].

Theorem 1. *Asymptotically (for a large number k of Boolean variables), all tautologies are simple i.e.*

$$\lim_{k \to \infty} \frac{\mu(G_k)}{\mu(Cl_k)} = 1.$$

Proof. We know that for any k the density of classical logic with k propositional variables $\mu(Cl_k)$ exists. Such a result is obtained by standard technics in analysis of algorithms; we skip the details and refer the interested reader to Flajolet and Sedgewick [3] or to [4].

Since $G_k \subset Cl_k \subset \mathcal{F}_k \setminus (SN_k \cup LN_k)$, and from the densities obtained in propositions 3, 4 and 5, we have

$$\frac{4k+1}{(2k+1)^2} = \mu(G_k) \leqslant \mu(Cl_k) \leqslant 1 - \left(\frac{k(k-1)}{(k+1)^2} + \frac{2k(k-1)^2}{(k+2)^4} \right).$$

The upper and lower bounds are asymptotically identical, equal to $1/k + O(1/k^2)$. □

Using the very same argument we can also obtain a result relating the asymptotic behavior of classical versus intuitionistic logics.

Corollary 1. *Asymptotically (for a large number k of Boolean variables), classical tautologies are intuitionistic i.e.*

$$\lim_{k\to\infty} \frac{\mu^-(Int_k)}{\mu(Cl_k)} = 1$$

where $\mu^-(Int_k) = \liminf_{n\to\infty} \frac{|Int_k^n|}{|\mathcal{F}_k^n|}$.

Proof. From the fact that $G_k \subset Int_k \subset Cl_k$, we have

$$\mu(G_k) = \lim_{n\to\infty} \frac{|G_k^n|}{|\mathcal{F}_k^n|} \leqslant \liminf_{n\to\infty} \frac{|Int_k^n|}{|\mathcal{F}_k^n|} \leqslant \limsup_{n\to\infty} \frac{|Int_k^n|}{|\mathcal{F}_k^n|} \leqslant \lim_{n\to\infty} \frac{|Cl_k^n|}{|\mathcal{F}_k^n|} = \mu(Cl_k).$$

The result follows from the fact that both $\mu(G_k)$ and $\mu(Cl_k)$ are equal to $1/k + O(1/k^2)$. □

This result also allows to estimate the size of the difference between classical and intuitionistic logics (so called Pierce formulas). Details are given in section 4.4.

4 Enumeration of Classes

We now compute the densities of the three sets SN_k, G_k and LN_k. The computation of these densities is done in a systematic way. First each set of expressions is defined recursively from simpler sets; this allows to build the generating functions enumerating the elements of these sets by their size (the number of leaves), and to obtain a generating function ϕ for the considered class. Then we extract the coefficient $[z^n]\phi(z)$ and obtain the density of the set under study as $\lim_{n\to\infty}[z^n]\phi(z)/[z^n]f(z)$ – we recall that f denotes the generating function of all formulas.

The last part deals with Pierce formulas. Although we don't know if this set of formulas has a density, we give some bounds and show that their order is $\Theta(1/k^2)$.

4.1 Simple Non-tautologies

We first consider the set SN_k of simple expressions that are non-tautologies. If $T \in SN_k$, then T is of the kind

$$T = A_1, \ldots, A_p \to r(T),$$

such that for all i, $r(A_i) \neq r(T)$. We first check that this is indeed not a tautology. Just consider the following valuation of propositional variables. Define $r(T)$ as *false* and all $r(A_i)$ as *true*. Under this valuation the whole expression is *false*. Let us next compute the generating function $SN(z)$ associated to SN_k.

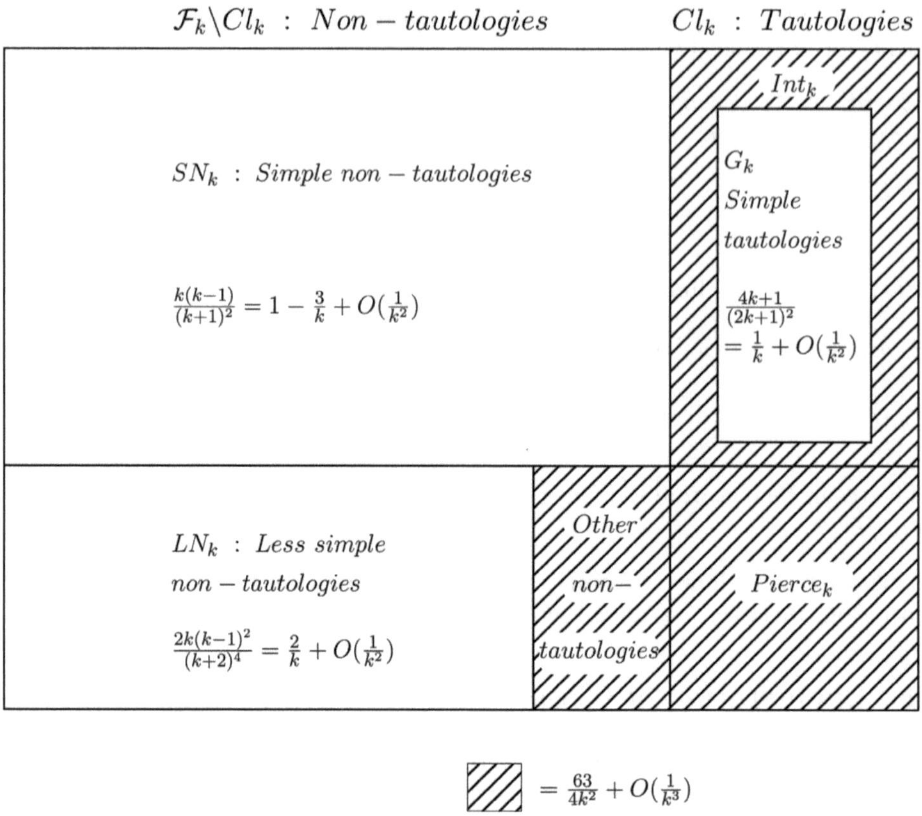

Fig. 2. Densities of simple tautologies, simple and less simple non-tautologies

First fix a Boolean variable α and consider all trees with $r(T) = \alpha$. Such a tree is a simple non-tautology if and only if all its premises A_i satisfy $r(A_i) \neq \alpha$. The generating function of all possible premises is $\frac{k-1}{k} f(z)$. As a simple non-tautology with goal α is a sequence of such premises followed by the leaf α, the generating function SN^α of simple non-tautologies with goal α is equal to

$$SN^\alpha(z) = \frac{z}{1 - \frac{k-1}{k} f(z)}.$$

Since α can be chosen arbitrarily among the k literals, we have $SN(z) = k \cdot SN^\alpha(z)$, which gives

$$SN(z) = \frac{kz}{1 - \frac{k-1}{k} f(z)}.$$

Proposition 3. *The density of simple non-tautologies exists and is equal to*

$$\mu(SN_k) = \frac{k(k-1)}{(k+1)^2}.$$

For large k, this density is $1 - 3/k + O(1/k^2)$.

Proof. This result was already given in the paper [9, page 586], with a different proof. We give an alternative proof here. If it exists, the density is given by the following formula:

$$\mu(SN_k) = \lim_{n\to\infty} \frac{|SN_k^n|}{|\mathcal{F}_k^n|} = \lim_{n\to\infty} \frac{[z^n]SN(z)}{[z^n]f(z)}.$$

After modifying the denominator of the generating function $SN(z)$, we obtain :

$$SN(z) = \frac{k(k+1)z + kz(1-k)\sqrt{1-4kz}}{2(1+z(k-1)^2)}.$$

The denominator of the rational fraction $SN(z)$ has a unique zero $\rho = -1/(k-1)^2$. However this value also cancels the numerator of the expression since

$$k(k+1)\rho + k(1-k)\rho\sqrt{(-\rho)((k-1)^2+4k)} = 0.$$

So ρ is not an actual pole. Hence the only singularity that matters asymptotically is $z = 1/4k$. Putting aside the error term, we obtain

$$[z^n]SN(z) = -\frac{2k^2(k-1)}{(k+1)^2}[z^{n-1}]\sqrt{1-4kz} = \frac{4k(k-1)}{(k+1)^2}k^n C_{n-2}.$$

This gives

$$\mu(SN_k) = \lim_{n\to\infty} \frac{|SN_k^n|}{|\mathcal{F}_k^n|} = \frac{4k(k-1)}{(k+1)^2} \lim_{n\to\infty} \frac{C_{n-2}}{C_{n-1}} = \frac{k(k-1)}{(k+1)^2},$$

hence the density of SN_k exists and is equal to $k(k-1)/(k+1)^2$. □

4.2 Simple Tautologies

If T is a simple tautology, then T can be written as

$$T = A_1, \ldots, A_p \to r(T),$$

with one of the A_i equal to $r(T)$. It is straightforward to check that T is indeed a tautology since it is logically equivalent with

$$T \sim \overline{A_1} \vee \ldots \vee \overline{r(T)} \vee \ldots \vee \overline{A_p} \vee r(T).$$

which obviously evaluates to *true*.

Let us now compute the generating function of simple tautologies. A tree T is not a simple tautology if and only if all its premises are different from $r(T)$ – see figure 3. The generating function for $\mathcal{F}_k \setminus G_k$ is therefore equal to $kz/(1-(f(z)-z))$. It follows that the generating function of G_k is

$$G(z) = f(z) - \frac{kz}{1+z-f(z)}.$$

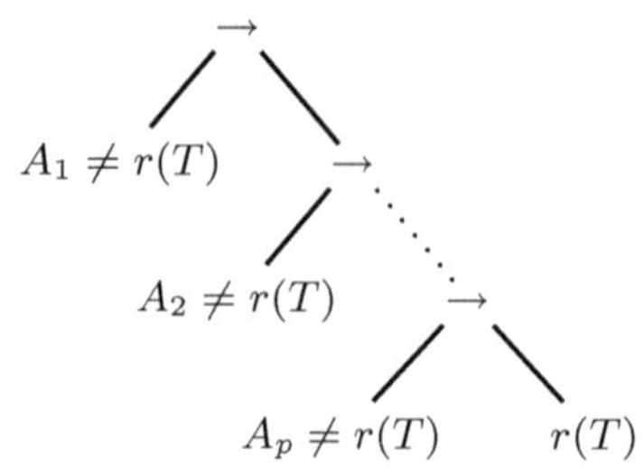

Fig. 3. Trees that are not simple tautologies

Proposition 4. *The limit density of simple tautologies on k variables exists and is equal to*

$$\mu(G_k) = \frac{4k+1}{(2k+1)^2}.$$

For large k, this density is asymptotically equal to $1/k - 3/4k^2 + O(1/k^3)$.

Proof. Another, earlier proof of this result is given in the paper [9, page 584]. We give here an alternative proof. The generating function $G(z)$ can be written as

$$G(z) = \frac{P(z) - (1+z)\sqrt{1-4kz}}{2(1+k+z)},$$

with $P(z)$ a suitable polynomial. Let ρ be its pole; $\rho = -1-k$. But ρ is larger that the algebraic singularity $1/(4k)$; hence $1/(4k)$ is the dominant singularity of $G(z)$. Finally we obtain (up to the error term)

$$\begin{aligned}[z^n]G(z) &= -\frac{2k}{(2k+1)^2}[z^n]\sqrt{1-4kz} - \frac{2k}{(2k+1)^2}[z^{n-1}]\sqrt{1-4kz} \\ &= \frac{4k}{(2k+1)^2}k^n C_{n-1} + \frac{4}{(2k+1)^2}k^n C_{n-2}.\end{aligned}$$

Let us prove the existence and compute the value of the density of G_k^n.

$$\begin{aligned}\mu(G_k) = \lim_{n\to\infty}\frac{|G_k^n|}{|F_k^n|} &= \lim_{n\to\infty}\left(\frac{4k}{(2k+1)^2}k^n C_{n-1} + \frac{4}{(2k+1)^2}k^n C_{n-2}\right)\cdot\frac{1}{k^n C_{n-1}} \\ &= \frac{4k}{(2k+1)^2} + \frac{4}{(2k+1)^2}\cdot\lim_{n\to\infty}\frac{C_{n-2}}{C_{n-1}}.\end{aligned}$$

Hence $\mu(G_k)$ does exist, and is equal to $(4k+1)/(2k+1)^2$. □

4.3 Less Simple Non-tautologies

In the family SN_k of simple non-tautologies, we did not allow any premise to have a rightmost leaf equal to $r(T)$. But here we will consider trees with exactly one such premise.

We recall that a tree T defines a less simple non-tautology if it is of the kind

$$T = B_1, \ldots, B_{i-1}, C, B_i, \ldots, B_p \to r(T),$$

where $C = C_1, \ldots, C_q \to r(C)$, with $r(C) = r(T)$, $q \geqslant 1$, and $C_1 = D_1, D_2, \ldots, D_s \to r(D)$ is such that $r(D) \neq r(T)$, $s \geqslant 0$, and the following holds: for all j, $r(B_j) \notin \{r(T), r(D)\}$ and $r(D_j) \notin \{r(T), r(D)\}$. See figure 4 for the general form of the tree and figure 5 for the subtree C; in these figures, if a subtree A is underlined, it means that it is subject to the constraint $r(A) \notin \{r(T), r(D)\}$.

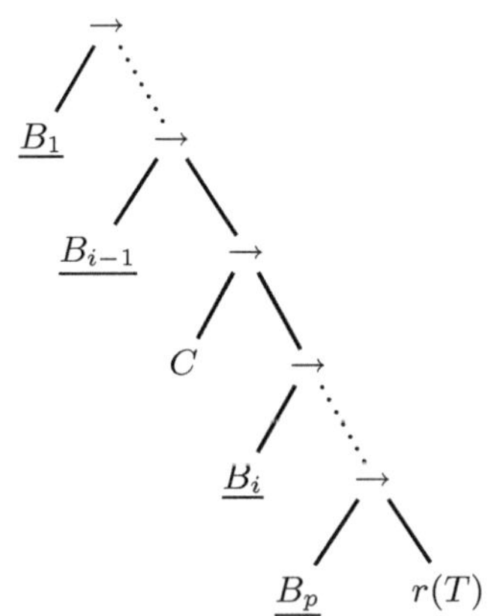

Fig. 4. Less simple non-tautologies

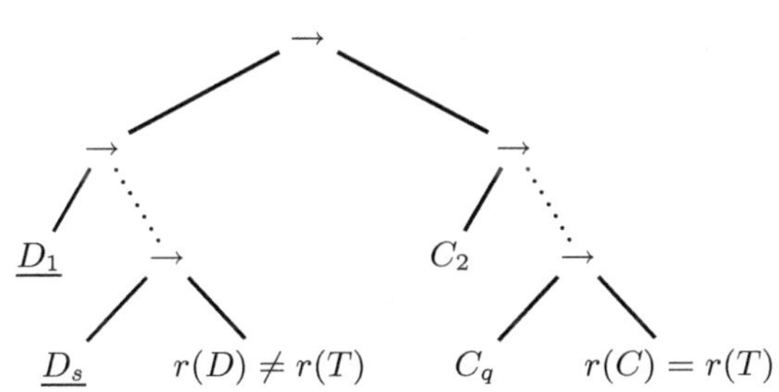

Fig. 5. Subtree C of a less simple non-tautology

Let us first prove that such a tree is not a tautology. For this, consider the assignement where all the variables are *true*, except $r(T)$ and $r(D)$ which are *false*; under this assignement, the whole expression evaluates to *false* – to check this, just notice that the function computed by such a tree can be developed into a conjuction of terms, one of them being $\bigvee_i \overline{r(B_i)} \vee r(T) \vee \bigvee_i \overline{r(D_i)} \vee r(D)$.

We shall now compute the generating function of LN_k. Let us fix α and β two distinct variables. We shall first compute $\psi(z)$ the generating functions of all trees $LN_k^{\alpha,\beta}$ from LN_k such that $r(T) = \alpha$ and $r(D) = \beta$. By symmetry, $\psi(z)$ is independent of the choice of α and β.

Let $b(z)$ be the generating function of all trees $T \in \mathcal{F}_k$ such that $r(T) \notin \{\alpha, \beta\}$. Of course $b(z) = (k-2)/k \cdot f(z)$. This generating function enumerates the possible subtrees B_j but also the possible subtrees D_j. Thus, the generating function of all possible trees for D is $d(z) = z/(1 - b(z))$, since it is a sequence of trees D_j such that $r(D_j) \notin \{\alpha, \beta\}$, followed by the leaf β. In the same way, the generating function for the subtree C is $c(z) = d(z) \cdot 1/(1 - f(z)) \cdot z$. Note that a tree of $LN_k^{\alpha,\beta}$ is built as a sequence of trees B_j with $r(B_j) \notin \{\alpha, \beta\}$, then a subtree C as described as above, another sequence of trees B_j with $r(B_j) \notin \{\alpha, \beta\}$, followed by the leaf α. Moreover, this decomposition is unique. The generating function for $LN_k^{\alpha,\beta}$ is thus equal to

$$\psi(z) = \frac{1}{1 - b(z)} c(z) \frac{1}{1 - b(z)} z.$$

Now it can be easily seen that LN_k is the disjoint union of the $LN_k^{\alpha,\beta}$. Indeed, given a tree $T \in LN_k$, then α is equal to $r(T)$ and the premise C of T is uniquely determined because it is the only premise of T with goal $r(T)$. Thus, β is uniquely determined as well since it is the goal of the first premise of C. It follows that $\phi(z) = k(k-1)\psi(z)$.

Proposition 5. *The density of less simple non-tautologies is equal to*

$$\mu(LN_k) = \frac{2k(k-1)^2}{(k+2)^4}.$$

For large k it is equal to $2/k + O(1/k^2)$.

Proof. After modifying the denominator of the generating function $\phi(z)$, we obtain :

$$\phi(z) = \frac{P(z) + k(k-1)(-k^2 + (2k^3 - 6k^2 + 8)z)z^2\sqrt{1 - 4kz}}{2(2 + (k-2)^2 z)^3},$$

where $P(z)$ is a suitable polynomial. The denominator of the rational fraction $\phi(z)$ has a zero $\rho = -2/(k-2)^2$. However this value also cancels the numerator (and its first two derivatives) of the expression, and is not an actual pole of ϕ. Hence the only singularity that matters asymptotically is $z = 1/4k$. Putting aside the error term, we obtain:

$$\begin{aligned}
[z^n]LN(z) &= -\frac{k^3(k-1)}{2(2 + \frac{(k-2)^2}{4k})^3}[z^{n-2}]\sqrt{1 - 4kz} \\
&\quad + \frac{k(k-1)(2k^3 - 6k^2 + 8)}{2(2 + \frac{(k-2)^2}{4k})^3}[z^{n-3}]\sqrt{1 - 4kz} \\
&= \frac{k^{n+1}(k-1)}{(2 + \frac{(k-2)^2}{4k})^3} C_{n-3} - \frac{k^{n-2}(k-1)(2k^3 - 6k^2 + 8)}{(2 + \frac{(k-2)^2}{4k})^3} C_{n-4}.
\end{aligned}$$

Let us prove the existence and compute the value of the density of LN_k^n:

$$\begin{aligned}
\mu(LN) &= \lim_{n\to\infty} \frac{|LN_k^n|}{|F_k^n|} \\
&= \lim_{n\to\infty} \Big(\frac{k^{n+1}(k-1)}{(2+\frac{(k-2)^2}{4k})^3} \frac{C_{n-3}}{k^n C_{n-1}} - \frac{k^{n-2}(k-1)(2k^3-6k^2+8)}{(2+\frac{(k-2)^2}{4k})^3} \frac{C_{n-4}}{k^n C_{n-1}} \Big) \\
&= \frac{64k^4(k-1)}{(k+2)^6} \cdot \lim_{n\to\infty} \frac{C_{n-3}}{C_{n-1}} - \frac{64k(k-1)(2k^3-6k^2+8)}{(k+2)^6} \cdot \lim_{n\to\infty} \frac{C_{n-4}}{C_{n-1}} \\
&= \frac{4k^4(k-1) - k(k-1)(2k^3-6k^2+8)}{(k+2)^6} = \frac{2k(k-1)^2}{(k+2)^4}.
\end{aligned}$$

This density does exist, and is equal to:

$$2k(k-1)^2/((k+2)^4).$$

For large k this is asymptotically equal to $2/k + O(1/k^2)$. □

4.4 Pierce Formulas

We are ready to estimate the number of Pierce formulas. Although we don't know if the set of Pierce formulas has a density, we shall give bounds on $\limsup_{n\to\infty} \frac{|Pierce_k^n|}{|\mathcal{F}_k^n|}$ and $\liminf_{n\to\infty} \frac{|Pierce_k^n|}{|\mathcal{F}_k^n|}$. A simple upper bound on $Pierce_k$ can be obtained from

$$Pierce_k = Cl_k \setminus Int_k \subset F_k \setminus (SN_k \cup LN_k \cup G_k).$$

Since SN_k, LN_k and G_k are disjoint we have a simple upper estimation based on propositions 3, 4 and 5:

$$\limsup_{n\to\infty} \frac{|Pierce_k^n|}{|\mathcal{F}_k^n|} \leqslant 1 - \frac{k(k-1)}{(k+1)^2} - \frac{2k(k-1)^2}{(k+2)^4} - \frac{4k+1}{(2k+1)^2} = \frac{63}{4k^2} + O(\frac{1}{k^3}).$$

However, we can obtain a sharper bound on the number of Pierce formulas. For this, we next bound the density of tautologies which are not simple – this density exists since we already know that both the density of all tautologies and the density of simple tautologies exist. Note that this result gives an alternative proof for Theorem 1.

Lemma 1. *The density of non simple tautologies T such that exactly one premise has a goal equal to $r(T)$ is bounded from above by $5/k^2 + O(1/k^3)$.*

Proof. Let A be a non simple tautology with goal $r(A) = \alpha$. Let p be the number of premises of A. We call B the premise of A whose goal is $r(A)$ and $\alpha_1, \ldots, \alpha_{p-1}$ the goal of the premises other than B. By hypothesis, $\alpha_i \neq \alpha$ for all $i \in \{1, \ldots, p-1\}$. Of course B cannot be reduced to a leaf (otherwise A would be a simple tautology). Let us decompose $B = (B_1, \ldots, B_m, \alpha)$, with $m \geqslant 1$. As

$\overline{B} = B_1 \wedge \ldots \wedge B_m \wedge \overline{\alpha}$, by developing the expression A, we obtain that necessarily, for all $j \in \{1, \ldots, m\}$,

$$B_j \vee \overline{\alpha}_1 \ldots \vee \overline{\alpha}_{p-1} \vee \alpha$$

computes *true*. Let us denote $\mathcal{C}_{(\alpha_1,\ldots,\alpha_{p-1},\alpha)}$ the set of trees such that

$$C \vee \overline{\alpha}_1 \ldots \vee \overline{\alpha}_{p-1} \vee \alpha$$

computes *true*. Let $C \in \mathcal{C}_{(\alpha_1,\ldots,\alpha_{p-1},\alpha)}$.

- If C is reduced to a leaf γ then necessarily $\gamma \in \{\alpha_1, \ldots, \alpha_{p-1}\}$.
- Otherwise, let us decompose $C = (C_1, \ldots, C_s, \gamma)$ with $s \geqslant 1$. Let $\gamma_i = r(C_i)$. Then

  $$\overline{\gamma_1} \vee \ldots \vee \overline{\gamma_s} \vee \gamma \vee \overline{\alpha_1} \ldots \vee \overline{\alpha_{p-1}} \vee \alpha$$

  has to evaluate to *true*. It follows that $\alpha \in \{\gamma_1, \ldots, \gamma_s\}$ or $\gamma \in \{\gamma_1, \ldots, \gamma_s, \alpha_1, \ldots, \alpha_{p-1}\}$.

We shall now compute a generating function $c_{(\alpha_1,\ldots,\alpha_{p-1},\alpha)}$ giving an upper bound on the number of trees of $\mathcal{C}_{(\alpha_1,\ldots,\alpha_{p-1},\alpha)}$. Let us define

$$c_{(\alpha_1,\ldots,\alpha_{p-1},\alpha)}(z) = (p-1)z + \frac{1}{1-((k-1)/k)f(z)} \cdot \frac{f(z)}{k} \cdot \frac{1}{1-f(z)} \cdot kz + \sum_{s=1}^{\infty} f(z)^s \cdot (s+p-1)z$$

the first term corresponding to the first point above, the second term corresponding to the case $\alpha \in \{\gamma_1, \ldots, \gamma_s\}$ and the third term to the case $\gamma \in \{\gamma_1, \ldots, \gamma_s, \alpha_1, \ldots, \alpha_{p-1}\}$. This generating function depends only on p; thus we shall now denote it by c_p. Let us now define

$$b_p(z) = \frac{c_p(z)}{1 - c_p(z)} \cdot z.$$

This function gives an upper bound on the number of trees B (for $p \geqslant 1$ and $\alpha, \alpha_1, \ldots, \alpha_{p-1}$ fixed) such that

$$B \vee \overline{\alpha}_1 \ldots \vee \overline{\alpha}_{p-1} \vee \alpha$$

computes *true*. Of course

$$b_p(z) \leqslant \tilde{b}_p(z) := c_p(z) + \frac{(c_p(z))^2}{1 - f(z)}.$$

We now define

$$a_p(z) = p \cdot ((k-1)/k \cdot f(z))^{p-1} \cdot \tilde{b}_p(z) \cdot z \cdot k.$$

The generating function a_p gives an upper bound on the number of non simple tautologies A with p premises, exactly one of them having a goal equal to $r(A)$. Indeed, z corresponds to $r(A) = \alpha$, k corresponds to the choice of α among the literals and p corresponds to the position of the unique premise with goal α.

We now define $a(z) = \sum_{p=1}^{\infty} a_p(z)$. This function bounds the number of non simple tautologies A with only one premise with goal $r(A)$. The computation based on the generating function defined above leads to an asymptotic density $5/k^2 + O(1/k^3)$. □

Lemma 2. *The density of non simple tautologies T such that exactly two premises have a goal equal to $r(T)$ is $O(1/k^3)$.*

Proof. Let us consider a non simple tautology A with exactly two premises B_1 and B_2 having a goal equal to $r(A)$. Let $\alpha_1, \ldots, \alpha_{p-2}$ the goals of the other premises. Since A is not simple, both B_1 and B_2 are not reduced to a leaf. Let C be the first premise of B_1, and D be the first premise of B_2. Let γ be the goal of C and $\gamma_1, \ldots, \gamma_s$ the goals of its premises (with $s \geqslant 0$). We define $\delta, \delta_1, \ldots, \delta_t$ the corresponding literals for the tree D. Since A is a tautology we can argue as in the previous lemma and we obtain that necessarily

$$\overline{\gamma_1} \vee \ldots \vee \overline{\gamma_s} \vee \gamma \vee \overline{\delta_1} \vee \ldots \vee \overline{\delta_t} \vee \delta \vee \overline{\alpha_1} \ldots \vee \overline{\alpha_{p-2}} \vee \alpha$$

evaluates to *true*. The same method as in the previous lemma (not detailed here) leads to a density $O(1/k^3)$. □

Lemma 3. *The asymptotic density of trees T such that at least three premises have a goal equal to $r(T)$ is $O(1/k^3)$.*

Proof. The generating function of this family of trees is equal to

$$\left(\frac{1}{1-(k/(k-1))f(z)} \cdot \frac{f(z)}{k}\right)^3 \cdot \frac{1}{1-f(z)} \cdot kz.$$

We obtain a density $O(1/k^3)$. □

Proposition 6. *The asymptotic density of non simple tautologies is bounded from above by $5/k^2 + O(1/k^3)$.*

Proof. A tautology is not reduced to a leaf. Moreover, a tautology T has (at least) a premise with goal $r(T)$: otherwise, it would be a simple non-tautology. The density of non simple tautologies is thus bounded from above by the sum of the three densities obtained in lemmas 1, 2 and 3. Hence it is bounded above by $5/k^2 + O(1/k^3)$. □

We can obtain a lower bound for Pierce formulas by the following argument. Consider special formulas from $\mathcal{F}_k$ of the form $((a \to T) \to a) \to a$ where $T = A_1, \ldots, A_p \to r(T)$ is a simple non-tautology taken from $\mathcal{F}_k$ (see section 4.1) and variable a differs from $r(T)$. We observe that $((a \to T) \to a) \to a$ must be a Pierce formula. It is obviously a classical tautology. Suppose $((a \to T) \to a) \to a$ is also an intuitionistic tautology. It means that there must exist a closed term of the type $((a \to T) \to a) \to a$. The long normal form of this term has the form $\lambda p_{(a\to T)\to a}.p(\lambda q_a.t)$ where t is a term of type T with only free variables p and q. Consider a closed term $\lambda p_{(a\to T)\to a}\lambda q_a.t$. The type of this term is the implicational formula

$$((a \to T) \to a) \to (a \to T).$$

But this type is again a simple non-tautology since the variables a and $r(T)$ are different. So the formula is unprovable classically and therefore intuitionistically too; contradiction. For more details about relation between intuitionistic logic and lambda calculus consult for example Sørensen, Urzyczyn [10].

Now we have to count this family. The number of such formulas is $(k-1) \cdot |SN_k^{n-3}|$. Thus the density of this special set of Pierce formulas exists and is equal to

$$\lim_{n\to\infty} \frac{(k-1)\cdot|SN_k^{n-3}|}{|\mathcal{F}_k^n|} = \lim_{n\to\infty} \frac{(k-1)\cdot|SN_k^{n-3}|}{|\mathcal{F}_k^{n-3}|} \cdot \frac{|\mathcal{F}_k^{n-3}|}{|\mathcal{F}_k^n|} = \frac{1}{64k^2}\frac{(k-1)^2}{(k+1)^2}$$

since $\lim_{n\to\infty} |\mathcal{F}_k^{n-3}|/|\mathcal{F}_k^n| = 1/(4k)^3$.

Proposition 7. *We have the following bounds on the number of Pierce formulas:*

$$\frac{1}{64k^2} - O\left(\frac{1}{k^3}\right) \leqslant \liminf_{n\to\infty} \frac{|Pierce_k^n|}{|\mathcal{F}_k^n|} \leqslant \limsup_{n\to\infty} \frac{|Pierce_k^n|}{|\mathcal{F}_k^n|} \leqslant \frac{5}{k^2} + O\left(\frac{1}{k^3}\right).$$

Proof. The lower bound comes from the previous discussion. Since Pierce formulas are non simple tautologies, the upper bound is a consequence of proposition 6. □

5 Final Remarks

We have shown that asymptotically, all tautologies over implication are simple, i.e. one of the premises is equal to the goal. The method developed in this paper extends to the logic of implication with both positive and negative literals. In this new setting again, we can prove that most of the tautologies, when the number of variables becomes large, exhibit a very simple structure; more precisely, most of the tautologies have one of their premises equal to the goal (as before), or have two of their premises which are opposite literals.

Some questions remain about the set of Pierce formulas. We conjecture that for any k, the densities $\mu(Int_k)$ and $\mu(Pierce_k)$ exist. If it is the case, it would be interesting to evaluate the asymptotic densities of these sets.

References

1. Chauvin, B., Flajolet, P., Gardy, D., Gittenberger, B.: And/Or trees revisited. Combinatorics, Probability and Computing 13(4-5), 475–497 (2004)
2. Flajolet, P., Sedgewick, R.: Analytic combinatorics: functional equations, rational and algebraic functions, INRIA, Number 4103 (2001)
3. Flajolet, P., Sedgewick, R.: Analytic combinatorics. Book in preparation (2007), available at `http://algo.inria.fr/flajolet/Publications/books.html`
4. Gardy, D.: Random Boolean expressions, Colloquium on Computational Logic and Applications. In: Proceedings in DMTCS, Chambery (France), June 2005, pp. 1–36 (2006)
5. Gardy, D., Woods, A.: And/or tree probabilities of Boolean function. Discrete Mathematics and Theoretical Computer Science , 139–146 (2005)

6. Kostrzycka, Z., Zaionc, M.: Statistics of intuitionnistic versus classical logic. Studia Logica 76(3), 307–328 (2004)
7. Lefmann, H., Savický, P.: Some typical properties of large And/Or Boolean formulas. Random Structures and Algorithms 10, 337–351 (1997)
8. Matecki, G.: Asymptotic density for equivalence. Electronic Notes in Theoretical Computer Science 140, 81–91 (2005)
9. Moczurad, M., Tyszkiewicz, J., Zaionc, M.: Statistical properties of simple types. Mathematical Structures in Computer Science 10(5), 575–594 (2000)
10. Sørensen, M., Urzyczyn, P.: Lectures on the Curry-Howard Isomorphism. Studies in Logic and the Foundations of Mathematics 149 (2006)
11. Wilf, H.: Generatingfunctionology, 2nd edn. Academic Press, Boston (1994)
12. Zaionc, M.: On the asymptotic density of tautologies in logic of implication and negation. Reports on Mathematical Logic 39, 67–87 (2005)
13. Zaionc, M.: Probability distribution for simple tautologies. Theoretical Computer Science 355(2), 243–260 (2006)

Qualitative Temporal and Spatial Reasoning Revisited

Manuel Bodirsky[1] and Hubie Chen[2]

[1] Department Algorithms and Complexity, Humboldt-Universität zu Berlin
bodirsky@informatik.hu-berlin.de
[2] Departament de Tecnologies de la Informació i les Comunicacions, Universitat Pompeu Fabra, Barcelona
hubie.chen@upf.edu

Abstract. Establishing local consistency is one of the main algorithmic techniques in temporal and spatial reasoning. In this area, one of the central questions for the various proposed temporal and spatial constraint languages is whether local consistency implies global consistency. Showing that a constraint language Γ has this "local-to-global" property implies polynomial-time tractability of the constraint language, and has further pleasant algorithmic consequences.

In the present paper, we study the "local-to-global" property by making use of a recently established connection of this property with universal algebra. Specifically, the connection shows that this property is equivalent to the presence of a so-called quasi near-unanimity polymorphism of the constraint language. We obtain new algorithmic results and give very concise proofs of previously known theorems. Our results concern well-known and heavily studied formalisms such as the point algebra and its extensions, Allen's interval algebra, and the spatial reasoning language RCC-5.

1 Introduction

Temporal and spatial reasoning is a subdiscipline in Artificial Intelligence that developed in the 1990s, and has many applications, for instance in natural language processing, geographic information systems, computational biology, and document interpretation; for references, see the monograph [FGV05] and the survey [RN07]. A common reasoning task in this field is to decide, given a set of relationships concerning temporal events or spatial regions, whether or not there exists a model fulfilling all of the relationships. It is well-acknowledged that instances of this reasoning task may be modelled using the *constraint satisfaction problem (CSP)*, a computational problem in which the input consists of a set of constraints on variables, and the question is whether or not there is an assignment to the variables satisfying all of the constraints. In this vein, a famous example from temporal reasoning is the CSP for *Allen's Interval Algebra*, where the variables denote intervals in time, and the constraints talk about relationships between intervals such as containment, overlap, and so forth [All83].

J. Duparc and T.A. Henzinger (Eds.): CSL 2007, LNCS 4646, pp. 194–207, 2007.

CSPs for temporal and spatial reasoning have been studied from a computational complexity perspective by restricting the sets of relationships that may be used. In the CSP formalism, this amounts to restricting the set of predicates that may be used to form constraints; we call such a restricted predicate set a *constraint language*, following the CSP literature.

A primary algorithmic technique for solving CSPs in spatial and temporal reasoning is the process of establishing k-consistency, an inferential process that yields a problem that is k-consistent: any partial solution on $(k-1)$ variables can be extended to any other variable. The notions of consistency that we employ in this paper are due to [Mac77, Fre82, DvB97]; formal definitions are provided later (see Section 2). For some constraint languages, it is known that (for some constant k) establishing k-consistency implies global consistency, the property of being i-consistent for all i. We refer to this property of constraint languages as the "local-to-global" property. Showing that a constraint language possesses this property implies that it is polynomial-time tractable, and has further desirable algorithmic consequences; for instance, we demonstrate a connection to a quantifier elimination algorithm for the *quantified* constraint satisfaction problem over the constraint language. One of the central questions for the various temporal and spatial constraint languages is to understand which such languages enjoy the "local-to-global" property.

In this paper, we study this question by making use of algebraic techniques for studying the complexity of constraint languages that have recently come into focus (see for instance the surveys [BJK05, CJ06]). Specifically, a fundamental result [BKJ05] associates to every constraint language an algebra in a way that permits the use of universal algebraic concepts and methods in the study of the complexity and algorithmic behavior of constraint languages. Utilizing this algebraic perspective, we both derive new algorithmic results and give very concise proofs of previously known theorems. We thus establish connections between two research areas–temporal and spatial reasoning, and algebraic techniques for constraint languages–that have up to the present seen little interaction, but indeed, as evidenced by our results, unite quite naturally. We hope that the present work serves to stimulate further interaction between these two areas.

Before giving more detail on our contributions, we describe how exactly we view temporal and spatial reasoning problems within the CSP framework, and which tools from CSP theory we employ. First, it should be pointed out that much of the work on CSP complexity has focused on constraint languages over *finite* domains. For such constraint languages, an exact algebraic characterization of the "local-to-global" property is known, namely, a constraint language has this property if and only if it has a so-called *near-unanimity polymorphism.* This characterization was presented in [JCC98] and in [FV99], and in part can be seen as reformulation of a classical result in universal algebra [BP74].

On the other hand, many temporal and spatial reasoning problems are naturally formulated as the CSP over constraint languages with *infinite* domains. Fortunately, a number of such problems can be formulated as the CSP over a language from the class of *ω-categorical constraint languages*, a class of languages that is known to be relatively manageable from the algebraic and logical viewpoints [BN06, Bod05]. Formulability in this class is well-known for the Point Algebra, and for Allen's interval algebra and all its fragments [Hir96]. Moreover, it has recently been shown that the mentioned algebraic characterization of the "local-to-global" property essentially remains valid for ω-categorical structures [BD06]. Precisely, it has been shown that an ω-categorical constraint language has this property if and only if it has a so-called *quasi near-unanimity polymorphism*.

Contributions and Outline. Sections 2 and 3 recall fundamental facts from the theory of constraint satisfaction that allow us to study temporal and spatial constraint languages algebraically. In Section 4 we prove results concerning the existence and properties of quasi near-unanimity polymorphisms for infinite posets. One of the applications of this result is a new and concise proof of the result of Koubarakis that $(2k+1)$-consistency (but not $2k$-consistency) implies global tractability for the Point Algebra with disjunctions of disequalities on at most k variables [Kou97] (discussed in Section 5). In Section 5, we provide characterizations of the fragments of the point algebra in terms of quasi near-unanimity polymorphisms. We also show that if we extend the constraint language for the Point Algebra to contain disjunctions of disequalities, the corresponding (uniform) *quantified CSP* can be solved in NL, giving a strict extension of the result of Koubarakis [Kou97].

In Section 6, we study the spatial reasoning formalism RCC-5. We first formulate the corresponding CSP with an ω-categorical constraint language, and then show that the so-called *basic relations* of RCC-5 possess the "local-to-global" property. In fact, similarly as we do for Koubarakis' result on the Point Algebra, we show that $(2k+1)$-consistency (but not $2k$-consistency) implies global consistency for the basic relations of RCC-5 with disjunctions of disequalities on at most k variables.

In the full version of this paper, we also present a general technique for establishing the "local-to-global" property for various temporal and spatial constraint languages, which is based on the model-theoretic concept of primitive positive interpretations. We apply this technique to the pointizable fragment of Allen's interval algebra: we show that the basic relations of the rectangular algebra have a 5-ary quasi near-unanimity polymorphism, and hence that they have the "local-to-global" property.

2 Preliminaries

Notation. A *relational signature* τ is a set of *relation symbols* R_i, each of which has an associated finite *arity* k_i. A *relational structure* Γ over the signature τ (also called τ-*structure*) is a set D_Γ (the *domain*) together with a relation $R_i \subseteq D_\Gamma^{k_i}$ for each relation symbol of arity k_i from τ. For simplicity, we use the same symbol for a relation symbol and the corresponding relation. If necessary, we write R^Γ to indicate that we are talking about the relation R belonging to the structure Γ. If a relational structure Δ can be obtained from a relational structure Γ by removing some of the relations from the structure and the signature of Γ, then Δ is called a *reduct* of Γ, and Γ is called an *expansion* of Δ.

For a subset S of D_Γ, we write $\Gamma[S]$ for the substructure of Γ induced by S. In this paper, substructure always means *induced substructure*, as in [Hod97]. An embedding of a τ-structure Γ in a τ-structure Δ is a mapping $f : D_\Gamma \to D_\Delta$ that is an isomorphism between Γ and $\Delta[f(D_\Gamma)]$.

The Constraint Satisfaction Problem. A *constraint language* is simply a relational structure; we typically refer to a relational structure Γ as a constraint language when we are interested in the constraint satisfaction or the quantified constraint satisfaction problem for Γ, which are defined below.

A first-order formula ϕ is called a *τ-formula* if all symbols in ϕ are either the standard logical symbols $\{\exists, \forall, \wedge, \vee, \neg, =\}$, variable symbols, or from τ. A first-order τ-formula ϕ is called a τ-sentence if ϕ has no free variables. A first-order τ-formula is called *primitive positive* (for short, *pp*) if it is of the form

$$\exists x_1, \ldots, x_n . \psi_1 \wedge \cdots \wedge \psi_m \quad ,$$

where each ψ_i is an atomic τ-formula (a formula $x = y$ or $R(x_1, \ldots, x_k)$ for $R \in \tau$) that can contain both free variables and quantified variables from $\{x_1, \ldots, x_n\}$. The *constraint satisfaction problem (CSP) for Γ* is the computational problem to determine for a given primitive positive τ-sentence Φ whether Φ is true in Γ.

A first-order τ-formula is *conjunctive positive* if it has the form

$$Q_1 v_1 \ldots Q_n v_n (\psi_1 \wedge \ldots \wedge \psi_m),$$

where each Q_i is a quantifier from $\{\forall, \exists\}$, and each ψ_i is an atomic τ-formula that can contain both free variables and quantified variables from $\{v_1, \ldots, v_n\}$. The *quantified constraint satisfaction problem* for Γ, denoted by $QCSP(\Gamma)$, is the problem of deciding for a given conjunctive positive τ-sentence whether or not the formula is true in Γ. Note that both the universal and existential quantification is understood to take place over the entire universe of Γ.

Let R be a k-ary relation and let Γ be a τ-structure. We say that R has a *pp-definition* (or is *pp-definable*) in Γ if there exists a pp-formula ϕ with free variables $x_1, \ldots, x_k$ such that $R(x_1, \ldots, x_k) = \phi(x_1, \ldots, x_k)$. Analogously, we define the concept of a *cp-definition.* The constraint language that contains all pp-definable relations in Γ is denoted by $\langle \Gamma \rangle$.

Amalgamation. As we have already mentioned in the introduction, it turns out that many CSPs in temporal and spatial reasoning (in particular, if they concern so-called *qualitative* formalisms [RN07]) can be formulated with constraint languages that are *ω-categorical.* A relational structure Γ is called *ω-categorical* if all countable models of the first-order theory[1] of Γ are isomorphic to Γ.

[1] The first-order theory of a τ-structure Γ is the set of all τ-sentences that are true in Γ.

To formulate computational problems for temporal and spatial calculi as constraint satisfaction problems with ω-categorical constraint languages, the following concept is very powerful. The *age* of a relational structure Γ is the set of finite structures that embed into Γ (this is terminology that goes back to Fraïssé [Fra86]). A class of finite relational structures $\mathcal{C}$ is an *amalgamation class* if $\mathcal{C}$ is nonempty, closed under isomorphisms and taking substructures, and has the *amalgamation property*, which says that for all $A, B_1, B_2 \in \mathcal{C}$ and embeddings $e_1 : A \to B_1$ and $e_2 : A \to B_2$ there exists $C \in \mathcal{C}$ and embeddings $f_1 : B_1 \to C$ and $f_2 : B_2 \to C$ such that $f_1 e_1 = f_2 e_2$.

A structure is *homogeneous* (sometimes called *ultra-homogeneous* [Hod97]) if every isomorphism between finite substructures of Γ can be extended to an automorphism.

Theorem 1 (Fraïssé [Fra86]). *A countable class $\mathcal{C}$ of finite relational structures with countable signature is the age of a countable homogeneous structure Γ if and only if $\mathcal{C}$ is an amalgamation class. In this case Γ is up to isomorphism unique and called the* Fraïssé-limit *of $\mathcal{C}$.*

Homogeneous structures provide a rich source of ω-categorical structures.

Proposition 2 (see e.g. [Hod97]). *A countable homogeneous structure Γ over a finite relational signature is ω-categorical.*

The following is well-known; a proof can be found in [BD06]. The second part of the proposition is not proven there, but can be shown analogously.

Proposition 3. *Let Δ be a countable structure and let Γ be ω-categorical. Then Δ homomorphically maps to Γ if and only if all finite (induced, or equivalently weak) substructures of Δ homomorphically map to Γ. The structure Δ injectively homomorphically maps to Γ if and only if all finite induced substructures of Δ injectively homomorphically map to Γ.*

Polymorphisms. The *(direct-, categorical-, or cross-) product* $\Gamma_1 \times \Gamma_2$ of two relational τ-structures Γ_1 and Γ_2 is a τ-structure on the domain $D_{\Gamma_1} \times D_{\Gamma_2}$. For all relations $R \in \tau$ the relation $R\big((x_1, y_1), \ldots, (x_k, y_k)\big)$ holds in $\Gamma_1 \times \Gamma_2$ iff $R(x_1, \ldots, x_k)$ holds in Γ_1 and $R(y_1, \ldots, y_k)$ holds in Γ_2. Homomorphisms from $\Gamma^k = \Gamma \times \ldots \times \Gamma$ to Γ are called *polymorphisms* of Γ. If $f : D^k \to D$ is a polymorphism of a relational structure (D, R), we also say that f *preserves* the relation R (and otherwise f *violates* the relation R).

The set of all polymorphisms of Γ gives rise to an algebra $\mathrm{Al}(\Gamma)$, defined as follows. The domain of the algebra equals the domain of Γ, and the algebra has a function (and an associated function symbol) for each polymorphisms of Γ. For finite domain constraint languages, Bulatov et al. [BKJ05] give a detailed exposition of this concept. A property of the algebra $\mathrm{Al}(\Gamma)$ is that it is *locally closed*, i.e., if f is a k-ary operation such that for every finite subset A of the domain there is a k-ary operation g in $\mathrm{Al}(\Gamma)$ such that $f(x) = g(x)$ for all $x \in A^k$, then f is also an operation in $\mathrm{Al}(\Gamma)$. (Note that this property is non-trivial only in the case that Γ has an infinite domain.)

We say that a polymorphism f of an ω-categorical structure Γ is *oligopotent* if the diagonal of f, that is, the function $f(x, \ldots, x)$, is contained in the locally closed clone generated by the automorphisms of Γ.

The importance of polymorphisms stems from the following powerful preservation theorems that characterize primitive positive and conjunctive positive definability over ω-categorical structures.

Theorem 4 (of [BN06]). *Let Γ be an ω-categorical structure. Then a relation R is pp-definable in Γ if and only if it is preserved by all polymorphisms of Γ.*

Theorem 5 (of [BC07b]). *Let Γ be an ω-categorical structure. A relation R is cp-definable in Γ if and only if it is preserved by all surjective polymorphisms of Γ.*

3 Consistency and QNUFs

Consistency. Establishing 2- and 3-consistency (defined below) are the most prominent algorithmic techniques for constraint satisfaction, due to their wide applicability in practical applications. On the other hand, the question whether a (quantified) constraint satisfaction problem can be solved in polynomial time by consistency techniques leads to challenging theoretical problems. We first introduce the basic definitions, and then present the mentioned connection to universal algebra.

Definition 6. *Let Γ be a relational structure. An instance A of CSP(Γ) is called* strongly k-consistent *if for every subset $S = \{v_1, \ldots, v_l\}$ with $l \leq k$ of the elements of A and every homomorphism h from $A[\{v_1, \ldots, v_{l-1}\}]$ to Γ there exists an extension of h that is a homomorphism from $A[S]$ to Γ.*

An important feature of k-consistency is that for every fixed k and for every finite or ω-categorical structure Γ there is an algorithm that *establishes strong k-consistency* for a given instance A of CSP(Γ), i.e., computes a strongly k-consistent instance B that is logically equivalent to A; to formalize this idea, we need the following definition.

Definition 7. *Let Γ be constraint language, and let Δ be an expansion of Γ by finitely many primitive positive definable relations of Γ (in particular, Γ and Δ are defined on the same domain D). We say that an instance B of CSP(Δ) is a* k-consistent variant *of an instance A of CSP(Γ) if B is k-consistent, has the same set of variables $V(A)$ as A, and every mapping from $V(A)$ to D is a solution for B if and only if it is a solution for A.*

In Proposition 8 below, we state a fact that is well-known for constraint languages over a finite domain; for ω-categorical constraint languages a proof can be found in [BD06]. We would like to remark that the algorithms that are used in the proof of Proposition 8 can be formulated as Datalog programs (Datalog programs can be seen as Prolog programs without function symbols, and are a well-studied concept in Database theory and finite model theory; see e.g. [AHV95, EF99]).

Proposition 8. *Let Γ be a finite or an ω-categorical structure over a finite relational language. Then for every k there is a polynomial-time algorithm that computes a k-consistent variant of a given instance A of CSP(Γ).*

Proof. This is a well-known fact for constraint languages Γ over a finite domain. For ω-categorical structures Γ, the statement follows from [BD06]. □

In artificial intelligence and in the theory of relation algebras, strong 3-consistency is usually called *path-consistency*, and the algorithm in Proposition 8 that computes for a given instance of the CSP a strongly 3-consistent variant is called the *path-consistency* algorithm.

Note that if a k-consistent variant B of an instance A of CSP(Γ) contains a constraint with a relation symbol that denotes the empty relation, then A does not have a solution. The converse need not be true in general.

Definition 9. *A constraint language Γ has* width k *when: a strongly $k+1$-consistent instance A of CSP(Γ) has a solution if and only if A does not contain a constraint with a relation symbol that denotes the empty relation.*

Global Consistency. Some constraint languages Γ have the strong property that every strongly k-consistent instance of CSP(Γ) is automatically *globally consistent* ([Fre82], see Definition 10 below).

Definition 10. *An instance A of CSP(Γ) is called* globally consistent *iff it is k-consistent for all $1 \leq k \leq |A|$.*

It follows easily from Proposition 8 that if Γ has the property that every strongly k-consistent instance of CSP(Γ) is globally consistent, then CSP(Γ) can be solved in polynomial time.

But note that Γ might have width k, while at the same time strong k-consistency does not establish global consistency. A well-known example over a two-element domain are boolean constraint languages that are preserved by the maximum operation (the relations in such a constraint language can be defined by Horn clauses). A well-known example over an infinite domain is CSP($\mathbb{Q}, \leq, \neq$). Vilain, Kautz and van Beek [KvBV90] have shown that this CSP has width 2, but establishing strong 3-consistency does not imply global consistency (however, establishing strong 5-consistency implies global consistency, due to a result by Koubarakis [Kou97]).

Quasi Near-Unanimity Functions. We now present the connection of the "local-to-global" property to universal algebra mentioned in the introduction.

Definition 11. *A function $f : D \to D$ is called a* quasi near-unanimity function *(short, a* QNUF*), if it satisfies $f(x,\ldots,x,y) = f(x,\ldots,x,y,x) = \cdots = f(y,x,\ldots,x) = f(x,\ldots,x)$ for all $x, y \in D$.*

As an example, consider the structure ($\mathbb{Q}, \leq$), and the operation *median*, which is the ternary function that returns the median of its three arguments. More precisely, for three elements x, y, z from $\mathbb{Q}$, suppose that $\{x, y, z\} = \{a, b, c\}$, where $a \leq b \leq c$. Then $median(x, y, z)$ is defined to have value b. It is easy to verify that *median* is a ternary quasi near-unanimity function, and that it is a polymorphism of $(\mathbb{Q}, \leq, <)$.

The following two results are of central importance in this paper, and they generalize well-known facts for finite structures Γ [BP74, FV99, JCC98].

Theorem 12 (of [BD06]). *An ω-categorical structure Γ has a k-ary oligopotent QNU-polymorphism if and only if every strongly k-consistent instance of CSP(Γ) is globally consistent.*

There is yet another characterization of constraint languages having a QNU-polymorphism. We say that a constraint language Γ is *k-decomposable* if every relation in Γ can be defined by a conjunction of at most k-ary primitive positive definable relations in Γ.

Theorem 13 (of [BC07a]). *An ω-categorical structure Γ has a $(k+1)$-ary oligopotent QNU-polymorphism if and only if Γ is k-decomposable.*

Innermost quantifier elimination. In this section, we show that if an ω-categorical constraint language has a surjective oligopotent QNU polymorphism, then $QCSP(\Gamma)$ can be solved in polynomial time. This was already known for constraint languages over finite domains, see e.g. [Che]. The same idea that was applied there can be applied for ω-categorical constraint languages.

Theorem 14. *Let Γ be an ω-categorical constraint language with a surjective oligopotent QNU polymorphism. Then QCSP(Γ) is in P.*

Proof (Sketch). The algorithm eliminates the variables of a given instance ϕ of $QCSP(\Gamma)$, starting from the innermost variable v_n, and computes a conjunctive-positive sentence ϕ' that is equivalent to ϕ, without the variable v_n. To do so, the algorithm first establishes k-consistency on the quantifier-free part of ϕ (viewing this part as an instance of the CSP; here we use Proposition 8 and the assumption that Γ is ω-categorical). If a constraint for the empty relation was derived during establishing k-consistency, then the algorithm reports that the instance is false over Γ. If v_n is existentially quantified, the algorithm simply removes all constraints of arity at least $k+1$ that contain v_n. This leads to an equivalent instance, essentially because Theorem 13 implies that Γ is k-decomposable. If v_n is universally quantified, then the universal quantifier distributes over all the conjunctively combined constraints ψ. Because the QNU polymorphism of Γ is surjective, it does not only preserve the primitive positive definable relations, but also the cp-definable relations in Γ (Theorem 5), and in particular it preserves the relation defined by $\forall v_n.\psi$. So we can without loss of generality assume that this relation is already a relation of Γ. Then, we replace ψ by a constraint for this relation.

Clearly, we can then proceed with the next variable v_{n-1} in the same fashion. This process can be iterated, and if the algorithm never report that the instance is false, it eventually shows that the sentence is true. □

4 Posets

In this section, we develop some general results on QNU polymorphisms of posets, which we will utilize in the following sections. We use $[k]$ to denote the first k natural numbers, $\{1, \ldots, k\}$.

Definition 15. *Relative to a poset $(D, \leq)$, we say that $b \in D$ is the* middle value *of a tuple $t = (t_1, \ldots, t_k)$ (with $k \geq 3$) if there exists a permutation $\pi : [k] \to [k]$ such that $t_{\pi(1)} \leq \cdots \leq t_{\pi(k)}$ and $b = t_{\pi(2)} = \cdots = t_{\pi(k-1)}$. In this situation, we call $t_{\pi(1)}$ the* low value *of t, and $t_{\pi(k)}$ the* high value.

Note that not every tuple has a middle value, but when a tuple does have a middle value, it has a unique middle value.

Definition 16. *Relative to a poset $(D, \leq)$, we say that $b \in D$ is the* main value *of a tuple $t = (t_1, \ldots, t_k)$ (with $k \geq 3$) if either b is the middle value of the tuple t, or b occurs in (at least) $k - 1$ coordinates of t.*

Again, not every tuple has a main value, but when a tuple has a main value, it is unique.

Using the definition of main value, we can now define an equivalence relation on the set of all tuples of length k. Let $(D, \leq)$ be a poset and let t, t' be two tuples of length k. We write $t \equiv_m t'$ if either $t = t'$ or t and t' have the same main value. We show that any QNUF on a poset must map equivalent tuples to the same point.

Proposition 17. *Let $\Gamma = (D, \leq)$ be a poset, and let $f : \Gamma^k \to \Gamma$ be a QNU polymorphism of Γ. For two tuples t, t' of length k, if $t \equiv_m t'$, then $f(t) = f(t')$. (Here, we assume $k \geq 3$).*

We now show that, under some mild assumptions, a poset has QNU polymorphisms of all arities which only identify together tuples that are equivalent under our equivalence relation $\equiv_m$.

Theorem 18. *Let $\Gamma = (D, \leq)$ be an ω-categorical poset such that there is an injective homomorphism from $(\mathbb{N}, \leq)$ to Γ. For all $k \geq 3$, there exists a k-ary QNU polymorphism $f : \Gamma^k \to \Gamma$ such that for all tuples $t, t' \in D^k$, it holds that $t \equiv_m t'$ if and only if $f(t) = f(t')$. Moreover, if $R \subseteq D^m$ is definable by a disjunction of disequalities over m variables and $k \geq 2m + 1$, the resulting polymorphism $f : \Gamma^k \to \Gamma$ preserves R.*

We say that a relation $R \subseteq D^m$ is definable by a disjunction of disequalities over m variables if there exists a disjunction of disequalities ϕ over variables $\{v_1, \ldots, v_m\}$ such that $R(v_1, \ldots, v_m) \equiv \phi(v_1, \ldots, v_m)$. As an example, consider the relation R such that $R(v_1, v_2, v_3, v_4, v_5) \equiv (v_1 \neq v_2 \vee v_2 \neq v_3 \vee v_4 \neq v_5)$. Here, we assume that each of the variables $v_1, \ldots, v_m$ appears in some disequality. Disjunctions of disequalities were studied in the context of temporal reasoning by Koubarakis [Kou01], and we may mention also that they appear in the classification result of [BC07b].

The following proposition complements Theorem 18, showing that the given arity bound for preserving disjunctions of disequalities is optimal.

Proposition 19. *Let $(D, \leq)$ be a poset with three distinct values $a, b, c \in D$ such that $a \leq b \leq c$. Let $R \subseteq D^m$ be a relation definable by a disjunction of disequalities. There is no QNUF of arity $2m$ preserving R.*

5 The Point Algebra, Fragments, and Extensions

The Point Algebra is one of the most fundamental formalisms for temporal reasoning. The corresponding constraint language is

$$\Gamma^{PA} = (\mathbb{Q}; \leq, <, >, \geq, \neq, =, \mathbb{Q}^2, \emptyset)$$

with the obvious interpretation to the eight binary relation symbols of this structure. We would like to remark that formally, the Point Algebra is a relation algebra [LM94, Due05], which has a natural representation by the relational structure Γ^{PA}. The notion of relation algebra and representations of relation algebras are not of importance in this paper, and it suffices to study Γ^{PA} for our purposes [LM94, Due05].

In this section, we study Γ^{PA}, its fragments, and its extensions from the algebraic viewpoint. We begin with Γ^{PA} itself. Koubarakis [Kou97] showed that strong 5-consistency implies global consistency for CSP(Γ^{PA}). In combination with Theorem 12, this implies the following.

Theorem 20. *The point algebra has an oligopotent QNU polymorphism of arity 5, but no oligopotent QNU polymorphism of arity 4.*

We observe that we may obtain an alternative proof of this theorem from results in the previous section. Note that via Theorem 12, this also leads to an alternative proof of the theorem of Koubarakis.

Proof (Theorem 20). It is clear that all relations of Γ^{PA} have a primitive positive definition in $(\mathbb{Q}, \leq, \neq)$, and we therefore work with this constraint language instead of Γ^{PA}. The statement then follows immediately from Theorem 18 and Proposition 19. □

It can be shown that the image of the constructed polymorphism is a dense subset of the rational numbers. It is then not hard to see from local closure arguments and the homogeneity of Γ^{PA} that Γ^{PA} also has a *surjective* 5-ary QNU polymorphism. Therefore, Theorem 14 shows that *QCSP*(Γ^{PA}) is in P. We will see a stronger algorithmic result for the QCSP over Γ^{PA} at the end of this section.

5.1 Fragments

Here, we give algebraic characterizations of fragments of the point algebra. If $f_1, f_2, \ldots$ are operations from $\mathbb{Q}^k \to \mathbb{Q}$, then $Inv_{\mathbb{Q}}(f_1, f_2, \ldots)$ denotes the set of relations with a first-order definition in $(\mathbb{Q}, <)$ that is preserved by all operations $f_1, f_2, \ldots$. The three binary relations $\mathbb{Q}^2$, $\emptyset$, and $=$ are trivial from our standpoint, as we study pp-definability; other than these relations, there are five binary first-order definable relations: $>$, $<$, $\geq$, $\leq$, and $\neq$. Note that the relation $<$ is the intersection of the relations $\leq$ and $\neq$.

We first consider two fragments with two binary relations.

Theorem 21. $\langle(\mathbb{Q}, <, \leq)\rangle = Inv_{\mathbb{Q}}(median)$.

Recall that *median* (defined in Section 3) is a QNUF and preserves $<$ and $\leq$.

Theorem 22. *There exists a QNUF* $f : \mathbb{Q}^3 \to \mathbb{Q}$ *of arity* 3 *such that* $\langle(\mathbb{Q}, <, \neq)\rangle = Inv_{\mathbb{Q}}(f)$.

Theorem 23. $\langle(\mathbb{Q}, \leq)\rangle = Inv_{\mathbb{Q}}(median, 0)$. *Here,* 0 *denotes the unary operation always equal to the constant* 0.

Theorem 24. $\langle(\mathbb{Q}, <)\rangle = Inv_{\mathbb{Q}}(median, f)$ *where* f *is the operation from Theorem 22.*

Theorem 25. $\langle(\mathbb{Q}, \neq)\rangle = Inv_{\mathbb{Q}}(-f)$ *where* f *is the operation from Theorem 22.*

5.2 Extensions

Koubarakis [Kou97] showed that establishing strong $2k+1$-consistency implies global consistency for the extension of Γ^{PA} by disjunctions of disequalities on at most k variables, but strong $2k$ consistency is not sufficient. By Theorem 18 and Proposition 19 of Section 4 in combination with Theorem 12, we have a new proof of this theorem.

Now let us consider the constraint language Γ that is the extension of the point algebra with all disjunctions of disequalities. This is a constraint language with infinitely many relations. What is the complexity of Γ? We can *not* derive the tractability of Γ using a QNUF, as it is immediate from Proposition 19 that Γ has no QNUF polymorphism. We show here that this extension of the point algebra is in fact tractable[2]; indeed, we show that the QCSP over this constraint language is solvable in NL.

Theorem 26. *Let* Γ *be the extension of the point algebra having arbitrary disjunctions of disequalities. The problem* $QCSP(\Gamma)$ *is in NL. (Note that we assume that constraints are presented syntactically, for instance, as "*$x \leq y$*" or "*$x \neq y \vee y \neq z$*".)*

---

[2] In the constraint satisfaction literature, this form of tractability of the CSP where the instances might contain arbitrary constraints built from an infinite constraint language is called *uniform* tractability.

6 Spatial Reasoning

In this section, we study constraint satisfaction problems that arise from the spatial calculus known under the name *RCC-5* [RCC92, Ben94, JD97, RN99].

We first introduce the constraint satisfaction problems as in [JD97], and later discuss how to formulate the same problems with ω-categorical templates. Many different but equivalent ways to introduce these problems appeared in the literature, see e.g. [JD97, LM94, Due05, RN99, Ben94]; the formulation with an ω-categorical template presented here appears to be new.

Let X be a countably infinite set. Consider the relational structure $B_0 = (2^X \setminus \{\emptyset\}, \mathtt{DR}, \mathtt{PO}, \mathtt{PP}, \mathtt{PPI}, \mathtt{EQ})$, whose elements are the non-empty subsets of X, and whose relations are defined as follows.

$$\begin{aligned}
\mathtt{DR}(X,Y) &\text{ iff } X \cap Y = \emptyset \\
\mathtt{PO}(X,Y) &\text{ iff } \exists a,b,c. a \in X \setminus Y, b \in X \cap Y, c \in Y \setminus X \\
\mathtt{PP}(X,Y) &\text{ iff } X \subset Y \\
\mathtt{PPI}(X,Y) &\text{ iff } X \supset Y \\
\mathtt{EQ}(X,Y) &\text{ iff } X = Y
\end{aligned}$$

Note that for each pair (U,V) of elements of B_0 exactly one of the relations $\mathtt{DR}, \mathtt{PO}, \mathtt{PP}, \mathtt{PPI}, \mathtt{EQ}$ holds.

The structure B_0 is not ω-categorical. To formulate the problem CSP(B_0) (and also the CSP for RCC-5) with a countably infinite ω-categorical structure F, we use Fraïssé's theorem. The resulting structure is known in model theory as the ω-categorical countably infinite atomless boolean ring without 1 [Abi72]. The approach to define this structure with Fraïssé-amalgamation is also not new (this is mentioned, for example, in [Eva94]). However, we give the amalgamation argument here, because it allows us to prove that the CSP of the resulting structure is indeed the same problem as CSP(B_0); moreover, it shows an interesting connection between amalgamation problems and solving certain CSPs.

Let $\mathcal{C}$ be the set of all finite induced substructures of B_0, considered up to isomorphism. Hence, $\mathcal{C}$ is a countable class of relational structures.

Proposition 27. *The class $\mathcal{C}$ is an amalgamation class.*

Corollary 28. *There exists an ω-categorical structure $\mathbb{B}_0$ such that CSP$(\mathbb{B}_0)$ equals CSP(B_0).*

We can easily obtain an ω-categorical constraint language for all of the RCC-5 relations by expanding the structure $\mathbb{B}_0$ by all binary relations that can be defined by boolean combinations of $\mathtt{DR}, \mathtt{PO}, \mathtt{PP}$, and $\mathtt{EQ}$. It is well-known (again, see [Eva94]) that there exists a countable set such that the elements of $\mathbb{B}_0$ can be seen as subsets of this set, and such that $\mathtt{DR}$ denotes disjointness, $\mathtt{PO}$ denotes proper overlap, and $\mathtt{PP}$ denotes strict containment between two subsets.

We now show that $\mathbb{B}_0$ has a QNU polymorphism.

Theorem 29. *For all $k \geq 5$, the structure $\mathbb{B}_0$ has an oligopotent QNU polymorphism of arity k such that for all $t, t' \in \Gamma^k$, it holds that $t \equiv_m t'$ if and only if $f(t) = f(t')$. The polymorphism of arity k preserves any disjunction of disequalities over m variables with $k \geq 2m+1$.*

References

[Abi72] Abian, A.: Categoricity of denumerable atomless boolean rings. Studia Logica 30(1), 63–67 (1972)

[AHV95] Abiteboul, S., Hull, R., Vianu, V.: Foundations of Databases. Addison-Wesley, Reading (1995)

[All83] Allen, J.F.: Maintaining knowledge about temporal intervals. Communications of the ACM 26(11), 832–843 (1983)

[BC07a] Bodirsky, M., Chen, H.: Oligomorphic clones. Algebra Universalis (to appear, 2007)

[BC07b] Bodirsky, M., Chen, H.: Quantified equality constraints. In: Proceedings of LICS'07 (to appear, 2007)

[BD06] Bodirsky, M., Dalmau, V.: Datalog and constraint satisfaction with infinite templates. In: Durand, B., Thomas, W. (eds.) STACS 2006. LNCS, vol. 3884, pp. 646–659. Springer, Heidelberg (2006) (A journal version is available from the webpage of the first author, 2006)

[Ben94] Bennett, B.: Spatial reasoning with propositional logics. In: International Conference on Knowledge Representation and Reasoning, Morgan Kaufmann, San Francisco (1994)

[BJK05] Bulatov, A., Jeavons, P., Krokhin, A.: The complexity of constraint satisfaction: An algebraic approach (a survey paper). In: Structural Theory of Automata, Semigroups and Universal Algebra (Montreal, 2003). NATO Science Series II: Mathematics, Physics, Chemistry, pp. 181–213 (2005)

[BKJ05] Bulatov, A., Krokhin, A., Jeavons, P.G.: Classifying the complexity of constraints using finite algebras. SIAM Journal on Computing 34, 720–742 (2005)

[BN06] Bodirsky, M., Nešetřil, J.: Constraint satisfaction with countable homogeneous templates. Journal of Logic and Computation 16(3), 359–373 (2006)

[Bod05] Bodirsky, M.: The core of a countably categorical structure. In: Diekert, V., Durand, B. (eds.) STACS 2005. LNCS, vol. 3404, pp. 100–110. Springer, Heidelberg (2005)

[BP74] Baker, K.A., Pixley, A.F.: Polynomial interpolation and the Chinese remainder theorem for algebraic systems. Math. Z. 143, 165–174 (1974)

[Che] Chen, H.: The computational complexity of quantified constraint satisfaction. Ph.D. thesis, Cornell University (August 2004)

[CJ06] Cohen, D., Jeavons, P.: The complexity of constraint languages. Appears in: Handbook of Constraint Programming (2006)

[Due05] Duentsch, I.: Relation algebras and their application in temporal and spatial reasoning. Artificial Intelligence Review 23, 315–357 (2005)

[DvB97] Dechter, R., van Beek, P.: Local and global relational consistency. TCS 173(1), 283–308 (1997)

[EF99] Ebbinghaus, H.-D., Flum, J.: Finite Model Theory, 2nd edn. Springer, Heidelberg (1999)

[Eva94] Evans, D.: Examples of aleph-zero categorical structures. Automorphisms of first-order structures, 33–72 (1994)

[FGV05] Fisher, M., Gabbay, D., Vila, L.: Handbook of Temporal Reasoning in Artificial Intelligence. Elsevier, Amsterdam (2005)

[Fra86] Fraïssé, R.: Theory of Relations. North-Holland, Amsterdam (1986)

[Fre82] Freuder, E.C.: A sufficient condition for backtrack-free search. Journal of the ACM 29(1), 24–32 (1982)

[FV99] Feder, T., Vardi, M.: The computational structure of monotone monadic SNP and constraint satisfaction: A study through Datalog and group theory. SIAM Journal on Computing 28, 57–104 (1999)

[Hir96] Hirsch, R.: Relation algebras of intervals. Artificial Intelligence Journal 83, 1–29 (1996)

[Hod97] Hodges, W.: A shorter model theory. Cambridge University Press, Cambridge (1997)

[JCC98] Jeavons, P., Cohen, D., Cooper, M.: Constraints, consistency and closure. AI 101(1-2), 251–265 (1998)

[JD97] Jonsson, P., Drakengren, T.: A complete classification of tractability in RCC-5. J. Artif. Intell. Res. 6, 211–221 (1997)

[Kou97] Koubarakis, M.: From local to global consistency in temporal constraint networks. Theor. Comput. Sci. 173(1), 89–112 (1997)

[Kou01] Koubarakis, M.: Tractable disjunctions of linear constraints: Basic results and applications to temporal reasoning. Theoretical Computer Science 266, 311–339 (2001)

[KvBV90] Kautz, H., van Beek, P., Vilain, M.: Constraint propagation algorithms: A revised report. Qualitative Reasoning about Physical Systems, 373–381 (1990)

[LM94] Ladkin, P.B., Maddux, R.D.: On binary constraint problems. Journal of the Association for Computing Machinery 41(3), 435–469 (1994)

[Mac77] Mackworth, A.K.: Consistency in networks of relations. AI 8, 99–118 (1977)

[RCC92] Randell, D.A., Cui, Z., Cohn, A.G.: A spatial logic based on regions and connection. In: Conference on Principles on Knowledge Representation and Reasoning (KR'92), pp. 165–176 (1992)

[RN99] Renz, J., Nebel, B.: On the complexity of qualitative spatial reasoning: A maximal tractable fragment of the region connection calculus. Artif. Intell. 108(1-2), 69–123 (1999)

[RN07] Renz, J., Nebel, B.: Qualitative spatial reasoning using constraint calculi. Handbook of Spatial Logics (to appear, 2007)

On Acyclic Conjunctive Queries and Constant Delay Enumeration

Guillaume Bagan[1], Arnaud Durand[2], and Etienne Grandjean[1]

[1] GREYC, Université de Caen, CNRS ; Campus 2, F-14032 Caen cedex, France
gbagan@info.unicaen.fr, grandjean@info.unicaen.fr
[2] Equipe de Logique Mathématique - CNRS UMR 7056, Université Paris 7 - Denis Diderot, 2 place Jussieu, 75251 Paris Cedex 05, France
durand@logique.jussieu.fr

Abstract. We study the enumeration complexity of the natural extension of acyclic conjunctive queries with disequalities. In this language, a number of **NP**-complete problems can be expressed. We first improve a previous result of Papadimitriou and Yannakakis by proving that such queries can be computed in time $c.|\mathcal{M}|.|\varphi(\mathcal{M})|$ where $\mathcal{M}$ is the structure, $\varphi(\mathcal{M})$ is the result set of the query and c is a simple exponential in the size of the formula φ. A consequence of our method is that, in the general case, tuples of such queries can be enumerated with a linear delay between two tuples.

We then introduce a large subclass of acyclic formulas called $\mathbf{CCQ}^{\neq}$ and prove that the tuples of a $\mathbf{CCQ}^{\neq}$ query can be enumerated with a linear time precomputation and a constant delay between consecutive solutions. Moreover, under the hypothesis that the multiplication of two $n \times n$ boolean matrices cannot be done in time $O(n^2)$, this leads to the following dichotomy for acyclic queries: either such a query is in $\mathbf{CCQ}^{\neq}$ or it cannot be enumerated with linear precomputation and constant delay. Furthermore we prove that testing whether an acyclic formula is in $\mathbf{CCQ}^{\neq}$ can be performed in polynomial time.

Finally, the notion of free-connex treewidth of a structure is defined. We show that for each query of free-connex treewidth bounded by some constant k, enumeration of results can be done with $O(|\mathcal{M}|^{k+1})$ precomputation steps and constant delay.

Introduction

From a complexity theoretical point of view, very few is known about enumeration problems and their complexity classes. In the context of logical or database query languages, enumeration amounts to produce one by one all the tuples of the result $\varphi(\mathcal{M})$ of a query $\varphi(\bar{x})$ over a structure $\mathcal{M}$. Producing solutions of a query with a regular delay or on user demand makes sense in that area. Although a lot of results are known about the complexity of query evaluation in the classical sense (from hard cases to islands of tractability) [6,9,12,14,16], only some recent works have started to investigate complexity issues of enumeration for query languages [8,10,1,11,2].

J. Duparc and T.A. Henzinger (Eds.): CSL 2007, LNCS 4646, pp. 208–222, 2007.

A large and well-studied class of queries is the class **ACQ** of acyclic conjunctive queries. A classical result of database theory, proved in [17], is that any acyclic conjunctive query φ can be evaluated in time $O(|\varphi|.|\mathcal{M}|.|\varphi(\mathcal{M})|)$. In [16], an extension $\mathbf{ACQ}^{\neq}$ is introduced in which disequalities between variables are allowed. In full generality, deciding a boolean $\mathbf{ACQ}^{\neq}$ query is **NP**-complete for combined complexity. However, it is shown in [16] that such a query can be evaluated in time $f(|\varphi|).|\mathcal{M}|.|\varphi(\mathcal{M})|.\log^2|\mathcal{M}|$ where f is some exponential function. Many **NP**-complete problems can be expressed by $\mathbf{ACQ}^{\neq}$ formulas: existence of some path of a given length k in a graph, multidimensional matching, color matching, etc.

The main goal of this paper is to revisit the complexity of the evaluation of conjunctive queries with or without disequalities from a dynamical point of view. As in [8,10,1,11,2], queries are seen as enumeration problems and the complexity is measured in terms of delay between two consecutive solutions. We consider the class of **F-ACQ** (resp $\mathbf{F\text{-}ACQ}^{\neq}$) formulas which are the "equivalent" of acyclic conjunctive formulas in functional structures. First, we prove that an elimination of quantifiers can be done on such formulas. More precisely, using a combinatorial argument, we show that given an $\mathbf{F\text{-}ACQ}^{\neq}$ formula φ and a structure $\mathcal{M}$ one can construct a formula φ' which is a disjunction of quantifier-free formulas of $\mathbf{F\text{-}ACQ}^{\neq}$ and a model $\mathcal{M}'$ such that $\varphi(\mathcal{M}) = \varphi'(\mathcal{M}')$. As a consequence, we prove that any query of $\mathbf{F\text{-}ACQ}^{\neq}$ can be evaluated with a linear time delay. This immediately implies that a boolean $\mathbf{ACQ}^{\neq}$ query can be evaluated in time $f(|\varphi|).|\mathcal{M}|$ and a general $\mathbf{ACQ}^{\neq}$ query can be evaluated in time $f(|\varphi|).|\mathcal{M}|.|\varphi(\mathcal{M})|$ where f is a simple exponential function which depends on the number of variables and the number of disequalities of the formula.

Constant-Delay$_{lin}$ introduced in [10] is the class of enumeration problems which are solved by algorithms that run in two phases : the first phase performs a precomputation in linear time and the second phase, using the precomputation, enumerates all the solutions with a constant delay between two consecutive ones. This class is regarded as the minimal (nontrivial) robust class for the complexity of enumeration. A natural question addressed is this paper is the following: is there a large subclass of $\mathbf{ACQ}^{\neq}$ queries whose enumeration belongs to Constant-Delay$_{lin}$? We introduce the class $\mathbf{CCQ}^{\neq}$ of formulas called *free-connex acyclic* formulas and defined by a connectivity condition between the free variables (in particular, it contains the quantifier-free acyclic formulas with disequalities). We show that such queries can be evaluated with linear precomputation and constant delay. Furthermore, it is proved that testing whether a formula is free-connex acyclic can be performed in polynomial time. This result is optimal in some sense because we prove the following dichotomy theorem: under the assumption that any two $n \times n$ boolean matrices *cannot* be multiplied in time $O(n^2)$ (a very reasonable conjecture), each acyclic conjunctive query (with or without disequalities) which is *simple* [1] either is free-connex acyclic or cannot be enumerated with linear precomputation and constant delay. Finally, we introduce

[1] This means that we assume that each relation symbol occurs once only in the conjunction (a slight restriction).

the notions of *free-connex tree-decomposition* and *free-connex treewidth*. We show that for any fixed query of free-connex treewidth bounded by some constant k, enumeration of solutions can be done with $O(|D|^{k+1} + |\mathcal{M}|)$ precomputation steps and constant delay (D is the domain of $\mathcal{M}$).

Notice that essentially all the results of this paper are new, to our knowledge, even if we consider only conjunctive queries (without disequalities). The paper is organized as follows. In Section 1, the main definitions about complexity, queries, hypergraphs and enumeration problems are given. Our results about quantifier elimination and enumeration are proved respectively in Sections 2 and 3. The notion of free-connex acyclic formula and the dichotomy theorem are exposed in Section 4. Finally, in Section 5, results about free-connex treewidth queries are presented. Note that by lack of space, many proofs are omitted.

1 Preliminaries

1.1 Problems

In order to handle problems in a general and uniform framework, we introduce the notion of *general problem* which is merely a binary relation $A \subseteq I \times O$ over two sets I and O respectively called *space of instances* (or *input space*) and *space of solutions* (or *output space*). For each input $x \in I$, the set $A(x) = \{y \in O : (x, y) \in A\}$ is assumed to be finite and is called the *set of solutions* of A on x. Let $<$ be a strict linear order on the space of solutions of A. For any input x and $A(x) = \{y_0, \ldots, y_{n-1}\}$ with $y_0 < y_1 < \ldots < y_{n-1}$, we denote by $\text{enum}(A, x, <) = y_0 y_1, \ldots y_{n-1}$ the increasing list of elements of $A(x)$. We are interested in the following problems.

EVAL(A)	ENUM($A, <$)
Input: An instance $x \in I$	*Input:* an instance $x \in I$
Output: $A(x)$	*Output:* $\text{enum}(A, x, <)$

In the case the order does not matter, we omit it and write ENUM(A).

1.2 Complexity Framework

The computation model used in this paper is the random access machine (RAM) with uniform cost measure and addition and substraction as the basic operations (see [13]). It has read-only input registers (containing the input x), read-write work memory registers and write-only output registers. Moreover, register values and memory addresses are assumed to be bounded by $c|x|$, for some constant c, where $|x|$ is the size of the input x (that is the number of registers it occupies).

An algorithm $\mathcal{A}$ is said to compute the problem ENUM($A, <$) if for any input x, $\mathcal{A}$ computes one by one the elements of the sequence *enum*($A, x, <$) and stops immediately after writing the last one. We define $\text{delay}_j(x) = \text{time}_j(x) - \text{time}_{j-1}(x)$ where $\text{time}_j(x)$ denotes the moment when $\mathcal{A}$ has completed the writing of the jth solution (by convention $\text{time}_0(x) = 0$). We say that $\mathcal{A}$ is a *constant delay* algorithm if for a constant c and any input x, we have $\text{delay}_j(x) \leq c$

and if furthermore, $\mathcal{A}$ uses space c (i.e. at most c memory registers) during its computation.

Definition 1. *A problem* $\text{ENUM}(A,<)$ *is computable* with constant delay and linear precomputation, *which is written* $\text{ENUM}(A,<) \in \text{CONSTANT-DELAY}_{lin}$ *if there is a function $r : x \mapsto r(x)$ (precomputation phase) computable in linear time and a constant delay algorithm $\mathcal{A}$ (enumeration phase) which, when applied to $r(x)$ for any input x, computes enum$(A, x, <)$.*

Note that the additional requirement that the enumeration phase uses constant space (uniform cost measure) appears to be a very strong condition. Nevertheless we have checked that it holds for all the constant delay algorithms we know (see [1,10,11,2]). However, this condition is not essential in this paper. We need a very precise notion of reduction for enumeration problems.

Definition 2. *Let $A \subseteq I_A \times O_A$ and $B \subseteq I_B \times O_B$ be two general problems. An* exact reduction *from* $\text{ENUM}(A)$ *to* $\text{ENUM}(B)$ *(resp. from* $\text{ENUM}(A,<_A)$ *to* $\text{ENUM}(B,<_B)$*) is a pair (r,t) of mappings $r : I_A \to I_B$ and $t : I_B \times O_B \to O_A$ which satisfy conditions 1 and 2*

1. *For every instance x of A, the correspondence $y \mapsto t(r(x), y)$ is a one-one (resp. strictly increasing) mapping from the set (resp $<_B$-ordered set) $B(r(x))$ onto the set (resp. $<_A$-ordered set) $A(x)$.*
2. *The instance $r(x)$ of B is computable from x in time and space $O(|x|)$ and the solution $t(r(x), y)$ is computable from $(r(x), y)$ in constant time and constant space.*

The following results are straightforward consequences of our definitions.

Lemma 3. *Assume there exists an exact reduction from* $\text{ENUM}(A)$ *to* $\text{ENUM}(B)$. *Then* $\text{ENUM}(B) \in \text{CONSTANT-DELAY}_{lin}$ *implies* $\text{ENUM}(A) \in \text{CONSTANT-DELAY}_{lin}$. *The similar result holds for* $\text{ENUM}(A,<_A)$ *and* $\text{ENUM}(B,<_B)$.

Lemma 4. *Let $A, B \subseteq I \times O$ be two general problems over the same input and output spaces. Let $<$ be a linear order on O. If both problems* $\text{ENUM}(A,<)$ *and* $\text{ENUM}(B,<)$ *belong to* $\text{CONSTANT-DELAY}_{lin}$, *then so does the union problem* $\text{ENUM}(A \cup B,<)$.

1.3 Logical Framework

The reader is supposed familiar with first-order logic on finite structures (see [15]). For a signature σ, a (finite) σ-structure consists of a domain D together with an interpretation of each symbol of σ over D (we do not distinguish between a symbol and its interpretation). For each σ-formula (or σ-query) $\varphi(\bar{x})$ with m variables $\bar{x} = (x_1, \ldots, x_m)$ and for each σ-structure $\mathcal{M}$ of domain D, we denote by $\varphi(\mathcal{M})$ the set $\{\bar{a} \in D^m : \mathcal{M} \models \varphi(\bar{a})\}$. Let $<$ be a linear order on D^m. We are interested in the evaluation and enumeration problems associated to $\varphi(\bar{x})$. Note that in most of our problems formula φ is considered as *fixed*.

EVAL(φ)		ENUM(φ, <)	
Input:	a σ-structure $\mathcal{M}$	*Input:*	a σ-structure $\mathcal{M}$
Output:	$\varphi(\mathcal{M})$	*Output:*	an <-ordered enumeration of $\varphi(\mathcal{M})$

1.4 Hypergraphs

Let us introduce some definitions on hypergraphs. A hypergraph is a pair (V, E) where V is a set of elements, called *vertices*, and E is a set of non-empty subsets of V called *hyperedges*. The *induced subhypergraph* of $H = (V, E)$ on a set of vertices $S \subseteq V$ is the hypergraph $H[S] = (S, \{e \cap S \neq \emptyset : e \in E\})$. A hypergraph is *acyclic* if there is a tree T, called a *tree-structure* of H, whose vertices are the hyperedges of H and having the following *connectivity property*: for each vertex v of H, the set of hyperedges that contain v consists of a subtree (connected subgraph) of T. For two hypergraphs $H_1 = (V_1, E_1)$ and $H_2 = (V_2, E_2)$, let us denote by $H_1 \cup H_2$ the hypergraph $(V_1 \cup V_2, E_1 \cup E_2)$. Given a graph $G = (V, E)$, the neighborhood of x denoted by $N(x)$ is $\{y \in V : \{x, y\} \in E\}$. For more details on hypergraphs (in particular, on paths) see [3].

1.5 Acyclic Conjunctive Queries

As usual, let **CQ** (resp. $\mathbf{CQ}^{\neq}$) denote the set of *conjunctive queries* (resp *conjunctive queries with disequalities*), i.e. relational σ-formulas of the form $\varphi(\bar{x}) \equiv \exists \bar{y} \psi(\bar{x}, \bar{y})$ where ψ is a conjunction of σ-atoms $R(\bar{v})$ for $\bar{v} \subseteq \{\bar{x}, \bar{y}\}$ (resp σ-atoms and disequalities $u \neq v$ for $u, v \in \{\bar{x}, \bar{y}\}$). A conjunctive query (or conjunctive formula) φ (with or without disequalities) is *acyclic* or **ACQ** (resp. $\mathbf{ACQ}^{\neq}$) if its hypergraph $H_\varphi = (V_\varphi, E_\varphi)$ is acyclic (Recall the set of vertices and the set of hyperedges of H_φ are respectively $V_\varphi = var(\varphi)$ and $E_\varphi = \{\{v_1, \ldots, v_{p_i}\} : R_i(v_1, \ldots, v_{p_i})$ is a σ-atom of $\varphi\})$. A tree-structure of H_φ is said to be a *join-tree* of φ.

Acyclic conjunctive functional queries: An acyclic conjunctive σ-query can be naturally translated in the functional framework, i.e., into a σ'-formula for a signature σ' that consists of unary fuction symbols and unary predicates.

Formula translation: Introduce for each σ-atom $A_i, 1 \leq i \leq q$, of the formula a new variable α_i.

Translation (1) of a quantifier-free ***ACQ*** *σ-formula:* $\psi(\bar{x}) \equiv \bigwedge_{1 \leq i \leq q} A_i$. Associate to ψ the following quantifier-free functional σ'-formula:

$$\psi'(\alpha_1, \ldots, \alpha_q) \equiv \bigwedge_{1 \leq i \leq q} D_{R_i}(\alpha_i) \wedge \bigwedge_{(A_i, A_j) \in T_\psi} \bigwedge_{\mathrm{var}_h(A_i) = \mathrm{var}_k(A_j)} f_h(\alpha_i) = f_k(\alpha_j) \tag{1}$$

where $\sigma' = \{D_R : R \in \sigma\} \cup \{f_1, \ldots, f_p\}, p = \text{arity}(\sigma) = \max_{R \in \sigma}(\text{arity}(R))$, T_ψ is a join-tree of ψ, $A_i \equiv R_i(v_i^1, \ldots, v_i^h, \ldots, v_i^{p_i})$, p_i is the arity of R_i and $\mathrm{var}_h(A_i)$ denotes the variable v_i^h that occurs at rank h in atom A_i.

Translation (2) of any **ACQ** *formula* $\varphi(\bar{x}) \equiv \exists\bar{y}\psi(\bar{x},\bar{y})$ where ψ is quantifier-free and $\bar{x} = (x_1, \ldots, x_m)$. Associate to φ the following functional σ'-formula:

$$\varphi''(\bar{x}) \equiv \exists\alpha_1 \ldots \exists\alpha_q(\psi'(\bar{\alpha}) \wedge \bigwedge_{1\leq i\leq m} f_{j_i}(\alpha_{k_i}) = x_i) \tag{2}$$

where ψ' is the above translation (1) of the quantifier-free part ψ of φ and A_{k_i} (represented by variable α_{k_i}) is a chosen atom of φ in which x_i occurs (at rank j_i).

Translation of the structure: Each relational σ-model $\mathcal{M} = (D; R_1, \ldots, R_s)$ is translated into the functional σ'-model $\mathcal{M}' = (D'; D, D_{R_1}, \ldots, D_{R_s}, f_1, \ldots, f_p)$:

- $D, D_{R_1}, \ldots, D_{R_s}$ are disjoint unary relations so that $D' = D \cup D_{R_1} \cup \ldots \cup D_{R_s} \cup \{\perp\}$ and, for each $i = 1, \ldots, s$, we have $D_{R_i} = R_i$, i.e. D_{R_i} is the set of tuples of R_i, and $\perp$ is an extra element.
- Each f_j $(1 \leq j \leq p)$ is a unary function: $D' \to D \cup \{\perp\}$ such that, for every $t = (a_1, \ldots, a_{p_i}) \in D_{R_i}$, we have $f_j(t) = a_j$ for $1 \leq j \leq p_i$ and $f_j(t) = \perp$ otherwise.

Example 5. Translation 1 transforms the quantifier-free **ACQ** formula $\psi(x, y, z) \equiv R(x, y) \wedge S(y, z)$ into the following (quantifier-free) formula: $\psi'(\alpha_1, \alpha_2) \equiv D_R(\alpha_1) \wedge D_S(\alpha_2) \wedge f_2(\alpha_1) = f_1(\alpha_2)$.

Example 6. Translation 2 transforms the **ACQ** formula $\varphi(x, z) \equiv \exists y(R(x, y) \wedge S(y, z))$ into the following formula (with the same free variables and the subformula ψ' as above): $\varphi''(x, z) \equiv \exists\alpha_1\exists\alpha_2(\psi'(\alpha_1, \alpha_2) \wedge f_1(\alpha_1) = x \wedge f_2(\alpha_2) = z)$.

Remark 7. Our translations (1) and (2) are easily extended to $\mathbf{ACQ}^{\neq}$ formulas.

Let us now introduce the notions of conjunctive and acyclic conjunctive functional queries.

Definition 8. *Let* **F-CQ** *(resp.* $\textbf{\textit{F-CQ}}^{\neq}$*) denote the set of conjunctive functional σ-formulas (for a unary functional signature σ) of the form $\varphi = \exists\bar{x}\psi$ where ψ is a conjunction of unary atoms $U(v)$ and equalities $\tau = \tau'$ (resp. equalities $\tau = \tau'$ and disequalities $\tau \neq \tau'$), each τ, τ' is a functional term of the form $f(v)$ or v, where v is a variable.*

Definition 9. *To each* **F-CQ** *or* $\textbf{\textit{F-CQ}}^{\neq}$ *formula φ, we naturally associate the undirected graph $G_\varphi = (V_\varphi, E_\varphi)$ defined as follows:*

- *$V_\varphi = var(\varphi)$ and*
- *for any pair of distinct variables v, v', set $\{v, v'\} \in E_\varphi$ iff an equality involving both v and v' occurs as a conjunct* [2] *in φ.*

An **F-CQ** *(resp.* $\textbf{\textit{F-CQ}}^{\neq}$*) formula φ is* acyclic *or* **F-ACQ** *(resp.* $\textbf{\textit{F-ACQ}}^{\neq}$*) if its graph G_φ is acyclic.*

[2] Here again, the disequalities of φ are not involved in the definition of G_φ.

Without loss of generality we assume that G_φ is also connex and hence is a tree T called a *join-tree* of φ . The following results are easy to prove.

Lemma 10. *1. Preservation of acyclicity: Let φ be a **CQ$^{\neq}$** formula. If φ is acyclic then its functional translations 1 (in the case φ is quantifier-free) and 2 are acyclic too.*

2. Linearity of the transformations:
- *For formulas: The sizes of the transformed formulas 1 and 2 are linear in the size of φ.*
- *For structures: Our transformation of a relational σ-structure $\mathcal{M}$ into a functional σ'-structure $\mathcal{M}'$ is computable in time $O(|\mathcal{M}|)$.*

3. Correctness of the reductions:
- *Translation 1: Let ψ' be the quantifier-free translation (**F-ACQ$^{\neq}$**) of a (quantifier-free) formula ψ (**ACQ$^{\neq}$**). There exists an exact reduction from problem* Enum(ψ) *to* Enum(ψ').
- *Translation 2: We have $\varphi(\mathcal{M}) = \varphi''(\mathcal{M}')$.*

2 Quantifier Elimination

2.1 Covers and Representative Sets

In all the definitions and results of this section, E and F are two finite sets and $\bar{f} = (f_1, \ldots, f_k)$ is a tuple of k unary functions from E to F.

Definition 11. *A* cover $\bar{c}$ *of $(E, \bar{f})$ is a tuple $(c_1, \ldots, c_k) \in F^k$ such that $\forall y \in E, \exists i : c_i = f_i(y)$. We denote by* covers$(E, \bar{f})$ *the set of covers of $(E, \bar{f})$.*

Definition 12. *A* representative set *of $(E, \bar{f})$ is a subset $E' \subseteq E$ such that* covers$(E, \bar{f})$ = covers$(E', \bar{f})$.

Lemma 13. *There is a representative set E' of $(E, \bar{f})$ of cardinality at most $O(k!)$. Such a set E' is called* a small representative set *of $(E, \bar{f})$ and can be computed in time $O(k!|E|)$.*

2.2 Quantifier Elimination

Lemma 14. *Let $\varphi(\bar{x})$ be an **F-ACQ$^{\neq}$** σ-formula with a join-tree T and of the form $\varphi(\bar{x}) = \exists z \psi(\bar{x}, z)$ where ψ is quantifier-free and z is a leaf of T. Let k be the number of disequalities in $\varphi(\bar{x})$ involving variable z and let $\mathcal{M}$ be a σ-structure. Then we can compute in time $O(k!.|\varphi|.|\mathcal{M}|)$ a quantifier-free σ'-formula $\varphi'(\bar{x})$ with $\sigma \subseteq \sigma'$ and a σ'-structure $\mathcal{M}'$ that expands $\mathcal{M}$ such that*

- *$\varphi(\mathcal{M}) = \varphi'(\mathcal{M}')$;*
- *$\varphi'(\bar{x})$ is a disjunction of at most $O(k!)$ formulas ψ_i of **F-ACQ$^{\neq}$**;*
- *for each disjunct ψ_i of φ', $|\psi_i| \leq |\varphi|$.*

Proof. Formula $\varphi(\bar{x})$ can be put under the form:

$$\varphi(\bar{x}) \equiv \psi_0(\bar{x}) \wedge \exists z(P(z) \wedge \bigwedge_{1\le i\le m} g_i(z) = g'_i(y) \wedge \bigwedge_{1\le i\le k} f_i(z) \neq f'_i(\bar{x}))$$

where $\psi_0(\bar{x})$ is a quantifier-free formula, $y \in \bar{x}$ is the parent of the leaf z in the tree T, $f'_i(\bar{x})$ denotes a term $f'_i(v_i)$ for $v_i \in \bar{x}$ and $P(z)$ is a quantifier-free formula on z. Let $\bar{g}(x) = (g_1(x), \dots g_m(x))$, $\bar{g'}(x) = (g'_1(x), \dots g'_m(x))$. One partitions the set $P = P(\mathcal{M})$ according to the values of $\bar{g}$; more precisely, for each $\bar{\alpha} \in \bar{g}(D)$, one defines the set $P_{\bar{\alpha}} = P \cap \bar{g}^{-1}(\bar{\alpha})$ and computes a small representative set of $(P_{\bar{\alpha}}, \bar{f})$ denoted $P'_{\bar{\alpha}}$ (of cardinality at most $h = O(k!)$). This can be done globally in time $\Sigma_{\bar{\alpha}\in\bar{g}(D)} O(k!|P_{\bar{\alpha}}|) = O(k!|P|) = O(k!|\mathcal{M}|)$. Clearly, $\varphi(\bar{x})$ can be rephrased as

$$\psi_0(\bar{x}) \wedge \exists z \in P_{\bar{g'}(y)} \bigwedge_{1\le i\le k} f_i(z) \neq f'_i(\bar{x}).$$

By definition of the representative sets, it is also equivalent to the following formula on $\mathcal{M}$:

$$\psi_0(\bar{x}) \wedge \exists z \in P'_{\bar{g'}(y)} \bigwedge_{1\le i\le k} f_i(z) \neq f'_i(\bar{x}).$$

Let $\mathcal{M}'$ be the σ'-expansion of $\mathcal{M}$ obtained by introducing the unary predicates E_j and the unary functions v_j, for $1 \le j \le h$, defined as follows: $y \in E_j \Leftrightarrow |P'_{\bar{g'}(y)}| \ge j$; $v_j(y)$ is the jth element of $P'_{\bar{g'}(y)}$ if $|P'_{\bar{g'}(y)}| \ge j$ and an arbitrary value otherwise. Finally, we obtain $\varphi(\mathcal{M}) = \varphi'(\mathcal{M}')$ for the following σ'-formula φ' of the required form:

$$\varphi'(\bar{x}) \equiv \bigvee_{1\le j\le h} (\psi_0(\bar{x}) \wedge E_j(y) \wedge \bigwedge_{1\le i\le k} f_i(v_j(y)) \neq f'_i(\bar{x}))$$

Using the result above, one can eliminate one by one all the variables but one in a bottom-up approach and derive the two following results. The iterative treatment of variables is similar in spirit to Yannakakis' algorithm [17].

Lemma 15. *Let $\varphi(x)$ be an **F-ACQ$^{\neq}$** σ-formula with only one free variable x and $\mathcal{M}$ be a σ-structure. Let l be the number of variables and k be the number of disequalities of $\varphi(x)$. One can compute in time $O(k!^l.|\varphi|.|\mathcal{M}|)$ a σ'-expansion $\mathcal{M}'$ of $\mathcal{M}$ ($\sigma \subseteq \sigma'$) and a σ'-formula φ' which is a disjunction of $O(k!^l)$ **F-ACQ$^{\neq}$** quantifier-free formulas ψ_i where $|\psi_i| \le |\psi|$ so that $\varphi(\mathcal{M}) = \varphi'(\mathcal{M}')$.*

Theorem 16. *Let φ be a fixed **F-ACQ$^{\neq}$** (resp. **ACQ$^{\neq}$**) query. Then* ENUM(φ) *can be evaluated with delay $O(|\mathcal{M}|)$.*

In the particular case of boolean queries then, surprisingly, the exponent due to the number of variables l can be avoided.

Proposition 17. *Let φ be an **F-ACQ$^{\neq}$** σ-formula without free variables and $\mathcal{M}$ be a σ-structure. Testing whether $\mathcal{M} \models \varphi$ can be decided in time $O(k!.|\varphi|.|\mathcal{M}|)$ where k is the number of disequalities of φ*

3 Constant Delay Enumeration

3.1 Combinatorial Tools

In this section, we will need a convenient notion of *two-phase algorithm* for a problem with two inputs : a *static input* x and a *dynamic input* y. It is defined by the following general scheme.

- *Static phase (or precomputation phase)*: it computes some data called $r(x)$ (only depending on x);
- *Dynamic phase*: from the input $(r(x), y)$, it computes the output of the problem.

LEASTNONCOVER
Static input : Two finite ordered sets E and F, a k-tuple of unary functions $\bar{f} = (f_1, \dots f_k)$ from E to F
Dynamic input : a vector $\bar{u} \in F^k$, an element $x \in E$
Parameter: k
Output: the least element $y \in E$ such that $y \geq x \wedge \bigwedge_{1 \leq i \leq k} f_i(y) \neq u_i$ if such an element exists, and $\perp$ otherwise.

Lemma 18. LEASTNONCOVER *is computable with a static phase of linear time and a dynamic phase of constant time.*

Proof. (sketch) In this proof, E, F and $\bar{f}$ are considered as implicit and we assume that $E = \{1, \dots, |E|\}$. Let x be an element of E, $\bar{u}$ be a k-tuple of elements of F and S be a subset of $[1, k]$. Then define Find$(x, \bar{u}, S)$ as the least element $y \in E$ if it exists ($\perp$ otherwise) such that $y \geq x \wedge \bigwedge_{i \in S} f_i(y) \neq u_i$. For each list $L = (a_1, \dots, a_l)$ of distinct elements of $[1, k]$, $0 \leq l \leq k$, we define the "pointer function" $\text{Ptr}_L : E \rightarrow E \cup \{\perp\}$ useful in the computation of Find as follows

- $\text{Ptr}_\epsilon(x) = x$;
- if $l \geq 1$ then $\text{Ptr}_{a_1,\dots,a_l}(x)$ is the least element $y \in E$ such that $y \geq x \wedge \forall j, 1 \leq j \leq l : f_{a_j}(y) \neq f_{a_j}(\text{Ptr}_{a_1,\dots a_{j-1}}(x))$ if it exists and $\text{Ptr}_{a_1,\dots a_{l-1}}(x) \neq \perp$, and is $\perp$ otherwise.

We assume that during the precomputation phase, we have computed the table PTR where $\text{PTR}_L[x] = \text{Ptr}_L(x)$ for each element $x \in E$ and each list L. We give an algorithm FIND$(x, \bar{u}, S)$ which computes Find$(x, \bar{u}, S)$. Clearly, the dynamic phase of LEASTNONCOVER consists in calling FIND with $S = [1, k]$. FIND$(x, \bar{u}, S)$ consists of the call FIND-aux$(x, \bar{u}, S, \epsilon)$.

Finally, Algorithm 2 is given which computes the PTR table for the precomputation phase.

The more elaborated problem below need to be defined.

Algorithm 1. FIND-aux($x, \bar{u}, S, L$)

```
y ← PTR_L[x]
if y = ⊥ ∨ ∀i ∈ S − L : f_i(y) ≠ u_i then
    return y
else
    choose i ∈ S − L such that f_i(y) = u_i
    return FIND-aux(x, ū, S, L.i)
end if
```

Algorithm 2. Precomputation phase

```
for x from |E| to 1 do
    PTR_ε[x] ← x
    for l from 1 to k do
        for each list L = (a_1, ..., a_l) of l distinct elements of [1, k] do
            if x = |E| or PTR_{a_1,...,a_{l-1}}[x] = ⊥ then
                PTR_L[x] ← ⊥
            else
                let ū = (u_1, ..., u_k) such that ∀j ≤ l : u_{a_j} = f_{a_j}(PTR_{a_1,...,a_{j-1}}[x])
                PTR_L[x] ← FIND(x + 1, ū, {a_1, ..., a_l})
            end if
        end for
    end for
end for
```

ENUMNONCOVER

Static input: Two finite ordered set E and F
a k-tuple of unary functions $\bar{f} = (f_1, \dots f_k)$ from E to F
Dynamic input: a vector $\bar{u} \in F^k$
Parameter: k
Output: the set $\{x \in E : \bigwedge_{1 \le i \le k} f_i(x) \neq u_i\}$ in increasing order

The following lemma is a consequence of Lemma 18

Lemma 19. ENUMNONCOVER *is computable with a static phase in linear time and a dynamic phase with a constant delay.*

3.2 Enumeration of Quantifier-Free and Free-Connex Acyclic Queries

Definition 20. *Let φ be a quantifier-free **F-ACQ**$^{\neq}$ formula and T a join-tree of φ. An elimination order (or T-order) of the variables of φ is an ordered list (may be empty) $x_1, \dots, x_p$ of its p variables such that (if $p > 0$) x_p is a leaf of T and $x_1, \dots, x_{p-1}$ is an elimination order of the tree $T - \{x_p\}$.*

In the following, we implicity assume that the variables of any quantifier-free **F-ACQ**$^{\neq}$ formula are numbered in some fixed T-order.

Theorem 21. *Let $\varphi(x_1, \ldots, x_p)$ be a quantifier-free σ-formula in $\boldsymbol{F}$-$\boldsymbol{ACQ}^{\neq}$. Then we have* ENUM$(\varphi, <_{lex}) \in$ CONSTANT-DELAY$_{lin}$ *for the lexicographical order* $<_{lex}$.

Proof. The theorem is proved by induction on the number of variables of φ. Without loss of generality, assume that φ is of the form $\varphi(\bar{x}, y) \equiv \varphi_0(\bar{x}) \wedge \Theta(\bar{x}, y)$ where

$$\Theta(\bar{x}, y) \equiv P(y) \wedge \bigwedge_{1 \leq i \leq m} f_i(y) = g_i(z) \wedge \bigwedge_{1 \leq j \leq k} h_j(y) \neq h'_j(\bar{x}).$$

In this formula, the variables y and $z \in \bar{x} = (x_1, \ldots, x_{p-1})$ are respectively the leaf x_p and its parent in the join-tree T of φ and $h'_j(\bar{x})$ denotes a term $h'_j(v)$ for some variable $v \in \bar{x}$. Trivially, φ is logically equivalent to the following formula $\varphi_1(\bar{x}, y) \equiv \psi(\bar{x}) \wedge \Theta(\bar{x}, y)$ where $\psi(\bar{x}) \equiv \exists y \varphi(\bar{x}, y)$. By Lemma 14, one can compute in linear time, for each σ-structure $\mathcal{M}$, a new σ'-structure $\mathcal{M}'$ (σ'-expansion of $\mathcal{M}$, $\sigma \subseteq \sigma'$) and a quantifier-free disjunction $\psi'(\bar{x}) \equiv \bigvee_{1 \leq i \leq N} \psi_i(\bar{x})$ of $\mathbf{ACQ}^{\neq}$ σ'-formulas ψ_i (each of join-tree $T - \{y\}$) so that we have $\psi(\mathcal{M}) = \psi'(\mathcal{M}')$. This yields $\varphi(\mathcal{M}) = \varphi'(\mathcal{M}')$ for the following quantifier-free formula: $\varphi'(\bar{x}, y) \equiv \psi'(\bar{x}) \wedge \Theta(\bar{x}, y)$. φ' is equivalent to the disjunction $\bigvee_{1 \leq i \leq N} \varphi_i(\bar{x}, y)$ where φ_i is the formula

$$\varphi_i(\bar{x}, y) \equiv \psi_i(\bar{x}) \wedge P(y) \wedge \bar{f}(y) = \bar{g}(z) \wedge \bigwedge_{1 \leq j \leq k} h_j(y) \neq h'_j(\bar{x})$$

By the induction hypothesis, we have for each $i \leq N$, ENUM$(\psi_i, <_{lex}) \in$ CONSTANT-DELAY$_{lin}$. By Lemma 4, it is sufficient to prove the same complexity result for ENUM$(\varphi_i, <_{lex}), 1 \leq i \leq N$. (Notice the essential fact that the linear order $<_{lex}$ is the same for each i.) Here is the required algorithm for ENUM$(\varphi_i, <_{lex})$.

Input: A σ'-structure $\mathcal{M}'$
Precomputation phase

- Perform the precomputation phase of ENUM$(\psi_i, <_{lex})$
- Set $P = \{y \in D : \mathcal{M}' \models P(y)\}$
- By sorting the elements $y \in P$ according to the values $\bar{f}(y)$, compute a partition of set P into non empty sets $P_{\bar{\alpha}} = \{y \in P : \bar{f}(y) = \alpha\}$
- Compute the set $A = \{\bar{\alpha} \in D^m : P_{\bar{\alpha}} \neq \emptyset\}$
- For each $\bar{\alpha} \in A$, execute the precomputation phase of the algorithm ENUM-NONCOVER on the (static) input $E = P_{\bar{\alpha}}$ and $(h_j)_{j \leq k}$.

Enumeration phase
For each $\bar{a}$ enumerated by ENUM$(\psi_i, <_{lex})$ do

- Let $\bar{\alpha} = \bar{g}(c)$ where c is the value of variable z in the tuple $\bar{a}$ (if z is the variable x_l, then $c = a_l$)
- Let $\bar{u} = (h'_j(\bar{a}))_{j \leq k}$
- For each b enumerated by ENUMNONCOVER$(P_{\bar{\alpha}}, (h_j)_{j \leq k}, \bar{u})$, output $(\bar{a}, b)$

Definition 22. *A query of* $\boldsymbol{F\text{-}ACQ}$ *(resp* $\boldsymbol{F\text{-}ACQ^{\neq}}$*) is* free-connex acyclic *if* φ *is acyclic and the set of free variables of* φ *is a connex subset of the join-tree of* φ*. We denote by* $\boldsymbol{F\text{-}CCQ}$ *(resp* $\boldsymbol{F\text{-}CCQ^{\neq}}$*) the class of free-connex acyclic queries of* $\boldsymbol{F\text{-}ACQ}$ *(resp* $\boldsymbol{F\text{-}ACQ^{\neq}}$*).*

Since free variables form a connex part of the tree decomposition, a repeated use of Lemma 14 permits to eliminate one by one all the quantified variables until only the free variables remain. Then, one applies Theorem 21 to obtain the following result.

Theorem 23. *For any* φ *of* $\boldsymbol{F\text{-}CCQ^{\neq}}$, $\textsc{Enum}(\varphi) \in \textsc{Constant-Delay}_{lin}$.

4 A Dichotomy Result

Definition 24. *Let* H *be a hypergraph, we say that* H' *is a P-extension of* H *if*

- $V(H) = V(H')$,
- $E(H) \subseteq E(H')$,
- $\forall e' \in E(H')\ \exists e \in E(H)\ e' \subseteq e$.

Definition 25. *Let* $H = (V, E)$ *be a hypergraph and let* $S \subseteq V$*. We say* (T, A) *is an* S-connex tree-structure *of* H *if* T *is a tree-structure of* H*, and* A *is a connex subset of the set of vertices of* T *such that* $\bigcup_{e \in A} e = S$.

- H *is* S-connex acyclic *if there is an* S*-connex tree-structure* (T, A) *of* H.
- H *is* e-S-connex acyclic *if there is a P-extension* H' *of* H *such that* H' *is* S*-connex acyclic.*
- *A conjunctive formula* φ *is* free-connex acyclic *if the hypergraph associed to* φ *is* S*-connex acyclic where* S *is the set free(*φ*) of free variables of* φ.

We denote by $\boldsymbol{CCQ}$ *(resp* $\boldsymbol{CCQ^{\neq}}$*) the class of free-connex acyclic queries* φ *of* $\boldsymbol{CQ}$ *(resp* $\boldsymbol{CQ^{\neq}}$*).*

Lemma 26. *Let* φ *be a formula of* $\boldsymbol{CCQ^{\neq}}$*. Then there exists a formula* φ' *of* $\boldsymbol{F\text{-}CCQ^{\neq}}$ *and an exact reduction from problem* $\textsc{Enum}(\varphi)$ *to* $\textsc{Enum}(\varphi')$

This immediately yields the following Theorem

Theorem 27. *Let* φ *be a query of* $\boldsymbol{CCQ^{\neq}}$*. Then* $\textsc{Enum}(\varphi) \in \textsc{Constant-Delay}_{lin}$.

We give now another characterization of S-connex acyclic hypergraphs. We need to introduce the notion of S-path.

Definition 28. *Let* $H = (V, E)$ *be a hypergraph and let* $S \subseteq V$*. An* S*-path is a path* $(x, a_1, \ldots, a_k, y)$ *such that*

- x *and* y *belong to* S,
- *vertices* $a_i, 1 \leq i \leq k$, *belong to* $V - S$, *and*

– *no hyperedge includes both x and y.*

Lemma 29. *Let $H = (V, E)$ be a hypergraph. H is e-S-connex acyclic if and only if H is acyclic and H admits no S-path.*

The previous lemma permits to obtain a polynomial (probably linear) algorithm to recognize **CCQ** formulas.

Lemma 30. *There is a polynomial time algorithm which, given an acyclic hypergraph $H = (V, E)$ and a set $S \subseteq V$ returns a P-extension H' of H and an S-connex tree-structure of H' if H is e-S-connex acyclic, and an S-path of H otherwise.*

We aim to show that the evaluation and enumeration problems of any acyclic query φ which is not **CCQ** are "hard" in some precise sense. For that purpose, we want to exhibit an exact reduction from the multiplication problem of $n \times n$ boolean matrices (a problem that is strongly conjectured to be not computable in $O(n^2)$ time) to $\text{Enum}(\varphi)$. We need to encode our matrix problem in the first-order framework. A Two-Matrix structure is a relational σ_{AB}-structure $\mathcal{M} = (D, A, B)$ where $D = [1, n]$, $\sigma_{AB} = \{A, B\}$, and A, B are binary relations. Clearly, the multiplication problem of two $n \times n$ boolean matrices is expressed by the following acyclic σ_{AB}-query: $\Pi(x, y) \equiv \exists z(A(x, z) \wedge B(z, y))$. More precisely, $\text{Enum}(\Pi)$ is the problem of enumerating, for each Two-Matrix structure $\mathcal{M} = ([1, n], A, B)$ given as input, the ordered pairs of indices (i, j) where 1 occurs in the matrix product $C = A \times B$, i.e., $C_{i,j} = 1, 1 \leq i, j \leq n$.

Definition 31. *A conjunctive formula is* simple *if each relation symbol appears in at most one atom.*

Remark 32. Given a formula φ of **ACQ** (resp **CCQ**, $\mathbf{ACQ}^{\neq}$, $\mathbf{CCQ}^{\neq}$) and a structure $\mathcal{M}$, one can compute in linear time a simple formula φ' of **ACQ** (resp **CCQ**, $\mathbf{ACQ}^{\neq}$, $\mathbf{CCQ}^{\neq}$) and a structure $\mathcal{M}'$ such that $\varphi(\mathcal{M}) = \varphi'(\mathcal{M}')$.

Lemma 33. *Let φ be a simple formula which is **ACQ** (resp $\mathbf{ACQ}^{\neq}$) but not **CCQ** (resp not $\mathbf{CCQ}^{\neq}$). Then there exists an exact reduction from the problem* $\text{Enum}(\Pi)$ *to* $\text{Enum}(\varphi)$.

Finally, one obtains the following result

Theorem 34. *(Dichotomy Theorem) Let φ be a simple **ACQ** (resp $\mathbf{ACQ}^{\neq}$) query. We have either (1) or (2):*

1. *$\varphi \in$ **CCQ** (resp $\mathbf{CCQ}^{\neq}$) and* $\text{Enum}(\varphi) \in \text{Constant-Delay}_{lin}$ *(and* $\text{Eval}(\varphi)$ *is computable in time $O(|\mathcal{M}| + |\varphi(\mathcal{M})|)$)*
2. *There is an exact reduction from* $\text{Enum}(\Pi)$ *to* $\text{Enum}(\varphi)$ *and then* $\text{Enum}(\varphi) \notin \text{Constant-Delay}_{lin}$ *(and* $\text{Eval}(\varphi)$ *is not computable in time $O(|\mathcal{M}|+|\varphi(\mathcal{M})|)$) under the hypothesis that the product of two $n \times n$ matrices cannot be computed in time $O(n^2)$.*

5 Free-Connex Treewidth

In this section, we introduce the notion of free-connex tree-decomposition of conjunctive queries: this is a variant of the tree-decomposition of graphs

Definition 35. *(see [4] for details) A* tree-decomposition *of a graph $G = (V, E)$ is a tree T whose vertices are subsets of V and are called the* bags *of T, so that*

- *T has the connectivity property (see above),*
- *each edge of G is included in some bag of T.*

The width *of a tree-decomposition T is $\max\{|B| : B$ is a bag of $T\} - 1$. The* treewidth *of G denoted by $tw(G)$ is the smallest width of any tree-decomposition of G.*

Definition 36. *Let $G = (V, E)$ be a graph and S be a subset of V. A S-*connex tree-decomposition *of G is a tuple (T, A) where T is a tree-decomposition of G and A is a connected subset of $V(T)$ such that $\bigcup_{B \in A} B = S$.*

*Given a graph $G = (V, E)$ and a set $S \subseteq V$, the S-*connex treewidth *of G denoted by $ctw(G, S)$ is the smallest width of any S-connex tree-decomposition of G.*
*Let φ be a formula of **CQ** (resp **CQ**$^{\neq}$). The* free-connex treewidth *of φ is the S-connex treewidth of the Gaifman graph [3] of φ for $S = free(\varphi)$.*

Lemma 37. *Let φ be a **CQ** (resp **CQ**$^{\neq}$) formula of free-connex treewidth k over a structure $\mathcal{M}$ of domain D. Then we can compute in time $O(|D|^{k+1} + |\mathcal{M}|)$ a formula of **CCQ** (resp **CCQ**$^{\neq}$) φ' (of size only depending on $|\varphi|$) and a model $\mathcal{M}'$ for φ' such that $\varphi(\mathcal{M}) = \varphi'(\mathcal{M}')$.*

By the previous Lemma and Theorem 27, we obtain the following theorem.

Theorem 38. *Given a fixed formula φ of **CQ**$^{\neq}$ of free-connex treewidth k, the problem* Enum(φ) *can be computed with $O(|D|^{k+1} + |\mathcal{M}|)$ precomputation and constant delay.*

We now give a polynomial algorithm to compute the free-connex treewidth of a formula. We need to introduce the notion of completed graph.

Definition 39. *Let G be a graph. The* completed graph *G' of (G, S) is built as follows*

- *$V(G') = V(G)$*
- *$E(G') = E(G) \cup \{\{x, y\} : x, y \in V$ and there is an S-path between x and $y\}$*

Theorem 40. *Let G be a graph, let $S \subseteq V(G)$ and let G' be the completed graph of (G, S). It holds*

[3] The Gaifman graph of φ is the graph $G_\varphi = (V, E)$ where $V = \text{var}(\varphi)$ and E is the set of pairs of variables $\{x, y\}$ such that there exists an atom of φ that contains both x and y.

1. *$ctw(G, S) = tw(G')$;*
2. *For any fixed k, there is a polynomial time algorithm which returns an S-connex tree-decomposition of G of width at most k if $ctw(G, S) \leq k$ and returns False otherwise.*

Notice that the algorithm of point (2) uses the k-tree decomposition procedure of [5] applied to the completed graph G'.

References

1. Bagan, G.: MSO queries on tree decomposable structures are computable with linear delay. In: Ésik, Z. (ed.) CSL 2006. LNCS, vol. 4207, pp. 167–181. Springer, Heidelberg (2006)
2. Bagan, G., Durand, A., Grandjean, E., Olive, F.: Computing the jth element of a first-order query (submitted 2007)
3. Berge, C.: Graphs and hypergraphs, 2nd edn. Amsterdam (1973)
4. Bodlaender, H.L.: A tourist guide through treewidth. Acta Cybernetica 11, 1–21 (1993)
5. Bodlaender, H.L.: A linear-time algorithm for finding tree-decompositions of small treewidth. SIAM J. Comput. 25(6), 1305–1317 (1996)
6. Chekuri, C., Rajaraman, A.: Conjunctive query containment revisited. Theor. Comput. Sci. 239(2), 211–229 (2000)
7. Cohn, H., Kleinberg, R., Szegedy, B., Umans, C.: Group-theoretic algorithms for matrix multiplication. In: FOCS, pp. 379–388 (2005)
8. Courcelle, B.: Linear delay enumeration and monadic second-order logic. Discrete Applied Maths (to appear, 2006)
9. Courcelle, B., Mosbah, M.: Monadic second-order evaluations on tree-decomposable graphs. Theor. Comput. Sci., 109(1&2), 49–82 (1993)
10. Durand, A., Grandjean, E.: First-order queries on structures of bounded degree are computable with constant delay. Transactions on Computational Logic (to appear)
11. Durand, A., Olive, F.: First-order queries over one unary function. In: Ésik, Z. (ed.) CSL 2006. LNCS, vol. 4207, pp. 334–348. Springer, Heidelberg (2006)
12. Flum, J., Frick, M., Grohe, M.: Query evaluation via tree-decompositions. J. ACM 49(6), 716–752 (2002)
13. Grandjean, E., Schwentick, T.: Machine-independent characterizations and complete problems for deterministic linear time. SIAM J. Comput., 32(1), 196–230 (2002)
14. Grohe, M., Schwentick, T., Segoufin, L.: When is the evaluation of conjunctive queries tractable? In: STOC, pp. 657–666 (2001)
15. Libkin, L.: Elements of finite model theory. Springer, Heidelberg (2004)
16. Papadimitriou, C.H., Yannakakis, M.: On the complexity of database queries. J. Comput. Syst. Sci. 58(3), 407–427 (1999)
17. Yannakakis, M.: Algorithms for acyclic database schemes, 82–94 (1981)

Integrating Linear Arithmetic into Superposition Calculus

Konstantin Korovin and Andrei Voronkov*

The University of Manchester
{korovin, voronkov}@cs.man.ac.uk

Abstract. We present a method of integrating linear rational arithmetic into superposition calculus for first-order logic. One of our main results is completeness of the resulting calculus under some finiteness assumptions.

1 Introduction

In this paper we consider superposition calculus extended with rules for rational linear arithmetic such as Gaussian Elimination for reasoning with equality and Fourier-Motzkin Elimination for reasoning with inequalities. These rules are similar to superposition and ordered chaining rules in first-order reasoning.

There are a number of approaches to integrate arithmetical reasoning into superposition calculus. Most of these approaches are based on approximation of arithmetical reasoning by considering an axiomatisable theory such as Abelian groups or divisible Abelian groups [4, 12–14]. Although this provides a sound approximation it is generally not complete w.r.t. reasoning in usual arithmetical structures such as rational numbers $\mathbb{Q}$. In our approach we consider $\mathbb{Q}$ as a fixed theory sort in the signature containing theory symbols $+, >, =$ together with non-theory sorts and function symbols. We present a sound Linear Arithmetic Superposition Calculus (LASCA) for this language based on a standard superposition calculus extended with rules for linear arithmetic. As we show, the validity problem for first-order formulas of linear arithmetic extended with non-theory function symbols is Π^1_1-complete even in the case when there are no variables over the theory sort. Therefore, there is no sound and complete calculus for this logic. Nevertheless, one of the main results of this paper is that under some finiteness assumptions it is possible to show completeness of our calculus. In particular, we can show that a finite saturated set of clauses (with variables over non-theory sorts) S is satisfiable if and only if S does not contain the empty clause. For this, we need to assume that a simplification ordering we use in our calculus is finite-based (a notion defined later in the paper). In this paper we also show how to construct such an ordering.

Our calculus LASCA is closely related to [4, 13], but here we are dealing directly with the structure $\mathbb{Q}$ rather than with axiomatisations. One of the differences with [13] is that we do not apply abstraction for theory terms. Such abstraction introduces new variables and can increase the number of inferences. On the other hand, in order to show our completeness result we impose additional restrictions on the ordering and variable occurrences. In our completeness proof we adapt the model generation technique

* This work is supported by EPSRC grant GR/T08760/01.

J. Duparc and T.A. Henzinger (Eds.): CSL 2007, LNCS 4646, pp. 223–237, 2007.

(see [2,9]). We use some ideas from normalised rewriting, symmetrisation [4,7,8] and many-sorted reasoning [3,5].

2 Preliminaries

We consider a many-sorted language. Let Σ be a signature consisting of a non-empty set of sorts $\mathcal{S}$, a set of function symbols $\mathcal{F}$, a set of predicate symbols $\mathcal{P}$ and an arity function $arity : \mathcal{F} \cup \mathcal{P} \rightarrow \mathcal{S}^+$, where $\mathcal{S}^+$ denotes the set of finite non-empty sequences of sorts. For a function symbol f with arity $arity(f) = \langle s_0, \ldots, s_n \rangle$, we call $s_0, \ldots, s_{n-1}$ *argument sorts* and s_n the *value sort* of f. In this paper we are mainly dealing with extensions of rational arithmetic. We write $\Sigma_{\mathbb{Q}}$ for a signature such that $\mathcal{S}_{\mathbb{Q}}$ consists of a designated *theory sort* $s_{\mathbb{Q}}$ of rationals, *theory predicate symbols* $\mathcal{P}_{\mathbb{Q}} = \{>, =\}$, and *theory function symbols* $\mathcal{F}_{\mathbb{Q}} = \{+\} \cup \{q, \cdot_q | q \in \mathbb{Q}\}$ where $\mathbb{Q}$ is the set of rationals. We assume that Σ extends $\Sigma_{\mathbb{Q}}$ with non-theory sorts and non-theory function symbols (note that non-theory functions can have arguments and values of the theory sort $s_{\mathbb{Q}}$). We assume that the only non-theory predicates in Σ are equalities on non-theory sorts, denoted as $\simeq_s$, we also write $\simeq$ if there is no confusion, and we use $=$ for equality over the theory sort $s_{\mathbb{Q}}$. Variables, terms, atoms, literals, clauses and first-order formulas are defined in the standard way. We use the standard semantics for many-sorted logic: a Σ-structure consists of a disjoint union of domains indexed by sorts with defined functions and predicates respecting their arities. In addition, we always assume that the domain of the theory sort $s_{\mathbb{Q}}$ is the rational numbers $\mathbb{Q}$ with the usual interpretation of $>, =, +$ and where elements of $\mathbb{Q}$ are also constants in our language and $\cdot_q$ is a unary function symbol interpreted as multiplication by q for each $q \in \mathbb{Q}$. We use convenient abbreviations qt for $\cdot_q(t)$ where t is a term of sort $s_{\mathbb{Q}}$ and $-t$ for $-1t$. We use $\bowtie$ to denote one of theory predicates $>$ or $=$.

We are interested in the question of whether a given first-order formula is (un)satisfiable in a Σ-structure. This question can be reformulated in a standard way as a question of (un)satisfiability of sets of clauses in a Herbrand interpretation which is defined later.

A non-variable term is called a *theory term* (*non-theory term*) if its top function symbol is a theory symbol (non-theory symbol respectively) and similarly for atoms. We assume that $>$ and $=$ occur only positively in clauses (for example $\neg(t > s)$ can be replaced by $s > t \vee s = t$ and $\neg(t = s)$ by $t > s \vee s > t$).

$\mathbb{Q}$-Normalised terms. Define a relation $=_{AC}$ on terms, called AC-congruence, as the least congruence relation generated by associativity and commutativity axioms for $+$. We assume $+$ to be variadic and define $\mathbb{Q}$-normalised terms as follows.

Definition 1. A term t is $\mathbb{Q}$-*normalised* if t is either:

1. a theory constant q, or
2. a non-theory term $f(t_1, \ldots, t_n)$ where $t_1, \ldots, t_n$ are $\mathbb{Q}$-normalised, or
3. $q_1t_1 + \cdots + q_nt_n$ where $n \geq 1$, and for each $1 \leq i \leq n$, the term t_i is a $\mathbb{Q}$-normalised non-theory term, $q_i \neq 0$ and $t_i \neq_{AC} t_j$ for $i \neq j$, and
4. $q_1t_1 + \cdots + q_nt_n + q$ where n and q_i, t_i for $1 \leq i \leq n$ are as in 3 above and $q \neq 0$.

It is not hard to argue that for every ground term t there is a unique, up to AC-congruence, $\mathbb{Q}$-equivalent term which is $\mathbb{Q}$-normalised. This term is called a $\mathbb{Q}$-*normal form* of t and denoted by $t\downarrow_{\mathbb{Q}}$. We say that s is an AC-subterm of a $\mathbb{Q}$-normalised term t if either: (i) $t =_{AC} s$, or (ii) $t = f(t_1, \ldots, t_n)$ and s is an AC-subterm of t_i for some $1 \leq i \leq n$, where f is a non-theory function symbol, or (iii) $t = qt'$ and s is an AC-subterm of t', or (iv) $t =_{AC} u + v$ and s is an AC-subterm of u or v. For example, $3d + 5a$ is an AC-subterm of $4f(5a + 2b + 3d)$.

In this paper we deal with orderings satisfying several properties defined below.

Definition 2. Let $\succ$ be an ordering on $\mathbb{Q}$-normalised terms. It is said to have a *subterm property* if $s[t] \succ t$ whenever t is a proper AC-subterm of $s[t]$. We say that $\succ$ is *AC-compatible* if it satisfies the following property: if $s \succ t$, $s =_{AC} s'$ and $t =_{AC} t'$, then $s' \succ t'$. We say that $\succ$ is $\mathbb{Q}$-*monotone* if for any $\mathbb{Q}$-normal form $t[s]$ where s is a non-theory term, from $s \succ s'$, it follows that $t[s] \succ (t[s'])\downarrow_{\mathbb{Q}}$.

An ordering $\succ$ is called $\mathbb{Q}$-total, if for all ground $\mathbb{Q}$-normal forms s, t, if $s \neq_{AC} t$, then either $s \succ t$ or $t \succ s$.

We say that an ordering $\succ$ has a *sum property* if for any non-theory term t and any finite family of non-theory terms $s_1, \ldots, s_n$ of sort $\mathbb{Q}$, such that $t \succ s_i$ for $1 \leq i \leq n$, it follows that $t \succ (q_1 s_1 + \cdots + q_n s_n + q)\downarrow_{\mathbb{Q}}$ for any coefficients $q_1, \ldots, q_n, q \in \mathbb{Q}$. □

From now on $\succ$ will denote an AC-compatible, $\mathbb{Q}$-monotone, $\mathbb{Q}$-total and well-founded ordering on $\mathbb{Q}$-normal forms which has sum and subterm properties. We show an example of such an ordering in Section 5. We use $\succeq$ for $\succ \cup =_{AC}$.

Let t be a $\mathbb{Q}$-normalised ground term of sort $\mathbb{Q}$, then the *leading monomial* m of t is defined as follows: if t is a theory constant then $m = t$, otherwise m is the greatest w.r.t. $\succ$ non-theory subterm of t. Let $\top$ denote the literal $0 = 0$ and $\bot$ the literal $0 > 1$. We call a ground literal L $\mathbb{Q}$-*normalised* if L is of one of the forms $l = s$, $l > s$, $-l > s$, $l \simeq r$, $l \not\simeq r$, $\top$, $\bot$ where l is a non-theory term and $l \succ r$, we also call l the *leading term* of L. A clause is $\mathbb{Q}$-normalised if all of its literals are $\mathbb{Q}$-normalised and the leading term of a clause is the greatest leading term of its literals. It is easy to see that every ground clause can be $\mathbb{Q}$-normalised into an equivalent clause. From now on we consider only $\mathbb{Q}$-normalised ground terms, literals and clauses.

In order to extend the ordering $\succ$ to literals we represent literals as multisets as follows $m(t = s) = \{t, s\}, m(t > s) = \{t, t, s, s\}, m(t \simeq s) = \{t, s\}, m(t \not\simeq s) = \{t, t, s, s\}$. Now we define $L \succ L'$ iff $m(L) \succ_m m(L')$ where $\succ_m$ is the multiset extension of $\succ$. We compare clauses in the two-fold multiset extension of $\succ$.

Herbrand Interpretation. An *evaluation function* is a mapping from ground non-theory terms of sort $s_{\mathbb{Q}}$ into $\mathbb{Q}$. Let ν be an evaluation function, then define $\bar{\nu}$ to be an extension of ν to the theory terms as follows: $\bar{\nu}(q_1 t_1 + \cdots + q_n t_n + q) = q_1 \nu(t_1) + \cdots + q_n \nu(t_n) + q$.

In order to define a Herbrand interpretation we need a congruence relation $\sim$ on ground $\mathbb{Q}$-normalised terms and an evaluation function ν, such that the following compatibility conditions are satisfied.

Compatibility Conditions:

1. If $t \sim s$ then $\nu(t) = \nu(s)$, for any non-theory terms t, s of sort $s_{\mathbb{Q}}$.
2. If $\bar{\nu}(u) = \bar{\nu}(v)$ then $u \sim v$, for any terms u, v of sort $s_{\mathbb{Q}}$.

We call a pair $\langle \nu, \sim \rangle$, satisfying Compatibility Conditions above, a *Herbrand interpretation*. A theory atom $t \bowtie s$ is true in $\langle \nu, \sim \rangle$ if $\mathbb{Q} \models \bar{\nu}(t) \bowtie \bar{\nu}(s)$, and otherwise false in $\langle \nu, \sim \rangle$. A non-theory atom $t \simeq s$ is true in $\langle \nu, \sim \rangle$ if $t \sim s$, and otherwise false in $\langle \nu, \sim \rangle$.

3 The Calculus for Ground Clauses

The inference rules of our Linear Arithmetic Superposition Calculus (LASCA) are presented in Table 1 (page 237). We assume that all inference rules are applied to $\mathbb{Q}$-normalised clauses and after application of an inference rule we implicitly $\mathbb{Q}$-normalise the conclusion. Note that if we write, e.g., $C \vee l = r$ then implicitly $l \succ r$, since the clause is assumed to be $\mathbb{Q}$-normalised. For a term t, we write $t \succeq C$ ($t \succ C$)if $t \succeq s$ ($t \succ s$) for any term s in C and similarly for literals. For a non-theory term l of sort $s_{\mathbb{Q}}$, we use $\pm l$ to denote l or $-l$, and assume that the choice of the sign is the same for a context, e.g., a rule and its conditions (we use $\mp$ to refer to the opposite sign).

Theorem 1. *Linear Arithmetic Calculus is sound: if the empty clause is derivable in LASCA from S then S is unsatisfiable.*

We say that a set of clauses S is saturated (w.r.t. LASCA) if S is closed under all inferences in LASCA. As we will see in Section 6 there is no sound and complete calculus for Linear Arithmetic extended with non-theory functions. Hence, our LASCA calculus is also incomplete in general: a saturated set of clauses S such that $\Box \notin S$ can still be unsatisfiable. Let us characterise some cases when from the fact that the set S is saturated and $\Box \notin S$ it follows that S is indeed satisfiable.

Definition 3. Let M be a set of terms or clauses. We say that M satisfies *Finiteness of Coefficients* condition if the following holds. There exists a finite set of coefficients P such that if a term qt or q is a subterm of a term in M then $q \in P$. $\Box$

In the sequel we impose the following assumption on sets of clauses.

Assumption 1 *Let S be a set of clauses. We assume that S satisfies Finiteness of Coefficients condition.*

Let us note that under Assumption 1, the number of occurrences of a non-theory term (or a theory constant) in S can be infinite. In Section 4 we show that the set of all ground instances of a finite set of clauses with variables over variable-safe sorts, satisfies Assumption 1. This will be used to show that if a finite set S of (possibly non-ground clauses) is saturated, then S is satisfiable if and only if $\Box \notin S$ (Theorem 3).

Definition 4. Consider a finite set of coefficients P, then T^P denotes the set of all $\mathbb{Q}$-normalised terms t such that any non-theory subterm of t of sort $s_{\mathbb{Q}}$ occurs in t with coefficients from P. An ordering on $\mathbb{Q}$-normalised terms is called *finite-based* if for any finite set of coefficients P and any ground term t the set of all terms in T^P less than t is finite. $\Box$

Assumption 2 *The ordering $\succ$ is finite-based.*

In Section 5 we show how to construct an appropriate ordering satisfying Assumption 2. Now we will show how to construct a candidate model $\langle \nu, \sim \rangle$ for a set of clauses S such that under Assumptions (1,2) if S is saturated and $\Box \notin S$ then S is true in $\langle \nu, \sim \rangle$.

Model Construction. For simplicity of exposition we consider the case when all functions have arguments and values in $\mathbb{Q}$. Let S be a set of ground clauses satisfying Finiteness of Coefficients Assumption 1. We consider terms modulo AC-congruence, and in particular all rewrite rules are implicitly applied modulo AC. Denote T_S the set of all AC-subterms of terms occurring in S and T_S^{nth} all non-theory AC-subterms of terms in S. Note that T_S and T_S^{nth} satisfy Finiteness of Coefficients condition. An equation $l = r$, where $l \succ r$ and l is a non-theory term, can be seen as a rewrite rule $l \rightarrow r$, replacing l with r (and applying $\mathbb{Q}$-normalisation to the resulting term). Any system R of such rules is terminating, and if the left-hand sides of any two rules in R are not overlapping then the system is also convergent. Let us construct a rewriting system R and an evaluation function ν for all terms in T_S^{nth}. The evaluation function ν will be represented via a convergent term rewriting system Υ such that the following holds: (i) Υ consists of rules of type $f(q_1, \dots, q_n) \rightarrow q$, where f is a non-theory function symbol $q_1, \dots, q_n, q \in \mathbb{Q}$, (ii) $R \cup \Upsilon$ is a convergent term rewriting system. We say that a term t is *evaluated* by Υ if $t \downarrow_\Upsilon \in \mathbb{Q}$. We construct R and Υ by induction on terms in T_S^{nth} ordered by $\succ$ as follows. For each term $l \in T_S^{nth}$ we define a set of rewrite rules ϵ_l and a set of evaluation rules δ_l. We define $R_l = \bigcup_{l \succ t \in T_S^{nth}} \epsilon_t$; $R^l = R_l \cup \epsilon_l$; $\Upsilon_l = \bigcup_{l \succ t \in T_S^{nth}} \delta_t$; $\Upsilon^l = \Upsilon_l \cup \delta_l$.

Consider a term l in T_S^{nth}. We inductively assume that we have constructed ϵ_t, δ_t for every $t \prec l, t \subset T_S^{nth}$ such that the following invariants hold.

Invariants (Inv):

1. either $\epsilon_t = \emptyset$, or $\epsilon_t = \{t \rightarrow r\}$ where $t \succ r, r \in T_S$, and
2. either $\delta_t = \emptyset$, or $\delta_t = \{f(q_1, \dots, q_n) \rightarrow q\}$ and $t = f(t_1, \dots, t_n)$, where $t_i \in T_S, q, q_i \in \mathbb{Q}$ for $1 \le i \le n, 0 \le n$, and
3. R^t, Υ^t and $R^t \cup \Upsilon^t$ are convergent term rewriting systems, and
4. t is evaluated by Υ^t, and
5. if t is R_t-irreducible then t is not evaluated by Υ_t, and
6. if t is R^t-irreducible then for any $u, v \in T_S$ such that t is the leading monomial of u, u is R^t-irreducible and $u \succ v$, we have $u \downarrow_{\Upsilon^t} \neq v \downarrow_{\Upsilon^t}$ (note $u \downarrow_{\Upsilon^t}, v \downarrow_{\Upsilon^t} \in \mathbb{Q}$ by *Inv* 4).

Let us note that since $\succ$ is finite-based, there are only a finite number of terms less than l in T_S^{nth}. Therefore $R_l = R^{l'}$ and $\Upsilon_l = \Upsilon^{l'}$ for some $l' \prec l$. We also have that R_l and Υ_l are finite. Now we show how to define ϵ_l, δ_l.

Consider the case when l can be reduced by R_l. If l is evaluated by Υ_l then we define $\epsilon_l = \delta_l = \emptyset$. If l is not evaluated by Υ_l, then $l = f(t_1, \dots, t_n)$ for a non-theory symbol f. We have $f(t_1 \downarrow_{\Upsilon_l}, \dots, t_n \downarrow_{\Upsilon_l}) = f(q_1, \dots, q_n)$ for some $q_i \in \mathbb{Q}$, $1 \le i \le n$, (since $l \succ t_i$ we have that all t_i are evaluated by Υ_l). Let us show that $f(q_1, \dots, q_n)$ does not occur in the left-hand sides of rules in R_l. Indeed, otherwise, $f(q_1, \dots, q_n) \in T_S^{nth}$ and $l \succ f(q_1, \dots, q_n)$, therefore $f(q_1, \dots, q_n)$ and l would be evaluated by Υ_l. Now we define $\epsilon_l = \emptyset$ and $\delta_l = \{f(q_1, \dots, q_n) \rightarrow q\}$ where $q \in \mathbb{Q}$ is selected arbitrary. It is straightforward to check that ϵ_l and δ_l satisfy all invariants above.

Now we assume that l is irreducible by R_l.

Claim. Let us show that l is not evaluated by Υ_l. Let $l = f(t_1, \dots, t_n)$, then $f(t_1 \downarrow_{\Upsilon_l}, \dots, t_n \downarrow_{\Upsilon_l}) = f(q_1, \dots, q_n)$. Assume that l is evaluated, then $f(q_1, \dots, q_n) \rightarrow q \in \Upsilon_l$

for some $q \in \mathbb{Q}$. Consider $s \in T_S^{nth}$, such that $l \succ s$ and $\delta_s = \{f(q_1, \dots, q_n) \to q\}$. We have $s = f(s_1, \dots, s_n)$ for some terms $s_i \in T_S$, $1 \leq i \leq n$ (see *Inv* 2). Since $l \succ s$, from monotonicity of $\succ$ it follows that $t_i \succ s_i$ for some $1 \leq i \leq n$. Let $t_i = \alpha_1 u_1 + \ldots + \alpha_k u_k + \alpha_{k+1}$ and $s_i = \beta_1 v_1 + \ldots + \beta_m v_m + \beta_{m+1}$ where we assume summands are ordered in a descending order (w.r.t. $\succ$). Let j be the smallest index such that $\alpha_j u_j \neq_{AC} \beta_j v_j$. If $j = k+1$ then $m = k$ and $\alpha_{k+1} \neq \beta_{m+1}$, we obtain a contradiction: $0 = t_i \downarrow_{\Upsilon_l} - s_i \downarrow_{\Upsilon_l} = \alpha - \beta \neq 0$. If $j \leq k$ then $\alpha_j u_j \succ \beta_p v_p$ for $j \leq p \leq m+1$. Since u_j is irreducible w.r.t. R_l (and therefore w.r.t. R^{u_j}) from *Inv* 6 it follows that $(\alpha_j u_j + \ldots + \alpha_k u_k + \alpha_{k+1}) \downarrow_{\Upsilon^{u_j}} \neq (\beta_j v_j + \ldots + \beta_m v_m + \beta_{m+1}) \downarrow_{\Upsilon^{u_j}}$. But then $t_i \downarrow_{\Upsilon_l} \neq s_i \downarrow_{\Upsilon_l}$, which is a contradiction.

We say that a literal $\pm l' \bowtie t$ with the leading term $l' \prec l$ is *true* w.r.t. Υ_l if $\mathbb{Q} \models \pm l' \downarrow_{\Upsilon_l} \bowtie t \downarrow_{\Upsilon_l}$ and *false* otherwise (note that $l' \downarrow_{\Upsilon_l}, t \downarrow_{\Upsilon_l} \in \mathbb{Q}$).

Let S^l be the set of all clauses in S with the leading term l. For a clause $C \in S^l$ define V_C^l, D_C^l such that $C = V_C^l \vee D_C^l$ and V_C^l consists of all literals in C with the leading term l (note D_C^l can be empty). We say that a clause $C \in S^l$, $C = C' \vee l = r$ *weakly produces* a rewrite rule $l \to r$, if the following holds.

- $l = r$ is a strictly maximal literal in C, and
- D_C^l is false w.r.t. Υ_l, and
- there is no $l = r' \in C'$ such that $\mathbb{Q} \models r \downarrow_{\Upsilon_l} = r' \downarrow_{\Upsilon_l}$.

If there is a clause in S^l weakly producing a rewrite rule then we take the smallest (w.r.t. $\succ$) such clause C. Let $l \to r$ be the rewrite rule weakly produced by C, then we say that $l \to r$ is *produced* by C. We define $\epsilon_l = \{l \to r\}$ and $\delta_l = \{l \downarrow_{\Upsilon_l} \to r \downarrow_{\Upsilon_l}\}$. Now we check that all *Inv* are satisfied. It follows immediately from the construction that *Inv* (1,2,4,5,6) are satisfied. Let us show that $R^l \cup \Upsilon^l$ is convergent. First we note that there are no critical pairs between $l \to r$ and R_l. Indeed, l is irreducible by R_l and l is greater (w.r.t. $\succ$) than all left-hand sides of rules in R_l. Likewise, from the Claim above it follows that that l is not evaluated by Υ_l and therefore there are no critical pairs between $l \downarrow_{\Upsilon_l} \to r \downarrow_{\Upsilon_l}$ and rules in Υ_l. The only new critical pairs possible are between $l \to r$ and rules in Υ^l, but they are joinable since $l \downarrow_{\Upsilon^l} = r \downarrow_{\Upsilon^l}$.

Now we assume that there is no clause in S^l producing a rewrite rule. We define $\epsilon_l = \emptyset$, and now we need to find an appropriate evaluation for l. Let us fix a numerical variable x_l. We say that a clause $C \in S^l$, $C = C' \vee \pm l > r$ *weakly produces* a bound $\pm x_l > r \downarrow_{\Upsilon_l}$, if the following holds.

- $\pm l > r$ is a strictly maximal literal in C, and
- D_C^l is false in Υ_l, and
- there is no literal $\pm l > r'$ in C', and
- if there is a literal $\mp l > r'$ in C', then $\mathbb{Q} \models r \downarrow_{\Upsilon_l} \geq -r' \downarrow_{\Upsilon_l}$.

Let B^l be the set of all bounds weakly produced by clauses in S^l, (B^l can be the empty set). It is not difficult to see that Assumptions 1 and 2 imply that B^l is finite. Let B_-^l be the set of lower bounds in B^l (i.e. bounds of the type $x_l > q$) an B_+^l be the set of upper bounds in B^l (i.e. bounds of the type $-x_l > q$). We have $B^l = B_-^l \cup B_+^l$. Since B^l is finite we have that each B_-^l and B_+^l are satisfiable. Let $x_l > q_{glb}$ be the greatest w.r.t. $>$ lower bound in B^l_-, (since B^l_- is finite such a bound always exists). Let U^l be the set of upper bounds $-x_l > q$ in B^l_+ such that $-q_{glb} > q$. Define $B^l_\pm = B^l_- \cup U^l$. We have $B^l_\pm$ is satisfiable and the set of solutions to $B^l_\pm$ is an open interval. Moreover, if $B^l_+ \neq U^l$ then B^l_- together with any bound from $B^l_+ \setminus U^l$ is unsatisfiable. Clauses weakly producing bounds in $B^l_\pm$ are called *productive*.

In order to satisfy *Inv* 6 we impose additional constraints on evaluation of l defined below. We say that a pair of terms $u, v \in T_S$, such that l is the leading monomial of u and $u \succ v$ produces a disequality constraint d_{uv} if the following holds. Assume that $u = \alpha l + u'$, $\alpha \neq 0$. If l is not a subterm of v and therefore $l \succ v$, then $d_{uv} = \{x_l \neq (v \downarrow_{\Upsilon_l} - u' \downarrow_{\Upsilon_l})/\alpha\}$ (note that $u' \downarrow_{\Upsilon_l}, v \downarrow_{\Upsilon_l} \in \mathbb{Q}$). If l is a subterm of v then $v = \beta l + v'$ and we need to consider the following possible cases. Case (i): $\beta = \alpha$. Then we have $u' \succ v'$ and we can apply *Inv* 6 to the leading term of u', obtaining $u' \downarrow_{\Upsilon_l} \neq v' \downarrow_{\Upsilon_l}$. In this case we have that under any evaluation of l, evaluation of u will be different from evaluation of v and therefore we define $d_{uv} = \emptyset$. Case (ii): $\beta \neq \alpha$. Then we define $d_{uv} = \{x_l \neq (v' \downarrow_{\Upsilon_l} - u' \downarrow_{\Upsilon_l})/(\alpha - \beta)\}$. We define D^l to be the union of all d_{uv} where $u, v \in T_S$, l is the leading monomial of u and $u \succ v$. From Assumptions (1, 2) it follows that D^l is finite, therefore D^l is satisfied by all but possible a finite number of rationals. We have $B^l_\pm \cup D^l$ is satisfiable. Define $\delta_l = \{l \downarrow_{\Upsilon_l} \to q\}$, where q is any rational satisfying $B^l_\pm \cup D^l$. It is straightforward to check that all *Inv* are satisfied by ϵ_l, δ_l.

We have shown how to construct ϵ_l, δ_l for every $l \in T^{nth}_S$. Now we define $R_S = \bigcup_{l \in T^{nth}_S} \epsilon_l$ and $\Upsilon_S = \bigcup_{l \in T^{nth}_S} \delta_l$. We have $R_S \cup \Upsilon_S$ is a convergent term rewriting system such that every term in T^{nth}_S is evaluated by Υ_S. Finally we need to extend evaluation Υ_S to all non-theory terms. We can do it by induction over all non-theory terms as follows. For each term t we define a set of evaluation rules κ_t as follows. Assume, by induction, that we have defined κ_s for non-theory terms $s \prec t$. Define $\Lambda_t = \Upsilon_S \bigcup \cup_{t \succ s} \kappa_s$. If t is evaluated by Λ_t then we define $\kappa_t = \emptyset$, otherwise we define $\kappa_t = \{t \downarrow_{\Lambda_t} \to q\}$ where $q \in \mathbb{Q}$ is selected arbitrary. Define $\Lambda^t = \Lambda_t \cup \kappa_t$ and $\Lambda_S = \cup \Lambda^t$. It is not difficult to check that $R_S \cup \Lambda_S$ is a convergent term rewriting system such that every non-theory term is evaluated by Λ_S.

Let us define a Herbrand interpretation $\langle \nu, \sim \rangle$, where $\nu(t) = t \downarrow_{\Lambda_S}$ and $t \sim s$ iff $t \downarrow_{\Lambda_S} = s \downarrow_{\Lambda_S}$. We call $\langle \nu, \sim \rangle$ the *candidate model* for S. □

Lemma 1. *In the Model Construction above if a clause C is productive then C is true in the candidate model $\langle \nu, \sim \rangle$.*

Proof. Immediately follows from the Model Construction. □

Lemma 2. *In the Model Construction above if a clause $C = C' \vee \pm l \bowtie r$ produces a rule or a bound $\pm l \bowtie r$ then C' is false in the candidate model $\langle \nu, \sim \rangle$.*

Proof. Consider first when $\pm l \bowtie r$ is $l = r$ and C generates the rule $l \to r$. We have $C' = V^l_{C'} \vee D^l_{C'}$ where $V^l_{C'}$ consists of all literals in C' with the leading term l. From the conditions on productiveness of C we have that $D^l_{C'}$ is false in $\langle \nu, \sim \rangle$. From the definition of the ordering on atoms $V^l_{C'}$ does not contain any atoms with $>$. If $V^l_{C'}$ contains an atom $l = r'$ then we have $r \downarrow_{\Upsilon_S} \neq r' \downarrow_{\Upsilon_S}$ and $l \downarrow_{\Upsilon_S} = r \downarrow_{\Upsilon_S}$ therefore $l = r'$ is false in $\langle \nu, \sim \rangle$.

Now we consider the case when $\pm l \bowtie r$ is $\pm l > r$ and C produces the bound $\pm x_l > r \downarrow_{\Upsilon_S}$. We have that $D^l_{C'}$ is false in $\langle \nu, \sim \rangle$. If $V^l_{C'}$ contains an atom $l = r'$, then by construction l is irreducible w.r.t. R^l. Since $l \succ r'$, *Inv* 6 implies that $l \downarrow_{\Upsilon_S} \neq r' \downarrow_{\Upsilon_S}$. If $V^l_{C'}$ contains an atom $\mp l > r'$ then we have $\pm l \downarrow_{\Upsilon_S} > r \downarrow_{\Upsilon_S} \geq -r' \downarrow_{\Upsilon_S}$ and therefore $r' \downarrow_{\Upsilon_S} > \mp l \downarrow_{\Upsilon_S}$ implying that $\mp l > r'$ is false. Also, by conditions on productiveness, there is no atom $\pm l > r'$ in $V^l_{C'}$. We have shown that all atoms in $V^l_{C'}$, and therefore in C', are false in $\langle \nu, \sim \rangle$. □

Theorem 2. *LASCA is complete under Assumptions (1,2). Let S be a set of ground clauses such that Assumptions (1,2) are satisfied. If S is saturated and $\square \notin S$ then S is true in the candidate model $\langle \nu, \sim \rangle$.*

Proof. Let S be a saturated set of clauses satisfying Assumption 1. We apply Model Construction above to obtain $R_S, \Upsilon_S, \Lambda_S$ and the candidate model $\langle \nu, \sim \rangle$. In order to show that $\langle \nu, \sim \rangle$ satisfies all clauses in S it is sufficient to show that Υ_S satisfies all clauses in S. Assume otherwise. Let C be the smallest clause in S that is false under Υ_S. Let $C = C' \vee \pm l \bowtie r$, where $\pm l \bowtie r$ be a maximal literal in C. First we show that l is irreducible by R_S. Indeed, assume that $l[l']$ is reducible by a rule $l' \to r'$. Consider the clause $D = D' \vee l' = r'$ producing $l' \to r'$. Then, there is an inference by Gaussian Elimination with the premise C, D and the conclusion $G = D' \vee C' \vee \pm l[r'] \bowtie r$. We have that $C \succ G$ and from Lemma 2 it follows that G is false in Υ_S. This contradicts minimality of C.

By Lemma 1 all productive clauses are true in Υ_S, therefore we assume that C is not productive. Consider possible cases.

Case (1): $C = C' \vee l = r$. If C is not weakly productive then $C' = C'' \vee l = r'$ and $r \downarrow_{\Upsilon_S} = r' \downarrow_{\Upsilon_S}$. Therefore, inference rule Theory Equality Factoring is applicable to C with the conclusion $D = C'' \vee r > r' \vee r' > r \vee l = r'$. We have $C \succ D$ and D is false in Υ_S, contradicting minimality of C. Now assume that C is weakly productive, then there is a clause $C' \preceq C$ which produces a rule $l \to r'$ to R_S. This contradicts that l is irreducible by R_S, which is shown above.

Case (2): $C = C' \vee -l > r$. If C is not weakly productive then either (i) there exists $D = D' \vee l = r'$ and D produces $l \to r'$, but this contradicts that l is irreducible by R_S, or (ii) there is a literal $-l > r'$ in C', or (iii) there is a literal $l > r'$ in C' such that $-r' \downarrow_{\Upsilon_S} > r \downarrow_{\Upsilon_S}$.

Case (2.ii). Assume that $C' = C'' \vee -l > r'$. Then, inference rules InF 1 and InF 2 are applicable to C with the conclusions $D_1 = C'' \vee r > r' \vee -l > r$ and $D_2 = C'' \vee r' > r \vee -l > r'$, respectively. Note that $D_1 \prec C$ and $D_2 \prec C$. Consider possible cases. If $r' \geq r$ is true in Υ_S then D_1 is false in Υ_S. If $r > r'$ is true in Υ_S then D_2 is false in Υ_S. In both cases we obtain a contradiction to the minimality of C.

Case (2.iii). Let us assume that there is a literal $l > r'$ in C' such that $-r' \downarrow_{\Upsilon_S} > r \downarrow_{\Upsilon_S}$. Since $l > r'$ and $-l > r$ are false in Υ_S we have $r' \downarrow_{\Upsilon_S} \geq l \downarrow_{\Upsilon_S}$ and $r \downarrow_{\Upsilon_S} \geq -l \downarrow_{\Upsilon_S}$, therefore $r \downarrow_{\Upsilon_S} \geq -r' \downarrow_{\Upsilon_S}$ which is a contradiction.

Case (2.iv). Now we assume that C is weakly productive. Let C weakly produces a bound $(-x_l > r \downarrow_{\Upsilon_S}) \in B^l_+$. If $(-x_l > r \downarrow_{\Upsilon_S}) \in U^l$ then $(-x_l > r \downarrow_{\Upsilon_S}) \in B^l_\pm$ implying C is productive which is a contradiction. If $(-x_l > r \downarrow_{\Upsilon_S}) \in B^l_+ \setminus U^l$ then we have the following. Let $D = D' \vee l > r_{glb}$ be the clause producing the greatest lower bound (w.r.t. $>$) into B^l_-. Then, the Fourier-Motzkin inference rule is applicable to C and D with the conclusion $K = C' \vee D' \vee -r_{glb} > r$. Let us show that K is false in Υ_S. Indeed, D' is false since D is productive (see Lemma 2), and $-r_{glb} > r$ is false in Υ_S since $(-x_l > r \downarrow_{\Upsilon_S}) \notin U^l$ (see definition of U^l). Now we show that $C \succ K$. Indeed, $(l > r_{glb}) \succ D'$ therefore $(-l > r) \succ D'$ and $(-l > r) \succ (-r_{glb} > r)$. These imply that $C \succ K$, obtaining a contradiction to the minimality of C.

Case (3): $C = C' \vee l > r$. Subcases (3.i-iii) are similar to (2.i-iii).

Case (3.iv). We assume that C is weakly productive. Since C weakly produces a bound $(l > r \downarrow_{\Upsilon_S}) \in B^l_- \subseteq B^l_\pm$ we have that C is also productive, which is a contradiction.

We have considered all possible cases arriving at a contradiction under the assumption that C is false in $\langle \nu, \sim \rangle$. Therefore all clauses in S are true in the candidate model $\langle \nu, \sim \rangle$. □

Let us note that our proof of the completeness theorem is based on the model generation technique, and therefore it is not difficult to adapt redundancy notions from the standard superposition calculus. For details we refer to [6].

4 Lifting

We now consider clauses with variables over variable-safe sorts defined below. It is convenient to define first the set of *variable-unsafe sorts* $\hat{S}$ (w.r.t. Σ) as the minimal set of sorts such that (i) $s_{\mathbb{Q}} \in \hat{S}$ and (ii) if there is a function symbol f in $\mathcal{F}$ with an argument of a sort in $\hat{S}$ then the value sort of f is also in $\hat{S}$. We define the set of *variable-safe sorts* as $\bar{S} = S \setminus \hat{S}$, (see Examples (1,2)).

Assumption 3 *For a set of clauses S, all variables in S are of variable-safe sorts.*

It is easy to see that if a finite set of clauses S satisfies Assumption 3, then the set of all ground instances of S satisfies Finiteness of Coefficients Assumption 1.

Our LASCA calculus for ground clauses works on $\mathbb{Q}$-normalised clauses. In order to lift LASCA calculus into non-ground case we need additional normalisation rules. In formulation of Normalisation rule below we assume that non-ground theory literals are in one-sided form $t \bowtie 0$.

For a pair of terms t, t' let $mgu_{AC}(t, t')$ be a minimal complete set of AC-unifiers.

Normalisation Rule:

$$\frac{C[qt + q't']}{C[(q + q')t]\sigma} \qquad \text{where } \sigma \in mgu_{AC}(t, t').$$

Equality Resolution:

$$\frac{C \vee t \not\simeq t'}{C\sigma} \qquad \text{where } \sigma \in mgu_{AC}(t, t').$$

Now lifting of LASCA calculus is straightforward and we show the corresponding rules only for Gaussian and Fourier-Motzkin elimination rules. We assume that $\succ$ is lifted to non-ground terms, literals and clauses, in such a way that $\succ$ is preserved under substitutions. As in the ground case we assume that before applying the LASCA rules, literals are represented in one of the form $l = r$, $l > r$, $-l > r$, $l \simeq r$, $l \not\simeq r$, $\top$, $\bot$ where there exists a grounding substitution σ such $l\sigma \succ r\sigma$ (there can be several choices for l and r in one literal).

Gaussian Elimination:

$$\frac{C \vee l = r \quad L[l']_p \vee D}{(C \vee D \vee L[r]_p)\sigma} \quad \begin{array}{l}\text{(i) } \sigma \in mgu_{AC}(l, l'),\\ \text{(ii) } (l = r)\sigma\theta \succ C\sigma\theta \text{ for some grounding } \theta.\end{array}$$

Fourier-Motzkin Elimination:

$$\frac{C \vee l > r \quad -l' > r' \vee D}{C \vee D \vee -r' > r}$$

(i) $\sigma \in mgu_{AC}(l, l')$,
and for some grounding substitution θ:
(ii) $(l > r)\sigma\theta \succ C\sigma\theta$,
(iii) there is no $l'' > r'' \in C$ such that $l''\sigma\theta =_{AC} l\sigma\theta$
(iv) $(-l' > r')\sigma\theta \succ D\sigma\theta$,
(v) there is no $-l'' > r'' \in D$ such that $l''\sigma\theta =_{AC} l\sigma\theta$.

Note that from the Assumption 3 it follows that l and l' are not variables in Gaussian and Fourier-Motzkin elimination rules.

Example 1. Let s be a non-theory sort. Let $f : \langle s, s_{\mathbb{Q}} \rangle$; $e : \langle s, s \rangle$; $g, h : \langle s_{\mathbb{Q}}, s_{\mathbb{Q}} \rangle$, assume that $g(x) \succ h(x)$. Consider set of clauses:

$$\begin{array}{rr} 2g(f(e(x))) + h(f(e(x))) = 0 & (1)\\ g(g(f(x)) + 1/2h(f(x))) > 2g(0) & (2)\\ g(0) > 0 & (3)\end{array}$$

We can prove unsatisfiability of the set of clauses $\{(1), (2), (3)\}$ by applying Gaussian Elimination between (1) and (2) obtaining $g(-1/2h(f(e(x))) + 1/2h(f(e(x)))) > 2g(0)$, then applying Normalisation obtaining $-g(0) > 0$ and Fourier-Motzkin with (3) obtaining $\bot$. Let us note that our next Theorem 3 implies that the set of clauses $\{(1), (2)\}$ is satisfiable, since the saturation process terminates.

Now we are ready to prove the following completeness theorem.

Theorem 3. *Let $\succ$ be finite-based and S be a finite saturated set of clauses satisfying Assumption 3. Then S is satisfiable if and only if $\Box \notin S$.*

Proof. Let S be a finite saturated set such that $\Box \notin S$. Let us show that S is satisfiable. Let S_{gr} be the set of all ground instances of clauses in S which are $\mathbb{Q}$-normalised. Since S is finite we have that S_{gr} satisfies Finiteness of Coefficients Assumption 1. Let $\langle \nu, \sim \rangle$ be the candidate model for S_{gr} (see Model Construction). Assume that $\langle \nu, \sim \rangle$ is not a model for S and let $C\sigma$ be the minimal w.r.t. $\succ$ instance of S false in $\langle \nu, \sim \rangle$. We can assume that $C\sigma$ is normalised. Indeed, if $C\sigma$ is not normalised, then we can apply Normalisation, or Equality Resolution rule to obtain a smaller clause false in $\langle \nu, \sim \rangle$.

Now we can proceed as in Theorem 2. The only additional case to consider is when $x\sigma = t$ is reducible by $R_{S_{gr}}$ for some variable in C. Let $l \rightarrow r$ in $R_{S_{gr}}$ and $t|_p = l$. Then $t \succ t[r]_p$. Define σ' to be a substitution such that $x\sigma' = t[r]_p$ and $y\sigma' = y\sigma$ for variables different from x. Then $C\sigma \succ C\sigma'$ and $C\sigma'$ is false in $\langle \nu, \sim \rangle$, contradicting minimality of $C\sigma$. □

Example 2. Let s be a non-theory sort. Consider $f : \langle s, s \rangle$, $h : \langle s, s_{\mathbb{Q}} \rangle$ and $c : \langle s_{\mathbb{Q}} \rangle$.

$$h(f(x)) > h(x)$$
$$c > h(y)$$

This set of clauses is saturated and therefore is satisfiable by Theorem 3. Note that after grounding this set of clauses we obtain an infinite number of inequalities, and the term c has infinitely many occurrences in different ground inequalities.

5 Finite-Based Ordering

In this section we present an ordering $\succ$ which satisfies all required properties: it is AC-compatible, $\mathbb{Q}$-monotone, $\mathbb{Q}$-total, finite-based, well-founded and satisfies sum and subterm properties. Without the condition to be finite-based, such orderings are well-known to exist by modifying the lexicographic path ordering (see e.g. [11]). Unfortunately these orderings are not finite-based which is a crucial condition for our completeness theorems. Here we show how to modify the Knuth-Bendix ordering to satisfy all requirements. Let $\succ_c$ be any well-founded total ordering on rationals such that $q \succ_c 1 \succ_c 0$ for any q different from $0, 1$. Let Σ_{nth} consists of all non-theory symbols. For an ordering $\succ$, let $\succ^{mul}$ denotes the multiset extension of $\succ$ and $\succ^{lex}$ denotes the lexicographic extension of $\succ$ defined over tuples of the same length.

Denote the set of natural numbers by $\mathbb{N}$. We call a *weight function* on Σ_{nth} any function $w : \Sigma_{nth} \rightarrow \mathbb{N}$ such that $w(e) > 0$ for every constant, or unary function symbol e. A *precedence relation* on Σ_{nth} is any linear order $\gg$ on Σ_{nth}. We call $w(g)$ the *weight* of g. The *weight* of any ground term t over signature Σ_{nth}, denoted $|t|$, is defined as follows: for any constant c we have $|c| = w(c)$ and for any function symbol g of a positive arity $|g(t_1, \ldots, t_n)| = w(g) + |t_1| + \ldots + |t_n|$.

First we define the Knuth-Bendix order on terms over Σ_{nth} in a usual way. The *Knuth-Bendix order induced by* w and $\gg$ is the binary relation $\succ_{KBO}$ on ground terms over Σ_{nth} defined as follows. For any ground terms $t = g(t_1, \ldots, t_n)$ and $s = h(s_1, \ldots, s_k)$ we have $t \succ_{KBO} s$ if one of the following conditions holds:

1. $|t| > |s|$;
2. $|t| = |s|$ and $g \gg h$;
3. $|t| = |s|$, $g = h$ and $(t_1, \ldots, t_n) \succ_{KBO}^{lex} (s_1, \ldots, s_n)$.

It is known that for every weight function w and precedence relation $\gg$ compatible with w, the Knuth-Bendix order induced by w and $\gg$ is a simplification order total on ground terms (see e. g. [1]). Let us note that for any term t there are only a finite number of terms less that t w.r.t. $\succ_{KBO}$. This property is crucial for defining a finite-based ordering.

Now we define an abstraction $abstr$ of terms over Σ into terms over Σ_{nth} as follows. Let $c_m \in \Sigma_{nth}$ be the least constant w.r.t. $\succ_{KBO}$. Then $abstr$ is defined by a structural induction on terms as follows: (i) $abstr(c) = c$ for a constant $c \in \Sigma_{nth}$, (ii) $abstr(f(t_1, \ldots, t_n)) = f(abstr(t_1), \ldots, abstr(t_n))$, for $f \in \Sigma_{nth}$, (iii) $abstr(q) = c_m$ for any constant $q \in \Sigma_{\mathbb{Q}}$, (iv) $abstr(q_1 t_1 + \ldots + q_n t_n + q) = abstr(t_j)$ where $abstr(t_j)$ is the greatest w.r.t. $\succ_{KBO}$ term among $abstr(t_1), \ldots, abstr(t_n)$, $1 \leq n$.

Given an ordering $\succ$ on non-theory terms we denote $\succ'$ an ordering extending $\succ$ to terms of the form qt where t is a non-theory term as follows. For non-theory terms t, s, we say that $qt \succ' q's$ iff either (i) $t \succ s$ or (ii) $t =_{AC} s$ and $q \succ_{c} q'$. Likewise we say $qt \succ' s$ iff $t \succeq s$, and $t \succ' qs$ iff $t \succ s$.

Finite-Based $\mathbb{Q}$*-KBO.* Now we define a $\mathbb{Q}$-Knuth-Bendix ordering ($\mathbb{Q}$-KBO) $\succ_{QKBO}$ on general terms as follows. Define $t \succ_{QKBO} s$ if one of the following conditions holds:

1. $abstr(t) \succ_{KBO} abstr(s)$, or
2. $abstr(t) = abstr(s)$ and
   (a) $t = f(t_1, \ldots, t_n)$ and $s = f(s_1, \ldots, s_n)$ for $f \in \Sigma_{nth}$ and
   $(t_1, \ldots, t_n) \succ_{QKBO}^{lex} (s_1, \ldots, s_n)$, or
   (b) $t = q_1 t_1 + \ldots + q_n t_n + q$ and $s = q'_1 s_1 + \ldots + q'_m s_m + q'$, and
      i. $\{t_1, \ldots, t_n\} \succ_{QKBO}^{mul} \{s_1, \ldots, s_m\}$ or,
      ii. $\{t_1, \ldots, t_n\} =_{AC} \{s_1, \ldots, s_m\}$ and
      $\{q_1 t_1, \ldots, q_n t_n, q\} \succ'^{mul}_{QKBO} \{q'_1 s_1, \ldots, q'_m s_m, q'\}$

Theorem 4. *$\mathbb{Q}$-Knuth-Bendix ordering is an AC-compatible, $\mathbb{Q}$-monotone, $\mathbb{Q}$-total, well-founded and finite-based ordering which satisfies sum and subterm properties.*

6 Negative Results

In this section we remark on complexity of the first-order theories for $\mathbb{Q}$ extended with non-theory function symbols. First we consider the structure $\mathbb{N}$ with theory symbols $\langle 0, S \rangle$ where S is interpreted as the successor function. Now, if we consider validity of sentences in an extended signature with non-theory function symbols then we are in the universal fragment of second-order arithmetic. Indeed, validity of a first-order sentence $\varphi(\bar{f})$ is equivalent to whether the second-order universal sentence $\forall \bar{f} \varphi(\bar{f})$ is true in $\mathbb{N}$. Therefore checking validity of formulas over $\mathbb{N}$ with non-theory function symbols is of the same complexity as checking validity of second-order universal sentences over $\mathbb{N}$ which is a Π_1^1-complete problem [10]. (Usually $\mathbb{N}$ is considered in the signature $\langle +, \cdot, 0, 1 \rangle$, but $+$ and $\cdot$ can be defined via S using standard inductive definitions.) It is easy to see that if $\mathbb{N}$ can be defined (up to isomorphism) in a language then the validity problem for such language extended with non-theory symbols is at least Π_1^1-hard (we can relativise formulas to $\mathbb{N}$).

Now we show that even if we consider formulas without quantifiers over variables of sort $s_{\mathbb{N}}$, still the validity problem is of the same complexity of being Π_1^1-complete. Indeed, consider a non-theory sort s and functions $0_s : \langle s \rangle$, $S_s : \langle s, s \rangle$, and $h : \langle s, s_{\mathbb{N}} \rangle$. Then, $\forall xy\, (h(x) = h(y) \rightarrow x \simeq y)$ axiomatises that h is an embedding of the domain of sort s into $\mathbb{N}$. Formulas $h(0_s) = 0$ and $\forall x\ (h(S_s(x)) = h(x) + 1)$, define $\mathbb{N}$ in the non-theory domain. Note that all variables in the above definition are of sort s.

Now we show that $\mathbb{N}$ is definable in $\mathbb{Q}$ extended with non-theory function symbols. Indeed, the following axioms define $\mathbb{N}$ in $\mathbb{Q}$, (for simplicity we consider N as a non-theory predicate symbol, but trivially $\mathbb{N}$ can be redefined using only function symbols).

$$\begin{array}{l} N(0) \\ \forall x\ (N(x) \rightarrow x = 0 \vee x > 0) \\ \forall x\ (N(x) \rightarrow N(x+1)) \\ \forall xy\ ((N(x) \wedge N(x+1) \wedge x + 1 > y > x) \rightarrow \neg N(y)) \\ \forall x\ (S(x) = x + 1) \end{array}$$

Since $\mathbb{Q}$ can be trivially coded in $\mathbb{N}$, we conclude that the validity problem for formulas in $\mathbb{Q}$ extended with non-theory function symbols is Π^1_1-complete.

Similar to the case of $\mathbb{N}$, we show that even if we consider formulas without quantifiers over variables of sort $\mathbb{Q}$, still the validity problem is Π^1_1-complete. The functions $h, 0_s, S_s$ are defined in the same way as for $\mathbb{N}$, but now h defines an embedding into $\mathbb{Q}$ rather than into $\mathbb{N}$. In order to define natural numbers N_s in the domain of sort s we need additional binary predicate $>_s$ over sort s and additional axioms:

$$\begin{array}{l} N_s(0_s) \\ \forall xy\ (x >_s y \leftrightarrow h(x) > h(y)) \\ \forall x\ (N_s(x) \rightarrow x \simeq 0_s \vee x >_s 0_s) \\ \forall x\ (N_s(x) \rightarrow N_s(S_s(x))) \\ \forall xy\ ((N_s(x) \wedge N_s(S_s(x)) \wedge S_s(x) >_s y >_s x) \rightarrow \neg N_s(y)) \end{array}$$

It is easy to check that these axioms define $\mathbb{N}$ in the domain of sort s. Therefore the validity problem for formulas without quantifiers over variables of sort $\mathbb{Q}$ is Π^1_1-complete. We summarise these results in the following theorem.

Theorem 5. *Consider $\mathbb{Q}$ in the signature $\langle 0, 1, +, > \rangle$ and $\mathbb{N}$ in the signature $\langle 0, S \rangle$. Then, the following problems are Π^1_1-complete.*

- *Unsatisfiability of sets of clauses, in a signature extending $\mathbb{Q}$ ($\mathbb{N}$) with non-theory function symbols.*
- *Unsatisfiability of sets of clauses with variables ranging over a non-theory sort s, in a signature extending $\mathbb{Q}$ ($\mathbb{N}$) with a non-theory sort s and non-theory function symbols.*

In particular, Theorem 5 implies that there is no sound and complete calculus for linear arithmetic extended with non-theory function symbols.

7 Conclusions

In this paper we have presented an extension of superposition calculus for first-order logic with rules for linear arithmetic. One of our main results is completeness of the resulting calculus under some finiteness assumptions. One of the possible applications of our results is to obtain new decision procedures for fragments of first-order logic extended with rational arithmetic.

References

1. Baader, F., Nipkow, T.: Term Rewriting and All That. Cambridge University press, Cambridge (1998)
2. Bachmair, L., Ganzinger, H.: Resolution theorem proving. In: Robinson, A., Voronkov, A. (eds.) Handbook of Automated Reasoning, vol. 1, pp. 19–100. Elsevier, Amsterdam (2001)
3. Bachmair, L., Ganzinger, H., Waldmann, U.: Refutational Theorem Proving for Hierarchic First-Order Theories. Applicable Algebra in Engineering, Communication and Computing 5(3/4), 193–212 (1994)
4. Godoy, G., Nieuwenhuis, R.: Superposition with completely built-in Abelian groups. J. Symb. Comput. 37(1), 1–33 (2004)
5. Hustadt, U., Motik, B., Sattler, U.: Reasoning in description logics with a concrete domain in the framework of resolution. In: ECAI, pp. 353–357 (2004)
6. Korovin, K., Voronkov, A.: Integrating linear arithmetic into superposition calculus. Journal version (in preparation)
7. Le Chenadec, P.: Canonical forms in finitely presented algebras. Research Notes in Theoretical Computer Science, Wiley, Chichester (1986)
8. Marché, C.: Normalized rewriting: An alternative to rewriting modulo a set of equations. J. Symb. Comput. 21(3), 253–288 (1996)
9. Nieuwenhuis, R., Rubio, A.: Paramodulation-based theorem proving. In: Robinson, A., Voronkov, A. (eds.) Handbook of Automated Reasoning, vol. I, pp. 371–443. Elsevier, Amsterdam (2001)
10. Rogers, H.: Theory of recursive functions and effective computability. MIT Press, Cambridge (1988)
11. Rubio, A., Nieuwenhuis, R.: A precedence-based total AC-compatible ordering. In: Kirchner, C. (ed.) Rewriting Techniques and Applications. LNCS, vol. 690, pp. 374–388. Springer, Heidelberg (1993)
12. Stuber, J.: Superposition theorem proving for Abelian groups represented as integer modules. Theor. Comput. Sci. 208(1-2), 149–177 (1998)
13. Waldmann, U.: Superposition and chaining for totally ordered divisible Abelian groups. In: International Joint Conference for Automated Reasoning, pp. 226–241 (2001)
14. Waldmann, U.: Cancellative Abelian monoids and related structures in refutational theorem proving (part I, II). Journal of Symbolic Computation 33(6), 777–829, 831–861 (2002)

Table 1. Linear Arithmetic Superposition Calculus (LASCA) for ground clauses

Ordered Paramodulation:

$$\frac{C \vee l \simeq r \quad L[l']_p \vee D}{C \vee D \vee L[r]_p} \qquad \begin{array}{l}\text{(i) } l =_{AC} l', \\ \text{(ii) } (l \simeq r) \succ C.\end{array}$$

Equality Factoring:

$$\frac{C \vee t' \simeq s' \vee t \simeq s}{C \vee s \not\simeq s' \vee t \simeq s'} \qquad \begin{array}{l}\text{(i) } t =_{AC} t', \\ \text{(ii) } (t \simeq s) \succeq C \vee t' \simeq s'.\end{array}$$

Gaussian Elimination:

$$\frac{C \vee l = r \quad L[l']_p \vee D}{C \vee D \vee L[r]_p} \qquad \begin{array}{l}\text{(i) } l =_{AC} l', \\ \text{(ii) } (l = r) \succ C.\end{array}$$

Theory Equality Factoring:

$$\frac{C \vee l' = r' \vee l = r}{C \vee r > r' \vee r' > r \vee l = r'} \qquad \begin{array}{l}\text{(i) } l =_{AC} l', \\ \text{(ii) } (l = r) \succeq C \vee l' = r'.\end{array}$$

Fourier-Motzkin Elimination:

$$\frac{C \vee l > r \quad -l' > r' \vee D}{C \vee D \vee -r' > r} \qquad \begin{array}{l}\text{(i) } l =_{AC} l', \\ \text{(ii) } (l > r) \succ C, \\ \text{(iii) there is no } l'' > r'' \in C \text{ such that } l'' =_{AC} l \\ \text{(iv) } (-l' > r') \succ D \\ \text{(v) there is no } -l'' > r'' \in D \text{ such that } l'' =_{AC} l.\end{array}$$

Inequality Factoring (InF1):

$$\frac{C \vee \pm l' > r' \vee \pm l > r}{C \vee r > r' \vee \pm l > r} \qquad \begin{array}{l}\text{(i) } l =_{AC} l', \\ \text{(ii) } (\pm l > r) \succeq C \vee \pm l' > r'.\end{array}$$

Inequality Factoring (InF2):

$$\frac{C \vee \pm l' > r' \vee \pm l > r}{C \vee r' > r \vee \pm l > r'} \qquad \begin{array}{l}\text{(i) } l =_{AC} l', \\ \text{(ii) } (\pm l > r) \succeq C \vee \pm l' > r'.\end{array}$$

$\perp$-Elimination:

$$\frac{C \vee \perp}{C} \qquad \text{(i) } C \text{ contains only } \top, \perp \text{ literals.}$$

The Theory of Calculi with Explicit Substitutions Revisited

Delia Kesner

PPS, Université Paris 7 and CNRS (UMR 7126), France

Abstract. Calculi with explicit substitutions (ES) are widely used in different areas of computer science. Complex systems with ES were developed these last 15 years to capture the good computational behaviour of the original systems (with meta-level substitutions) they were implementing.

In this paper we first survey previous work in the domain by pointing out the motivations and challenges that guided the development of such calculi. Then we use very simple technology to establish a general theory of explicit substitutions for the lambda-calculus which enjoys fundamental properties such as simulation of one-step beta-reduction, confluence on metaterms, preservation of beta-strong normalisation, strong normalisation of typed terms and full composition. The calculus also admits a natural translation into Linear Logic's proof-nets.

1 Introduction

This paper is about *explicit substitutions* (ES), an intermediate formalism that - by decomposing the *higher-order* substitution operation into more atomic steps - allows a better understanding of the execution models of complex languages.

Indeed, higher-order substitution is a *meta-level* operation used in higher-order languages (such as functional, logic, concurrent and object-oriented programming), while ES is an *object-level* notion internalised and handled by symbols and reduction rules belonging to their own worlds. However, the two formalisms are still very close, this can be easily seen for example in the case of the λ-calculus whose reduction rule is given by $(\lambda x.t)\ u \rightarrow_\beta t\{x/u\}$, where the operation $t\{x/v\}$ denotes the result of substituting all the *free* occurrences of x in t by u, a notion that can be formally defined *modulo α-conversion* [1] as follows:

$$\begin{array}{ll} x\{x/u\} := u & (t_1\, t_2)\{x/u\} := (t_1\{x/u\}t_2\{x/u\}) \\ y\{x/u\} := y\ (x \neq y) & (\lambda y.v)\{x/u\} := \lambda y.v\{x/u\} \end{array}$$

Then, the simplest way to specify a λ-calculus with ES is to incorporate substitutions into the language, then to transform the equalities of the previous specification into reduction rules (so that one still works modulo α-conversion), thus yielding the following reduction system known as λx [36, 37, 44, 10].

[1] Definition of substitution modulo α-conversion avoids to explicitly deal with the variable capture case. Thus, for example $(\lambda x.y)\{y/x\} =_\alpha (\lambda z.y)\{y/x\} =_{def} \lambda z.y\{y/x\} = \lambda z.x$.

J. Duparc and T.A. Henzinger (Eds.): CSL 2007, LNCS 4646, pp. 238–252, 2007.

$$\begin{array}{lll}
(\lambda x.t)\,u & \rightarrow t[x/u] & \\
x[x/u] & \rightarrow u & \\
y[x/u] & \rightarrow y & (x \neq y) \\
(t_1\,t_2)[x/u] & \rightarrow (t_1[x/u]\,t_2[x/u]) & \\
(\lambda y.v)[x/u] & \rightarrow \lambda y.v[x/u] &
\end{array}$$

The λx-calculus corresponds to the minimal behaviour [2] that can be found in most of the calculi with ES appearing in the literature. More sophisticated treatments of substitutions also consider a composition operator allowing much more interactions between them. This is exactly the source of the problems that we discuss below.

Related Work. In these last years there has been a growing interest in λ-calculi with ES. They can be defined either with unary [44, 35] or n-ary [2, 23] substitutions, by using de Bruijn notation [11, 12, 32, 27], or levels [39], or combinators [20], or director strings [46], or ... simply by named variables as in λx. Also, a calculus with ES can be seen as a term notation for a logical system where the reduction rules behave like cut elimination transformations [22, 29, 16].

In any case, all these calculi were introduced as a bridge between formal higher-order calculi and their concrete implementations. However, implementing an atomic substitution operation by several elementary explicit steps comes at a price. Indeed, while λ-calculus is perfectly *orthogonal* (does not have critical pairs), calculi with ES such as λx suffer at least from the following well-known diverging example:

$$t[y/v][x/u[y/v]] \;{}^*\!\!\leftarrow ((\lambda x.t)\,u)[y/v] \rightarrow^* t[x/u][y/v]$$

Different solutions were adopted in the literature to close this diagram. If no new rewriting rule is added to those of λx, then reduction turns out to be confluent on terms but not on *metaterms* (terms with metavariables used to represent incomplete programs and proofs). If naive rules for composition are considered, then one recovers confluence on metaterms but loses normalisation: there exist terms which are strongly normalisable in λ-calculus but not in the corresponding ES version. This phenomenon, known as Melliès' counter-example [40], shows a flaw in the design of ES calculi in that they are supposed to implement their underlying calculus (in our case the λ-calculus) without losing its good properties. More precisely, let us call λ_{Z}*-calculus* an arbitrary set of (λ_{Z}-)terms together with a set of (λ_{Z}-)reduction rules. Also, let us consider a mapping $\mathtt{to}_{\mathrm{Z}}$ from λ-terms to λ_{Z}-terms. The following list of properties can be identified:

- **(C)** The λ_{Z}-reduction relation is confluent on λ_{Z}-terms: If $u \;{}^*_{\lambda_{\mathrm{Z}}}\!\!\leftarrow t \rightarrow^*_{\lambda_{\mathrm{Z}}} v$, then there is t' such that $u \rightarrow^*_{\lambda_{\mathrm{Z}}} t' \;{}^*_{\lambda_{\mathrm{Z}}}\!\!\leftarrow v$.
- **(MC)** The λ_{Z}-reduction relation is confluent on λ_{Z}-metaterms.
- **(PSN)** The λ_{Z}-reduction relation preserves β-strong normalisation: If the λ-term t is in $\mathcal{SN}_\beta$, then $\mathtt{to}_{\mathrm{Z}}(t)$ is in $\mathcal{SN}_{\lambda_{\mathrm{Z}}}$.
- **(SN)** Strong normalisation holds for λ_{Z}-typed terms: If the λ_{Z}-term t is typed, then t is in $\mathcal{SN}_{\lambda_{\mathrm{Z}}}$.
- **(SIM)** Any evaluation step in λ-calculus can be implemented by λ_{Z}: If $t \rightarrow_\beta t'$, then $\mathtt{to}_{\mathrm{Z}}(t) \rightarrow^*_{\lambda_{\mathrm{Z}}} \mathtt{to}_{\mathrm{Z}}(t')$.

[2] Some presentations replace the rule $y[x/u] \rightarrow y$ by the more general one $t[x/u] \rightarrow t$ $(x \notin \overline{t})$.

(FC) Full composition can be implemented by λ_Z: The λ_Z-term $t[x/u]$ λ_Z-reduces to $t\{x/u\}$ for an appropriate notion of (meta)substitution on λ_Z-terms.

In particular, (MC) implies (C) and (PSN) usually implies (SN).

The result of Melliès appeared as a challenge to find a calculus having all the properties mentioned above. There are already several propositions in the literature giving (partial) answers to this challenge; they are summarised in the following table, where we just write one representative calculus for each line, even if there are currently many more references available in the literature (by lack of space we cannot cite all of them).

Calculus	C	MC	PSN	SN	SIM	FC
λ_x [44]	Yes	No	Yes	Yes	Yes	No
λ_σ [2]	Yes	No	No	No	Yes	Yes
$\lambda_{\sigma\Uparrow}$ [23]	Yes	Yes	No	No	Yes	Yes
λ_ς [41]	Yes	Yes	Yes	Yes	No	No
λ_{ws} [14]	Yes	Yes	Yes	Yes	Yes	No
$\lambda\mathtt{lxr}$ [29]	Yes	?	Yes	Yes	Yes	Yes

In other words, there are many ways to avoid Melliès' counter-example in order to recover the PSN property. More precisely, one can forbid the substitution operators to cross lambda-abstractions [38, 18] or avoid composition of substitutions [6]. One can also impose a simple strategy on the calculus with ES to mimic exactly the calculus without ES. The first solution leads to *weak* lambda calculi, not able to express *strong* beta-equality (used for example in implementations of proof-assistants). The second solution is drastic when composition of substitutions is needed for implementations of HO unification [15] or functional abstract machines [24]. The last one does not take advantage of the notion of ES because they can be neither composed nor even delayed.

In order to cope with this problem David and Guillaume [14] defined a calculus with *labels* called λ_{ws}, which allows *controlled* composition of ES without losing PSN and SN. But the λ_{ws}-calculus has a complicated syntax and its named version [13] is even less intelligible. However, the strong normalisation proof for λ_{ws} given in [13] reveals a natural semantics for composition of ES via Linear Logic's proof-nets [19], suggesting that weakening (explicit erasure) and contraction (explicit duplication) can be added to the calculus without losing strong normalisation.

Explicit weakening and contraction are the starting points of the $\lambda\mathtt{lxr}$-calculus [29], which is in some sense a (complex) precursor of the $\lambda\mathtt{es}$-calculus that we present in this paper. However, while λ-syntax could be seen as a particular case of $\lambda\mathtt{es}$-syntax, a special encoding is needed to incorporate weakening and contraction operators to λ-terms in order to verify the so-called linearity constraints of $\lambda\mathtt{lxr}$. Moreover, the reduction system of $\lambda\mathtt{lxr}$ contains 6 equations and 19 rewriting rules, thus requiring an important amount of combinatorial reasoning. This is notably discouraging when one needs to check properties by cases on the reduction step; a reason why confluence on metaterms for $\lambda\mathtt{lxr}$ is just conjectured but not still proved.... Also, whereas $\lambda\mathtt{lxr}$ gives the evidence that explicit weakening and contraction are *sufficient* to verify all the properties one expects from a calculus with ES, there is no justified reason to think that they are also *necessary*.

We choose here to introduce the λes-calculus by using concise and simple syntax in named variable notation style (as in λx) in order to dissociate all the renaming details which are necessary to specify higher-order substitution on first-order terms (such as for example terms in de Bruijn notation). Even if this choice implies the use of α-equivalence, we think that this presentation is more appropriate to focus on the fundamental computational properties of the calculus. Moreover, this can also be justified by the fact that it is now perfectly well-understood in the literature how to translate terms with named variables into equivalent terms in first-order notation. Another important choice made in this paper is the use of minimal equational reasoning (just one equation) to specify commutation of independent substitutions. This will turn out to be essential to obtain a *safe* notion of (full)composition which does not need the complex use of explicit operators for contraction and weakening. Also, simultaneous substitution (also called n-ary substitution), can be simply expressed within our framework.

We thus achieve the definition of a simple language being easy to understand, and enjoying a useful set of properties: confluence on metaterms (and thus on terms), simulation of one-step β-reduction, strong normalisation of typed terms, preservation of β-strong normalisation, simulation of one-step β-reduction and full composition. Moreover, these properties can be proved using very simple proof techniques while this is not the case for other calculi axiomatising commutation of substitutions. Thus for example, the calculus proposed in [45] specifies commutation of independent substitutions by a *non-terminating* rewriting system (instead of an equation), thus leading to complicated notions and proofs of its underlying normalisation properties.

The λes-calculus admits a natural translation into Linear Logic's proof-nets, thus providing an alternative proof of strong normalisation. Also, a more implementation oriented calculus based on λes could be specified by means of de Bruijn notation and n-ary substitutions. These two last topics are however omitted in this paper because of lack of space, we refer the interested reader to [28].

The rest of the paper is organised as follows. Section 2 introduces syntax for Λes-terms and appropriate notions of equivalence and reduction. In Section 3 we develop a proof of confluence for metaterms. Preservation of β-strong normalisation is studied and proved in Section 4. The typing system for λes is presented in Section 5 as well as the subject reduction property and the relation between typing derivations in λes and λ-calculus. Finally, strong normalisation based on PSN is proved in this same section.

We refer the reader to [28] for detailed proofs and to [9, 47] for standard notions from rewriting that we will use throughout the paper.

2 Syntax

A Λes-term is inductively defined by a *variable* x, an *application* $t\ u$, an *abstraction* $\lambda x.t$ or a *substituted term* $t[x/u]$, when t and u are Λes-terms. The syntactic object $[x/u]$, which is not a term itself, is called an *explicit substitution*.

The terms $\lambda x.t$ and $t[x/u]$ bind x in t. The sets of *free* and *bound* variables of a term t, denoted $\overline{t}$ and $\underline{t}$ respectively, can be defined as usual. Thus, the standard notion of α-conversion on higher-order terms is obtained so that one may assume, when *necessary*, that two bound variables have different names, and no variable is free and bound at the same time. Indeed, when using different symbols x and y to talk about two *nested* bound variables, as for example in the terms $(\lambda y.t)[x/u]$ and $t[x/u][y/v]$, we implicitly mean $x \neq y$. The use of the same name for bound variables appearing in *parallel/disjoint* positions, as for example in $t[x/u]\ v[x/u]$ or $(\lambda x.x)\ (\lambda x.x)$ is not problematic.

Besides α-conversion the following equations and reduction rules are considered.

Equations	**Reduction Rules**			
$t[x/u][y/v] =_{\mathtt{C}} t[y/v][x/u]$	$(\lambda x.t)\ u$	$\rightarrow_{\mathtt{B}}$	$t[x/u]$	
$(y \notin \overline{u}\ \&\ x \notin \overline{v})$	**The (sub)set of rules** $\mathtt{s}$**:**			
	$x[x/u]$	$\rightarrow_{\mathtt{Var}}$	u	
	$t[x/u]$	$\rightarrow_{\mathtt{Gc}}$	t	$(x \notin \overline{t})$
	$(t\ u)[x/v]$	$\rightarrow_{\mathtt{App}_1}$	$t[x/v]\ u[x/v]$	$(x \in \overline{t}\ \&\ x \in \overline{u})$
	$(t\ u)[x/v]$	$\rightarrow_{\mathtt{App}_2}$	$t\ u[x/v]$	$(x \notin \overline{t}\ \&\ x \in \overline{u})$
	$(t\ u)[x/v]$	$\rightarrow_{\mathtt{App}_3}$	$t[x/v]\ u$	$(x \in \overline{t}\ \&\ x \notin \overline{u})$
	$(\lambda y.t)[x/v]$	$\rightarrow_{\mathtt{Lamb}}$	$\lambda y.t[x/v]$	
	$t[x/u][y/v]$	$\rightarrow_{\mathtt{Comp}_1}$	$t[y/v][x/u[y/v]]$	$(y \in \overline{u}\ \&\ y \in \overline{t})$
	$t[x/u][y/v]$	$\rightarrow_{\mathtt{Comp}_2}$	$t[x/u[y/v]]$	$(y \in \overline{u}\ \&\ y \notin \overline{t})$

It is appropriate to point out here that α-conversion is necessary in order to avoid capture of variables. Thus for example the left-hand side of the Lamb-rule $(\lambda y.t)[x/v]$ implicitly assumes $y \neq x$ and $y \notin \overline{v}$. See also Sections 4.2 and 6 for a a discussion about the minimality of the subset $\mathtt{s}$ w.r.t its number of rules.

The *higher-order rewriting system* containing the rules $\{\mathtt{B}\} \cup \mathtt{s}$ is called $\mathtt{Bs}$. The *equivalence relation* generated by the conversions $\mathtt{E_s} = \{\alpha, \mathtt{C}\}$ is denoted by $=_{\mathtt{E_s}}$. The *reduction relation* generated by the *rewriting rules* $\mathtt{s}$ (resp. $\mathtt{Bs}$) *modulo the equivalence relation* $=_{\mathtt{E_s}}$ is denoted by $\rightarrow_{\mathtt{es}}$ (resp. $\rightarrow_{\lambda\mathtt{es}}$), the $\mathtt{e}$ means equational and the $\mathtt{s}$ substitution. More precisely,

$$\begin{array}{ll} t \rightarrow_{\mathtt{es}} t' & \text{iff there are } u, u' \text{ s.t. } t =_{\mathtt{E_s}} u \rightarrow_{\mathtt{s}} u' =_{\mathtt{E_s}} t' \\ t \rightarrow_{\lambda\mathtt{es}} t' & \text{iff there are } u, u' \text{ s.t. } t =_{\mathtt{E_s}} u \rightarrow_{\mathtt{Bs}} u' =_{\mathtt{E_s}} t' \end{array}$$

The notation $\rightarrow^*_{\lambda\mathtt{es}}$ (resp. $\rightarrow^+_{\lambda\mathtt{es}}$) is used for the reflexive and transitive (resp. transitive) closure of $\rightarrow_{\lambda\mathtt{es}}$.

Remark that any simultaneous (n-ary) substitution can now be thought as a sequence of consecutive *independent* unary substitutions representing the same mapping. Thus for example $[x/u, y/v]$ can be expressed as $[x/u][y/v]$ (or $[y/v][x/u]$) where $y \notin \overline{u}$ and $x \notin \overline{v}$. The use of the equation C to make a list of independent substitutions behave like a simultaneous one is essential. We leave to the reader the verification that composition of simultaneous substitution can be expressed within our $\lambda\mathtt{es}$-reduction relation.

The equivalence relation preserves free variables and the reduction relation either preserves or decreases them. Thus, $t \rightarrow_{\lambda\mathtt{es}} u$ implies $\overline{u} \subseteq \overline{t}$.

Also, the (sub)calculus $\mathtt{es}$, which is intended to implement (meta-level) substitution, can be shown to be terminating by associating to each $\Lambda\mathtt{es}$-term t a measure which does not change by $\mathtt{E_s}$ but strictly decreases by $\rightarrow_s$ (details can be found in [28]).

We now address the property of full composition. For that, we extend the standard notion of (meta-level)substitution on λ-terms given in the introduction to all the $\Lambda\mathtt{es}$-terms by adding the new case $t[y/u]\{x/v\} := t\{x/v\}[y/u\{x/v\}]$, where we implicitly mean $x \neq y\ \&\ y \notin \overline{v}$. Remark that $t\{x/u\} = t$ if $x \notin \overline{t}$, thus we can prove:

Lemma 1 (Full Composition). *Let t and u be Λes-terms. Then* $t[x/u] \rightarrow^*_{\lambda\mathtt{es}} t\{x/u\}$.

We now establish basic connections between λ and λes-reduction. As expected, β-reduction can be implemented by the more atomic notion of λes-reduction while this one can be projected into β.

Lemma 2 (Simulating β-reduction). *Let t be a λ-term s.t. $t \rightarrow_\beta t'$. Then* $t \rightarrow^+_{\lambda\mathtt{es}} t'$.

Proof. By induction on β-reduction using Lemma 1.

Λes-terms are encoded into λ-terms as follows: $\mathtt{L}(x) := x$, $\mathtt{L}(\lambda x.t) := \lambda x.\mathtt{L}(t)$, $\mathtt{L}(t\,u) := \mathtt{L}(t)\,\mathtt{L}(u)$ and $\mathtt{L}(t[x/u]) := \mathtt{L}(t)\{x/\mathtt{L}(u)\}$. Thus, projection is obtained:

Lemma 3 (Projecting into β-reduction). *If $t \rightarrow_{\lambda\mathtt{es}} u$, then* $\mathtt{L}(t) \rightarrow^*_\beta \mathtt{L}(u)$.

Proof. First prove that $t =_{\mathtt{E_s}} u$ implies $\mathtt{L}(t) = \mathtt{L}(u)$ by the well-known substitution lemma [4] of λ-calculus. Remark that $t \rightarrow_{\mathtt{s}} u$ trivially implies $\mathtt{L}(t) = \mathtt{L}(u)$. Finally, prove that $t \rightarrow_{\mathtt{B}} u$ implies $\mathtt{L}(t) \rightarrow^*_\beta \mathtt{L}(u)$ by induction on the reduction step $t \rightarrow_{\mathtt{B}} u$.

3 Confluence on Metaterms

Metaterms are terms containing *metavariables* denoting *incomplete* programs/proofs in a higher-order unification framework [25]. Metavariables should come with a minimal amount of information to guarantee that some basic operations such as instantiation (replacement of metavariables by metaterms) are sound in a typing context. However, known formalisms in the literature for the specification of higher-order metaterms, such as Combinatory Reduction Systems (CRS) [30] or Expression Reduction Systems (ERS) [26], do not allow, at least in a simpler way, to specify the precise set of free variables which is expected from a (sound)instantiation. Thus for example, a CRS metaterm like $M(x, y)$ specifies that x and y *may* occur in the instantiation of M, but M can also be further instantiated by any other term not containing x and y at all. Another example is given by the (raw) ERS metaterm $t = \lambda y.y\ \mathbb{X}\ (\lambda z.\mathbb{X})$ because the instantiation of $\mathbb{X}$ by a term containing a free occurrence of z would be unsound (see [41, 15, 17] for details).

We thus propose to specify incomplete proofs as follows. We consider a countable set of *raw* metavariables $\mathbb{X}, \mathbb{Y}, \ldots$ associated to sets of variables $\Gamma, \Delta, \ldots$, thus yielding *decorated* metavariables denoted by $\mathbb{X}_\Gamma, \mathbb{Y}_\Delta$, etc. This decoration says nothing about the *structure* of the incomplete proof itself but is sufficient to guarantee that different occurrences of the same metavariable inside a metaterm are never instantiated by different metaterms.

The grammar for Λes-terms is extended to generate Λes-metaterms as follows:

$$t ::= x \mid \mathbb{X}_\Delta \mid t\,t \mid \lambda x.t \mid t[x/t]$$

We extend the notion of *free variables* to *metaterms* by $\overline{X_\Delta} = \Delta$.

Reduction on metaterms must be understood in the same way reduction on terms: the λes-relation is generated by the Bs-relation on $\mathtt{E_s}$-equivalence classes of *metaterms*.

In contrast to the ERS notion of metaterm, α-conversion turns out to be perfectly well-defined on λes-metaterms by extending the renaming of bound variables to the decoration sets. Thus for example $\lambda x.Y_x =_\alpha \lambda z.Y_z$.

It is well-known that confluence on metaterms fails for calculi *without* composition for ES as for example the following critical pair in λx shows

$$s = t[x/u][y/v] \;^*\!\!\leftarrow ((\lambda x.t)\ u)[y/v] \rightarrow^* t[y/v][x/u[y/v]] = s'$$

Indeed, while this diagram can be closed in λx for terms *without metavariables* [10], there is no way to find a common reduct between s and s' whenever t is (or contains) metavariables: no λx-reduction rule is able to mimic composition on raw or decorated metavariables. This can be fortunately recovered in the case of the λes-calculus.

3.1 The Confluence Proof

This section develops a confluence proof for reduction on λes-metaterms based on Tait and Martin-Löf's technique: define a simultaneous reduction relation denoted $\Rrightarrow_{\mathtt{es}}$; prove that $\Rrightarrow^*_{\mathtt{es}}$ and $\rightarrow^*_{\mathtt{es}}$ are the same relation; show that $\Rrightarrow^*_{\mathtt{es}}$ is confluent; and finally conclude. While many steps in this proof are similar to those appearing in other proofs of confluence for the λ-calculus, some special considerations are to be used here in order to accommodate correctly the substitution calculus as well as the equational part of our notion of reduction (see in particular Lemma 6).

A first interesting property of the system es is that it can be used as a function on $\mathrm{E_s}$-equivalence classes:

Lemma 4. *The* es*-normal forms of metaterms are unique modulo* $\mathrm{E_s}$ *so that* $t =_{\mathrm{E_s}} u$ *implies* $\mathtt{es}(t) =_{\mathrm{E_s}} \mathtt{es}(u)$.

The simultaneous reduction relation $\Rrightarrow_{\mathtt{es}}$ on es-normal forms is now defined in terms of a simpler relation $\Rrightarrow$ working on $\mathrm{E_s}$-equivalence classes.

Definition 1 (The relations $\Rrightarrow$ and $\Rrightarrow_{\mathtt{es}}$). *Simultaneous reduction is defined on metaterms in* es*-normal form as follows:* $t \Rrightarrow_{\mathtt{es}} t'$ *iff* $\exists\, u, u'$ *s.t.* $t =_{\mathrm{E_s}} u \Rrightarrow u' =_{\mathrm{E_s}} t'$, *where*

- $x \Rrightarrow x$
- *If* $t \Rrightarrow t'$, *then* $\lambda x.t \Rrightarrow \lambda x.t'$
- *If* $t \Rrightarrow t'$ *and* $u \Rrightarrow u'$, *then* $t\ u \Rrightarrow t'\ u'$
- *If* $t \Rrightarrow t'$ *and* $u \Rrightarrow u'$, *then* $(\lambda x.t)\ u \Rrightarrow \mathtt{es}(t'[x/u'])$
- *If* $u_i \Rrightarrow u'_i$ *and* $x_i \notin \overline{u_j}$ *for all* $i, j \in [1, n]$, *then* $\mathbb{X}_\Delta[x_1/u_1]\ldots[x_n/u_n] \Rrightarrow \mathbb{X}_\Delta[x_1/u'_1]\ldots[x_n/u'_n]$

The simultaneous relation is stable in the following sense.

Lemma 5. *If* $t \Rrightarrow_{\mathtt{es}} t'$ *and* $u \Rrightarrow_{\mathtt{es}} u'$, *then* $\mathtt{es}(t[x/u]) \Rrightarrow_{\mathtt{es}} \mathtt{es}(t'[x/u'])$.

It can be now shown that the relation $\Rrightarrow_{\mathtt{es}}$ has the *diamond property*.

Lemma 6. *If* $t_1\ {}_{\mathtt{es}}\!\Lleftarrow t \Rrightarrow_{\mathtt{es}} t_2$, *then* $\exists t_3$ *s.t.* $t_1 \Rrightarrow_{\mathtt{es}} t_3\ {}_{\mathtt{es}}\!\Lleftarrow t_2$.

Proof. 1. First prove that $t \Lleftarrow u =_{\mathrm{E_s}} u'$ implies $t =_{\mathrm{E_s}} t' \Lleftarrow u'$ for some t' by induction on $t \Lleftarrow u$. Thus conclude that $v\ {}_{\mathtt{es}}\!\Lleftarrow v' =_{\mathrm{E_s}} u'$ implies $v =_{\mathrm{E_s}} t' \Lleftarrow u'$ for some t'.

2. Prove that $t_1 \Lleftarrow t \Rrightarrow t_2$ implies $t_1 \Rrightarrow_{\mathtt{es}} t_3 \;{}_{\mathtt{es}}\!\Lleftarrow t_2$ for some t_3 by induction on $\Rrightarrow$ using Lemma 5.
3. Finally prove the diamond property as follows. Let $t_1 \;{}_{\mathtt{es}}\!\Lleftarrow t =_{\mathtt{E_s}} u \Rrightarrow u' =_{\mathtt{E_s}} t_2$. By point (1) there is u_1 such that $t_1 =_{\mathtt{E_s}} u_1 \Lleftarrow u$ and by point (2) there is t_3 such that $u_1 \Rrightarrow_{\mathtt{es}} t_3 \;{}_{\mathtt{es}}\!\Lleftarrow u'$. Conclude $t_1 \Rrightarrow_{\mathtt{es}} t_3 \;{}_{\mathtt{es}}\!\Lleftarrow t_2$.

We thus obtain the main result of this section:

Corollary 1. *The reduction relation $\rightarrow^*_{\mathtt{es}}$ is confluent.*

Proof. The relation $\Rrightarrow^*_{\mathtt{es}}$ enjoys the diamond property (Lemma 6) so that it turns out to be confluent [9]. Since $\Rrightarrow^*_{\mathtt{es}}$ and $\rightarrow^*_{\lambda\mathtt{es}}$ can be shown (using Lemmas 4 and 5) to be the same relation, then conclude that $\rightarrow^*_{\lambda\mathtt{es}}$ is also confluent.

Although this confluence result guarantees that all the critical pairs in $\lambda\mathtt{es}$ can be closed, let us analyse a concrete example being the source of interesting diverging diagrams in calculi with ES (c.f. Section 1), giving by the following case:

$$\begin{array}{ccccc} s_3 & {}^{*}_{\lambda\mathtt{es}}\!\leftarrow & s_1 & \rightarrow_{\mathtt{B}} & s_2 \\ ? & & ((\lambda x.t)\ u)[y/v] & & t[x/u][y/v] \end{array}$$

The metaterm s_3 as well as the one used to close the diagram can be determined by the following four different cases:

$y \in \overline{t}$	$y \in \overline{u}$	s_3	Close the diagram by
Yes	Yes	$t[y/v][x/u[y/v]]$	$s_3 \;{}_{\mathtt{Comp_1}}\!\leftarrow s_2$
Yes	No	$t[y/v][x/u]$	$s_3 =_{\mathtt{E_s}} s_2$
No	Yes	$t[x/u[y/v]]$	$s_3 \;{}_{\mathtt{Comp_2}}\!\leftarrow s_2$
No	No	$(\lambda x.t)\ u$	$s_3 \rightarrow_{\mathtt{B}} t[x/u] \;{}_{\mathtt{Gc}}\!\leftarrow s_2$

4 Preservation of β-Strong Normalisation

Preservation of β-strong normalisation (PSN) in calculi with ES received a lot of attention (see for example [2, 6, 10, 32]), starting from an unexpected result given by Melliès [40] who has shown that there are β-strongly normalisable terms in λ-calculus that are not strongly normalisable when evaluated by the reduction rules of an explicit version of the λ-calculus. This is for example the case for $\lambda\sigma$ [2] and $\lambda\sigma_{\Uparrow}$ [23].

Since then, different notions of safe composition where introduced, even if PSN becomes more difficult to prove ([8, 14, 1, 29, 31]). This is mainly because the so-called *decent* terms are not stable by reduction : a term t is said to be *decent* in the calculus $\lambda_{\mathtt{Z}}$ if every subterm v appearing in some substituted subterm $u[x/v]$ of t is $\lambda_{\mathtt{Z}}$-strongly normalising. As an example, the term $x[x/(y\ y)][y/\lambda w.w\ w]$ is decent in $\lambda\mathtt{es}$ since $y\ y$ and $\lambda w.w\ w$ are both $\lambda\mathtt{es}$-strongly normalising, but its $\mathtt{Comp_2}$-reduct $x[x/(y\ y)[y/\lambda w.w\ w]]$ is not.

This section proves that $\lambda\mathtt{es}$ preserves β-strong normalisation. For that, we use a simulation proof technique based on the following steps. We first define a calculus $\lambda\mathtt{esw}$ (Section 4.1). We then give a translation $\mathtt{K}$ from $\Lambda\mathtt{es}$-terms (and thus also from λ-terms) into $\lambda\mathtt{esw}$ s.t. $t \in \mathcal{SN}_{\beta}$ implies $\mathtt{K}(t) \in \mathcal{SN}_{\lambda\mathtt{esw}}$ (Corollary 4) and $\mathtt{K}(t) \in \mathcal{SN}_{\lambda\mathtt{esw}}$ implies $t \in \mathcal{SN}_{\lambda\mathtt{es}}$ (Corollary 2).

4.1 The λesw-Calculus

A Λesw-term is inductively defined by x, $t\ u$, $\lambda x.t$, $t[x/u]$ or $\mathcal{W}_x(t)$ (an *explicit weakening*). We extend the notion of free variables to explicit weakenings by adding the case $\overline{\mathcal{W}_x(t)} = \{x\} \cup \overline{t}$. The notion of *strict* term will be essential: every subterm $\lambda x.t$ and $t[x/u]$ is such that $x \in \overline{t}$ and every subterm $\mathcal{W}_x(t)$ is such that $x \notin \overline{t}$.

Besides equations and rules in λes, those in the following table are also considered.

Additional Equations	**Additional Reduction Rules**
$\mathcal{W}_x(\mathcal{W}_y(t)) =_{\mathtt{WC}} \quad \mathcal{W}_y(\mathcal{W}_x(t))$	$\mathcal{W}_x(t)[x/u] \to \mathcal{W}_{\overline{u}\setminus\overline{t}}(t)$
$\mathcal{W}_y(t)[x/u] =_{\mathtt{Weak1}} \mathcal{W}_y(t[x/u])\ (x \neq y\ \&\ y \notin \overline{u})$	$\mathcal{W}_y(t)\ u \to t\ u \quad (y \in \overline{u})$
$\mathcal{W}_y(\lambda x.t) \quad =_{\mathtt{WAbs}} \lambda x.\mathcal{W}_y(t) \quad (x \neq y)$	$\mathcal{W}_y(t)\ u \to \mathcal{W}_y(t\ u) \quad (y \notin \overline{u})$
	$t\ \mathcal{W}_y(u) \to t\ u \quad (y \in \overline{t})$
	$t\ \mathcal{W}_y(u) \to \mathcal{W}_y(t\ u) \quad (y \notin \overline{t})$
	$\mathcal{W}_y(t)[x/u] \to t[x/u] \quad (y \in \overline{u})$
	$t[x/\mathcal{W}_y(u)] \to \mathcal{W}_y(t[x/u])\ (y \notin \overline{t})$
	$t[x/\mathcal{W}_y(u)] \to t[x/u] \quad (y \in \overline{t})$

Given a set of variables $\Gamma = \{x_1, \ldots, x_n\}$, the use of the abbreviation $\mathcal{W}_\Gamma(t)$ for $\mathcal{W}_{x_1}(\ldots \mathcal{W}_{x_n}(t))$ in the first reduction rule is justified by the equation WC. In the particular case $\Gamma = \emptyset$, we define $\mathcal{W}_\emptyset(t) = t$. It is suitable again to recall that we work modulo α-conversion. Thus for example the terms $\mathcal{W}_y(\lambda x.t)$ and $t[x/\mathcal{W}_y(u)]$ have to be always understood as $x \neq y$. However, this is not the case for example for $\lambda x.\mathcal{W}_y(t)$ or $\mathcal{W}_y(t)[x/u]$ where the variables x and y may be equal or different, that's the reason to explicitly add the side-condition $x \neq y$ in some of the previous equations and rules.

The rewriting system containing all the reduction rules in the previous table plus those in system s is called sw. The notation Bsw is used for the system $\{\mathtt{B}\} \cup \mathtt{sw}$. The equivalence relation generated by all the equations in the previous table plus those in $\mathrm{E_s}$ is denoted by $=_{\mathrm{E_{sw}}}$. The relation generated by the reduction rules sw (resp. Bsw) modulo the equivalence relation $=_{\mathrm{E_{sw}}}$ is denoted by $\to_{\mathtt{esw}}$ (resp. $\to_{\lambda\mathtt{esw}}$). More precisely,

$$
\begin{array}{ll}
t \to_{\mathtt{esw}} t' & \text{iff there are } u, u' \text{ s.t. } t =_{\mathrm{E_{sw}}} u \to_{\mathtt{sw}} u' =_{\mathrm{E_{sw}}} t' \\
t \to_{\lambda\mathtt{esw}} t' & \text{iff there are } u, u' \text{ s.t. } t =_{\mathrm{E_{sw}}} u \to_{\mathtt{Bsw}} u' =_{\mathrm{E_{sw}}} t'
\end{array}
$$

From now on, we only work with strict terms, a choice that is justified by the fact that λesw-reduction relation preserves strict terms.

In order to infer normalisation of λes from that of λesw, a relation between both notions of reduction is needed. For that, a translation K from Λes-terms (and thus also from λ-terms) to (strict) Λesw-terms is defined as follows:

$$
\begin{array}{llll}
\mathtt{K}(x) := x & & \mathtt{K}(u\ v) := \mathtt{K}(u)\,\mathtt{K}(v) & \\
\mathtt{K}(\lambda x.t) := \lambda x.\mathtt{K}(t) & \text{If } x \in \overline{t} & \mathtt{K}(\lambda x.t) := \lambda x.\mathcal{W}_x(\mathtt{K}(t)) & \text{If } x \notin \overline{t} \\
\mathtt{K}(u[x/v]) := \mathtt{K}(u)[x/\mathtt{K}(v)] & \text{If } x \in \overline{t} & \mathtt{K}(u[x/v]) := \mathcal{W}_x(\mathtt{K}(u))[x/\mathtt{K}(v)] & \text{If } x \notin \overline{t}
\end{array}
$$

Remark that $\overline{\mathtt{K}(t)} = \overline{t}$. Also, λesw-reduction can be used to push out useless weakening constructors as follows:

Lemma 7. *If* $u \to_{\lambda es} v$*, then* $\mathtt{K}(u) \to^+_{\lambda esw} \mathcal{W}_{\overline{u}\setminus\overline{v}}(\mathtt{K}(v))$.

Proof. The proof is by induction on $\to_{\lambda es}$ and it accurately puts in evidence the fact that `Weak1` and `WAbs` are needed as equations and not as rewriting rules.

The previous lemma allows us to conclude with the following preservation result:

Corollary 2. *If* $\mathtt{K}(t) \in \mathcal{SN}_{\lambda esw}$*, then* $t \in \mathcal{SN}_{\lambda es}$.

4.2 The Λ_I-Calculus

The Λ_I-calculus is another intermediate language used as technical tool to prove PSN. The set of Λ_I-terms [30] is defined by the grammar:

$$M ::= x \mid M\ M \mid \lambda x.M \mid [M, M]$$

We consider the extended notions of free variables and (meta)level substitution on Λ_I-terms. We restrict again the syntax to *strict* terms (every subterm $\lambda x.M$ satisfies $x \in \overline{M}$). The following two reduction rules will be used:

$$\boxed{\begin{array}{l} (\lambda x.M)\ N \to_\beta M\{x/N\} \\ [M, N]\ L \quad \to_\pi [M\ L, N] \end{array}}$$

Strict Λ_I-terms turn out to be stable by reduction since they do not lose free variables during reduction.

A binary relation (and not a function) $\mathcal{I}$ is used to relate λesw and Λ_I-terms, this because Λesw-terms are translated into Λ_I-syntax by adding some *garbage* information which is not uniquely determined. Thus, each Λesw-term can be projected into different Λ_I-terms, and this will be essential in the simulation property (Theorem 1).

Definition 2. *The relation* $\mathcal{I}$ *between* strict Λesw*-terms and* strict Λ_I*-terms is inductively given by the following rules:*

$$\frac{}{x\ \mathcal{I}\ x} \qquad \frac{t\ \mathcal{I}\ T}{\lambda x.t\ \mathcal{I}\ \lambda x.T} \qquad \frac{t\ \mathcal{I}\ T \quad u\ \mathcal{I}\ U}{t\ u\ \mathcal{I}\ T\ U} \qquad \frac{t\ \mathcal{I}\ T \quad u\ \mathcal{I}\ U}{t[x/u]\ \mathcal{I}\ T\{x/U\}}$$

$$\frac{t\ \mathcal{I}\ T\ \&\ M\ \textit{strict}}{t\ \mathcal{I}\ [T, M]} \qquad \frac{t\ \mathcal{I}\ T\ \&\ x \in \overline{T}}{\mathcal{W}_x(t)\ \mathcal{I}\ T}$$

The relation $\mathcal{I}$ enjoys the following properties.

Lemma 8. *Let* $t\ \mathcal{I}\ M$*. Then* $\overline{t} \subseteq \overline{M}$*,* $M \in \Lambda_I$ *and* $x \notin \overline{t}\ \&\ N \in \Lambda_I$ *implies* $t\ \mathcal{I}\ M\{x/N\}$.

Remark however that $t\ \mathcal{I}\ M$ implies $\overline{t} \subseteq \overline{M}$ only on *strict* terms. This can be seen as a proof technical argument to exclude from our calculus rewriting rules not preserving strict terms like

$$\begin{array}{lll} (\mathtt{App}) & (t\ u)[x/v] & \to t[x/v]\ u[x/v] \\ (\mathtt{Comp}) & t[x/u][y/v] & \to t[y/v][x/u[y/v]] \quad (y \in \overline{u}) \end{array}$$

Reduction in λesw can be related to reduction in Λ_I by means of the following simulation property (proved by induction on the reduction/equivalence step).

Theorem 1. *Let $s \in \Lambda\texttt{esw}$ and $S \in \Lambda_I$.*

1. *If $s \, \mathcal{I} \, S$ and $s =_{\texttt{E}_{\texttt{sw}}} t$, then $t \, \mathcal{I} \, S$.*
2. *If $s \, \mathcal{I} \, S$ and $s \to_{\texttt{sw}} t$, then $t \, \mathcal{I} \, S$.*
3. *If $s \, \mathcal{I} \, S$ and $s \to_{\texttt{B}} t$, then there is $T \in \Lambda_I$ s.t. $t \, \mathcal{I} \, T$ and $S \to^{+}_{\beta\pi} T$.*

The second preservation result can be now stated as follows:

Corollary 3. *If $s \, \mathcal{I} \, S$ and $S \in \mathcal{SN}_{\beta\pi}$, then $s \in \mathcal{SN}_{\lambda\texttt{esw}}$.*

Proof. Suppose $s \notin \mathcal{SN}_{\lambda\texttt{esw}}$. As $\to_{\texttt{esw}}$ can easily be show to be well-founded (see [28] for details), then an infinite $\lambda\texttt{esw}$-reduction sequence starting at s is necessarily projected by the previous Theorem into an infinite $\beta\pi$-reduction sequence starting at S. This leads to a contradiction with the hypothesis.

4.3 Solving the Puzzle

All the parts of the puzzle together give a PSN argument for $\lambda\texttt{es}$. The starting point is the following encoding from λ to Λ_I-terms:

$$\begin{array}{llll} \texttt{I}(x) & := x & \texttt{I}(\lambda x.t) := \lambda x.\texttt{I}(t) & x \in \overline{t} \\ \texttt{I}(t\,u) & := \texttt{I}(t)\;\texttt{I}(u) & \texttt{I}(\lambda x.t) := \lambda x.[\texttt{I}(t), x] & x \notin \overline{t} \end{array}$$

Now, starting from a λ-term u, which is also a $\Lambda\texttt{es}$-term, one computes its $\texttt{K}$-image - a $\lambda\texttt{esw}$-term - so that some Λ_I-term will be in $\mathcal{I}$-relation with it. More precisely, a straightforward induction on u gives:

Theorem 2. *For any λ-term u, $\texttt{K}(u) \, \mathcal{I} \, \texttt{I}(u)$.*

Preservation of β-strong-normalisation, which is one of the main results of the paper, can be finally stated:

Corollary 4 (PSN). *If $t \in \mathcal{SN}_{\beta}$, then $t \in \mathcal{SN}_{\lambda\texttt{es}}$.*

Proof. If $t \in \mathcal{SN}_{\beta}$, then $\texttt{I}(t) \in WN_{\beta\pi}$ [34] and thus $\texttt{I}(t) \in SN_{\beta\pi}$ [42]. As $\texttt{K}(t) \, \mathcal{I} \, \texttt{I}(t)$ by Theorem 2, then $\texttt{K}(t) \in \mathcal{SN}_{\lambda\texttt{esw}}$ by Corollary 3 so that $t \in \mathcal{SN}_{\lambda\texttt{es}}$ by Corollary 2.

5 The Typed $\lambda\texttt{es}$-Calculus

Simply types are built over a countable set of atomic symbols (base types) and the type constructor $\to$ (functional types). An *environment* is a finite set of pairs of the form $x : A$. Two environments Γ and Δ are said to be *compatible* iff for all $x : A \in \Gamma$ and $y : B \in \Delta$, $x = y$ implies $A = B$. The *union of compatible contexts* is written $\Gamma \uplus \Delta$. Thus for example $(x : A, y : B) \uplus (x : A, z : C) = (x : A, y : B, z : C)$. The following properties on compatible environments will be used:

Lemma 9.

1. *If $\Gamma \subseteq \Gamma'$ and $\Delta \subseteq \Delta'$, then $\Gamma \uplus \Delta \subseteq \Gamma' \uplus \Delta'$.*
2. $\Gamma \uplus (\Delta \uplus \Pi) = (\Gamma \uplus \Delta) \uplus \Pi$.

Typing judgements have the form $\Gamma \vdash t : A$ where t is a term, A is a type and Γ is an environment. *Derivations* of typing judgements, written $\Gamma \vdash_{\lambda es} t : A$, can be obtained by application of the (*multiplicative*) rules in the following table.

$\dfrac{}{x : A \vdash x : A}$ (axiom)	$\dfrac{\Gamma \vdash t : A \to B \qquad \Delta \vdash u : A}{\Gamma \uplus \Delta \vdash (t\ u) : B}$ (app)
$\dfrac{\Gamma \vdash t : B}{\Gamma \setminus \{x : A\} \vdash \lambda x.t : A \to B}$ (abs)	$\dfrac{\Gamma \vdash u : B \qquad \Delta \vdash t : A}{\Gamma \uplus (\Delta \setminus \{x : B\}) \vdash t[x/u] : A}$ (subs)

The axiom rule types a variable in a minimal environment but variables not appearing free may be introduced by binder symbols by means of the rules abs and subs. Thus for example starting from the derivable typing judgement $x : B \vdash x : B$ one can derive judgements like $\vdash \lambda x.x : B \to B$ or $x : B \vdash \lambda z.x : A \to B$. Remark that when $\Gamma \uplus \Delta$ appears in the conclusion of some rule, then by definition, Γ and Δ are compatible.

The typing rules for λes ensure that every environment Γ contains *exactly* the set of free variables of the term t. Thus, $\Gamma \vdash_{\lambda es} t : A$ implies $\Gamma = \overline{t}$.

The typed calculus enjoys *local* subject reduction in the sense that no meta-theorem stating *weakening* or *thinning* is needed to show preservation of types.

Lemma 10 (Subject Reduction). *Let $\Gamma \vdash_{\lambda es} s : A$. Then $s =_{E_s} s'$ implies $\Gamma \vdash_{\lambda es} s' : A$ and $s \to_{\lambda es} s'$ implies $\Pi' \vdash_{\lambda es} s' : A$ for some $\Pi' \subseteq \Pi$.*

The connexion between *typed* derivations in λ-calculus (written $\vdash_\lambda$) and *typed* derivations in λes-calculus is stated as follows, where $\Gamma|_S$ denotes the environment Γ restricted to the set of variables S.

Lemma 11. *If $\Gamma \vdash_\lambda t : A$, then $\Gamma|_{\overline{t}} \vdash_{\lambda es} t : A$ and if $\Gamma \vdash_{\lambda es} t : A$, then $\Gamma \vdash_\lambda \mathtt{L}(t) : A$.*

We now prove strong-normalisation for λes-typed terms by using PSN. Another proof of strong-normalisation based on a translation of typed λes-terms into Linear Logic's proof-nets is also developed in [28].

Theorem 3 (Strong Normalisation). *Every typable Λes-term M is in $SN_{\lambda es}$.*

Proof. First define a translation $\mathtt{C}$ from λes to λ as follows: $\mathtt{C}(x) := x$, $\mathtt{C}(t\ u) := \mathtt{C}(t)\ \mathtt{C}(u)$, $\mathtt{C}(\lambda x.t) := \lambda x.\mathtt{C}(t)$ and $\mathtt{C}(t[x/u]) := (\lambda x.\mathtt{C}(t))\ \mathtt{C}(u)$. Thus for example, $\mathtt{C}((x[x/y]\ z)[w/(w_1\ w_2)]) = (\lambda w.((\lambda x.x)\ y)\ z)(w_1\ w_2)$.

We remark that for every Λes-term one has $\mathtt{C}(t) \to^*_{\lambda es} t$. Also, when t is typable in λes, then also $\mathtt{C}(t)$ is typable in λes (just change the use of subs by abs followed by app). By Lemma 11 the term $\mathtt{L}(\mathtt{C}(t)) = \mathtt{C}(t)$ is also typable in simply typed λ-calculus and thus $\mathtt{C}(t) \in SN_\beta$ [5]. We get $\mathtt{C}(t) \in SN_{\lambda es}$ by Corollary 4 so that $t \in SN_{\lambda es}$.

This proof technique, which is very simple in the case of the λes-calculus, needs some additional work to be applied to other (de Bruijn) calculi [43, 3].

6 Conclusion

In this paper we survey some properties concerning ES calculi and we describe work done in the domain during these last 15 years. We propose simple syntax and simple equations and rewriting rules to model a formalism enjoying good properties, specially confluence on metaterms, preservation of β-strong normalisation, strong normalisation of typed terms and implementation of full composition.

We believe however that some of our proofs can be simplified. In particular, PSN might be proved directly without using translations of $\lambda\mathtt{es}$ to other formalisms. We leave this for future work.

Another interesting issue is the extension of Pure Type Systems (PTS) with ES in order to improve the understanding of logical systems used in theorem-provers. Work done in this direction is based on sequent calculi [33] or natural deduction [41]. The main contribution of $\lambda\mathtt{es}$ w.r.t the formalisms previously mentioned would be our *safe* notion of composition.

It is also legitimate to ask whether $\lambda\mathtt{es}$ is minimal w.r.t. the number of rewriting rules. Indeed, it is really tempted to gather the rules $\{\mathtt{App}_1, \mathtt{App}_2, \mathtt{App}_3\}$ (resp. $\{\mathtt{Comp}_1, \mathtt{Comp}_2\}$) into the single rule App for application (resp. Comp for composition) given just after Lemma 8. While this change seems to be sound w.r.t. the properties of the calculus[3], the translation of $\Lambda\mathtt{es}$-terms into Λ_I-terms (c.f. Section 4.2), respectively into proof-nets (c.f. [28]), does not work anymore. We thus leave this question as an open problem. Note however that $\lambda\mathtt{es}$-reduction can be translated to the correspondent notion of reduction in this calculus : thus for example $\mathtt{App}_1$ can be obtained by App followed by Gc.

As far as implementation is concerned, it would be preferable from a practical point of view to avoid the systematic use of the equivalence classes generated by the axioms α and C. In other words, it would be more efficient to work with a pure rewriting system (without equations) verifying the same properties than $\lambda\mathtt{es}$. We believe that simultaneous substitutions will be needed to avoid axiom C while some technology like de Bruijn notation will be needed to avoid axiom α (as in the $\lambda_{\sigma_\Uparrow}$-calculus). We leave this topic for future investigations, but we refer the interested reader to [28] for a concrete proposition of such a calculus.

Acknowledgements

This work has benefited from fruitful discussions with E. Bonelli, R. David, R. Di Cosmo, J-P. Jouannaud, S. Lengrand, C. Muñoz and V. van Oostrom.

References

[1] Arbiser, A., Bonelli, E., Ríos, A.: Perpetuality in a lambda calculus with explicit substitutions and composition. In: WAIT (2000)

[2] Abadi, M., Cardelli, L., Curien, P.L., Lévy, J.-J.: Explicit substitutions. JFP 4(1), 375–416 (1991)

[3] While the weaker rule for composition given by $t[x/u][y/v] \to t[x/u[y/v]]$ $(y \notin \overline{t})$, is well-known [7] to affect strong normalisation and preservation of β-strong normalisation.

[3] Arbiser, A.: Explicit Substitution Systems and Subsystems. PhD thesis, Universidad Buenos Aires (2006)

[4] Barendregt, H.: The Lambda Calculus: Its Syntax and Semantics. North-Holland, Amsterdam (1984)

[5] Barendregt, H.: Lambda calculus with types. Handbook of Logic in Computer Science 2 (1992)

[6] Benaissa, Z.-E.-A., Briaud, D., Lescanne, P., Rouyer-Degli, J.: $\lambda\upsilon$, a calculus of explicit substitutions which preserves strong normalisation. JFP (1996)

[7] Bloo, R., Geuvers, H.: Explicit substitution: on the edge of strong normalization. TCS 6(5), 699–722 (1999)

[8] Bloo, R.: Preservation of Termination for Explicit Substitution. PhD thesis, Eindhoven University of Technology (1997)

[9] Baader, F., Nipkow, T.: Term Rewriting and All That. Cambridge University Press, Cambridge (1998)

[10] Bloo, R., Rose, K.: Preservation of strong normalization in named lambda calculi with explicit substitution and garbage collection. In: Computer Science in the Netherlands (1995)

[11] de Bruijn, N.: Lambda-calculus notation with nameless dummies, a tool for automatic formula manipulation, with application to the church-rosser theorem. Indag. Mat. 5(35), 381–392 (1972)

[12] de Bruijn, N.: Lambda-calculus notation with namefree formulas involving symbols that represent reference transforming mappings. Indag. Mat. 40, 356–384 (1978)

[13] Di Cosmo, R., Kesner, D., Polonovski, E.: Proof nets and explicit substitutions. In: Tiuryn, J. (ed.) ETAPS 2000 and FOSSACS 2000. LNCS, vol. 1784, Springer, Heidelberg (2000)

[14] David, R., Guillaume, B.: A λ-calculus with explicit weakening and explicit substitution. MSCS 11, 169–206 (2001)

[15] Dowek, G., Hardin, T., Kirchner, C.: Higher-order unification via explicit substitutions. I&C 157, 183–235 (2000)

[16] Dyckhoff, R., Urban, C.: Strong normalisation of Herbelin's explicit substitution calculus with substitution propagation. In: WESTAPP 2001 (2001)

[17] de Flavio Moura, M.A.-R., Kamareddine, F.: Higher order unification: A structural relation between Huet's method and the one based on explicit substitution. Available from `http://www.macs.hw.ac.uk/~fairouz/papers/`

[18] Forest, J.: A weak calculus with explicit operators for pattern matching and substitution. In: Tison, S. (ed.) RTA 2002. LNCS, vol. 2378, Springer, Heidelberg (2002)

[19] Girard, J.-Y.: Linear logic. TCS 50(1), 1–101 (1987)

[20] Goubault-Larrecq, J.: Conjunctive types and SKInT. In: Altenkirch, T., Naraschewski, W., Reus, B. (eds.) TYPES 1998. LNCS, vol. 1657, Springer, Heidelberg (1999)

[21] Hardin, T.: Résultats de confluence pour les règles fortes de la logique combinatoire catégorique et liens avec les lambda-calculs. Thèse de doctorat, Université de Paris VII (1987)

[22] Herbelin, H.: A λ-calculus structure isomorphic to sequent calculus structure. In: Pacholski, L., Tiuryn, J. (eds.) CSL 1994. LNCS, vol. 933, Springer, Heidelberg (1995)

[23] Hardin, T., Lévy, J.-J.: A confluent calculus of substitutions. In: France-Japan Artificial Intelligence and Computer Science Symposium (1989)

[24] Hardin, T., Maranget, L., Pagano, B.: Functional back-ends within the lambda-sigma calculus. In: ICFP (1996)

[25] Huet, G.: Résolution d'équations dans les langages d'ordre $1, 2, \ldots, \omega$. Thèse de doctorat d'état, Université Paris VII (1976)

[26] Khasidashvili, Z.: Expression reduction systems. In: Proceedings of IN Vekua Institute of Applied Mathematics, Tbilisi, vol. 36 (1990)

[27] Kesner, D.: Confluence properties of extensional and non-extensional λ-calculi with explicit substitutions. In: Ganzinger, H. (ed.) Rewriting Techniques and Applications. LNCS, vol. 1103, Springer, Heidelberg (1996)
[28] Kesner, D.: The theory of calculi with explicit substitutions revisited (2006), Available as `http://hal.archives-ouvertes.fr/hal-00111285/`
[29] Kesner, D., Lengrand, S.: Extending the explicit substitution paradigm. In: Giesl, J. (ed.) RTA 2005. LNCS, vol. 3467, Springer, Heidelberg (2005)
[30] Klop, J.-W.: Combinatory Reduction Systems. PhD thesis, Mathematical Centre Tracts 127, CWI, Amsterdam (1980)
[31] Khasidashvili, Z., Ogawa, M., van Oostrom, V.: Uniform Normalization Beyond Orthogonality. In: Middeldorp, A. (ed.) RTA 2001. LNCS, vol. 2051, Springer, Heidelberg (2001)
[32] Kamareddine, F.: A λ-calculus à la de Bruijn with explicit substitutions. In: Swierstra, S.D. (ed.) PLILP 1995. LNCS, vol. 982, Springer, Heidelberg (1995)
[33] Lengrand, S., Dyckhoff, R., McKinna, J.: A sequent calculus for type theory. In: Ésik, Z. (ed.) CSL 2006. LNCS, vol. 4207, Springer, Heidelberg (2006)
[34] Lengrand, S.: Normalisation and Equivalence in Proof Theory and Type Theory. PhD thesis, University Paris 7 and University of St Andrews (2006)
[35] Lescanne, P.: From λ_σ to λ_υ, a journey through calculi of explicit substitutions. In: POPL (1994)
[36] Lins, R.: A new formula for the execution of categorical combinators. In: Siekmann, J.H. (ed.) 8th International Conference on Automated Deduction. LNCS, vol. 230, Springer, Heidelberg (1986)
[37] Lins, R.: Partial categorical multi-combinators and Church Rosser theorems. Technical Report 7/92, Computing Laboratory, University of Kent at Canterbury (1992)
[38] Lévy, J.-J., Maranget, L.: Explicit substitutions and programming languages. In: Pandu Rangan, C., Raman, V., Ramanujam, R. (eds.) Foundations of Software Technology and Theoretical Computer Science. LNCS, vol. 1738, Springer, Heidelberg (1999)
[39] Lescanne, P., Rouyer-Degli, J.: Explicit substitutions with de Bruijn levels. In: Hsiang, J. (ed.) Rewriting Techniques and Applications. LNCS, vol. 914, Springer, Heidelberg (1995)
[40] Melliès, P.-A.: Typed λ-calculi with explicit substitutions may not terminate. In: Dezani-Ciancaglini, M., Plotkin, G. (eds.) TLCA 1995. LNCS, vol. 902, Springer, Heidelberg (1995)
[41] Muñoz, C.: Un calcul de substitutions pour la représentation de preuves partielles en théorie de types. PhD thesis, Université Paris 7 (1997)
[42] Nederpelt, R.: Strong Normalization in a Typed Lambda Calculus with Lambda Structured Types. PhD thesis, Eindhoven University of Technology (1973)
[43] Polonovski, E.: Substitutions explicites, logique et normalisation. Thèse de doctorat, Université Paris 7 (2004)
[44] Rose, K.: Explicit cyclic substitutions. In: Rusinowitch, M., Remy, J.-L. (eds.) Conditional Term Rewriting Systems. LNCS, vol. 656, Springer, Heidelberg (1993)
[45] Sakurai, T.: Strong normalizability of calculus of explicit substitutions with composition. Available on
`http://www.math.s.chiba-u.ac.jp/~sakurai/papers.html`
[46] Sinot, F.-R., Fernández, M., Mackie, I.: Efficient reductions with director strings. In: Nieuwenhuis, R. (ed.) RTA 2003. LNCS, vol. 2706, Springer, Heidelberg (2003)
[47] Terese.: Term Rewriting Systems. Cambridge Tracts in Theoretical Computer Science, vol. 55. Cambridge University Press, Cambridge (2003)

A Soft Type Assignment System for λ-Calculus

Marco Gaboardi and Simona Ronchi Della Rocca

Dipartimento di Informatica
Università degli Studi di Torino
Corso Svizzera 185, 10149 Torino, Italy
gaboardi,ronchi@di.unito.it

Abstract. Soft Linear Logic (SLL) is a subsystem of second-order linear logic with restricted rules for exponentials, which is correct and complete for PTIME. We design a type assignment system for the λ-calculus (STA), which assigns to λ-terms as types (a proper subset of) SLL formulas, in such a way that typable terms inherit the good complexity properties of the logical system. Namely STA enjoys subject reduction and normalization, and it is correct and complete for PTIME and FPTIME.

1 Introduction

The light logics, Light Linear Logic (LLL) [1] and Soft Linear Logic (SLL) [2], were both introduced as logical counterpart of the polynomial complexity. Namely proofs of both logics normalize in a polynomial number of cut-elimination steps, if their depth is fixed, and moreover they are complete for FPTIME and PTIME respectively. So they can be used for the design of programming languages with an intrinsically polynomial computational bound. This can be done in a straightforward way by a complete decoration of the logical proofs, but for both these logics, since the presence of modalities, the resulting languages have a very complex syntactical structure, and they cannot be reasonably proposed for programming (an example of complete decoration of SLL is in [3]). A different approach is to fix as starting points:

1. The use of λ-calculus as an abstract paradigm of programming languages.
2. The use of types to characterize program properties.

In this line, the aim becomes the design of a type assignment system for λ-calculus, where types are formulae of a light logic, in such a way that the logical properties are inherited by the well typed terms. Then types can be used for checking, beside the usual notion of correctness, also the property of having polynomial complexity. A type assignment for λ-calculus correct and complete with respect to polynomial time computations has been designed by Baillot and Terui [4], using as types formulae of Light Affine Logic (LAL), a simplified version of LLL defined in [5,6]. The aim of this paper is to design a type assignment system for λ-calculus, enjoying the same properties, but using as types formulae

J. Duparc and T.A. Henzinger (Eds.): CSL 2007, LNCS 4646, pp. 253–267, 2007.

of SLL. The motivation for doing it is twofold. The design of λ-calculi with implicit complexity properties is a very new research line, and so exploring different approaches is necessary in order to compare them. In particular, SLL formulas look simpler than LAL ones, since they have just one modality while LAL has two modalities, and this could in principle give rise to a system easier to deal with. Moreover, SLL has been proved to be complete just for PTIME, while we want a type assignment system complete for FPTIME: so we want to explore at the same time if it is possible, when switching from formulae to types, to prove this further property.

We start from the original version of second order SLL with the only connectives $\multimap$ and ! and the quantifier $\forall$, given in sequent calculus style. The principal problem in designing the desired type assignment system is that, in general, in a modal logic setting, the good properties of proofs are not easily inherited by λ-terms. In particular, there is a mismatch between β-reduction in the λ-calculus and cut-elimination in logical systems, which makes it difficult both getting the subject reduction property and inheriting the complexity properties from the logic, as discussed in [4]. To solve this problem we exploit the fact that, in the decorated sequent calculus, there is a redundancy of proofs, in the sense that the same λ-term can arise from different proofs. So we propose a restricted system, called Soft Type Assignment (STA), where the set of derivations corresponds to a proper subset of proofs in the affine version of SLL. Types are a subset of SLL formulae, where, in the same spirit of [4], the modality ! is not allowed in the right hand of an arrow, and the application of some rules (in particular the *cut* rule) is restricted to linear types. STA enjoys subject reduction, and the set of typable terms characterizes PTIME. Moreover we prove that the language of typable terms is complete both for PTIME and FPTIME. The completeness we are speaking about is a functional completeness, in the sense that we prove (through a simulation of polynomial time Turing Machines) that all polynomial functions can be computed by terms typable in STA. The algorithmic expressivity of the language has been not studied here.

This paper is a first step: we are actually working on a natural deduction version of STA, and on the type inference problem. The type inference problem for STA seems undecidable, but we think it is possible to build decidable restrictions that do not lose the complexity completeness. Moreover a comparison with respect to DLAL, the type assignment system in [4], is in order. The two systems have incomparable typability power, in the sense that there exist terms typable in STA but not in DLAL and vice versa. STA has an easier treatment of contexts, and no notion of discharged assumptions. This, together with the fact that types in STA have just one modality, can help in designing a simpler natural deduction and a more efficient type inference algorithm.

The paper is organized as follows. In Section 2 SLL is introduced and a discussion is made about the problem of subject reduction. In Section 3 STA is defined, and the proof of subject reduction is given. Section 4 contains the results about the complexity bound: polynomial bound is proved in Subsection 4.1, and the completeness for both PTIME and FPTIME is proved in Subsection 4.2.

2 Soft Linear Logic and λ-Calculus

In [7] a decoration of SLL by terms of λ-calculus has been presented, as technical tool for studying the expressive power of the system. The decoration is presented in Table 1, where $\Gamma\#\Delta$ denotes the fact that the domain of contexts Γ and Δ are disjoint sets of variables. In such system, which we call SLL_λ, we have the failure of subject reduction. Let us give a detailed analysis of the problem. In a decorated sequent calculus system, β-reduction is the counterpart, in terms, of the *cut* rule of the logic. The *cut* rule in Table 1 can be split into three different rules, according to the shape of U, of the contexts and of the derivation:

$$\frac{\Gamma \vdash_l M : U \quad \Delta, x : U \vdash_l N : V \quad U \text{ not modal}}{\Gamma, \Delta \vdash_l N[M/x] : V}\ (L\ cut)$$

$$\frac{\Pi \triangleright !\Gamma \vdash_l M : !U \quad \Delta, x : !U \vdash_l N : V \quad \Pi \text{ duplicable}}{!\Gamma, \Delta \vdash_l N[M/x] : V}\ (D\ cut)$$

$$\frac{\Pi \triangleright \Gamma \vdash_l M : !U \quad \Delta, x : !U \vdash_l N : V \quad \Pi \text{ not duplicable}}{\Gamma, \Delta \vdash_l N[M/x] : V}\ (S\ cut)$$

where $\Pi \triangleright !\Gamma \vdash_l M : !U$ is duplicable if it corresponds to a !-box in the respective proof-net of SLL [2], and L, D, S are short for *Linear*, *Duplication* and *Sharing*. The problem is that, while (L *cut*) and (D *cut*) both correspond to β-reduction, (S *cut*) is not necessarily reflected into a β-reduction.
Let us show it by an example. Consider the term $M \equiv y((\lambda z.sz)w)((\lambda z.sz)w)$ and let $Z = U \multimap U \multimap V, S = V \multimap !U$. It has the typing $y : Z, s : S, w : V \vdash_l y((\lambda z.sz)w)((\lambda z.sz)w) : V$ as proved by the following (incomplete) derivation:

$$\dfrac{\dfrac{s : S \vdash_l \lambda z.sz : S \quad t : S, w : V \vdash_l tw : !U}{s : S, w : V \vdash_l (\lambda z.sz)w : !U}\ (cut) \qquad \dfrac{y : Z, r : U, l : U \vdash_l yrl : V}{y : Z, x : !U \vdash_l yxx : V}\ (m)}{y : Z, s : S, w : V \vdash_l y((\lambda z.sz)w)((\lambda z.sz)w) : V}\ (cut)$$

Table 1. SLL_λ

$$\frac{}{x : U \vdash_l x : U}\ (Id) \qquad \frac{\Gamma \vdash_l M : U \quad x : V, \Delta \vdash_l N : Z \quad \Gamma\#\Delta \quad y \text{ fresh}}{\Gamma, y : U \multimap V, \Delta \vdash_l N[yM/x] : Z}\ (\multimap L)$$

$$\frac{\Gamma \vdash_l M : U \quad \Delta, x : U \vdash_l N : V \quad \Gamma\#\Delta}{\Gamma, \Delta \vdash_l N[M/x] : V}\ (cut) \qquad \frac{\Gamma, x : U \vdash_l M : V}{\Gamma \vdash_l \lambda x.M : U \multimap V}\ (\multimap R)$$

$$\frac{\Gamma \vdash_l M : U}{!\Gamma \vdash_l M : !U}\ (sp) \qquad \frac{\Gamma, x_0 : U, ..., x_n : U \vdash_l M : V}{\Gamma, x : !U \vdash_l M[x/x_0, ..., x/x_n] : V}\ (m)$$

$$\frac{\Gamma \vdash_l M : U}{\Gamma \vdash_l M : \forall\alpha.U}\ (\forall R) \qquad \frac{\Gamma, x : U[V/\alpha] \vdash_l M : Z}{\Gamma, x : \forall\alpha.U \vdash_l M : Z}\ (\forall L)$$

where the *cut* is clearly a $(S\ cut)$. It is easy to check that $y((\lambda z.sz)w)((\lambda z.sz)w) \rightarrow_\beta y(sw)((\lambda z.sz)w)$ but unfortunately there is no derivation with conclusion: $y : Z, s : S, w : V \vdash_l y((sw)((\lambda z.sz)w) : V$.

The technical reason is that, in the previous derivation, there is a mismatch between the term and the derivation: in M there are two copies of $(\lambda z.sz)w$, while in the derivation there is just one subderivation with subject $(\lambda z.sz)w$, and this subderivation is not duplicable, since in particular S is not modal. Then the two copies of this subterm cannot be treated in a non uniform way.

The problem is not new, and all the type assignment systems for λ-calculus derived from Linear Logic need to deal with it. Until now the proposed solutions follow three different paths, all based on a natural deduction definition of the type assignment. The first one, proposed in [8], explicitly checks the duplicability condition before to perform a normalization step. So in the resulting language (which is a fully typed λ-calculus) the set of redexes is a proper subset of the set of classical β-redexes. In the light logics setting, in [9], a type assignment for the λ-calculus, based on EAL, is designed. There the authors use the call-by-value λ-calculus, where the restricted definition of reduction corresponds exactly to linear substitution and duplication. In the type assignment for λ-calculus based on LAL, made in [4], a syntax of types is used, where the modality ! is no more present, and there are two arrows, a linear and an intuitionistic one, whose elimination reflects the linear and the duplication cut respectively. The set of types corresponds to a restriction of the set of logical formulas. All the approaches need a careful control of the context, which technically has been realized by splitting it in different parts, collecting respectively the linear and modal assumptions (in case of [9] a further context is needed). Here we want to explore a different approach. A type derivation for λ-calculus based on sequent calculus is in some sense redundant, since the same typing can be proved by a plethora of different derivations, corresponding to different ways of building the term. A (cut) rule and a $(\multimap L)$ rule both correspond to a modification of the subject through a substitution. Our key observation is that a term can always be built using linear substitutions. As example, take the term before $y((\lambda z.sz)w)((\lambda z.sz)w)$. It can be seen as $yxx[(\lambda z.sz)w/x]$ but also as $yx_1x_2[(\lambda z.s_1z)w_1/x_1, (\lambda z.s_2z)w_2/x_2][s/s_1, s/s_2, w/w_1,\ w/w_2]$, where all substitutions are linear. Then we want to restrict the set of proofs, in such a way that both sharing and duplication are forbidden, and terms are built by means of linear substitutions only. Such restriction preserves subject reduction since duplication is a derived rule. In order to do this, we start from an affine version of SLL and restrict the formulae, using as types just a subset of the SLL formulae which are, in some sense, recursively linear. The gain is that the splitting of the context is no more necessary.

3 The Soft Type Assignment System

In this section we will present a type assignment system, which assigns to λ-terms a proper subset of SLL formulae, and we will prove that it enjoys subject

Table 2. Soft Type Assignment system

$$\frac{}{x : A \vdash x : A}\ (Id) \qquad \frac{\Gamma \vdash M : \tau \quad x : A, \Delta \vdash N : \sigma \quad \Gamma\#\Delta \quad y \text{ fresh}}{\Gamma, y : \tau \multimap A, \Delta \vdash N[yM/x] : \sigma}\ (\multimap L)$$

$$\frac{\Gamma, x : \sigma \vdash M : A}{\Gamma \vdash \lambda x.M : \sigma \multimap A}\ (\multimap R) \qquad \frac{\Gamma \vdash M : A \quad \Delta, x : A \vdash N : \sigma \quad \Gamma\#\Delta}{\Gamma, \Delta \vdash N[M/x] : \sigma}\ (cut)$$

$$\frac{\Gamma \vdash M : \sigma}{\Gamma, x : A \vdash M : \sigma}\ (w) \qquad \frac{\Gamma \vdash M : \sigma}{!\Gamma \vdash M : !\sigma}\ (sp) \qquad \frac{\Gamma, x : A[B/\alpha] \vdash M : \sigma}{\Gamma, x : \forall\alpha.A \vdash M : \sigma}\ (\forall L)$$

$$\frac{\Gamma, x_1 : \tau, ..., x_n : \tau \vdash M : \sigma}{\Gamma, x : !\tau \vdash M[x/x_1, ..., x/x_n] : \sigma}\ (m) \qquad \frac{\Gamma \vdash M : A}{\Gamma \vdash M : \forall\alpha.A}\ (\forall R)$$

reduction. Types correspond in some sense to "recursively linear" formulae and quantification can be applied only to linear types. The type assignment system is such that all rules dealing with substitution (($\multimap L$) and (cut)) are applicable only if the type of the replaced variable is linear. So all terms are built through linear substitutions.

Definition 1. *i) The syntax of λ-terms is the usual one. Let α range over a countable set of type variables. The set* **T** *of* soft types *is defined as follows:*

$$A ::= \alpha \mid \sigma \multimap A \mid \forall\alpha.A \ \textit{(Linear Types)}$$
$$\sigma ::= A \mid !\sigma$$

Type variables are ranged over by α, β, linear types by A, B, C, and types by σ, τ, ζ. $\equiv$ denotes the syntactical equality both for types and terms (modulo renaming of bound variables).

ii) A context is a set of assumptions of the shape $x : \sigma$, where all variables are different. Contexts are ranged over by Γ, Δ. $dom(\Gamma) = \{x \mid \exists x : \sigma \in \Gamma\}$ and $\Gamma\#\Delta$ means $dom(\Gamma) \cap dom(\Delta) = \emptyset$.

iii) STA proves sequents of the shape $\Gamma \vdash M : \sigma$ where Γ is a context, M is a λ-term, and σ is a soft type. The rules are given in Table 2, where as usual the rule ($\forall R$) has the side condition that α must not be free in Γ.

iv) Derivations are denoted by $\Pi, \Sigma, \Phi, \Psi, \Theta$. $\Pi \triangleright \Gamma \vdash M : \sigma$ denotes a derivation Π with conclusion $\Gamma \vdash M : \sigma$. $\vdash M : \sigma$ is short for $\emptyset \vdash M : \sigma$.

We will use the following:

Notation. *$FV(M)$ denotes the set of free variables of M, $n_o(x, M)$ the number of free occurrences of the variable x in M. $M\{z\}$ ($M\{zQ\}$) denotes that z (zQ) occurs once in M. $!^n\sigma$ is an abbreviation for $!...!\sigma$ n-times. $!^0\sigma \equiv \sigma$. As usual $\multimap$ associates to the right and has precedence on $\forall$, while $!$ has precedence on everything else. $\sigma[A/\alpha]$ denotes the capture free substitution in σ of all occurrences of the type variable α by the linear type A: note that this kind of substitution* *preserves the correct syntax of types.* $\boldsymbol{\sigma}$ *denotes a sequence of types, and* $\sigma[\mathbf{A}/\boldsymbol{\alpha}]$ *denotes the replacement of the i-th type variable in* $\boldsymbol{\alpha}$ *by the i-th type in* $\mathbf{A}$. $|\boldsymbol{\sigma}|$ *denotes the length of the sequence* $\boldsymbol{\sigma}$. $\Gamma[\mathbf{A}/\boldsymbol{\alpha}]$ *is the context* Γ*, where all type* τ *has been replaced by* $\tau[\mathbf{A}/\boldsymbol{\alpha}]$. $\forall\boldsymbol{\alpha}.A$ *is an abbreviation for* $\forall\alpha_1....\forall\alpha_n.A$*, for some* $n \geq 0$. *Note that each type is of the shape* $!^n\forall\boldsymbol{\alpha}.A$

Some comments on STA are needed. The (*cut*) is a linear cut, according to the classification given in the previous section, and rule ($\multimap L$) requires that the type of the replaced variable is linear, so in particular all substitutions in the subject act on a variable occurring at most once (as we will see in the following lemma). Moreover both the (*Id*) rule and (*w*) rule can introduce only linear types.

Note for example that the term M considered in Section 2. is typable with typing: $y : A \multimap A \multimap B, s :!(A \multimap A), w :!A \vdash y((\lambda z.sz)w)((\lambda z.sz)w) : B$

Lemma 2. *i)* $\Gamma, x : A \vdash M : \nu$ *implies* $n_o(x, M) \leq 1$.
ii) $\Gamma, x : A \vdash M :!\sigma$ *implies* $x \notin FV(M)$.

The system enjoys the classical substitution properties.

Lemma 3. *i)* $\Gamma \vdash M : \sigma$ *implies* $\Gamma[\boldsymbol{\zeta}/\boldsymbol{\alpha}] \vdash M : \sigma[\boldsymbol{\zeta}/\boldsymbol{\alpha}]$.
ii) $\Gamma \vdash M : \forall\boldsymbol{\alpha}.A$ *implies* $\Gamma \vdash M : A$.

Let $\Pi \triangleright \Gamma, x : \tau \vdash M : \sigma$. The notion of chain of x in Π will be used to remember the successive renaming of the assumptions which give rise to the assumption $x : \tau$. The notion of s-chain is necessary for dealing with the particular case of the replacements of a variable by another one in a cut rule.

Definition 4. *Let* $\Pi \triangleright \Gamma, x : \tau \vdash M : \sigma$.

- *A* chain *of* x *in* Π *is a sequence of variables inductively defined as follows:*
  - *Let the last applied rule of* Π *be:*

$$\frac{}{x : A \vdash x : A} \;,\quad \frac{\Gamma' \vdash P : \sigma}{\Gamma', x : A \vdash P : \sigma} \quad or \quad \frac{\Delta \vdash P : \tau \quad z : A, \Gamma \vdash N : \sigma}{x : \tau \multimap A, \Gamma, \Delta \vdash N[xP/z] : \sigma}$$

    Then the only chain of x *is* x *itself.*
  - *Let the last applied rule of* Π *be:*

$$\frac{\Pi' \triangleright \Gamma', x_1 : \tau, ..., x_k : \tau \vdash N : \sigma}{\Gamma', x :!\tau \vdash N[x/x_1, ..., x/x_k] : \sigma}\ (m)$$

    Then a chain of x *in* Π *is every sequence* $x\mathbf{c}$*, where* $\mathbf{c}$ *is a chain of* x_i *in* Π' *for* $1 \leq i \leq k$.
  - *In every other case there is an assumption with subject* x *both in the conclusion of the rule and in one of its premises* Σ. *Then a chain of* x *in* Π *is every sequence* $x\mathbf{c}$*, where* $\mathbf{c}$ *is a chain of* x *in* Σ.
- *Let* $\mathbf{c}$ *be a chain of* x *in* Π *and let* x_k *be the last variable of* $\mathbf{c}$. *Then* x_k *is an* ancestor *of* x *in* Π. $n_a(x, \Pi)$ *denotes the number of ancestors of* x *in* Π.

- *y is an* effective ancestor *of x if and only if it is an ancestor of x, and moreover every variable z in the chain of x ending with y occurs in sequent where z belongs to the free variable set of the subject of the sequent. $n_e(x, \Pi)$ denotes the number of effective ancestors of x in Π.*
- *A* s-chain *of x in Π is inductively defined analogously to the chain of x in Π but the (cut) case:*

$$\frac{\Gamma \vdash M : A \quad \Sigma \triangleright \Delta, z : A \vdash N : \sigma}{\Gamma, \Delta \vdash N[M/z] : \sigma} \ (cut)$$

If $M \not\equiv x$ then an s-chain of x in Π is defined as a chain of x in Π, otherwise an s-chain of x in Π is every sequence $z\mathbf{c}$, where $\mathbf{c}$ is a chain of z in Σ. An s-ancestor *is defined analogously to ancestor, but with respect to s-chains.*

The following lemma follows easily from the previous definition.

Lemma 5. *Let $\Pi \triangleright \Gamma \vdash M : \sigma$. Then $\forall x : \ n_o(x, M) \leq n_e(x, \Pi) \leq n_a(x, \Pi)$.*

The notion of effective ancestor will be used in Section 4. The Generation Lemma connects the shape of a term with its possible typings, and will be useful in the sequel. Let Π and Π' be two derivations in STA, proving the same conclusion: $\Pi \rightsquigarrow \Pi'$ denotes the fact that Π' is obtained from Π by commuting some rule applications, by erasing m rule applications, by inserting $n \leq m$ applications of rule (w), for some $n, m \geq 0$, and by renaming some variables.

Lemma 6 (Generation Lemma)

1. *$\Gamma \vdash \lambda x.M : \sigma$ implies $\sigma \equiv !^i(\forall \boldsymbol{\alpha}.\tau \multimap A)$, for some τ, A and $i, |\boldsymbol{\alpha}| \geq 0$;*
2. *$\Pi \triangleright \Gamma \vdash \lambda x.P : \sigma \multimap A$ implies $\Pi \rightsquigarrow \Pi'$, whose last rule is $(\multimap R)$.*
3. *$\Pi \triangleright \Gamma, x : \forall \boldsymbol{\alpha}.\tau \multimap A \vdash M\{xQ\} : \sigma$ implies that Π is composed by a subderivation Σ, followed by a sequence δ of rules not containing rule (sp), and the last rule of Σ is:*

$$\frac{\Gamma' \vdash Q' : \tau' \quad \Gamma'', z : A' \vdash P\{z\} : \sigma'}{\Gamma', \Gamma'', t : \tau' \multimap A' \vdash P\{z\}[tQ'/z] : \sigma'} \ (\multimap L)$$

where $\tau' \equiv \tau[\boldsymbol{B}/\boldsymbol{\alpha}]$, $A' \equiv A[\boldsymbol{B}/\boldsymbol{\alpha}]$, t is the only s-ancestor of x in Π, for some $P, Q', \Gamma', \Gamma'', \boldsymbol{B}, \boldsymbol{\alpha}, \sigma'$.
4. *$\Pi \triangleright \Gamma \vdash M :!\sigma$ implies $\Pi \rightsquigarrow \Pi'$ where Π' is composed by a subderivation, ending with the rule (sp) proving $!\Gamma' \vdash M :!\sigma$, followed by a sequence of rules (w), $(\multimap L)$, (m), $(\forall L)$, (cut), all dealing with variables not occurring in M.*
5. *$\Pi \triangleright !\Gamma \vdash M :!\sigma$ implies $\Pi \rightsquigarrow \Pi'$, whose last rule is (sp).*
6. *$\Pi \triangleright \Gamma \vdash \lambda x.M : \forall \alpha.A$ implies $\Pi \rightsquigarrow \Pi'$ where the last rule of Π' is $(\forall R)$.*

3.1 Subject Reduction

STA enjoys the subject reduction property. The proof is based on the fact that, while formally the (*cut*) rule in it is linear, the duplication cut is a derived rule. To prove this, we need a further lemma.

Lemma 7. *$\Pi \triangleright \Gamma, x :!^n \forall \boldsymbol{\alpha}.A \vdash M : \sigma$ where $A \not\equiv \forall \boldsymbol{\beta}.B$ and y is an ancestor of x in Π imply that y has been introduced with type $\forall \boldsymbol{\alpha}'.A[\boldsymbol{B}/\boldsymbol{\alpha}'']$, for some $\boldsymbol{\alpha}', \boldsymbol{\alpha}''$ (possible empty) such that $\boldsymbol{\alpha} = \boldsymbol{\alpha}'', \boldsymbol{\alpha}'$*

Lemma 8. *The following rule is derivable in STA:*

$$\frac{\Pi \triangleright !^n \Gamma \vdash M :!^n B \quad \Sigma \triangleright x :!^n B, \Delta \vdash N : \sigma \quad !^n \Gamma \# \Delta}{!^n \Gamma, \Delta \vdash N[M/x] : \sigma} \ (dup)$$

Proof. In case $n = 0$ the proof is obvious, since (dup) coincides with the (cut) rule. Otherwise let $B \equiv \forall \boldsymbol{\alpha}.A$. By Lemma 6.5, Π can be transformed into a derivation Π'', which is $\Pi' \triangleright \Gamma \vdash M : A$ followed by n applications of rule (sp). Let the ancestors of x in Σ be $x_1, ..., x_m$. By Lemma 7, x_j has been introduced with type $\forall \boldsymbol{\alpha}_j'.A[\boldsymbol{C_j}/\boldsymbol{\alpha}_j'']$, for some $\boldsymbol{C_j}, \boldsymbol{\alpha}_j', \boldsymbol{\alpha}_j''$ $(1 \leq j \leq m)$. By Lemma 3, there are disjoint derivations $\Pi_j' \triangleright \Gamma_j \vdash M_j : \forall \boldsymbol{\alpha}_j'.A[\boldsymbol{C_j}/\boldsymbol{\alpha}_j'']$, where M_j and Γ_j are fresh copies of M and Γ $(1 \leq j \leq m)$. Then, for every x_j introduced by an (Id) rule, replace such rule by Π_j'. For every x_j introduced by a $(\multimap L)$ rule as:

$$\frac{\Gamma' \vdash R : \tau[\boldsymbol{C_j}/\boldsymbol{\alpha}_j''] \quad \Delta', z : B'[\boldsymbol{C_j}/\boldsymbol{\alpha}_j''] \vdash T : \sigma}{\Gamma', x_j : (\tau \multimap B')[\boldsymbol{C_j}/\boldsymbol{\alpha}_j''], \Delta' \vdash T[x_j R/z] : \sigma} \ (\multimap L)$$

where $\forall \boldsymbol{\alpha}_j'.A[\boldsymbol{C_j}/\boldsymbol{\alpha}_j''] \equiv (\tau \multimap B')[\boldsymbol{C_j}/\boldsymbol{\alpha}_j'']$, $|\boldsymbol{\alpha}_j'| = 0$, after this insert the rule:

$$\frac{\Pi_j' \quad \Gamma', x_j : A[\boldsymbol{C_j}/\boldsymbol{\alpha}_j''], \Delta' \vdash T[x_j R/z] : \sigma}{\Gamma', \Delta', \Gamma_j \vdash T[M_j R/z] : \sigma} \ (cut).$$

For every x_j introduced by a (w) rule, just erase the rule. Moreover arrange the context and the subject in all rules according with these modifications. Then replace every application of a rule (m), when applied on variables in a chain of x in Σ, by a sequence of rules (m) applied to the free variables of the corresponding copy of M. So the resulting derivation Φ proves $!^n \Gamma, \Delta \vdash N[M/y] : \sigma$. □

Note that the rule (dup) cannot represent a $(S\ cut)$ since by Lemma 6.4 a derivation where both the context and the type are modal always corresponds to a !-box. The above lemma allow us to freely use (dup) rule in what follows.

Theorem 9 (Subject Reduction). *If $\Gamma \vdash M : \sigma$ and $M \rightarrow_\beta M'$ then $\Gamma \vdash M' : \sigma$*

Proof. By induction on the derivation. The only interesting case is when the last rule is (cut), creating the redex $(\lambda y.P)Q$ reduced in M.

$$\frac{\Psi \triangleright \Gamma \vdash \lambda y.P : A \quad \Sigma \triangleright \Delta, x : A \vdash N\{xQ\} : \sigma \quad \Gamma \# \Delta}{\Gamma, \Delta \vdash N\{xQ\}[\lambda y.P/x] : \sigma} \ (cut).$$

By Lemma 6.1, and by the constraint on the (cut) rule, $A \equiv \forall \boldsymbol{\alpha}.\tau \multimap B$. By Lemma 6.3, Σ is composed by a subderivation Σ_2, followed by a sequence δ of rules not containing (sp). Σ_2 is:

$$\frac{\Theta_1 \triangleright \Delta'' \vdash Q' : \tau' \quad \Theta_2 \triangleright \Delta', z : B' \vdash N'\{z\} : \sigma'}{\Delta', t : \tau' \multimap B', \Delta'' \vdash N'\{z\}[tQ'/z] : \sigma'}$$

where $\tau' = \tau[\boldsymbol{C}/\boldsymbol{\alpha}]$, $B' = B[\boldsymbol{C}/\boldsymbol{\alpha}]$, t is the only s-ancestor of x, for some $\Delta', \Delta'', N', Q', \sigma', \boldsymbol{C}$. By Lemma 6.2,6, $\Psi \rightsquigarrow \Psi'$ ending as:

$$\dfrac{\dfrac{\Gamma, y : \tau \vdash P : B}{\Gamma \vdash \lambda y.P : \tau \multimap B}\,(\multimap R)}{\Gamma \vdash \lambda y.P : \forall \boldsymbol{\alpha}.\tau \multimap B}\,(\forall R)^*$$

hence by Lemma 3.ii), there is a derivation $\Phi \triangleright \Gamma, y : \tau' \vdash P : B'$, since $\boldsymbol{\alpha}$ are not free in Γ.

Let $\tau' \equiv !^n A'$ $(n \geq 0)$ for some A'. By Lemma 6.4,5, Θ_1 can be transformed in a derivation composed by a subderivation $\Theta_4 \triangleright !^n \Delta''' \vdash Q' :!^n A'$, followed by a sequence δ' of applications of rules (w), $(\multimap L)$,(m), $(\forall L)$, (cut), all dealing with variables not occurring in Q. So we can apply the derived rule (dup) obtaining:

$$\dfrac{\Theta_4 :!^n \Delta''' \vdash Q' :!^n A' \quad \Phi \triangleright \Gamma, y :!^n A' \vdash P : B'}{\Gamma, !^n \Delta''' \vdash P[Q'/y] : B'}$$

Then, by applying δ' we have $\Theta_3 \triangleright \Gamma, \Delta'' \vdash P[Q'/y] : B'$. Hence,

$$\dfrac{\Theta_3 \quad \Theta_2}{\Gamma, \Delta', \Delta'' \vdash N'\{z\}[P[Q'/y]/z] : \sigma}\,(cut)$$

and by applying δ the desired derivation can be built. □

The proof of subject reduction gives evidence to the fact that β-reduction, while corresponding to the cut elimination in case the bound variable occurs at most once, is reflected into a global transformation of the derivation, in case a duplication of the argument is necessary.

4 Complexity

In this section we will show that STA is correct and complete for polynomial complexity. Namely, in Subsection 4.1, we will prove that, if a term M can be typed in STA by a derivation Π, then it reduces to normal form in a number of β-reduction steps which is bounded by $|M|^{\mathtt{d}(\Pi)+1}$, where $|M|$ is the number of symbols of M and $\mathtt{d}(\Pi)$ is the number of nested applications of rule (sp) in Π. So working with terms typed by derivations of fixed degree assures us to keep only a polynomial number of computation steps. Then we show that this polystep result extends to a polytime result. In the Subsection 4.2, we prove that STA is complete both for PTIME and FPTIME, using a representation of Turing machine working in polynomial time by λ-terms, typable in STA through derivations obeying suitable constraints. The key idea is that data can be represented by terms typable by linear types using derivations of degree 0. Obviously programs can duplicate their data, so a derivation typing an application of a program to its data can have degree greater than 0, but this degree depends only on the program, so it does not affect the complexity measure.

4.1 Complexity of Reductions in STA

SLL enjoys a polynomial time bound, namely the cut-elimination procedure is polynomial in the size of the proof. This result holds obviously also for STA. But we need something more, namely to relate the polynomial bound to the size of the λ-terms. So we prove this result by defining measures of both terms and proofs, which are an adaptation of those given by Lafont, but they do not take into account the subderivations that do not contribute to the term formation (which we will call "erasing").

Definition 10.

- *The size* $|M|$ *of a term* M *is defined as* $|x| = 1$, $|\lambda x.M| = |M| + 1$, $|MN| = |M| + |N| + 1$. *The size* $|\Pi|$ *of a proof* Π *is the number of rules in* Π.
- *In the rules* ($\multimap L$) *and* (*cut*), *the subderivation proving* $\Gamma \vdash M : \tau$ *and* $\Gamma \vdash M : A$ *respectively is* erasing *if* $x \notin FV(N)$ *(using notation of Table 2).*
- *The* rank *of a rule* (*m*), *as defined in Table 2, is the number* $k \leq n$ *of variables* x_i *such that* $x_i \in FV(M)$ $(1 \leq i \leq n$ *). Let* r *be the the maximum rank of a rule* (*m*) *in* Π, *not considering erasing subderivations. The rank* $\mathtt{rk}(\Pi)$ *of* Π *is the maximum between* 1 *and* r.
- *The degree* $\mathtt{d}(\Pi)$ *of* Π *is the maximum nesting of applications of rule* (*sp*) *in* Π, *not considering erasing subderivations;*
- *The weight* $\mathtt{W}(\Pi, r)$ *of* Π *with respect to* r *is defined inductively as follows.*
  - *If the last applied rule is* (*Id*) *then* $\mathtt{W}(\Pi, r) = 1$.
  - *If the last applied rule is* ($\multimap R$) *with premise a derivation* Σ, *then* $\mathtt{W}(\Pi, r) = \mathtt{W}(\Sigma, r) + 1$.
  - *If the last applied rule is* (*sp*) *with premise a derivation* Σ, *then* $\mathtt{W}(\Pi, r) = r\mathtt{W}(\Sigma, r)$.
  - *If the last applied rule is:*

$$\frac{\Sigma \triangleright \Gamma \vdash M : A \quad \Phi \triangleright x : A, \Delta \vdash N : \sigma}{\Gamma, \Delta \vdash N[M/x] : \sigma}\ (cut)$$

    then $\mathtt{W}(\Pi, r) = \mathtt{W}(\Sigma, r) + \mathtt{W}(\Phi, r) - 1$ *if* $x \in FV(N)$, $\mathtt{W}(\Pi, r) = \mathtt{W}(\Phi, r)$ *otherwise.*
  - *In every other case* $\mathtt{W}(\Pi, r)$ *is the sum of the weights of the premises with respect to* r, *not counting erasing subderivations.*

Lemma 11. *Let* $\Pi \triangleright \Gamma \vdash M : \sigma$. *Then:*
1. $\mathtt{rk}(\Pi) \leq |M| \leq |\Pi|$. 2. $\mathtt{W}(\Pi, 1) \leq |M|$. 3. $\mathtt{W}(\Pi, r) \leq r^{\mathtt{d}(\Pi)}\mathtt{W}(\Pi, 1)$
4. $x :!^q A \in \Gamma$ *implies* $n_o(x, M) \leq n_e(x, \Pi) \leq \mathtt{rk}(\Pi)^q$.
5. $\Pi \rightsquigarrow \Pi'$ *implies* $\mathtt{W}(\Pi', r) \leq \mathtt{W}(\Pi, r)$.

The normalization of proofs is based on the notion of substitution. The following lemma extends the weight definition to the derived rule (*dup*) by taking into account the weight of the involved proofs.

Lemma 12. *Let* Φ *be a derivation ending with the rule* (*dup*), *with premises* Σ *and* Π. *Then, if* $r \geq \mathtt{rk}(\Phi)$, $\mathtt{W}(\Phi, r) \leq \mathtt{W}(\Sigma, r) + \mathtt{W}(\Pi, r)$.

Proof. It suffices to verify how the weights are modified in the proof of Lemma 8, using Lemma 11. We will use exactly the same notations as in the lemma.

The transformation of Π in Π'' leaves the weight unchanged. In particular $\mathtt{W}(\Pi, r) = r^n \mathtt{W}(\Pi', r)$ and $\mathtt{W}(\Pi'_j, r) = \mathtt{W}(\Pi', r)$.

By Lemma 6.4 and Lemma 6.5 Σ can be transformed in a derivation Σ' followed by $k \leq n$ applications of rule (sp) and by a sequence of rules dealing with variables not occurring in N. In particular $\mathtt{W}(\Sigma, r) = r^k \mathtt{W}(\Sigma', r)$.

Then, for every ancestor x_j in Σ the replacement by Π'_j increases the weight $\mathtt{W}(\Sigma', r)$ of at most a quantity $\mathtt{W}(\Pi', r)$. By definition of weight we are interested only in effective ancestors, hence by Lemma 11.4:

$$\mathtt{W}(\Phi, r) = r^k(\mathtt{W}(\Sigma', r) + n_e(x, \Sigma')\mathtt{W}(\Pi', r)) \leq r^k \mathtt{W}(\Sigma', r) + r^k r^{n-k} \mathtt{W}(\Pi', r)$$
$$\leq r^k \mathtt{W}(\Sigma', r) + r^n \mathtt{W}(\Pi', r) = \mathtt{W}(\Sigma, r) + \mathtt{W}(\Pi, r).$$ □

Now we can show that the weight decreases when a β-reduction is performed.

Lemma 13. *Let $\Pi \triangleright \Gamma \vdash M : \sigma$ and $M \to_\beta M'$. There is a derivation $\Pi' \triangleright \Gamma \vdash M' : \sigma$, with $\mathtt{rk}(\Pi) \geq \mathtt{rk}(\Pi')$, such that if $r \geq \mathtt{rk}(\Pi')$ then $\mathtt{W}(\Pi', r) < \mathtt{W}(\Pi, r)$.*

Proof. It suffices to verify how the weights are modified in the proof of Theorem 9, using Lemma 11. We will use exactly the same notations as in the theorem. Note that if a derivation $\tilde{\Phi}$ is composed by a derivation $\tilde{\Phi}_1$, followed either by the sequence δ or δ', then either $\mathtt{W}(\tilde{\Phi}, r) = \mathtt{W}(\tilde{\Phi}_1, r) + c$ or $\mathtt{W}(\tilde{\Phi}, r) = \mathtt{W}(\tilde{\Phi}_1, r)$, where c is a constant depending only on δ, since δ does not contain rule (sp). Then looking at the definition of weight and to the theorem we can state the following (in)equalities:

$$\mathtt{W}(\Pi', r) = \mathtt{W}(\Theta_3, r) + \mathtt{W}(\Theta_2, r) + c - 1 \leq \mathtt{W}(\Phi, r) + \mathtt{W}(\Theta_4, r) + \mathtt{W}(\Theta_2, r) + c - 1$$
$$= \mathtt{W}(\Phi, r) + \mathtt{W}(\Theta_1, r) + \mathtt{W}(\Theta_2, r) + c - 1 = \mathtt{W}(\Phi, r) + \mathtt{W}(\Sigma, r) - 1$$
$$< \mathtt{W}(\Psi, r) + \mathtt{W}(\Sigma, r) - 1 = \mathtt{W}(\Pi, r).$$ □

Finally the desired results can be obtained.

Theorem 14 (Strong Polystep Soundness). *Let $\Pi \triangleright \Gamma \vdash M : \sigma$, and M β-reduces to M' in m steps. Then:*

$$(1) \quad m \leq |M|^{\mathtt{d}(\Pi)+1} \qquad (2) \quad |M'| \leq |M|^{\mathtt{d}(\Pi)+1}$$

Proof. (1) By repeatedly using Lemma 13 and by Lemma 11.3, since $|M| \geq \mathtt{rk}(\Pi)$. (2) By repeatedly using Lemma 13 and by Lemma 11.2. □

Theorem 15 (Polytime Soundness). *Let $\Pi \triangleright \Gamma \vdash M : \sigma$, then M can be evaluated to normal form on a Turing Machine in time $O(|M|^{3(\mathtt{d}(\Pi)+1)})$.*

Proof. Clearly, as pointed in [10], a β reduction step $N \to_\beta N'$ can be simulated in time $O(|N|^2)$ on a Turing Machine. Let $M \equiv M_0 \to_\beta M_1 \to_\beta \cdots \to_\beta M_n$ be a reduction of M to normal form M_n. By Theorem 14.2 $|M_i| \leq |M|^{\mathtt{d}(\Pi)+1}$ for $0 \leq i \leq n$, hence each step in the reduction takes time $O(|M|^{2(\mathtt{d}(\Pi)+1)})$. Furthermore since by Theorem 14.1 n is $O(|M|^{\mathtt{d}(\Pi)+1})$, the conclusion follows. □

Clearly Theorem 15 implies that a strong polytime soundness holds, considering Turing Machine with an oracle for strategies.

4.2 Polynomial Time Completeness

In order to prove polynomial time completeness of STA we need to encode Turing Machines (TM) configurations, transitions between configurations and iterators. We achieve such results considering the usual notion of lambda definability, given in [11], generalized to different kinds of data. We encode input data in the usual way, TM configurations and transitions following the lines of [7] and iterators as usual by Church numerals. We stress that we allow a liberal typing, the only constraint we impose in order to respect Theorem 14 is that each input data must be typable through derivations with degree 0.

Some syntactic sugar. To keep notation concise we add some syntactic sugar. Let $M \circ N$ stand for $\lambda z.M(Nz)$, $M_1 \circ M_2 \circ \cdots \circ M_n$ stand for $\lambda z.M_1(M_2(\cdots(M_n z)))$. As usual I stands for $\lambda x.x$, and $\vdash I : \forall\alpha.\alpha \multimap \alpha$.

Tensor product is definable by second order as $\sigma \otimes \tau \doteq \forall\alpha.(\sigma \multimap \tau \multimap \alpha) \multimap \alpha$. The constructors and destructors for this data type are $\langle M, N\rangle \doteq \lambda x.xMN$, $\texttt{let}\ z\ \texttt{be}\ x, y\ \texttt{in}\ N \doteq z(\lambda x.\lambda y.N)$. n-ary tensor product can be easily defined through the binary one and we use σ^n to denote $\sigma \otimes \cdots \otimes \sigma$ n-times.

Note that, since STA is an affine system, tensor product enjoys some properties of the additive conjunction, as to allow the projectors.

Polynomials To encode natural numbers we will use Church numerals, i.e. $\underline{n} \doteq \lambda s.\lambda z.s^n(z)$. Successor, addition and multiplication are λ-definable in the usual way: $\underline{succ} \doteq \lambda psz.s(psz)$, $\underline{add} \doteq \lambda pqsz.ps(qsz)$ and $\underline{mul} \doteq \lambda pqs.p(qs)$. Note that the above terms are not typable by using the usual type for natural numbers $\mathbf{N} \doteq \forall\alpha.!(\alpha \multimap \alpha) \multimap \alpha \multimap \alpha$. For this reason we define *indexed types*, one for each $i \in \mathbb{N}$: $\mathbf{N}_i \doteq \forall\alpha.!^i(\alpha \multimap \alpha) \multimap \alpha \multimap \alpha$ in particular we have $\mathbf{N}_1 \equiv \mathbf{N}$. Clearly for each Church numerals $\underline{n}$ and for each $i > 0 \in \mathbb{N}$ it holds $\vdash \underline{n} : \mathbf{N}_i$. Using indexed type it is easy to derive the following typing: $\vdash \underline{succ} : \mathbf{N}_i \multimap \mathbf{N}_{i+1}$, $\vdash \underline{add} : \mathbf{N}_i \multimap \mathbf{N}_j \multimap \mathbf{N}_{max(i,j)+1}$, $\vdash \underline{mul} : \mathbf{N}_j \multimap !^j\mathbf{N}_i \multimap \mathbf{N}_{i+j}$.

Note that Church numerals behave as iterators only on terms typable with type $\mu \multimap \mu$ for some linear type μ. For this reason the above terms cannot be iterated. Nevertheless they can be composed. In fact, if we want to multiply two natural numbers we can use $\underline{mul}$ typed as $\vdash \underline{mul} : \mathbf{N} \multimap !\mathbf{N} \multimap \mathbf{N}_2$ while if we want to multiply three natural numbers, through the term $\lambda xyz.\underline{mul}(\underline{mul}\,x\,y)z$, the two occurrences of $\underline{mul}$ in it need to be typed by different types, for example $\mathbf{N} \multimap !\mathbf{N} \multimap \mathbf{N}_2$ the innermost one and $\mathbf{N}_2 \multimap !!\mathbf{N} \multimap \mathbf{N}_3$ the outermost. Such change of type, in particular on the number of modalities, does not depend on the values of data.

Lemma 16. *Let P be a polynomial in the variable X and $deg(P)$ its degree. Then there is a term $\underline{P}$ λ-defining P typable as $x :!^{deg(P)}\mathbf{N} \vdash \underline{P} : \mathbf{N}_{2deg(P)+1}$.*

Proof. Consider P in Horner normal form, i.e., $P = a_0 + X(a_1 + X(\cdots(a_{n-1} + Xa_n)\cdots)$. By induction on $deg(P)$ we show something stronger, i.e., for $i > 0$, it is derivable $x_0 : \mathbf{N}_i, x_1 :!^i\mathbf{N}_i, \ldots, x_n :!^{i(deg(P^*)-1)}\mathbf{N}_i \vdash \underline{P^*} : \mathbf{N}_{i(deg(P^*))+deg(P^*)+1}$ where $P^* = a_0 + X_0(a_1 + X_1(\cdots(a_{n-1} + X_n a_n)\cdots)$ so conclusion follows using (m) rule and taking $i = 1$.

Base case is trivial, so consider $P^* = a_0 + X_0(P')$. By induction hypothesis $x_1 : \mathbf{N}_i, \ldots, x_n :!^{i(deg(P')-1)}\mathbf{N}_i \vdash \underline{P'} : \mathbf{N}_{i(deg(P'))+deg(P')+1}$.
Take $\underline{P^*} \equiv \underline{add}(\underline{a_0}, \underline{mul}(x_0, \underline{P'}))$, clearly we have $x_0 : \mathbf{N}_i, x_1 :!^i\mathbf{N}_i, \ldots, x_n : !^{i(deg(P')-1)+i}\mathbf{N}_i \vdash \underline{P^*} : \mathbf{N}_{i(deg(P')+1)+deg(P')+1+1}$. Since $deg(P^*) = deg(P') + 1$: it follows $x_0 : \mathbf{N}_i, x_1 :!^i\mathbf{N}_i, \ldots, x_n :!^{i(deg(P^*)-1)}\mathbf{N}_i \vdash \underline{P^*} : \mathbf{N}_{i(deg(P^*))+deg(P^*)+1}$.
Now by taking $i = 1$ and repeatedly applying (m) rule we conclude $x :!^{deg(P)}\mathbf{N} \vdash \underline{P} \equiv \underline{P^*}[x/x_1, \cdots, x/x_n] : \mathbf{N}_{2deg(P)+1}$ □

Booleans. Let us encode booleans, as usual, by: $\mathbf{0} \doteq \lambda xy.x$, $\mathbf{1} \doteq \lambda xy.y$ and $\mathtt{if}\ x\ \mathtt{then}\ M\ \mathtt{else}\ N \doteq xMN$. They can be typed in STA using as type for booleans: $\mathbf{B} \doteq \forall\alpha.\alpha \multimap \alpha \multimap \alpha$. In particular it holds $\vdash b : \mathbf{B}$ for $b \in \{\mathbf{0}, \mathbf{1}\}$. Moreover assuming $\Gamma \vdash M : \sigma$ and $\Delta \vdash N : \sigma$ and $\Gamma\#\Delta$ it follows $\Gamma, \Delta, x : \mathbf{B} \vdash \mathtt{if}\ x\ \mathtt{then}\ M\ \mathtt{else}\ N : \sigma$.

With the help of the conditional we can define all the usual boolean functions $\vdash \underline{And} : \mathbf{B} \multimap \mathbf{B} \multimap \mathbf{B}$, $\vdash \underline{Or} : \mathbf{B} \multimap \mathbf{B} \multimap \mathbf{B}$, $\vdash \underline{Not} : \mathbf{B} \multimap \mathbf{B}$. In particular we have contraction on booleans: $\vdash \underline{Cnt} \doteq \lambda b.\ \mathtt{if}\ b\ \mathtt{then}\ \langle \mathbf{0}, \mathbf{0}\rangle\ \mathtt{else}\ \langle \mathbf{1}, \mathbf{1}\rangle : \mathbf{B} \multimap \mathbf{B} \otimes \mathbf{B}$. The above functions and weakening are useful to prove the following.

Lemma 17. *Each boolean total function $f : \mathbb{B}^n \to \mathbb{B}^m$, where $n, m \geq 1$, can be λ-defined by a term $\underline{f}$ typable in STA as $\vdash \underline{f} : \mathbf{B}^n \multimap \mathbf{B}^m$.*

Strings String of booleans can be encoded by: $[\,] \doteq \lambda cz.z$ and $[b_0, b_1, \ldots, b_n] \doteq \lambda cz.cb_0(\cdots(cb_nz)\cdots)$ where $b_i \in \{\mathbf{0}, \mathbf{1}\}$. Boolean strings are typable in STA with the indexed type $\mathbf{S}_i \doteq \forall\alpha.!^i(\mathbf{B} \multimap \alpha \multimap \alpha) \multimap (\alpha \multimap \alpha)$. In particular for each $n, i > 0 \in \mathbb{N}$ it holds $b_0 : \mathbf{B}, ..., b_n : \mathbf{B} \vdash [b_0, \ldots, b_n] : \mathbf{S}_i$. The term $\mathbf{len} \doteq \lambda cs.c(\lambda xy.sy)$ λ-defines the function returning the length of an input string and is typable in STA with typing $\vdash \mathbf{len} : \mathbf{S}_i \multimap \mathbf{N}_i$.

Turing Machine. We can λ-define Turing Machine configurations by terms of the shape $\lambda c.\langle cb^l_0 \circ \cdots \circ cb^l_n, cb^r_0 \circ \cdots \circ cb^r_m, Q\rangle$ where $cb^l_0 \circ \cdots \circ cb^l_n$ and $cb^r_0 \circ \cdots \circ cb^r_m$ represent respectively the left and the right part of the tape, while $Q \equiv \langle b_1, \cdots, b_n\rangle$ is a n-ary tensor product of boolean values representing the current state. We assume without loss of generality that by convention the left part of the tape is represented in a reversed order, that the alphabet is composed of only the two symbols $\mathbf{0}$ and $\mathbf{1}$, that the scanned symbol is the first symbol of the right part and that states are divided in accepting and rejecting. The indexed type of a Turing Machine configuration in STA is $\mathbf{TM}_i \doteq \forall\alpha.!^i(\mathbf{B} \multimap \alpha \multimap \alpha) \multimap ((\alpha \multimap \alpha)^2 \otimes \mathbf{B}^q)$. In fact, $\vdash \lambda c.\langle cb^l_0 \circ \cdots \circ cb^l_n, cb^r_0 \circ \cdots \circ cb^r_m, Q\rangle : \mathbf{TM}_i$. The term $\mathbf{Init} \doteq \lambda tc.\langle \lambda z.z, \lambda z.t(c\mathbf{0})z, \underline{q_0}\rangle$ λ-defines the function that, taking as input $\underline{Q}$, defining a polynomial Q, gives as output a Turing Machine with tape of length Q filled by $\mathbf{0}$'s in the initial state q_0 and with the head at the beginning of the tape. As expected $\vdash \mathbf{Init} : \mathbf{N}_i \multimap \mathbf{TM}_i$. In what follows ID_i will denote:

$$\forall\alpha.!^i(\mathbf{B} \multimap \alpha \multimap \alpha) \multimap ((\alpha \multimap \alpha)^2 \otimes (\mathbf{B} \multimap \alpha \multimap \alpha) \otimes \mathbf{B} \otimes (\mathbf{B} \multimap \alpha \multimap \alpha) \otimes \mathbf{B} \otimes \mathbf{B}^q)$$

Following [7], the work of showing that Turing Machine transitions are λ-definable can be decomposed in two steps. Firstly we have a term

$$\mathbf{Dec} \doteq \lambda sc.\ \mathtt{let}\ s(F(c))\ \mathtt{be}\ l, r, q\ \mathtt{in}\ \mathtt{let}\ l\langle I, \lambda x.I, \mathbf{0}\rangle$$
$$\mathtt{be}\ t_l, c_l, b^l_0\ \mathtt{in}\ \mathtt{let}\ r\langle I, \lambda x.I, \mathbf{0}\rangle\ \mathtt{be}\ t_r, c_r, b^r_0\ \mathtt{in}\ \langle t_l, t_r, c_l, b^l_0, c_r, b^r_0, q\rangle$$

where $F(c) \doteq \lambda bz.\ \texttt{let}\ z\ \texttt{be}\ g,h,i\ \texttt{in}\ \langle hi \circ g, c, b\rangle$. Such a term is typable as $\vdash \mathbf{Dec} : \mathbf{TM}_i \multimap ID_i$ and its behaviour is to decompose a configuration as:

$$\mathbf{Dec}(\lambda c.\langle cb_0^l \circ \cdots \circ cb_n^l, cb_0^r \circ \cdots \circ cb_m^r, Q\rangle) \to_\beta^* \\ \lambda c.\langle cb_1^l \circ \cdots \circ cb_n^l, cb_1^r \circ \cdots \circ cb_m^r, c, b_0^l, c, b_0^r, Q\rangle$$

Then we can define a term

$$\mathbf{Com} \doteq \lambda sc.\ \texttt{let}\ sc\ \texttt{be}\ l,r,c_l,b_l,c_r,b_r,q\ \texttt{in} \\ \texttt{let}\ \underline{\delta}\langle b_r,q\rangle\ \texttt{be}\ b',q',m\ \texttt{in}\ (\texttt{if}\ m\ \texttt{then}\ \mathbf{R}\ \texttt{else}\ \mathbf{L})b'q'\langle l,r,c_l,b_l,c_r\rangle$$

where $\mathbf{R} \doteq \lambda b'q's.\ \texttt{let}\ s\ \texttt{be}\ l,r,c_l,b_l,c_r\ \texttt{in}\ \langle c_rb' \circ c_lb_l \circ l, r, q'\rangle$ and $\mathbf{L} \doteq \lambda b'q's.\ \texttt{let}\ s\ \texttt{be}\ l,r,c_l,b_l,c_r\ \texttt{in}\ \langle l, c_lb_l \circ c_rb' \circ r, q'\rangle$. Such a term is typable as $\vdash \mathbf{Com} : ID_i \multimap \mathbf{TM}_i$ and depending on the δ transition function, it combines the symbols returning a configuration as:

$$\mathbf{Com}\ (\lambda c.\langle cb_1^l \circ \cdots \circ cb_n^l, cb_1^r \circ \cdots \circ cb_m^r, c, b_0^l, c, b_0^r, Q\rangle) \\ \to_\beta^* \lambda c.\langle cb' \circ cb_0^l \circ cb_1^l \circ \cdots \circ cb_n^l, cb_1^r \circ \cdots \circ cb_m^r, Q'\rangle \ \text{ if } \delta(b_0^r, Q) = (b', Q', \text{Right}) \\ \text{or} \\ \to_\beta^* \lambda c.\langle cb_1^l \circ \cdots \circ cb_n^l, cb_0^l \circ cb' \circ cb_1^r \circ \cdots \circ cb_m^r, Q'\rangle \ \text{ if } \delta(b_0^r, Q) = (b', Q', \text{Left})$$

Checking the typing and the behavior for **Dec** and **Com** is boring but easy.

By combining the above terms we obtain the term $\mathbf{Tr} \doteq \mathbf{Com} \circ \mathbf{Dec}$ which λ-defines a Turing Machine transition and as expected admits the typing $\vdash \mathbf{Tr} : \mathbf{TM}_i \multimap \mathbf{TM}_i$.

Let the open term $\mathbf{T}(b)$ be $\lambda sc.\ \texttt{let}\ sc\ \texttt{be}\ l,r,c_l,b_l,c_r,b_r,q\ \texttt{in}\ \mathbf{R}bq\langle l,r,c_l,b_l,c_r\rangle$ where $\mathbf{R}$ is defined as above. Such term is typable as $b : \mathbf{B} \vdash \mathbf{T}(b) : ID_i \multimap \mathbf{TM}_i$, and it is useful to define the term $\mathbf{In} \doteq \lambda sm.s(\lambda b.\mathbf{T}(b) \circ \mathbf{Dec})m$ that, when supplied by a boolean string and a Turing Machine, writes the input string on the tape of the Turing Machine. Such a term is typable as $\vdash \mathbf{In} : \mathbf{S} \multimap \mathbf{TM}_i \multimap \mathbf{TM}_i$.

The term $\mathbf{Ext} \doteq \lambda s.\ \texttt{let}\ s(\lambda b.\lambda c.c)\ \texttt{be}\ l,r,q\ \texttt{in}\ \underline{f}(q)$ permits to extract the information that the machine is in an accepting or non-accepting state. Since Lemma 17 assures the existence of $\underline{f}$, it holds $\vdash \mathbf{Ext} : \mathbf{TM}_i \multimap \mathbf{B}$. Now we can finally show that STA is complete for PTIME.

Theorem 18 (PTIME Completeness). *Let a decision problem $\mathfrak{P}$ be decided in* polynomial time *P, where $deg(P) = m$, and in* polynomial space *Q, where $deg(Q) = l$, by a Turing Machine $\mathcal{M}$. Then it is λ-definable by a term $\underline{M}$ typable in STA as $s :!^{max(l,m,1)+1}\mathbf{S} \vdash \underline{M} : \mathbf{B}$.*

Proof. By Lemma 16: $s_p :!^m\mathbf{S} \vdash P[\mathbf{lens}_p/x] : \mathbf{N}_{2m+1}$ and $s_q :!^l\mathbf{S} \vdash Q[\mathbf{lens}_q/x] : \mathbf{N}_{2l+1}$. Furthermore by composition: $s : \mathbf{S}, q : \mathbf{N}_{2l+1}, p : \mathbf{N}_{2m+1} \vdash \mathbf{Ext}(p\mathbf{Tr}(\mathbf{In}s(\mathbf{Init}(q)))) : \mathbf{B}$ so by (cut) and some applications of (m) rule the conclusion follows. □

Without loss of generality we can assume that a Turing machine stops on accepting states with the head at the begin of the tape. Hence the term $\mathbf{Ext_F} \doteq \lambda sc.\ \texttt{let}\ sc\ \texttt{be}\ l,r,q\ \texttt{in}\ r$ extracts the result from the tape. It is easy to verify that the typing $\vdash \mathbf{Ext_F} : \mathbf{TM}_i \multimap \mathbf{S}_i$ holds. So we can conclude that STA is also complete for FPTIME.

Theorem 19 (FPTIME Completeness). *Let a function $\mathcal{F}$ be computed in* polynomial time *P, where $deg(P) = m$, and in* polynomial space *Q, where $deg(Q) = l$, by a Turing Machine $\mathcal{M}$. Then it is λ-definable by a term $\underline{M}$ typable in STA as $!^{max(l,m,1)+1}\mathbf{S} \vdash \underline{M} : \mathbf{S}_{2m+1}$.*

In the work of Lafont [2], in the proof of PTIME completeness of SLL, an important role is played by the notions of generic and homogeneous proofs, being respectively proofs without multiplexors (rule (m)) and with multiplexors of a fixed rank. Lafont uses homogeneous proofs for representing data and generic proofs for representing programs. We do not use this classification for proving the completeness of STA.

References

1. Girard, J.Y.: Light Linear Logic. Information and Computation 143(2), 175–204 (1998)
2. Lafont, Y.: Soft linear logic and polynomial time. Theoretical Computer Science 318(1-2), 163–180 (2004)
3. Baillot, P., Mogbil, V.: Soft lambda-calculus: a language for polynomial time computation. In: Walukiewicz, I. (ed.) FOSSACS 2004. LNCS, vol. 2987, pp. 27–41. Springer, Heidelberg (2004)
4. Baillot, P., Terui, K.: Light types for polynomial time computation in lambda-calculus. In: Proceedings of LICS 2004, pp. 266–275. IEEE Computer Society Press, Los Alamitos (2004)
5. Asperti, A.: Light Affine Logic. In: Proceedings of LICS 1998, pp. 300–308. IEEE Computer Society Press, Los Alamitos (1998)
6. Asperti, A., Roversi, L.: Intuitionistic Light Affine Logic. ACM Transactions on Computational Logic 3(1), 137–175 (2002)
7. Mairson, H.G., Terui, K.: On the Computational Complexity of Cut-Elimination in Linear Logic. In: Blundo, C., Laneve, C. (eds.) ICTCS 2003. LNCS, vol. 2841, pp. 23–36. Springer, Heidelberg (2003)
8. Ronchi Della Rocca, S., Roversi, L.: Lambda calculus and Intuitionistic Linear Logic. Studia Logica 59(3) (1997)
9. Coppola, P., Dal Lago, U., Ronchi Della Rocca, S.: Elementary Affine Logic and the Call by Value Lambda Calculus. In: Urzyczyn, P. (ed.) TLCA 2005. LNCS, vol. 3461, pp. 131–145. Springer, Heidelberg (2005)
10. Terui, K.: Light Affine Lambda Calculus and Polytime Strong Normalization. In: Proceedings of LICS 2001, pp. 209–220. IEEE Computer Society Press, Los Alamitos (2001)
11. Barendregt, H.P.: The Lambda Calculus: Its Syntax and Semantics. Elsevier/North-Holland, Amsterdam, London, New York (1984)

Lambda Theories of Effective Lambda Models

Chantal Berline[1], Giulio Manzonetto[1,2], and Antonino Salibra[2]

[1] Laboratoire PPS, CNRS-Université Paris 7,
2, place Jussieu (case 7014), 75251 Paris Cedex 05, France
[2] Università Ca'Foscari di Venezia,
Dipartimento di Informatica Via Torino 155, 30172 Venezia, Italy
chantal.berline@pps.jussieu.fr, {gmanzone, salibra}@dsi.unive.it

Abstract. A longstanding open problem is whether there exists a non-syntactical model of the untyped λ-calculus whose theory is exactly the least λ-theory $\lambda\beta$. In this paper we investigate the more general question of whether the equational/order theory of a model of the untyped λ-calculus can be recursively enumerable (r.e. for brevity). We introduce a notion of *effective model* of λ-calculus, which covers in particular all the models individually introduced in the literature. We prove that the order theory of an effective model is never r.e.; from this it follows that its equational theory cannot be $\lambda\beta$, $\lambda\beta\eta$. We then show that no effective model living in the stable or strongly stable semantics has an r.e. equational theory. Concerning Scott's semantics, we investigate the class of graph models and prove that no order theory of a graph model can be r.e., and that there exists an effective graph model whose equational/order theory is the minimum one. Finally, we show that the class of graph models enjoys a kind of downwards Löwenheim-Skolem theorem.

Keywords: Lambda calculus, Effective lambda models, Recursively enumerable lambda theories, Graph models, Löwenheim-Skolem theorem.

1 Introduction

Lambda theories are equational extensions of the untyped λ-calculus closed under derivation. They arise by syntactical or semantic considerations. Indeed, a λ-theory may correspond to a possible operational (observational) semantics of λ-calculus, as well as it may be induced by a model of λ-calculus through the kernel congruence relation of the interpretation function. Although researchers have mainly focused their interest on a limited number of them, the class of λ-theories constitutes a very rich and complex structure (see [1,4,5]).

Topology is at the center of the known approaches to giving models of the untyped λ-calculus. After the first model, found by Scott in 1969 in the category of complete lattices and Scott continuous functions, a large number of mathematical models for λ-calculus, arising from syntax-free constructions, have been introduced in various categories of domains and were classified into semantics

J. Duparc and T.A. Henzinger (Eds.): CSL 2007, LNCS 4646, pp. 268–282, 2007.

according to the nature of their representable functions, see e.g. [1,4,19]. Scott continuous semantics [22] is given in the category whose objects are complete partial orders and morphisms are Scott continuous functions. The stable semantics (Berry [7]) and the strongly stable semantics (Bucciarelli-Ehrhard [8]) are refinements of the continuous semantics, introduced to capture the notion of "sequential" Scott continuous function. In each of these semantics it is possible to build up $2^{\aleph_0}$ models inducing pairwise distinct λ-theories [16,17]. Nevertheless, all are equationally *incomplete* (see [15,2,20,21]) in the sense that they do not represent all possible consistent λ-theories. It is interesting to note that there are very few known equational theories of λ-models living in these semantics that can be described syntactically: namely, the theory of Böhm trees and variants of it. None of these theories is r.e.

Berline has raised in [4] the natural question of whether, given a class of models of λ-calculus, there is a minimum λ-theory represented by it. This question relates to the longstanding open problem proposed by Barendregt about the existence of a continuous model or, more generally, of a non-syntactical model of $\lambda\beta$ ($\lambda\beta\eta$). Di Gianantonio, Honsell and Plotkin [12] have shown that Scott continuous semantics admits a minimum theory, at least if we restrict to extensional models. Another result of [12], in the same spirit, is the construction of an extensional model whose theory is $\lambda\beta\eta$, a fortiori minimal, in a weakly-continuous semantics. However, the construction of this model starts from the term model of $\lambda\beta\eta$, and hence it cannot be seen as having a purely non syntactical presentation. More recently, Bucciarelli and Salibra [9,10] have shown that the class of graph models admits a minimum λ-theory different from $\lambda\beta$. Graph models, isolated in the seventies by Plotkin, Scott and Engeler (see e.g. [1]) within the continuous semantics, have proved useful for showing the consistency of extensions of λ-calculus and for studying operational features of λ-calculus (see [4]).

In this paper we investigate the related question of whether the equational theory of a model can be recursively enumerable (r.e. for brevity). As far as we know, this problem was first raised in [5], where it is conjectured that no graph model can have an r.e. theory. But we expect that this could indeed be true for all models living in the continuous semantics, and its refinements.

We find it natural to concentrate on models with built-in effectivity properties. It seems indeed reasonable to think that, if effective models do not even succeed to have an r.e. theory, then the other ones have no chance to succeed. Another justification for considering effective models comes from a previous result obtained for typed λ-calculus. Indeed, it was proved in [3] that there exists a non-syntactical model of Girard's system F whose theory is $\lambda\beta\eta$. This model lives in Scott's continuous semantics, and can easily be checked to be "effective" in the same spirit as in the present paper (see [3, Appendix C] for a sketchy presentation of the model).

Starting from the known notion of an effective domain, we introduce a general notion of an *effective model* of λ-calculus and we study the main properties of these models. Effective models are omni-present in the continuous, stable and strongly stable semantics. In particular, all the models which have been introduced individually in the literature can easily be proved effective[1].

The following are the main results of the paper:

1. Let $\mathcal{D}$ be an effective model of λ-calculus. Then:
   (i) The order theory $\mathrm{Ord}(\mathcal{D})$ of $\mathcal{D}$ is not r.e.
   (ii) The equational theory $\mathrm{Eq}(\mathcal{D})$ of $\mathcal{D}$ is not the theory $\lambda\beta$ ($\lambda\beta\eta$).
   (iii) If for some λ-term M there are only finitely many λ-definable elements below the interpretation of M (e.g. if $\bot \in \mathcal{D}$ is λ-definable), then $\mathrm{Eq}(\mathcal{D})$ is not r.e.

Concerning the existence of a non-syntactical effective model with an r.e. equational theory, we are able to give a definite negative answer for all (effective) stable and strongly stable models:

2. No effective model living in the stable or strongly stable semantics has an r.e. equational theory.

Concerning Scott continuous semantics, the problem looks much more difficult. We concentrate here on the class of graph models (see [5,6,9,10,11] for earlier investigation of this class) and show the following results:

3. Let $\mathcal{D}$ be an arbitrary graph model. Then:
   (i) The order theory $\mathrm{Ord}(\mathcal{D})$ of $\mathcal{D}$ is not r.e.
   (ii) If $\mathcal{D}$ is freely generated by a finite "partial model", then the equational theory $\mathrm{Eq}(\mathcal{D})$ of $\mathcal{D}$ is not r.e.
4. There exists an effective graph model whose equational/order theory is minimal among all theories of graph models.
5. (Löwenheim-Skolem theorem for graph models) Every equational/order graph theory (where "graph theory" means "theory of a graph model") is the theory of a graph model having a carrier set of minimal cardinality.

The last result positively answers Question 3 in [4, Section 6.3] for the class of graph models.

The central technical device used in this paper is Visser's result [25] stating that the complements of β-closed r.e. sets of λ-terms enjoy the finite intersection property (see Theorem 2).

2 Preliminaries

To keep this article self-contained, we summarize some definitions and results concerning λ-calculus that we need in the subsequent part of the paper. With regard to the lambda calculus we follow the notation and terminology of [1].

---

[1] As far as we know, only Giannini and Longo [13] have introduced a notion of an effective model; but their definition is *ad hoc* for two particular models (Scott's P_ω and Plotkin's T_ω) and their results depend on the fact that these models have a very special (and well known) common theory.

We denote by $\mathbb{N}$ the set of natural numbers. A set $A \subseteq \mathbb{N}$ is *recursively enumerable* (r.e. for short) if it is the domain of a partial recursive function. The complement of a recursively enumerable set is called a *co-r.e.* set. If both A and its complement are r.e., A is called *decidable*. We will denote by $\mathcal{RE}$ the collection of all r.e. subsets of $\mathbb{N}$.

A numeration of a set A is a map from $\mathbb{N}$ onto A. $\mathcal{W} : \mathbb{N} \to \mathcal{RE}$ denotes the usual numeration of r.e. sets (i.e., $\mathcal{W}_n$ is the domain of the n-th computable function ϕ_n).

2.1 Lambda Calculus and Lambda Models

Λ and Λ^o are, respectively, the set of λ-terms and of closed λ-terms. Concerning specific λ-terms we set:

$$\mathbf{I} \equiv \lambda x.x; \quad \mathbf{T} \equiv \lambda xy.x; \quad \mathbf{F} \equiv \lambda xy.y; \quad \Omega \equiv (\lambda x.xx)(\lambda x.xx).$$

A set X of λ-terms is *trivial* if either $X = \emptyset$ or $X = \Lambda$.

We denote $\alpha\beta$-conversion by $\lambda\beta$. A λ*-theory* $\mathcal{T}$ is a congruence on Λ (with respect to the operators of abstraction and application) which contains $\lambda\beta$. We write $M =_\mathcal{T} N$ for $(M, N) \in \mathcal{T}$. If $\mathcal{T}$ is a λ-theory, then $[M]_\mathcal{T}$ denotes the set $\{N : N =_\mathcal{T} M\}$. A λ-theory $\mathcal{T}$ is: *consistent* if $\mathcal{T} \neq \Lambda \times \Lambda$; *extensional* if it contains the equation $\mathbf{I} = \lambda xy.xy$; *recursively enumerable* if the set of Gödel numbers of all pairs of $\mathcal{T}$-equivalent λ-terms is r.e. Finally, $\lambda\beta\eta$ is the least extensional λ-theory.

Solvable λ-terms can be characterized as follows: a λ-term M is solvable if, and only if, it has a *head normal form*, that is, $M =_{\lambda\beta} \lambda x_1 \dots x_n.yM_1 \dots M_k$ for some $n, k \geq 0$ and λ-terms $M_1, \dots, M_k$. $M \in \Lambda$ is *unsolvable* if it is not solvable.

The λ-theory $\mathcal{H}$, generated by equating all the unsolvable λ-terms, is consistent by [1, Theorem 16.1.3]. A λ-theory $\mathcal{T}$ is *sensible* if $\mathcal{H} \subseteq \mathcal{T}$, while it is *semi-sensible* if it contains no equations of the form $U = S$ where S is solvable and U unsolvable. Consistent sensible theories are semi-sensible (see [1, Cor. 4.1.9]) and are never r.e. (see [1, Section 17.1]).

It is well known [1, Chapter 5] that a model of λ-calculus (λ*-model*, for short) can be defined as a reflexive object in a ccc (Cartesian closed category) $\mathbf{C}$, that is to say a triple $(D, \mathcal{F}, \lambda)$ such that D is an object of $\mathbf{C}$ and $\mathcal{F} : D \to [D \to D]$, $\lambda : [D \to D] \to D$ are morphisms such that $\mathcal{F} \circ \lambda = id_{[D \to D]}$. In the following we will mainly be interested in Scott's ccc of cpos and Scott continuous functions (*continuous semantics*), but we will also draw conclusions for Berry's ccc of *DI*–domains and stable functions (*stable semantics*), and for Ehrhard's ccc of *DI*-domains with coherence and strongly stable functions between them (*strongly stable semantics*). We recall that *DI*-domains are special Scott domains, and that Scott domains are special cpos (see, e.g., [24]).

Let D be a cpo. The partial order of D will be denoted by $\sqsubseteq_D$. We let Env_D be the set of environments ρ mapping the set Var of variables of λ-calculus into D. For every $x \in Var$ and $d \in D$ we denote by $\rho[x := d]$ the environment ρ' which coincides with ρ, except on x, where ρ' takes the value d. A reflexive cpo

D generates a λ-model $\mathcal{D} = (D, \mathcal{F}, \lambda)$ with the interpretation of a λ-term defined as follows:

$$x^{\mathcal{D}}_{\rho} = \rho(x);\ (MN)^{\mathcal{D}}_{\rho} = \mathcal{F}(M^{\mathcal{D}}_{\rho})(N^{\mathcal{D}}_{\rho});\ (\lambda x.M)^{\mathcal{D}}_{\rho} = \lambda(f),$$

where f is defined by $f(d) = M^{\mathcal{D}}_{\rho[x:=d]}$ for all $d \in D$. We write $M^{\mathcal{D}}$ for $M^{\mathcal{D}}_{\rho}$ if M is a closed λ-term. In the following $\mathcal{F}(d)(e)$ will also be written $d \cdot e$ or de.

Each λ-model $\mathcal{D}$ induces a λ-theory, denoted here by $\mathrm{Eq}(\mathcal{D})$, and called *the equational theory of* $\mathcal{D}$. Thus, $M = N \in \mathrm{Eq}(\mathcal{D})$ if, and only if, M and N have the same interpretation in $\mathcal{D}$. A reflexive cpo $\mathcal{D}$ induces also an *order theory* $\mathrm{Ord}(\mathcal{D}) = \{M \sqsubseteq N : M^{\mathcal{D}}_{\rho} \sqsubseteq_D N^{\mathcal{D}}_{\rho}$ for all environments $\rho\}$.

2.2 Effective Domains

A triple $\mathcal{D} = (D, \sqsubseteq_D, d)$ is called an *effective domain* if $(D, \sqsubseteq_D)$ is a Scott domain and d is a numeration of the set $K(\mathcal{D})$ of its compact elements such that the relations "d_m and d_n have an upper bound" and "$d_n = d_m \sqcup d_k$" are both decidable (see, e.g., [24, Chapter 10]).

We recall that an element v of an effective domain $\mathcal{D}$ is said *r.e.* (*decidable*) if the set $\{n : d_n \sqsubseteq_D v\}$ is r.e. (decidable); we will write $\mathcal{D}^{r.e.}$ ($\mathcal{D}^{dec}$) for the set of r.e. (decidable) elements of $\mathcal{D}$. The set $K(\mathcal{D})$ of compact elements is included within $\mathcal{D}^{dec}$. Using standard techniques of recursion theory it is possible to get in a uniform way a numeration $\xi : \mathbb{N} \to \mathcal{D}^{r.e.}$ which is *adequate* in the sense that the relation $d_k \sqsubseteq_D \xi_n$ is r.e. in (k, n) and the inclusion mapping $\iota : K(\mathcal{D}) \to \mathcal{D}^{r.e.}$ is computable w.r.t. d, ξ.

The full subcategory **ED** of the category of Scott-domains with effective domains as objects and continuous functions as morphisms is a ccc.

A continuous function $f : D \to D'$ is an r.e. element in the effective domain of Scott continuous functions (i.e., $f \in [\mathcal{D} \to \mathcal{D}']^{r.e.}$) if, and only if, its restriction $f\restriction : \mathcal{D}^{r.e.} \to \mathcal{D}'^{r.e.}$ is *computable w.r.t.* ξ, ξ', i.e., there is a computable map $g : \mathbb{N} \to \mathbb{N}$ such that $f(\xi_n) = \xi'_{g(n)}$. In such a case we say that g *tracks* f.

2.3 Graph Models

The class of graph models belongs to Scott continuous semantics (see [5] for a complete survey on this class of models). Historically, the first graph model was Scott's P_ω, which is also known in the literature as "the graph model". "Graph" referred to the fact that the continuous functions were encoded in the model via (a sufficient fragment of) their graph.

As a matter of notation, for every set G, G^* is the set of all finite subsets of G, while $\mathcal{P}(G)$ is the powerset of G.

Definition 1. *A graph model $\mathcal{G}$ is a pair $(G, c_{\mathcal{G}})$, where G is an infinite set, called the* carrier set *of $\mathcal{G}$, and $c_{\mathcal{G}} : G^* \times G \to G$ is an injective total function.*

Such pair $\mathcal{G}$ generates the reflexive cpo $(\mathcal{P}(G), \subseteq, \lambda, \mathcal{F})$, where λ and $\mathcal{F}$ are defined as follows, for all $f \in [\mathcal{P}(G) \to \mathcal{P}(G)]$ and $X, Y \subseteq G$: $\lambda(f) = \{c_{\mathcal{G}}(a, \alpha) :$

$\alpha \in f(a)$ and $a \in G^*\}$ and $\mathcal{F}(X)(Y) = \{\alpha \in G : (\exists a \subseteq Y)\ c_{\mathcal{G}}(a,\alpha) \in X\}$. For more details we refer the reader to Berline [4].

The interpretation of a λ-term M into a λ-model has been defined in Section 2.1. However, in this context we can make explicit the interpretation $M^{\mathcal{G}}_\rho$ of a λ-term M as follows:

$$(MN)^{\mathcal{G}}_\rho = \{\alpha : (\exists a \subseteq N^{\mathcal{G}}_\rho)\ c_{\mathcal{G}}(a,\alpha) \in M^{\mathcal{G}}_\rho\}; (\lambda x.M)^{\mathcal{G}}_\rho = \{c_{\mathcal{G}}(a,\alpha) : \alpha \in M^{\mathcal{G}}_{\rho[x:=a]}\}.$$

We turn now to the interpretation of Ω in graph models (the details of the proof are, for example, worked out in [6, Lemma 4]).

Lemma 1. *$\alpha \in \Omega^{\mathcal{G}}$ if, and only if, there is $a \subseteq (\lambda x.xx)^{\mathcal{G}}$ such that $c_{\mathcal{G}}(a,\alpha) \in a$.*

In the following we use the terminology "*graph theory*" as a shorthand for "theory of a graph model". It is well known that the equational graph theories are never extensional and that there exists a continuum of them (see [16]). In [9,10] the existence of a minimum equational graph theory was proved and it was also shown that this minimum theory is different from $\lambda\beta$.

The completion method for building graph models from "partial pairs" was initiated by Longo in [18] and developed on a wide scale by Kerth in [16,17].

Definition 2. *A* partial pair $\mathcal{A}$ *is given by a set A and by a partial, injective function $c_{\mathcal{A}} : A^* \times A \to A$.*

A partial pair is *finite* if A is finite, and is a graph model if $c_{\mathcal{A}}$ is total.

The interpretation of a λ-term in a partial pair $\mathcal{A}$ is defined in the obvious way: $(MN)^{\mathcal{A}}_\rho = \{\alpha \in A : (\exists a \subseteq N^{\mathcal{A}}_\rho)\ [(a,\alpha) \in dom(c_{\mathcal{A}}) \wedge c_{\mathcal{A}}(a,\alpha) \in M^{\mathcal{A}}_\rho]\}$; $(\lambda x.M)^{\mathcal{A}}_\rho = \{c_{\mathcal{A}}(a,\alpha) \in \mathcal{A} : (a,\alpha) \in dom(c_{\mathcal{A}}) \wedge \alpha \in M^{\mathcal{A}}_{\rho[x:=a]}\}$.

Definition 3. *Let $\mathcal{A}$ be a partial pair. The* completion *of $\mathcal{A}$ is the graph model $\mathcal{E}_{\mathcal{A}} = (E_{\mathcal{A}}, c_{\mathcal{E}_{\mathcal{A}}})$ defined as follows:*

- *$E_{\mathcal{A}} = \bigcup_{n\in\mathbb{N}} E_n$, where $E_0 = A$ and $E_{n+1} = E_n \cup ((E_n^* \times E_n) - dom(c_{\mathcal{A}}))$.*
- *Given $a \in E^*_{\mathcal{A}}$, $\alpha \in E_{\mathcal{A}}$,*

$$c_{\mathcal{E}_{\mathcal{A}}}(a,\alpha) = \begin{cases} c_{\mathcal{A}}(a,\alpha) & \text{if } c_{\mathcal{A}}(a,\alpha) \text{ is defined} \\ (a,\alpha) & \text{otherwise} \end{cases}$$

A notion of *rank* can be naturally defined on the completion $\mathcal{E}_{\mathcal{A}}$ of a partial pair $\mathcal{A}$. The elements of A are the elements of rank 0, while an element $\alpha \in E_{\mathcal{A}} - A$ has rank n if $\alpha \in E_n$ and $\alpha \notin E_{n-1}$.

Let $\mathcal{A}$ and $\mathcal{B}$ be two partial pairs. A *morphism* from $\mathcal{A}$ into $\mathcal{B}$ is a map $f : A \to B$ such that $(a,\alpha) \in dom(c_{\mathcal{A}})$ implies $(fa, f\alpha) \in dom(c_{\mathcal{B}})$ and, in such a case $f(c_{\mathcal{A}}(a,\alpha)) = c_{\mathcal{B}}(fa, f\alpha)$. Isomorphisms and automorphisms can be defined in the obvious way. $Aut(\mathcal{A})$ denotes the group of automorphisms of the partial pair $\mathcal{A}$.

Lemma 2. *Let $\mathcal{G}, \mathcal{G}'$ be graph models and $f : \mathcal{G} \to \mathcal{G}'$ be a morphism. If $M \in \Lambda$ and $\alpha \in M^{\mathcal{G}}_\rho$, then $f\alpha \in M^{\mathcal{G}'}_{f\circ\rho}$.*

2.4 Co-r.e. Sets of Lambda Terms

In this section we recall the main properties of recursion theory concerning λ-calculus that will be applied in the following sections.

An r.e. (co-r.e.) set of λ-terms closed under β-conversion will be called a *β-r.e.* (*β-co-r.e.*) set.

The following theorem is due to Scott (see [1, Thm. 6.6.2]).

Theorem 1. *A set of λ-terms which is both β-r.e. and β-co-r.e. is trivial.*

Definition 4. *A family $X = (X_i : i \in I)$ of sets has the* FIP *(finite intersection property) if $X_{i_1} \cap \cdots \cap X_{i_n} \neq \emptyset$ for all $i_1, \dots, i_n \in I$.*

Visser (see [1, Ch. 17] and [25, Thm. 2.5]) has shown that the topology on Λ generated by the β-co-r.e. sets of λ-terms is hyperconnected (i.e., the intersection of two non-empty open sets is non-empty). In other words:

Theorem 2. *The family of all non-empty β-co-r.e. subsets of Λ has the FIP.*

Corollary 1. *Every non-empty β-co-r.e. set of λ-terms contains a non-empty β-co-r.e. set of unsolvable λ-terms.*

Proof. The set of all unsolvable λ-terms is β-co-r.e. The conclusion follows from Theorem 2.

3 Effective Lambda Models

In this section we introduce the notion of an effective λ-model and we study the main properties of these models. We show that the order theory of an effective λ-model is not r.e. and that its equational theory is different from $\lambda\beta, \lambda\beta\eta$. Effective λ-models are omni-present in the continuous, stable and strongly stable semantics (see Section 4). In particular, all the λ-models which have been introduced individually in the literature, to begin with Scott's $\mathcal{D}_\infty$, can easily be proved effective.

The following natural definition is enough to force the interpretation function of λ-terms to be computable from Λ^o into $\mathcal{D}^{r.e.}$. However, other results of this paper will need a more powerful notion. That is the reason why we only speak of "weak effectivity" here.

Definition 5. *A λ-model is called* weakly effective *if it is a reflexive object $(\mathcal{D}, \mathcal{F}, \lambda)$ in the category* **ED** *and, $\mathcal{F} \in [\mathcal{D} \to [\mathcal{D} \to \mathcal{D}]]$ and $\lambda \in [[\mathcal{D} \to \mathcal{D}] \to \mathcal{D}]$ are r.e. elements.*

In the following a weakly effective λ-model $(\mathcal{D}, \mathcal{F}, \lambda)$ will be denoted by $\mathcal{D}$.

We fix bijective effective numerations $\nu_\Lambda : \mathbb{N} \to \Lambda$ of the set of λ-terms and $\nu_{var} : \mathbb{N} \to Var$ of the set of variables of λ-calculus. In particular this gives to the set Env_D of all environments a structure of effective domain. $\Lambda_\perp = \Lambda \cup \{\perp\}$ is the usual flat domain of λ-terms. The element $\perp$ is always interpreted as $\perp_D$ in a cpo $(D, \sqsubseteq_D)$.

Proposition 1. *Let $\mathcal{D}$ be a weakly effective λ-model. Then the function f mapping $(\rho, M) \mapsto M^{\mathcal{D}}_{\rho}$ is an element of $[Env_{D} \times \Lambda_{\bot} \to \mathcal{D}]^{r.e.}$.*

Proof. **(Sketch)** By structural induction on M it is possible to show the existence of a partial computable map tracking f. The only difficult case is $M \equiv \lambda x.N$. Since λ is r.e. it is sufficient to prove that the function $g : e \mapsto N^{\mathcal{D}}_{\rho[x:=e]}$ is also r.e. Once shown that $h : (\rho, x, e) \mapsto \rho[x := e]$ is r.e., from the induction hypothesis it follows that the function $g'(\rho, x, e) = f(h(\rho, x, e), N)$ is r.e. Then by applying the s-m-n theorem of recursion theory to the computable function tracking g' we obtain a computable function tracking g, which is then r.e.

Notation 1 *We define for any $e \in D$ and $M \in \Lambda^o$:*
(i) $e^- \equiv \{P \in \Lambda^o : P^{\mathcal{D}} \sqsubseteq_D e\}$;
(ii) $M^- \equiv \{P \in \Lambda^o : P^{\mathcal{D}} \sqsubseteq_D M^{\mathcal{D}}\}$.

Corollary 2. *If $e \in \mathcal{D}^{dec}$, then e^- is a β-co-r.e. set of λ-terms.*

Proof. Let $\rho \in (Env_{D})^{r.e.}$ be an environment. By Proposition 1 there is a computable map ϕ tracking the interpretation function $M \mapsto M^{\mathcal{D}}_{\rho}$ of λ-terms from Λ into $\mathcal{D}^{r.e.}$ with respect to the effective numeration ν_{Λ} of Λ and an adequate numeration ξ of $\mathcal{D}^{r.e.}$. From $e \in \mathcal{D}^{dec}$ it follows that the set $X = \{n : \xi_n \sqsubseteq_{\mathcal{D}} e\}$ is co-r.e. This implies that the set $\phi^{-1}(X)$, which is the set of the codes of the elements of $\{M \in \Lambda : M^{\mathcal{D}}_{\rho} \sqsubseteq_{\mathcal{D}} e\}$, is also co-r.e. We get the conclusion because Λ^o is a decidable subset of Λ.

Definition 6. *A weakly effective λ-model $\mathcal{D}$ is called* effective *if it satisfies the following two further conditions:*

(i) If $d \in K(\mathcal{D})$ and $e_i \in \mathcal{D}^{dec}$, then $de_1 \dots e_n \in \mathcal{D}^{dec}$.
(ii) If $f \in [\mathcal{D} \to \mathcal{D}]^{r.e.}$ and $f(e) \in \mathcal{D}^{dec}$ for all $e \in K(\mathcal{D})$, then $\lambda(f) \in \mathcal{D}^{dec}$.

An environment ρ is *compact* in the effective domain Env_{D} (i.e., $\rho \in K(Env_{D})$) if $\rho(x) \in K(\mathcal{D})$ for all variables x and $\{x : \rho(x) \neq \bot_D\}$ is finite.

Notation 2 *We define: $\Lambda^{dec}_{\mathcal{D}} \equiv \{M \in \Lambda : M^{\mathcal{D}}_{\rho} \in \mathcal{D}^{dec}$ for all $\rho \in K(Env_{D})\}$.*

Theorem 3. *Suppose $\mathcal{D}$ is an effective λ-model. Then the set $\Lambda^{dec}_{\mathcal{D}}$ is closed under the following rules:*

1. *$x \in \Lambda^{dec}_{\mathcal{D}}$ for every variable x.*
2. *$M_1, \dots, M_k \in \Lambda^{dec}_{\mathcal{D}} \;\Rightarrow\; yM_1 \dots M_k \in \Lambda^{dec}_{\mathcal{D}}$.*
3. *$M \in \Lambda^{dec}_{\mathcal{D}} \;\Rightarrow\; \lambda x.M \in \Lambda^{dec}_{\mathcal{D}}$.*

In particular, $\Lambda^{dec}_{\mathcal{D}}$ contains all the β-normal forms.

Proof. Let $\rho \in K(Env_{D})$. We have three cases.
(1) $x^{\mathcal{D}}_{\rho} = \rho(x)$ is compact, hence it is decidable.
(2) By definition $(yM_1 \dots M_k)^{\mathcal{D}}_{\rho} = \rho(y)(M_1)^{\mathcal{D}}_{\rho} \dots (M_k)^{\mathcal{D}}_{\rho}$. Hence the result follows from Definition 6(i), $\rho(y) \in K(\mathcal{D})$ and $(M_i)^{\mathcal{D}}_{\rho} \in \mathcal{D}^{dec}$.
(3) By definition we have that $(\lambda x.M)^{\mathcal{D}}_{\rho} = \lambda(f)$, where $f(e) = M^{\mathcal{D}}_{\rho[x:=e]}$ for all $e \in D$. Note that $\rho[x := e]$ is also compact for all $e \in K(D)$. Hence the conclusion follows from $M^{\mathcal{D}}_{\rho[x:=e]} \in D^{dec}$ $(e \in K(\mathcal{D}))$, Definition 6(ii) and $f \in [\mathcal{D} \to \mathcal{D}]^{r.e.}$.

Recall that Eq($\mathcal{D}$) and Ord($\mathcal{D}$) are respectively the equational theory and the order theory of $\mathcal{D}$.

Theorem 4. *Let $\mathcal{D}$ be an effective λ-model, and let $M_1, \dots M_k \in \Lambda^{dec}_{\mathcal{D}}$ $(k \geq 1)$ be closed terms. Then we have:*

(i) $M_1^- \cap \dots \cap M_k^-$ *is a β-co-r.e. set, which contains a non-empty β-co-r.e. set of unsolvable terms.*
(ii) *If* $e \in \mathcal{D}^{dec}$ *and* e^- *is non-empty and finite modulo Eq($\mathcal{D}$), then Eq($\mathcal{D}$) is not r.e. (in particular, if* $\bot_{\mathcal{D}}^- \neq \emptyset$ *then Eq($\mathcal{D}$) is not r.e.).*
(iii) *Ord($\mathcal{D}$) is not r.e.*
(iv) *Eq($\mathcal{D}$)* $\neq \lambda\beta, \lambda\beta\eta$.

Proof. (i) By Theorem 3, Corollary 1, Corollary 2 and the FIP.

(ii) By Corollary 2 we have that e^- is a β-co-r.e. set of closed λ-terms. The conclusion follows because e^- is non-empty and finite modulo Eq($\mathcal{D}$).

(iii) Let $M \in \Lambda^{dec}_{\mathcal{D}}$ be a closed term. If Ord($\mathcal{D}$) were r.e., then we could enumerate the set M^-. However, by (i) this set is non-empty and β-co-r.e. By Theorem 1 it follows that $M^- = \Lambda^o$. By the arbitrariness of M, it follows that $\mathbf{T}^- = \mathbf{F}^-$. Since $\mathbf{F} \in \mathbf{T}^-$ and conversely we get $\mathbf{F} = \mathbf{T}$ in $\mathcal{D}$, contradiction.

(iv) Because of (iii), if Eq($\mathcal{D}$) is r.e. then Ord($\mathcal{D}$) strictly contains Eq($\mathcal{D}$). Hence the conclusion follows from Selinger's result stating that in any partially ordered λ-model, whose theory is $\lambda\beta$, the interpretations of distinct closed terms are incomparable [23, Corollary 4]. Similarly for $\lambda\beta\eta$.

4 Can Effective λ-Models Have an r.e. Theory?

In this section we give a sufficient condition for a wide class of graph models to be effective and show that no effective graph model generated freely by a partial pair, which is finite modulo its group of automorphisms, can have an r.e. equational theory. Finally, we show that no effective λ-model living in the stable or strongly stable semantics can have an r.e. equational theory.

In Section 5 we will show that every equational/order graph theory is the theory of a graph model $\mathcal{G}$ whose carrier set is the set $\mathbb{N}$ of natural numbers. In the next theorem we characterize the effectivity of these models.

Theorem 5. *Let $\mathcal{G}$ be a graph model such that, after encoding, $G = \mathbb{N}$ and $c_{\mathcal{G}}$ is a computable map. Then $\mathcal{G}$ is weakly effective. Moreover, $\mathcal{G}$ is effective under the further hypothesis that $c_{\mathcal{G}}$ has a decidable range.*

Proof. It is easy to check, using the definitions given in Section 2.3, that $\mathcal{F}, \lambda$ are r.e. in their respective domains and that condition (i) of Definition 6 is satisfied. Then $\mathcal{G}$ is weakly effective. Moreover, Definition 6(ii) holds under the hypothesis that the range of $c_{\mathcal{G}}$ is decidable.

Completions of partial pairs have been extensively studied in the literature. They are useful for solving equational and inequational constraints (see [4,5,10,11]). In

[11] Bucciarelli and Salibra have recently proved that the theory of the completion of a partial pair which is not a graph model is semi-sensible. The following theorem shows, in particular, that the theory of the completion of a finite partial pair is not r.e.

Theorem 6. *Let $\mathcal{A}$ be a partial pair such that A is finite or equal to $\mathbb{N}$ after encoding, and $c_{\mathcal{A}}$ is a computable map with a decidable domain. Then we have:*

(i) The completion $\mathcal{E}_{\mathcal{A}}$ of $\mathcal{A}$ is weakly effective;
(ii) If the range of $c_{\mathcal{A}}$ is decidable, then $\mathcal{E}_{\mathcal{A}}$ is effective;
(iii) If $\mathcal{A}$ is finite modulo its group of automorphisms (in particular, if A is finite), then $Eq(\mathcal{E}_{\mathcal{A}})$ is not r.e.

Proof. Since A is finite or equal to $\mathbb{N}$ we have that $E_{\mathcal{A}}$ is also decidable (see Definition 3). Moreover, the map $c_{\mathcal{E}_{\mathcal{A}}} : E^*_{\mathcal{A}} \times E_{\mathcal{A}} \to E_{\mathcal{A}}$ is computable, because it is an extension of a computable function $c_{\mathcal{A}}$ with decidable domain, and it is the identity on the decidable set $(E^*_{\mathcal{A}} \times E_{\mathcal{A}}) - dom(c_{\mathcal{A}})$. Then (i)-(ii) follow from Theorem 5.

Clearly A is a decidable subset of $E_{\mathcal{A}}$; then by Corollary 2 the set A^- is a β-co-r.e. set of λ-terms. We now show that this set is non-empty because $\Omega^{\mathcal{E}_{\mathcal{A}}} \subseteq A$. By Lemma 1 we have that $\alpha \in \Omega^{\mathcal{E}_{\mathcal{A}}}$ implies that $c_{\mathcal{E}_{\mathcal{A}}}(a, \alpha) \in a$ for some $a \in E^*_{\mathcal{A}}$. Immediate considerations on the rank show that this is only possible if $(a, \alpha) \in dom(c_{\mathcal{A}})$, which forces $\alpha \in A$.

The orbit of $\alpha \in A$ modulo $Aut(\mathcal{A})$ is defined by $O(\alpha) = \{\theta(\alpha) : \theta \in Aut(\mathcal{A})\}$.

We now show that, if the set of orbits of $\mathcal{A}$ has cardinality k for some $k \in \mathbb{N}$, then the cardinality of A^- modulo $\mathrm{Eq}(\mathcal{E}_{\mathcal{A}})$ is less than or equal to 2^k. Assume $p \in M^{\mathcal{E}_{\mathcal{A}}} \subseteq A$. Then by Lemma 2 the orbit of p modulo $Aut(\mathcal{A})$ is included within $M^{\mathcal{E}_{\mathcal{A}}}$. By hypothesis the number of the orbits is k; hence, the number of all possible values for $M^{\mathcal{E}_{\mathcal{A}}}$ cannot overcome 2^k.

In conclusion, A^- is non-empty, β-co-r.e. and modulo $\mathrm{Eq}(\mathcal{E}_{\mathcal{A}})$ is finite. Then (iii) follows from Theorem 4.

All the material developed in Section 3 could be adapted to the stable semantics (Berry's ccc of DI–domains and stable functions) and strongly stable semantics (Ehrhard's ccc of DI-domains with coherence and strongly stable functions). We recall that the notion of an effectively given DI-domain has been introduced by Gruchalski in [14], where it is shown that the category having effective DI-domains as objects and stable functions as morphisms is a ccc. There are also many effective models in the stable and strongly stable semantics. Indeed, the stable semantics contains a class which is analogous to the class of graph models (see [4]), namely Girard's class of *reflexive coherent spaces* called *G-models* in [4]. The results shown in Theorem 5 and in Theorem 6 for graph models could also be adapted for G-models, even if it is more delicate to complete partial pairs in this case (see [17]). It could also be developed for Ehrhard's class of strongly stable H-models (see [4]) even though working in the strongly stable semantics certainly adds technical difficulties.

Theorem 7. *Let $\mathcal{D}$ be an effective λ-model in the stable or strongly stable semantics. Then $Eq(\mathcal{D})$ is not r.e.*

Proof. Since $\bot_D \in \mathcal{D}^{dec}$ and the interpretation function is computable, then $\bot_D^- = \{M \in \Lambda^o : M^\mathcal{D} = \bot_D\}$ is co-r.e. If we show that this set is non-empty, then Eq($\mathcal{D}$) cannot be r.e. Since $\mathcal{D}$ is effective, then by Theorem 4(i) $\mathbf{F}^- \cap \mathbf{T}^-$ is a non-empty and co-r.e. set of λ-terms. Let $N \in \mathbf{F}^- \cap \mathbf{T}^-$ and let $f, g, h : \mathcal{D} \to \mathcal{D}$ be three (strongly) stable functions such that $f(x) = \mathbf{T}^\mathcal{D} \cdot x$, $g(x) = \mathbf{F}^\mathcal{D} \cdot x$ and $h(x) = N^\mathcal{D} \cdot x$ for all $x \in D$. By monotonicity we have $h \leq_s f, g$ in the stable ordering. Now, g is the constant function taking value $\mathbf{I}^\mathcal{D}$, and $f(\bot_D) = \mathbf{T}^\mathcal{D} \cdot \bot_D$. The first assertion forces h to be a constant function, because in the stable ordering all functions under a constant map are also constant, while the second assertion together with the fact that h is pointwise smaller than f forces the constant function h to satisfy $h(x) = \mathbf{T}^\mathcal{D} \cdot \bot_D$ for all x. Then an easy computation provides that $(NPP)^\mathcal{D} = \bot_D$ for every closed term P. In conclusion, we have that $\{M \in \Lambda^o : M^\mathcal{D} = \bot_D\} \neq \emptyset$ and the theory of $\mathcal{D}$ is not r.e.

5 The Löwenheim-Skolem Theorem

In this section we show that for each graph model $\mathcal{G}$ there is a countable graph model $\mathcal{P}$ with the same equational/order theory. This result is a kind of downwards Löwenheim-Skolem theorem for graph models which positively answers Question 3 in [4, Section 6.3]. Note that we cannot apply directly the classical Löwenheim-Skolem theorem since graph models are not first-order structures.

Let $\mathcal{A}, \mathcal{B}$ be partial pairs. We say that $\mathcal{A}$ is a *subpair* of $\mathcal{B}$, and we write $\mathcal{A} \leq \mathcal{B}$, if $A \subseteq B$ and $c_\mathcal{B}(a, \alpha) = c_\mathcal{A}(a, \alpha)$ for all $(a, \alpha) \in dom(c_\mathcal{A})$.

As a matter of notation, if ρ, σ are environments and C is a set, we let $\sigma = \rho \cap C$ mean $\sigma(x) = \rho(x) \cap C$ for every variable x, and $\rho \subseteq \sigma$ mean $\rho(x) \subseteq \sigma(x)$ for every variable x.

The proof of the following lemma is straightforward. Recall that the definition of interpretation with respect to a partial pair is defined in Section 2.3.

Lemma 3. *Suppose $\mathcal{A} \leq \mathcal{B}$, then $M_\rho^\mathcal{A} \subseteq M_\sigma^\mathcal{B}$ for all environments $\rho : Var \to \mathcal{P}(A)$ and $\sigma : Var \to \mathcal{P}(B)$ such that $\rho \subseteq \sigma$.*

Lemma 4. *Let M be a λ-term, $\mathcal{G}$ be a graph model and $\alpha \in M_\rho^\mathcal{G}$ for some environment ρ. Then there exists a finite subpair $\mathcal{A}$ of $\mathcal{G}$ such that $\alpha \in M_{\rho\cap A}^\mathcal{A}$.*

Proof. The proof is by induction on M.

If $M \equiv x$, then $\alpha \in \rho(x)$, so that we define $A = \{\alpha\}$ and $dom(c_\mathcal{A}) = \emptyset$.

If $M \equiv \lambda x.P$, then $\alpha \equiv c_\mathcal{G}(b, \beta)$ for some b and β such that $\beta \in P^\mathcal{G}_{\rho[x:=b]}$. By the induction hypothesis there exists a finite subpair $\mathcal{B}$ of $\mathcal{G}$ such that $\beta \in P^\mathcal{B}_{\rho[x:=b]\cap B}$. We define another finite subpair $\mathcal{A}$ of $\mathcal{G}$ as follows: $A = B \cup b \cup \{\beta, \alpha\}$; $dom(c_\mathcal{A}) = dom(c_\mathcal{B}) \cup \{(b, \beta)\}$. Then we have that $\mathcal{B} \leq \mathcal{A}$ and $\rho[x := b] \cap B \subseteq \rho[x := b] \cap A$. From $\beta \in P^\mathcal{B}_{\rho[x:=b]\cap B}$ and from Lemma 3 it follows that $\beta \in P^\mathcal{A}_{\rho[x:=b]\cap A} = P^\mathcal{A}_{(\rho\cap A)[x:=b]}$. Then we have that $\alpha \equiv c_\mathcal{A}(b, \beta) \in (\lambda x.P)^\mathcal{A}_{\rho\cap A}$.

If $M \equiv PQ$, then there is $a = \{\alpha_1, \dots, \alpha_n\}$ such that $c_\mathcal{G}(a, \alpha) \in P_\rho^\mathcal{G}$ and $a \subseteq Q_\rho^\mathcal{G}$. By the induction hypothesis there exist finite subpairs $\mathcal{A}_0, \mathcal{A}_1, \dots, \mathcal{A}_n$

of $\mathcal{G}$ such that $c_{\mathcal{G}}(a,\alpha) \in P^{\mathcal{A}_0}_{\rho\cap A_0}$ and $\alpha_k \in Q^{\mathcal{A}_k}_{\rho\cap A_k}$ for $k = 1,\dots,n$. We define another finite subpair $\mathcal{A}$ of $\mathcal{G}$ as follows: $A = \cup_{0\le k\le n} A_k \cup a \cup \{\alpha\}$ and $dom(c_{\mathcal{A}}) = (\cup_{0\le k\le n} dom(c_{\mathcal{A}_k})) \cup \{(a,\alpha)\}$. The conclusion follows from Lemma 3.

Proposition 2. *Let $\mathcal{G}$ be a graph model, and suppose $\alpha \in M^{\mathcal{G}} - N^{\mathcal{G}}$ for some $M, N \in \Lambda^o$. Then there exists a finite $\mathcal{A} \le \mathcal{G}$ such that: for all pairs $\mathcal{C} \ge \mathcal{A}$, if there is a morphism $f : \mathcal{C} \to \mathcal{G}$ such that $f(\alpha) = \alpha$, then $\alpha \in M^{\mathcal{C}} - N^{\mathcal{C}}$.*

Proof. By Lemma 4 there is a finite pair $\mathcal{A}$ such that $\alpha \in M^{\mathcal{A}}$. By Lemma 3 we have $\alpha \in M^{\mathcal{C}}$. Now, if $\alpha \in N^{\mathcal{C}}$ then, by Lemma 2 $\alpha = f(\alpha) \in N^{\mathcal{G}}$, which is a contradiction.

Corollary 3. *Let $\mathcal{G}$ be a graph model, and suppose $\alpha \in M^{\mathcal{G}} - N^{\mathcal{G}}$ for some $M, N \in \Lambda^o$. Then there exists a finite $\mathcal{A} \le \mathcal{G}$ such that: for all pairs $\mathcal{B}$ satisfying $\mathcal{A} \le \mathcal{B} \le \mathcal{G}$ we have $\alpha \in M^{\mathcal{B}} - N^{\mathcal{B}}$.*

Let $\mathcal{G}$ be a graph model. A graph model $\mathcal{P}$ is called a *sub graph model* of $\mathcal{G}$ if $\mathcal{P} \le \mathcal{G}$. It is easy to check that the class of sub graph models of $\mathcal{G}$ is closed under (finite and infinite) intersection. If $\mathcal{A} \le \mathcal{G}$ is a partial pair, then the *sub graph model generated by* $\mathcal{A}$ is defined as the intersection of all graph models $\mathcal{P}$ such that $\mathcal{A} \le \mathcal{P} \le \mathcal{G}$.

Theorem 8. (Löwenheim-Skolem Theorem for graph models) *For every graph model $\mathcal{G}$ there exists a sub graph model $\mathcal{P}$ of $\mathcal{G}$ with a countable carrier set and such that Ord($\mathcal{P}$) = Ord($\mathcal{G}$), and hence Eq($\mathcal{P}$) = Eq($\mathcal{G}$).*

Proof. We will define an increasing sequence of countable subpairs $\mathcal{A}_n$ of $\mathcal{G}$, and take for $\mathcal{P}$ the sub graph model of $\mathcal{G}$ generated by $\mathcal{A} \equiv \cup\mathcal{A}_n$.

First we define $\mathcal{A}_0$. Let I be the countable set of inequations between closed λ-terms which fail in $\mathcal{G}$. Let $e \in I$. By Corollary 3 there exists a finite partial pair $\mathcal{A}_e \le \mathcal{G}$ such that e fails in every partial pair $\mathcal{B}$ satisfying $\mathcal{A}_e \le \mathcal{B} \le \mathcal{G}$. Then we define $\mathcal{A}_0 = \cup_{e\in I}\mathcal{A}_e \le \mathcal{G}$. Assume now that $\mathcal{A}_n$ has been defined. We define $\mathcal{A}_{n+1}$ as follows. For each inequation $e \equiv M \sqsubseteq N$ which holds in $\mathcal{G}$ and fails in the sub graph model $\mathcal{P}_n \le \mathcal{G}$ generated by $\mathcal{A}_n$, we consider the set $L_e = \{\alpha \in P_n : \alpha \in M^{\mathcal{P}_n} - N^{\mathcal{P}_n}\}$. Let $\alpha \in L_e$. Since $\mathcal{P}_n \le \mathcal{G}$ and $\alpha \in M^{\mathcal{P}_n}$, then by Lemma 3 we have that $\alpha \in M^{\mathcal{G}}$. By $\mathcal{G} \models M \sqsubseteq N$ we also obtain $\alpha \in N^{\mathcal{G}}$. By Lemma 4 there exists a partial pair $\mathcal{F}_{\alpha,e} \le \mathcal{G}$ such that $\alpha \in N^{\mathcal{F}_{\alpha,e}}$. We define $\mathcal{A}_{n+1}$ as the union of the partial pair $\mathcal{A}_n$ and the partial pairs $\mathcal{F}_{\alpha,e}$ for every $\alpha \in L_e$.

Finally take for $\mathcal{P}$ the sub graph model of $\mathcal{G}$ generated by $\mathcal{A} \equiv \cup\mathcal{A}_n$. By construction we have, for every inequation e which fails in $\mathcal{G}$: $\mathcal{A}_e \le \mathcal{P}_n \le \mathcal{P} \le \mathcal{G}$. Now, Ord($\mathcal{P}$) $\subseteq$ Ord($\mathcal{G}$) follows from Corollary 3 and from the choice of $\mathcal{A}_e$.

Let now $M \sqsubseteq N$ be an inequation which fails in $\mathcal{P}$ but not in $\mathcal{G}$. Then there is an $\alpha \in M^{\mathcal{P}} - N^{\mathcal{P}}$. By Corollary 3 there is a finite partial pair $\mathcal{B} \le \mathcal{P}$ satisfying the following condition: for every partial pair $\mathcal{C}$ such that $\mathcal{B} \le \mathcal{C} \le \mathcal{P}$, we have $\alpha \in M^{\mathcal{C}} - N^{\mathcal{C}}$. Since B is finite, we have that $\mathcal{B} \le \mathcal{P}_n$ for some n. This implies that $\alpha \in M^{\mathcal{P}_n} - N^{\mathcal{P}_n}$. By construction of $\mathcal{P}_{n+1}$ we have that $\alpha \in N^{\mathcal{P}_{n+1}}$; this implies $\alpha \in N^{\mathcal{P}}$. Contradiction.

6 The Minimum Order Graph Theory

In this section we show one of the main theorems of the paper: the minimum order graph theory exists and it is the theory of an effective graph model. This result has the interesting consequence that no order graph theory can be r.e.

Lemma 5. *Suppose $\mathcal{A} \leq \mathcal{G}$ and let $f : E_\mathcal{A} \to G$ be defined by induction over the rank of $x \in E_\mathcal{A}$ as follows:*

$$f(x) = \begin{cases} x & \text{if } x \in A \\ c_\mathcal{G}(fa, f\alpha) & \text{if } x \notin A \text{ and } x \equiv (a, \alpha). \end{cases}$$

Then f is a morphism from $\mathcal{E}_\mathcal{A}$ into $\mathcal{G}$.

Lemma 6. *Suppose $\alpha \in M^\mathcal{G} - N^\mathcal{G}$ for some $M, N \in \Lambda^o$. Then there exists a finite $\mathcal{A} \leq \mathcal{G}$ such that: for all pairs $\mathcal{B}$ satisfying $\mathcal{A} \leq \mathcal{B} \leq \mathcal{G}$, we have $\alpha \in M^{\mathcal{E}_\mathcal{B}} - N^{\mathcal{E}_\mathcal{B}}$.*

Proof. By Proposition 2 and Lemma 5.

Theorem 9. *There exists an effective graph model whose order/equational theory is the minimum order/equational graph theory.*

Proof. It is not difficult to define an effective bijective numeration $\mathcal{N}$ of all finite partial pairs whose carrier set is a subset of $\mathbb{N}$. We denote by $\mathcal{N}_k$ the k-th finite partial pair with $N_k \subseteq \mathbb{N}$. We now make the carrier sets N_k ($k \in \mathbb{N}$) disjoint. Let p_k be the k-th prime natural number. Then we define another finite partial pair $\mathcal{P}_k$ as follows: $P_k = \{p_k^{x+1} : x \in N_k\}$ and $c_{\mathcal{P}_k}(\{p_k^{\alpha_1+1}, \dots, p_k^{\alpha_n+1}\}, p_k^{\alpha+1}) = p_k^{c_{\mathcal{N}_k}(\{\alpha_1,\dots,\alpha_n\},\alpha)+1}$ for all $(\{\alpha_1, \dots, \alpha_n\}, \alpha) \in dom(c_{\mathcal{N}_k})$. In this way we get an effective bijective numeration of all finite partial pairs $\mathcal{P}_k$. Finally, we take $\mathcal{P} \equiv \cup_{k\in\mathbb{N}}\mathcal{P}_k$. It is an easy matter to prove that P is a decidable subset of $\mathbb{N}$ and that, after encoding, $c_\mathcal{P} = \cup_{k\in\mathbb{N}} c_{\mathcal{P}_k}$ is a computable map with a decidable domain and range. Then by Theorem 6(ii) $\mathcal{E}_\mathcal{P}$ is an effective graph model. Notice that $\mathcal{E}_\mathcal{P}$ is also isomorphic to the completion of the union $\cup_{k\in\mathbb{N}}\mathcal{E}_{\mathcal{P}_k}$, where $\mathcal{E}_{\mathcal{P}_k}$ is the completion of the partial pair $\mathcal{P}_k$.

We now prove that the order theory of $\mathcal{E}_\mathcal{P}$ is the minimum one. Let $e \equiv M \sqsubseteq N$ be an inequation which fails in some graph model $\mathcal{G}$. By Lemma 6 e fails in the completion of a finite partial pair $\mathcal{A}$. Without loss of generality, we may assume that the carrier set of $\mathcal{A}$ is a subset of $\mathbb{N}$, and then that $\mathcal{A}$ is one of the partial pairs $\mathcal{P}_k$. For such a $\mathcal{P}_k$, e fails in $\mathcal{E}_{\mathcal{P}_k}$. Now, it was shown by Bucciarelli and Salibra in [9, Proposition 2] that, if a graph model $\mathcal{G}$ is the completion of the disjoint union of a family of graph models $\mathcal{G}_i$, then $Q^{\mathcal{G}_i} = Q^\mathcal{G} \cap G_i$ for any closed λ-term Q. Then we can conclude the proof as follows: if the inequation e holds in $\mathcal{E}_\mathcal{P}$, then by [9, Proposition 2] we get a contradiction: $M^{\mathcal{E}_{\mathcal{P}_k}} = M^{\mathcal{E}_\mathcal{P}} \cap E_{P_k} \subseteq N^{\mathcal{E}_\mathcal{P}} \cap E_{P_k} = N^{\mathcal{E}_{\mathcal{P}_k}}$.

Theorem 10. *Let $\mathcal{T}_{min}$ and $\mathcal{O}_{min}$ be, respectively, the minimum equational graph theory and the minimum order graph theory. We have:*

(i) $\mathcal{O}_{min}$ *is not r.e.*
(ii) $\mathcal{T}_{min}$ *is an intersection of a countable set of non-r.e. equational graph theories.*

Proof. (i) follows from Theorem 9 and from Theorem 4(iii), because $\mathcal{O}_{min}$ is the theory of an effective λ-model.

(ii) By the proof of Theorem 9 we have that $\mathcal{T}_{min}$ is an intersection of a countable set of graph theories, which are theories of completions of finite partial pairs. By Theorem 6(iii) these theories are not r.e.

Corollary 4. *For all graph models* $\mathcal{G}$*, Ord*($\mathcal{G}$) *is not r.e.*

Proof. If Ord($\mathcal{G}$) is r.e. and M is closed and β-normal, then $M^- = \{N \in \Lambda^o : N^{\mathcal{G}} \subseteq M^{\mathcal{G}}\}$ is a β-r.e. set, which contains the β-co-r.e. set $\{N \in \Lambda^o : \mathcal{O}_{min} \vdash N \sqsubseteq M\}$. By the FIP $M^- = \Lambda^o$. By the arbitrariness of M, it follows that $\mathbf{T}^- = \mathbf{F}^-$. Since $\mathbf{F} \in \mathbf{T}^-$ and conversely we get $\mathbf{F} = \mathbf{T}$ in $\mathcal{G}$, contradiction.

Corollary 5. *Let* $\mathfrak{G}$ *be the class of all graph models. For any finite sequence* $M_1, \ldots, M_n$ *of closed* β*-normal forms, there exists a non-empty* β*-closed co-r.e. set* $\mathcal{U}$ *of closed unsolvable terms such that*

$$(\forall \mathcal{G} \in \mathfrak{G})(\forall U \in \mathcal{U})\ U^{\mathcal{G}} \subseteq M_1^{\mathcal{G}} \cap \cdots \cap M_n^{\mathcal{G}}.$$

Proof. By Theorem 4(i) applied to any effective graph model with minimum theory, we have $(\forall U \in \mathcal{U})\ \mathcal{O}_{min} \vdash U \sqsubseteq M_1 \wedge \cdots \wedge \mathcal{O}_{min} \vdash U \sqsubseteq M_n$. The conclusion follows.

The authors do not know any example of unsolvable satisfying the above condition.

References

1. Barendregt, H.P.: The lambda calculus: Its syntax and semantics. North-Holland Publishing, Amsterdam (1984)
2. Bastonero, O., Gouy, X.: Strong stability and the incompleteness of stable models of λ-calculus. Annals of Pure and Applied Logic 100, 247–277 (1999)
3. Berardi, S., Berline, C.: $\beta\eta$-complete models for system F. Mathematical Structure in Computer Science 12, 823–874 (2002)
4. Berline, C.: From computation to foundations via functions and application: The λ-calculus and its webbed models. Theoretical Computer Science 249, 81–161 (2000)
5. Berline, C.: Graph models of λ-calculus at work, and variations. Math. Struct. in Comp. Science 16, 185–221 (2006)
6. Berline, C., Salibra, A.: Easiness in graph models. Theoretical Computer Science 354, 4–23 (2006)
7. Berry, G.: Stable models of typed lambda-calculi. In: Ausiello, G., Böhm, C. (eds.) Automata, Languages and Programming. LNCS, vol. 62, Springer, Heidelberg (1978)
8. Bucciarelli, A., Ehrhard, T.: Sequentiality and strong stability. In: Sixth Annual IEEE Symposium on Logic in Computer Science (LICS 1991), pp. 138–145. IEEE Computer Society Press, Los Alamitos (1991)

9. Bucciarelli, A., Salibra, A.: The minimal graph model of lambda calculus. In: Rovan, B., Vojtáš, P. (eds.) MFCS 2003. LNCS, vol. 2747, pp. 300–307. Springer, Heidelberg (2003)
10. Bucciarelli, A., Salibra, A.: The sensible graph theories of lambda calculus. In: 19th Annual IEEE Symposium on Logic in Computer Science (LICS 2004), IEEE Computer Society Press, Los Alamitos (2004)
11. Bucciarelli, A., Salibra, A.: Graph lambda theories. Mathematical Structures in Computer Science (to appear)
12. Di Gianantonio, P., Honsell, F., Plotkin, G.D.: Uncountable limits and the lambda calculus. Nordic J. Comput. 2, 126–145 (1995)
13. Giannini, P., Longo, G.: Effectively given domains and lambda-calculus models. Information and Control 62, 36–63 (1984)
14. Gruchalski, A.: Computability on dI-Domains. Information and Computation 124, 7–19 (1996)
15. Honsell, F., Ronchi della Rocca, S.: An approximation theorem for topological lambda models and the topological incompleteness of lambda calculus. Journal of Computer and System Sciences 45, 49–75 (1992)
16. Kerth, R.: Isomorphism and equational equivalence of continuous lambda models. Studia Logica 61, 403–415 (1998)
17. Kerth, R.: On the construction of stable models of λ-calculus. Theoretical Computer Science 269, 23–46 (2001)
18. Longo, G.: Set-theoretical models of λ-calculus: theories, expansions and isomorphisms. Ann. Pure Applied Logic 24, 153–188 (1983)
19. Plotkin, G.D.: Set-theoretical and other elementary models of the λ-calculus. Theoretical Computer Science 121, 351–409 (1993)
20. Salibra, A.: A continuum of theories of lambda calculus without semantics. In: 16th Annual IEEE Symposium on Logic in Computer Science (LICS 2001), pp. 334–343. IEEE Computer Society Press, Los Alamitos (2001)
21. Salibra, A.: Topological incompleteness and order incompleteness of the lambda calculus. ACM Transactions on Computational Logic 4, 379–401 (2003) (LICS'01 Special Issue)
22. Scott, D.S.: Continuous lattices. In: Toposes, Algebraic geometry and Logic. LNM, vol. 274, Springer, Heidelberg (1972)
23. Selinger, P.: Order-incompleteness and finite lambda reduction models. Theoretical Computer Science 309, 43–63 (2003)
24. Stoltenberg-Hansen, V., Lindström, I., Griffor, E.R.: Mathematical theory of domains. In: Cambridge Tracts in Theoretical Computer Science, vol. 22, Cambridge University Press, Cambridge (1994)
25. Visser, A.: Numerations, λ-calculus and arithmetic. In: Hindley, J.R., Seldin, J.P. (eds.) To H.B. Curry: Essays on Combinatory Logic, Lambda-Calculus and Formalism, pp. 259–284. Academic Press, New York (1980)

Typed Normal Form Bisimulation

Soren B. Lassen[1] and Paul Blain Levy[2]

[1] Google, Inc.
soren@google.com
[2] University of Birmingham, U.K.
pbl@cs.bham.ac.uk

Abstract. Normal form bisimulation is a powerful theory of program equivalence, originally developed to characterize Lévy-Longo tree equivalence and Boehm tree equivalence. It has been adapted to a range of untyped, higher-order calculi, but types have presented a difficulty. In this paper, we present an account of normal form bisimulation for types, including recursive types. We develop our theory for a continuation-passing style calculus, Jump-With-Argument (JWA), where normal form bisimilarity takes a very simple form. We give a novel congruence proof, based on insights from game semantics. A notable feature is the seamless treatment of eta-expansion. We demonstrate the normal form bisimulation proof principle by using it to establish a syntactic minimal invariance result and the uniqueness of the fixed point operator at each type.

1 Introduction

1.1 Background

Normal form bisimulation—also known as open (applicative) bisimulation—originated as a coinductive way of describing Lévy-Longo tree equivalence for the lazy λ-calculus [1], and has subsequently been extended to call-by-name, call-by-value, nondeterminism, aspects, storage, and control [2–7].

Suppose we have two functions $V, V' : A \to B$. When should they be deemed equivalent? Here are two answers:

- when VW and $V'W$ behave the same for every *closed* $W : A$
- when $V\mathtt{y}$ and $V'\mathtt{y}$ behave the same for *fresh* $\mathtt{y} : A$.

The first answer leads to the theory of *applicative bisimulation* [8, 9], the second to that of *normal form bisimulation.* The first answer requires us to run closed terms only, the second to run non-closed terms.

To illustrate the difference[1], let $G(p, q)$ be the following function

$$\lambda\mathtt{x}.\mathtt{if}\ (\mathtt{x}\,p)\ \mathtt{then}\ \mathtt{x}\,q\ \mathtt{else}\ \mathtt{not}(\mathtt{x}\,q)$$

and let V be $G(\mathtt{true}, \mathtt{false})$ and V' be $G(\mathtt{false}, \mathtt{true})$, both of type $(\mathtt{bool} \to \mathtt{bool}) \to \mathtt{bool}$. Assuming the language is free of effects besides divergence,

[1] This example works in both call-by-value and call-by-name.

J. Duparc and T.A. Henzinger (Eds.): CSL 2007, LNCS 4646, pp. 283–297, 2007.

they cannot be distinguished by applying to closed arguments. But let us apply them to a fresh identifier `y` and evaluate the resulting open terms, symbolically. Then the first begins by applying `y` to `true` with the continuation `if – then x false else not(x true)`. The second applies `y` to `false` with the continuation `if – then x true else not(x false)`.

Normal form bisimilarity requires that two (nondivergent) programs end up *either* applying the same identifier to equivalent arguments with equivalent continuations, *or* returning equivalent values. In our example, the arguments are different (`true` and `false` respectively) and so are the continuations. Therefore V and V' are not normal form bisimilar.

This is a situation where normal form bisimulation gives an equivalence that is finer than contextual equivalence, in contrast with applicative bisimulation. However, the addition of state and suitable control effects makes normal form bisimulation coincide with contextual equivalence [2, 7].

Compared to applicative bisimulation, the absence of the universal quantification over closed arguments to functions in the definition of normal form bisimulation makes certain program equivalence proofs possible that have not been accomplished using applicative bisimulation. An example is our proof of syntactic minimal invariance in Section 4. See also the examples in [2, 4, 7].

But all the results on normal form bisimulation are in untyped settings only. The adaptation to typed calculi has presented difficulties. For example, in call-by-value, if A is an empty type, such as $\mu\mathtt{X}.(\mathtt{bool} \times \mathtt{X})$, then all functions of type $A \to B$ should be equivalent—without appealing to the absence of any closed arguments, as we would for applicative bisimulation.

Another problem that has slowed progress on normal form bisimulation is that the congruence proofs given in the literature (especially [7]) are complicated: they establish congruence for η-long terms, and separately prove the validity of the η-law.

1.2 Contributions

This paper makes three important contributions to the development of normal form bisimulation. First, we extend normal form bisimulation to types, inclusive product, sum, and higher-order types, empty types and recursive types. Second, we give a new, lucid congruence proof that highlights connections with game semantics. Third, we present a seamless treatment of η-expansion—we do not need to mention it in our definitions or proofs.

To illustrate the power of our theory, we use it to prove

- uniqueness of the fixpoint operator at each type
- syntactic minimal invariance.

It is not known how to prove the latter using applicative bisimulation, so this shows a definite advantage of normal form bisimulation as an operational reasoning technique.

Our work is part of a larger programme to explore the scope of normal form bisimulation, both (1) as a syntactic proof principle for reasoning about program equivalence and (2) as an operational account of pointer game semantics. Space constraints prevent us from giving a formal account of the second point but we mention some connections in our discussion of related work below and in the technical exposition of our theory we give some hints to the benefit of readers familiar with game semantics.

For simplicity, we develop our theory in a continuation-passing style calculus, called Jump-With-Argument (JWA), where it takes a very simple form. It is then trivial to adapt it to a direct-style calculus (with or without control operators), by applying the appropriate CPS transform. The pointer game semantics for JWA, and the relationship with direct-style programs, is given in [10, 11]; but here we do not assume familiarity with pointer games.

In this paper, we look at JWA without storage. In game terminology, the strategies we are studying are innocent. It is clear from the game literature that, when we extend JWA with storage, normal form bisimilarity will coincide with contextual equivalence, but we leave such an analysis to future work.

1.3 Related Work

In Section 1.1 we descibed the origins of and previous work on normal form bisimulation.

We refer the reader to [7] for a survey of other syntactic theories for reasoning about program equivalence: equational theories, context lemmas, applicative bisimulation, environmental bisimulation, and syntactic logical relations. Given that our calculus, JWA, is a CPS calculus, note that applicative bisimulation has been studied recently in a CPS setting by Merro and Biasi [12].

On the other hand, let us discuss some relevant literature on game semantics, because its relationship with normal form bisimulation has not been surveyed before.

Pointer game semantics is a form of denotational model in which a term denotes a strategy for a game with pointers between moves. It was introduced in [13], and has been used to model typed and untyped languages, call-by-name, call-by-value, recursion, storage, control operators and much more [14–19]. In general, denotational equality is finer than contextual equivalence, but in the presence of storage and control, they coincide [20]. The simplest models, for terms without local state, use *innocent* strategies, which correspond to Böhm trees [21], or variants thereof such as PCF trees, Nakajima trees and Lévy-Longo trees.

Pointer games are analyzed operationally in [22, 10], relating a term's denotation to its behaviour in an abstract machine. But these abstract machines are complex to describe and reason about. Moreover, the only terms studied in these accounts are η-long: thus [22] states "in the sequel, we will consider terms up to η-equality". This is a limitation, especially in the presence of recursive types at which terms cannot be fully η-expanded.

In [23], Sect. 1–2, a quite different operational account of a game model is given, without abstract machines or η-expansion—albeit in the limited setting of a first-order language. It is defined in terms of an operational semantics for *open* terms (unlike traditional operational semantics, defined on closed terms only).

Clearly there are many similarities between normal form bisimulation and pointer game semantics: the connection with Böhm trees and Lévy-Longo trees, the completeness in the presence of storage and control, the use of operational semantics for open terms. Readers familiar with game semantics will immediately see that V and V' in the example in Section 1.1 denote different strategies.

An immediate precursor for our definition of typed normal form bisimulation is the labelled transition system and pointer game semantics in [24]. This work also describes an extension to mutable references, which is something we plan to explore in future work. Recently and independently, Laird [25] developed a very similar labelled transition system semantics for a typed functional language with mutable references.

The representation of strategies as π-calculus processes by Hyland, Ong, Fiore, and Honda [26, 27] leads to a bisimulation approach to equality of strategies that is analogous to normal form bisimulation. However, because of the detour via π-calculus encodings, the resulting π-calculus-based bisimulation proof principles for program equivalence are not as direct as normal form bisimulation, for proving specific program equivalences between terms.

2 Jump-With-Argument

2.1 Syntax and Semantics

Jump-With-Argument is a continuation-passing style calculus, extending the CPS calculus in [28]. Its types are given (including recursive types) by

$$A ::= \ \neg A \mid \textstyle\sum_{i \in I} A_i \mid 1 \mid A \times A \mid \mathtt{X} \mid \mu \mathtt{X}.A$$

where I is any finite set. The type $\neg A$ is the type of functions that take an argument of type A and do not return. JWA has two judgements: *values* written $\Gamma \vdash^{\mathsf{v}} V$ and *nonreturning commands* written $\Gamma \vdash^{\mathsf{n}} M$. The syntax[2] is shown in Fig. 1. We write `pm` as an abbreviation for "pattern-match", and write `let` to make a binding. We omit typing rules, etc., for 1, since 1 is analogous to $\times$.

From the cpo viewpoint, a JWA type denotes an (unpointed) cpo. In particular, $\neg A$ denotes $[\![A]\!] \to R$, where R is a chosen pointed cpo.

Operational semantics. To evaluate a command $\Gamma \vdash^{\mathsf{n}} M$, simply apply the transitions (β-reductions) in Fig. 2 until a terminal command is reached. Every command M is either a redex or terminal; by determinism, either $M \leadsto^* T$ for unique terminal T, or else $M \leadsto^\infty$. This operational semantics is called the *C-machine.*

We define a fixed point combinator Y as follows

$$Y \stackrel{\text{def}}{=} \Phi(\mathtt{fold}\ \lambda\langle \mathtt{x}, \langle \mathtt{u}, \mathtt{f}\rangle\rangle.\mathtt{f}\langle \mathtt{u}, \lambda \mathtt{v}.\Phi(\mathtt{x})\langle \mathtt{v}, \mathtt{f}\rangle\rangle)$$

$$\Phi(x) \stackrel{\text{def}}{=} \lambda \mathtt{z}.\mathtt{pm}\ x\ \mathtt{as}\ \mathtt{fold}\ \mathtt{y}.\mathtt{y}\langle \mathtt{fold}\ \mathtt{y}, \mathtt{z}\rangle$$

[2] In earlier works e.g. [10], $\gamma\mathtt{x}.M$ was written for $\lambda\mathtt{x}.M$ and $W \nearrow V$ for VW.

$$\frac{}{\Gamma \vdash^{\mathsf{v}} \mathtt{x} : A}\ (\mathtt{x} : A) \in \Gamma \qquad \frac{\Gamma \vdash^{\mathsf{v}} V : A \quad \Gamma, \mathtt{x} : A \vdash^{\mathsf{n}} M}{\Gamma \vdash^{\mathsf{n}} \mathtt{let}\ V\ \mathtt{be}\ \mathtt{x}.\ M}$$

$$\frac{\hat{\imath} \in I \quad \Gamma \vdash^{\mathsf{v}} V : A_{\hat{\imath}}}{\Gamma \vdash^{\mathsf{v}} \langle \hat{\imath}, V \rangle : \sum_{i \in I} A_i} \qquad \frac{\Gamma \vdash^{\mathsf{v}} V : \sum_{i \in I} A_i \quad \Gamma, \mathtt{x}_i : A_i \vdash^{\mathsf{n}} M_i\ (\forall i \in I)}{\Gamma \vdash^{\mathsf{n}} \mathtt{pm}\ V\ \mathtt{as}\ \{\langle i, \mathtt{x}_i \rangle . M_i\}_{i \in I}}$$

$$\frac{\Gamma \vdash^{\mathsf{v}} V : A \quad \Gamma \vdash^{\mathsf{v}} V' : A'}{\Gamma \vdash^{\mathsf{v}} \langle V, V' \rangle : A \times A'} \qquad \frac{\Gamma \vdash^{\mathsf{v}} V : A \times A' \quad \Gamma, \mathtt{x} : A, \mathtt{y} : A' \vdash^{\mathsf{n}} M}{\Gamma \vdash^{\mathsf{n}} \mathtt{pm}\ V\ \mathtt{as}\ \langle \mathtt{x}, \mathtt{y} \rangle . M}$$

$$\frac{\Gamma, \mathtt{x} : A \vdash^{\mathsf{n}} M}{\Gamma \vdash^{\mathsf{v}} \lambda \mathtt{x}.M : \neg A} \qquad \frac{\Gamma \vdash^{\mathsf{v}} V : \neg A \quad \Gamma \vdash^{\mathsf{v}} W : A}{\Gamma \vdash^{\mathsf{n}} VW}$$

$$\frac{\Gamma \vdash^{\mathsf{v}} V : A[\mu \mathtt{X}.A/\mathtt{X}]}{\Gamma \vdash^{\mathsf{v}} \mathtt{fold}\ V : \mu \mathtt{X}.A} \qquad \frac{\Gamma \vdash^{\mathsf{v}} V : \mu \mathtt{X}.A \quad \Gamma, \mathtt{x} : A[\mu \mathtt{X}.A/\mathtt{X}] \vdash^{\mathsf{n}} M}{\Gamma \vdash^{\mathsf{n}} \mathtt{pm}\ V\ \mathtt{as}\ \mathtt{fold}\ \mathtt{x}.\ M}$$

Fig. 1. Syntax of JWA, with type recursion

Transitions

$$\begin{array}{lcl}
\mathtt{pm}\ \langle \hat{\imath}, V \rangle\ \mathtt{as}\ \{\langle i, \mathtt{x} \rangle . M_i\}_{i \in I} & \rightsquigarrow & M_{\hat{\imath}}[V/\mathtt{x}] \\
\mathtt{pm}\ \langle V, V' \rangle\ \mathtt{as}\ \langle \mathtt{x}, \mathtt{y} \rangle . M & \rightsquigarrow & M[V/\mathtt{x}, V'/\mathtt{y}] \\
(\lambda \mathtt{x}.M)V & \rightsquigarrow & M[V/\mathtt{x}] \\
\mathtt{pm}\ \mathtt{fold}\ V\ \mathtt{as}\ \mathtt{fold}\ \mathtt{x}.\ M & \rightsquigarrow & M[V/\mathtt{x}] \\
\mathtt{let}\ V\ \mathtt{be}\ \mathtt{x}.M & \rightsquigarrow & M[V/\mathtt{x}]
\end{array}$$

Terminal Commands

$$\begin{array}{l}
\mathtt{pm}\ \mathtt{z}\ \mathtt{as}\ \{\langle i, \mathtt{x} \rangle .\ M_i\}_{i \in I} \\
\mathtt{pm}\ \mathtt{z}\ \mathtt{as}\ \langle \mathtt{x}, \mathtt{y} \rangle .\ M \\
\mathtt{z}V \\
\mathtt{pm}\ \mathtt{z}\ \mathtt{as}\ \mathtt{fold}\ \mathtt{x}.\ M
\end{array}$$

Fig. 2. C-machine

where notation $\lambda \langle \mathtt{x}, \langle \mathtt{u}, \mathtt{f} \rangle \rangle .M$ is short for $\lambda \mathtt{t}.\mathtt{pm}\ \mathtt{t}\ \mathtt{as}\ \langle \mathtt{x}, \mathtt{p} \rangle .\mathtt{pm}\ \mathtt{p}\ \mathtt{as}\ \langle \mathtt{u}, \mathtt{f} \rangle .M$.

Using type recursion, we assign type $\neg(A \times \neg(A \times \neg A))$ to Y, for every type A, by giving the argument to Φ type $\mu \mathtt{X}.\neg(\mathtt{X} \times (A \times \neg A))$.

Let $V = \lambda \langle \mathtt{x}, \langle \mathtt{u}, \mathtt{f} \rangle \rangle .\mathtt{f}\langle \mathtt{u}, \lambda \mathtt{v}.\Phi(\mathtt{x})\langle \mathtt{v}, \mathtt{f} \rangle \rangle$, then we calculate:

$$\begin{aligned}
Y\langle \mathtt{u}, \mathtt{f} \rangle &\rightsquigarrow \mathtt{pm}\ \mathtt{fold}\ V\ \mathtt{as}\ \mathtt{fold}\ \mathtt{y}.\mathtt{y}\langle \mathtt{fold}\ \mathtt{y}, \langle \mathtt{u}, \mathtt{f} \rangle \rangle \\
&\rightsquigarrow V\langle \mathtt{fold}\ V, \langle \mathtt{u}, \mathtt{f} \rangle \rangle \\
&\rightsquigarrow^3 \mathtt{f}\langle \mathtt{u}, \lambda \mathtt{v}.\Phi(\mathtt{fold}\ V)\langle \mathtt{v}, \mathtt{f} \rangle \rangle \\
&= \mathtt{f}\langle \mathtt{u}, \lambda \mathtt{v}.Y\langle \mathtt{v}, \mathtt{f} \rangle \rangle
\end{aligned}$$

That is, Y is a solution to the fixed point equation

$$\vdash^{\mathsf{v}} Y =_{\beta} \lambda\langle \mathtt{u}, \mathtt{f}\rangle.\mathtt{f}\langle \mathtt{u}, \lambda \mathtt{v}.Y\langle \mathtt{v}, \mathtt{f}\rangle\rangle : \neg(A \times \neg(A \times \neg A))$$

2.2 Ultimate Pattern Matching

To describe normal form bisimulation, we need to decompose a value into an *ultimate pattern* (the tags) and a value sequence (the rest). Take, for example, the value

$$\langle \hat{\imath}_0, \langle\langle\langle \lambda \mathtt{w}.M, \mathtt{x}\rangle, \mathtt{y}\rangle, \langle \hat{\imath}_1, \mathtt{x}\rangle\rangle\rangle$$

Provided $\mathtt{x}$ and $\mathtt{y}$ have $\neg$ type, we can decompose this as the ultimate pattern

$$\langle \hat{\imath}_0, \langle\langle\langle -_{\neg A}, -_{\neg B}\rangle, -_{\neg C}\rangle, \langle \hat{\imath}_1, -_{\neg B}\rangle\rangle\rangle$$

(for appropriate types A, B, C), where the holes are filled with the value sequence

$$\lambda \mathtt{w}.M, \mathtt{x}, \mathtt{y}, \mathtt{x}$$

As this example shows, an ultimate pattern is built up out of tags and holes; the holes are to be filled by values of $\neg$ type. For each type A, we define the set $\mathsf{ult}^{\mathsf{v}}(A)$ of ultimate patterns of type A, by mutual induction:

- $-_{\neg A} \in \mathsf{ult}^{\mathsf{v}}(\neg A)$
- if $p \in \mathsf{ult}^{\mathsf{v}}(A)$ and $p' \in \mathsf{ult}^{\mathsf{v}}(A')$ then $\langle p, p'\rangle \in \mathsf{ult}^{\mathsf{v}}(A \times A')$
- if $\hat{\imath} \in I$ and $p \in \mathsf{ult}^{\mathsf{v}}(A_{\hat{\imath}})$ then $\langle \hat{\imath}, p\rangle \in \mathsf{ult}^{\mathsf{v}}(\sum_{i \in I} A_i)$
- if $p \in \mathsf{ult}^{\mathsf{v}}(A[\mu \mathtt{X}.A/\mathtt{X}])$ then $\mathtt{fold}\, p \in \mathsf{ult}^{\mathsf{v}}(\mu \mathtt{X}.A)$.

For $p \in \mathsf{ult}^{\mathsf{v}}(A)$, we write $H(p)$ for the list of types (all $\neg$ types) of holes of p. Given a value sequence $\Gamma \vdash^{\mathsf{v}} \overrightarrow{V} : H(p)$, we obtain a value $\Gamma \vdash^{\mathsf{v}} p(\overrightarrow{V}) : A$ by filling the holes of p with $\overrightarrow{V}$. We can now state our decomposition theorem.

Proposition 1. *Let* $\overrightarrow{\mathtt{x} : \neg A} \vdash V : B$ *be a value. Then there is a unique ultimate pattern* $p \in \mathsf{ult}^{\mathsf{v}}(B)$ *and value sequence* $\overrightarrow{\mathtt{x} : \neg A} \vdash^{\mathsf{v}} \overrightarrow{W} : H(p)$ *such that* $V = p(\overrightarrow{W})$.

Proof. Induction on V.

3 Normal Form Bisimulation

In this section, we define normal form bisimulation. Some readers may like to see this as a way of characterizing when two terms have the same Böhm tree[3], or

[3] The Böhm trees for JWA are given by the following classes of commands and values, defined coinductively:

$$\begin{aligned} M &::= \mathtt{diverge} \mid \mathtt{x}_i p(\overrightarrow{V}) \\ V &::= \lambda\{p(\overrightarrow{\mathtt{y}}).M_p\}_{p \in \mathsf{ult}^{\mathsf{v}}(A)} \end{aligned}$$

These trees are not actually (infinite) JWA terms, because ultimate pattern matching is not part of the syntax of JWA.

(equivalently) denote the same innocent strategy, but neither of these concepts will be used in the paper.

Notation. For any $n \in \mathbb{N}$, we write $\$n$ for the set $\{0, \ldots, n-1\}$. For a sequence $\overrightarrow{a}$, we write $|\overrightarrow{a}|$ for its length.

Any terminal command $\overrightarrow{\mathtt{x} : \neg A} \vdash^{\mathsf{n}} M$ must be of the form $\mathtt{x}_i p(\overrightarrow{V})$. The core of our definition is that we regard (i, p) as an observable action, so we write

$$M \overset{ip}{\rightsquigarrow} \overrightarrow{\mathtt{x} : \neg A} \vdash^{\mathsf{v}} \overrightarrow{V} : H(p)$$

(This action is called a "Proponent move". Interchangeably, we write it as $\mathtt{x}_i p$ when it is more convenient to name the identifier than its index.) Thus, for any command $\overrightarrow{\mathtt{x} : A} \vdash N$, we have either

$$N \rightsquigarrow^* \overset{ip}{\rightsquigarrow} \overrightarrow{V}$$

for unique $i, p, \overrightarrow{V}$, or else $N \rightsquigarrow^\infty$.

Suppose we are given a value sequence $\overrightarrow{\mathtt{x} : \neg A} \vdash^{\mathsf{v}} \overrightarrow{V : \neg B}$. For each $j \in \$|\overrightarrow{\neg B}|$ and $q \in \mathsf{ult}^{\mathsf{v}}(B_j)$, we define $(\overrightarrow{V : \neg B}) : jq$ to be the command

$$\overrightarrow{\mathtt{x} : \neg A}, \overrightarrow{\mathtt{y}} : H(q) \vdash^{\mathsf{n}} V_j q(\overrightarrow{y})$$

where $\overrightarrow{\mathtt{y}}$ are fresh. (We call this operation an "Opponent move".)

Definition 1. *Let $\mathcal{R}$ be a set of pairs of commands $\overrightarrow{\mathtt{x} : \neg A} \vdash^{\mathsf{n}} M, M'$.*

1. *Let $\overrightarrow{\mathtt{x} : \neg A} \vdash^{\mathsf{v}} \overrightarrow{V}, \overrightarrow{V'} : \overrightarrow{\neg B}$ be two value sequences. We say $\overrightarrow{V} \mathcal{R}^{\mathsf{v}} \overrightarrow{V'}$ when for any $j \in \$|\overrightarrow{\neg B}|$ and $q \in \mathsf{ult}^{\mathsf{v}}(B_j)$, we have $(\overrightarrow{V} : jq) \mathcal{R} (\overrightarrow{V'} : jq)$.*
2. *$\mathcal{R}$ is a* normal form bisimulation *when $\overrightarrow{\mathtt{x} : \neg A} \vdash^{\mathsf{n}} N \mathcal{R} N'$ implies either*
   - *$N \rightsquigarrow^\infty$ and $N' \rightsquigarrow^\infty$, or*
   - *$N \rightsquigarrow^* \overset{ip}{\rightsquigarrow} \overrightarrow{V}$ and $N' \rightsquigarrow^* \overset{ip}{\rightsquigarrow} \overrightarrow{V'}$ and $\overrightarrow{V} \mathcal{R}^{\mathsf{v}} \overrightarrow{V'}$.*

We write $\approx$ for normal form bisimilarity, *i.e. the greatest normal form bisimulation.*

Proposition 2. *(preservation under renaming)*
For any renaming $\overrightarrow{\mathtt{x} : \neg A} \xrightarrow{\theta} \overrightarrow{\mathtt{y} : \neg A}$, we have $\overrightarrow{\mathtt{x} : \neg A} \vdash M \approx M'$ implies $M\theta \approx M'\theta$, and $\overrightarrow{\mathtt{x} : \neg A} \vdash^{\mathsf{v}} \overrightarrow{V} \approx^{\mathsf{v}} \overrightarrow{V'} : \overrightarrow{\neg C}$ implies $\overrightarrow{V\theta} \approx^{\mathsf{v}} \overrightarrow{V'\theta}$.

Proof. Let $\mathcal{R}$ be the set of pairs $(M\theta, M'\theta)$ where $\overrightarrow{\mathtt{x} : \neg A} \vdash M \approx M'$ and $\overrightarrow{\mathtt{x} : \neg A} \xrightarrow{\theta} \overrightarrow{\mathtt{y} : \neg A}$ is a renaming. Then, for such θ, $(\overrightarrow{V\theta}, \overrightarrow{V'\theta}) \in \mathcal{R}^{\mathsf{v}}$ whenever $\overrightarrow{\mathtt{x} : \neg A} \vdash^{\mathsf{v}} \overrightarrow{V} \approx^{\mathsf{v}} \overrightarrow{V'} : \overrightarrow{\neg C}$. It is easy to show $\mathcal{R}$ is a normal form bisimulation.

Proposition 3. *(preservation under substitution)*
Suppose $\overrightarrow{\mathtt{y} : \neg B} \vdash^{\mathsf{v}} \overrightarrow{W} \approx^{\mathsf{v}} \overrightarrow{W'} : \overrightarrow{\neg A}$. Then $\overrightarrow{\mathtt{x} : \neg A} \vdash^{\mathsf{n}} M \approx M'$ implies $M[\overrightarrow{W/\mathtt{x}}] \approx M'[\overrightarrow{W'/\mathtt{x}}]$ and $\overrightarrow{\mathtt{x} : \neg A} \vdash^{\mathsf{v}} \overrightarrow{V} \approx^{\mathsf{v}} \overrightarrow{V'} : \overrightarrow{\neg C}$ implies $\overrightarrow{V[\overrightarrow{W/\mathtt{x}}]} \approx^{\mathsf{v}} \overrightarrow{V'[\overrightarrow{W'/\mathtt{x}}]}$.

This is proved in the next section.

Next we have to extend normal form bisimilarity to arbitrary commands and values (conceptually following the categorical construction in [14]).

Definition 2. – *Given commands* $\overrightarrow{\mathtt{x}:A} \vdash^{\mathsf{n}} M, M'$*, we say* $M \approx M'$ *when for each* $\overrightarrow{p \in \mathsf{ult}^{\mathsf{v}}(A)}$ *we have*

$$\overrightarrow{\overrightarrow{\mathtt{y}} : H(p)} \vdash^{\mathsf{n}} M[\overrightarrow{p(\overrightarrow{\mathtt{y}})/\mathtt{x}}] \approx M'[\overrightarrow{p(\overrightarrow{\mathtt{y}})/\mathtt{x}}]$$

– *Given values* $\overrightarrow{\mathtt{x}:A} \vdash^{\mathsf{v}} V, V' : B$*, we say* $V \approx V'$ *when for each* $\overrightarrow{p \in \mathsf{ult}^{\mathsf{v}}(A)}$*, decomposing* $V[\overrightarrow{p(\overrightarrow{\mathtt{y}})/\mathtt{x}}]$ *as* $q(\overrightarrow{W})$ *and* $V'[\overrightarrow{p(\overrightarrow{\mathtt{y}})/\mathtt{x}}]$ *as* $q'(\overrightarrow{W'})$*, we have* $q = q'$ *and*

$$\overrightarrow{\overrightarrow{\mathtt{y}} : H(p)} \vdash^{\mathsf{v}} \overrightarrow{W} \approx^{\mathsf{v}} \overrightarrow{W'} : H(q)$$

It is easy to see that normal form bisimilarity for JWA is an equivalence relation and validates all the β and η laws [10].

Proposition 4. $\approx$ *is a substitutive congruence.*

Proof. Substitutivity follows from Prop. 3. For each term constructor, we prove it preserves $\approx$ using substitutivity, as in [7].

As a corollary, we get that normal form bisimilar terms are contextually equivalent, for any reasonable definition of contextual equivalence. The opposite is not true. For example, this equation holds for contextual equivalence:

$$\mathtt{x} : \neg\neg 1, \mathtt{y} : \neg 1 \vdash^{\mathsf{n}} \mathtt{x}\,\mathtt{y} \cong_{\mathrm{ctx}} \mathtt{x}\,(\lambda \mathtt{z}.\mathtt{x}\,\mathtt{y}) \tag{1}$$

The intuition is that either x ignores its argument, and then the equivalence holds trivially, or else x invokes its argument at some point, and from that point onward $\mathtt{x}\,(\lambda\mathtt{z}.\mathtt{x}\,\mathtt{y})$ emulates $\mathtt{x}\,\mathtt{y}$ from the beginning.[4] But $\mathtt{x}\,\mathtt{y}$ and $\mathtt{x}\,(\lambda\mathtt{z}.\mathtt{x}\,\mathtt{y})$ are not normal form bisimilar: $\mathtt{x}\,\mathtt{y} \overset{\mathtt{x}(-_{\neg 1})}{\rightsquigarrow} \mathtt{y}$ and $\mathtt{x}\,(\lambda\mathtt{z}.\mathtt{x}\,\mathtt{y}) \overset{\mathtt{x}(-_{\neg 1})}{\rightsquigarrow} \lambda\mathtt{z}.\mathtt{x}\,\mathtt{y}$ but y and $\lambda\mathtt{z}.\mathtt{x}\,\mathtt{y}$ are clearly not bisimilar since, given an argument z, the labelled transitions $\mathtt{y}\,\mathtt{z} \overset{\mathtt{y}(-_1)}{\rightsquigarrow} \mathtt{z}$ and $(\lambda\mathtt{z}.\mathtt{x}\,\mathtt{y})\,\mathtt{z} \overset{\mathtt{x}(-_{\neg 1})}{\rightsquigarrow} \mathtt{y}$ mismatch.

3.1 Alternating Substitution

To prove Prop. 3, it turns out to be easier to show preservation by a more general operation on terms, *alternating substitution*, which is applied to an *alternating table* of terms. They are defined in Fig. 3. A table provides the following information:

- the context of each term is the identifiers to the left of it
- the type of each identifier, and of each term, is given in the top row.

[4] We omit both the definition of contextual equivalence and the proof of (1) but the reasoning is analogous to Thielecke's proof of Filinski's equation $M \cong_{\mathrm{ctx}} M; M$ in a direct-style calculus with exception and continuations but without state [29]. Indeed, (1) is essentially derived from Filinski and Thielecke's example by a CPS transform.

For example, the table

$\overrightarrow{\neg A}_{\mathsf{out}}$	$\overrightarrow{\neg B}_{\mathsf{out}}$	$\overrightarrow{\neg A}_0$	$\overrightarrow{\neg B}_0$	$\overrightarrow{\neg A}_1$	$\overrightarrow{\neg B}_1$	$\vdash^{\mathsf{n}}$
$\vec{\mathbf{x}}_{\mathsf{out}}$		$\vec{V}_0$	$\vec{\mathbf{x}}_0$	$\vec{V}_1$	$\vec{\mathbf{x}}_1$	M
	$\vec{\mathbf{z}}_{\mathsf{out}}$	$\vec{\mathbf{z}}_0$	$\vec{W}_0$	$\vec{\mathbf{z}}_1$	$\vec{W}_1$	

consists of the following value-sequences and commands:

$$\vec{\mathbf{x}}_{\mathsf{out}} : \overrightarrow{\neg A}_{\mathsf{out}} \vdash^{\mathsf{v}} \vec{V}_0 : \overrightarrow{\neg A}_0$$
$$\vec{\mathbf{z}}_{\mathsf{out}} : \overrightarrow{\neg B}_{\mathsf{out}}, \vec{\mathbf{z}}_0 : \overrightarrow{\neg A}_0 \vdash^{\mathsf{v}} \vec{W}_0 : \overrightarrow{\neg B}_0$$
$$\vec{\mathbf{x}}_{\mathsf{out}} : \overrightarrow{\neg A}_{\mathsf{out}}, \vec{\mathbf{x}}_0 : \overrightarrow{\neg B}_0 \vdash^{\mathsf{v}} \vec{V}_1 : \overrightarrow{\neg A}_1$$
$$\vec{\mathbf{z}}_{\mathsf{out}} : \overrightarrow{\neg B}_{\mathsf{out}}, \vec{\mathbf{z}}_0 : \overrightarrow{\neg A}_0, \vec{\mathbf{z}}_1 : \overrightarrow{\neg A}_1 \vdash^{\mathsf{v}} \vec{W}_1 : \overrightarrow{\neg B}_1$$
$$\vec{\mathbf{x}}_{\mathsf{out}} : \overrightarrow{\neg A}_{\mathsf{out}}, \vec{\mathbf{x}}_0 : \overrightarrow{\neg B}_0, \vec{\mathbf{x}}_1 : \overrightarrow{\neg B}_1 \vdash^{\mathsf{n}} M$$

We define transitions between tables in Fig. 4. The key result is the following:

A *command table* T is a collection of terms, either

of the form (2)

$\overrightarrow{\neg A}_{\mathsf{out}}$	$\overrightarrow{\neg B}_{\mathsf{out}}$	$\overrightarrow{\neg A}_0$	$\overrightarrow{\neg B}_0$	$\cdots$	$\overrightarrow{\neg A}_{n-1}$	$\overrightarrow{\neg B}_{n-1}$	$\vdash^{\mathsf{n}}$
$\vec{\mathbf{x}}_{\mathsf{out}}$		$\vec{V}_0$	$\vec{\mathbf{x}}_0$	$\cdots$	$\vec{V}_{n-1}$	$\vec{\mathbf{x}}_{n-1}$	M
	$\vec{\mathbf{z}}_{\mathsf{out}}$	$\vec{\mathbf{z}}_0$	$\vec{W}_0$	$\cdots$	$\vec{\mathbf{z}}_{n-1}$	$\vec{W}_{n-1}$	

or of the form (3)

$\overrightarrow{\neg A}_{\mathsf{out}}$	$\overrightarrow{\neg B}_{\mathsf{out}}$	$\overrightarrow{\neg A}_0$	$\overrightarrow{\neg B}_0$	$\cdots$	$\overrightarrow{\neg A}_{n-1}$	$\overrightarrow{\neg B}_{n-1}$	$\overrightarrow{\neg A}_n$	$\vdash^{\mathsf{n}}$
$\vec{\mathbf{x}}_{\mathsf{out}}$		$\vec{V}_0$	$\vec{\mathbf{x}}_0$	$\cdots$	$\vec{V}_{n-1}$	$\vec{\mathbf{x}}_{n-1}$	$\vec{V}_n$	
	$\vec{\mathbf{z}}_{\mathsf{out}}$	$\vec{\mathbf{z}}_0$	$\vec{W}_0$	$\cdots$	$\vec{\mathbf{z}}_{n-1}$	$\vec{W}_{n-1}$	$\vec{\mathbf{z}}_n$	M

We define the command $\vec{\mathbf{x}}_{\mathsf{out}} : \overrightarrow{\neg A}_{\mathsf{out}}, \vec{\mathbf{z}}_{\mathsf{out}} : \overrightarrow{\neg B}_{\mathsf{out}} \vdash^{\mathsf{n}}$ subst T to be

$$M[\vec{\mathbf{x}}_{n-1}\backslash\vec{W}_{n-1}][\vec{\mathbf{z}}_{n-1}\backslash\vec{V}_{n-1}]\cdots[\vec{\mathbf{x}}_0\backslash\vec{W}_0][\vec{\mathbf{z}}_0\backslash\vec{V}_0] \quad \text{in case (2)}$$
$$M[\vec{\mathbf{z}}_n\backslash\vec{V}_n][\vec{\mathbf{x}}_{n-1}\backslash\vec{W}_{n-1}][\vec{\mathbf{z}}_{n-1}\backslash\vec{V}_{n-1}]\cdots[\vec{\mathbf{x}}_0\backslash\vec{W}_0][\vec{\mathbf{z}}_0\backslash\vec{V}_0] \quad \text{in case (3)}$$

A *value-sequence table* R is a collection of terms, either

of the form (4)

$\overrightarrow{\neg A}_{\mathsf{out}}$	$\overrightarrow{\neg B}_{\mathsf{out}}$	$\overrightarrow{\neg A}_0$	$\overrightarrow{\neg B}_0$	$\cdots$	$\overrightarrow{\neg A}_{n-1}$	$\overrightarrow{\neg B}_{n-1}$	$\vdash^{\mathsf{v}} \overrightarrow{\neg C}$
$\vec{\mathbf{x}}_{\mathsf{out}}$		$\vec{V}_0$	$\vec{\mathbf{x}}_0$	$\cdots$	$\vec{V}_{n-1}$	$\vec{\mathbf{x}}_{n-1}$	$\vec{U}$
	$\vec{\mathbf{z}}_{\mathsf{out}}$	$\vec{\mathbf{z}}_0$	$\vec{W}_0$	$\cdots$	$\vec{\mathbf{z}}_{n-1}$	$\vec{W}_{n-1}$	

or of the form (5)

$\overrightarrow{\neg A}_{\mathsf{out}}$	$\overrightarrow{\neg B}_{\mathsf{out}}$	$\overrightarrow{\neg A}_0$	$\overrightarrow{\neg B}_0$	$\cdots$	$\overrightarrow{\neg A}_{n-1}$	$\overrightarrow{\neg B}_{n-1}$	$\overrightarrow{\neg A}_n$	$\vdash^{\mathsf{v}} \overrightarrow{\neg C}$
$\vec{\mathbf{x}}_{\mathsf{out}}$		$\vec{V}_0$	$\vec{\mathbf{x}}_0$	$\cdots$	$\vec{V}_{n-1}$	$\vec{\mathbf{x}}_{n-1}$	$\vec{V}_n$	
	$\vec{\mathbf{z}}_{\mathsf{out}}$	$\vec{\mathbf{z}}_0$	$\vec{W}_0$	$\cdots$	$\vec{\mathbf{z}}_{n-1}$	$\vec{W}_{n-1}$	$\vec{\mathbf{z}}_n$	$\vec{U}$

We define the value-sequence $\vec{\mathbf{x}}_{\mathsf{out}} : \overrightarrow{\neg A}_{\mathsf{out}}, \vec{\mathbf{z}}_{\mathsf{out}} : \overrightarrow{\neg B}_{\mathsf{out}} \vdash^{\mathsf{v}}$ subst $R : \overrightarrow{\neg C}$ just as for command tables.

Fig. 3. Alternating tables and substitution

Proposition 5. *1. There is no infinite chain of consecutive switching transitions* $T_0 \rightsquigarrow_{\mathrm{switch}} T_1 \rightsquigarrow_{\mathrm{switch}} T_2 \rightsquigarrow_{\mathrm{switch}} \cdots$

2. *If* $T_0 \rightsquigarrow_{\mathrm{inner}} T_1$, *then* $\mathrm{subst}\ T_0 \rightsquigarrow \mathrm{subst}\ T_1$.
3. *If* $T_0 \rightsquigarrow_{\mathrm{switch}} T_1$, *then* $\mathrm{subst}\ T_0 = \mathrm{subst}\ T_1$.
4. *If* $T_0 \overset{ip}{\rightsquigarrow} T_1$ *then* $\mathrm{subst}\ T_0 \overset{ip}{\rightsquigarrow} \mathrm{subst}\ T_1$.
5. *Let* R *be a value-sequence table of type* $\overrightarrow{\neg C}$ *(in the rightmost column), and let* $j \in \$|\overrightarrow{\neg C}|$ *and* $q \in \mathsf{ult}^{\mathsf{v}}(C_j)$. *Then* $\mathrm{subst}\ (R : jq) = (\mathrm{subst}\ R) : jq$.

Proof. (1) In a command-table T that is *inner-terminal* (i.e. the command is terminal), the command will be of the form $\mathtt{x}p(\overrightarrow{U})$, where $\mathtt{x}$ is declared in some column of T. We call this column $\mathrm{col}(T)$. If $T_0 \rightsquigarrow_{\mathrm{switch}} T_1$, and T_1 is itself inner-terminal, then $\mathrm{col}(T_1)$ must be to the left of $\mathrm{col}(T_0)$. To see this, suppose that T_0 is of the form (2). Then the command of T_0 is $\mathtt{x}_{m,i}p(\overrightarrow{V}_n)$, and the command of T_1 is $W_{m,i}p(\overrightarrow{\mathtt{z}}_n)$. Since T_1 is inner-terminal, $W_{m,i}$ must be an identifier, declared in the context of $W_{m,i}$ i.e. somewhere to the left of column $\mathrm{col}(T_0)$. Similarly if T_0 is of the form (3).

Hence, if there are N columns to the left of $\mathrm{col}(T_0)$ (counting the two "outer" columns as one) then there are at most N switching transitions from T_0.

(2)–(5) are trivial.

We say that two tables are *componentwise bisimilar* when they have the same types and identifiers, the corresponding value sequences are related by $\approx^{\mathsf{v}}$, and the commands (if they are command tables) are related by $\approx$.

Proposition 6. *If* T, T' *are command tables that are componentwise bisimilar, then* $\mathrm{subst}\ T \approx \mathrm{subst}\ T'$. *If* R, R' *are value-sequence tables that are componentwise bisimilar, then* $\mathrm{subst}\ R \approx^{\mathsf{v}} \mathrm{subst}\ R'$.

Proof. Let $\mathcal{R}$ be the set of pairs $(\mathrm{subst}\ T, \mathrm{subst}\ T')$, where T, T' are command tables that are componentwise bisimilar. Then $(\mathrm{subst}\ R, \mathrm{subst}\ R') \in \mathcal{R}^{\mathsf{v}}$ whenever R, R' are value-sequence tables that are componentwise bisimilar. We wish to show that $\mathcal{R}$ is a normal form bisimulation.

We show, by induction on n, that if $\mathrm{subst}\ T \rightsquigarrow^n \overset{ip}{\rightsquigarrow} U$, then $T(\rightsquigarrow_{\mathrm{inner}} \cup \rightsquigarrow_{\mathrm{switch}})^* \overset{ip}{\rightsquigarrow} R$ where $\mathrm{subst}\ R = U$. For this, Prop. 5 gives us $T \rightsquigarrow^*_{\mathrm{switch}} T_1 \not\rightsquigarrow_{\mathrm{switch}}$ with $\mathrm{subst}\ T = \mathrm{subst}\ T_1$ and the rest is straightforward.

We next show, by induction on n, that if $(\mathrm{subst}\ T, \mathrm{subst}\ T') \in \mathcal{R}$ and $T(\rightsquigarrow^*_{\mathrm{inner}}\rightsquigarrow_{\mathrm{switch}})^n \rightsquigarrow^*_{\mathrm{inner}} \overset{ip}{\rightsquigarrow} R$ then $T'(\rightsquigarrow^*_{\mathrm{inner}}\rightsquigarrow_{\mathrm{switch}})^n \rightsquigarrow^*_{\mathrm{inner}} \overset{ip}{\rightsquigarrow} R'$ for some R' componentwise bisimilar to R; and hence $\mathrm{subst}\ T' \rightsquigarrow^* \overset{ip}{\rightsquigarrow} \mathrm{subst}\ R'$. The inductive step uses Prop. 2.

These two facts give us the required property of $\mathcal{R}$.

The first part of Prop. 3 is now given by

$$M[\overrightarrow{W/\mathtt{x}}] = \mathrm{subst}\ \begin{array}{|c|c||c||c|}\hline \epsilon & \overrightarrow{\neg B} & \overrightarrow{\neg A} & \vdash^{\mathsf{n}} \\ \hline & & \overrightarrow{\mathtt{x}} & M \\ & \overrightarrow{\mathtt{y}} & \overrightarrow{W} & \\ \hline \end{array} \approx \mathrm{subst}\ \begin{array}{|c|c||c||c|}\hline \epsilon & \overrightarrow{\neg B} & \overrightarrow{\neg A} & \vdash^{\mathsf{n}} \\ \hline & & \overrightarrow{\mathtt{x}} & M' \\ & \overrightarrow{\mathtt{y}} & \overrightarrow{W'} & \\ \hline \end{array} = M'[\overrightarrow{W'/\mathtt{x}}]$$

and the second part is similar.

Let T be of the form (2). There are 3 possibilities for M:

– If $M \rightsquigarrow M'$, we have an *inner transition*

$T \quad \rightsquigarrow_{\text{inner}}$

$\overrightarrow{\neg A}_{\text{out}}$	$\overrightarrow{\neg B}_{\text{out}}$	$\overrightarrow{\neg A}_0$	$\overrightarrow{\neg B}_0$	$\cdots$	$\overrightarrow{\neg A}_{n-1}$	$\overrightarrow{\neg B}_{n-1}$	$\vdash^{\mathsf{n}}$
$\overrightarrow{\mathtt{x}}_{\text{out}}$		$\overrightarrow{V}_0$	$\overrightarrow{\mathtt{x}}_0$	$\cdots$	$\overrightarrow{V}_{n-1}$	$\overrightarrow{\mathtt{x}}_{n-1}$	M'
	$\overrightarrow{\mathtt{z}}_{\text{out}}$	$\overrightarrow{\mathtt{z}}_0$	$\overrightarrow{W}_0$	$\cdots$	$\overrightarrow{\mathtt{z}}_{n-1}$	$\overrightarrow{W}_{n-1}$	

– If $M = \mathtt{x}_{m,i}p(\overrightarrow{U})$, we have a *switching transition*

$T \quad \rightsquigarrow_{\text{switch}}$

$\overrightarrow{\neg A}_{\text{out}}$	$\overrightarrow{\neg B}_{\text{out}}$	$\overrightarrow{\neg A}_0$	$\overrightarrow{\neg B}_0$	$\cdots$	$\overrightarrow{\neg A}_{n-1}$	$\overrightarrow{\neg B}_{n-1}$	$H(p)$	$\vdash^{\mathsf{n}}$
$\overrightarrow{\mathtt{x}}_{\text{out}}$		$\overrightarrow{V}_0$	$\overrightarrow{\mathtt{x}}_0$	$\cdots$	$\overrightarrow{V}_{n-1}$	$\overrightarrow{\mathtt{x}}_{n-1}$	$\overrightarrow{U}$	
	$\overrightarrow{\mathtt{z}}_{\text{out}}$	$\overrightarrow{\mathtt{z}}_0$	$\overrightarrow{W}_0$	$\cdots$	$\overrightarrow{\mathtt{z}}_{n-1}$	$\overrightarrow{W}_{n-1}$	$\overrightarrow{\mathtt{z}}_n$	$W_{m,i}p(\overrightarrow{\mathtt{z}}_n)$

– If $M = \mathtt{x}_{\text{out},i}p(\overrightarrow{U})$, we have an *outer Proponent move*

$T \quad \overset{ip}{\rightsquigarrow}$

$\overrightarrow{\neg A}_{\text{out}}$	$\overrightarrow{\neg B}_{\text{out}}$	$\overrightarrow{\neg A}_0$	$\overrightarrow{\neg B}_0$	$\cdots$	$\overrightarrow{\neg A}_{n-1}$	$\overrightarrow{\neg B}_{n-1}$	$\vdash^{\mathsf{v}} H(p)$
$\overrightarrow{\mathtt{x}}_{\text{out}}$		$\overrightarrow{V}_0$	$\overrightarrow{\mathtt{x}}_0$	$\cdots$	$\overrightarrow{V}_{n-1}$	$\overrightarrow{\mathtt{x}}_{n-1}$	$\overrightarrow{U}$
	$\overrightarrow{\mathtt{z}}_{\text{out}}$	$\overrightarrow{\mathtt{z}}_0$	$\overrightarrow{W}_0$	$\cdots$	$\overrightarrow{\mathtt{z}}_{n-1}$	$\overrightarrow{W}_{n-1}$	

The case where T is of the form (3) is similar, with $T \overset{ip}{\rightsquigarrow}$ replaced by $T \overset{(|\overrightarrow{\mathtt{x}}|+i)p}{\rightsquigarrow}$.
Let R be of the form (4). For $j \in \$|\overrightarrow{\neg C}|$ and $q \in \mathsf{ult}^{\mathsf{v}}(E_j)$, we define

$R : jq \quad \overset{\text{def}}{=}$

$\overrightarrow{\neg A}_{\text{out}}, H(q)$	$\overrightarrow{\neg B}_{\text{out}}$	$\overrightarrow{\neg A}_0$	$\overrightarrow{\neg B}_0$	$\cdots$	$\overrightarrow{\neg A}_{n-1}$	$\overrightarrow{\neg B}_{n-1}$	$\vdash^{\mathsf{n}}$
$\overrightarrow{\mathtt{x}}_{\text{out}}, \overrightarrow{\mathtt{y}}$		$\overrightarrow{V}_0$	$\overrightarrow{\mathtt{x}}_0$	$\cdots$	$\overrightarrow{V}_{n-1}$	$\overrightarrow{\mathtt{x}}_{n-1}$	$U_j q(\overrightarrow{\mathtt{y}})$
	$\overrightarrow{\mathtt{z}}_{\text{out}}$	$\overrightarrow{\mathtt{z}}_0$	$\overrightarrow{W}_0$	$\cdots$	$\overrightarrow{\mathtt{z}}_{n-1}$	$\overrightarrow{W}_{n-1}$	

where $\overrightarrow{\mathtt{y}}$ is fresh (this is called an *outer Opponent move*). The case where R is of the form (5) is similar.

Fig. 4. Transitions between alternating tables

4 Examples

4.1 Fixed Point Combinators Are Unique

To illustrate the use of normal form bisimulation, we now prove that at each type, there is a unique fixpoint combinator up to normal form bisimilarity. The proof is similar to the classical result that all λ-calculus fixed point combinators have the same Böhm tree [30].

Theorem 1. *All solutions* $\vdash^{\mathsf{v}} U : \neg(A \times \neg(A \times \neg A))$ *to the fixed point equation*

$$\vdash^{\mathsf{v}} U \approx^{\mathsf{v}} \lambda\langle \mathtt{u}, \mathtt{f}\rangle.\mathtt{f}\langle \mathtt{u}, \lambda\mathtt{v}.U\langle \mathtt{v}, \mathtt{f}\rangle\rangle : \neg(A \times \neg(A \times \neg A))$$

are normal form bisimilar.

Proof. Suppose U_1 and U_2 are both solutions. We show they are bisimilar

$$\vdash^{\mathsf{v}} U_1 \precsim^{\mathsf{v}} U_2 : \neg(A \times \neg(A \times \neg A)) \tag{6}$$

by exhibiting a bisimulation relation that relates

$$\overrightarrow{\mathtt{y}} : H(q) \vdash^{\mathsf{n}} U_1 q(\overrightarrow{\mathtt{y}}), U_2 q(\overrightarrow{\mathtt{y}}) \tag{7}$$

for all $q \in \mathsf{ult}^{\mathsf{v}}(A \times \neg(A \times \neg A))$, that is, all $q = \langle p, -_{\neg(A\times\neg A)}\rangle$ where $p \in \mathsf{ult}^{\mathsf{v}}(A)$.

We define $\mathcal{R}$ to be the relation between all commands $\Gamma \vdash^{\mathsf{n}} M_1, M_2$ where

$$\Gamma \vdash^{\mathsf{n}} M_i \precsim U_i\langle p(\overrightarrow{\mathtt{x}}), \mathtt{f}\rangle, \quad \text{for } i \in \{1,2\},$$

and $p \in \mathsf{ult}^{\mathsf{v}}(A)$ and $\Gamma = \overrightarrow{\mathtt{x}} : H(p), \mathtt{f} : \neg(A \times \neg A), \overrightarrow{\mathtt{y} : \neg B}$. Then $\mathcal{R} \cup \precsim$ is a normal form bisimulation. To see this, let $p' = \langle p, -_{\neg A}\rangle$ and observe that since $\Gamma \vdash^{\mathsf{n}} M_i \precsim U_i\langle p(\overrightarrow{\mathtt{y}}), \mathtt{f}\rangle$ and U_i is a fixed point, there exist $\overrightarrow{W_i}, V_i$ such that

$$M_i \leadsto^* \mathtt{f}\langle p(\overrightarrow{W_i}), V_i\rangle = \mathtt{f}p'(\overrightarrow{W_i}, V_i) \overset{\mathtt{f}p'}{\leadsto} \overrightarrow{W_i}, V_i$$
$$\Gamma \vdash^{\mathsf{v}} \overrightarrow{W_i} \precsim^{\mathsf{v}} \overrightarrow{\mathtt{x}} : H(p)$$
$$\Gamma \vdash^{\mathsf{v}} V_i \precsim^{\mathsf{v}} \lambda \mathtt{v}.U_i\langle \mathtt{v}, \mathtt{f}\rangle : \neg A$$

for $i \in \{1,2\}$. Hence $\Gamma \vdash^{\mathsf{v}} \overrightarrow{W_1} \precsim^{\mathsf{v}} \overrightarrow{W_2} : H(p)$ and $\Gamma \vdash^{\mathsf{v}} V_1 \mathcal{R}^{\mathsf{v}} V_2 : \neg A$. We conclude that $\mathcal{R} \cup \precsim$ is a bisimulation and thus (6), since $\mathcal{R}$ relates (7).

4.2 Syntactic Minimal Invariance

Finally, we prove a syntactic version of the domain-theoretic minimal invariance property [31] for JWA. Our proof is greatly facilitated by the normal form bisimulation proof principle and is simpler than other syntactic minimal invariance proofs in the literature for typed and untyped calculi [32, 33].

For every closed type A, we define the type $A^\dagger \overset{\text{def}}{=} A \times \neg A$ and we will define a closed term $\vdash^{\mathsf{v}} h(A) : \neg A^\dagger$. More generally, to deal with recursive types, we define in Figure 4.2, by structural induction on A, open terms:

$$\Gamma \vdash^{\mathsf{v}} h(\Gamma \vdash A) : \neg A[\Gamma]^\dagger,$$

where

- $\Gamma = \overrightarrow{\mathtt{X} : \neg B^\dagger}$ (we take the liberty to use $\overrightarrow{\mathtt{X}}$ as term identifiers in Γ and $h(\Gamma \vdash A)$ and as type identifiers in A),
- the types in $\overrightarrow{B}$ are closed,
- A is an open type: $\overrightarrow{\mathtt{X}} \vdash A$ type, and
- $[\Gamma]$ denotes the type substitution $[\overrightarrow{B/\mathtt{X}}]$.

When A is closed, $h(A) \overset{\text{def}}{=} h(\vdash A)$.

$$
\begin{aligned}
h(\Gamma \vdash \neg A_0) &= \lambda\langle \mathtt{x}, \mathtt{k}\rangle.\mathtt{k}\,\lambda \mathtt{x}_0.h(\Gamma \vdash A_0)\langle \mathtt{x}_0, \mathtt{x}\rangle \\
h(\Gamma \vdash \textstyle\sum_{i\in I} A_i) &= \lambda\langle \mathtt{x}, \mathtt{k}\rangle.\mathtt{pm\ x\ as}\ \{\langle i, \mathtt{x}_i\rangle.h(\Gamma \vdash A_i)\langle \mathtt{x}_i, \lambda \mathtt{y}_i.\mathtt{k}\langle i, \mathtt{y}_i\rangle\rangle\}_{i\in I} \\
h(\Gamma \vdash 1) &= \lambda\langle \mathtt{x}, \mathtt{k}\rangle.\mathtt{k}\,\mathtt{x} \\
h(\Gamma \vdash A_1 \times A_2) &= \lambda\langle \mathtt{x}, \mathtt{k}\rangle.\mathtt{pm\ x\ as}\ \langle \mathtt{x}_1, \mathtt{x}_2\rangle. \\
&\qquad h(\Gamma \vdash A_1)\langle \mathtt{x}_1, \lambda \mathtt{y}_1.h(\Gamma \vdash A_2)\langle \mathtt{x}_2, \lambda \mathtt{y}_2.\mathtt{k}\langle \mathtt{y}_1, \mathtt{y}_2\rangle\rangle\rangle \\
h(\Gamma \vdash \mathtt{X}) &= \mathtt{X} \\
h(\Gamma \vdash \mu\mathtt{X}.A) &= \lambda \mathtt{u}.Y\langle \mathtt{u}, \lambda\langle \mathtt{v}, \mathtt{X}\rangle. \\
&\qquad \mathtt{pm\ v\ as}\ \langle \mathtt{x}, \mathtt{k}\rangle. \\
&\qquad \mathtt{pm\ x\ as\ fold\ x}_0. \\
&\qquad h(\Gamma, \mathtt{X} : \neg(\mu\mathtt{X}.A)^\dagger \vdash A)\langle \mathtt{x}_0, \lambda \mathtt{y}_0.\mathtt{k}(\mathtt{fold\ y}_0)\rangle\rangle
\end{aligned}
$$

Fig. 5. Definition of $h(\Gamma \vdash A)$

Proposition 7. $h(A[\Gamma]) = h(\Gamma \vdash A)[\overrightarrow{h(B)/\mathtt{X}}]$, *if* $\Gamma = \overrightarrow{\mathtt{X} : \neg B^\dagger}$.

Proof. By structural induction on A.

In particular, if $\mu\mathtt{X}.A$ is closed,

$$h(A[\mu\mathtt{X}.A/\mathtt{X}]) = h(\mathtt{X} : \neg(\mu\mathtt{X}.A)^\dagger \vdash A)[h(\mu\mathtt{X}.A)/\mathtt{X}] \tag{8}$$

Lemma 1. *Let* $g(V : \neg A) = \lambda \mathtt{y}.h(A)\langle \mathtt{y}, V\rangle$ *and, by extension, let* $g(\overrightarrow{V : \neg A})$ *be the value sequence where* $g(\overrightarrow{V : \neg A})_i = g(V_i : \neg A_i)$*. Then, for all closed* A *and* $p \in \mathsf{ult}^{\mathsf{v}}(A)$*,* $h(A)\langle p(\overrightarrow{V}), \mathtt{k}\rangle \rightsquigarrow^* \mathtt{k}(p(g(\overrightarrow{V : H(p)})))$.

Proof. By structural induction on p. For illustration, we show the induction step for the case when $A = \mu\mathtt{X}.A_0$ and $p = \mathtt{fold}\ p_0$. Let $K = \lambda \mathtt{y}_0.\mathtt{k}(\mathtt{fold\ y}_0)$.

$$
\begin{aligned}
h(A)\langle p(\overrightarrow{V}), \mathtt{k}\rangle &\rightsquigarrow^* h(\mathtt{X} : \neg A^\dagger \vdash A_0)[h(A)/\mathtt{X}]\langle p_0(\overrightarrow{V}), K\rangle \\
&= h(A_0[h(A)/\mathtt{X}])\langle p_0(\overrightarrow{V}), K\rangle, \text{ by (8)} \\
&\rightsquigarrow^* K(p_0(\overrightarrow{V : H(p_0)})), \text{ by the induction hypothesis} \\
&\rightsquigarrow \mathtt{k}(\mathtt{fold}\ p_0(\overrightarrow{V : H(p_0)})) = \mathtt{k}(p(\overrightarrow{V : H(p)}))
\end{aligned}
$$

Theorem 2 (Syntactic minimal invariance). *For all closed types* A*,* $\vdash^{\mathsf{v}} h(A) \approx^{\mathsf{v}} \lambda\langle \mathtt{x}, \mathtt{k}\rangle.\mathtt{k}\,\mathtt{x} : \neg A^\dagger$.

Proof. We need to show, for all $p \in \mathsf{ult}^{\mathsf{v}}(A)$,

$$\mathtt{k} : \neg A, \overrightarrow{\mathtt{x} : H(p)} \vdash^{\mathsf{n}} h(\Gamma \vdash A)\langle p(\overrightarrow{\mathtt{x}}), \mathtt{k}\rangle \approx (\lambda\langle \mathtt{x}, \mathtt{k}\rangle.\mathtt{k}\,\mathtt{x})\langle p(\overrightarrow{\mathtt{x}}), \mathtt{k}\rangle$$

By Lemma 1, the left hand side β-reduces to $\mathtt{k}(p(g(\overrightarrow{\mathtt{x} : H(p)})))$ and the right hand side β-reduces to $\mathtt{k}(p(\overrightarrow{\mathtt{x}}))$. It remains to show that

$$\mathtt{k} : \neg A, \overrightarrow{\mathtt{x} : H(p)} \vdash^{\mathsf{v}} g(\overrightarrow{\mathtt{x} : H(p)}) \precsim^{\mathsf{v}} \overrightarrow{\mathtt{x}} : H(p)$$

This follows because the relation that relates, for all closed A and $p \in \mathsf{ult}^{\mathsf{v}}(A)$,

$$\overrightarrow{\mathtt{z} : \neg B}, \mathtt{x} : \neg A, \overrightarrow{\mathtt{y} : H(p)} \vdash^{\mathsf{n}} g(\mathtt{x} : \neg A)(p(\overrightarrow{\mathtt{y}})), \mathtt{x}(p(\overrightarrow{\mathtt{y}}))$$

is a bisimulation, which is immediate from the calculation (using Lemma 1)

$$g(\mathtt{x} : \neg A)(p(\overrightarrow{\mathtt{y}})) \rightsquigarrow h(A)\langle p(\overrightarrow{\mathtt{y}}), \mathtt{x}\rangle \rightsquigarrow^{*} \mathtt{x}(p(g(\overrightarrow{\mathtt{y} : H(p)})))$$

References

1. Sangiorgi, D.: The lazy lambda calculus in a concurrency scenario. Information and Computation 111(1), 120–153 (1994)
2. Jagadeesan, R., Pitcher, C., Riely, J.: Open bisimulation for aspects. In: Intl. Conf. on Aspect-Oriented Software Development, ACM, pp. 209–224. ACM Press, New York (2007)
3. Lassen, S.B.: Bisimulation in untyped lambda calculus: Böhm trees and bisimulation up to context. In: MFPS XV. ENTCS, vol. 20, pp. 346–374. Elsevier, Amsterdam (1999)
4. Lassen, S.B.: Eager normal form bisimulation. In: 20th LICS, pp. 345–354. IEEE Computer Society Press, Los Alamitos (2005)
5. Lassen, S.B.: Normal form simulation for McCarty's amb. In: MFPS XXI. ENTCS, vol. 155, pp. 445–465. Elsevier, Amsterdam (2005)
6. Lassen, S.B.: Head normal form bisimulation for pairs and the $\lambda\mu$-calculus. In: 21st LICS, pp. 297–306. IEEE Computer Society Press, Los Alamitos (2006) (extended abstract)
7. Støvring, K., Lassen, S.B.: A complete, co-inductive syntactic theory of sequential control and state. In: 34th POPL, pp. 63–74. ACM Press, New York (2007)
8. Abramsky, S.: The lazy λ-calculus. In: Turner, D. (ed.) Research Topics in Functional Programming, pp. 65–116. Addison-Wesley, Reading (1990)
9. Gordon, A.D.: Functional Programming and Input/Output. CUP (1994)
10. Levy, P.B.: Call-By-Push-Value. A Functional/Imperative Synthesis. In: Semantic Struct. in Computation, Springer, Heidelberg (2004)
11. Levy, P.B.: Adjunction models for call-by-push-value with stacks. Theory and Applications of Categories 14(5), 75–110 (2005)
12. Merro, M., Biasi, C.: On the observational theory of the CPS calculus. In: MFPS XXII. ENTCS, vol. 158, pp. 307–330. Elsevier, Amsterdam (2006)
13. Hyland, J.M.E., Ong, C.H.L.: On full abstraction for PCF: I, II, and III. Information and Computation 163(2), 285–408 (2000)
14. Abramsky, S., McCusker, G.: Call-by-value games. In: Nielsen, M. (ed.) CSL 1997. LNCS, vol. 1414, pp. 1–17. Springer, Heidelberg (1998)
15. Abramsky, S., Honda, K., McCusker, G.: A fully abstract game semantics for general references. In: 13th LICS, pp. 334–344. IEEE Computer Society Press, Los Alamitos (1998)

16. Laird, J.: Full abstraction for functional languages with control. In: 12th LICS, pp. 58–67. IEEE Computer Society Press, Los Alamitos (1997)
17. Ker, A.D., Nickau, H., Ong, C.H.L.: Innocent game models of untyped lambda-calculus. Theoretical Computer Science 272(1-2), 247–292 (2002)
18. Ker, A.D., Nickau, H., Ong, C.H.L.: Adapting innocent game models for the Böhm tree λ-theory. Theoretical Computer Science 308(1-3), 333–366 (2003)
19. Ong, C.H.L., Gianantonio, P.D.: Games characterizing Lévy-Longo trees. Theoretical Computer Science 312(1), 121–142 (2004)
20. Laird, J.: A categorical semantics of higher-order store. In: 9th Conference on Category Theory and Computer Science. ENTCS, vol. 69, Elsevier, Amsterdam (2003)
21. Curien, P.L., Herbelin, H.: Computing with abstract Böhm trees. In: Fuji International Symposium on Functional and Logic Programming, pp. 20–39 (1998)
22. Danos, V., Herbelin, H., Regnier, L.: Game semantics and abstract machines. In: 11th LICS, pp. 394–405. IEEE Computer Society Press, Los Alamitos (1996)
23. Levy, P.B.: Infinite trace equivalence. In: MFPS XXI. ENTCS, vol. 155, pp. 467–496. Elsevier, Amsterdam (2006)
24. Levy, P.B.: Game semantics using function inventories.Talk given at Geometry of Computation 2006, Marseille (2006)
25. Laird, J.: A fully abstract trace semantics for general references. In: 34th ICALP. LNCS, vol. 4596, Springer, Heidelberg (2007)
26. Hyland, J.M.E., Ong, C.H.L.: Pi-calculus, dialogue games and PCF. In: 7th FPCA, pp. 96–107. ACM Press, New York (1995)
27. Fiore, M.P., Honda, K.: Recursive types in games: Axiomatics and process representation. In: 13th LICS, pp. 345–356. IEEE Computer Society Press, Los Alamitos (1998)
28. Thielecke, H.: Categorical Structure of Continuation Passing Style. PhD thesis, University of Edinburgh (1997)
29. Thielecke, H.: Contrasting exceptions and continuations. Unpublished (October 2001)
30. Barendregt, H.P.: The Lambda Calculus: Its Syntax and Semantics. Revised edn., North-Holland, Amsterdam (1984)
31. Pitts, A.M.: Relational properties of domains. Information and Computation 127, 66–90 (1996)
32. Birkedal, L., Harper, R.: Operational interpretations of recursive types in an operational setting (summary). In: Ito, T., Abadi, M. (eds.) TACS 1997. LNCS, vol. 1281, Springer, Heidelberg (1997)
33. Lassen, S.B.: Relational reasoning about contexts. In: Gordon, A.D., Pitts, A.M. (eds.) Higher Order Operational Techniques in Semantics.CUP, pp. 91–135 (1998)

Not Enough Points Is Enough

Antonio Bucciarelli[1], Thomas Ehrhard[1,2], and Giulio Manzonetto[1,3]

[1] Laboratoire PPS, CNRS UMR 7126-Université Paris 7,
2, place Jussieu (case 7014), 75251 Paris Cedex 05, France
[2] Institut de Mathématiques de Luminy CNRS UMR 6206,
163, avenue de Luminy, case 907 13288 Marseille, France
[3] Università Ca'Foscari di Venezia, Dipartimento di Informatica,
Via Torino 155, 30172 Venezia, Italia
{antonio.bucciarelli, thomas.ehrhard, giulio.manzonetto}@pps.jussieu.fr

Abstract. Models of the untyped λ-calculus may be defined either as applicative structures satisfying a bunch of first-order axioms (λ-models), or as reflexive objects in cartesian closed categories (categorical models). In this paper we show that any categorical model of λ-calculus can be presented as a λ-model, even when the underlying category does not have enough points. We provide an example of an extensional model of λ-calculus in a category of sets and relations which has not enough points. Finally, we present some of its algebraic properties which make it suitable for dealing with non-deterministic extensions of λ-calculus.

Keywords: λ-calculus, cartesian closed categories, λ-models, relational model, non-determinism.

1 Introduction

In 1969 Scott constructed the first mathematical model of the untyped λ-calculus [24], but only at the end of the seventies, researchers were able to provide general definitions of λ-calculus model. Barendregt, inspired by proof theoretical considerations such as ω-incompleteness [22], introduced two classes of models: the λ-algebras and the λ-models [4, Ch. 5].

Taking for granted the definition of *combinatory algebra* $(A, \cdot, \mathbf{k}, \mathbf{s})$, we recall that:

- A λ-*algebra* is a combinatory algebra satisfying the five combinatory axioms of Curry [4, Thm. 5.2.5].
- A λ-*model* is a λ-algebra satisfying the Meyer-Scott (or *weak extensionality*) axiom: $\forall x\ (a\cdot x = b\cdot x) \Rightarrow \varepsilon\cdot a = \varepsilon\cdot b$ where ε is the combinator $\mathbf{s}\cdot(\mathbf{k}\cdot((\mathbf{s}\cdot\mathbf{k})\cdot\mathbf{k}))$.

All the other known notions of model coincide essentially with λ-models except for categorical models which have been proved equivalent to λ-algebras. More precisely, given a λ-model, it is always possible to define a cartesian closed category (ccc, for short) where its carrier set is a reflexive object with enough points [3, Sec. 9.5]. On the other side, by applying a construction due to Koymans [19] and based on work of Scott, arbitrary reflexive objects in ccc's give rise

J. Duparc and T.A. Henzinger (Eds.): CSL 2007, LNCS 4646, pp. 298–312, 2007.

to λ-algebras and to all of them. Moreover, using this method, the λ-models are exactly those λ-algebras that come from reflexive objects with enough points.

The class of λ-algebras is not sound for λ-theories since, with the failure of the Meyer-Scott axiom, it is no longer guaranteed that (under the interpretation) $M = N$ implies $\lambda x.M = \lambda x.N$. Nevertheless, λ-algebras can be desirable since they satisfy all the provable equations of λ-calculus and constitute an equational class. Hence, Koymans had as *aim* to provide λ-algebras, but this led to a common belief that only the reflexive objects having enough points give rise to λ-models.

The point of our paper, in its first part, is to describe a simple method for turning *any* reflexive object $\mathcal{U}$ of a ccc into a λ-model in such a way that one can easily switch from categorical to algebraic interpretation of λ-terms and *vice versa*. It turns out that the resulting λ-model is isomorphic to the λ-model obtained by freely adjoining the variables of λ-calculus as indeterminates to the λ-algebra associated with $\mathcal{U}$ by Koymans' construction. See [25] for more details.

Our approach asks for countable products. In practice, this hypothesis does not seem to be very restrictive. Nevertheless, we do claim full generality for this method and in Section 3.2 we sketch an alternative, but less simple and natural, construction which does not need this additional hypothesis. Before going further, let us remark that our construction does not give anything new for the categories of domains generally used to solve the domain inequality $U \Rightarrow U \lhd U$ (see, e.g., [24,23,18]), which do have enough points. Once set up the framework allowing to associate a reflexive object (without enough points) of a ccc with a λ-model, we discuss in Section 5 a paradigmatic example to which it can be applied.

In denotational semantics, ccc's without enough points arise naturally when morphisms are not simply functions, but carry some "intensional" information, like for instance sequential algorithms or strategies in various categories of games [6,1,16]. The original motivation for these constructions was the semantic characterization of sequentiality, in the simply typed case. As far as we know, most often the study of reflexive objects in the corresponding ccc's has not been undertaken. Notable exceptions are [11] and [17], where reflexive objects in categories of games yielding the λ-theories $\mathcal{H}^*$ and $\mathcal{B}$, respectively, are defined.

This probably deserves a short digression, from the perspective of the present work: there is of course no absolute need of considering the combinatory algebra associated with a reflexive object, in order to study the λ-theory thereof; it is often a matter of taste whether to use categorical or algebraic notations. What we are proposing here is simply an algebraic counterpart of any categorical model which satisfies weak extensionality.

A framework simpler than game semantics, where reflexive objects cannot have enough points is the following: given the category **Rel** of sets and relations, consider the comonad $\mathcal{M}_f(-)$ of "finite multisets". **MRel**, the Kleisli category of $\mathcal{M}_f(-)$, is a ccc which has been studied in particular as a semantic framework for linear logic [12,2,7].

An even simpler framework, based on **Rel**, would be provided by the functor "finite sets" instead of "finite multisets". The point is that the former is not a comonad. Nevertheless, a ccc may eventually be obtained in this case too, via a "quasi Kleisli" construction [15]. Interestingly, from the perspective of the present work, these Kleisli categories over **Rel** are advocated in [15] as the "natural" framework in which standard models of the λ-calculus like Engeler's model, and graph models [5] in general, *should* live.

In Section 5 we define a relational version, in **MRel**, of another classical model: Scott's $\mathcal{D}_\infty$. Instead of the inverse limit construction, we get our reflexive object $\mathcal{D}$ by an iterated completion operation similar to the canonical completion of graph models. In this case D is isomorphic to $D \Rightarrow D$ by construction.

Finally, in Section 6 we show that the λ-model $\mathcal{M}_\mathcal{D}$ associated with $\mathcal{D}$ by our construction has a rich algebraic structure. In particular, we define two operations of sum and product making the carrier set of $\mathcal{M}_\mathcal{D}$ a commutative semiring, which are left distributive with respect to the application. This opens the way to the interpretation of conjunctive-disjunctive λ-calculi [9] in the relational framework.

2 Preliminaries

To keep this article self-contained, we summarize some definitions and results used in the paper. With regard to the λ-calculus we follow the notation and terminology of [4]. Our main reference for category theory is [3].

2.1 Generalities

Let S be a set. We denote by $\mathcal{P}(S)$ (resp. $\mathcal{P}_\mathrm{f}(S)$) the collection of all subsets (resp. finite subsets) of S and we write $A \subset_\mathrm{f} S$ to express that A is a finite subset of S. A *multiset* m over S can be defined as an unordered list $m = [a_1, a_2, \dots]$ with repetitions such that $a_i \in S$ for all i. For each $a \in S$ the *multiplicity of* a *in* m is the number of occurrences of a in m. Given a multiset m over S, its *support* is the set of elements of S belonging to m. A multiset m is called *finite* if it is a finite list, we denote by $[]$ the empty multiset. Given two multisets $m_1 = [a_1, a_2, \dots]$ and $m_2 = [b_1, b_2, \dots]$ the *multiset union* of m_1, m_2 is defined by $m_1 \uplus m_2 = [a_1, b_1, a_2, b_2, \dots]$. We will write $\mathcal{M}_f(S)$ for the set of all finite multisets over S.

We denote by $\mathbb{N}$ the set of natural numbers. A $\mathbb{N}$-indexed sequence $\sigma = (m_1, m_2, \dots)$ of multisets is *quasi-finite* if $m_i = []$ holds for all but a finite number of indices i; σ_i stands for the i-th element of σ. If S is a set, we denote by $\mathcal{M}_f(S)^{(\omega)}$ the set of all quasi-finite $\mathbb{N}$-indexed sequences of multisets over S. We write $*$ for the $\mathbb{N}$-indexed family of empty multisets, i.e., $*$ is the only inhabitant of $\mathcal{M}_f(\emptyset)^{(\omega)}$.

2.2 Cartesian Closed Categories

Throughout the paper, **C** is a small *cartesian closed category* (ccc, for short). Let A, B, C be arbitrary objects of **C**. We denote by $A \,\&\, B$ the *product*[1] of A

[1] We use the symbol & instead of × because, in the example we are interested in, the categorical product is the disjoint union. The usual notation is kept to denote the set-theoretical product.

and B, by $\pi_1 \in \mathbf{C}(A\&B, A)$, $\pi_2 \in \mathbf{C}(A\&B, B)$ the associated *projections* and, given a pair of arrows $f \in \mathbf{C}(C, A)$ and $g \in \mathbf{C}(C, B)$, by $\langle f, g\rangle \in \mathbf{C}(C, A\&B)$ the unique arrow such that $\pi_1 \circ \langle f, g\rangle = f$ and $\pi_2 \circ \langle f, g\rangle = g$. We write $A \Rightarrow B$ for the *exponential object* and $ev_{AB} \in \mathbf{C}((A \Rightarrow B)\&A, B)$ for the *evaluation morphism*[2]. Moreover, for any object C and arrow $f \in \mathbf{C}(C\&A, B)$, $\Lambda(f) \in \mathbf{C}(C, A \Rightarrow B)$ stands for the (unique) morphism such that $ev_{AB} \circ (\Lambda(f) \times Id_A) = f$. Finally, $\mathbb{1}$ denotes the terminal object and $!_A$ the only morphism in $\mathbf{C}(A, \mathbb{1})$.
We recall that in every ccc the following equalities hold:

$$\text{(pair)} \quad \langle f, g\rangle \circ h = \langle f \circ h, g \circ h\rangle \qquad \Lambda(f) \circ g = \Lambda(f \circ (g \times Id)) \quad \text{(Curry)}$$
$$\text{(beta)} \quad ev \circ \langle \Lambda(f), g\rangle = f \circ \langle Id, g\rangle \qquad \Lambda(ev) = Id \quad \text{(Id-Curry)}$$

We say that $\mathbf{C}$ *has enough points* if, for all $f, g \in \mathbf{C}(A, B)$, whenever $f \neq g$, there exists a morphism $h \in \mathbf{C}(\mathbb{1}, A)$ such that $f \circ h \neq g \circ h$.

2.3 The Untyped λ-Calculus and Its Models

The set Λ of λ-terms over a countable set Var of variables is constructed as usual: every variable is a λ-term; if P and Q are λ-terms, then also are PQ and $\lambda z.P$ for each variable z. As a matter of notation, given a λ-term M, we will denote by $FV(M)$ the set of its free variables. A λ-term M is *closed* if $FV(M) = \emptyset$.

It is well known [4, Ch. 5] that there are, essentially, two tightly linked notions of λ-calculus models. The former is connected with category theory (*categorical models*) and the latter is related to combinatory algebras (λ-*models*).

Categorical Models. A *categorical model* of λ-calculus is a *reflexive object* in a ccc $\mathbf{C}$, that is, a triple $\mathcal{U} = (U, \mathrm{Ap}, \lambda)$ such that U is an object of $\mathbf{C}$, and $\lambda \in \mathbf{C}(U \Rightarrow U, U)$ and $\mathrm{Ap} \in \mathbf{C}(U, U \Rightarrow U)$ satisfy $\mathrm{Ap} \circ \lambda = Id_{U \Rightarrow U}$. In this case we write $U \Rightarrow U \lhd U$. When moreover $\lambda \circ \mathrm{Ap} = Id_U$, the model $\mathcal{U}$ is called *extensional.*

In the sequel we suppose that $\overline{x} = (x_1, \ldots, x_n)$ is a finite ordered sequence of variables without repetitions of length n. Given an arbitrary λ-term M and a sequence $\overline{x}$, we say that $\overline{x}$ is *adequate for M* if $\overline{x}$ contains all the free variables of M. We simply say that $\overline{x}$ is adequate whenever M is clear from the context.

Given a categorical model $\mathcal{U} = (U, \mathrm{Ap}, \lambda)$, for all $M \in \Lambda$ and for all adequate $\overline{x}$, the *interpretation of M* (in $\overline{x}$) is a morphism $|M|_{\overline{x}} \in \mathbf{C}(U^n, U)$ defined by structural induction on M as follows:

- If $M \equiv z$, then $|z|_{\overline{x}} = \pi_i$, if z occurs in i-th position in the sequence $\overline{x}$.
- If $M \equiv PQ$, then by inductive hypothesis we have defined $|P|_{\overline{x}}, |Q|_{\overline{x}} \in \mathbf{C}(U^n, U)$. So we set $|PQ|_{\overline{x}} = ev \circ \langle \mathrm{Ap} \circ |P|_{\overline{x}}, |Q|_{\overline{x}}\rangle \in \mathbf{C}(U^n, U)$.
- If $M \equiv \lambda z.P$, then by inductive hypothesis we have defined $|P|_{\overline{x},z} \in \mathbf{C}(U^n \& U, U)$, where we assume that z does not occur in $\overline{x}$. So we set $|\lambda z.P|_{\overline{x}} = \lambda \circ \Lambda(|P|_{\overline{x},z})$.

It is routine to check that, if M and N are β-equivalent, then $|M|_{\overline{x}} = |N|_{\overline{x}}$ for all $\overline{x}$ adequate for M and N. If the reflexive object $\mathcal{U}$ is extensional, then $|M|_{\overline{x}} = |N|_{\overline{x}}$ holds as soon as M and N are $\beta\eta$-equivalent.

[2] We simply write ev when A and B are clear from the context.

Combinatory Algebras and λ-Models. An *applicative structure* $\mathcal{A} = (A, \cdot)$ is an algebra where $\cdot$ is a binary operation on A called *application*. We may write it infix as $s \cdot t$, or even drop it and write st. Application associates to the left.

A *combinatory algebra* $\mathcal{C} = (C, \cdot, \mathbf{k}, \mathbf{s})$ is an applicative structure with distinguished elements $\mathbf{k}$ and $\mathbf{s}$ satisfying $\mathbf{k}xy = x$ and $\mathbf{s}xyz = xz(yz)$ for all x, y, and z. See, e.g., [8] for a full treatment.

We call $\mathbf{k}$ and $\mathbf{s}$ the *basic combinators*. In the equational language of combinatory algebras the derived combinators $\mathbf{i}$ and ε are defined as $\mathbf{i} \equiv \mathbf{skk}$ and $\varepsilon \equiv \mathbf{s}(\mathbf{ki})$. It is not hard to verify that every combinatory algebra satisfies the identities $\mathbf{i}x = x$ and $\varepsilon xy = xy$.

We say that $c \in C$ *represents* a function $f : C \to C$ (and that f is *representable*) if $cz = f(z)$ for all $z \in C$. Two elements $c, d \in C$ are *extensionally equal* when they represent the same function in $\mathcal{C}$. For example, c and εc are always extensionally equal.

The axioms of an elementary subclass of combinatory algebras, called λ-*models*, were expressly chosen to make coherent the definition of interpretation of λ-terms (see [4, Def. 5.2.1]). The *Meyer-Scott axiom* is the most important axiom in the definition of a λ-model. In the first-order language of combinatory algebras it becomes: $\forall x \forall y (\forall z (xz = yz) \Rightarrow \varepsilon x = \varepsilon y)$.

The combinator ε becomes an inner choice operator, that makes coherent the interpretation of an abstraction λ-term. A λ-model is said *extensional* if, moreover, we have that $\forall x \forall y (\forall z (xz = yz) \Rightarrow x = y)$.

3 From Reflexive Objects to λ-Models

In the common belief, a reflexive object $\mathcal{U}$ in a ccc $\mathbf{C}$ may be turned into a λ-model only when U has *enough points*, i.e., for all $f, g \in \mathbf{C}(U, U)$, if $f \neq g$ then there exists a morphism $h \in \mathbf{C}(\mathbb{1}, U)$ such that $f \circ h \neq g \circ h$. This holds *a fortiori* if $\mathbf{C}$ has enough points.

In the main result of this section we show that this hypothesis is unnecessary if we choose appropriately the associated λ-model and in Section 5 we will also provide a concrete example.

3.1 Building a Syntactical λ-Model

We give now the definition of "syntactical λ-models" [13]. Recall that, by [4, Thm. 5.3.6], λ-models are equal to syntactical λ-models, up to isomorphism.

Given an applicative structure $\mathcal{A}$, we let $Env_{\mathcal{A}}$ be the set of environments ρ mapping the set Var of variables of λ-calculus into A. For every $x \in$ Var and $a \in A$ we denote by $\rho[x := a]$ the environment ρ' which coincides with ρ, except on x, where ρ' takes the value a.

Definition 1. *A* syntactical λ-model *is a pair* $(\mathcal{A}, [\![-]\!])$ *where,* $\mathcal{A}$ *is an applicative structure and* $[\![-]\!] : \Lambda \times Env_{\mathcal{A}} \to A$ *satisfies the following conditions:*

(i) $[\![z]\!]_\rho = \rho(z)$,

(ii) $[\![PQ]\!]_\rho = [\![P]\!]_\rho \cdot [\![Q]\!]_\rho$,
(iii) $[\![\lambda z.P]\!]_\rho \cdot a = [\![P]\!]_{\rho[z:=a]}$,
(iv) $\rho\restriction_{FV(M)} = \rho'\restriction_{FV(M)} \Rightarrow [\![M]\!]_\rho = [\![M]\!]_{\rho'}$,
(v) $\forall a \in A,\ [\![M]\!]_{\rho[z:=a]} = [\![N]\!]_{\rho[z:=a]} \Rightarrow [\![\lambda z.M]\!]_\rho = [\![\lambda z.N]\!]_\rho$

A syntactical λ*-model is* extensional *if, moreover,* $\forall a \forall b (\forall x (a \cdot x = b \cdot x) \Rightarrow a = b)$.

Let us fix a categorical model $\mathcal{U} = (U, \mathrm{Ap}, \lambda)$ living in a ccc $\mathbf{C}$ *having countable products*[3]. The set $\mathbf{C}(U^{\mathrm{Var}}, U)$ can be naturally seen as an applicative structure whose application is defined by $u \bullet v = ev \circ \langle \mathrm{Ap} \circ u, v \rangle$. Moreover, the categorical interpretation $|M|_{\overline{x}}$ of a λ-term M, can be intuitively viewed as a morphism in $\mathbf{C}(U^{\mathrm{Var}}, U)$ only depending from a finite number of variables.

In order to capture this informal idea, we now focus our attention on the set $A_{\mathcal{U}}$ whose elements are the "finitary" morphisms in $\mathbf{C}(U^{\mathrm{Var}}, U)$.

A morphism $f \in \mathbf{C}(U^{\mathrm{Var}}, U)$ is *finitary* if there exist a finite set J of variables and a morphism $f_J \in \mathbf{C}(U^J, U)$ such that $f = f_J \circ \pi_J$, where π_J denotes the canonical projection from U^{Var} onto U^J. In this case we say that the pair (f_J, J) is *adequate* for f, and we write $(f_J, J) \in \mathrm{Ad}(f)$.

Given two finitary morphisms f, g it is easy to see that $f \bullet g$ is also finitary and that, if $(f_J, J) \in \mathrm{Ad}(f)$ and $(g_I, I) \in \mathrm{Ad}(g)$, then $((f_J \circ \pi_J) \bullet (g_I \circ \pi_I), J \cup I) \in \mathrm{Ad}(f \bullet g)$. Hence, we can define the following applicative structure.

Definition 2. *Let* $\mathcal{U}$ *be a reflexive object in a ccc* $\mathbf{C}$ *having countable products. The applicative structure associated with* $\mathcal{U}$ *is defined by* $\mathcal{A}_{\mathcal{U}} = (A_{\mathcal{U}}, \bullet)$*, where:*

- $A_{\mathcal{U}} = \{ f \in \mathbf{C}(U^{\mathrm{Var}}, U) : f$ *is finitary* $\}$,
- $a \bullet b = ev \circ \langle \mathrm{Ap} \circ a, b \rangle$.

We are going to define a syntactical λ-model $\mathcal{M}_{\mathcal{U}} = (\mathcal{A}_{\mathcal{U}}, [\![-]\!])$. In order to prove that $[\![-]\!]$ satisfies conditions $(i)-(v)$ of Definition 1 we need the technical lemma below. As a matter of fact, in this presentation we use Lemma 1 to define $[\![-]\!]$ itself (since our proof of Lemma 2 relies on Lemma 1). We remark that this is just a shortcut: indeed, the definition of $[\![-]\!]$ is sound independently from Lemma 1.

Lemma 1. *Let* $f_1, \ldots, f_n \in A_{\mathcal{U}}$ *and* $(f'_k, J_k) \in \mathrm{Ad}(f_k)$ *for all* $1 \le k \le n$*. Given* $z \in \mathrm{Var}$ *such that* $z \notin \bigcup_{k \le n} J_k$*, and* $\eta_z \in \mathbf{C}(U^{\mathrm{Var}} \& U, U^{\mathrm{Var}})$ *defined by:*

$$\eta_z^x = \begin{cases} \pi_2 & \text{if } x = z, \\ \pi_x \circ \pi_1 & \text{otherwise,} \end{cases}$$

the following diagram commutes:

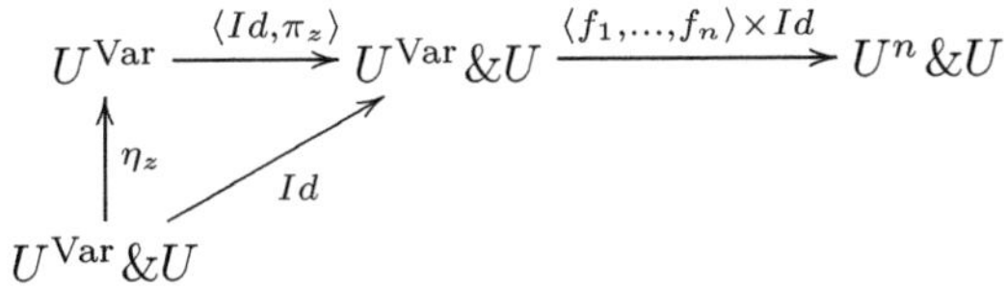

[3] Note that all the underlying categories of the models present in the literature, including those without enough points, satisfy this requirement.

Proof. Starting by $(\langle f_1,\dots,f_n\rangle \times Id)\circ\langle Id,\pi_z\rangle\circ\eta_z$, we get $\langle\langle f_1,\dots,f_n\rangle\circ\eta_z,\pi_2\rangle$ via easy calculations. Hence, it is sufficient to prove that $\langle f_1,\dots,f_n\rangle\circ\eta_z = \langle f_1,\dots,f_n\rangle\circ\pi_1$. We show that this equality holds componentwise. By hypothesis, we have that, for all $1\le k\le n$, $f_k\circ\eta_z = f'_k\circ\pi_{J_k}\circ\eta_z$. Since $z\notin J_k$, we have that $\pi_{J_k}\circ\eta_z=\pi_{J_k}\circ\pi_1$ (computing componentwise in π_{J_k} and applying the definition of η_z). To conclude, we note that $f'_k\circ\pi_{J_k}\circ\eta_z = f'_k\circ\pi_{J_k}\circ\pi_1 = f_k\circ\pi_1$. □

As a matter of notation, given a sequence $\overline{x}$ of variables and an environment $\rho\in Env_{\mathcal{A}_\mathcal{U}}$, we denote by $\rho(\overline{x})$ the morphism $\langle\rho(x_1),\dots,\rho(x_n)\rangle\in\mathbf{C}(U^{\mathrm{Var}},U^n)$.

Lemma 2. *For all λ-terms M, environments ρ and sequences $\overline{x},\overline{y}$ adequate for M, we have that $|M|_{\overline{x}}\circ\rho(\overline{x}) = |M|_{\overline{y}}\circ\rho(\overline{y})$.*

Proof. The proof is by structural induction on M.
If $M\equiv z$, then z occurs in, say, i-th position in $\overline{x}$ and j-th position in $\overline{y}$. Then $|z|_{\overline{x}}\circ\rho(\overline{x}) = \pi_i\circ\rho(\overline{x}) = \rho(z) = \pi_j\circ\rho(\overline{y}) = |z|_{\overline{y}}\circ\rho(\overline{y})$.
If $M\equiv PQ$, then $|PQ|_{\overline{x}}\circ\rho(\overline{x}) = ev\circ\langle\mathrm{Ap}\circ|P|_{\overline{x}},|Q|_{\overline{x}}\rangle\circ\rho(\overline{x})$. By (pair), this is equal to $ev\circ\langle\mathrm{Ap}\circ|P|_{\overline{x}}\circ\rho(\overline{x}),|Q|_{\overline{x}}\circ\rho(\overline{x})\rangle$ which is, by inductive hypothesis, $ev\circ\langle\mathrm{Ap}\circ|P|_{\overline{y}}\circ\rho(\overline{y}),|Q|_{\overline{y}}\circ\rho(\overline{y})\rangle = |PQ|_{\overline{y}}\circ\rho(\overline{y})$.
If $M\equiv\lambda z.N$, then $|\lambda z.N|_{\overline{x}}\circ\rho(\overline{x}) = \lambda\circ\Lambda(|N|_{\overline{x},z})\circ\rho(\overline{x})$ and by (Curry), we obtain $\lambda\circ\Lambda(|N|_{\overline{x},z}\circ(\rho(\overline{x})\times Id))$. Let $(\rho_1,J_1)\in\mathrm{Ad}(\rho(x_1)),\dots,(\rho_n,J_n)\in\mathrm{Ad}(\rho(x_n))$. By α-conversion we can suppose that $z\notin\bigcup_{k\le n}J_k$, hence by Lemma 1 we obtain $\lambda\circ\Lambda(|N|_{\overline{x},z}\circ(\rho(\overline{x})\times Id)\circ\langle Id,\pi_z\rangle\circ\eta_z) = \lambda\circ\Lambda(|N|_{\overline{x},z}\circ\rho[z:=\pi_z](\overline{x},z)\circ\eta_z)$. This is equal, by inductive hypothesis, to $\lambda\circ\Lambda(|N|_{\overline{y},z}\circ\rho[z:=\pi_z](\overline{y},z)\circ\eta_z) = |\lambda z.N|_{\overline{y}}$. □

As a consequence of Lemma 2 we have that the following definition is sound.

Definition 3. $\mathcal{M}_\mathcal{U} = (\mathcal{A}_\mathcal{U},[\![-]\!])$, *where* $[\![M]\!]_\rho = |M|_{\overline{x}}\circ\rho(\overline{x})$ *for some adequate sequence* $\overline{x}$.

We are going to prove that $\mathcal{M}_\mathcal{U}$ is a syntactical λ-model, which is extensional exactly when $\mathcal{U}$ is an extensional categorical model.

For this second property we need another categorical lemma. Remark that the morphism $\iota_{J,x}\in\mathbf{C}(U^{J\cup\{x\}},U^{\mathrm{Var}})$ defined below is a sort of canonical injection and that, in particular, the morphism $\lambda\circ\Lambda(Id_U)\circ!_{U^{J\cup\{x\}}}$ does not play any role in the rest of the argument.

Lemma 3. *Let $f\in A_\mathcal{U}$, $(f_J,J)\in\mathrm{Ad}(f)$, $x\notin J$ and $\iota_{J,x}$ defined as follows:*

$$\iota^z_{J,x} = \begin{cases}\pi_z & \text{if } z\in J\cup\{x\},\\ \lambda\circ\Lambda(Id_U)\circ!_{U^{J\cup\{x\}}} & \text{otherwise.}\end{cases}$$

Then the following diagram commutes:

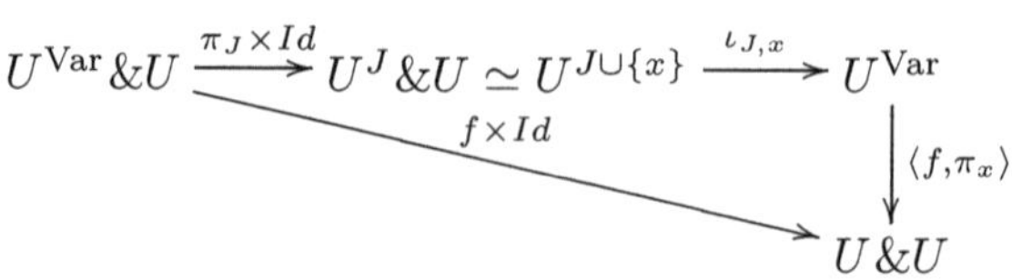

Proof. Since by hypothesis $f = f_J \circ \pi_J$, this is equivalent to ask that the following diagram commutes, and this is obvious from the definition of $\iota_{J,x}$.

$$\begin{array}{ccccc} U^{\mathrm{Var}} \& U & \xrightarrow{\pi_J \times Id} & U^J \& U \simeq U^{J\cup\{x\}} & \xrightarrow{\iota_{J,x}} & U^{\mathrm{Var}} \\ & & \searrow^{f_J \times Id} & & \searrow^{\langle \pi_J, \pi_x \rangle} \\ & & & U \& U \xleftarrow[f_J \times Id]{} U^{J\cup\{x\}} & \end{array}$$ □

Theorem 1. *Let $\mathcal{U}$ be a reflexive object in a ccc* **C** *having countable products. Then:*

1) $\mathcal{M}_{\mathcal{U}}$ is a syntactical λ-model,
2) $\mathcal{M}_{\mathcal{U}}$ is extensional if, and only if, $\mathcal{U}$ is.

Proof. 1) In the following $\overline{x}$ is any adequate sequence and the items correspond to those in Definition 1.
(i) $[\![z]\!]_\rho = |z|_{\overline{x}} \circ \rho(\overline{x}) = \pi_z \circ \rho(\overline{x}) = \rho(z)$.
(ii) $[\![PQ]\!]_\rho = |PQ|_{\overline{x}} \circ \rho(\overline{x}) = (|P|_{\overline{x}} \bullet |Q|_{\overline{x}}) \circ \rho(\overline{x}) = ev \circ \langle \mathrm{Ap} \circ |P|_{\overline{x}}, |Q|_{\overline{x}} \rangle \circ \rho(\overline{x})$. By (pair) this is equal to $ev \circ \langle \mathrm{Ap} \circ |P|_{\overline{x}} \circ \rho(\overline{x}), |Q|_{\overline{x}} \circ \rho(\overline{x}) \rangle = [\![P]\!]_\rho \bullet [\![Q]\!]_\rho$.
(iii) $[\![\lambda z.P]\!]_\rho \bullet a = (|\lambda z.P|_{\overline{x}} \circ \rho(\overline{x})) \bullet a = ev \circ \langle \mathrm{Ap} \circ \lambda \circ \Lambda(|P|_{\overline{x},z}) \circ \rho(\overline{x}), a \rangle$. Since $\mathrm{Ap} \circ \lambda = Id_{U \Rightarrow U}$ and by applying the rules (Curry) and (beta) we obtain $|P|_{\overline{x},z} \circ (\rho(\overline{x}) \times Id) \circ \langle Id, a \rangle$. Finally, by (pair) we get $|P|_{\overline{x},z} \circ \langle \rho(\overline{x}), a \rangle = [\![P]\!]_{\rho[z:=a]}$.
(iv) Obvious since, by Lemma 2, $[\![M]\!]_\rho = |M|_{\overline{x}} \circ \rho(\overline{x})$ where $\overline{x}$ are exactly the free variables of M.
(v) $[\![\lambda z.M]\!]_\rho = |\lambda z.M|_{\overline{x}} \circ \rho(\overline{x}) = \lambda \circ \Lambda(|M|_{\overline{x},z} \circ (\rho(\overline{x}) \times Id))$. Let $(\rho_1, J_1) \in \mathrm{Ad}(\rho(x_1)), \ldots, (\rho_n, J_n) \in \mathrm{Ad}(\rho(x_n))$. Without loss of generality we can suppose that $z \notin \bigcup_{k \leq n} J_k$. Hence, by Lemma 1 we obtain $\lambda \circ \Lambda(|M|_{\overline{x},z} \circ (\rho(\overline{x}) \times Id) \circ \langle Id, \pi_z \rangle \circ \eta_z)$. By (pair), this is $\lambda \circ \Lambda(|M|_{\overline{x},z} \circ \langle \rho(\overline{x}), \pi_z \rangle \circ \eta_z) = \lambda \circ \Lambda([\![M]\!]_{\rho[z:=\pi_z]} \circ \eta_z)$ which is equal to $\lambda \circ \Lambda([\![N]\!]_{\rho[z:=\pi_z]} \circ \eta_z)$ since, by hypothesis, $[\![M]\!]_{\rho[z:=a]} = [\![N]\!]_{\rho[z:=a]}$ for all $a \in A_{\mathcal{U}}$. It is, now, routine to check that $\lambda \circ \Lambda([\![N]\!]_{\rho[z:=\pi_z]} \circ \eta_z) = [\![\lambda z.N]\!]_\rho$.
2) $(\Rightarrow)$ Let $x \in \mathrm{Var}$ and $\pi_x \in \mathbf{C}(U^{\mathrm{Var}}, U)$. For all $a \in A_{\mathcal{U}}$ we have $(\lambda \circ \mathrm{Ap} \circ \pi_x) \bullet a = ev \circ \langle \mathrm{Ap} \circ \lambda \circ \mathrm{Ap} \circ \pi_x, a \rangle = ev \circ \langle \mathrm{Ap} \circ \pi_x, a \rangle = \pi_x \bullet a$. If $\mathcal{M}_{\mathcal{U}}$ is extensional, this implies $\lambda \circ \mathrm{Ap} \circ \pi_x = \pi_x$. Since π_x is an epimorphism, we get $\lambda \circ \mathrm{Ap} = Id_U$.
$(\Leftarrow)$ Let $a, b \in A_{\mathcal{U}}$, then there exist $(a_J, J) \in \mathrm{Ad}(a)$ and $(b_I, I) \in \mathrm{Ad}(b)$ such that $I = J$. Let us set $\varphi = \iota_{J,x} \circ (\pi_J \times Id)$ where $x \notin J$ and $\iota_{J,x}$ is defined in Lemma 3. Suppose that for all $c \in A_{\mathcal{U}}$ we have $(a \bullet c = b \bullet c)$ then, in particular, $ev \circ \langle \mathrm{Ap} \circ a, \pi_x \rangle = ev \circ \langle \mathrm{Ap} \circ b, \pi_x \rangle$ and this implies that $\langle \mathrm{Ap} \circ a, \pi_x \rangle \circ \varphi = \langle \mathrm{Ap} \circ b, \pi_x \rangle \circ \varphi$. By applying Lemma 3, we get $\langle \mathrm{Ap} \circ a, \pi_x \rangle \circ \varphi = (\mathrm{Ap} \circ a) \times Id$ and $\langle \mathrm{Ap} \circ b, \pi_x \rangle \circ \varphi = (\mathrm{Ap} \circ b) \times Id$. Then $\mathrm{Ap} \circ a = \mathrm{Ap} \circ b$ which implies $\lambda \circ \mathrm{Ap} \circ a = \lambda \circ \mathrm{Ap} \circ b$. We conclude since $\lambda \circ \mathrm{Ap} = Id_U$. □

Note that, by using a particular environment $\hat{\rho}$, it is possible to "recover" the categorical interpretation $|M|_{\overline{x}}$ from the interpretation $[\![M]\!]_\rho$ in the syntactical λ-model. Let us fix the environment $\hat{\rho}(x) = \pi_x$ for all $x \in \mathrm{Var}$. Then

$$[\![M]\!]_{\hat{\rho}} = |M|_{\overline{x}} \circ \langle \pi_{x_1}, \dots, \pi_{x_n} \rangle,$$

i.e., it is the morphism $|M|_{\overline{x}}$ "viewed" as element of $\mathbf{C}(U^{\mathrm{Var}}, U)$.

3.2 Working Without Countable Products

The construction provided in the previous section works if the underlying category $\mathbf{C}$ has countable products. We remark, once again, that this hypothesis is not really restrictive since all the categories used in the literature in order to obtain models of λ-calculus satisfy this requirement. Nevertheless, the discussions on the relation between categorical and algebraic models of λ-calculus in [20,21,25] can help us get rid of this additional hypothesis. We give here the basic ideas of this approach.

In [25], Selinger implicitly suggests that every λ-algebra $\mathcal{A}$ can be embedded into a λ-model $\mathcal{A}[\mathrm{Var}]$, which is obtained from $\mathcal{A}$ by freely adjoining the variables of λ-calculus as indeterminates. More precisely, he shows that, under the interpretation in $\mathcal{A}[x_1, \dots, x_n]$, $M = N$ implies $\lambda z.M = \lambda z.N$ holds as soon as M, N are λ-terms with free variables among $x_1, \dots, x_n$. Moreover, if $\mathcal{A}$ is the λ-algebra associated with a categorical model $\mathcal{U}$ by Koymans' construction, then for all $I \subset_{\mathrm{f}} \mathrm{Var}$ the free extension $\mathcal{A}[I]$ is isomorphic (in the category of combinatory algebras and homomorphisms between them) to $\mathbf{C}(U^I, U)$ endowed with the natural structure of combinatory algebra.

Since there exist canonical homomorphisms $\mathcal{A}[I] \mapsto \mathcal{A}[J]$ and $\mathbf{C}(U^I, U) \mapsto \mathbf{C}(U^J, U)$ which are one-to-one if $I \subseteq J \subset_{\mathrm{f}} \mathrm{Var}$, we can construct the inductive limit of both $\mathcal{P}_{\mathrm{f}}(\mathrm{Var})$-indexed diagrams.

From one side we obtain a λ-model isomorphic to $\mathcal{A}[\mathrm{Var}]$ and from the other side we get $A' = \bigcup_{I \subset_{\mathrm{f}} \mathrm{Var}} \mathbf{C}(U^I, U)/_{\sim}$, where $\sim$ is the equivalence relation defined as follows: if $f \in \mathbf{C}(U^J, U)$ and $g \in \mathbf{C}(U^I, U)$, then $f \sim g$ if, and only if, $f \circ \pi_J = g \circ \pi_I$ where $\pi_J \in \mathbf{C}(U^{I \cup J}, U^J)$ and $\pi_I \in \mathbf{C}(U^{I \cup J}, U^I)$. The above isomorphism is obviously preserved at the limit; hence A', endowed with a suitable application operator on the equivalence classes, is also a λ-model. This approach, although less simple and natural, also works in the case that the underlying category $\mathbf{C}$ does not have countable products. Finally, it is easy to check that if $\mathbf{C}$ does have countable products then A' is isomorphic to the set of finitary morphisms in $\mathbf{C}(U^{\mathrm{Var}}, U)$.

4 A Cartesian Closed Category of Sets and Relations

It is quite well known [12,2,15,7] that, by endowing the monoidal closed category **Rel** with a suitable comonad, one gets a ccc via the co-Kleisli construction. In this section we present the ccc **MRel** obtained by using the comonad $\mathcal{M}_f(-)$, without explicitly going through the monoidal structure of **Rel**.

Hence we directly define the category **MRel** as follows:

- The objects of **MRel** are all the sets.
- Given two sets S and T, a morphism from S to T is a relation from $\mathcal{M}_f(S)$ to T. In other words, $\mathbf{MRel}(S, T) = \mathcal{P}(\mathcal{M}_f(S) \times T)$.

– The identity morphism of S is the relation:
$$Id_S = \{([a], a) : a \in S\} \in \mathbf{MRel}(S, S)\,.$$
– Given two morphisms $s \in \mathbf{MRel}(S,T)$ and $t \in \mathbf{MRel}(T,U)$, we define:
$$t \circ s = \{(m, c) : \exists (m_1, b_1), \ldots, (m_k, b_k) \in s \text{ such that } m = m_1 \uplus \ldots \uplus m_k \text{ and } ([b_1, \ldots, b_k], c) \in t\}.$$

It is easy to check that this composition law is associative, and that the identity morphisms defined above are neutral for this composition.

Theorem 2. *The category* **MRel** *is cartesian closed.*

Proof. The terminal object $\mathbb{1}$ is the empty set $\emptyset$, and the unique element of $\mathbf{MRel}(S, \emptyset)$ is the empty relation.

Given two sets S_1 and S_2, their categorical product $S_1\&S_2$ in **MRel** is their disjoint union:
$$S_1\&S_2 = (\{1\} \times S_1) \cup (\{2\} \times S_2)$$
and the projections π_1, π_2 are given by:
$$\pi_i = \{([(i, a)], a) : a \in S_i\} \in \mathbf{MRel}(S_1\&S_2, S_i), \text{ for } i = 1, 2.$$
It is easy to check that this is actually the categorical product of S_1 and S_2 in **MRel**; given $s \in \mathbf{MRel}(U, S_1)$ and $t \in \mathbf{MRel}(U, S_2)$, the corresponding morphism $\langle s, t\rangle \in \mathbf{MRel}(U, S_1\&S_2)$ is given by:
$$\langle s, t\rangle = \{(m, (1, a)) : (m, a) \in s\} \cup \{(m, (2, b)) : (m, b) \in t\}\,.$$

We will consider the canonical bijection between $\mathcal{M}_f(S_1) \times \mathcal{M}_f(S_2)$ and $\mathcal{M}_f(S_1 \,\&\, S_2)$ as an equality, hence we will still denote by (m_1, m_2) the corresponding element of $\mathcal{M}_f(S_1\&S_2)$.

Given two objects S and T the exponential object $S \Rightarrow T$ is $\mathcal{M}_f(S) \times T$ and the evaluation morphism is given by:
$$ev_{ST} = \{(([(m, b)], m), b) : m \in \mathcal{M}_f(S) \text{ and } b \in T\} \in \mathbf{MRel}((S \Rightarrow T)\&S, T)\,.$$
Again, it is easy to check that in this way we defined an exponentiation. Indeed, given any set U and any morphism $s \in \mathbf{MRel}(U \,\&\, S, T)$, there is exactly one morphism $\Lambda(s) \in \mathbf{MRel}(U, S \Rightarrow T)$ such that:
$$ev_{ST} \circ (\Lambda(s) \times Id_S) = s.$$
where $\Lambda(s) = \{(p, (m, b)) : ((p, m), b) \in s\}$. □

Here, the points of an object S, i.e., the elements of $\mathbf{MRel}(\mathbb{1}, S)$, are relations between $\mathcal{M}_f(\emptyset)$ and S. These are, up to isomorphism, the subsets of S.

In the next section we will present an extensional model of λ-calculus living in **MRel**, which is a "strongly" non extensional ccc: not only it has not enough points but there exists *no* object $U \neq \mathbb{1}$ of **MRel** having enough points.

Indeed, we can always find $t_1, t_2 \in \mathbf{MRel}(U, U)$ such that $t_1 \neq t_2$ and, for all $s \in \mathbf{MRel}(\mathbb{1}, U)$, $t_1 \circ s = t_2 \circ s$. Recall that, by definition of composition, $t_1 \circ s = \{([], b) : \exists a_1, \ldots, a_n \in U \; ([], a_i) \in s \quad ([a_1, \ldots, a_n], b) \in t_1\} \in \mathbf{MRel}(\mathbb{1}, U)$, and similarly for $t_2 \circ s$. Hence it is sufficient to choose $t_1 = \{(m_1, b)\}$ and $t_2 = \{(m_2, b)\}$ such that m_1 and m_2 are different multisets with the same support.

5 An Extensional Relational Model of λ-Calculus

In this section we build a reflexive object in **MRel**, which is extensional by construction.

5.1 Constructing an Extensional Reflexive Object

We build a family of sets $(D_n)_{n\in\mathbb{N}}$ as follows[4]:

- $D_0 = \emptyset$,
- $D_{n+1} = \mathcal{M}_f(D_n)^{(\omega)}$.

Since the operation $S \mapsto \mathcal{M}_f(S)^{(\omega)}$ is monotonic on sets, and since $D_0 \subseteq D_1$, we have $D_n \subseteq D_{n+1}$ for all $n \in \mathbb{N}$. Finally, we set $D = \bigcup_{n\in\mathbb{N}} D_n$.

So we have $D_0 = \emptyset$ and $D_1 = \{*\} = \{([],[],\ldots)\}$. The elements of D_2 are quasi-finite sequences of multisets over a singleton, i.e., quasi-finite sequences of natural numbers. More generally, an element of D can be represented as a finite tree which alternates two kinds of layers:

- ordered nodes (the quasi-finite sequences), where immediate subtrees are indexed by a possibly empty finite set of natural numbers,
- unordered nodes where subtrees are organised in a *non-empty* multiset.

In order to define an isomorphism in **MRel** between D and $D \Rightarrow D$ (which is equal to $\mathcal{M}_f(D) \times D$) just remark that every element $\sigma \in D$ stands for the pair $(\sigma_0, (\sigma_1, \sigma_2, \ldots))$ and *vice versa*. Given $\sigma \in D$ and $m \in \mathcal{M}_f(D)$, we write $m \cdot \sigma$ for the element $\tau \in D$ such that $\tau_1 = m$ and $\tau_{i+1} = \sigma_i$. This defines a bijection between $\mathcal{M}_f(D) \times D$ and D, and hence an isomorphism in **MRel** as follows:

Proposition 1. *The triple $\mathcal{D} = (D, \mathrm{Ap}, \lambda)$ where:*
- $\lambda = \{(([(m,\sigma)], m\cdot\sigma) : m \in \mathcal{M}_f(D), \sigma \in D\} \in \mathbf{MRel}(D\Rightarrow D, D)$,
- $\mathrm{Ap} = \{([m\cdot\sigma], (m,\sigma)) : m \in \mathcal{M}_f(D), \sigma \in D\} \in \mathbf{MRel}(D, D\Rightarrow D)$,

is an extensional categorical model of λ-calculus.

Proof. It is easy to check that $\lambda \circ \mathrm{Ap} = Id_D$ and $\mathrm{Ap} \circ \lambda = Id_{D\Rightarrow D}$. □

5.2 Interpreting the Untyped λ-Calculus in $\mathcal{D}$

In Section 2.3, we have recalled how a λ-term is interpreted when a reflexive object is given, in any ccc. We provide the result of the corresponding computation, when it is performed in the present structure $\mathcal{D}$.

Given a λ-term M and a sequence $\overline{x}$ of length n, which is adequate for M, the interpretation $|M|_{\overline{x}}$ is an element of $\mathbf{MRel}(D^n, D)$, where $D^n = D \,\&\, \ldots \,\&\, D$, i.e., a subset of $\mathcal{M}_f(D)^n \times D$. This set is defined by structural induction on M.

[4] Note that, in greater generality, we can start from a set A of "atoms" and take: $D_0 = \emptyset$, $D_{n+1} = \mathcal{M}_f(D_n)^{(\omega)} \times A$. Nevertheless the set of atoms A is not essential to produce a non-trivial model of λ-calculus.

- $|x_i|_{\overline{x}} = \{(([],\ldots,[],[\sigma],[],\ldots,[]),\sigma) : \sigma \in D\}$, where the only non-empty multiset stands in i-th position.
- $|NP|_{\overline{x}} = \{((m_1,\ldots,m_n),\sigma) : \exists k \in \mathbb{N}$
  $\exists(m_1^j,\ldots,m_n^j) \in \mathcal{M}_f(D)^n$ for $j = 0,\ldots,k$
  $\exists\sigma_1,\ldots,\sigma_k \in D$ such that
  $m_i = m_i^0 \uplus \ldots \uplus m_i^k$ for $i = 1,\ldots,n$
  $((m_1^0,\ldots,m_n^0),[\sigma_1,\ldots,\sigma_k]\cdot\sigma) \in |N|_{\overline{x}}$
  $((m_1^j,\ldots,m_n^j),\sigma_j) \in |P|_{\overline{x}}$ for $j = 1,\ldots,k\}$.
- $|\lambda z.P|_{\overline{x}} = \{((m_1,\ldots,m_n),m\cdot\sigma) : ((m_1,\ldots,m_n,m),\sigma) \in |P|_{\overline{x},z}\}$, where we assume that z does not occur in $\overline{x}$.

Since $\mathcal{D}$ is extensional, if $M =_{\beta\eta} N$ then M and N have the same interpretation in the model. Note that if M is a closed λ-terms then it is simply interpreted, in the empty sequence, by a subset of D. If M is moreover a solvable term, i.e., if it is β-convertible to a term of the shape $\lambda x_1\ldots x_n.x_iM_1\cdots M_k$ $(n,k\geq 0)$, then its interpretation is non-empty. Indeed, it is quite clear that $[]\cdots[]\cdot[*]\cdot * \in |M|$ (where $[*]$ stands in i-th position).

6 Modelling Non-determinism

Since **MRel** has countable products, the construction given in Subsection 3.1 provides an applicative structure $\mathcal{A}_\mathcal{D} = (A_\mathcal{D},\bullet)$, whose elements are the finitary morphisms in $\mathbf{MRel}(D^{\mathrm{Var}},D)$, and the associated λ-model $\mathcal{M}_\mathcal{D} = (\mathcal{A}_\mathcal{D},[\![-]\!])$. This λ-model is extensional by Theorem 1*(2)*.

We are going to define two operations of sum and product on $A_\mathcal{D}$; in order to show easily that these operations are well defined, we provide a characterization of the finitary elements of $\mathbf{MRel}(D^{\mathrm{Var}},D)$.

Proposition 2. *Let* $f \in \mathbf{MRel}(D^{\mathrm{Var}},D)$ *and* $J \subset_{\mathrm{f}} \mathrm{Var}$. *Then there exists* f_J *such that* $(f_J,J) \in \mathrm{Ad}(f)$ *if, and only if, for all* $(m,\sigma) \in f$ *and for all* $x \notin J$, $\pi_x(m) = []$.

Proof. Straightforward.

Hence, the union of finitary elements is still a finitary element. As a matter of notation, we will write $a \oplus b$ for $a \cup b$.

We now recall the definition of semilinear applicative structure given in [10].

Definition 4. *A* semilinear applicative structure *is a pair* $((A,\cdot),+)$ *such that:*

(i) $(A,\cdot)$ *is an applicative structure.*
(ii) $+ : A^2 \to A$ *is an idempotent, commutative and associative operation.*
(iii) $\forall x,y,z \in A \;\; (x+y)\cdot z = (x\cdot z)+(y\cdot z)$.

Straightforwardly, the union operation makes $\mathcal{A}_\mathcal{D}$ semilinear.

Proposition 3. $(\mathcal{A}_\mathcal{D},\oplus)$ *is a semilinear applicative structure.*

Moreover, the syntactic interpretation of Definition 1 may be extended to the non-deterministic λ-calculus $\Lambda_\oplus$ of [10], by stipulating that $[\![M \oplus N]\!]_\rho = [\![M]\!]_\rho \oplus [\![N]\!]_\rho$. Hence, we get that $(\mathcal{A}_\mathcal{D}, \oplus, [\![-]\!])$ is an extensional syntactical model of $\Lambda_\oplus$ in the sense of [10]. The operation $\oplus$ can be seen intuitively as a non-deterministic choice.

We define another binary operation on $A_\mathcal{D}$, which can be thought of as parallel composition.

Definition 5.

- *Given* $\sigma, \tau \in D$, *we set* $\sigma \odot \tau = (\sigma_1 \uplus \tau_1, \dots, \sigma_n \uplus \tau_n, \dots)$.
- *Given* $a, b \in A_\mathcal{D}$, *we set* $a \odot b = \{(m_1 \uplus m_2, \sigma \odot \tau) : (m_1, \sigma) \in a, (m_2, \tau) \in b\}$.

Once again, it is easy to see that $\odot$ produces finitary elements when applied to finitary elements.

Note that $\mathcal{A}_\mathcal{D}$, equipped with $\odot$, is *not* a semilinear applicative structure, simply because the operator $\odot$ is not idempotent. Nevertheless, the left distributivity with respect to the application is satisfied.

Proposition 4. *For all* $a, b, c \in A_\mathcal{D}$, $(a \odot b) \bullet c = (a \bullet c) \odot (b \bullet c)$.

Proof. Straightforward.

The units of the operations $\oplus$ and $\odot$ are $0 = \emptyset$ and $1 = \{([], *)\}$, respectively; $(A_\mathcal{D}, \oplus, 0)$ and $(A_\mathcal{D}, \odot, 1)$ are commutative monoids. Moreover 0 annihilates $\odot$, and multiplication distributes over addition. Summing up, the following proposition holds.

Proposition 5. *1.* $(A_\mathcal{D}, \oplus, \odot, 0, 1)$ *is a commutative semiring.*
2. $\oplus$ *and* $\odot$ *are left distributive over* $\bullet$.
3. $\oplus$ *is idempotent.*

7 Conclusions and Further Works

We have proposed a general method for getting a λ-model out of a reflexive object of a ccc, which does not rely on the fact that the object has enough points. We have applied this construction to an extensional reflexive object $\mathcal{D}$ of **MRel**, the Kleisli category of the comonad "finite multisets" on **Rel**, and showed some algebraic properties of the resulting λ-model $\mathcal{M}_\mathcal{D}$. A first natural question about $\mathcal{M}_\mathcal{D}$ concerns its theory. We know that it is extensional, and that $\mathcal{M}_\mathcal{D}$ can be "stratified" following the construction of $D = \bigcup_{n\in\mathbb{N}} D_n$ given in Subsection 5.1. Not surprisingly, the theory of $\mathcal{M}_\mathcal{D}$ turns out to be $\mathcal{H}^*$, the maximal consistent sensible λ-theory. In a forthcoming paper, we show how the proof method based on the approximation theorem, due to Hyland [14], can be adapted to all suitably defined "stratified λ-models" in order to prove that their theory is $\mathcal{H}^*$.

Proposition 5 shows that $\mathcal{M}_\mathcal{D}$ has a quite rich algebraic structure. In order to interpret conjunctive-disjunctive λ-calculi, endowed with both "non-deterministic choice" and "parallel composition", a notion of λ-lattice has been introduced in [9]. It is interesting to notice that our structure $(A_{\mathcal{D}}, \subseteq, \bullet, \oplus, \odot)$ does not give rise to a real λ-lattice essentially because $\odot$ is not idempotent. Roughly speaking, this means that in the model $\mathcal{M}_{\mathcal{D}}$ of the conjunctive-disjunctive calculus $[\![M||M]\!] \neq [\![M]\!]$, i.e., that the model is "resource sensible". We aim to investigate full abstraction results for must/may semantics in $\mathcal{M}_{\mathcal{D}}$.

A concluding remark: for historical reasons, most of the work on models of untyped λ-calculus, and its extensions, has been carried out in subcategories of **CPO**. *A posteriori*, we can propose two motivations:

(i) because of the seminal work of Scott, the Scott-continuity of morphisms has been seen as *the* canonical way of getting $U \Rightarrow U \lhd U$;
(ii) the classic result relating algebraic and categorical models of untyped λ-calculus asks for reflexive objects with enough points.

Our proposal allows to overcome (ii). It remains to be proved that working in categories like **MRel** allows to get interesting classes of models.

References

1. Abramsky, S., Jagadeesan, R., Malacaria, P.: Full abstraction for PCF. Information and Computation 163(2), 409–470 (2000)
2. Amadio, R.M., Curien, P.-L.: Domains and lambda-calculi. Cambridge University Press, New York (1998)
3. Asperti, A., Longo, G.: Categories, types and structures. Categories, types and structures, MIT Press, Cambridge (1991)
4. Barendregt, H.P.: The lambda calculus: Its syntax and semantics. North-Holland Publishing Co, Amsterdam (1984)
5. Berline, C.: From computation to foundations via functions and application: The λ-calculus and its webbed models. Theoretical Computer Science 249, 81–161 (2000)
6. Berry, G., Curien, P.-L.: Sequential algorithms on concrete data structures. Theoretical Computer Science 20, 265–321 (1985)
7. Bucciarelli, A., Ehrhard, T.: On phase semantics and denotational semantics: the exponentials. Ann. Pure Appl. Logic 109(3), 205–241 (2001)
8. Curry, H.B., Feys, R.: Combinatory Logic, vol. I. North-Holland, Amsterdam (1958)
9. Dezani Ciancaglini, M., de'Liguoro, U., Piperno, A.: A filter model for concurrent λ-calculus. SIAM Journal of Computing 27(5), 1376–1419 (1998)
10. de'Liguoro, U., Piperno, A.: Non deterministic extensions of untyped λ-calculus. Information and Computation 109, 149–177 (1955)
11. Di Gianantonio, P., Franco, G., Honsell, F.: Game semantics for untyped lambda calculus. In: Duke, D.J., Marshall, M.S., Herman, I. (eds.) PREMO: A Framework for Multimedia Middleware. LNCS, vol. 1591, pp. 114–128. Springer, Heidelberg (1999)
12. Girard, J.-Y.: Normal functors, power series and the λ-calculus. Annals of pure and applied logic 37, 129–177 (1988)
13. Hindley, J.R., Longo, G.: Lambda calculus models and extensionality. Z. Math. Logik Grundlag. Math. 26, 289–310 (1980)

14. Hyland, M.: A syntactic characterization of the equality in some models for the lambda calculus. J. London Math. Soc. 12(3), 361–370 (1975/76)
15. Hyland, M., Nagayama, M., Power, J., Rosolini, G.: A category theoretic formulation for Engeler-style models of the untyped λ-calculus. Electronic notes in theoretical computer science 161, 43–57 (2006)
16. Hyland, M., Ong, C.-H.L.: On full abstraction for PCF: I. Models, observables and the full abstraction problem, II. Dialogue games and innocent strategies, III. A fully abstract and universal game model. Inf. and Comp 163, 285–408 (2000)
17. Ker, A.D., Nickau, H., Ong, C.-H.L.: A universal innocent game model for the Boehm tree lambda theory. In: Flum, J., Rodríguez-Artalejo, M. (eds.) CSL 1999. LNCS, vol. 1683, pp. 405–419. Springer, Heidelberg (1999)
18. Kerth, R.: On the construction of stable models of λ-calculus. Theoretical Computer Science 269, 23–46 (2001)
19. Koymans, C.P.J.: Models of the lambda calculus. Information and Control 52(3), 306–332 (1982)
20. Lambek, J.: From lambda calculus to Cartesian closed categories. In: Seldin, J.P., Hindley, J. (eds.) To H. B. Curry: Essays on Combinatory Logic, Lambda Calculus and Formalism, Academic Press, London (1980)
21. Lambek, J., Scott, P.J.: Introduction to Higher Order Categorical Logic. The Journal of Symbolic Logic 54(3) (1989)
22. Plotkin, G.D.: The λ-calculus is ω-incomplete. J. Symb. Log. 39(2), 313–317 (1974)
23. Plotkin, G.D.: T_ω as a Universal Domain. J. Comput. System Sci. 17, 209–236 (1978)
24. Scott, D.S.: Continuous lattices. In: Toposes, Algebraic geometry and Logic. LNM, vol. 274, Springer, Heidelberg (1972)
25. Selinger, P.: The lambda calculus is algebraic. J. Funct. Program. 12(6), 549–566 (2002)

Classical Program Extraction in the Calculus of Constructions

Alexandre Miquel

PPS & Université Paris 7
Case 7014, 2 Place Jussieu
75251 PARIS Cedex 05 – France
alexandre.miquel@pps.jussieu.fr

Abstract. We show how to extract classical programs expressed in Krivine λ_c-calculus from proof-terms built in a proof-irrelevant and classical version of the calculus of constructions with universes. For that, we extend Krivine's realisability model of classical second-order arithmetic to the calculus of constructions with universes using a structure of Π-set which is reminiscent of ω-sets, and show that our realisability model validates Peirce's law and proof-irrelevance. Finally, we extend the extraction scheme to a primitive data-type of natural numbers in a way which preserves the whole compatibility with the classical realisability interpretation of second-order arithmetic.

Introduction

Program extraction has been a major concern from the early development of the calculus of constructions (CC) [3] to its more recent extensions [13,17] implemented in proof assistants such as Coq [18], LEGO or Plastic. The first extraction scheme implemented in Coq [16] was based on the dependency erasing translation from CC to $F\omega$ [4], with a facility allowing to distinguish computationally relevant parts of the proof from the purely logical parts. (This facility relied on a distinction between two impredicative sorts Prop and Set.) However, as the system grew up, the initial mechanism became obsolete, so that in 2002 the extraction mechanism of Coq was completely redesigned [11]. The currently implemented mechanism now extracts the constructive skeleton of terms (corresponding to the parts built in the predicative universes Type_i) while removing their purely logical part (corresponding to the parts built in Prop).

In this paper we present another extraction mechanism, that extends the extraction mechanism used in system AF2 [6] to the whole type system of Coq. Moreover, this mechanism is able to extract programs from classical proofs (using the control operator call/cc), and it is actually compatible with Krivine's recent results about realising different forms of the axiom of choice [9,10].

The richness of the type system of CC naturally raises difficulties which do not exist when program extraction is performed in second-order arithmetic only. The first difficulty comes from the fact that in CC, types (and propositions) may depend on proofs. The traditional way to solve this problem is to add an axiom

J. Duparc and T.A. Henzinger (Eds.): CSL 2007, LNCS 4646, pp. 313–327, 2007.

of *proof-irrelevance* (or to modify the conversion rule) in order to make all the proofs of a given proposition equal. For a long time it has been believed that proof-irrelevance was incompatible with an extraction *à la* AF2, where programs are extracted from purely logical proofs (i.e. built in Prop).[1] As we shall see, this is not the case. Proof-irrelevance is not only compatible with classical program extraction, but it also removes the need of introducing an equational theory for each constant implementing classical reasoning, simply because such constants become transparent in the conversion rule.

From ω-sets to Π-sets. The classical extraction presented here is based on a realisability model constructed with Π-sets, a structure which is directly inspired from ω-sets [5,12], D-sets [19] and saturated λ-sets [2,14]. Historically, ω-sets (and their generalisation to all partial combinatory algebras: D-sets) have been used to define a realisability model of CC that provides a non-trivial interpretation of the impredicative sort Prop. (Such a model is extended to ECC in [13].) The next improvement came with Altenkirch [2], who noticed that by adding saturation conditions to λ-sets, the λ-set model of CC could be turned into a strong normalisation model. (The structure of saturated λ-set was later reused to extend this construction to a larger class of systems [14].)

However, recent works about normalisation stressed on the importance of definitions by orthogonality in the design of reducibility candidates. Since classical realisability [8,9,10] also deeply relies on definitions by orthogonality, it is natural to shift the point of view of realisability from the player (the λ-term) to the opponent (the stack it is applied to). In this move, the realisability relation becomes an orthogonality relation, and the structure of λ-set is turned into a new structure: the structure of Π-set which is described in section 3.

Which target (classical) λ-calculus?. Since the beginning of the 90's, many λ-calculi have been designed to extend the Curry-Howard correspondence to classical logic. Although we are convinced that most (if not all) of them could be used successfully as the target calculus of classical extraction, we focus here to the λ_c-calculus for several reasons.

The first reason, which is pedagogical, is that it illustrates the combination of two formalisms that are technically very different: on one side the calculus of constructions, whose conversion rule is based on strong evaluation and whose meta-theory deeply relies on the Church-Rosser property; on the other side the λ_c-calculus, that only performs weak head evaluation and for which the notion of normal form and the notion of confluence are simply irrelevant.

The second reason, and actually the main reason, is that λ_c can be extended with extra instructions allowing several forms of the axiom of choice to be realised—and in particular the axiom of dependent choice [9,10].[2] By taking λ_c as the target calculus—and provided we ensure that the full realisability model of CC is compatible with Krivine's realisability model of second-order

[1] Notice that since the currently extraction of Coq erases all proof-terms (in Prop), it is *de facto* compatible with proof-irrelevance.

[2] These results have not been ported yet to other classical calculi.

arithmetic—we thus allow the extraction mechanism to deal with axioms such as the axiom of the dependent choice (now expressed in CC) and more generally to benefit from the most advanced results of classical realisability.

1 A Proof-Irrelevant Calculus of Constructions

1.1 Syntax

The *proof-irrelevant calculus of constructions with universes* ($\mathrm{CC}_\omega^{\mathrm{irr}}$) is built from the same syntax as the calculus of constructions with universes (CC_ω):

$$\begin{array}{llcll} \textbf{Sorts} & s \in \mathscr{S} & ::= & \mathsf{Prop} \mid \mathsf{Type}_i & (i \geq 1) \\ \textbf{Terms} & M, N, T, U & ::= & x \mid s \mid \Pi x : T \,.\, U \mid \lambda x : T \,.\, M \mid MN & \end{array}$$

Here, Prop denotes the sort of *propositions* (seen as the types of their proofs) whereas Type_i $(i \geq 1)$ denotes the ith *predicative universe*.

The set $\mathscr{S}$ of sorts is equipped with a set of axioms $\mathscr{A} \subset \mathscr{S}^2$ and with a set of rules $\mathscr{R} \subset \mathscr{S}^3$ defined by

$$\begin{array}{rcl} \mathscr{A} & = & \{(\mathsf{Prop} : \mathsf{Type}_1);\ (\mathsf{Type}_i : \mathsf{Type}_{i+1}) \mid i \geq 1\} \\ \mathscr{R} & = & \{(\mathsf{Prop}, \mathsf{Prop}, \mathsf{Prop});\ (\mathsf{Type}_i, \mathsf{Prop}, \mathsf{Prop}); \\ & & \ (\mathsf{Prop}, \mathsf{Type}_i, \mathsf{Type}_i);\ (\mathsf{Type}_i, \mathsf{Type}_i, \mathsf{Type}_i) \mid i \geq 1\} \end{array}$$

as well as a (total) order $s_1 \leq s_2$ (the *cumulative order*) which is generated from the relations $\mathsf{Prop} \leq \mathsf{Type}_1$ and $\mathsf{Type}_i \leq \mathsf{Type}_{i+1}$ $(i \geq 1)$.

In both constructions $\lambda x : T \,.\, M$ and $\Pi x : T \,.\, U$, the symbols λ and Π are binders, which bind all the free occurrences of the variable x in M and U, but no occurrence of x in T. The set of free variables of M is written $FV(M)$. As usual we write $T \to U \equiv \Pi x : T \,.\, U$ (when $x \notin FV(U)$) the non-dependent product. The external operation of substitution, written $M\{x := N\}$, is defined as expected (taking care of renaming bound variables to avoid variable capture). In what follows, we work with terms up to α-conversion.

Terms of $\mathrm{CC}_\omega^{\mathrm{irr}}$ come with an untyped notion of β-reduction (defined as expected) which is confluent and enjoys Church and Rosser's property. However, we will not consider the untyped notion of β-reduction of $\mathrm{CC}_\omega^{\mathrm{irr}}$ in what follows, since we will identify β-equivalent terms (and more) in the conversion/subsumption rule using a typed equality judgement $\Gamma \vdash M = M' : T$ à la Martin-Löf.

1.2 Typing

A *typing context* (for short: a *context*) is a finite list of declarations of the form

$$\Gamma \ \equiv\ x_1 : T_1, \ldots, x_n : T_n$$

where $x_1, \ldots, x_n$ are pairwise distinct variables and where $T_1, \ldots, T_n$ are arbitrary terms. The *domain* of a context $\Gamma = x_1 : T_1, \ldots, x_n : T_n$ is written dom(Γ) and defined by dom(Γ) $= \{x_1; \ldots; x_n\}$. Given two contexts Γ and Γ', we write $\Gamma \subseteq \Gamma'$ when all the declarations that appear in Γ also appear in Γ', not necessarily in the same order.

The type system of $\mathrm{CC}_\omega^{\mathrm{irr}}$ is defined from four forms of judgements, namely:

$\vdash \Gamma\ \mathbf{ctx}$	'Γ is a well-formed context'
$\Gamma \vdash M : T$	'in the context Γ, the term M has type T'
$\Gamma \vdash T_1 \leq T_2$	'in the context Γ, T_1 is a subtype of T_2
$\Gamma \vdash M_1 = M_2 : T$	'in Γ, M_1 and M_2 are equal terms of type T'

The corresponding rules of inference are given in Fig. 1.

The main syntactic properties of this type system are the following (writing J any of $M : T$ or $T_1 \leq T_2$ or $M_1 = M_2 : T$). We do not indicate the proofs, that all proceed by induction on the suitable derivation.

Lemma 1 (Context well-formedness). *From any derivation of $\Gamma \vdash J$, one can extract a sub-derivation of $\vdash \Gamma$ **ctx**.*

Lemma 2 (Weakening). *If $\Gamma \vdash J$ and $\Gamma \subseteq \Gamma'$ and $\vdash \Gamma'$ **ctx**, then $\Gamma' \vdash J$.*

Lemma 3 (Substitutivity). *If $\Gamma, x : T, \Delta \vdash J$ and $\Gamma \vdash N : T$, then $\Gamma, \Delta\{x := N\} \vdash J\{x := N\}$.*

Lemma 4 (Type of types).

- *If $\Gamma \vdash M : T$ or $\Gamma \vdash M_1 = M_2 : T$, then $\Gamma \vdash T : s$ for some $s \in \mathscr{S}$.*
- *If $\Gamma \vdash T \leq T'$, then $\Gamma \vdash T : s$ and $\Gamma \vdash T' : s'$ for some $s, s' \in \mathscr{S}$.*

2 The Language of Realisers

2.1 Terms, Stacks and Processes

Terms of λ_c [7,10] are simply the pure λ-terms enriched with infinitely many constants taken in a denumerable set $\mathcal{C}$:

Terms $\qquad t, u \quad ::= \quad x \quad | \quad \lambda x\,.\,t \quad | \quad tu \quad | \quad c \qquad (c \in \mathcal{C})$

The notion of free and bound variable is defined as usual, as well as the external operation of substitution. In what follows, we will only consider closed terms, and write Λ for the set of all closed terms.

Stacks are built from a denumerable set $\mathcal{B}$ of *stack constants* (a.k.a. *stack bottoms*). Formally, stacks are defined as lists of closed terms terminated by a stack constant:

Stacks $\qquad \pi \quad ::= \quad b \quad | \quad t \cdot \pi \qquad (b \in \mathcal{B},\ t \in \Lambda)$

(writing $t \cdot \pi$ the 'consing' operation). The set of stacks is written Π.

Context formation rules

$$\frac{}{\vdash [\,]\ \mathbf{ctx}} \qquad \frac{\Gamma \vdash T : s \qquad x \notin \mathrm{dom}(\Gamma)}{\vdash \Gamma, x : T\ \mathbf{ctx}}$$

Typing rules

$$\frac{\vdash \Gamma\ \mathbf{ctx}}{\Gamma \vdash x : T}\ {\scriptstyle (x:T)\in\Gamma} \qquad \frac{\vdash \Gamma\ \mathbf{ctx}}{\Gamma \vdash s_1 : s_2}\ {\scriptstyle (s_1,s_2)\in\mathscr{A}} \qquad \frac{\Gamma \vdash M : T \qquad \Gamma \vdash T \leq T'}{\Gamma \vdash M : T'}$$

$$\frac{\Gamma \vdash T : s_1 \qquad \Gamma, x : T \vdash U : s_2}{\Gamma \vdash \Pi x{:}T\,.\,M : s_3}\ {\scriptstyle (s_1,s_2,s_3)\in\mathscr{R}}$$

$$\frac{\Gamma \vdash \Pi x{:}T\,.\,U : s \qquad \Gamma, x : T \vdash M : U}{\Gamma \vdash \lambda x{:}T\,.\,M : \Pi x{:}T\,.\,U} \qquad \frac{\Gamma \vdash M : \Pi x{:}T\,.\,U \qquad \Gamma \vdash N : T}{\Gamma \vdash MN : U\{x := N\}}$$

Subtyping rules

$$\frac{\Gamma \vdash T = T' : s}{\Gamma \vdash T \leq T'} \qquad \frac{\Gamma \vdash T \leq T' \qquad \Gamma \vdash T' \leq T''}{\Gamma \vdash T \leq T''}$$

$$\frac{\vdash \Gamma\ \mathbf{ctx} \qquad s_1 \leq s_2}{\Gamma \vdash s_1 \leq s_2} \qquad \frac{\Gamma \vdash T = T' : s \qquad \Gamma \vdash U \leq U'}{\Gamma \vdash \Pi x{:}T\,.\,U \leq \Pi x{:}T'\,.\,U'}$$

Equality rules

$$\frac{\Gamma \vdash M : T}{\Gamma \vdash M = M : T} \qquad \frac{\Gamma \vdash M = M' : T \qquad \Gamma \vdash M' = M'' : T}{\Gamma \vdash M = M'' : T}$$

$$\frac{\Gamma \vdash M = M' : T}{\Gamma \vdash M' = M : T} \qquad \frac{\Gamma \vdash M = M' : T \qquad \Gamma \vdash T \leq T'}{\Gamma \vdash M = M' : T'}$$

$$\frac{\Gamma \vdash T = T' : s_1 \qquad \Gamma, x : T \vdash U = U' : s_2 \qquad (s_1, s_2, s_3) \in \mathscr{R}}{\Gamma \vdash \Pi x{:}T\,.\,U = \Pi x{:}T'\,.\,U' : s_3}$$

$$\frac{\Gamma \vdash T = T' : s \qquad \Gamma, x : T \vdash M = M' : U}{\Gamma \vdash \lambda x{:}T\,.\,M = \lambda x{:}T'\,.\,M' : \Pi x{:}T\,.\,U}$$

$$\frac{\Gamma \vdash M = M' : \Pi x{:}T\,.\,U \qquad \Gamma \vdash N = N' : T}{\Gamma \vdash MN = M'N' : U\{x := N\}}$$

$$\frac{\Gamma \vdash \Pi x{:}T\,.\,U : s \qquad \Gamma, x : T \vdash M : U \qquad \Gamma \vdash N : T}{\Gamma \vdash (\lambda x{:}T\,.\,M)N = M\{x := N\} : U\{x := N\}}$$

$$\frac{\Gamma \vdash T : \mathsf{Prop} \qquad \Gamma \vdash M : T \qquad \Gamma \vdash M' : T}{\Gamma \vdash M = M' : T}$$

Fig. 1. Typing rules of CC_ω

To each stack $\pi \in \Pi$ we associate a constant $\mathsf{k}_\pi \in \mathcal{C}$ in such a way that (1) the correspondence $\pi \mapsto \mathsf{k}_\pi$ is injective, and (2) the set of all $c \neq \mathsf{k}_\pi$ (for all $\pi \in \Pi$) is still a denumerably infinite subset of $\mathcal{C}$.

In what follows, we call a *quasi-proof* any closed term containing none of the constants k_π ($\pi \in \Pi$). Finally, we take a fresh constant $\mathsf{c\!c} \in \mathcal{C}$ ('call/cc')[3] such that $\mathsf{c\!c} \neq \mathsf{k}_\pi$ for all $\pi \in \Pi$, which we reserve to realise Peirce's law.

Processes are then defined as pairs formed by a closed term and a stack:

Processes $\qquad p, q \quad ::= \quad t \star \pi \qquad (t \in \Lambda,\ \pi \in \Pi)$

2.2 Evaluation and Realisability

Processes are equipped with a binary relation of one step evaluation, written $p \succ p'$, which is defined by the following rules:

$$\begin{array}{rcl} \lambda\xi\,.\,t \star u \cdot \pi & \succ & t\{\xi := u\} \star \pi \\ tu \star \pi & \succ & t \star u \cdot \pi \end{array} \qquad\qquad \begin{array}{rcl} \mathsf{c\!c} \star t \cdot \pi & \succ & t \star \mathsf{k}_\pi \cdot \pi \\ \mathsf{k}_\pi \star t \cdot \pi' & \succ & t \star \pi \end{array}$$

The definition of a realisability model based on the language λ_c (for second-order arithmetic or for CC_ω) is parameterised by a fixed set of processes $\perp\!\!\!\perp$ that we assume to be *saturated*, in the sense that:

$$p \succ p' \quad \text{and} \quad p' \in \perp\!\!\!\perp \quad \text{imply} \quad p \in \perp\!\!\!\perp \qquad\qquad (\text{for all } p, p')$$

Intuitively, $\perp\!\!\!\perp$ represents a set of accepting processes (w.r.t. some correctness criterion), and the condition of saturation expresses that each processes that evaluates to an accepting process is itself an accepting process. A typical candidate for $\perp\!\!\!\perp$ is the set $\perp\!\!\!\perp_0$ of all terminating processes defined by:

$$\perp\!\!\!\perp_0 \;=\; \{p \mid \exists p'\ (p \succ^* p' \ \wedge\ p' \not\succ)\}\,.$$

In classical realisability, sets of stacks are used as *falsity values* (that is, as sets of potential refutations). Each falsity value $S \subset \Pi$ defines by orthogonality a *truth value* written $S^{\perp\!\!\!\perp}$ and defined by

$$S^{\perp\!\!\!\perp} \;=\; \{t \in \Lambda \mid \forall \pi \in S \ \ t \star \pi \in \perp\!\!\!\perp\}\,.$$

In section 4, we will construct a model where all the objects are annotated with falsity values, using a structure of Π-set.

3 The Π-Set Structure

3.1 Definition

Definition 1 (Π-set). *A* Π-set *is a pair $X = \langle |X|, \perp_X \rangle$ formed by a set $|X|$ (called the* carrier *of X) equipped with a binary relation $\perp_X \subset |X| \times \Pi$ (called the* local orthogonality relation *of X).*

[3] From the Scheme operator 'call-cc' (call with current continuation).

Intuitively, the binary relation $x \perp_X \pi$ expresses that the stack π is an attempt to refute (or to attack, or to falsify) the denotation $x \in |X|$. Given an element $x \in |X|$, we write $x^{\perp_X} = \{\pi \mid x \perp_X \pi\}$ the orthogonal of x w.r.t. X. From this we define the *local realisability relation* $t \Vdash_X x$ by setting

$$\begin{array}{lll} t \Vdash_X x & \text{iff} & \forall\pi\ (x \perp_X \pi \Rightarrow t \star \pi \in \perp\!\!\!\perp) \\ & \text{iff} & t \in (x^{\perp_X})^{\perp\!\!\!\perp} \end{array}$$

for all $t \in \Lambda$ and $x \in |X|$.

Remark. Unlike ω-sets, we do not require that each element of X is realised by at least a quasi-proof—we do not even require that each element of X has a realiser. However, all the elements of the carrier have realisers as soon as the set $\perp\!\!\!\perp$ is inhabited: given a fixed process $t_0 \star \pi_0 \in \perp\!\!\!\perp$, it is easy to check that the term $\mathsf{k}_{\pi_0} t_0$ is orthogonal to any stack (w.r.t. $\perp\!\!\!\perp$), and thus realises any denotation.

Coarse Π-sets. We say that a Π-set X is *coarse* when $\perp_X = \emptyset$ (i.e. when the orthogonality relation on X is empty). By duality, we get $t \Vdash_X x$ for all $t \in \Lambda$ and $x \in |X|$, which means that any term realises any element of the carrier of X.

Notice that any set X can be given the structure of a coarse Π-set simply by taking $|X| = X$ and $\vdash_X = \emptyset$.

Pointed Π-sets. In many situations, it is desirable to exclude the empty Π-set and to work only with Π-sets whose carrier is inhabited. To avoid the cost of introducing the axiom of choice in the meta-theory (typically to ensure that the Cartesian product of a family of inhabited Π-sets is inhabited), we will only consider Π-sets with a distinguished element of the carrier, that is: *pointed Π-sets.* Formally, a pointed Π-set is a triple $X = \langle |X|, \perp_X, \varepsilon_X \rangle$ where $\langle |X|, \perp_X \rangle$ is a Π-set and where $\varepsilon_X \in |X|$.

3.2 Cartesian Product of a Family of Π-Sets

Let $(Y_x)_{x \in |X|}$ be a family of Π-sets indexed by the carrier of a Π-set X. The *Cartesian product* of the family $(Y_x)_{x \in |X|}$ is the Π-set written $\Pi x : X . Y_x$ and defined by:

$$|\Pi x : X . Y_x| = \prod_{x \in |X|} |Y_x| \quad \text{and} \quad f^{\perp_{\Pi x:X.Y_x}} = \bigcup_{x \in |X|} \left((x^{\perp_X})^{\perp\!\!\!\perp} \cdot (f(x)^{\perp_{Y_x}}) \right)$$

for all $f \in |\Pi x : X . Y_x|$. Moreover if Y_x is a pointed Π-set for all $x \in |X|$, then the product $\Pi x : X . Y_x$ can be given the structure of a pointed Π-set by setting

$$\varepsilon_{\Pi x : X . Y_x} = (x \in |X| \mapsto \varepsilon_{Y_x}) .$$

Fact 1. *If Y_x is a coarse Π-set (resp. a coarse pointed Π-set) for all $x \in |X|$, then $\Pi x : X . Y_x$ is a coarse Π-set (resp. a coarse pointed Π-set).*

In section 4 we will interpret the sorts Prop, Type_i by coarse pointed Π-sets; hence the fact above will automatically imply that more generally, all types of predicates will be interpreted by coarse pointed Π-sets.

3.3 Degenerated Π-Sets

A Π-set is said to be *degenerated* when its carrier is a singleton: $|X| = \{\varepsilon_X\}$. (In this case we can always consider X as a pointed Π-set by taking ε_X as the unique element of its carrier.) A degenerated Π-set X is characterised by its unique element ε_X and by the set of stacks $\varepsilon_X^{\perp_X}$ that are orthogonal to this element, which set of stacks will be written $X^\perp\,(=\varepsilon_X^{\perp_X})$.

Fact 2 (Product of degenerated Π-sets). *If Y_x is a degenerated Π-set for all $x \in |X|$, then the Π-set $\Pi x\!:\!X\,.\,Y_x$ is degenerated too, and we have*

$$(\Pi x\!:\!X\,.\,Y_x)^\perp \;=\; \bigcup_{x\in|X|} \left((x^{\perp_X})^{\perp\!\!\perp} \cdot Y_x^\perp\right)$$

Moreover, if the Π-set X is degenerated, then the Cartesian product $\Pi x\!:\!X\,.\,Y_x$ is non-dependent and $(\Pi x\!:\!X\,.\,Y_x)^\perp \;=\; (X \to Y)^\perp \;=\; (X^\perp)^{\perp\!\!\perp} \cdot Y^\perp$.

In what follows, degenerated Π-sets will be used to interpret propositions.

3.4 Subtyping

Given two Π-sets X and X', we write $X \le X'$ (X is a *subtype* of X') when

$$|X| \subseteq |X'| \qquad \text{and} \qquad x^{\perp_X} \supseteq x^{\perp_{X'}} \quad \text{for all } x \in |X|\,.$$

(Notice that the reverse inclusion above is necessary to ensure that $t \Vdash_X x$ implies $t \Vdash_{X'} x$ for all $t \in \Lambda$ and $x \in |X|$.) When both X and X' are pointed Π-sets, we also require that: $\varepsilon_{X'} = \varepsilon_X$.

4 Constructing the Model

In what follows, we work in ZF set theory extended with an axiom expressing the existence of infinitely many inaccessible cardinals to interpret the hierarchy of predicative universes.

4.1 An Alternative Encoding of Functions

To achieve proof-irrelevance in the model, we will interpret all proof-objects by a single value written $\bullet$ and all propositions by degenerated Π-sets based on the singleton $\{\bullet\}$. Since we want to keep the property of closure under Cartesian products (Fact 2), it is necessary to identify all constant functions $(x \in D \mapsto \bullet)$ for the proof-object $\bullet$ itself. For that, we adopt a set-theoretic encoding of functions (proposed by [1] and inspired from the notion of trace in domain theory) in which functions are represented not as set of pairs $\langle x, y\rangle$ such that $y = f(x)$, but as set of pairs $\langle x, z\rangle$ such that $z \in f(x)$.

Formally, we introduce the following abbreviations:

$$\begin{array}{rcl} f \text{ function} & \equiv & \forall p \in f\ \exists x\ \exists y\ \ p = \langle x, y\rangle \\ (x \in D \mapsto E_x) & = & \{\langle x, z\rangle \mid x \in D\ \wedge\ z \in E_x\} \\ f(x) & = & \{z \mid \exists x\ \langle x, z\rangle \in f\} \end{array}$$

This encoding is sound in the sense that given a function $f = (x \in D \mapsto E_x)$, we have $f(d) = E_d$ for all $d \in D$. However, the domain information is partially lost since the encoding keeps no track of elements of the domain mapped to the empty set. We can only define the *support* of a function

$$\text{supp}(f) \ = \ \{x \mid \exists z\ \ \langle x, z\rangle \in f\}\,.$$

Apart from this (minor) difference, this alternative encoding of functions can be used the same way as the traditional encoding. From now on we consider that functions in the model are represented in this way, and we take $\bullet = \emptyset$ so that the equality $(x \in D \mapsto \bullet) = \bullet$ now holds for all D.

4.2 Interpreting Sorts

Let $(\lambda_i)_{i\geq 1}$ be an increasing sequence of inaccessible cardinals and set:

$$\mathcal{U}_0 = \{\{\bullet\}\} \times \mathfrak{P}(\{\bullet\} \times \Pi) \times \{\bullet\}$$

$$\mathcal{U}_i = \bigcup_{\substack{S \in V_{\lambda_i} \\ s_0 \in S}} \{S\} \times \mathfrak{P}(S \times \Pi) \times \{s_0\} \qquad (\subset V_{\lambda_i})$$

By definition, $\mathcal{U}_0$ is the set of all degenerated pointed Π-sets based on the singleton $\{\bullet\}$ whereas $\mathcal{U}_i$ $(i \geq 1)$ is the set of all pointed Π-sets whose (nonempty) carrier belongs to V_{λ_i} (i.e. the ith ZF-universe). Each set of Π-sets $\mathcal{U}_i$ $(i \geq 0)$ can be given in turn the structure of a coarse pointed Π-set $\mathcal{U}'_i$ by setting:

$$\mathcal{U}'_0 = \langle \mathcal{U}_0,\ \emptyset,\ \langle\{\bullet\},\ \emptyset,\ \bullet\rangle\rangle \qquad \text{and} \qquad \mathcal{U}'_i = \langle \mathcal{U}_i,\ \emptyset,\ \mathcal{U}'_{i-1}\rangle \quad \text{for } i \geq 1\,.$$

Finally, the domain of all denotations $\mathcal{M}$ is taken as the union of all carriers of the Π-sets in the universes $\mathcal{U}_i$: $\mathcal{M} = \bigcup_{i\in\omega} \bigcup_{X\in\mathcal{U}_i} |X|$.

Fact 3 (Closure under Cartesian product). *For all $i \geq 1$:*

1. *If $X \in \mathcal{U}_i$ and $Y_x \in \mathcal{U}_0$ for all $x \in |X|$, then $\Pi x : X\,.\,Y_z \in \mathcal{U}_0$;*
2. *If $X \in \mathcal{U}_i$ and $Y_x \in \mathcal{U}_i$ for all $x \in |X|$, then $\Pi x : X\,.\,Y_z \in \mathcal{U}_i$.*

4.3 The Interpretation Function

A *valuation* in $\mathcal{M}$ is a partial function $\rho : \mathcal{X} \to \mathcal{M}$ (writing $\mathcal{X}$ the set of all variables) whose domain $\text{dom}(\rho) \subset \mathcal{X}$ is finite. (Here, it is more convenient to keep the traditional set-theoretic encoding of functions to represent valuations.)

The set of all valuations is written $\mathrm{Val}_{\mathcal{M}}$. Given a valuation ρ, a variable $x \in \mathcal{X}$ and a value $v \in \mathcal{M}$, we write $(\rho, x \leftarrow v)$ the valuation defined by

$$(\rho, x \leftarrow v)(x) = v \qquad \text{and} \qquad (\rho, x \leftarrow v)(y) = \rho(y) \qquad (y \in \mathrm{dom}(\rho) \setminus \{x\})$$

To each term M we associate a partial function $[\![M]\!]_{\_} : \mathrm{Val}_{\mathcal{M}} \rightharpoonup \mathcal{M}$ which is inductively defined on M by the equations:

$$\begin{array}{rlrl} [\![x]\!]_\rho &= \rho(x) & [\![\Pi x : T \,.\, U]\!]_\rho &= \Pi v : [\![T]\!]_\rho \,.\, [\![U]\!]_{\rho; x \leftarrow v} \quad \text{(product of } \Pi\text{-sets)} \\ [\![\mathsf{Prop}]\!]_\rho &= \mathcal{U}'_0 & [\![\lambda x : T \,.\, M]\!]_\rho &= (v \in |[\![T]\!]_\rho| \mapsto [\![M]\!]_{\rho; x \leftarrow v}) \\ [\![\mathsf{Type}_i]\!]_\rho &= \mathcal{U}'_i & [\![MN]\!]_\rho &= [\![M]\!]_\rho([\![N]\!]_\rho) \end{array}$$

Since the function $\rho \mapsto [\![M]\!]_\rho$ is partial, it is important to precise when the denotation $[\![M]\!]\rho$ is defined:

- $[\![x]\!]_\rho$ is defined when $x \in \mathrm{dom}(\rho)$;
- $[\![\mathsf{Prop}]\!]_\rho$ and $[\![\mathsf{Type}_i]\!]_\rho$ are always defined;
- $[\![\Pi x : T \,.\, U]\!]_\rho$ is defined when
  - $[\![T]\!]_\rho$ is defined, and it is a pointed Π-set,
  - For all $v \in |[\![T]\!]_\rho|$, $[\![U]\!]_{\rho, x \leftarrow v}$ is defined, and it is a pointed Π-set,
  - $\Pi v : [\![T]\!]_\rho \,.\, [\![U]\!]_{\rho; x \leftarrow v}$ is an element of $\mathcal{M}$;
- $[\![\lambda x : T \,.\, M]\!]_\rho$ is defined when
  - $[\![T]\!]_\rho$ is defined, and it is a pointed Π-set,
  - For all $v \in |[\![T]\!]_\rho|$, $[\![M]\!]_{\rho, x \leftarrow v}$ is defined,
  - $(v \in |[\![T]\!]_\rho| \mapsto [\![U]\!]_{\rho; x \leftarrow v})$ is an element of $\mathcal{M}$;
- $[\![MN]\!]_\rho$ is defined when
  - $[\![M]\!]_\rho$ and $[\![N]\!]_\rho$ are defined,
  - $[\![M]\!]_\rho$ is a function, and $[\![M]\!]_\rho([\![N]\!]_\rho)$ is an element of $\mathcal{M}$.

The interpretation function is extended to all contexts by setting:

$$[\![\Gamma]\!] \quad = \quad \{\rho \in \mathrm{Val}_{\mathcal{M}} \mid \forall (x : T) \in \Gamma \, \rho(x) \in |[\![T]\!]_\rho|\}$$

4.4 Soundness

Definition 2 (Soundness conditions).

1. *A typing judgement $\Gamma \vdash M : T$ is sound w.r.t. $\mathcal{M}$ if for all $\rho \in [\![\Gamma]\!]$:*
   - *The denotations $[\![M]\!]_\rho$ and $[\![T]\!]_\rho$ are defined;*
   - *$[\![T]\!]_\rho$ is a Π-set; and*
   - *$[\![M]\!]_\rho \in |[\![T]\!]_\rho|$.*
2. *A subtyping judgement $\Gamma \vdash T \leq T'$ is sound w.r.t. $\mathcal{M}$ if for all $\rho \in [\![\Gamma]\!]$:*
   - *The denotations $[\![T]\!]_\rho$ and $[\![T']\!]_\rho$ are defined;*
   - *$[\![T]\!]_\rho$ and $[\![T']\!]_\rho$ are Π-sets;*
   - *$[\![T]\!]_\rho \leq [\![T']\!]_\rho$.*
3. *An equality judgement $\Gamma \vdash M = M' : T$ is sound w.r.t. $\mathcal{M}$ if for all $\rho \in [\![\Gamma]\!]$:*
   - *The denotations $[\![M]\!]_\rho$, $[\![M']\!]_\rho$ and $[\![T]\!]_\rho$ are defined;*
   - *$[\![T]\!]_\rho$ is a Π-set; and*
   - *$[\![M]\!]_\rho = [\![M']\!]_\rho \in |[\![T']\!]_\rho|$.*

Proposition 1 (Soundness). *If a typing judgement, a subtyping judgement or an equality judgement is derivable (Fig. 1), then it is sound w.r.t. $\mathcal{M}$.*

Proof. By induction of the derivation of the judgement. □

4.5 Adequacy

The basic extraction scheme To each (raw-)term M of CC_ω we associate a term M^* of λ_c which is inductively defined from the equations

$$x^* = x \qquad s^* = (\Pi x : T \,.\, U)^* = \lambda z \,.\, z \qquad \text{(or any quasi-proof)}$$
$$(\lambda x : T \,.\, M)^* = \lambda x \,.\, M^* \qquad (MN)^* = M^* N^*$$

Intuitively, this extraction function erases all non-computational information related to types, but preserves all the computational contents of proof-terms.

Substitutions. A *substitution* is a partial function $\sigma : \mathcal{X} \mapsto \Lambda$ whose domain $\mathrm{dom}(\sigma) \subset \mathcal{X}$ is finite. Given an open term t of the λ_c-calculus and a substitution σ, we write $t[\sigma]$ the result of applying the substitution σ to t.

Let Γ be a context, ρ a valuation and σ a substitution. We say that σ *realises* ρ *in* Γ and write $\sigma \Vdash_\Gamma \rho$ when

1. $\mathrm{dom}(\sigma) = \mathrm{dom}(\Gamma)$
2. $\rho \in [\![\Gamma]\!]$
3. For all $(x : T) \in \Gamma$: $\sigma(x) \Vdash_{[\![T]\!]_\rho} \rho(x)$

We can now extend the property of adequacy of second-order arithmetic [10] to CC_ω as follows:

Proposition 2 (Adequacy). *If $\Gamma \vdash M : T$, then for all valuations $\rho \in \Gamma$ and for all substitutions σ such that $\sigma \Vdash_\Gamma \rho$, we have*

$$M^*[\sigma] \Vdash_{[\![T]\!]_\rho} [\![M]\!]_\rho \,.$$

Proof. By induction of the derivation of the judgement. □

In particular, when the judgement $\vdash M : T$ is derivable in the empty context, the extracted term M^* realises the denotation of M: $M^* \Vdash_{[\![T]\!]} [\![M]\!]$.

5 Extensions of the Formalism

5.1 Peirce's Law and the Excluded Middle

Let us now extend CC_ω^{irr} with a new constant

$$\mathsf{peirce} \quad : \quad \Pi A, B : \mathsf{Prop} \,.\, (((A \to B) \to A) \to A)$$

that we interpret in the model $\mathcal{M}$ as $[\![\mathsf{peirce}]\!]_\rho = \bullet$. We then extend the basic extraction scheme by setting $\mathsf{peirce}^* = \lambda\_ \,.\, \lambda\_ \,.\, \mathbb{c}$ and check that this extension is adequate in the sense of Prop. 2:

Fact 4. $\mathsf{peirce}^* \in ([\![\Pi A, B : \mathsf{Prop} \,.\, (((A \to B) \to A) \to A)]\!]^\perp)^{\perp\!\!\perp}$.

From this extension of the calculus, it is easy to derive the law of excluded middle $\Pi A : \mathsf{Prop} \,.\, (A \vee \neg A)$ at the level of propositions (defining disjunction and negation by the mean of standard second-order encodings).

Remark. In the source calculus (CC_ω^{irr}), it is not necessary to endow the extra constant peirce with specific equality rules, since the rule of proof-irrelevance already performs all possible identifications at the level of proof-terms. In the target calculus (λ_c), the constant peirce is extracted to the λ_c-term $\lambda\_.\,\lambda\_.\,\mathbf{c}$ that evaluates as expected, by consuming two computationally irrelevant arguments (corresponding to types) before capturing the current continuation.

5.2 Decomposing the Propositional Dependent Product

In intuitionistic and classical realisability [10], (non relativised) first- and second-order quantification is usually interpreted parametrically, that is, as an intersection (or as a union on the side of stacks). In CC_ω^{irr}, universal quantification is represented by a dependent-product $\Pi x:T\,.\,U(x)$, that is, by a type of functions taking a value $x : T$ and returning a proof of $U(x)$.

To bridge both interpretations of universal quantification, we first extend the formalism with three new syntactic constructs, namely:

- An intersection type binder $\forall x:T\,.\,U$ corresponding to the parametric interpretation of the universal quantification. This construction is exactly the implicit dependent product of the implicit calculus of constructions [15], but here restricted to propositions.
- A construction $M \in T$ representing the *propositional contents* of the typing judgement $M : T$. As we shall see, the construction $M \in T$ represents the proposition whose proofs are the realisers of the term M in the type T.
- A constant $\top$ representing the proposition realised by all λ-terms.

Terms $\quad T, U \quad ::= \quad \cdots \quad | \quad \forall x:T\,.\,U \quad | \quad M \in T \quad | \quad \top$

These new syntactic constructs that we interpret in $\mathcal{M}$ by

$$[\![\forall x:T\,.\,U]\!]_\rho \;=\; \Big\langle \{\bullet\},\ \{\bullet\} \times \bigcup_{v \in [\![T]\!]_\rho} [\![U]\!]^{\perp}_{\rho, x\leftarrow v},\ \bullet \Big\rangle$$

$$[\![M \in T]\!]_\rho \;=\; \langle \{\bullet\},\ \{\bullet\} \times [\![M]\!]_\rho^{\perp [\![T]\!]_\rho},\ \bullet \rangle \qquad\qquad [\![\top]\!]_\rho \;=\; \langle \{\bullet\},\ \emptyset,\ \bullet \rangle$$

come with typing, subtyping and equality rules that are given in Fig. 2.

Fact 5. *The typing rules, subtyping rules and equality rules of Fig. 2 are sound w.r.t. the interpretation of the constructs* $\forall x:T\,.\,U$, $M \in T$ *and* $\top$ *in* $\mathcal{M}$.

In the extended formalism, the propositional dependent product can now be decomposed in terms of $\forall$ and $\in$ as

$$\Pi x:T\,.\,U \;=\; \forall x:T\,.\,((x \in T) \to U)$$

using the decomposition rule of Fig. 2. Intuitively, this equality rule expresses that in CC_ω^{irr}, the propositional dependent product corresponds exactly to the relativised universal quantification in the sense of AF2. In subsection 5.3, we

Typing rules

$$\frac{\Gamma, x:T \vdash U : \mathsf{Prop}}{\Gamma \vdash \forall x{:}T\,.\,U : \mathsf{Prop}} \qquad \frac{\Gamma \vdash M : T}{\Gamma \vdash M \in T : \mathsf{Prop}} \qquad \frac{\vdash \Gamma\ \textbf{ctx}}{\Gamma \vdash \top : \mathsf{Prop}}$$

Subtyping rules

$$\frac{\Gamma \vdash T = T' : s \qquad \Gamma \vdash U \leq U' \qquad \Gamma \vdash U' : \mathsf{Prop}}{\Gamma \vdash \forall x{:}T\,.\,U \leq \forall x{:}T'\,.\,U'}$$

$$\frac{\Gamma \vdash M : T \qquad \Gamma \vdash T \leq T'}{\Gamma \vdash (M \in T) \leq (M \in T')} \qquad \frac{\Gamma \vdash T : \mathsf{Prop}}{\Gamma \vdash T \leq \top}$$

Equality rules

$$\frac{\Gamma \vdash T = T' : s \qquad \Gamma, x:T \vdash U = U' : \mathsf{Prop}}{\Gamma \vdash \forall x{:}T\,.\,U = \forall x{:}T'\,.\,U' : \mathsf{Prop}}$$

$$\frac{\Gamma \vdash M = M' : T \qquad \Gamma \vdash T = T' : s}{\Gamma \vdash (M \in T) = (M' \in T') : \mathsf{Prop}}$$

(Decomposition of Π)

$$\frac{\Gamma, x:T \vdash U : \mathsf{Prop}}{\Gamma \vdash \Pi x{:}T\,.\,U = \forall x{:}T\,.\,(x \in T \to U) : \mathsf{Prop}}$$

$$\frac{\Gamma \vdash T : s}{\Gamma \vdash \Pi x{:}T\,.\,\top = \top : \mathsf{Prop}}$$

(Simplification of $\in$)

$$\frac{\Gamma \vdash T : s}{\Gamma \vdash (T \in s) = \top : \mathsf{Prop}} \qquad \frac{\Gamma \vdash M : \top}{\Gamma \vdash (M \in \top) = \top : \mathsf{Prop}}$$

$$\frac{\Gamma \vdash M : \Pi x{:}T\,.\,U}{\Gamma \vdash (M \in \Pi x{:}T\,.\,U) = \forall x{:}T\,.\,((x \in T) \to Mx \in U) : \mathsf{Prop}}$$

$$\frac{\Gamma \vdash M : \forall x{:}T\,.\,U}{\Gamma \vdash (M \in \forall x{:}T\,.\,U) = \forall x{:}T\,.\,(M \in U) : \mathsf{Prop}}$$

Fig. 2. Typing, subtyping and equality rules of $\forall$, $\in$ and $\top$

will exploit this fact in order to recover the usual interpretation of the numeric quantification in classical realisability.

5.3 Adding a Primitive Type of Natural Numbers

Let us now extend $\mathrm{CC}_\omega^{\mathrm{irr}}$ with the following set of typed constants ($i \geq 1$):

$$\begin{array}{lcl} \mathsf{nat} & : & \mathsf{Type}_1 \qquad \mathsf{0} : \mathsf{nat} \qquad \mathsf{s} : \mathsf{nat} \to \mathsf{nat} \\ \mathsf{nat\_ind} & : & \Pi X : \mathsf{nat} \to \mathsf{Prop}\,.\,(X\mathsf{0} \to \Pi y : \mathsf{nat}\,.\,(Xy \to X(\mathsf{s}\,y)) \to \Pi x : \mathsf{nat}\,.\,Xx) \\ \mathsf{nat\_rec}_i & : & \Pi X : \mathsf{nat} \to \mathsf{Type}_i\,.\,(X\mathsf{0} \to \Pi y : \mathsf{nat}\,.\,(Xy \to X(\mathsf{s}\,y)) \to \Pi x : \mathsf{nat}\,.\,Xx) \end{array}$$

The constants $\mathsf{nat\_ind}$ and $\mathsf{nat\_rec}_i$ ($i \geq 1$) respectively implement the induction principle and (dependently-typed) recursion in Type_i.

Interpreting nat. The constant nat is interpreted as the pointed Π-set defined by $|[\![\mathsf{nat}]\!]| = \mathbb{N}$, $\varepsilon_{[\![\mathsf{nat}]\!]} = 0$, and whose orthogonality relation is given by

$$n^{\perp_{[\![\mathsf{nat}]\!]}} = [\![\forall X : \mathsf{nat} \to \mathsf{Prop}\,.\,(X\mathsf{0} \to \forall y : \mathsf{nat}\,.\,(Xy \to X(\mathsf{s}\,y)) \to Xx)]\!]^{\perp}_{x \leftarrow n}$$

for all $n \in \mathbb{N}$. Notice that the definition above is not circular, since the r.h.s. only depends on the definition of the carrier of nat, but not on its orthogonality relation. The constants 0 and s are then interpreted as expected.

The interest of this definition is that the proposition $x \in \mathsf{nat}$ is interpreted exactly as the relativisation predicate which is traditionally used in second-order arithmetic to define numeric quantification:

$$\mathsf{Nat}(x) \equiv \forall X : \mathsf{nat} \to \mathsf{Prop}\,.\,(X\mathsf{0} \to \forall y : \mathsf{nat}\,.\,(Xy \to X(\mathsf{s}\,y)) \to Xx)$$

Fact 6. *The following equality rule is sound in $\mathcal{M}$:*

$$\frac{\Gamma \vdash M : \mathsf{nat}}{\Gamma \vdash (M \in \mathsf{nat}) = \mathsf{Nat}(M) : \mathsf{Prop}}$$

Combining the latter with the decomposition of the dependent product (cf subsection 5.2) we get the equality $\Pi x : \mathsf{nat}\,.\,P(x) = \forall x : \mathsf{nat}\,.\,(\mathsf{Nat}(x) \to P(x))$ expressing that the PTS-style quantification $\Pi x : \mathsf{nat}\,.\,P(x)$ is interpreted in $\mathcal{M}$ exactly the same way as the numeric quantification in classical realisability [10].

Interpreting $\mathsf{nat\_ind}$ *and* $\mathsf{nat\_rec}_i$. The constant nat_ind is interpreted as the proof object • whereas the constants $\mathsf{nat\_rec}_i$ are interpreted the obvious way (i.e. as the corresponding set-theoretic recursors in the universes $\mathcal{U}_i$). From the latter definition, it is immediate that:

Fact 7. *The following equality rules are sound in $\mathcal{M}$:*

$$\frac{\begin{array}{ll} \Gamma \vdash P : \mathsf{nat} \to \mathsf{Type}_i & \Gamma \vdash N : \mathsf{nat} \\ \Gamma \vdash M_0 : P\,\mathsf{0} & \Gamma \vdash M_1 : \Pi p : \mathsf{nat}\,.\,P\,p \to P(\mathsf{s}\,p) \end{array}}{\begin{array}{l} \Gamma \vdash \mathsf{nat\_rec}_i\,P\,M_0\,M_1\,\mathsf{0} = M_0 : P\,\mathsf{0} \\ \Gamma \vdash \mathsf{nat\_rec}_i\,P\,M_0\,M_1\,(\mathsf{s}\,N) = M_1\,N\,(\mathsf{nat\_rec}_i\,P\,M_0\,M_1\,N) : P\,(\mathsf{s}\,N) \end{array}}$$

Extraction. We finally extend the extraction mechanism to the new constants nat, 0, s, nat_ind and $\mathsf{nat\_rec}_i$ by setting:

$$\mathsf{nat}^* = \lambda z\,.\,z \qquad \text{(or any quasi-proof)}$$
$$0^* = \lambda xf\,.\,x \qquad \mathsf{s}^* = \lambda nxf\,.\,f(nxf) \qquad \mathsf{nat\_rec}_i^* = \mathsf{nat\_ind}^*$$
$$\mathsf{nat\_ind}^* = \lambda\_xfn\,.\,n\,(\lambda z\,.\,z\,0^*x)\,(\lambda p\,.\,p\,(\lambda myz\,.\,z\,(s^*m)\,(fmy)))\,(\lambda xy\,.\,y)$$

Proposition 3. *The extraction scheme extended to the constants* nat, 0, s, nat_ind *and* $\mathsf{nat\_rec}_i$ *is adequate w.r.t.* $\mathcal{M}$ *(in the sense of Prop. 2).*

References

1. Aczel, P.: On relating type theories and set theories. In: Altenkirch, Naraschewski, Reus (eds.) Proceedings of Types'98 (1999)
2. Altenkirch, T.: Constructions, Inductive Types and Strong Normalization. PhD thesis, University of Edinburgh (November 1993)
3. Coquand, T., Huet, G.: The calculus of constructions. Information and Computation 120(76), 95 (1988)
4. Geuvers, J.H., Nederhof, M.J.: A modular proof of strong normalization for the calculus of constructions. Journal of Functional Programming 1,2(1991), 155–189 (1991)
5. Hyland, J.M.E.: The effective topos. In: Troelstra, A.S., van Dalen, D. (eds.) The L. E. J. Brouwer Centenary Symposium, North Holland, Amsterdam (1982)
6. Krivine, J.-L.: Lambda-calcul, types et modèles. Masson (1991)
7. Krivine, J.-L.: A general storage theorem for integers in call-by-name lambda-calculus. Th. Comp. Sc. 129, 79–94 (1994)
8. Krivine, J.-L.: Typed lambda-calculus in classical Zermelo-Fraenkel set theory. Arch. Math. Log. 40(3), 189–205 (2001)
9. Krivine, J.-L.: Dependent choice, 'quote' and the clock. Th. Comp. Sc. 308, 259–276 (2003)
10. Krivine, J.-L.: Realizability in classical logic. Unpublished lecture notes (available on the author's web page) (2005)
11. Letouzey, P.: A new extraction for Coq. In: Geuvers, H., Wiedijk, F. (eds.) TYPES 2002. LNCS, vol. 2646, pp. 200–219. Springer, Heidelberg (2003)
12. Longo, G., Moggi, E.: A category-theoretic characterization of functional completeness. Theor. Comput. Sci. 70(2), 193–211 (1990)
13. Luo, Z.: Computation and Reasoning: A Type Theory for Computer Science. Oxford University Press, Oxford (1994)
14. Melliès, P.-A., Werner, B.: A generic normalisation proof for pure type systems. In: Giménez, E., Paulin-Mohring, C. (eds.) TYPES 1996. LNCS, vol. 1512, pp. 254–276. Springer, Heidelberg (1998)
15. Miquel, A.: Le calcul des constructions implicite: syntaxe et sémantique. PhD thesis, Université Paris 7 (2001)
16. Paulin-Mohring, C.: Extracting $F\omega$'s programs from proofs in the calculus of constructions. In: POPL'89, pp. 89–104 (1989)
17. Paulin-Mohring, C.: Définitions Inductives en Théorie des Types d'Ordre Supérieur. Habilitation à diriger les recherches, Université Claude Bernard Lyon I (1996)
18. The Coq Development Team (LogiCal Project). The Coq Proof Assistant Reference Manual - Version 8.1. Technical report, INRIA (2006)
19. Streicher, T.: Semantics of Type Theory. Birkhäuser (1991)

Building Decision Procedures in the Calculus of Inductive Constructions

Frédéric Blanqui[1], Jean-Pierre Jouannaud[2], and Pierre-Yves Strub[2]

[1] LORIA*, Equipe Protheo, Campus Scientifique
BP 239, 54506 Vandoeuvre-lès-Nancy Cedex
`blanqui@loria.fr`

[2] LogiCal**, LIX, UMR CNRS 7161
École Polytechnique, 91128 Plaiseau, France
`{jouannaud,strub}@lix.polytechnique.fr`

Abstract. It is commonly agreed that the success of future proof assistants will rely on their ability to incorporate computations within deduction in order to mimic the mathematician when replacing the proof of a proposition P by the proof of an equivalent proposition P' obtained from P thanks to possibly complex calculations.

In this paper, we investigate a new version of the calculus of inductive constructions which incorporates arbitrary decision procedures into deduction via the conversion rule of the calculus. The novelty of the problem in the context of the calculus of inductive constructions lies in the fact that the computation mechanism varies along proof-checking: goals are sent to the decision procedure together with the set of user hypotheses available from the current context. Our main result shows that this extension of the calculus of constructions does not compromise its main properties: confluence, subject reduction, strong normalization and consistency are all preserved.

Keywords: Calculus of Inductive Constructions, Decision procedures, Theorem provers.

1 Introduction

Background. It is commonly agreed that the success of future proof assistants will rely on their ability to incorporate computations within deduction in order to mimic the mathematician when replacing the proof of a proposition P by the proof of an equivalent proposition P' obtained from P thanks to possibly complex calculations.

Proof assistants based on the Curry-Howard isomorphism such as Coq [1] allow to build the proof of a proposition by applying appropriate proof tactics generating a proof term that can be checked with respect to the rules of logic.

* UMR 7503 CNRS-INPL-INRIA-Nancy2-UHP.

** Project LogiCal, Pôle Commun de Recherche en Informatique du Plateau de Saclay, CNRS, École Polytechnique, INRIA, Univ. Paris-Sud.

J. Duparc and T.A. Henzinger (Eds.): CSL 2007, LNCS 4646, pp. 328–342, 2007.

The proof-checker, also called the *kernel* of the proof assistant, implements the inference and deduction rules of the logic on top of a term manipulation layer. Trusting the kernel is vital since the mathematical correctness of a proof development relies entirely on the kernel.

The (intuitionist) logic on which Coq is based is the Calculus of Constructions (CC) of Coquand and Huet [2], an impredicative type theory incorporating polymorphism, dependent types and type constructors. As other logics, CC enjoys a computation mechanism called cut-elimination, which is nothing but the β-reduction rule of the underlying λ-calculus. But unlike logics without dependent types, CC enjoys also a powerful type-checking rule, called *conversion*, which incorporates computations within deduction, making decidability of type-checking a non-trivial property of the calculus.

The traditional view that computations coincide with β-reductions suffers several drawbacks. A methodological one is that the user must encode other forms of computations as deduction, which is usually done by using appropriate, complex tactics. A practical one is that proofs become much larger than necessary, up to a point that they cannot be type-checked anymore. These questions become extremely important when carrying out complex developments involving a large amount of computation as the formal proof of the four colour (now proof-checked) theorem completed by Gonthier and Werner using Coq [3].

The Calculus of Inductive Constructions of Coquand and Paulin was a first attempt to solve this problem by introducing inductive types and the associated elimination rules [4]. The recent versions of Coq are based on a slight generalization of this calculus [5]. Besides the β-reduction rule, they also include the so-called ι-reductions which are recursors for terms and types. While the kernel of CC is extremely compact and simple enough to make it easily readable -hence trustable-, the kernel of CIC is much larger and quite complex. Trusting it would require a formal proof, which was done once [6]. Updating that proof for each new release of the system is however unrealistic. CIC does not solve our problem, though, since such a simple function as *reverse* of a *dependent list* cannot be defined in CIC because $a :: l$ and $l :: a$, assuming :: is list concatenation and the element a can be coerced to a list of length 1, have non-convertible types $list(n+1)$ and $list(1+n)$.

A more general attempt was carried out since the early 90's, by adding user-defined computations as rewrite rules, resulting in the Calculus of Algebraic Constructions [7]. Although conceptually quite powerful, since CAC captures CIC [8], this paradigm does not yet fulfill all needs, because the set of user-defined rewrite rules must satisfy several strong assumptions. No implementation of CAC has indeed been released because making type-checking efficient would require compiling the user-defined rules, a complex task resulting in a kernel too large to be trusted anymore.

The proof assistant PVS uses a potentially stronger paradigm than Coq by combining its deduction mechanism[1] with a notion of computation based on the

[1] PVS logic is not based on Curry-Howard and proof-checking is not even decidable making both frameworks very different and difficult to compare.

powerful Shostak's method for combining decision procedures [9], a framework dubbed *little proof engines* by Shankar [10]: the *little proof engines* are the decision procedures, required to be convex, combined by Shostak's algorithm. A given decision procedure encodes a fixed set of axioms P. But an important advantage of the method is that the relevant assumptions A present in the context of the proof are also used by the decision procedure to prove a goal G, and become therefore part of the notion of computation. For example, in the case where the little proof engines is the congruence closure algorithm, the fixed set of axioms P is made of the axioms for equality, A is the set of algebraic ground equalities declared in the context, while the goal G is an equality $s = t$ between two ground expressions. The congruence closure algorithm will then process A and $s = t$ together in order to decide whether or not $s = t$ follows from $P \cup A$. In the Calculus of Constructions, this proof must be constructed by a specific tactic called by the user, which applies the inference rules of CC to the axioms in P and the assumptions in A, and becomes then part of the proof term being built. Reflexion techniques allow to omit checking this proof term by proving the decision procedure itself, but the soundness of the entire mechanism cannot be guaranteed [11].

Two further steps in the direction of integrating decision procedures into the Calculus of Constructions are Stehr's Open Calculus of Constructions OCC [12] and Oury's Extensional Calculus of Constructions [13]. Implemented in Maude, OCC allows for the use of an arbitrary equational theory in conversion. ECC can be seen as a particular case of OCC in which all provable equalities can be used in conversion, which can also be achieved by adding the extensionality and Streicher's axiom [14] to CIC, hence the name of this calculus. Unfortunately, strong normalization and decidability of type checking are lost in ECC (and OCC), which shows that we should look for more restrictive extensions. In a preliminary work, we also designed a new, quite restrictive framework, the Calculus of Congruent Constructions (CCC), which incorporates the congruence closure algorithm in CC's conversion [15], while preserving the good properties of the calculus, including the decidability of type checking.

Problem. The main question investigated in this paper is the incorporation of a general mechanism calling a decision procedure for solving conversion-goals in the Calculus of Inductive Constructions which uses the relevant information available from the current context of the proof.

Contributions. Our main contribution is the definition and the meta-theoretical investigation of the Calculus of Congruent Inductive Constructions (CCIC), which incorporates arbitrary *first-order theories* for which entailment is decidable into deduction via an abstract conversion rule of the calculus. A major technical innovation of this work lies in the computation mechanism: goals are sent to the decision procedure together with the set of user hypotheses available from the current context. Our main result shows that this extension of CIC does not compromise its main properties: confluence, strong normalization, coherence and decidability of proof-checking are all preserved. Unlike previous calculi, the main difficulty here is confluence, which led to a complex definition of conversion as a fixpoint. As a consequence of this definition, decidability of type checking becomes itself difficult.

Finally, we explain why the new system is still trustable, by leaving decision procedures *out* of its kernel, assuming that each procedure delivers a checkable *certificate* which becomes part of the proof. Certificate checkers become themselves part of the kernel, but are usually quite small and efficient and can be added one by one, making this approach a good compromise between CIC and the aforementioned extensions.

We assume some familiarity with typed lambda calculi [16] and the Calculus of Inductive Constructions.

2 The Calculus

For ease of the presentation, we restrict ourselves to $\mathrm{CC}_{\mathbb{N}}$, a calculus of constructions with a type nat of natural numbers generated by its two constructors $\mathbf{0}$ and $\mathbf{S}$ and equipped with its weak recursor $\mathrm{Rec}_{\mathbb{N}}^{\mathcal{W}}$. The calculus is also equipped with a polymorphic equality symbol $\doteq$ for which we use here a mixfix notation, writing $t \doteq_T u$ (or even $t \doteq u$ when T is not relevant) instead of $\doteq T\, t\, u$.

Let $\mathcal{S} = \{\star, \Box, \triangle\}$ the set of $\mathrm{CC}_{\mathbb{N}}$ *sorts*. For $s \in \{\star, \Box\}$, $\mathcal{X}^s$ denotes a countably infinite set of *s-sorted variables* s.t. $\mathcal{X}^\star \cap \mathcal{X}^\Box = \emptyset$. The union $\mathcal{X}^\star \cup \mathcal{X}^\Box$ will be written $\mathcal{X}$. For $x \in \mathcal{X}$, we write s_x the sort of x. Let $\mathcal{A} = \{\mathbf{u}, \mathbf{r}\}$ a set of two constants called *annotations*, totally ordered by $\mathbf{u} \prec_{\mathcal{A}} \mathbf{r}$, where $\mathbf{r}$ stands for *restricted* and $\mathbf{u}$ for *unrestricted*. We use a for an arbitrary annotation.

Definition 1 (Pseudo-terms of $\mathrm{CC}_{\mathbb{N}}$). *We define the* pseudo-terms *of* $\mathrm{CC}_{\mathbb{N}}$ *by the grammar rules:*

$$t, T := x \in \mathcal{X} \mid s \in \mathcal{S} \mid \mathrm{nat} \mid \doteq \mid \mathbf{0} \mid \mathbf{S} \mid \dot{+} \mid \mathrm{Eq}(t) \mid t\, u$$
$$\mid\ \lambda[x :^a T]t \mid \forall(x :^a T).\, t \mid \mathrm{Rec}_{\mathbb{N}}^{\mathcal{W}}(t, T)\{t_0, t_S\}$$

We use $\mathrm{FV}(t)$ *for the set of free variables of* t.

Definition 2 (Pseudo-contexts of $\mathrm{CC}_{\mathbb{N}}$). *The* typing environments *of* $\mathrm{CC}_{\mathbb{N}}$ *are defined as* $\Gamma, \Delta := [] \mid \Gamma, [x :^a T]$ *s.t. a variable cannot appear twice. We use* $\mathrm{dom}(\Gamma)$ *for the domain of* Γ *and* $x\Gamma$ *for the type associated to* x *in* Γ.

Remark that in our calculus, assumptions stored in the proof context always come along with an annotation used to control whether they can be used (in case the annotation is $\mathbf{r}$) or not in a conversion goal. We will later point out why this is necessary.

Definition 3 (Syntactic classes). *The pairwise disjoint syntactic classes of* $\mathrm{CC}_{\mathbb{N}}$*, called* objects *($\mathcal{O}$),* predicates *or* types *($\mathcal{P}$),* kinds *($\mathcal{K}$),* externs *($\mathcal{E}$) and* $\triangle$ *are defined in Figure 1.*

This enumeration defines a postfixed successor function +1 on classes ($\mathcal{O}+1 = \mathcal{P}$, $\mathcal{P}+1 = \mathcal{K}$, ... $\Delta + 1 = \perp$) . We also define $\mathrm{Class}(t) = \mathcal{D}$ *if* $t \in \mathcal{D}$ *and* $\mathcal{D} \in \{\mathcal{O}, \mathcal{P}, \mathcal{K}, \mathcal{E}, \triangle\}$ *and* $\mathrm{Class}(t) = \perp$ *otherwise.*

$$\begin{aligned}
\mathcal{O} &:= \mathcal{X}^\star \mid \mathbf{0} \mid \mathbf{S} \mid \dot{+} \mid \mathcal{O}\,\mathcal{O} \mid \mathcal{O}\,\mathcal{P} \mid [\lambda\mathcal{X}^\star :^a \mathcal{P}]\mathcal{O} \mid \\
&:= [\lambda\mathcal{X}^\Box :^a \mathcal{K}]\mathcal{O} \mid \mathrm{Rec}_{\mathbb{N}}^{\mathcal{W}}(\mathcal{O},\cdot)\{\mathcal{O},\mathcal{O}\} \\
\mathcal{P} &:= \mathcal{X}^\Box \mid \mathrm{nat} \mid \mathcal{P}\,\mathcal{O} \mid \mathcal{P}\,\mathcal{P} \mid [\lambda\mathcal{X}^\star :^a \mathcal{P}]\mathcal{P} \mid \doteq \mid \\
&:= [\lambda\mathcal{X}^\Box :^a \mathcal{K}]\mathcal{P} \mid (\forall\mathcal{X}^\star :^a \mathcal{P})\mathcal{P} \mid (\forall\mathcal{X}^\Box :^a \mathcal{K})\mathcal{P} \\
\mathcal{K} &:= \star \mid \mathcal{K}\,\mathcal{O} \mid \mathcal{K}\,\mathcal{P} \mid [\lambda\mathcal{X}^\star :^a \mathcal{P}]\mathcal{K} \mid \\
&:= [\lambda\mathcal{X}^\Box :^a \mathcal{K}]\mathcal{K} \mid (\forall\mathcal{X}^\star :^a \mathcal{P})\mathcal{K} \mid (\forall\mathcal{X}^\Box :^a \mathcal{K})\mathcal{K} \\
\mathcal{E} &:= \Box \mid (\forall\mathcal{X}^\star :^a \mathcal{P})\mathcal{E} \mid (\forall\mathcal{X}^\Box :^a \mathcal{K})\mathcal{E} \\
\triangle &:= \triangle
\end{aligned}$$

Fig. 1. $\mathrm{CC}_{\mathbb{N}}$ terms classes

Our typing judgments are classically written $\Gamma \vdash t : T$, meaning that the *well formed term* t is a proof of the proposition T under the assumptions in the *well-formed* environment Γ. *Typing rules* are those of CIC restricted to the single inductive type of natural numbers, with one exception, [Conv], based on an equality relation called *conversion* defined in section 2.1.

Definition 4 (Typing). *Typing rules of* $\mathrm{CC}_{\mathbb{N}}$ *are defined in Figure 2.*

2.1 Computation by Conversion

Our calculus has a complex notion of computation reflecting its rich structure made of three different ingredients, the typed lambda calculus, the type nat with its weak recursor and the Presburger arithmetic.

Our typed lambda calculus comes along with the β-rule. The η-rule raises known technical difficulties, see [17].

The type nat is generated by the two constructors $\mathbf{0}$ and $\mathbf{S}$ whose typing rules are given in Figure 2. We use $\mathrm{Rec}_{\mathbb{N}}^{\mathcal{W}}$ for its weak recursor whose typing rule is given in Figure 2 as well. Following CIC's tradition, we separate their arguments into two groups, using parentheses for the first two, and curly brackets for the two branches. The computation rules of nat are given below:

Definition 5 (ι-reduction). *The ι-reduction is defined by the following rewriting system:*

$$\begin{aligned}
\mathrm{Rec}_{\mathbb{N}}^{\mathcal{W}}(\mathbf{0},Q)\{t_0,t_S\} &\to_\iota t_0 \\
\mathrm{Rec}_{\mathbb{N}}^{\mathcal{W}}(\mathbf{S}\,t,Q)\{t_0,t_S\} &\to_\iota t_S\ t\ (\mathrm{Rec}_{\mathbb{N}}^{\mathcal{W}}(t,Q)\{t_0,t_S\})
\end{aligned}$$

where $t_0, t_S \in \mathcal{O}$.

These rules are going to be part of the conversion $\sim_\Gamma$. Of course, we do not want to type-check terms at each single step of conversion, we want to type-check only the starting two terms forming the equality goal in [Conv]. But intermediate terms could then be non-typable and strong normalization be lost.

$$[\textsc{Axiom-1}]\ \frac{}{\vdash \star : \square} \qquad [\textsc{Axiom-2}]\ \frac{}{\vdash \square : \triangle}$$

$$[\doteq\textsc{-Intro}]\ \frac{}{\vdash \doteq : \forall(T :^{u} \star).\, T \rightarrow T \rightarrow \star}$$

$$[\textsc{Product}]\ \frac{\Gamma \vdash T : s_T \quad \Gamma, [x :^{a} T] \vdash U : s_U}{\Gamma \vdash \forall(x :^{a} T).\, U : s_U}$$

$$[\textsc{Lamda}]\ \frac{\Gamma \vdash \forall(x :^{a} T).\, U : s \quad \Gamma, [x :^{a} T] \vdash u : U}{\Gamma \vdash \lambda[x :^{a} T]u : \forall(x :^{a} T).\, U}$$

$$[\textsc{Weak}]\ \frac{\Gamma \vdash V : s \quad \Gamma \vdash t : T \qquad s \in \{\star, \square\} \quad x \in \mathcal{X}^{s} - \mathrm{dom}(\Gamma)}{\Gamma, [x :^{a} V] \vdash t : T}$$

$$[\textsc{Var}]\ \frac{x \in \mathrm{dom}(\Gamma) \quad \Gamma \vdash x\Gamma : s_x}{\Gamma \vdash x : x\Gamma}$$

$$[\textsc{App}]\ \frac{\begin{array}{c}\Gamma \vdash t : \forall(x :^{a} U).\, V \quad \Gamma \vdash u : U \\ \text{if } a = \mathbf{r} \text{ and } U \rightarrow_{\beta}^{*} t_1 \doteq_T t_2 \text{ with } t_1, t_2 \in \mathcal{O} \\ \text{then } t_1 \sim_{\Gamma} t_2 \text{ must hold}\end{array}}{\Gamma \vdash t\, u : V\{x \mapsto u\}}$$

$$[\mathbf{0}\textsc{-Intro}]\ \frac{}{\vdash \mathbf{0} : \mathrm{nat}} \qquad [\mathbf{S}\textsc{-Intro}]\ \frac{}{\vdash \mathbf{S} : \mathrm{nat} \rightarrow \mathrm{nat}}$$

$$[\textsc{Nat}]\ \frac{}{\vdash \mathrm{nat} : \star} \qquad [\dot{+}\textsc{-Intro}]\ \frac{}{\vdash \dot{+} : \mathrm{nat} \rightarrow \mathrm{nat} \rightarrow \mathrm{nat}}$$

$$[\textsc{Eq-Intro}]\ \frac{\begin{array}{c}\Gamma \vdash t_1 : T \quad \Gamma \vdash t_2 : T \\ \Gamma \vdash p : \forall(P : T \rightarrow \star).\, P\, t_1 \rightarrow P\, t_2\end{array}}{\Gamma \vdash \mathrm{Eq}(p) : t_1 \doteq_T t_2}$$

$$[\iota\textsc{-Elim}]\ \frac{\begin{array}{c}\Gamma \vdash t : \mathrm{nat} \quad \Gamma \vdash Q : \mathrm{nat} \rightarrow \star \quad \Gamma \vdash f_0 : \mathrm{nat} \\ \Gamma \vdash f_S : \forall(n :^{\mathrm{u}} \mathrm{nat}).\, Q\, n \rightarrow Q\, (\mathbf{S}\, n)\end{array}}{\Gamma \vdash \mathrm{Rec}_{\mathbb{N}}^{\mathcal{W}}(t, Q)\{f_0, f_S\} : Q\, t}$$

$$[\textsc{Conv}]\ \frac{\Gamma \vdash t : T \quad \Gamma \vdash T' : s' \quad T \sim_{\Gamma} T'}{\Gamma \vdash t : T'}$$

Fig. 2. Typing judgment of $\mathrm{CC}_{\mathbb{N}}$

The constructors **0** and **S**, as well as the additional first-order constant $\dot{+}$ are *also* used to build up expressions in the algebraic world of Presburger arithmetic, in which function symbols have arities. We therefore have two different possible views of terms of type nat, either as a term of the calculus of inductive constructions, or as an algebraic term of Presburger arithmetic. We now define precisely this algebraic world and explain in detail how to extract algebraic information from arbitrary terms of $CC_{\mathbb{N}}$.

Let $\mathcal{T}$ be the theory of *Presburger arithmetic* defined on the signature $\Sigma = \{0, S(\_), \_+\_\}$ and $\mathcal{Y}$ a set of variables distinct from $\mathcal{X}$. Note that we syntactically distinguish the algebraic symbols from the $CC_{\mathbb{N}}$ symbols by using a different font (0 and S for the algebraic symbols, **0** and **S** for the constructors).

We write $\mathcal{T} \models F$ if F is a valid formula in $\mathcal{T}$, and $\mathcal{T}, E \models F$ for $\mathcal{T} \models E \Rightarrow F$.

Definition 6 (Algebraic terms). *The set* **Alg** *of* $CC_{\mathbb{N}}$ algebraic terms *is the smallest subset of* $\mathcal{O}$ *s.t. i)* $\mathcal{X}^\star \subseteq \mathbf{Alg}$, *ii)* $\mathbf{0} \in \mathbf{Alg}$, *iii)* $\forall t \in CC_{\mathbb{N}}. \mathbf{S}\, t \in \mathbf{Alg}$, *iv)* $\forall t, u \in CC_{\mathbb{N}}. t \dot{+} u \in \mathbf{Alg}$.

Definition 7 (Algebraic cap and aliens). *Given a relation R on* $CC_{\mathbb{N}}$*, let $\mathcal{R}$ be the smallest congruence on* $CC_{\mathbb{N}}$ *containing R, and π_R a function from* $CC_{\mathbb{N}}$ *to $\mathcal{Y} \cup \mathcal{X}^\star$ such that $t \,\mathcal{R}\, u \iff \pi_R(t) = \pi_R(u)$.*

The algebraic cap of t modulo R, $\mathrm{cap}_R(t)$*, is defined by:*

- $\mathrm{cap}_R(\mathbf{0}) = 0$, $\mathrm{cap}_R(\mathbf{S}\, u) = S(\mathrm{cap}_R(u))$, $\mathrm{cap}_R(u \dot{+} v) = \mathrm{cap}_R(u) + \mathrm{cap}_R(v)$,
- *otherwise,* $\mathrm{cap}_R(t) = t$ *if* $t \in \mathcal{X}^*$ *and else* $\pi_R(t)$.

We call aliens *the subterms of t abstracted by a variable in $\mathcal{Y}$.*

Observe that a term not headed by an algebraic symbol is abstracted by a variable from our new set of variables $\mathcal{Y}$ in such a way that $\mathcal{R}$-equivalent terms are abstracted by the same variable.

We can now glue things together to define *conversion.*

Definition 8 (Conversion relation). *The family* $\{\sim_\Gamma\}_\Gamma$ *of* Γ-conversions *is defined by the rules of Figure 3.*

This definition is technically complex.

Being a congruence, $\sim_\Gamma$ includes congruence rules. However, all these rules are not quite congruence rules since crossing a binder increases the current context Γ by the new assumption made inside the scope of the binding construct, resulting in a family of congruences. More questions are raised by the three different kinds of basic conversions.

First, $\sim_\Gamma$ includes the rules $\rightarrow_\beta$ and $\rightarrow_\iota$ of $CC_{\mathbb{N}}$. Unlike the beta rule, $\rightarrow_\iota$ interacts with first-order rewriting, and therefore the CONV rule of Figure 2 cannot be expressed by $T \leftrightarrow^*_{\beta\iota} \sim_\Gamma \leftrightarrow^*_{\beta\iota} T'$ as one would expect.

Second, $\sim_\Gamma$ includes the relevant assumptions grabbed from the context, this is rule EQ. These assumptions must be of the form $[x :^{\mathbf{r}} T]$, with the appropriate annotation r, and T must be an equality assumption or otherwise *reduce* to an equality assumption. Note that we use only $\rightarrow_\beta$ here. Using $\sim_\Gamma$ recursively instead is indeed an equivalent formulation under our assumptions. Without annotations, $CC_{\mathbb{N}}$ does not enjoy subject reduction. Generating appropriate annotations is discussed in section 4.

$$[\beta\iota]\ \frac{t \leftrightarrow^*_{\beta\iota} u}{t \sim_\Gamma u} \qquad [\text{EQ}]\ \frac{[x :^{\mathrm{r}} T] \in \Gamma \quad T \to^*_\beta t_1 \doteq t_2 \quad t_1, t_2 \in \mathcal{O}}{t_1 \sim_\Gamma t_2}$$

$$[\text{DED}]\ \frac{t_1, t_2 \in \mathcal{O} \quad \mathcal{T}, \{\text{cap}_{\sim_\Gamma}(u_1) = \text{cap}_{\sim_\Gamma}(u_2) \mid u_1 \sim_\Gamma u_2\} \models \text{cap}_{\sim_\Gamma}(t_1) = \text{cap}_{\sim_\Gamma}(t_2)}{t_1 \sim_\Gamma t_2}$$

$$[\text{SYM}]\ \frac{t \sim_\Gamma u}{u \sim_\Gamma t} \qquad [\text{TRANS}]\ \frac{t \sim_\Gamma u \quad u \sim_\Gamma v}{t \sim_\Gamma v}$$

$$[\text{CC}_{\mathbb{N}}\text{-EQ}]\ \frac{t \sim_\Gamma u}{\text{Eq}(t) \sim_\Gamma \text{Eq}(u)} \qquad [\text{APP}]\ \frac{t_1 \sim_\Gamma t_2 \quad u_1 \sim_\Gamma u_2}{t_1\, u_1 \sim_\Gamma t_2\, u_2}$$

$$[\text{PROD}]\ \frac{T \sim_\Gamma U \quad t \sim_{\Gamma,[x:^a T]} u \quad b \preceq a}{\forall(x :^b T).\, t \sim_\Gamma \forall(x :^b U).\, u} \qquad [\text{LAM}]\ \frac{T \sim_\Gamma U \quad t \sim_{\Gamma,[x:^a T]} u \quad b \preceq a}{\lambda[x :^b T]t \sim_\Gamma \lambda[x :^b U]u}$$

$$[\text{ELIM-}\mathcal{W}]\ \frac{t \sim_\Gamma u \quad P \sim_\Gamma Q \quad t_0 \sim_\Gamma u_0 \quad t_S \sim_\Gamma u_S}{\text{Rec}^{\mathcal{W}}_{\mathbb{N}}(t, P)\{t_0, t_S\} \sim_\Gamma \text{Rec}^{\mathcal{W}}_{\mathbb{N}}(u, Q)\{u_0, u_S\}}$$

Fig. 3. Conversion relation $\sim_\Gamma$

Third, with rule [DED], we can also generate new assumptions by using Presburger arithmetic. This rule here uses the property that two algebraic terms are equivalent in $\sim_\Gamma$ if their caps relative to $\sim_\Gamma$ are equivalent in $\sim_\Gamma$ (the converse being false). This is so because the abstraction function $\pi_{\sim_\Gamma}$ abstracts equivalent aliens by the same variable taken from $\mathcal{Y}$. It is therefore the case that deductions on caps made in Presburger arithmetic can be lifted to deductions on arbitrary terms via the abstraction function. As a consequence, the two definitions of the abstraction function $\pi_{\sim_\Gamma}$ and of the congruence $\sim_\Gamma$ are mutually inductive: our conversion relation is defined as a least fixpoint.

2.2 Two Simple Examples

More automation - smaller proofs. We start with a simple example illustrating how the equalities extracted from a context Γ can be use to deduce new equalities in $\sim_\Gamma$.

$$\begin{aligned}\Gamma = {} & [x\, y\, t :^{\mathrm{u}} \text{nat}], [f :^{\mathrm{u}} \text{nat} \to \text{nat}], \\ & [p_1 :^{\mathrm{r}} t \doteq 2], [p_2 :^{\mathrm{r}} f\,(x \dotplus 3) \doteq x \dotplus 2], \\ & [p_3 :^{\mathrm{r}} f\,(y \dotplus t) \dotplus 2 \doteq y], [p_4 :^{\mathrm{r}} y \dotplus 1 \doteq x \dotplus 2]\end{aligned}$$

From p_1 and p_4 (extracted from the context by [EQ]), [DED] will deduce that $y \dotplus t \sim_\Gamma x \dotplus 3$, and by congruence, $f\,(y \dotplus t) \sim_\Gamma f\,(x \dotplus 3)$. Therefore, $\pi_{\sim_\Gamma}$ will abstract $f(x \dot{+} 3)$ and $f(y \dot{+} t)$ by the same variable z, resulting in two new equations available for [DED]: $z = x + 2$ and $z + 2 = y$. Now, $z = x + 2$, $z + 2 = y$ and $y + 1 = x + 2$ form a set of unsatisfiable equations and we deduce $0 \sim_\Gamma 1$ by the DED rule: contradiction has been obtained. This shows that we can easily carry out a proof by contradiction in $\mathcal{T}$.

More typable terms. We continue with a second example showing that the new calculus can type more terms than CIC. For the sake of this example we assume that the calculus is extended by dependent lists on natural numbers. We denote by **list** (of type nat $\to \star$) the type of dependent lists and by **nil** (of type $\mathbf{list}\,\mathbf{0}$) and **cons** (of type $\forall(n : \text{nat}).\, \mathbf{list}\, n \to \text{nat} \to \mathbf{list}\,(\mathbf{S}\, n))$ the lists constructors. We also add a weak recursor $\text{Rec}_{\mathbb{L}}^{\mathcal{W}}$ such that, given $P : \forall(n : \text{nat}).\, \mathbf{list}\, n \to \star$, $l_0 : P\,\mathbf{0}\,\mathbf{nil}$ and $l_S : \forall(n : \text{nat})(l : \mathbf{list}\, n).\, P\, n\, l \to \forall(x : \text{nat}).\, P\,(\mathbf{S}\, n)\,(\mathbf{cons}\, n\, x\, l)$, then $\text{Rec}_{\mathbb{L}}^{\mathcal{W}}(l, P)\{l_0, l_S\}$ has type $P\, n\, l$ for any list l of type $\mathbf{list}\, n$.

Assume now given a dependent reverse function (of type $\forall(n : \text{nat}).\, \mathbf{list}\, n \to \mathbf{list}\, n)$ and the list concatenation function @ (of type $\forall(n\, n' : \text{nat}), \mathbf{list}\, n \to \mathbf{list}\, n' \to \mathbf{list}\,(n \dot{+} n'))$. We can simply express that a list l is a palindrome: l is a palindrome if reverse $l \doteq l$.

Suppose now that one wants to prove that palindromes are closed under substitution of letters by palindromes. To make it easier, we will simply consider a particular case: the list $l_1 l_2 l_2 l_1$ is a palindrome if l_1 and l_2 are palindromes. The proof sketch is simple: it suffices to apply as many times as needed the lemma reverse(ll') = reverse(l')@ reverse(l) $(*)$. What can be quite surprising is that Lemma $(*)$ is rejected by Coq. Indeed, if l and l' are of length n and n', it is easy to check that reverse(ll') is of type $\mathbf{list}\,(n \dot{+} n')$ and reverse(l') :: reverse(l) of type $\mathbf{list}\,(n' \dot{+} n)$ which are clearly not $\beta\iota$-convertible. This is not true in our system: $n \dot{+} n'$ will of course be convertible to $n' \dot{+} n$ and lemma $(*)$ is therefore well-formed. Proving the more general property needs of course an additional induction on natural numbers to apply lemma $(*)$ the appropriate number of times, which can of course be carried out in our system.

Note that, although possible, writing a reverse function for dependent lists in Coq is not that simple. Indeed, a direct inductive definition of reverse will define reverse$(\mathbf{cons}\, n\, a\, l)$, of type $\mathbf{list}\,(1 \dot{+} n)$, as reverse$(l)$ @ a, of type $\mathbf{list}\,(n \dot{+} 1)$. Coq will reject such a definition since $\mathbf{list}\,(1 \dot{+} n)$ and $\mathbf{list}\,(n \dot{+} 1)$ are not convertible. Figure 4 shows how reverse can be defined in Coq.

3 Metatheorical Properties

Most basic properties of Pure Type Systems (see [18]) are not too difficult. Those using substitution instances are more delicate. They rely on the annotations decorating the abstractions and products which were introduced for that purpose.

3.1 Stability by Substitution

Assume that Γ is a typing environment of the form $\Gamma_1, [p :^{\mathsf{r}} a \doteq b], \Gamma_2$ (a and b being two variables of type nat in Γ). The stability by substitution claims that if

```
Coq < Definition reverse: forall (n: nat), (list n) -> (list n) .
Coq <   assert (reverse_acc : forall (n m : nat),
Coq <              list n -> list m -> list (m+n)) .
Coq <   refine (fix reverse_acc (n m : nat) (from : list n) (to : list m)
Coq <              {struct from} : list (m+n) := _) .
Coq <   destruct from as [ | n' v rest ] .
Coq <     rewrite <- plus_n_O_transparent; exact to .
Coq <     rewrite <- plus_n_Sm_transparent;
Coq <       exact (reverse_acc n' (S m) rest (cons _ v to)) .
Coq <   intros n l . exact (reverse_acc _ _ l nil) .
Coq < Defined .
```

Fig. 4. reverse function is Coq

we have a typing derivation $\Gamma \vdash t : T$, then we can substitute p by a term P (of type $a \doteq b$ under Γ_1) in this derivation and obtain a proof of $\Gamma_1, \Gamma_2\theta \vdash t\theta : T\theta$, where θ is the substitution $\{p \mapsto P\}$. This property can easily be proved for Pure Type Systems as soon as the conversion relation is itself stable by substitution. In our example one can easily check that $a \sim_\Gamma b$, but $a \sim_{\Gamma_1,\Gamma_2\theta} b$ will not hold in general: the assumption $a \doteq b$ has been inlined and thus is no more extractable by the conversion relation. As a result, we need to strengthen the formulation of stability by substitution:

Lemma 1. *Let $\Gamma = \Gamma_1, [z :^a W], \Gamma_2$ and assume that i) $T \sim_\Gamma T'$, ii) if $a = \mathbf{r}$ and $W \to_\beta^* t_1 \doteq t_2$ then $t_1 \sim_{\Gamma_1} t_2$. Then, $T\theta \sim_\Delta T'\theta$ where $\theta = \{z \mapsto w\}$ and $\Delta = \Gamma_1, \Gamma_2\theta$*

Corollary 1 (Stability by substitution). *Let $\Gamma = \Gamma_1, [z :^a W], \Gamma_2$ and assume that i) $T \sim_\Gamma T'$ ii) if $a = \mathbf{r}$ and $W \to_\beta^* t_1 \doteq t_2$ then $t_1 \sim_{\Gamma_1} t_2$. Then, $\Delta \vdash t\theta : T\theta$ where $\theta = \{z \mapsto w\}$, $\Gamma_1 \vdash w : W$ and $\Delta = \Gamma_1, \Gamma_2\theta$.*

As usual, the substitutivity lemma is to be used in the proof of subject reduction (for $\to_{\beta\iota}$) to come later. Because it requires a specific typing property for the equality assumptions annotated by $\mathbf{r}$, we need to ensure this property in the application case of the coming subject reduction proof. This is indeed the origin of the similar condition arising in the typing rule [App].

3.2 Conversion as Rewriting

We now turn conversion into a rewriting relation in order to prove that our system is logically consistent by analyzing a proof in normal form of $\forall(x :^{\mathbf{u}} \star). x$. The notion of a normal proof is of course more complicated than in CIC, since we must account for the congruence $\sim_\Gamma$ associated with an arbitrary context Γ. The difficulty is that the set of equalities assumed in a given environment Γ together with the axioms of the theory $\mathcal{T}$ may be inconsistent, making all first-order terms equal in $\sim_\Gamma$ which could break strong normalization of our rewriting relation.

Definition 9 ($\mathcal{T}$-consistent environment). *A typing environment Γ is $\mathcal{T}$-*consistent *if there exist two terms $t, u \in \mathcal{O}$ s.t. $\neg(t \sim_\Gamma u)$.*

Lemma 2. *If Γ is $\mathcal{T}$-consistent then $\neg(\mathbf{0} \sim_\Gamma \mathbf{S}\, t)$ for any term t.*

Definition 10 (Weak conversion). *We inductively define a family of* weak conversion relations $\{\cong_\Gamma\}_\Gamma$ *as the smallest congruent relation s.t. $t \cong_\Gamma u$ if $\mathcal{T}, \mathrm{Eq}(\Gamma) \models \mathrm{cap}_\emptyset(t) = \mathrm{cap}_\emptyset(u)$, where $\mathrm{Eq}(\Gamma) = \{\mathrm{cap}_\emptyset(w_1) = \mathrm{cap}_\emptyset(w_2) \mid w_1, w_2 \in \mathcal{O}, [x :^{\mathrm{r}} w_1 \doteq w_2] \in \Gamma\}$.*

Definition 11. *We inductively define a family $\{\rightarrow_\Gamma\}_\Gamma$ of rewriting relations modulo weak-conversion as the smallest rewriting relations satisfying the rules of Figure 5.*

The first rule shows that rewriting is modulo weak conversion in a consistent environment. The second equates all object terms when the environment is inconsistent, replacing them by the new constant $\bullet$. The others are as expected.

$$[\textsc{Rw-Mod}]\ \frac{\Gamma \text{ is } \mathcal{T}\text{-consistent} \quad t \cong_\Gamma t' \rightarrow_\Gamma u' \cong_\Gamma u}{t \rightarrow_\Gamma u}$$

$$[\textsc{Rw-}\bullet]\ \frac{\Gamma \text{ is } \mathcal{T}\text{-inconsistent} \quad t \in \mathcal{O} \quad t \neq \bullet}{t \rightarrow_\Gamma \bullet}$$

$$[\textsc{Rw-}\beta\iota]\ \frac{t \rightarrow_{\beta\iota} u}{t \rightarrow_\Gamma u} \qquad [\textsc{Rw-Fwd}]\ \frac{t \rightarrow_\Delta u \quad \Gamma \rightarrow_\beta \Delta}{t \rightarrow_\Gamma u}$$

$$[\text{W-}\forall]\ \frac{t \rightarrow_{\Gamma,[x:^a T]} u \quad b \preceq a}{\forall(x :^b T).\, t \rightarrow_\Gamma \forall(x :^b T).\, u}$$

$$[\text{W-}\lambda]\ \frac{t \rightarrow_{\Gamma,[x:^a T]} u \quad b \preceq a}{\lambda[x :^b T].\, t \rightarrow_\Gamma \lambda[x :^b T].\, u}$$

Fig. 5. Conversion as a rewriting system

Lemma 3. *1. The rewriting relation $\rightarrow_\Gamma$ is confluent.*
2. If $t \sim_\Gamma u$ then $t \leftrightarrow^*_\Gamma u$.
3. If $t \leftrightarrow^*_\Gamma u$ with $\bullet \notin t$ and $\bullet \notin u$ then $t \sim_\Gamma u$.
4. If $\Gamma \vdash t : T$ with Γ $\mathcal{T}$-consistent and $t \cong_\Gamma u$, then $\Gamma \vdash u : T$.

Lemma 4. *If $\Gamma \vdash t : T$ and $t \rightarrow_\Gamma u$ with $\bullet \notin u$, then $\Gamma \vdash u : T$.*

Proof. The proof is standard, by induction on the type derivation of the left-hand side. The interesting case is when a β-reduction applies to the top of a term of the form $(\lambda[x :^a U]v)\ w$ and the typing rule is [App]: we then conclude by using Lemma 1. Note that the side condition of rule [App] provides us with the property needed for using Lemma 1.

Lemma 5. *The rewriting relation* $\rightarrow_{\Gamma}$ *is strongly normalizing for well formed terms.*

Proof. The proof is a direct application of proof irrelevance [19], because $\sim_{\Gamma}$ is a congruence generated by equalities between object terms, apart from beta-reduction. What makes this true is that $\mathrm{Rec}_{\mathbb{N}}^{\mathcal{W}}$ is a weak recursor, working at the object level. Including strong elimination rules invalidates this argument. □

We finally conclude that $\mathrm{CC}_{\mathbb{N}}$ is consistent:

Theorem 1. *There is no proof of* $\vdash t : \forall(x :^{\mathbf{u}} \star).\, x$.

Proof. Assume that $\vdash t : \mathbf{0} \doteq \mathbf{S}\,\mathbf{0}$ where t is $\rightarrow_{\Gamma}$-normal. Since $\mathbf{0} \doteq \mathbf{S}\,\mathbf{0}$ is not convertible to a sort, t cannot be equal to nat, or a sort, or a product. Since t is necessarily closed, t is not a variable. Moreover, t cannot be of the form $\mathrm{Rec}_{\mathbb{N}}^{\mathcal{W}}(u, Q)\{t_0, t_S\}$ since t is closed and in $\rightarrow_{\iota}$-normal form.

If t is an application, it is necessarily of the form $c\,\vec{u}$ with $c \in \{\mathbf{0}, \mathbf{S}, \dot{+}, \doteq\}$. By using inversion it suffices to check that in all these cases, t has a type T which is not convertible to $\mathbf{0} \doteq \mathbf{S}\,\mathbf{0}$.

If $t = \mathrm{Eq}(u)$, then t has type $u \doteq u$ with u of type nat and $u \doteq u$ convertible to $\mathbf{0} \doteq \mathbf{S}\,\mathbf{0}$. Thus $\mathbf{0} \sim_{[]} \mathbf{S}\,\mathbf{0}$, and $\mathcal{T} \models 0 = 1$, which is impossible. □

3.3 Decidability of Type Checking

Theorem 2. *Type checking of* $\mathrm{CC}_{\mathbb{N}}$ *is decidable.*

Decidability of type checking needs two ingredients. First-of-all, eliminating [Conv], which is non-structural, by incorporating it to [App]. This is classical, and it is easy to prove decidability of the transformed set of rules for type-checking, assuming $\sim_{\Gamma}$ is decidable.

Deciding $\sim_{\Gamma}$ is more complex. We cannot use the rewrite system $\rightarrow_{\Gamma}$ for that purpose since the first two rules use the $\mathcal{T}$-consistency of Γ as a prerequisite. We use instead a saturation based algorithm. The method resembles very much the one used for combining first-order decision procedures operating on disjoint alphabets [20,21]. Basic ingredients are: purification of formulas (here equations) by abstracting aliens by new variables; deriving new equalities among variables by using the appropriate decision procedure for pure formulas; propagating these new equalities to the other formulas.

4 Conclusion and Discussion

$\mathrm{CC}_{\mathbb{N}}$ is an extension of CIC (restricted to the weak elimination rules of the inductive type nat) by a fragment of Presburger arithmetic (without the natural strict order $\mathbb{N}$) in which conversion incorporates Presburger arithmetic, β-reduction and higher-order primitive recursion into a single mechanism. We now discuss in more details how this can be generalized to full CIC, how this can be used in practice, how useful that is, and whether the obtained kernel is trustable.

Relevance. Our second example shows very clearly the expressivity of our calculus with respect to CIC. However, what is done here by a typing rule could be done alternatively in CIC by a tactic. Besides, if one wants to avoid building a proof term which can be quite large and slow down the type-checker, it is possible to prove the tactic and then use a reflexion mechanism in order to avoid type-checking the proof each time the tactic is called. In both cases, however, the user must call the tactic explicitly. In our approach, this is completely transparent, and would remain transparent in case of a succession of uses of the decision procedure separated by eliminations, since conversion incorporates both, or in case of different decision procedures called successively.

Extension to CIC. Building decision procedures in a type-theoretic framework is not that easy. The main difficulty lies in the adequate definition of the congruence $\sim_\Gamma$. Once the definition is obtained, carrying out the technical development is not too difficult in the case of the pure Calculus of Constructions (the congruence becomes quite simpler in this case), difficult in the present case of $CC_{\mathbb{N}}$ (because of the presence of the weak recursors for nat), no more difficult when other decidable theories are introduced such as lists with their associated recursors, but much harder when including strong elimination rules which interact with the first-order theories. In this case, it is necessary to block the congruence below the strong recursor in order to avoid lifting an incoherence from the object level to the predicate level, which would immediately yield paradoxes [22].

Annotations restriction. One may wonder how annotations can be handled in practice. As seen, annotations are used to forbid *inlining* (when a β-redex is contracted) of equational assumptions which are used by conversion. This could be seen as a restriction since our calculus, in order to avoid the creation of problematic β-redexes, forbids in most cases applications of symbols of type $\forall(p :^{\mathbf{r}} t \doteq u). T$.

This restriction can be removed by using the notion of *opaque definitions* (as opposed to *transparent definitions*) of Coq which allows the user to define symbols that the system cannot inline. In most cases, definitions having a computational behavior (like $\dot{+}$) are transparent whereas definitions representing lemmas (like the associativity of $\dot{+}$) are opaque. This convention is used in the standard library of Coq.

Returning to our previous example, if the user needs to prove a lemma of the form $\forall(p :^{\mathbf{r}} t \doteq u). T$, he or she should declare it as an opaque definition $P := \lambda[p :^{\mathbf{r}} t \doteq u]q$. The application of P to a term v should then be allowed: the term $P\,v$ cannot reduce to $q\{p \mapsto v\}$. Of course, if P is defined transparently, the application $P\,v$ has to be forbidden.

Moreover, this gives us a simple heuristic to automatically tag products and abstractions: **r** annotation should by used by default when the user is defining an opaque symbol, whereas **u** annotation should be used everywhere else.

Arbitrary decision procedures. So far, we have considered only decidable equational theories. But it is well-known that a decidable theory can always be transformed into a decidable equational theory over the type Bool of truth values equipped with its usual operations. This is so because of the decidability assumption.

Type levels equalities. One may wonder whether the conversion relation of $CC_{\mathbb{N}}$ could use type level equalities (or hypotheses of the form $P \leftrightarrow Q$). The answer seems to be negative: extracting type levels equalities breaks subject reduction and β-strong normalization (see [13]), two properties needed for the decidability of our calculus.

Trusting the kernel. Decision procedures require complex coding. It took a lot of time to get a correct tactic for Presburger arithmetic in Coq. Including a tactic into the kernel of the system is therefore unrealistic, unless it is itself proved correct with a trustable proof assistant. On the other hand, most decision procedures can provide a *certificate* that is quite compact and can be verified by a *certificate-checker* which is usually small, and easy to write and read, and is therefore a trustable piece of code. The reason is that the procedure *searches* for a proof while the certificate-checker *verifies* that the certificate is correct. A certificate checker looks indeed like a proof-checker. It is then easy to modify the conversion rule so as to output a certificate each time a decision procedure is used. The kernel of $CC_{\mathbb{N}}$ should therefore include a certificate-checker for Presburger arithmetic. In case of CCIC with several decision procedures, the kernel would include one proof-checker for each decision procedure. Besides, the process is incremental: the procedures and the associated proof-checkers can be included one by one, because decision procedures for different inductive types operate on disjoint vocabularies, hence can be combined [20,21].

An implementation of CCIC has started and should be available soon as a prototype in a version without certificate generation and checking.

References

1. Coq-Development-Team: The Coq Proof Assistant Reference Manual - Version 8.0. INRIA, INRIA Rocquencourt, France (2004), `http://coq.inria.fr/`
2. Coquand, T., Huet, G.: The Calculus of Constructions. Information and Computation 76(2-3), 95–120 (1988)
3. Gonthier, G.: The four color theorem in coq. In: Filliâtre, J.-C., Paulin-Mohring, C., Werner, B. (eds.) TYPES 2004. LNCS, vol. 3839, Springer, Heidelberg (2006)
4. Coquand, T., Paulin-Mohring, C.: Inductively defined types. In: Martin-Löf, P., Mints, G. (eds.) COLOG-88. LNCS, vol. 417, pp. 50–66. Springer, Heidelberg (1990)
5. Giménez, E.: Structural recursive definitions in type theory. In: Larsen, K.G., Skyum, S., Winskel, G. (eds.) ICALP 1998. LNCS, vol. 1443, pp. 397–408. Springer, Heidelberg (1998)
6. Barras, B.: Auto-validation d'un systéme de preuves avec familles inductives. PhD thesis, Université de Paris VII (1999)
7. Blanqui, F.: Definitions by rewriting in the calculus of constructions. Mathematical Structures in Computer Science 15(1), 37–92 (2005) (Journal version of LICS'01)
8. Blanqui, F.: Inductive types in the calculus of algebraic constructions. Fundamenta Informaticae 65(1-2), 61–86 (2005) (Journal version of TLCA'03)

9. Shostak, R.E.: An efficient decision procedure for arithmetic with function symbols. J. of the Association for Computing Machinery 26(2), 351–360 (1979)
10. Shankar, N.: Little engines of proof. In: Plotkin, G. (ed.) Proceedings of the Seventeenth Annual IEEE Symp. on Logic in Computer Science, LICS 2002, IEEE Computer Society Press, Los Alamitos (2002) (Invited Talk)
11. Corbineau, P.: Démonstration automatique en Théorie des Types. PhD thesis, University of Paris IX (2005)
12. Stehr, M.: The Open Calculus of Constructions: An equational type theory with dependent types for programming, specification, and interactive theorem proving (part I and II). Fundamenta Informaticae (to appear, 2007)
13. Oury, N.: Extensionality in the calculus of constructions. In: Hurd, J., Melham, T. (eds.) TPHOLs 2005. LNCS, vol. 3603, pp. 278–293. Springer, Heidelberg (2005)
14. Hofmann, M., Streicher, T.: The groupoid interpretation of type theory. In: Twenty-five years of constructive type theory. Oxford Logic Guides, vol. 36, pp. 83–111. Oxford University Press, Oxford (1998)
15. Blanqui, F., Jouannaud, J.P., Strub, P.Y.: A Calculus of Congruent Constructions. Unpublished draft (2005)
16. Barendregt, H.: Lambda calculi with types. Handbook of logic in computer science, vol. 2. Oxford University Press, Oxford (1992)
17. Werner, B.: Une Théorie des Constructions Inductives. PhD thesis, University of Paris VII (1994)
18. Blanqui, F.: Type Theory and Rewriting. PhD thesis, Université de Paris XI, Orsay, France (2001)
19. Barthe, G.: The relevance of proof irrelevance. In: Larsen, K.G., Skyum, S., Winskel, G. (eds.) ICALP 1998. LNCS, vol. 1443, Springer, Heidelberg (1998)
20. Schmidt-Schauß, M.: Unification in a combination of arbitrary disjoint equational theories. J. Symbolic Computation 8, 51–99 (1989) (Special issue on Unification)
21. Baader, F., Schulz, K.: Unification in the union of disjoint equational theories: Combining decision procedures. In: Kapur, D. (ed.) Automated Deduction - CADE-11. LNCS, vol. 607, Springer, Heidelberg (1992)
22. Strub, P.Y.: Type Theory and Decision Procedures. PhD thesis, École Polytechnique, Palaiseau, France (Work in progress)

Structure Theorem and Strict Alternation Hierarchy for FO^2 on Words*

Philipp Weis and Neil Immerman

Department of Computer Science
University of Massachusetts, Amherst
140 Governors Drive, Amherst, MA 01003, USA
{pweis,immerman}@cs.umass.edu
http://www.cs.umass.edu/~{pweis,immerman}

Abstract. It is well-known that every first-order property on words is expressible using at most three variables. The subclass of properties expressible with only two variables is also quite interesting and well-studied. We prove precise structure theorems that characterize the exact expressive power of first-order logic with two variables on words. Our results apply to $FO^2[<]$ and $FO^2[<, \mathrm{Suc}]$, the latter of which includes the binary successor relation in addition to the linear ordering on string positions.

For both languages, our structure theorems show exactly what is expressible using a given quantifier depth, n, and using m blocks of alternating quantifiers, for any $m \leq n$. Using these characterizations, we prove, among other results, that there is a strict hierarchy of alternating quantifiers for both languages. The question whether there was such a hierarchy had been completely open. As another consequence of our structural results, we show that satisfiability for $FO^2[<]$, which is NEXP-complete in general, becomes NP-complete once we only consider alphabets of a bounded size.

1 Introduction

It is well-known that every first-order property on words is expressible using at most three variables [7, 8]. The subclass of properties expressible with only two variables is also quite interesting and well-studied (Fact 1).

In this paper we prove precise structure theorems that characterize the exact expressive power of first-order logic with two variables on words. Our results apply to $FO^2[<]$ and $FO^2[<, \mathrm{Suc}]$, the latter of which includes the binary successor relation in addition to the linear ordering on string positions.

For both languages, our structure theorems show exactly what is expressible using a given quantifier depth, n, and using m blocks of alternating quantifiers, for any $m \leq n$. Using these characterizations, we prove that there is a strict hierarchy of alternating quantifiers for both languages. The question whether there was such a hierarchy had been completely open since it was asked in [3,4].

* Supported in part by NSF grant CCF-0514621.

J. Duparc and T.A. Henzinger (Eds.): CSL 2007, LNCS 4646, pp. 343–357, 2007.

As another consequence of our structural results, we show that satisfiability for $FO^2[<]$, which is NEXP-complete in general [4], becomes NP-complete once we only consider alphabets of a bounded size.

Our motivation for studying FO^2 on words comes from the desire to understand the trade-off between formula size and number of variables. This is of great interest because, as is well-known, this is equivalent to the trade-off between parallel time and number of processors [6]. Adler and Immerman introduced a game that can be used to determine the minimum size of first-order formulas with a given number of variables needed to express a given property. These games, which are closely related to the communication complexity games of Karchmer and Wigderson [9], were used to prove two optimal size bounds for temporal logics [1]. Later Grohe and Schweikardt used similar methods to study the size versus variable trade-off for first-order logic on unary words [5]. They proved that all first-order expressible properties of unary words are already expressible with two variables and that the variable-size trade-off between two versus three variables is polynomial whereas the trade-off between three versus four variables is exponential. They left open the trade-off between k and $k+1$ variables for $k \geq 4$. While we do not directly address that question here, our classification of FO^2 on words is a step towards the general understanding of the expressive power of FO needed for progress on such trade-offs.

Our characterization of $FO^2[<]$ and $FO^2[<, \mathrm{Suc}]$ on words is based on the very natural notion of n-ranker (Definition 5). Informally, a ranker is the position of a certain combination of letters in a word. For example, $\triangleright_{\mathtt{a}}$ and $\triangleleft_{\mathtt{b}}$ are 1-rankers where $\triangleright_{\mathtt{a}}(w)$ is the position of the first $\mathtt{a}$ in w (from the left) and $\triangleleft_{\mathtt{b}}(w)$ is the position of the first $\mathtt{b}$ in w from the right. Similarly, the 2-ranker $r_2 = \triangleright_{\mathtt{a}}\triangleright_{\mathtt{c}}$ denotes the position of the first $\mathtt{c}$ to the right of the first $\mathtt{a}$, and the 3-ranker, $r_3 = \triangleright_{\mathtt{a}} \triangleright_{\mathtt{c}} \triangleleft_{\mathtt{b}}$ denotes the position of the first $\mathtt{b}$ to the left of r_2. If there is no such letter then the ranker is undefined. For example, $r_3(\mathtt{cababcba}) = 5$ and $r_3(\mathtt{acbbca})$ is undefined.

Our first structure theorem (Theorem 11) says that the properties expressible in $FO^2_n[<]$, i.e. first-order logic with two variables and quantifier depth n, are exactly boolean combinations of statements of the form, "r is defined", and "r is to the left (right) of r'" for k-rankers, r, and k'-rankers, r', with $k \leq n$ and $k' < n$. A non-quantitative version of this theorem was previously known [13].[1] Furthermore, a quantitative version in terms of iterated block products of the variety of semilattices is presented in [16], based on work by Straubing and Thérien [14].

Surprisingly, Theorem 11 can be generalized in almost exactly the same form to characterize $FO^2_{m,n}[<]$ where there are at most m blocks of alternating quantifiers, $m \leq n$. This second structure theorem (Theorem 17) uses the notion of (m, n)-ranker where there are m blocks of $\triangleright$'s or $\triangleleft$'s, that is, changing direction in rankers corresponds exactly to alternation of quantifiers. Using Theorem 17 we prove that there is a strict alternation hierarchy for $FO^2_n[<]$ (Theorem 20)

---

[1] See item 7 in Fact 1: a "turtle language" is a language of the form "r is defined", for some ranker, r.

but that exactly at most $|\Sigma| + 1$ alternations are useful, where $|\Sigma|$ is the size of the alphabet (Theorem 18).

The language $FO^2[<, \mathrm{Suc}]$ is more expressive than $FO^2[<]$ because it allows us to talk about consecutive strings of symbols[2]. For $FO^2[<, \mathrm{Suc}]$, a straightforward generalization of n-ranker to n-successor-ranker allows us to prove exact analogs of Theorems 11 and 17. We use the latter to prove that there is also a strict alternation hierarchy for $FO^2_n[<, \mathrm{Suc}]$ (Theorem 26). Since in the presence of successor we can encode an arbitrary alphabet in binary, no analog of Theorem 18 holds for $FO^2[<, \mathrm{Suc}]$.

The expressive power of first-order logic with three or more variables on words has been well-studied. The languages expressible are of course the star-free regular languages [10]. The dot-depth hierarchy is the natural hierarchy of these languages. This hierarchy is strict [2] and identical to the first-order quantifier alternation hierarchy [18, 19].

Many beautiful results on FO^2 on words were also already known. The main significant outstanding question was whether there was an alternating hierarchy. The following is a summary of the main previously known characterizations of $FO^2[<]$ on words. For a nice survey that discusses all of these characterizations, and even more, see [15].

Fact 1. [3, 4, 11, 12, 17, 13] *Let $R \subseteq \Sigma^\star$. The following conditions are equivalent:*

1. $R \in FO^2[<]$
2. *R is expressible in unary temporal logic*
3. $R \in \Sigma_2 \cap \Pi_2[<]$
4. *R is an unambiguous regular language*
5. *The syntactic semi-group of R is a member of* **DA**
6. *R is recognizable by a partially-ordered 2-way automaton*
7. *R is a boolean combination of "turtle languages"*

The proofs of our structure theorems are self-contained applications of Ehrenfeucht-Fraïssé games. All of the above characterizations follow from these results. Furthermore, we have now exactly connected quantifier and alternation depth to the picture, thus adding tight bounds and further insight to the above results.

For example, one can best understand item 4 above – that $FO^2[<]$ on words corresponds to the unambiguous regular languages – via Theorem 15 which states that any $FO^2_n[<]$ formula with one free variable that is always true of at most one position in any string, necessarily denotes an n-ranker.

In the conclusion of [13], the authors define the subclasses of rankers with one and two blocks of alternation. They write that, "...turtle languages might turn out to be a helpful tool for further studies in algebraic language theory." We feel that the present paper fully justifies that prediction. Turtle languages — aka rankers — do provide an exceptionally clear and precise understanding of the expressive power of FO^2 on words, with and without successor.

[2] With three variables we can express $\mathrm{Suc}(x, y)$ using the ordering: $x < y \wedge \forall z(z \leq x \vee y \leq z)$.

In summary, our structure theorems provide a complete classification of the expressive power of FO^2 on words in terms of both quantifier depth and alternation. They also tighten several previous characterizations and lead to the alternation hierarchy results.

We begin the remainder of this extended abstract with a brief review of logical background including Ehrenfeucht-Fraïssé games, our main tool. In Sect. 3 we formally define rankers and present our structure theorem for $FO^2_n[<]$. The structure theorem for $FO^2_{m,n}[<]$ is covered in Sect. 4, including our alternation hierarchy result that follows from it. Sect. 5 extends our structure theorems and the alternation hierarchy result to $FO^2[<, Suc]$. Finally, we discuss applications of our structural results to satisfiability for $FO^2[<]$ in Sect. 6. The full version [20] of this paper includes all proofs that had to be left out here due to space constraints.

2 Background and Definitions

We recall some notation concerning strings, first-order logic, and Ehrenfeucht-Fraïssé games. See [6] for more details, including the proof of Facts 2 and 3.

Σ will always denote a finite alphabet and ε the empty string. For $w \in \Sigma^\ell$ and $i \in [1, \ell]$, let w_i be the i-th letter of w; and for $[a, b]$ a subinterval of $[1, \ell]$, let $w_{[a,b]}$ be the substring $w_a \dots w_b$. We identify a word, $w \in \Sigma^\ell$ with the logical structure, $w = (\{1, \dots, \ell\}, Q_\sigma, \sigma \in \Sigma)$, where $(w, i/x) \models Q_\sigma(x)$ iff $w_i = \sigma$.

We use FO[<] to denote first-order logic with a binary linear order predicate $<$, and FO = FO[<, Suc] for first-order logic with an additional binary successor predicate. FO^k_n refers to the restriction of first-order logic to use at most k distinct variables, and quantifier depth n. $FO^k_{m,n}$ is the further restriction to formulas such that any path in their parse tree has at most m blocks of alternating quantifiers, and $FO^k\text{–ALT}\,[m] = \bigcup_{n \geq m} FO^k_{m,n}$. We write $u \equiv^2_n v$ to mean that u and v agree on all formulas from FO^2_n, and $u \equiv^2_{m,n}$ if they agree on $FO^2_{m,n}$.

We assume that the reader is familiar with our main tool: the Ehrenfeucht-Fraïssé game. In each of the n moves of the game $FO^2_n(u, v)$, Samson places one of the two pebbles pairs, x or y on a position in one of the two words and Delilah then answers by placing that pebble's mate on a position of the other word. Samson wins if after any move, the map from the chosen points in u to those in v, i.e., $x(u) \mapsto x(v)$, $y(u) \mapsto y(v)$ is not an isomorphism of the induced substructures; and Delilah wins otherwise. The fundamental theorem of Ehrenfeucht-Fraïssé games is the following:

Fact 2. *Let $u, v \in \Sigma^\star$, $n \in \mathbb{N}$. Delilah has a winning strategy for the game $FO^2_n(u, v)$ iff $u \equiv^2_n v$.*

Thus, Ehrenfeucht-Fraïssé games are a perfect tool for determining what is expressible in first-order logic with a given quantifier-depth and number of variables. The game $FO^2_{m,n}(u, v)$ is the restriction of the game $FO^2_n(u, v)$ in which Samson may change which word he plays on at most $m - 1$ times.

Fact 3. *Let $u, v \in \Sigma^\star$ and let $m, n \in \mathbb{N}$ with $m \leq n$. Delilah has a winning strategy for the game $\mathrm{FO}^2_{m,n}(u, v)$ iff $u \equiv^2_{m,n} v$.*

3 Structure Theorem for $\mathrm{FO}^2[<]$

We define boundary positions that point to the first or last occurrences of a letter in a word, and define an n-ranker as a sequence of n boundary positions. In terms of [13], boundary positions are turtle instructions and n-rankers are turtle programs of length n. The following three lemmas show that basic properties about the definedness and position of these rankers can be expressed in $\mathrm{FO}^2[<]$, and we use these results to prove our structure theorem.

Definition 4. *A* boundary position *denotes the first or last occurrence of a letter in a given word. Boundary positions are of the form d_a where $d \in \{\triangleright, \triangleleft\}$ and $a \in \Sigma$. The interpretation of a boundary position d_a on a word $w = w_1 \ldots w_{|w|} \in \Sigma^\star$ is defined as follows.*

$$d_a(w) = \begin{cases} \min\{i \in [1, |w|] \mid w_i = a\} & \text{if } d = \triangleright \\ \max\{i \in [1, |w|] \mid w_i = a\} & \text{if } d = \triangleleft \end{cases}$$

Here we set $\min\{\}$ and $\max\{\}$ to be undefined, thus $d_a(w)$ is undefined if a does not occur in w. A boundary position can also be specified with respect to a position $q \in [1, |w|]$.

$$d_a(w, q) = \begin{cases} \min\{i \in [q+1, |w|] \mid w_i = a\} & \text{if } d = \triangleright \\ \max\{i \in [1, q-1] \mid w_i = a\} & \text{if } d = \triangleleft \end{cases}$$

Definition 5. *Let n be a positive integer. An n-*ranker *r is a sequence of n boundary positions. The interpretation of an n-ranker $r = (p_1, \ldots, p_n)$ on a word w is defined as follows.*

$$r(w) := \begin{cases} p_1(w) & \text{if } r = (p_1) \\ \text{undefined} & \text{if } (p_1, \ldots, p_{n-1})(w) \text{ is undefined} \\ p_n(w, (p_1, \ldots, p_{n-1})(w)) & \text{otherwise} \end{cases}$$

Instead of writing n-rankers as a formal sequence $(p_1, \ldots, p_n)$, we often use the simpler notation $p_1 \ldots p_n$. We denote the set of all n-rankers by R_n, and the set of all n-rankers that are defined over a word w by $R_n(w)$. Furthermore, we set $R^\star_n := \bigcup_{i \in [1,n]} R_i$ and $R^\star_n(w) := \bigcup_{i \in [1,n]} R_i(w)$.

Definition 6. *Let r be an n-ranker. As defined above, we have $r = (p_1, \ldots, p_n)$ for boundary positions p_i. The k-*prefix ranker *of r for $k \in [1, n]$ is $r_k := (p_1, \ldots, p_k)$.*

Definition 7. *Let $i, j \in \mathbb{N}$. The* order type *of i and j is defined as*

$$\mathit{ord}(i, j) = \begin{cases} < & \text{if } i < j \\ = & \text{if } i = j \\ > & \text{if } i > j \end{cases}$$

Lemma 8 (distinguishing points on opposite sides of a ranker). *Let n be a positive integer, let $u, v \in \Sigma^\star$ and let $r \in R_n(u) \cap R_n(v)$. Samson wins the game $\mathrm{FO}^2_n(u, v)$ where initially $ord(x(u), r(u)) \neq ord(x(v), r(v))$.*

Proof. We only look at the case where $x(u) \geq r(u)$ and $x(v) < r(v)$ since all other cases are symmetric to this one. For $n = 1$ Samson has a winning strategy: If r is the first occurrence of a letter, then Samson places y on $r(u)$ and Delilah cannot reply. If r marks the last occurrence of a letter in the whole word, then Samson places y on $r(v)$. Again, Delilah cannot reply with any position and thus loses.

For $n > 1$, we look at the prefix ranker r_{n-1} of r. One of the following two cases applies.

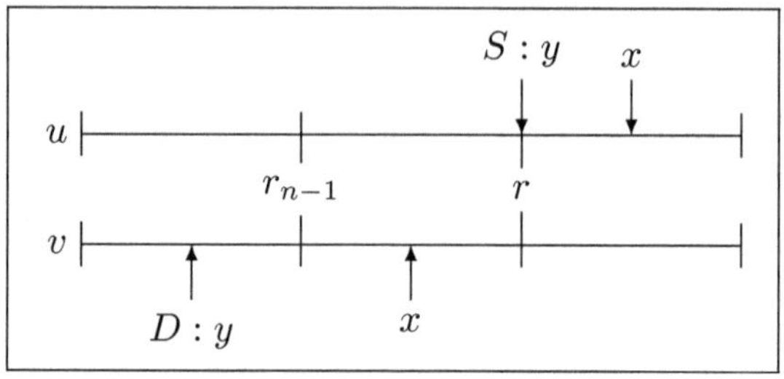

Fig. 1. The case $r_{n-1}(u) < r(u)$

(1) $r_{n-1}(u) < r(u)$, as shown in Fig. 1. Samson places pebble y on $r(u)$, and Delilah has to reply with a position to the left of $x(v)$. She cannot choose a position in the interval $(r_{n-1}(v), r(v))$, because this section does not contain the letter $u_{r(u)}$. Thus she has to choose a position left of or equal to $r_{n-1}(v)$. By induction Samson wins the remaining game.

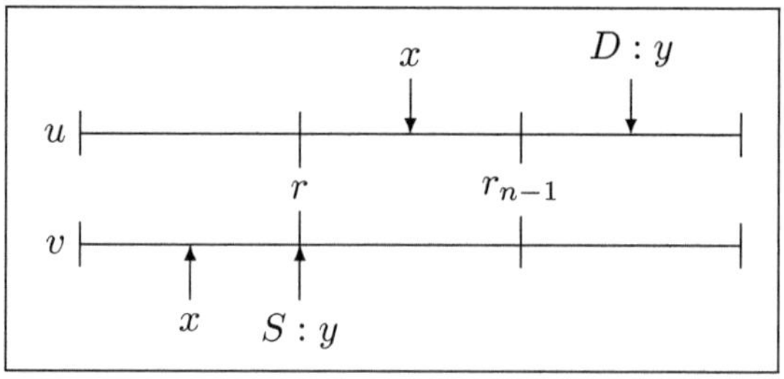

Fig. 2. The case $r(u) < r_{n-1}(u)$

(2) $r(u) < r_{n-1}(u)$, as shown in Fig. 2. Samson places y on $r(v)$, and Delilah has to reply with a position right of $x(u)$ and thus right of $r(u)$. She cannot choose any position in $(r(u), r_{n-1}(u))$, because this interval does not contain the letter $v_{r(v)}$, thus Delilah has to choose a position right of or equal to $r_{n-1}(u)$. By induction Samson wins the remaining game. □

Lemma 9 (expressing the definedness of a ranker). *Let n be a positive integer, and let $r \in R_n$. There is a formula $\varphi_r \in \mathrm{FO}^2_n[<]$ such that for all $w \in \Sigma^\star$, $w \models \varphi_r \iff r \in R_n(w)$.*

Proof. Let $u, v \in \Sigma^\star$ such that $r \in R_n(u)$ and $r \notin R_n(v)$. We show that Samson wins the game $\mathrm{FO}^2_n(u, v)$. If r_1, the shortest prefix ranker of r, is not defined over v, the letter referred to by r_1 occurs in u but does not occur in v. Thus Samson easily wins in one move.

Otherwise we let $r_i = (p_1, \ldots, p_i)$ be the shortest prefix ranker of r that is undefined over v. Thus r_{i-1} is defined over both words. Without loss of generality we assume that $p_i = \lhd_{\mathtt{a}}$. This situation is illustrated in Fig. 3. Notice that v does not contain any a's to the left of $r_{i-1}(v)$, otherwise r_i would be defined over v. Samson places x in u on $r_i(u)$, and Delilah has to reply with a position right of or equal to $r_{i-1}(v)$. Now Lemma 8 applies and Samson wins in $i-1$ more moves. □

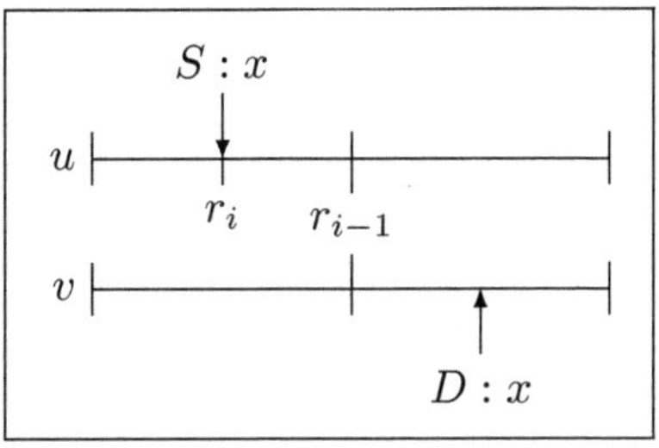

Fig. 3. $r_i(v)$ is undefined

Lemma 10 (position of a ranker). *Let n be a positive integer and let $r \in R_n$. There is a formula $\varphi_r \in FO^2_n[<]$ such that for all $w \in \Sigma^\star$ and for all $i \in [1, |w|]$, $(w, i/x) \models \varphi_r \iff i = r(w)$.*

Proof. Let $u, v \in \Sigma^\star$. We show that Samson wins the game $FO^2_n(u, v)$ where initially $x(u) = r(u)$ and $x(v) \neq r(v)$. If $r(v)$ is defined over v, then we can apply Lemma 8 immediately to get the desired strategy for Samson. Otherwise we use the strategy from Lemma 9. □

Theorem 11 (structure of $FO^2_n[<]$). *Let u and v be finite words, and let $n \in \mathbb{N}$. The following two conditions are equivalent.*

(i) (a) $R_n(u) = R_n(v)$, and,
(b) for all $r \in R^\star_n(u)$ and $r' \in R^\star_{n-1}(u)$, ord$(r(u), r'(u))$ = ord$(r(v), r'(v))$
(ii) $u \equiv^2_n v$

Notice that condition (i)(a) is equivalent to $R^\star_n(u) = R^\star_n(v)$. Instead of proving Theorem 11 directly, we prove the following more general version on words with two interpreted variables.

Theorem 12. *Let u and v be finite words, let $i_1, i_2 \in [1, |u|]$, let $j_1, j_2 \in [1, |v|]$, and let $n \in \mathbb{N}$. The following two conditions are equivalent.*

(i) (a) $R_n(u) = R_n(v)$, and,
(b) for all $r \in R^\star_n(u)$ and $r' \in R^\star_{n-1}(u)$, ord$(r(u), r'(u))$ = ord$(r(v), r'(v))$, and,
(c) $(u, i_1/x, i_2/y) \equiv^2_0 (v, j_1/x, j_2/y)$, and,
(d) for all $r \in R^\star_n(u)$, ord$(i_1, r(u))$ = ord$(j_1, r(v))$ and ord$(i_2, r(u))$ = ord$(j_2, r(v))$
(ii) $(u, i_1/x, i_2/y) \equiv^2_n (v, j_1/x, j_2/y)$

Proof. For $n = 0$, (i)(a), (i)(b) and (i)(d) are vacuous, and (i)(c) is equivalent to (ii). For $n \geq 1$, we prove the two implications individually using induction on n.

We first show "$\neg$(i) $\Rightarrow$ $\neg$(ii)". Assuming that (i) holds for $n \in \mathbb{N}$ but fails for $n+1$, we show that $(u, i_1/x, i_2/y) \not\equiv^2_n (v, j_1/x, j_2/y)$ by giving a winning strategy for Samson in the FO^2_n game on the two structures. If (i)(c) does not hold, then Samson wins immediately. If (i)(d) does not hold for $n+1$, then Samson wins by Lemma 8. If (i)(a) or (i)(b) do not hold for $n+1$, then one of the following three cases applies.

(1) There are two n-rankers that don't agree on their ordering in u and v.
(2) There is an $(n+1)$-ranker that is defined over one word but not over the other.
(3) There is an $(n+1)$-ranker that does not appear in the same order on both structures with respect to a k-ranker where $k \leq n$.

We first look at case (1) where there are two rankers $r, r' \in R_n^\star(u)$ such that $\mathrm{ord}(r(u), r'(u)) \neq \mathrm{ord}(r(v), r'(v))$. Without loss of generality we assume that $r(u) \leq r'(u)$ and $r(v) > r'(v)$, and present a winning strategy for Samson in the FO^2_{n+1} game. In the first move he places x on $r(u)$ in u. Delilah has to reply with $r(v)$ in v, otherwise she would lose the remaining n-move game as shown in Lemma 8. Let r'_{n-1} be the $(n-1)$-prefix-ranker of r'. We look at two different cases depending on the ordering of r'_{n-1} and r'.

For $r'_{n-1}(u) < r'(u)$, the situation is illustrated in Fig. 4. In his second move, Samson places y on $r'(v)$. Delilah has to reply with a position to the left of $x(u)$, but she cannot choose any position from the interval $(r'_{n-1}(u), r'(u))$ because it does not contain the letter $v_{y(v)}$. So she has to reply with a position left of or equal to $r'_{n-1}(u)$, and Samson wins the remaining FO^2_{n-1} game as shown in Lemma 8.

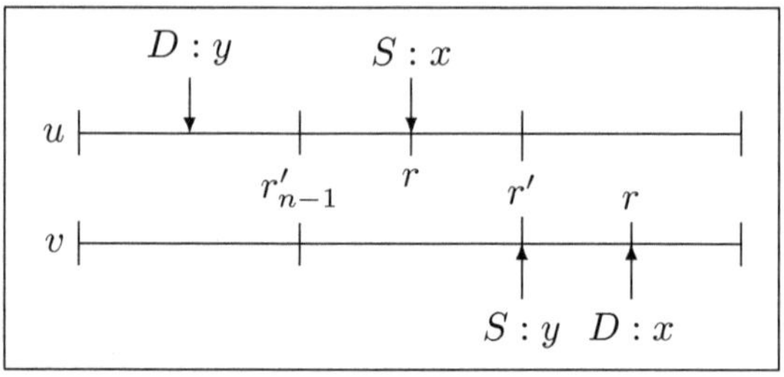

Fig. 4. Two n-rankers appear in different order and r' ends with $\triangleright$

For $r'_{n-1}(u) > r'(u)$, the situation is illustrated in Fig. 5. In his second move, Samson places pebble y on $r'(u)$, and Delilah has to reply with a position to the right of $x(v)$, but she cannot choose anything from the interval $(r'(v), r'_{n-1}(v))$ because this section does not contain the letter $u_{y(u)}$. Thus she has to reply with a position right of or equal to $r'_{n-1}(v)$, and Samson wins the remaining FO^2_{n-1} game as shown in Lemma 8.

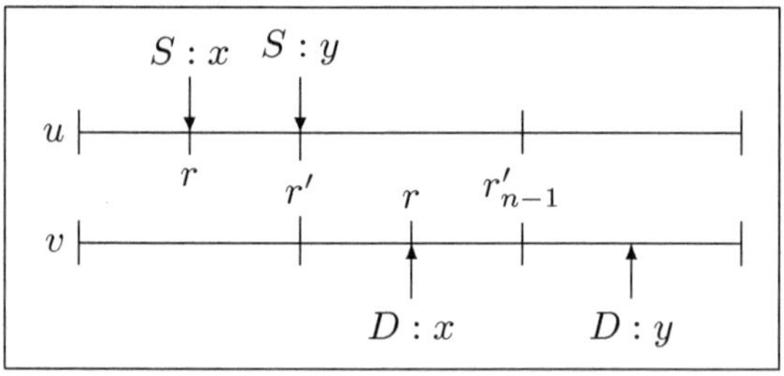

Fig. 5. Two n-rankers appear in different order and r' ends with $\triangleleft$

If (i) fails but all n-rankers agree on their ordering, then there are two consecutive n-rankers $r, r' \in R_n(u)$ with $r(u) < r'(u)$ and a letter $a \in \Sigma$ such that without loss of generality a occurs in the segment $u_{((r(u),r'(u))}$ but not in the segment $v_{(r(v),r'(v))}$. We describe a winning strategy for Samson in the game $\mathrm{FO}^2_{n+1}(u, v)$. He places x on an a in the segment $(r(u), r'(u))$ of u, as shown in Fig. 6. Delilah cannot reply with any- thing in the interval $(r(v), r'(v))$. If she replies with a position left of or equal to $r(v)$, then x is on different sides of the

n-ranker r in the two words. Thus Lemma 8 applies and Samson wins the remaining n-move game. If Delilah replies with a position right of or equal to $r'(v)$, then we can apply Lemma 8 to r' and get a winning strategy for the remaining game as well. This concludes the proof of "$\neg$(i) $\Rightarrow$ $\neg$(ii)".

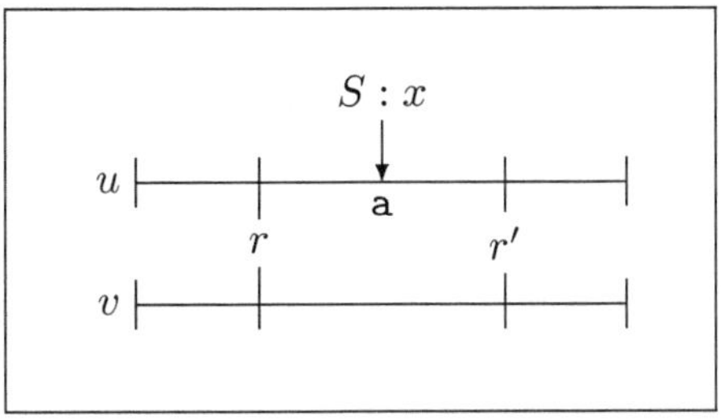

Fig. 6. A letter a occurs between n-rankers r, r' in u but not in v

To show (i) $\Rightarrow$ (ii), we assume (i) for $n+1$, and present a winning strategy for Delilah in the FO^2_{n+1} game on the two structures. In his first move Samson picks up one of the two pebbles, and places it on a new position. Without loss of generality we assume that Samson picks up x and places it on u in his first move. If $x(u) = r(u)$ for any ranker $r \in R^\star_{n+1}(u)$, then Delilah replies with $x(v) = r(v)$. This establishes (i)(c) and (i)(d) for n, and thus Delilah has a winning strategy for the remaining FO^2_n game by induction.

If Samson does not place $x(u)$ on any ranker from $R^\star_{n+1}(u)$, then we look at the closest rankers from $R^\star_n(u)$ to the left and right of $x(u)$, denoted by r_ℓ and r_r, respectively. Let $\mathtt{a} := u_{x(u)}$ and define the $(n+1)$-ranker $s = (r_\ell, \triangleright_\mathtt{a})$. On u we have $r_\ell(u) < s(u) < r_r(u)$. Because of (i)(a) s is defined on v as well, and because of (i)(b), we have $r_\ell(v) < s(v) < r_r(v)$. If $y(u)$ is not contained in the interval $(r_\ell(u), r_r(u))$, then Delilah places x on $s(v)$, which establishes (i)(c) and (i)(d) for n. Thus by induction Delilah has a winning strategy for the remaining FO^2_n game.

If both pebbles $x(u)$ and $y(u)$ are in the interval $(r_\ell(u), r_r(u))$, then we have to be more careful. Without loss of generality we assume $y(u) < x(u)$ as illustrated in Fig. 7. Thus Delilah has to place x somewhere in the segment $(y(v), r_r(v))$ and at a position with letter $\mathtt{a} := u_{x(u)}$. We define the $n+1$-ranker $s = (r_r, \triangleleft_\mathtt{a})$. From (i)(d) we know that s appears on the same side of y in both structures, thus we have $y(v) < s(v) < r_r(v)$. Delilah places her pebble x on $s(v)$, and thus establishes (i)(c) and (i)(d) for n. By induction, Delilah has a winning strategy for the remaining FO^2_n game. □

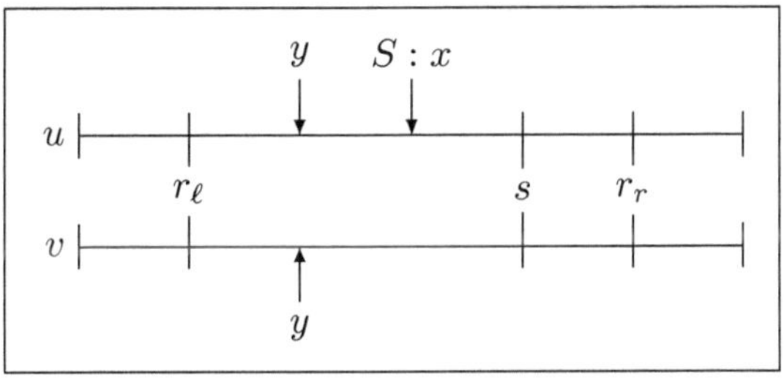

Fig. 7. x and y are in the same section

A fundamental property of an n-ranker is that it uniquely describes a position in a given word. Now we show that the converse holds as well: any unique position in a word can be described by a ranker.

Definition 13 (unique position formula). *A formula $\varphi \in FO^2[<]$ with x as a free variable is a* unique position formula *if for all $w \in \Sigma^\star$ there is at most one $i \in [1, |w|]$ such that $(w, i/x) \models \varphi$.*

Lemma 14. *Let n be a positive integer and let $\varphi \in \mathrm{FO}^2_n[<]$ be a unique position formula. Let $u \in \Sigma^\star$ and let $i \in [1, |u|]$ such that $(u, i/x) \models \varphi$. Then $i = r(u)$ for some ranker $r \in R^\star_n$.*

Theorem 15. *Let n be a positive integer and let $\varphi \in \mathrm{FO}^2_n[<]$ be a unique position formula. There is a $k \in \mathbb{N}$, and there are mutually exclusive formulas $\alpha_i \in \mathrm{FO}^2_n[<]$ and rankers $r_i \in R^\star_n$ such that*

$$\varphi \equiv \bigvee_{i \in [1,k]} (\alpha_i \wedge \varphi_{r_i})$$

where $\varphi_{r_i} \in \mathrm{FO}^2_n[<]$ is the formula from Lemma 10 that uniquely describes the ranker r_i.

4 Alternation Hierarchy for $\mathrm{FO}^2[<]$

We define alternation rankers and prove our structure theorem (Theorem 17) for $\mathrm{FO}^2_{m,n}[<]$. Surprisingly the number of alternating blocks of $\lhd$ and $\rhd$ in the rankers corresponds exactly to the number of alternating quantifier blocks. The main ideas from our proof of Theorem 11 still apply here, but keeping track of the number of alternations does add complications.

Definition 16 (m-alternation n-ranker). *Let $m, n \in \mathbb{N}$ with $m \leq n$. An* m-alternation n-ranker, *or (m, n)-ranker, is an n-ranker with exactly m blocks of boundary positions that alternate between $\rhd$ and $\lhd$.*

We use the following notation for alternation rankers.

$$R_{m,n}(w) := \{r \mid r \text{ is an } m\text{-alternation } n\text{-rankers and defined over the word } w\}$$
$$R_{m\rhd,n}(w) := \{r \in R_{m,n}(w) \mid r \text{ ends with } \rhd\}$$
$$R^\star_{m,n}(w) := \bigcup_{i \in [1,m], j \in [1,n]} R_{i,j}(w)$$
$$R^\star_{m\rhd,n}(w) := R^\star_{m-1,n}(w) \cup \bigcup_{i \in [1,n]} R_{m\rhd,i}(w)$$

Theorem 17 (structure of $\mathrm{FO}^2_{m,n}[<]$). *Let u and v be finite words, and let $m, n \in \mathbb{N}$ with $m \leq n$. The following two conditions are equivalent.*

(i) (a) $R_{m,n}(u) = R_{m,n}(v)$, and,
(b) for all $r \in R^\star_{m,n}(u)$ and for all $r' \in R^\star_{m-1,n-1}(u)$, we have $ord(r(u), r'(u)) = ord(r(v), r'(v))$, and,
(c) for all $r \in R^\star_{m,n}(u)$ and $r' \in R^\star_{m,n-1}(u)$ such that r and r' end with different directions, $ord(r(u), r'(u)) = ord(r(v), r'(v))$
(ii) $u \equiv^2_{m,n} v$

Using Theorem 17, we show that for any fixed alphabet Σ, at most $|\Sigma| + 1$ alternations are useful. Intuitively, each boundary position in a ranker says that a certain letter does not occur in some part of a word. Alternations are only useful if they visit one of these previous parts again. Once we visited one part of a word $|\Sigma|$ times, this part cannot contain any more letters and thus is empty.

Theorem 18. *Let Σ be a finite alphabet, let $u, v \in \Sigma^\star$ and $n \in \mathbb{N}$. If $u \equiv^2_{|\Sigma|+1,n} v$, then $u \equiv^2_n v$.*

In order to prove that the alternation hierarchy for FO^2 is strict, we define example languages that can be separated by a formula of a given alternation depth m, but that cannot be separated by any formula of lower alternation depth. As Theorem 18 shows, we need to increase the size of the alphabet with increasing alternation depth. We inductively define the example words $u_{m,n}$ and $v_{m,n}$ and the example languages K_m and L_m over finite alphabets $\Sigma_m = \{a_0, \ldots, a_{m-1}\}$. Here i, m and n are positive integers.

$$\begin{aligned} u_{1,n} &:= a_0 & v_{1,n} &:= \varepsilon \\ u_{2,n} &:= a_0(a_1a_0)^{2n} & v_{2,n} &:= (a_1a_0)^{2n} \\ u_{2i+1,n} &:= (a_0 \ldots a_{2i})^n\; u_{2i,n} & v_{2i+1,n} &:= (a_0 \ldots a_{2i})^n\; v_{2i,n} \\ u_{2i+2,n} &:= u_{2i+1,n}\; (a_{2i+1} \cdots a_0)^n & v_{2i+2,n} &:= v_{2i+1,n}\; (a_{2i+1} \cdots a_0)^n \end{aligned}$$

Notice that $u_{m,n}$ and $v_{m,n}$ are almost identical – if we delete a_0 in the center of $u_{m,n}$, we get $v_{m,n}$. Finally, we set $K_m := \bigcup_{n\geq 1}\{u_{m,n}\}$ and $L_m := \bigcup_{n\geq 1}\{v_{m,n}\}$.

Definition 19. *A formula φ* separates *two languages $K, L \subseteq \Sigma^\star$ if for all $w \in K$ we have $w \models \varphi$ and for all $w \in L$ we have $w \not\models \varphi$ or vice versa.*

Our example words are constructed such that for $m \geq 3$, $u_{m,n}$ and $v_{m,n}$ can be distinguished by the ordering of two $(m-1)$-rankers. In the case $m = 3$ for example, we can use the two rankers $r_3 := \lhd_{a_2}\rhd_{a_0}$ and $s_3 := \lhd_{a_2}\rhd_{a_1}$. A formal argument for all m is given in [20]. There we also argue that the example words $u_{m,n}$ and $v_{m,n}$ agree on the definedness of all $(m-1,n)$-rankers, and that these rankers appear in exactly the same order with respect to shorter rankers. Thus the two languages K_m and L_m cannot be separated by any $\mathrm{FO}^2[<]$–ALT $[m-1]$ formula. Thus we have the following theorem.

Theorem 20 (alternation hierarchy for $\mathrm{FO}^2[<]$). *For any positive integer m, there is a $\varphi_m \in \mathrm{FO}^2[<]$–ALT $[m]$ and there are two languages K_m, L_m such that φ_m separates K_m and L_m, but no $\psi \in \mathrm{FO}^2[<]$–ALT $[m-1]$ separates K_m and L_m.*

Theorem 20 resolves an open question from [3, 4].

5 Structure Theorem and Alternation Hierarchy for $\mathrm{FO}^2[<, \mathrm{Suc}]$

We extend our definitions of boundary positions and rankers from Sect. 3 to include the substrings of a given length that occur immediately before and after the position of the ranker.

Definition 21. *A* (k,ℓ)-neighborhood boundary position *denotes the first or last occurrence of a substring in a word. More precisely, a* (k,ℓ)*-neighborhood boundary position is of the form* $d_{(s,a,t)}$ *with* $d \in \{\triangleright, \triangleleft\}$, $s \in \Sigma^k$, $a \in \Sigma$ *and* $t \in \Sigma^\ell$. *The interpretation of a* (k,ℓ)*-neighborhood boundary position* $p = d_{(s,a,t)}$ *on a word* $w = w_1 \dots w_{|w|}$ *is defined as follows.*

$$p(w) = \begin{cases} \min\{i \in [k+1, |w| - \ell] \mid w_{i-k} \dots w_{i+\ell} = sat\} & \text{if } d = \triangleright \\ \max\{i \in [k+1, |w| - \ell] \mid w_{i-k} \dots w_{i+\ell} = sat\} & \text{if } d = \triangleleft \end{cases}$$

Notice that $p(w)$ *is undefined if the sequence sat does not occur in* w. *A* (k,ℓ)*-neighborhood boundary position can also be specified with respect to a position* $q \in [1, |w|]$.

$$p(w,q) = \begin{cases} \min\{i \in [\max\{q+1, k+1\}, |w| - \ell] \mid w_{i-k} \dots w_{i+\ell} = sat\} & \text{if } d = \triangleright \\ \max\{i \in [k+1, \min\{q-1, |w| - \ell\}] \mid w_{i-k} \dots w_{i+\ell} = sat\} & \text{if } d = \triangleleft \end{cases}$$

Observe that (0,0)-neighborhood boundary positions are identical to the boundary positions from Definition 4. As before in the case without successor, we build rankers out of these boundary positions.

Definition 22. *An* n-successor-ranker r *is a sequence of* n *neighborhood boundary positions,* $r = (p_1, \dots, p_n)$, *where* p_i *is a* (k_i, ℓ_i)*-neighborhood boundary position and* $k_i, \ell_i \in [0, (i-1)]$. *The interpretation of an* n*-successor-ranker* r *on a word* w *is defined as follows.*

$$r(w) := \begin{cases} p_1(w) & \text{if } r = (p_1) \\ \text{undefined} & \text{if } (p_1, \dots, p_{n-1})(w) \text{ is undefined} \\ p_n(w, (p_1, \dots, p_{n-1})(w)) & \text{otherwise} \end{cases}$$

We denote the set of all n*-successor-rankers that are defined over a word* w *by* $SR_n(w)$, *and set* $SR_n^\star(w) := \bigcup_{i \in [1,n]} SR_i(w)$.

Because we now have the additional atomic relation Suc, we need to extend our definition of order type as well.

Definition 23. *Let* $i, j \in \mathbb{N}$. *The* successor order type *of* i *and* j *is defined as*

$$ord_S(i,j) = \begin{cases} \ll & \text{if } i < j-1 \\ -1 & \text{if } i = j-1 \\ = & \text{if } i = j \\ +1 & \text{if } i = j+1 \\ \gg & \text{if } i > j+1 \end{cases}$$

With this new definition of n-successor-rankers, our proofs for Lemmas 8, 9, 10 and Theorem 11 go through with only minor modifications. Instead of working through all the details again, we simply point out the differences.

First we notice that 1-successor-rankers are simply 1-rankers, so the base case of all inductions remains unchanged. In the proofs of Lemmas 8, 9 and 10, and in the proof of (ii) ⇒ (i) from Theorem 11, we argued that Delilah cannot reply with a position in a given section because it does not contain a certain ranker and therefore it does not contain the symbol used to define this ranker. Now we need to know more – we need to show that Delilah cannot reply with a certain letter in a given section that is surrounded by a specified neighborhood, given that this section does not contain the corresponding successor-ranker. Whenever Samson's winning strategy depends on the fact that an n-successor-ranker does not occur in a given section, he has $n-1$ additional moves left. So if Delilah does not reply with a position with the same letter and the same neighborhood, Samson can point out a difference in the neighborhood with at most $(n-1)$ additional moves.

For the other direction of Theorem 11, we need to make sure that Delilah can reply with a position that is contained in the correct interval, has the same symbol and is surrounded by the same neighborhood. Where we previously defined the n-ranker $s := (r_\ell, \triangleright_{\mathtt{a}})$ or $s := (r_r, \triangleleft_{\mathtt{a}})$, we now include the $(n-1)$-neighborhood of the respective positions chosen by Samson. Thus we make sure that Samson cannot point out a difference in the two words, and Delilah still has a winning strategy. Thus we have the following three theorems for $FO^2[<, Suc]$.

Theorem 24 (structure of $FO^2_n[<, Suc]$). *Let u and v be finite words, and let $n \in \mathbb{N}$. The following two conditions are equivalent.*

(i) (a) $SR_n(u) = SR_n(v)$, and,
(b) for all $r \in SR^\star_n(u)$ and for all $r' \in SR^\star_{n-1}(u)$, $ord_S(r(u), r'(u)) = ord_S(r(v), r'(v))$

(ii) $u \equiv^2_n v$

Theorem 25 (structure of $FO^2_{m,n}[<, Suc]$). *Let u and v be finite words, and let $m, n \in \mathbb{N}$ with $m \leq n$. The following two conditions are equivalent.*

(i) (a) $SR_{m,n}(u) = SR_{m,n}(v)$, and,
(b) for all $r \in SR^\star_{m,n}(u)$ and for all $r' \in SR^\star_{m-1,n-1}(u)$, $ord_S(r(u), r'(u)) = ord_S(r(v), r'(v))$, and,
(c) for all $r \in SR^\star_{m,n}(u)$ and $r' \in SR^\star_{m,n-1}(u)$ such that r and r' end with different directions, $ord_S(r(u), r'(u)) = ord_S(r(v), r'(v))$

(ii) $u \equiv^2_{m,n} v$

Theorem 26 (alternation hierarchy for $FO^2[<, Suc]$). *Let m be a positive integer. There is a $\varphi_m \in FO^2[<, Suc]$-ALT $[m]$ and there are two languages $K_m, L_m \subseteq \Sigma^\star$ such that φ_m separates K_m and L_m, but there is no $\psi \in FO^2[<, Suc]$-ALT $[m-1]$ that separates K_m and L_m.*

The proof of Theorem 26 is given in [20], and mostly identical to the proof of Theorem 20. We use n copies of an extra letter between any two letters in our example words, and thus ensure that the successor predicate is not useful.

6 Small Models and Satisfiability for $FO^2[<]$

The complexity of satisfiability for $FO^2[<]$ was investigated in [4]. There it is shown that any satisfiable $FO^2_n[<]$ formula has a model of size at most exponential in n. It follows that satisfiability for $\mathrm{FO}^2[<]$ is in NEXP, and a reduction from TILING shows that satisfiability for $\mathrm{FO}^2[<]$ is NEXP-complete. Using our characterization of $\mathrm{FO}^2[<]$, Wilke observed that satisfiability becomes NP-complete if we look at binary alphabets only [21]. We generalize this observation and show that satisfiability for $\mathrm{FO}^2[<]$ is NP-complete for any fixed alphabet size. In contrast to this, satisfiability for $\mathrm{FO}^2[<, \mathrm{Suc}]$ is NEXP-complete even for binary alphabets [4], since in the presence of a successor predicate we can encode an arbitrary alphabet in binary.

Theorem 27. *Let $n \in \mathbb{N}$ and let $\varphi \in \mathrm{FO}^2_n[<]$ be a formula over a k-letter alphabet. If φ is satisfiable, then φ has a model of size $O(n^k)$.*

The proof of Theorem 27 is presented in [20]. We argue that any fixed word has as most $O(n^k)$ positions that can be reached with n-rankers, and thus we have a word of size $O(n^k)$ that satisfies the given formula.

Theorem 28. *Satisfiability for $\mathrm{FO}^2[<]$ where the size of the alphabet is bounded by some fixed $k \geq 2$ is NP-complete.*

Proof. Membership in NP follows immediately from Theorem 27 – we nondeterministically guess a model of size $O(n^k)$ where n is the quantifier depth of the given formula, and verify that it is a model of the formula. Now we give a reduction from SAT. Let α be a boolean formula in conjunctive normal form over the variables $X_1, \ldots, X_n$. We construct a $\mathrm{FO}^2[<]$ formula $\varphi = \varphi_n \wedge \alpha[\xi_i/X_i]$, where φ_n says that every model has size exactly n, and where we replace every occurrence of X_i in α with a formula ξ_i of length $O(n)$ which says that the i-th letter is a 1. The total length of φ is $O(|\alpha| \cdot n)$, and φ is satisfiable iff α is satisfiable. □

7 Conclusion

We proved precise structure theorems for FO^2, with and without the successor predicate, that completely characterize the expressive power of the respective logics, including exact bounds on the quantifier depth and on the alternation depth. Using our structure theorems, we show that the quantifier alternation hierarchy for FO^2 is strict, settling an open question from [3, 4]. Both our structure theorems and the alternation hierarchy results add further insight to and simplify previous characterizations of FO^2. We also hope that the insights gained in our study of FO^2 on words will be useful in future investigations of the trade-off between formula size and number of variables.

Acknowledgment

We would like to thank Thomas Wilke for pointing out the consequences of our structural results to the satisfiability problem for $\mathrm{FO}^2[<]$.

References

1. Adler, M., Immerman, N.: An n! lower bound on formula size. ACM Transactions on Computational Logic 4(3), 296–314 (2003)
2. Brzozowski, J., Knast, R.: The dot-depth hierarchy of star-free languages is infinite. Journal of Computer and System Science 16, 37–55 (1978)
3. Etessami, K., Vardi, M.Y., Wilke, T.: First-order logic with two variables and unary temporal logic. In: IEEE Symposium on Logic in Computer Science (1997)
4. Etessami, K., Vardi, M.Y., Wilke, T.: First-order logic with two variables and unary temporal logic. Information and Computation 179(2), 279–295 (2002)
5. Grohe, M., Schweikardt, N.: The succinctness of first-order logic on linear orders. Logical Methods in Computer Science 1, 1–6 (2005)
6. Immerman, N.: Descriptive complexity. Springer, Heidelberg (1999)
7. Immerman, N., Kozen, D.: Definability with bounded number of bound variables. Information and Computation 83(2), 121–139 (1989)
8. Kamp, J.A.: Tense logic and the theory of linear order. PhD thesis, University of California, Los Angeles (1968)
9. Karchmer, M., Wigderson, A.: Monotone circuits for connectivity require super-logarithmic depth. SIAM Journal of Discrete Mathematics 3(2), 255–265 (1990)
10. McNaughton, R., Papert, S.A.: Counter-free automata. MIT Press, Cambridge (1971)
11. Pin, J.-E., Weil, P.: Polynomial closure and unambiguous product. Theory of Computing Systems 30, 1–39 (1997)
12. Schützenberger, M.P.: Sur le produit de concatenation non ambigu. Semigroup Forum 13, 47–75 (1976)
13. Schwentick, T., Thérien, D., Vollmer, H.: Partially-ordered two-way automata: a new characterization of DA. In Developments in Language Theory (2001)
14. Straubing, H., Thérien, D.: Weakly iterated block products. In: 5th Latin American Theoretical Informatics Conference (2002)
15. Tesson, P., Thérien, D.: Diamonds are forever: the variety DA. In Semigroups, Algorithms, Automata and Languages (2001)
16. Tesson, P., Thérien, D.: Algebra meets logic: the case of regular languages. Logical Methods in Computer Science 3, 1–4 (2007)
17. Thérien, D., Wilke, T.: Over words, two variables are as powerful as one quantifier alternation. In: ACM Symposium on Theory of Computing (1998)
18. Thomas, W.: Classifying regular events in symbolic logic. Journal of Computer and System Science 25, 360–376 (1982)
19. Thomas, W.: An application of the Ehrenfeucht-Fraïssé game in formal language theory. Mémoires de la S.M.F 16, 11–21 (1984)
20. Weis, P., and Immerman, N.: Structure theorem and strict alternation hierarchy for FO^2 on words (2007), `http://www.cs.umass.edu/~immerman/pub/FO2_on_words.pdf`
21. Wilke, T.: Personal communication (2007)

On the Complexity of Reasoning About Dynamic Policies

Stefan Göller*

Universität Stuttgart, FMI, Germany
goeller@informatik.uni-stuttgart.de

Abstract. We study the complexity of satisfiability for $\mathrm{DLP}^+_{\mathrm{dyn}}$, an expressive logic introduced by Demri that allows to reason about dynamic policies. $\mathrm{DLP}^+_{\mathrm{dyn}}$ extends the logic $\mathrm{DLP}_{\mathrm{dyn}}$ of Pucella and Weissman, which in turn extends van der Meyden's Dynamic Logic of Permission (DLP). $\mathrm{DLP}^+_{\mathrm{dyn}}$ generously enhances DLP and $\mathrm{DLP}_{\mathrm{dyn}}$ by allowing to update the policy set by adding or removing policy transitions, which are defined as a direct product of two sets, each specified by a formula of the logic itself. It is proven that satisfiability for $\mathrm{DLP}^+_{\mathrm{dyn}}$ is complete for deterministic exponential time. Our results close the complexity gap of satisfiability for $\mathrm{DLP}^+_{\mathrm{dyn}}$ from 2EXP, and for $\mathrm{DLP}_{\mathrm{dyn}}$ from NEXP, to EXP respectively, matching the EXP lower bound both inherit from Propositional Dynamic Logic (PDL). To prove the EXP upper bound for $\mathrm{DLP}^+_{\mathrm{dyn}}$, we first proceed by accurately identifying a suitable generalization of PDL, which allows to use compressed programs and then find a satisfiability preserving translation from $\mathrm{DLP}^+_{\mathrm{dyn}}$ to this extension of PDL. To finally show the EXP upper bound for $\mathrm{DLP}^+_{\mathrm{dyn}}$, we prove that satisfiability of our extension of PDL lies in EXP.

1 Introduction

Numerous applications contain a set of policies that, roughly speaking, describe what is prohibited and what is permitted. Policies arise in many contexts. For one, they can comprise control policies, thus specifying which agents are permitted to access resources. They can as well be legal policies, describing actions that are legally permitted. In the course of time, the policies of an application may change. This dynamic behaviour originates from the interaction between the application and its user(s). When changing policies, the system has to guarantee that unintentional side-effects do not occur. Furthermore, it is often not straightforward to decide whether to modify the current policy set or not, not to mention the duty of creating it. Various examples of typical practical scenarios are given in [10]. In order to allow comparison of different policies and reasoning about them, a variety of languages have been introduced, an overview is given in [15]. Van der Meyden's Dynamic Logic of Permission (DLP) is an example of an expressive logic that allows to reason about dynamic policies. Formally, it extends test-free Propositional Dynamic Logic (PDL) and allows to reason about a fixed policy set that describes the set of all permitted transitions of a system which is in turn modeled by a Kripke structure. In addition to test-free PDL, the logic DLP allows to ask queries

* The author is supported by the DFG project GELO.

J. Duparc and T.A. Henzinger (Eds.): CSL 2007, LNCS 4646, pp. 358–373, 2007.

of the kind 'does there exist a sequence of solely permitted transitions to some world where the property φ holds' and 'does every sequence of transitions to worlds satisfying φ solely consist of permitted transitions'. Extending DLP, the logic $\mathrm{DLP}_{\mathrm{dyn}}$ of Pucella and Weissman [10] additionally allows to update the policy set by removing and adding transitions. These removed or added transitions are defined as the direct product of two sets of worlds, each specified by boolean combinations of atomic propositions. In [10], numerous applications are stated that demand for the possibility to update the policy set within the logic. The even more general logic $\mathrm{DLP}^+_{\mathrm{dyn}}$, introduced by Demri [3], allows to update the policy set by adding (via the `grant`-operator) and removing (via the `revoke`-operator) a direct product of world sets, but each specified by an *arbitrary* formula of the logic itself. In this paper, we focus on the computational complexity of an important algorithmic problem for $\mathrm{DLP}^+_{\mathrm{dyn}}$, namely satisfiability. For its fragment $\mathrm{DLP}_{\mathrm{dyn}}$, it is pointed out in [10] that the complexity of satisfiability is in NEXP. Focusing on naturalness, Demri gives a satisfiability preserving translation from $\mathrm{DLP}^+_{\mathrm{dyn}}$ to PDL [3]. By the presence of an exponential blowup in formula size in this translation, a 2EXP upper bound for satisfiability of $\mathrm{DLP}^+_{\mathrm{dyn}}$ was derived. However, if the depth of applied `grant` and `revoke` operators is bounded by some constant, an EXP upper bound was shown. Since the latter is the case for formulas of DLP, satisfiability for DLP was shown to be in EXP as a corollary. In this paper, we close the complexity gap for $\mathrm{DLP}^+_{\mathrm{dyn}}$ from 2EXP, and for $\mathrm{DLP}_{\mathrm{dyn}}$ from NEXP, to EXP respectively, matching the EXP lower bound both inherit from PDL. An approach proposed in [3] to improve the complexity status of full $\mathrm{DLP}^+_{\mathrm{dyn}}$ is a polynomial time computable reduction to PDL enhanced with an operator $\oplus$ on programs (we call the resulting logic PDL$\oplus$ from now on) and then proving that satisfiability of PDL$\oplus$ lies in EXP. A program $\oplus(\pi, \varphi_1, \varphi_2)$, where π is a program and both φ_1 and φ_2 are formulas, relates pairs of worlds (x, y) of a Kripke structure, that are related via π such that additionally φ_1 holds in x or φ_2 holds in y. As we will remark later, PDL$\oplus$ is definable in PDL, but the size of the programs of the resulting PDL formula may grow exponentially in the size of the programs of the original PDL$\oplus$ formula. Alas, it turns out that a translation from $\mathrm{DLP}^+_{\mathrm{dyn}}$ to PDL$\oplus$ will not lead to an improvement of the complexity of $\mathrm{DLP}^+_{\mathrm{dyn}}$ – we prove that PDL$\oplus$ is 2EXP-complete. Thus, PDL$\oplus$ identifies a translatable fragment of both PDL with intersection [2] and of PDL, where programs are represented as dags, that is already hard for 2EXP. Yet to prove an EXP upper bound for $\mathrm{DLP}^+_{\mathrm{dyn}}$, we accurately identify a fragment PDL$\oplus[\mathbb{A}]$ of PDL$\oplus$, into which we translate $\mathrm{DLP}^+_{\mathrm{dyn}}$ and that we prove to lie in EXP. Our translation from $\mathrm{DLP}^+_{\mathrm{dyn}}$ to PDL$\oplus[\mathbb{A}]$ consists of a concise examination of how applied `grant` and `revoke` operators influence the truth of subformulas. For proving that PDL$\oplus[\mathbb{A}]$ lies in EXP, we translate an input formula of PDL$\oplus[\mathbb{A}]$ φ into an alternating Büchi tree automaton over infinite trees $\mathcal{A}(\varphi)$ and check the tree language of $\mathcal{A}(\varphi)$ for non-emptiness.

Firstly, our main contribution is to prove that $\mathrm{DLP}^+_{\mathrm{dyn}}$ and thus $\mathrm{DLP}_{\mathrm{dyn}}$ is EXP-complete, which solves two open problems stated in [3]. The various applications of DLP and $\mathrm{DLP}_{\mathrm{dyn}}$, as listed in [12,10], and the technical difficulties of reasoning about dynamic policies that arise, motivate an exact examination of the complexity of satisfiability of the more general $\mathrm{DLP}^+_{\mathrm{dyn}}$. Secondly, we believe that PDL with (restrictions on) the operator $\oplus$ is an interesting logic to study w.r.t. complexity, situated between EXP and 2EXP.

The paper is organized as follows. After introducing preliminaries in Section 2, we formally define $\mathrm{DLP}^+_{\mathrm{dyn}}$ and related logics in Section 3. Known complexity results for comparable logics and the difficulty of handling $\mathrm{DLP}^+_{\mathrm{dyn}}$ are summarized and discussed in Section 4. In Section 5, we give a satisfiability preserving translation from $\mathrm{DLP}^+_{\mathrm{dyn}}$ to $\mathrm{PDL}\oplus[\mathbb{A}]$. An EXP upper bound for satisfiability of $\mathrm{PDL}\oplus[\mathbb{A}]$ is proven in Section 6. Finally, in Section 7, we show that satifiability for $\mathrm{PDL}\oplus$ is 2EXP-complete.

2 Preliminaries

If A and B are sets and $f : A \to B$ is a mapping, then for every subset $C \subseteq A$, we define $f(C) = \{f(c) \mid c \in C\}$. If $w = a_1 a_2 \cdots a_n$ is a string over the alphabet Σ and $a_i \in \Sigma$ for each $1 \leq i \leq n$, then $w^{(j)} = a_1 \cdots a_j$ denotes the *j-th prefix* of w for every $0 \leq j \leq n$, where $w^{(0)} = \varepsilon$. For a string w, let $|w|$ denote the *length* of w. If X is a set, $R \subseteq X \times X$ is a binary relation over X and $A, B \subseteq X$, then $\oplus(R, A, B) = R \cap ((A \times X) \cup (X \times B))$. If X and Y are sets with $X \cap Y = \emptyset$, then $X \uplus Y$ denotes the union of X and Y and recalls the fact that X and Y are disjoint. For every $l, k \in \mathbb{N}$, define $[k] = \{1, \ldots, k\}$, and $[l, k] = \{l, l+1, \ldots, k\}$. Let us introduce NFAs. An *NFA* is a tuple $A = (Q, \Sigma, q_0, \delta, F)$, where (i) Q is a finite set of *states*, (ii) Σ is a finite *alphabet*, (iii) $q_0 \in Q$ is an *initial state*, (iv) $F \subseteq Q$ is a set of *final states*, and (v) $\delta : Q \times \Sigma \to 2^Q$ is a *transition function*. We abbreviate $q' \in \delta(q, a)$ by $q \overset{a}{\Rightarrow}_A q'$. Next, we extend $\Rightarrow_A$ to words over Σ. For all $q \in Q$ we have $q \overset{\varepsilon}{\Rightarrow}_A q$. If $w \in \Sigma^*$, $a \in \Sigma$, $q \overset{w}{\Rightarrow}_A q'$, and $q' \overset{a}{\Rightarrow}_A q''$, then $q \overset{wa}{\Rightarrow}_A q''$. Let $L(A) = \{w \in \Sigma^* \mid q_0 \overset{w}{\Rightarrow}_A q \text{ for some } q \in F\}$ denote the *language* of A.

3 Logic

For the rest of the paper, fix some countable set of *atomic propositions* $\mathbb{P}$ and some countable set of *atomic programs* $\mathbb{A}$.

3.1 $\mathrm{DLP}^+_{\mathrm{dyn}}$ and Its Fragments $\mathrm{DLP}_{\mathrm{dyn}}$ and DLP

Formulas φ and *programs* π of the logic $\mathrm{DLP}^+_{\mathrm{dyn}}$ are given by the following grammar, where p ranges over $\mathbb{P}$ and a ranges over $\mathbb{A}$:

$$\begin{aligned} \varphi \quad &::= \quad p \mid \neg\varphi \mid \varphi_1 \vee \varphi_2 \mid \langle\pi\rangle\,\varphi \mid \mathtt{perm}(\pi)\varphi \mid \mathtt{fperm}(\pi)\varphi \mid \\ &\qquad \mathtt{grant}(\varphi_1,\varphi_2)\varphi_3 \mid \mathtt{revoke}(\varphi_1,\varphi_2)\varphi_3 \\ \pi \quad &::= \quad a \mid \pi_1 \cup \pi_2 \mid \pi_1 \circ \pi_2 \mid \pi^* \mid \varphi? \end{aligned}$$

We introduce the following abbreviations: $\varphi_1 \wedge \varphi_2 = \neg(\neg\varphi_1 \vee \neg\varphi_2)$, $\mathtt{true} = p \vee \neg p$ for some $p \in \mathbb{P}$, $\mathtt{false} = \neg\mathtt{true}$, and $[\pi]\varphi = \neg\langle\pi\rangle\neg\varphi$. Let Φ denote the set of all $\mathrm{DLP}^+_{\mathrm{dyn}}$ formulas and let $\mathrm{Test} = \{\varphi? \mid \varphi \in \Phi\}$ denote the set of all $\mathrm{DLP}^+_{\mathrm{dyn}}$ *test programs*. The logic $\mathrm{DLP}_{\mathrm{dyn}}$ is the syntactic fragment of $\mathrm{DLP}^+_{\mathrm{dyn}}$, where the first two components of `grant` and `revoke` formulas are restricted to be boolean combinations of atomic propositions and where test programs do not occur. The logic DLP is the fragment of DLP_{dyn}, where the operators `grant` and `revoke` do not occur. For every $\text{DLP}^+_{\text{dyn}}$ program π let $L(\pi)$ denote the *regular language* of π interpreted as a regular expression over some finite subset from $\mathbb{A} \cup \text{Test}$. A *Kripke structure* is a tuple $(X, \{\rightarrow_a \mid a \in \mathbb{A}\}, \varrho)$, where X is a set of *worlds*, $\rightarrow_a \subseteq X \times X$ is a binary relation for each $a \in \mathbb{A}$, and $\varrho : X \rightarrow 2^{\mathbb{P}}$ assigns to each world $x \in X$ a set of atomic propositions $\varrho(x)$. An *extended Kripke* structure is a tuple $(X, \{\rightarrow_a \mid a \in \mathbb{A}\}, \varrho, P)$, where $(X, \{\rightarrow_a \mid a \in \mathbb{A}\}, \varrho)$ is a Kripke structure and $P \subseteq X \times X$ is a binary relation that we call *policy set*. If $\text{op} \in \{\cup, \setminus\}$, $\mathcal{K} = (X, \{\rightarrow_a \mid a \in \mathbb{A}\}, \varrho, P)$ is an extended Kripke structure, and $A, B \subseteq X$, then $\mathcal{K} \upharpoonright (A, B, \text{op}) = (X, \{\rightarrow_a \mid a \in \mathbb{A}\}, \varrho, P \text{ op } (A \times B))$. Fix some extended Kripke structure $\mathcal{K} = (X, \{\rightarrow_a \mid a \in \mathbb{A}\}, \varrho, P)$. Then, for each atomic/test program π, we define a binary relation $[\![\pi]\!]_{\mathcal{K}} \subseteq X \times X$ and for each formula $\varphi \in \Phi$ we define a subset $[\![\varphi]\!]_{\mathcal{K}} \subseteq X$ inductively as follows, where $p \in \mathbb{P}$ and $a \in \mathbb{A}$:

$$
\begin{aligned}
[\![a]\!]_{\mathcal{K}} &= \rightarrow_a \\
[\![\varphi?]\!]_{\mathcal{K}} &= \{(x, x) \mid x \in [\![\varphi]\!]_{\mathcal{K}}\} \\
[\![p]\!]_{\mathcal{K}} &= \{x \in X \mid p \in \varrho(x)\} \\
[\![\neg\varphi]\!]_{\mathcal{K}} &= X \setminus [\![\varphi]\!]_{\mathcal{K}} \\
[\![\varphi_1 \vee \varphi_2]\!]_{\mathcal{K}} &= [\![\varphi_1]\!]_{\mathcal{K}} \cup [\![\varphi_2]\!]_{\mathcal{K}}
\end{aligned}
$$

$$
\begin{aligned}
[\![\langle\pi\rangle\varphi]\!]_{\mathcal{K}} = \{x \in X \mid{} & \text{there exist } x_0, x_1, \ldots, x_n \in X \text{ with} \\
& x = x_0, x_n \in [\![\varphi]\!]_{\mathcal{K}}, \text{ and there exist} \\
& A_1 \cdots A_n \in L(\pi) \text{ s.t. for all } 1 \leq i \leq n \\
& \text{we have } (x_{i-1}, x_i) \in [\![A_i]\!]_{\mathcal{K}}\}
\end{aligned}
$$

$$
\begin{aligned}
[\![\texttt{perm}(\pi)\varphi]\!]_{\mathcal{K}} = \{x \in X \mid{} & \text{there exist } x_0, x_1, \ldots, x_n \in X \text{ with} \\
& x = x_0, x_n \in [\![\varphi]\!]_{\mathcal{K}}, \text{ and there exist} \\
& A_1 \cdots A_n \in L(\pi) \text{ s.t. for all } 1 \leq i \leq n \\
& \text{we have } (x_{i-1}, x_i) \in [\![A_i]\!]_{\mathcal{K}} \text{ and} \\
& A_i \in \mathbb{A} \text{ implies } (x_{i-1}, x_i) \in P\}
\end{aligned}
$$

$$
\begin{aligned}
[\![\texttt{fperm}(\pi)\varphi]\!]_{\mathcal{K}} = \{x \in X \mid{} & \text{there does not exist} \\
& x_0, x_1, \ldots, x_n \in X \text{ and } A_1 \cdots A_n \in L(\pi) \text{ with} \\
& x = x_0, x_n \in [\![\varphi]\!]_{\mathcal{K}} \text{ and } (x_{i-1}, x_i) \in [\![A_i]\!]_{\mathcal{K}} \\
& \text{for all } 1 \leq i \leq n \text{ such that } A_j \in \mathbb{A} \text{ and} \\
& (x_{j-1}, x_j) \notin P \text{ for some } 1 \leq j \leq n\}
\end{aligned}
$$

$$
\begin{aligned}
[\![\texttt{grant}(\varphi_1, \varphi_2)\varphi_3]\!]_{\mathcal{K}} &= [\![\varphi_3]\!]_{\mathcal{K}\upharpoonright([\![\varphi_1]\!]_{\mathcal{K}}, [\![\varphi_2]\!]_{\mathcal{K}}, \cup)} \\
[\![\texttt{revoke}(\varphi_1, \varphi_2)\varphi_3]\!]_{\mathcal{K}} &= [\![\varphi_3]\!]_{\mathcal{K}\upharpoonright([\![\varphi_1]\!]_{\mathcal{K}}, [\![\varphi_2]\!]_{\mathcal{K}}, \setminus)}
\end{aligned}
$$

We abbreviate $x \in [\![\varphi]\!]_{\mathcal{K}}$ by $(\mathcal{K}, x) \models \varphi$. If for some state $x \in X$ we have $(\mathcal{K}, x) \models \varphi$ then $\mathcal{K}$ is a *model* for φ. We say that a formula φ is *satisfiable* if there exists some model for φ. The size $|\varphi|$ of a $\text{DLP}^+_{\text{dyn}}$ formula φ and the size $|\pi|$ of a $\text{DLP}^+_{\text{dyn}}$ program π is inductively defined as follows: $|p| = |a| = 1$ for every $p \in \mathbb{P}$ and for every $a \in \mathbb{A}$, $|\varphi_1 \vee \varphi_2| = |\varphi_1| + |\varphi_2| + 1$, $|\neg\varphi| = |\varphi| + 1$, $|\langle\pi\rangle\varphi| = |\pi| + |\varphi| + 1$, $|\texttt{perm}(\pi)\varphi| = |\texttt{fperm}(\pi)\varphi| = |\pi| + |\varphi| + 1$, $|\texttt{grant}(\varphi_1, \varphi_2)\varphi_3| = |\texttt{revoke}(\varphi_1, \varphi_2)\varphi_3| = |\varphi_1| + |\varphi_2| + |\varphi_3| + 1$, $|\pi_1 \cup \pi_2| = |\pi_1 \circ \pi_2| = |\pi_1| + |\pi_2| + 1$, $|\pi^*| = |\pi| + 1$, and $|\varphi?| = |\varphi| + 1$. The set $\text{subf}(\varphi)$ of *subformulas* of a formula φ and the set of *subformulas* $\text{subf}(\pi)$ of a program π is inductively defined as follows: (i) $\text{subf}(p) = \{p\}$ for all $p \in \mathbb{P}$, (ii) $\text{subf}(\neg\varphi) = \{\neg\varphi\} \cup \text{subf}(\varphi)$, (iii) $\text{subf}(\varphi_1 \vee \varphi_2) = \{\varphi_1 \vee \varphi_2\} \cup \text{subf}(\varphi_1) \cup \text{subf}(\varphi_2)$, (iv) if $\varphi = \langle\pi\rangle\psi$, $\varphi = \texttt{perm}(\pi)\psi$, or $\varphi = \texttt{fperm}(\pi)\psi$, then $\text{subf}(\varphi) = \{\varphi\} \cup \text{subf}(\pi) \cup \text{subf}(\psi)$, (v) if $\varphi = \texttt{grant}(\varphi_1, \varphi_2)\varphi_3$ or $\varphi = \texttt{revoke}(\varphi_1, \varphi_2)\varphi_3$, then $\text{subf}(\varphi) = \{\varphi\} \cup \bigcup_{i=1}^{3} \text{subf}(\varphi_i)$, (vi) $\text{subf}(a) = \emptyset$ for all $a \in \mathbb{A}$, (vii) $\text{subf}(\pi_1 \cup \pi_2) = \text{subf}(\pi_1 \circ \pi_2) = \text{subf}(\pi_1) \cup \text{subf}(\pi_2)$, $\text{subf}(\pi^*) = \text{subf}(\pi)$, and finally (viii) $\text{subf}(\varphi?) = \text{subf}(\varphi)$.

3.2 PDL and Its Compressed Variants PDL⊕ and PDL⊕[$\mathbb{A}$]

Formulas φ and *programs* π of the logic PDL⊕ are given by the following grammar, where p ranges over $\mathbb{P}$ and a ranges over $\mathbb{A}$:

$$\begin{array}{lll} \varphi & ::= & p \mid \neg\varphi \mid \varphi_1 \vee \varphi_2 \mid \langle\pi\rangle\,\varphi \\ \pi & ::= & a \mid \pi_1 \cup \pi_2 \mid \pi_1 \circ \pi_2 \mid \pi^* \mid \varphi? \mid \oplus(\pi, \varphi_1, \varphi_2) \end{array}$$

Formulas and programs of PDL⊕ are interpreted in Kripke structures (hence policy sets do not occur). If $\mathcal{K} = (X, \{\rightarrow_a \mid a \in \mathbb{A}\}, \varrho)$ is a Kripke structure, then for each program π, we can assign a binary relation $[\![\pi]\!]_{\mathcal{K}} \subseteq X \times X$, which is defined homomorphic for $\cup$, $\circ$, and $*$, and where the semantics of the program operator $\oplus$ is defined as $[\![\oplus(\pi, \varphi_1, \varphi_2)]\!]_{\mathcal{K}} = \oplus([\![\pi]\!]_{\mathcal{K}}, [\![\varphi_1]\!]_{\mathcal{K}}, [\![\varphi_2]\!]_{\mathcal{K}})$. The size of $\oplus(\pi, \varphi_1, \varphi_2)$ is defined as $|\oplus(\pi, \varphi_1, \varphi_2)| = 1 + |\pi| + |\varphi_1| + |\varphi_2|$. If $\pi = \oplus(\pi', \varphi_1, \varphi_2)$, then the set of subformulas $\text{subf}(\pi)$ is defined as $\text{subf}(\pi) = \text{subf}(\pi') \cup \text{subf}(\varphi_1) \cup \text{subf}(\varphi_2)$. Similarly as above, a Kripke structure $(X, \{\rightarrow_a \mid a \in \mathbb{A}\}, \varrho)$ is a *model* for a PDL⊕ formula φ, if for some state $x \in X$ we have $(\mathcal{K}, x) \models \varphi$. A PDL⊕ formula φ is *satisfiable*, if there exists some model for φ. The logic PDL⊕[$\mathbb{A}$] is the syntactic fragment of PDL⊕, where the program arguments of a $\oplus$ program must either be an atomic program or a $\oplus$ program itself. More formally, *formulas* φ, *basic programs* α and *programs* π of PDL⊕[$\mathbb{A}$] are given by the following grammar, where p ranges over $\mathbb{P}$ and a ranges over $\mathbb{A}$:

$$\begin{array}{lll} \varphi & ::= & p \mid \neg\varphi \mid \varphi_1 \vee \varphi_2 \mid \langle\pi\rangle\,\varphi \\ \alpha & ::= & a \mid \oplus(\alpha, \varphi_1, \varphi_2) \\ \pi & ::= & \alpha \mid \pi_1 \cup \pi_2 \mid \pi_1 \circ \pi_2 \mid \pi^* \mid \varphi? \end{array}$$

The logic PDL is the syntactic fragment of PDL⊕ without the $\oplus$ operator.

Remark 1. A PDL⊕ program $\oplus(\pi, \varphi_1, \varphi_2)$ can be translated into an equivalent PDL program $\|\oplus(\pi, \varphi_1, \varphi_2)\|$ as follows, where $\|\pi\|$, $\|\varphi_1\|$, and $\|\varphi_2\|$ are inductively the translations for π, φ_1, and φ_2 respectively:

$$\|\oplus(\pi, \varphi_1, \varphi_2)\| \quad = \quad (\|\varphi_1\|? \circ \|\pi\| \cup \|\pi\| \circ \|\varphi_2\|?)$$

Hence, the logics PDL$\oplus$, PDL$\oplus[\mathbb{A}]$, and PDL are equi-expressive. However, by the presence of $\oplus$ programs, the above translation might cause an exponential blowup. Thus, the three logics may differ in succintness. In fact, our 2EXP-completeness result for PDL$\oplus$ shown in Section 7 implies that there does not exist a satisfiability preserving translation from PDL$\oplus$ to PDL that is computable in polynomial time.

4 Known Results and Difficulties of Reasoning About DLP$^+_{\text{dyn}}$ and Related Logics

In this section, we state known results and explain some difficulties of determining the complexity of reasoning about DLP$^+_{\text{dyn}}$ and logics related to it.

The *satisfiability problem* asks, given a formula φ of any of the logics introduced above, whether φ is satisfiable. By Fischer/Ladner and Pratt it is known:

Theorem 1 ([4,9]). *Satisfiability for PDL is* EXP*-complete.*

Focusing on a natural translation from DLP$^+_{\text{dyn}}$ to PDL, Demri has recently shown:

Theorem 2 ([3]). *There exists a satisfiability preserving reduction from DLP$^+_{\text{dyn}}$ to PDL computable in polynomial time.*

Since the size of the resulting PDL formula in the proof of Theorem 2 is exponential in the size of the original DLP$^+_{\text{dyn}}$ formula, the following theorem holds.

Theorem 3 ([3]). *Satisfiability for DLP$^+_{\text{dyn}}$ is in* 2EXP.

For its fragment DLP$_{\text{dyn}}$, the following upper bound is known.

Theorem 4 ([10]). *Satisfiability for DLP$_{\text{dyn}}$ is in* NEXP .

Let us summarize the difficulties of identifying the exact complexity of DLP$^+_{\text{dyn}}$. By the presence of the operators `grant` and `revoke`, the truth of a formula may depend on the truth of a subformula in some modified extended Kripke structure. This behaviour is very much in the flavor of sabotage modal logic of van Benthem [11]. The latter is basically modal logic enhanced with a sabotage operator $\langle - \rangle$. A formula $\langle - \rangle\varphi$ holds in a world x of a Kripke structure $\mathcal{K}$ with state set X, if φ holds in x in $\mathcal{K}'$, where $\mathcal{K}'$ is a Kripke structure that emerges from $\mathcal{K}$ by removing some world from $X \setminus \{x\}$. The decidability status for sabotage modal logic is unknown so far. Yet, a variant of it, in which the sabotage operator requires to remove some labeled transition instead of some world, has been proven undecidable by Löding and Rohde [7]. At first glance, one could think that the situation for DLP$^+_{\text{dyn}}$ is even worse, since transitions can be removed *and* added and since moreover these transitions can be specified in the logic itself. Interestingly, it turns out, that precisely the fact that the updated transitions are specified in the logic itself, allows translations to other logics that are more manageable w.r.t. satisfiability. So, one promising approch to decide DLP$^+_{\text{dyn}}$ in EXP was a translation into PDL with intersection and negation of atomic programs given in [3]. This translation has the property that the width of nested intersection of the resulting formulas is bounded by some constant. Firstly, a precise analysis of [5] yields that satisfiability of PDL with the intersection on programs, where formulas have bounded intersection width, is in EXP. Secondly, a result from [8] states that PDL with negation of atomic programs is in EXP too. But unfortunately, PDL with intersection *and* negation of atomic programs has proven to be undecidable recently [5], even when restricting to formulas of constant intersection width. A proposal of Demri [3] to improve the 2EXP upper bound for $\mathrm{DLP}^+_{\mathrm{dyn}}$ is to give a polynomial translation from $\mathrm{DLP}^+_{\mathrm{dyn}}$ to PDL$\oplus$ and to try to decide satisfiability of PDL$\oplus$ in EXP. However, we will prove in Section 7, that PDL$\oplus$ is complete for 2EXP.

Wrapping up, all mentioned translations have the drawback that either the target logic was too hard w.r.t. complexity or the translations have a blowup in formula size. Nevertheless, our solution to decide $\mathrm{DLP}^+_{\mathrm{dyn}}$ in deterministic exponential time is to find a tricky translation into PDL$\oplus$'s adequate fragment PDL$\oplus[\mathbb{A}]$ and to show that PDL$\oplus[\mathbb{A}]$ is in EXP. By combining the EXP-hardness $\mathrm{DLP}^+_{\mathrm{dyn}}$ inherits from PDL [4], we state the main result of this paper.

Theorem 5. *Satisfiability for $\mathrm{DLP}^+_{\mathrm{dyn}}$ is* EXP*-complete.*

As stated above, for proving Theorem 5, we first give a satisfiability preserving translation from $\mathrm{DLP}^+_{\mathrm{dyn}}$ to PDL$\oplus[\mathbb{A}]$, that can be computed in polynomial time in Section 5. An EXP upper bound for PDL$\oplus[\mathbb{A}]$ is proven in Section 6.

5 A Translation from $\mathrm{DLP}^+_{\mathrm{dyn}}$ to PDL$\oplus[\mathbb{A}]$

In this section, we give a satisfiability preserving translation from $\mathrm{DLP}^+_{\mathrm{dyn}}$ to PDL$\oplus[\mathbb{A}]$ that is computable in polynomial time. In this translation, a precise analysis of how applied `grant` and `revoke` operators influence the truth of subformulas, allows to handle $\mathrm{DLP}^+_{\mathrm{dyn}}$. In parts, it combines some ideas from [3] and [6].

First, we introduce a notion of certain modified Kripke structures. Recall that Φ denotes the set of all $\mathrm{DLP}^+_{\mathrm{dyn}}$ formulas. Let $\Sigma = (\Phi \times \Phi \times \{\cup, \setminus\})^*$. For each $\sigma = (\theta_1, \theta'_1, \mathrm{op}_1) \cdots (\theta_k, \theta'_k, \mathrm{op}_k) \in \Sigma$, where $k \geq 0$, let $U_\sigma = \{m \in [k] \mid \mathrm{op}_m = \cup\}$ and $M_\sigma = \{m \in [k] \mid \mathrm{op}_m = \setminus\}$. If $\mathcal{K}$ is an extended Kripke structure, then for every $\sigma \in \Sigma$ define the extended Kripke structure $\mathcal{K} \restriction \sigma$, by the length of σ, inductively as follows: $\mathcal{K} \restriction \varepsilon = \mathcal{K}$ and $\mathcal{K} \restriction \sigma(\psi_1, \psi_2, \mathrm{op}) = (\mathcal{K} \restriction \sigma) \restriction ([\![\psi_1]\!]_{\mathcal{K}\restriction\sigma}, [\![\psi_2]\!]_{\mathcal{K}\restriction\sigma}, \mathrm{op})$ for all $\psi_1, \psi_2 \in \Phi$ and $\mathrm{op} \in \{\cup, \setminus\}$.

Remark 2. If $\mathcal{K}$ is an extended Kripke structure and $\sigma \in \Sigma$, then the extended Kripke structures $\mathcal{K}$ and $\mathcal{K} \restriction \sigma$ can only differ in their policy set. Thus, for all $a \in \mathbb{A}$ we have $[\![a]\!]_{\mathcal{K}} = [\![a]\!]_{\mathcal{K}\restriction\sigma}$ and for all $p \in \mathbb{P}$ we have $[\![p]\!]_{\mathcal{K}} = [\![p]\!]_{\mathcal{K}\restriction\sigma}$.

For a formula $\varphi \in \Phi$, let $\mathrm{Occ}(\varphi)$ denote the set of all occurrences of subformulas of φ and for each $\psi \in \mathrm{Occ}(\varphi)$, define the unique sequence $\sigma(\psi) \in \Sigma$ that we get by considering the `grant` and `revoke` operators that occur along the path from φ to ψ in the syntax tree of φ. We define, in a top down manner, $\sigma : \mathrm{Occ}(\varphi) \to \Sigma$ as follows:

- $\sigma(\varphi) = \varepsilon$
- If $\psi = \neg\chi$, then $\sigma(\chi) = \sigma(\psi)$.
- If $\psi = \psi_1 \vee \psi_2$, then $\sigma(\psi_1) = \sigma(\psi_2) = \sigma(\psi)$.

- If $\psi = \langle\pi\rangle\chi$, $\psi = \mathtt{perm}(\pi)\chi$, or $\psi = \mathtt{fperm}(\pi)\chi$, then $\sigma(\chi) = \sigma(\psi') = \sigma(\psi)$ for every test program $\psi'?$ of π.
- If $\psi = \mathtt{grant}(\psi_1, \psi_2)\chi$, then $\sigma(\psi_1) = \sigma(\psi_2) = \sigma(\psi)$, and $\sigma(\chi) = \sigma(\psi)(\psi_1, \psi_2, \cup)$.
- If $\psi = \mathtt{revoke}(\psi_1, \psi_2)\chi$, then $\sigma(\psi_1) = \sigma(\psi_2) = \sigma(\psi)$, and $\sigma(\chi) = \sigma(\psi)(\psi_1, \psi_2, \setminus)$.

Before giving the translation from $\mathrm{DLP}^+_{\mathrm{dyn}}$ to $\mathrm{PDL}\oplus[\mathbb{A}]$, the following lemma characterizes the policy set of $\mathcal{K} \restriction \sigma$, for every $\sigma \in \Sigma$. Its proof is by induction on k.

Lemma 1. *Let $\mathcal{K} = (X, \{\rightarrow_a \mid a \in \mathbb{A}\}, \varrho, P)$ be an extended Kripke structure, $\sigma = (\theta_1, \theta'_1, \mathrm{op}_1) \cdots (\theta_k, \theta'_k, \mathrm{op}_k) \in \Sigma$ (with $k \geq 0$), $\mathcal{K} \restriction \sigma = (X, \{\rightarrow_a \mid a \in \mathbb{A}\}, \varrho, P_\sigma)$ and $a \in \mathbb{A}$.*

Then, for all $(x, y) \in X \times X$, we have $(x, y) \in [\![a]\!]_{\mathcal{K}\restriction\sigma} \cap P_\sigma$ if and only if there exists some $u \in \{0\} \uplus U_\sigma$ such that for all $m \in M_\sigma \cap [u, k]$ we have $(x, y) \in [\![\oplus(a, \neg\theta_m, \neg\theta'_m)]\!]_{\mathcal{K}\restriction\sigma^{(m-1)}}$ and either (i) $u = 0$ and $(x, y) \in P \cap [\![a]\!]_{\mathcal{K}}$, or (ii) $u \in U_\sigma$ and $(x, y) \in [\![a]\!]_{\mathcal{K}} \cap ([\![\theta_u]\!]_{\mathcal{K}\restriction\sigma^{(u-1)}} \times [\![\theta'_u]\!]_{\mathcal{K}\restriction\sigma^{(u-1)}})$.

Conversely, for all $(x, y) \in X \times X$, we have $(x, y) \in [\![a]\!]_{\mathcal{K}\restriction\sigma} \setminus P_\sigma$ if and only if there exists some $m \in \{0\} \uplus M_\sigma$ such that for all $u \in U_\sigma \cap [m, k]$ we have $(x, y) \in [\![\oplus(a, \neg\theta_u, \neg\theta'_u)]\!]_{\mathcal{K}\restriction\sigma^{(u-1)}}$ and either (i) $m = 0$ and $(x, y) \in [\![a]\!]_{\mathcal{K}} \setminus P$, or (ii) $m \in M_\sigma$ and $(x, y) \in [\![a]\!]_{\mathcal{K}} \cap ([\![\theta_m]\!]_{\mathcal{K}\restriction\sigma^{(m-1)}} \times [\![\theta'_m]\!]_{\mathcal{K}\restriction\sigma^{(m-1)}})$.

Let us turn to our translation. For this, fix some $\mathrm{DLP}^+_{\mathrm{dyn}}$ formula φ over atomic programs A and over atomic propositions P for the rest of this section. Let $\{\psi_1, \ldots, \psi_n\}$ be an enumeration of $\mathrm{Occ}(\varphi)$ and assume $\psi_n = \varphi$. Let $\mathsf{A}' = \mathsf{A} \uplus \overline{\mathsf{A}}$ and $\mathsf{P}' = \mathsf{P} \uplus \{p_1, \ldots, p_n\}$. Below, we also write $p(\psi_i)$ for p_i whenever $\psi_i \in \mathrm{Occ}(\varphi)$. If $\sigma = (\theta_1, \theta'_1, \mathrm{op}_1) \cdots (\theta_k, \theta'_k, \mathrm{op}_k) \in \sigma(\mathrm{Occ}(\varphi))$, then for every subset $S = \{j_1, \ldots, j_l\} \subseteq [k]$, where $j_1 < j_2 < \cdots < j_l$, and every $\mathrm{PDL}\oplus[\mathbb{A}]$ program ζ, define the $\mathrm{PDL}\oplus$ program $\zeta^S = \zeta^S_l$, where $\zeta^S_0 = \zeta$ and $\zeta^S_h = \oplus(\zeta^S_{h-1}, \neg p(\theta_{j_h}), \neg p(\theta'_{j_h}))$ for all $1 \leq h \leq l$. We will build a $\mathrm{PDL}\oplus[\mathbb{A}]$ formula φ' over the atomic programs A' and over the atomic propositions P' such that φ is satisfiable if and only if φ' is satisfiable. Intuitively, if $\mathcal{K}$ is a model for φ with policy set P and $\mathcal{K}'$ is a model of φ', think of the relation $[\![a]\!]_{\mathcal{K}'}$ as $[\![a]\!]_{\mathcal{K}} \cap P$ and of $[\![\overline{a}]\!]_{\mathcal{K}'}$ as $[\![a]\!]_{\mathcal{K}} \setminus P$. Let us first, for every $\psi_i \in \mathrm{Occ}(\varphi)$, define the $\mathrm{PDL}\oplus[\mathbb{A}]$ formula $\|\psi_i\|$ over the atomic propositions P' and over the atomic programs A' inductively as follows:

- If $\psi_i = p$ for some $p \in \mathsf{P}$, then $\|\psi_i\| = p$.
- If $\psi_i = \neg\psi_j$, then $\|\psi_i\| = \neg p_j$.
- If $\psi_i = \psi_j \vee \psi_k$, then $\|\psi_i\| = p_j \vee p_k$.
- If $\psi_i = \mathtt{grant}(\psi_j, \psi_k)\psi_l$ or $\psi_i = \mathtt{revoke}(\psi_j, \psi_k)\psi_l$, then $\|\psi_i\| = p_l$.
- If $\psi_i = \langle\pi\rangle\psi_j$, then $\|\psi_i\| = \langle T(\pi)\rangle p_j$, where $T(\pi)$ is homomorphic on $\cup, \circ$, and on $*$, $T(\psi_k?) = p_k?$ for every test program $\psi_k?$, and $T(a) = a \cup \overline{a}$ for every $a \in \mathsf{A}$.
- Assume $\psi_i = \mathtt{perm}(\pi)\psi_j$ and $\sigma = \sigma(\psi_i) = (\theta_1, \theta'_1, \mathrm{op}_1) \cdots (\theta_k, \theta'_k, \mathrm{op}_k)$. Recall that $U_\sigma = \{u \in [k] \mid \mathrm{op}_u = \cup\}$ and $M_\sigma = \{m \in [k] \mid \mathrm{op}_m = \setminus\}$. Then, we define $\|\psi_i\| = \langle T^\forall(\pi)\rangle p_j$, where $T^\forall(\pi)$ is homomorphic on $\cup, \circ$, and on $*$,

$T^{\forall}(\psi_k?) = p_k?$ for every test program $\psi_k?$. For every $a \in \mathsf{A}$, we define [1]:

$$T^{\forall}(a) \quad = \quad a^{M_\sigma} \cup \bigcup_{u \in U_\sigma} p(\theta_u)? \circ (a \cup \overline{a})^{M_\sigma \cap [u,k]} \circ p(\theta'_u)?$$

Via application of Lemma 1, the intention behind $T^{\forall}(a)$ is to describe the relation $[\![a]\!]_{\mathcal{K}\restriction\sigma} \cap P_\sigma$, where P_σ is the policy set of $\mathcal{K} \restriction \sigma$.

– Assume $\psi_i = \mathtt{fperm}(\pi)\psi_j$ and $\sigma = \sigma(\psi_i) = (\theta_1, \theta'_1, \mathrm{op}_1) \cdots (\theta_k, \theta'_k, \mathrm{op}_k)$. Then, we define $\|\psi_i\| = \neg\langle T^{\exists}(\pi)\rangle p_j$, where $T^{\exists}(\pi)$ is inductively defined as follows:
  - $T^{\exists}(\pi_1 \cup \pi_2) = T^{\exists}(\pi_1) \cup T^{\exists}(\pi_2)$
  - $T^{\exists}(\pi_1 \circ \pi_2) = T^{\exists}(\pi_1) \circ T(\pi_2) \ \cup \ T(\pi_1) \circ T^{\exists}(\pi_2)$, where T is defined as above.
  - $T^{\exists}(\pi^*) = T(\pi^*) \circ T^{\exists}(\pi) \circ T(\pi^*)$, where T is defined as above.
  - $T^{\exists}(\psi_k?) = \mathtt{false}?$
  - For every $a \in \mathsf{A}$, we define[1]:

$$T^{\exists}(a) \quad = \quad \overline{a}^{U_\sigma} \cup \bigcup_{m \in M_\sigma} p(\theta_m)? \circ (a \cup \overline{a})^{U_\sigma \cap [m,k]} \circ p(\theta'_m)?$$

Via application of Lemma 1, the intention behind $T^{\exists}(a)$ is to describe the relation $[\![a]\!]_{\mathcal{K}\restriction\sigma} \setminus P_\sigma$, where P_σ is the policy set of $\mathcal{K} \restriction \sigma$.

Recall $\varphi = \psi_n$. By identifying, for every subset $X \subseteq \mathsf{A} \cup \overline{\mathsf{A}}$ the program X with $\bigcup_{x \in X} x$, we define

$$\varphi' \quad = \quad p_n \wedge [(\mathsf{A} \cup \overline{\mathsf{A}})^*] \bigwedge_{i \in [n]} (p_i \leftrightarrow \|\psi_i\|).$$

Note that expressing $\leftrightarrow$ with $\neg$ and $\vee$ only roughly doubles the size of the formula. The following upper bound on the size of φ' is not difficult to see.

Lemma 2. $|\varphi'| \in \mathcal{O}(|\varphi|^3)$.

The correctness of the above translation follows from the following lemma.

Lemma 3. *The formula φ is satisfiable if and only if φ' is satisfiable.*

By combining Lemma 2 and Lemma 3, we can deduce the following theorem.

Theorem 6. *There exists a satisfiability preserving translation from DLP^+_{dyn} to PDL$\oplus[\mathbb{A}]$, which is computable in polynomial time.*

---

[1] To avoid lengthy notations, the programs $T^{\forall}(a)$ and $T^{\exists}(a)$ for $a \in \mathsf{A}$ are not PDL$\oplus[\mathbb{A}]$ programs (since the $\oplus$ operator is applied to a non-atomic programs of the kind $a \cup \overline{a}$, where $a \in \mathsf{A}$). However, the program $T^{\forall}(a)$ can be rewritten as

$$a^{M_\sigma} \cup \bigcup_{u \in U_\sigma} p(\theta_u)? \circ a^{M_\sigma \cap [u,k]} \circ p(\theta'_u)? \cup \bigcup_{u \in U_\sigma} p(\theta_u)? \circ \overline{a}^{M_\sigma \cap [u,k]} \circ p(\theta'_u)?$$

which is a PDL$\oplus[\mathbb{A}]$ program. We can proceed analogously for $T^{\exists}(a)$.

6 Satisfiability for PDL⊕[$\mathbb{A}$]

In this section, our goal is to prove that satisfiability of PDL⊕[$\mathbb{A}$] is in EXP. So for the rest of this section, fix some input PDL⊕[$\mathbb{A}$] formula φ over atomic propositions P and over atomic programs A. To decide satisfiability of φ, we translate φ into an alternating Büchi tree automaton $\mathcal{A}(\varphi)$ that accepts exactly the set of so called *Hintikka trees* for φ. Roughly speaking, Hintikka trees for φ summarize relevant information of models of φ. So the satisfiability of φ reduces to non-emptiness of the tree language of $\mathcal{A}(\varphi)$. We follow a similar approach as in [8]. Although in the following the $\neg$-operator may occur in front of non-atomic formulas, we implictly assume w.l.o.g. that every occurring PDL⊕[$\mathbb{A}$] formula is in *negation normal form*, i.e. negations occur only in front of atomic propositions. This can be achieved by introducing the operators $\wedge$ and [ ] and by abbreviating $\neg(\varphi_1 \vee \varphi_2)$ by $\neg\varphi_1 \wedge \neg\varphi_2$ and $\neg\langle\pi\rangle\psi$ by $[\pi]\neg\psi$. Thus, we associate $\neg\neg\psi$ with ψ. We call formulas of the kind $\langle\pi\rangle\psi$ *diamond formulas* and formulas of the kind $[\pi]\psi$ *box formulas*. The definition of subf is straightforward. As for $\mathrm{DLP}^+_{\mathrm{dyn}}$, for every program π that occurs in φ, we can associate a regular language $L(\pi)$ over the alphabet $\Sigma(\pi)$, where the latter consists of the basic programs and the test programs that occur in π. Moreover, we assume that the programs π, that occur in φ, are given by NFAs $A(\pi)$ over the alphabet $\Sigma(\pi)$, i.e. $L(A(\pi)) = L(\pi)$. Moreover, if $\mathcal{K}$ is a Kripke structure and $w = w_1 \cdots w_n$ with $w_i \in \Sigma(A(\pi))$ for all $i \in [n]$, then $[\![w]\!]_{\mathcal{K}} = [\![w_1]\!]_{\mathcal{K}} \circ \cdots \circ [\![w_n]\!]_{\mathcal{K}}$. If $A = (Q, \Sigma, q_0, \delta, F)$ is an NFA and $q \in Q$, then $A_q = (Q, \Sigma, q, \delta, F)$ denotes the same NFA as A, but with initial state q. If we do not explicitly define A, then $Q(A)$ denotes the state set of A, $\Sigma(A)$ denotes the alphabet of A, $q_0(A)$ denotes the initial state of A, and $F(A)$ denotes the set of final states of A respectively.
We start by introducing the *closure* of φ analogously as in [8,14,4].

Definition 1. *The* closure $\mathsf{cl}(\varphi)$ *of* φ *is the smallest set such that:*

- $\varphi \in \mathsf{cl}(\varphi)$,
- *if* $\chi \in \mathrm{subf}(\psi)$ *for some* $\psi \in \mathsf{cl}(\varphi)$, *then* $\chi \in \mathsf{cl}(\varphi)$,
- *if* $\psi \in \mathsf{cl}(\varphi)$, *then* $\neg\psi \in \mathsf{cl}(\varphi)$,
- *if* $\langle A\rangle\psi \in \mathsf{cl}(\varphi)$, *then* $\chi \in \mathsf{cl}(\varphi)$ *for all* $\chi? \in \Sigma(A)$,
- *if* $\langle A\rangle\psi \in \mathsf{cl}(\varphi)$, $\alpha \in \Sigma(A)$ *is a basic program, and* $\chi \in \mathrm{subf}(\alpha)$, *then* $\chi \in \mathsf{cl}(\varphi)$,
- *if* $\langle A\rangle\psi \in \mathsf{cl}(\varphi)$, *then for all* $q \in Q(A)$ *we have* $\langle A_q\rangle\psi \in \mathsf{cl}(\varphi)$,
- *if* $[A]\psi \in \mathsf{cl}(\varphi)$, *then* $\chi \in \mathsf{cl}(\varphi)$ *for all* $\chi? \in \Sigma(A)$,
- *if* $[A]\psi \in \mathsf{cl}(\varphi)$, $\alpha \in \Sigma(A)$ *is a basic program, and* $\chi \in \mathrm{subf}(\alpha)$, *then* $\chi \in \mathsf{cl}(\varphi)$, *and finally*
- *if* $[A]\psi \in \mathsf{cl}(\varphi)$, *then for all* $q \in Q(A)$ *we have* $[A_q]\psi \in \mathsf{cl}(\varphi)$.

It is straightforward to verify, that the size of $\mathsf{cl}(\varphi)$ is polynomial in the size of φ. Let us now introduce Hintikka sets for φ. These are subsets of $\mathsf{cl}(\varphi)$, that satisfy certain closure properties.

Definition 2. *A subset* $M \subseteq \mathsf{cl}(\varphi)$ *is a* Hintikka set for φ, *if the following five closure properties are satisfied:*

(C1) *if* $\chi_1 \wedge \chi_2 \in M$*, then* $\chi_1, \chi_2 \in M$*,*
(C2) *if* $\chi_1 \vee \chi_2 \in M$*, then* $\chi_1 \in M$ *or* $\chi_2 \in M$*,*
(C3) $\psi \in M$ *if and only if* $\neg\psi \notin M$ *for all* $\psi \in \mathsf{cl}(\varphi)$*,*
(C4) *if* $[A]\chi \in M$ *and* $q_0(A) \in F(A)$*, then* $\chi \in M$*,*
(C5) *if* $[A]\chi \in M$*, then for all* $q \in Q(A)$ *and all* $\chi'? \in \Sigma(A)$ *with* $q_0(A) \overset{\chi'?}{\Rightarrow}_A q$*, we have* $\neg\chi' \in M$ *or* $[A_q]\chi \in M$*.*

Recall that in any Hintikka set M for φ, by the presence of (C3), for all $\psi \in \mathsf{cl}(\varphi)$, we either have $\psi \in M$ or $\neg\psi \in M$ (but not both). Let $\mathcal{B}$ denote the set of all basic programs that occur in some diamond formula or in some box formula from $\mathsf{cl}(\varphi)$. For each $\alpha \in \mathcal{B}$ inductively define $\mathrm{At}(\alpha) = a$ if $\alpha = a \in \mathbb{A}$ and $\mathrm{At}(\alpha) = \mathrm{At}(\beta)$ if $\alpha = \oplus(\beta, \varphi_1, \varphi_2)$. For handling nestings of applied $\oplus$-operators in basic programs, we introduce, for basic programs $\alpha \in \mathcal{B}$, an appropriate notion of whether α holds between two subsets of $\mathsf{cl}(\varphi)$.

Definition 3. *For all* $\alpha \in \mathcal{B}$ *and all subsets* $S, T \subseteq \mathsf{cl}(\varphi)$ *for* φ*, we define the relation* $(S,T) \models \alpha$ *inductively by the syntactic structure of* α *as follows:*

- $(S,T) \models a$ *for all subsets* $S, T \subseteq \mathsf{cl}(\varphi)$*, and all* $a \in \mathsf{A}$*.*
- *If* $\alpha = \oplus(\beta, \chi_1, \chi_2)$*, then* $(S,T) \models \alpha$ *if and only if (*$\chi_1 \in S$ *or* $\chi_2 \in T$*) and* $(S,T) \models \beta$*.*

Let us introduce infinite trees and infinite paths in them. If Γ and Υ are sets, then a *Γ-labeled Υ-tree* is a mapping $\mathcal{T} : D \to \Gamma$ for some non-empty prefix-closed subset $D \subseteq \Upsilon^*$. An *infinite path of $\mathcal{T}$* is a mapping $\tau : \omega \to \Upsilon$ such that $\tau(1) \cdots \tau(n) \in D$ for all $n \geq 1$. If $k \in \omega$, then a *Γ-labeled k-tree* is a Γ-labeled $[k]$-tree $\mathcal{T} : D \to \Gamma$ such that $D = [k]^*$.

Fix an enumeration $\psi_1, \ldots, \psi_k$ of all diamond formulas in $\mathsf{cl}(\varphi)$. Let us now define a concrete labeling set Γ, that we will comment on below in more detail. Define the labeling set $\Gamma = 2^{\mathsf{cl}(\varphi)} \times (\mathcal{B} \uplus \{\bot\}) \times [0, k]$. Intuitively, each model $\mathcal{K}$ of φ corresponds to some Γ-labeled k-tree $\mathcal{T}$ that contains the necessary information about how diamond formulas are satisfied in $\mathcal{K}$. We can think of it as assigning each $u \in [k]^*$ some state x of $\mathcal{K}$. The first component of $\mathcal{T}(u)$ contains exactly the set of formulas of $\mathsf{cl}(\varphi)$ that hold in x. The second component of $\mathcal{T}(u)$ either contains the information by which witnessing basic program the state x was reached or whether this information is not important (then it equals $\bot$). If, on the one hand, the third component of $\mathcal{T}(u)$ contains the information $i \in [k]$, then u has the obligation, in its subtree, to show that the diamond formula ψ_i is true in x. If, on the other hand, the third component of $\mathcal{T}(u)$ equals 0, then u is a witness that some diamond formula $\psi_j = \langle A \rangle \chi \in \mathsf{cl}(\varphi)$ holds in some world $y \in X$ – more precisely, we have $(\mathcal{K}, x) \models \chi$ and $(y, x) \in [\![w]\!]_{\mathcal{K}}$ for some $w \in L(A)$. For every $\gamma \in \Gamma$ and every $j \in \{1, 2, 3\}$, let γ_j denote the j-th component of γ.

Before defining what a Hintikka tree is, let us first, for Γ-labeled k-trees, introduce a notion of local consistency of the labeling $\mathcal{T}(u) \in \Gamma$ of worlds $u \in [k]^*$. For a string w over some alphabet Σ let $\mathrm{Occ}(w) \subseteq \Sigma$ denote the set of all letters that occur in w.

Definition 4. *If* $\mathcal{T} : [k]^* \to \Gamma$ *is a* Γ*-labeled* k*-tree, and* $u \in [k]^*$ *is some world, we say that* $\mathcal{T}$ *is* locally consistent in u*, if the following two conditions hold:*

(L1) *Whenever* $\psi_i = \langle A\rangle\chi \in \mathcal{T}(u)_1$, *then for some sequence of test programs* $w \in (\Sigma(A)\backslash\mathcal{B})^*$ *and some state* $q' \in Q(A)$ *such that* $\theta? \in \mathrm{Occ}(w)$ *implies* $\theta \in \mathcal{T}(u)_1$, $q_0(A) \overset{w}{\Rightarrow}_A q'$, *and at least one of the following two conditions holds:*

(a)
- $q' \in F(A)$,
- $\chi \in \mathcal{T}(u)_1$,
- $\mathcal{T}(ui)_2 = \bot$, *and*
- $\mathcal{T}(ui)_3 = 0$, *or*

(b) *there exists some basic program* $\alpha \in \Sigma(A) \cap \mathcal{B}$ *and some state* $q \in Q(A)$ *such that*
- $q' \overset{\alpha}{\Rightarrow}_A q$,
- $\psi_j = \langle A_q\rangle\chi \in \mathcal{T}(ui)_1$,
- $\mathcal{T}(ui)_2 = \alpha$,
- $\mathcal{T}(ui)_3 = j$, *and*
- $(\mathcal{T}(u)_1, \mathcal{T}(ui)_1) \models \alpha$.

(L2) *Whenever* $[A]\chi \in \mathcal{T}(u)_1$ *and* $q_0(A) \overset{\alpha}{\Rightarrow}_A q$ *for some basic program* $\alpha \in \Sigma(A)\cap\mathcal{B}$ *and some* $q \in Q(A)$, *then the following implication holds for all* $j \in [k]$ *with* $\mathcal{T}(uj)_2 = \beta \in \mathcal{B}$ *and* $\mathrm{At}(\alpha) = \mathrm{At}(\beta)$*:*

$$(\mathcal{T}(u)_1, \mathcal{T}(uj)_1) \models \alpha \qquad \Rightarrow \qquad [A_q]\chi \in \mathcal{T}(uj)_1$$

Let us summarize the intention of local consistency of a Γ-labeled k-tree in a world $u \in [k]^*$. As indicated above, condition (L1) ensures that whenever u obliges to prove some diamond formula ψ_i, then this proof is provided in the subtree of u: Either we directly prove that ψ_i holds in u ((L1)(a)), or we delay this proof by obliging an appropriate successor of u to prove an appropriate diamond formula ψ_j ((L1)(b)). Conditon (L2), on the other hand, ensures that if u obliges to prove some box formula, then all relevant successors of u must oblige to prove all relevant box formulas. We call a diamond formula ψ_i *infinitely delaying in a world* $u \in [k]^*$ of some Γ-labeled k-tree $\mathcal{T}$, if there exists an infinite path $\tau : \omega \to [k]$ such that $\tau(1) = i$ and $\mathcal{T}(u\tau(1)\cdots\tau(n))_3 = \tau(n+1)$ for all $n \geq 1$.

Definition 5. *A* Hintikka tree for φ *is a* Γ*-labeled* k*-tree* $\mathcal{T}$ *such that*

(H1) $\varphi \in \mathcal{T}(\varepsilon)_1$,
(H2) $\mathcal{T}(u)_1$ *is a Hintikka set for* φ *for all* $u \in [k]^*$,
(H3) $\mathcal{T}$ *is locally consistent in all* $u \in [k]^*$, *and*
(H4) *for all* $u \in [k]^*$ *and all* $i \in [k]$ *such that* $\psi_i \in \mathcal{T}(u)_1$, *the diamond formula* ψ_i *is not infinitely delaying in* u.

The following lemma can be shown.

Lemma 4. *The formula* φ *is satisfiable if and only if there exists a Hintikka tree for* φ.

Let us introduce alternating Büchi tree automata over infinite trees. Our goal is to construct an alternating Büchi tree automaton $\mathcal{A}(\varphi)$ which accepts exactly the set of all Hintikka trees for φ. Thus, satisfiability of φ reduces to non-emptiness of the tree language of $\mathcal{A}(\varphi)$. If X is a set, let $\mathbb{B}^+(X)$ denote the set of all *positive boolean formulas* over the set X. An *alternating Büchi tree automaton* over Λ-labeled l-trees is a tuple $\mathcal{A} = (S, s_0, \delta, F)$, where (1) S is a finite set of *states*, (2) $s_0 \in S$ is the *initial state*, (3) $\delta : S \times \Lambda \to \mathbb{B}^+(S \times ([l] \cup \{\varepsilon\}))$ maps every pair $(s, \lambda) \in S \times \Lambda$ to a positive boolean formula $\delta(s, \lambda)$ over $(S \times ([l] \cup \{\varepsilon\}))$, and (4) $F \subseteq S$ is a set of *final states*. Let $\mathcal{T} : [l]^* \to \Lambda$ be a Λ-labeled l-tree. A *run* of $\mathcal{A}$ on $\mathcal{T}$ is an $S \times [l]^*$-labeled ω-tree $\mathcal{R} : D \to S \times [l]^*$ (i.e. D is a non-empty prefix-closed subset of ω^*), which satisfies the following: (1) $\mathcal{R}(\varepsilon) = (s_0, \varepsilon)$, and (2) for all $u \in D$ such that $\mathcal{R}(u) = (s, x)$ and $\delta(s, \mathcal{T}(x)) = \theta$ there exists some set $Y = \{(s_1, i_1), \ldots, (s_n, i_n)\} \subseteq S \times ([l] \cup \{\varepsilon\})$ such that (i) Y satisfies the formula θ, and (ii) for all $1 \leq j \leq n$, we have $uj \in D$ and $\mathcal{R}(uj) = (s_j, xi_j)$. We call a run $\mathcal{R}$ of $\mathcal{A}$ on $\mathcal{T}$ *accepting*, if all infinite paths $\tau : \omega \to \omega$ of $\mathcal{R}$ satisfy the Büchi acceptance condition, i.e. $\{s \in S \mid \mathcal{R}(\tau(1) \cdots \tau(n)) \in (\{s\} \times [l]^*) \text{ for infinitely many } n \in \omega\} \cap F \neq \emptyset$. Let $L(\mathcal{A}) = \{\mathcal{T} \mid \text{there exists an accepting run of } \mathcal{A} \text{ on } \mathcal{T}\}$ denote the *language* accepted by $\mathcal{A}$. From [13], the following complexity for non-emptiness of the language of alternating Büchi automata over infinite trees can be derived .

Theorem 7 ([13]). *The language non-emptiness for alternating Büchi tree automata $\mathcal{A} = (S, s_0, \delta, F)$ over Λ-labeled l-trees is decidable in time exponential in $|S|$ and l, and in time polynomial in $|\Lambda|$ and in $|\delta|$.*

Note that in several definitions, alternating Büchi tree automata are not allowed to use ε-transitions. It can easily be verified that Theorem 7 still holds, if we allow ε-transitions. By translating Definition 1, 2, 3, 4, and 5 into an alternating Büchi tree automaton, we can derive the following lemma.

Lemma 5. *There exists an alternating Büchi tree automaton $\mathcal{A} = \mathcal{A}(\varphi) = (S, s_0, \delta, F)$ over Γ-labeled k-trees that accepts exactly the set of Hintikka trees for φ. Moreover, the size of S is polynomial in $|\varphi|$ and the size of δ is exponential in $|\varphi|$.*

Note that the Büchi acceptance condition suffices to check whether some diamond formula is not infinitely delaying in some world. Clearly, we need alternation to check $(S, T) \models \alpha$ for subsets $S, T \subseteq \mathsf{cl}(\varphi)$ and $\alpha \in \mathcal{B}$ and thus for checking local consistency (H3). By Theorem 7 and Lemma 5, the existence of a Hintikka tree for φ can be decided in time exponential in $|\varphi|$. Thus, by applying Lemma 4, we get the following theorem.

Theorem 8. *Satisfiability for PDL$\oplus[\mathbb{A}]$ is in* EXP.

7 2EXP-Completeness of PDL$\oplus$

In this section, we prove that satisfiability for PDL$\oplus$ is complete for 2EXP. Let us first introduce alternating Turing machines. An alternating Turing machine (ATM) is a tuple $\mathcal{M} = (Q, \Sigma, \Gamma, q_0, \delta, \square)$, where (i) $Q = Q_{\text{acc}} \uplus Q_{\text{rej}} \uplus Q_\exists \uplus Q_\forall$ is a finite *set of states*, which is partitioned into *accepting* (Q_{acc}), *rejecting* (Q_{rej}), *existential* ($Q_\exists$), and *universal* ($Q_\forall$) states, (ii) Γ is a finite *tape alphabet*, (iii) $\Sigma \subseteq \Gamma$ is the *input alphabet*, (iv) $q_0 \in Q$ is the *initial state*, (v) $\square \in \Gamma \setminus \Sigma$ is the *blank symbol*, and (vi) the mapping $\delta : (Q_\exists \cup Q_\forall) \times \Gamma \to \text{Moves} \times \text{Moves}$ with $\text{Moves} = Q \times \Gamma \times \{\leftarrow, \rightarrow\}$, assigns to every pair $(q, \gamma) \in (Q_\exists \cup Q_\forall) \times \Gamma$ a pair of moves. So we assume that a configuration in a current state from $Q_\exists \cup Q_\forall$ has exactly two successors. We call a configuration C of $\mathcal{M}$ in current state $q \in Q$ *accepting*, if either (i) $q \in Q_{\mathrm{acc}}$,(ii) $q \in Q_\exists$ and there exists an accepting successor configuration of C, or (iii) $q \in Q_\forall$ and all successor configurations of C are accepting. Let $L(\mathcal{M}) = \{w \in \Sigma^* \mid \text{ configuration } q_0 w \text{ is accepting}\}$ denote the *language* of $\mathcal{M}$.

Theorem 9. *Satisfiability for PDL⊕ is* 2EXP*-complete.*

Proof (sketch). The upper bound follows easily by a translation from PDL⊕ into PDL with an exponential blowup in formula size (cf. Remark 1) and by the EXP upper bound of PDL (cf. Theorem 1). The proof of the lower bound is an adaption of the 2EXP lower bound for PDL with intersection from [6]. Fix some ATM $\mathcal{M} = (Q, \Sigma, \Gamma, q_0, \delta, \Box)$ with a 2EXP-hard acceptance problem, that, on an input $w = w_1 \cdots w_n \in \Sigma^*$ where $w_i \in \Sigma$ for each $i \in [n]$, uses at most $2^{p(n)}$ space, where p is some polynomial. By [1], such a Turing machine $\mathcal{M}$ exists. We will give a reduction, computable in polynomial time in n, that translates w and $\mathcal{M}$ into a PDL⊕ formula $\varphi = \varphi(\mathcal{M}, w)$ such that $w \in L(\mathcal{M})$ if and only if φ is satisfiable. Let $N = p(n)$ and assume that Q and Γ are disjoint. A *configuration* of $\mathcal{M}$ is a word from the language $\bigcup_{i=0}^{2^N-1} \Gamma^i Q \Gamma^{2^N-i}$. Let $C = \gamma_0 \cdots \gamma_{i-1} q \gamma_i \cdots \gamma_{2^N-1}$ be a configuration of $\mathcal{M}$, where $0 \le i \le 2^N - 1$, $q \in Q$, and $\gamma_j \in \Gamma$ for each $0 \le j \le 2^N - 1$. Figure 1 illustrates how we will represent C.

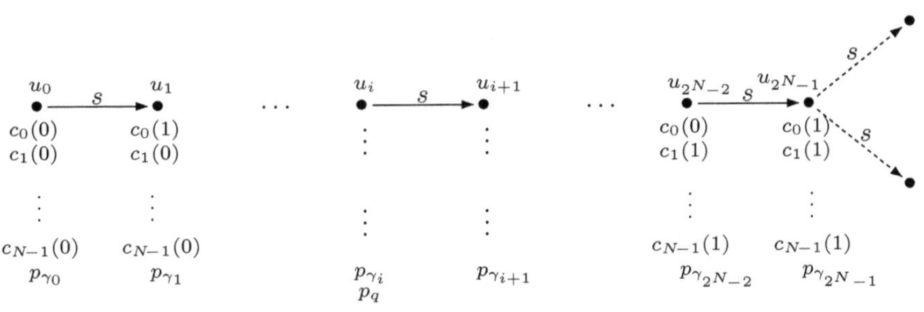

Fig. 1. Representation of a configuration of $\mathcal{M}$

So each world u_j ($0 \le j \le 2^N - 1$) represents the cell on position j of the configuration C (starting to count from 0) and the symbols below each world are the atomic propositions that hold in u_j. The atomic propositions $c_l(0)$ and $c_l(1)$, where $0 \le l \le N - 1$, represent the binary encoding of j. For each $0 \le j \le 2^N - 1$, there exists exactly one $\gamma \in \Gamma$ such that the atomic proposition p_γ holds in u_j, expressing that the content of the cell on position j of C is γ. On the other hand, for exactly one $0 \le j < 2^N - 1$ and exactly one $q \in Q$, the atomic proposition p_q holds in u_j, expressing that $\mathcal{M}$ is currently in state q and scans cell j. Furthermore, worlds u_j and u_{j+1} are connected by the atomic program s for all $0 \le j < 2^N - 1$. Moreover, the world u_{2^N-1} may be connected to the first cell of one or both successor configuration(s) of C via the atomic program s. Let $\Lambda = Q \uplus \Gamma$. Formally, the formula φ will be over the atomic propositions $\mathsf{P} = \{c_j(0), c_j(1) \mid 0 \le j \le N-1\} \cup \{p_\lambda \mid \lambda \in \Lambda\}$ and over the atomic program set $\mathsf{A} = \{s\}$. The crucial part of the formula φ is to find a program match that relates the current cell to the cell of some successor configuration at same positions. Let $\varphi_{\text{first}} = \bigwedge_{1 \leq j \leq N-1} c_j(0)$, which expresses that the current world represents some cell at position 0. The auxiliary program $\pi = (s \circ \neg\varphi_{\text{first}}?)^* \circ s \circ \varphi_{\text{first}}? \circ (s \circ \neg\varphi_{\text{first}}?)^*$ relates a cell c with a cell c' such that c' is in some successor configuration of the configuration of c, not necessarily at same positions. Let $\alpha_{-1} = \pi$ and, for all $0 \leq i \leq N-1$, define $\alpha_i = \oplus(\oplus(\alpha_{i-1}, c_i(0), c_i(1)), c_i(1), c_i(0))$. We put $\text{match} = \alpha_{N-1}$. Since we enforced that each reachable world in a model of φ satisfies exactly one of the atomic propositions $c_i(0)$ and $c_i(1)$ for all $0 \leq i \leq N-1$, the program match relates only worlds that agree on the same atomic propositions from $\{c_i(0), c_i(1) \mid 0 \leq i \leq N-1\}$. Since the program π relates cells in consecutive configurations, we relate cells in consecutive configurations at same positions. □

Acknowledgments. The author thanks the anonymous referees for interesting and substantial comments. Moreover the author thanks Markus Lohrey and Carsten Lutz for fruitful discussions on the topic as well as Florian Student for reading a draft version of this paper.

References

1. Chandra, A.K., Kozen, D.C., Stockmeyer, L.J.: Alternation. Journal of the Association for Computing Machinery 28(1), 114–133 (1981)
2. Danecki, R.: Nondeterministic Propositional Dynamic Logic with Intersection is decidable. In: Skowron, A. (ed.) Computation Theory. LNCS, vol. 208, pp. 34–53. Springer, Heidelberg (1985)
3. Demri, S.: A reduction from DLP to PDL. Journal of Logic and Computation 15(5), 767–785 (2005)
4. Fischer, M.J., Ladner, R.E.: Propositional Dynamic Logic of Regular Programs. Journal of Computer and System Sciences 18(2), 194–211 (1979)
5. Göller, S., Lohrey, M., Lutz, C.: PDL with Intersection and Converse is 2EXP-Complete. In: Proc. of FoSSaCS 2007, vol. 4423, pp. 198–212. Springer, Heidelberg (2007)
6. Lange, M., Lutz, C.: 2-ExpTime Lower Bounds for Propositional Dynamic Logics with Intersection. Journal of Symbolic Logic 70(4), 1072–1086 (2005)
7. Löding, C., Rohde, P.: Model Checking and Satisfiability for Sabotage Modal Logic. In: Pandya, P.K., Radhakrishnan, J. (eds.) FST TCS 2003: Foundations of Software Technology and Theoretical Computer Science. LNCS, vol. 2914, pp. 302–313. Springer, Heidelberg (2003)
8. Lutz, C., Walther, D.: PDL with Negation of Atomic Programs. Journal of Applied Non-Classical Logic 15(2), 189–214 (2005)
9. Pratt, V.R.: A Practical Decision Method for Propositional Dynamic Logic: Preliminary Report. In: Proc. of STOC 1980, pp. 326–337. ACM Press, New York (1978)
10. Pucella, R., Weissman, V.: Reasoning about Dynamic Policies. In: Walukiewicz, I. (ed.) FOSSACS 2004. LNCS, vol. 2987, pp. 453–467. Springer, Heidelberg (2004)
11. van Benthem, J.: An Essay on Sabotage and Obstruction. In: Hutter, D., Stephan, W. (eds.) Mechanizing Mathematical Reasoning. LNCS (LNAI), vol. 2605, pp. 268–276. Springer, Heidelberg (2005)
12. van der Meyden, R.: The Dynamic Logic of Permission. Journal of Logic and Computation 6(3), 465–479 (1996)

13. Vardi, M.Y.: Reasoning about the Past with Two-Way Automata. In: Larsen, K.G., Skyum, S., Winskel, G. (eds.) ICALP 1998. LNCS, vol. 1443, pp. 628–641. Springer, Heidelberg (1998)
14. Vardi, M.Y., Wolper, P.: Automata-Theoretic Techniques for Modal Logics of Programs. Journal of Computer and System Sciences 32(2), 183–221 (1986)
15. Wieringa, R.J., Meyer, J.-J.C.: Applications of Deontic Logic in Computer Science: A Concise Overview. In: Deontic Logic in Computer Science, pp. 17–40 (1993)

Relativizing Small Complexity Classes and Their Theories

Klaus Aehlig*, Stephen Cook, and Phuong Nguyen

University of Toronto

Abstract. Existing definitions of the relativizations of $\mathbf{NC}^1$, $\mathbf{L}$ and $\mathbf{NL}$ do not preserve the inclusions $\mathbf{NC}^1 \subseteq \mathbf{L}$, $\mathbf{NL} \subseteq \mathbf{AC}^1$. We start by giving the first definitions that preserve them. Here for $\mathbf{L}$ and $\mathbf{NL}$ we define their relativizations using Wilson's stack oracle model, but limit the height of the stack to a constant (instead of $\log(n)$). We show that the collapse of any two classes in $\{\mathbf{AC}^0(m), \mathbf{TC}^0, \mathbf{NC}^1, \mathbf{L}, \mathbf{NL}\}$ implies the collapse of their relativizations. Next we develop theories that characterize the relativizations of subclasses of $\mathbf{P}$ by modifying theories previously defined by the second two authors. A function is provably total in a theory iff it is in the corresponding relativized class. Finally we exhibit an oracle α that makes $\mathbf{AC}^k(\alpha)$ a proper hierarchy. This strengthens and clarifies the separations of the relativized theories in [Takeuti, 1995]. The idea is that a circuit whose nested depth of oracle gates is bounded by k cannot compute correctly the $(k+1)$ compositions of every oracle function.

1 Introduction

Oracles that separate $\mathbf{P}$ from $\mathbf{NP}$ and oracles that collapse $\mathbf{NP}$ to $\mathbf{P}$ have both been constructed. This rules out the possibility of proofs of the separation or collapse of $\mathbf{P}$ and $\mathbf{NP}$ by methods that relativize. However, similar results have not been established for subclasses of $\mathbf{P}$ such as $\mathbf{L}$ and $\mathbf{NL}$. Indeed, prior to this work there has not been a satisfying definition of the relativized version of $\mathbf{NL}$ that preserves simultaneously the inclusions

$$\mathbf{NC}^1 \subseteq \mathbf{L} \subseteq \mathbf{NL} \subseteq \mathbf{AC}^1 \tag{1}$$

For example [LL76], if the Turing machines are allowed to be nondeterministic when writing oracle queries, then there is an oracle α so that $\mathbf{NL}(\alpha) \not\subseteq \mathbf{P}(\alpha)$. Later definitions of $\mathbf{NL}(\alpha)$ adopt the requirement specified in [RST84] that the nondeterministic oracle machines be deterministic whenever the oracle tape (or oracle stack) is nonempty. Then the inclusion $\mathbf{NL}(\alpha) \subseteq \mathbf{P}(\alpha)$ relativizes, but not all inclusions in (1).

Because the nesting depth of oracle gates in an oracle $\mathbf{NC}^1$ circuit can be bigger than one, the model of relativization that preserves the inclusion $\mathbf{NC}^1 \subseteq \mathbf{L}$ must allow an oracle logspace Turing machine to have access to more than one oracle query tape [Orp83, Bus86, Wil88]. For the model defined by Wilson [Wil88],

* Supported by DFG grant Ae 102/1-1.

J. Duparc and T.A. Henzinger (Eds.): CSL 2007, LNCS 4646, pp. 374–388, 2007.

the partially constructed oracle queries are stored in a stack. The machine can write queries only on the oracle tape at the top of the stack. It can start a new query on an empty oracle tape (thus *pushing* down the current oracle tape, if there is any), or query the content of the top tape which then becomes empty and the stack is *popped*.

Following Cook [Coo85], the circuits accepting languages in relativized $\mathbf{NC}^1$ are those with logarithmic depth where the Boolean gates have bounded fanin and an oracle gate of m inputs contributes $\log(m)$ to the depths of its parents. Then in order to relativize the inclusion $\mathbf{NC}^1 \subseteq \mathbf{L}$, the oracle logspace machines defined by Wilson [Wil88] are required to satisfy the condition that at any time,

$$\sum_{i=1}^{k} max\{\log(|q_i|), 1\} = \mathcal{O}(\log(n))$$

where $q_1, q_2, \ldots, q_k$ are the contents of the stack and $|q_i|$ are their lengths. For the simulation of an oracle $\mathbf{NC}^1$ circuit by such an oracle logspace machine the upper bound $\mathcal{O}(\log(n))$ cannot be improved.

Although the above definition of $\mathbf{L}(\alpha)$ (and $\mathbf{NL}(\alpha)$) ensures that $\mathbf{NC}^1(\alpha) \subseteq \mathbf{L}(\alpha)$, unfortunately we know only that $\mathbf{NL}(\alpha) \subseteq \mathbf{AC}^2(\alpha)$ [Wil88]; the inclusion $\mathbf{NL}(\alpha) \subseteq \mathbf{AC}^1(\alpha)$ is left open.

We observe that if the height of the oracle stack is bounded by a constant (while the lengths of the queries are still bounded by a polynomial in the length of the inputs), then an oracle $\mathbf{NL}$ machine can be simulated by an oracle $\mathbf{AC}^1$ circuit, i.e., $\mathbf{NL}(\alpha) \subseteq \mathbf{AC}^1(\alpha)$. In fact, $\mathbf{NL}(\alpha)$ can then be shown to be the $\mathbf{AC}^0(\alpha)$ closure of the Reachability problem for directed graphs. Similarly, $\mathbf{L}(\alpha)$ is the $\mathbf{AC}^0(\alpha)$ closure of the Reachability problem for directed graphs whose outdegree is at most one.

The $\mathbf{AC}^0(\alpha)$ closure of the Boolean Sentence Value problem (which is $\mathbf{AC}^0$ complete for $\mathbf{NC}^1$) turns out to be the languages computable by uniform oracle $\mathbf{NC}^1$ circuits (defined as before) where the nesting depth of oracle gates is now bounded by a constant. We redefine $\mathbf{NC}^1(\alpha)$ using this new restriction on the oracle gates; the new definition is more suitable in the context of $\mathbf{AC}^0(\alpha)$ reducibility (the previous definition of $\mathbf{NC}^1(\alpha)$ seems suitable when one considers $\mathbf{NC}^1(\alpha)$ reducibility). Consequently, we obtain the first definition of $\mathbf{NC}^1(\alpha)$, $\mathbf{L}(\alpha)$ and $\mathbf{NL}(\alpha)$ that preserves the inclusions in (1).

Furthermore, the $\mathbf{AC}^0$-complete problems for $\mathbf{NC}^1$, $\mathbf{L}$, and $\mathbf{NL}$ (as well as $\mathbf{AC}^0(m)$, $\mathbf{TC}^0$) become $\mathbf{AC}^0(\alpha)$-complete for the corresponding relativized classes. Therefore the existence of any oracle that separates two of the mentioned classes implies the separation of the respective nonrelativized classes. (If the non-relativised classes would be equal, their complete problems would be equivalent under $\mathbf{AC}^0$-reductions, hence even more under $\mathbf{AC}^0(\alpha)$-reductions and therefore the relativised classes would coincide as well.) Separating the relativized classes is as hard as separating their nonrelativized counterparts. This nicely generalizes known results [Wil88, Sim77, Wil89].

On the other hand, oracles that separate the classes $\mathbf{AC}^k$ (for $k = 0, 1, 2, \ldots$) and $\mathbf{P}$ have been constructed [Wil89]. Here we prove a sharp separation between relativized circuit classes whose nesting depths of oracle gates differ by one. More precisely, we show that a family of uniform circuits with nesting depth of oracle gates bounded by k cannot compute correctly the $(k+1)$ iterated compositions

$$f(f(\dots f(0)\dots)) \tag{2}$$

for all oracle function f. (Clearly (2) can be computed correctly by a circuit with oracle gates having nesting depth $(k+1)$.) As a result, there is an oracle α such that

$$\mathbf{NL}(\alpha) \subsetneq \mathbf{AC}^1(\alpha) \subsetneq \mathbf{AC}^2(\alpha) \subsetneq \dots \subsetneq \mathbf{P}(\alpha) \tag{3}$$

The idea of using (2) to separate relativized circuit classes is already present in the work of Takeuti [Tak95] where it is used to separate the relativized versions of first-order theories $TLS(\alpha)$ and $TAC^1(\alpha)$. Here TLS and TAC are (single sorted) theories associated with $\mathbf{L}$ and $\mathbf{AC}^1$, respectively. Thus with simplified arguments we strengthen his results.

Finally, building up from the work of the second two authors [CN06, NC05] we develop relativized two-sorted theories that are associated with the newly defined classes $\mathbf{NC}^1(\alpha), \mathbf{L}(\alpha), \mathbf{NL}(\alpha)$ as well as other relativized circuit classes.

The paper is organized as follows. In Section 2 we define the relativized classes and prove the inclusions mentioned above. In Section 3 we define the associated theories. An oracle that separates classes in (3) is shown in Section 4.

2 Definitions of Small Relativized Classes

2.1 Relativized Circuit Classes

Throughout this paper, α denotes a unary relation on binary strings.

A problem is in $\mathbf{AC}^k$ if it can be solved by a polynomial size family of Boolean circuits whose depth is bounded by $\mathcal{O}((\log n)^k)$ (n is the number of the inputs), where $\wedge$ and $\vee$ gates are allowed unbounded fanin. The relativized class $\mathbf{AC}^k(\alpha)$ generalizes this by allowing, in addition to (unbounded fanin) Boolean gates $(\neg, \wedge, \vee)$, oracle gates that output 1 if and only if the inputs to the gates (viewed as binary strings) belong to α (these gates are also called α gates).

In this paper we always require circuit families to be *uniform*. Our default definition of uniform is DLOGTIME, a robust notion of uniformity that has a number of equivalent definitions [BIS90, Imm99]. In particular, a language $L \subseteq \{0,1\}^*$ is in (uniform) $\mathbf{AC}^0$ iff it represents the set of finite models $\{1,\dots,n\}$ of some fixed first-order formula with an uninterpreted unary predicate symbol and ternary predicates which are interpreted as addition and multiplication.

Recall that $\mathbf{TC}^0$ (resp. $\mathbf{AC}^0(m)$) is defined in the same way as $\mathbf{AC}^0$, except the circuits allow unbounded fanin threshold (resp. mod$m$) gates.

Definition 1 ($\mathbf{AC}^k(\alpha)$, $\mathbf{AC}^0(m,\alpha)$, $\mathbf{TC}^0(\alpha)$). *For $k \geq 0$, the class $\mathbf{AC}^k(\alpha)$ (resp. $\mathbf{AC}^0(m,\alpha)$, $\mathbf{TC}^0(\alpha)$) is defined as uniform $\mathbf{AC}^k$ (resp. $\mathbf{AC}^0(m)$, $\mathbf{TC}^0$) except that unbounded fan-in α gates are allowed.*

The class $\mathbf{NC}^k$ is the subclass of $\mathbf{AC}^k$ defined by restricting the $\wedge$ and $\vee$ gates to have fanin 2. Defining $\mathbf{NC}^k(\alpha)$ is more complicated. In [Coo85] the depth of an oracle gate with m inputs is defined to be $\log(m)$. A circuit is an $\mathbf{NC}^k(\alpha)$-circuit provided that it has polynomial size and the total depth of all gates along any path from the output gate to an input gate is $\mathcal{O}((\log n)^k)$. Note that if there is a mix of large and small oracle gates, the number of oracle gates may not be $\mathcal{O}((\log n)^{k-1})$.

Here we restrict the definition further, requiring that the nested depth of oracle gates is $\mathcal{O}((\log n)^{k-1})$.

Definition 2 ($\mathbf{NC}^k(\alpha)$). *For $k \geq 1$, a language is in $\mathbf{NC}^k(\alpha)$ if it is computable by a uniform family of $\mathbf{NC}^k(\alpha)$ circuits, i.e., $\mathbf{AC}^k(\alpha)$ circuits where the $\wedge$ and $\vee$ gates have fanin 2, and the nested depth of α gates is $\mathcal{O}((\log n)^{k-1})$.*

2.2 Relativized Logspace Classes

To define oracle logspace classes, we use a modification of Wilson's stack model [Wil88]. An advantage is that the relativized classes defined here are closed under $\mathbf{AC}^0$-reductions. This is not true for the non-stack model.

A Turing machine M with a stack of oracle tapes can write 0 or 1 onto the top oracle tape if it already contains some symbols, or it can start writing on an empty oracle tape. In the latter case, the new oracle tape will be at the top of the stack, and we say that M performs a *push* operation. The machine can also *pop* the stack, and its next action and state depend on $\alpha(Q)$, where Q is the content of the top oracle tape. Note that here the oracle tapes are write-only.

Instead of allowing an arbitrary number of oracle tapes, we modify Wilson's model by allowing only a stack of constant height (hence the prefix "cs" in cs$\mathbf{L}(\alpha)$ and cs$\mathbf{NL}(\alpha)$). This places the relativized classes in the same order as the order of their unrelativized counterparts.

In the definition of cs$\mathbf{NL}(\alpha)$, we also use the restriction [RST84] that the machine is deterministic when the oracle stack is non empty.

Definition 3 (cs$\mathbf{L}(\alpha)$, cs$\mathbf{NL}(\alpha)$). *For a unary relation α on strings, cs$\mathbf{L}(\alpha)$ is the class of languages computable by logspace, polytime Turing machines using an α-oracle stack whose height is bounded by a constant. cs$\mathbf{NL}(\alpha)$ is defined as cs$\mathbf{L}(\alpha)$ but the Turing machines are allowed to be nondeterministic when the oracle stack is empty.*

Theorem 4. $\mathbf{NC}^1(\alpha) \subseteq$ *cs*$\mathbf{L}(\alpha) \subseteq$ *cs*$\mathbf{NL}(\alpha) \subseteq \mathbf{AC}^1(\alpha)$.

Proof. The second inclusion is immediate from the definitions, the first can be proved as in the standard proof of the fact that $\mathbf{NC}^1 \subseteq \mathbf{L}$ (see also [Wil88]). The last inclusion can actually be strengthened, as shown in the next theorem. □

Theorem 5. *Each language in cs$\mathbf{NL}(\alpha)$ can be computed by a uniform family of $\mathbf{AC}^1(\alpha)$ circuits whose nested depth of oracle gates is bounded by a constant.*

A proof of the theorem is given in the full version of this paper [ACN07].

2.3 csL(α) Reducibility

A cs$\mathbf{L}(\alpha)$ function is defined by allowing the cs$\mathbf{L}(\alpha)$ machine to write on a write-only output tape. Then the notion of many-one cs$\mathbf{L}(\alpha)$ reducibility is defined as usual. Using this notion, the next lemma can be used to show that the Immerman-Szelepcsényi Theorem and Savitch's Theorem relativize. Recall that **STCONN** is the problem: given (G, s, t), where s, t are two designated vertices of a directed graph G, decide whether there is a path from s to t. A proof of the next lemma is given in [ACN07].

Lemma 6. *A language is in cs*$\mathbf{NL}(\alpha)$ *iff it is many-one cs*$\mathbf{L}(\alpha)$ *reducible to* **STCONN**.

Corollary 7 (Relativized Immerman-Szelepcsényi Theorem). *cs*$\mathbf{NL}(\alpha)$ *is closed under complementation.*

Proof. Any language in *co*-cs$\mathbf{NL}(\alpha)$ is cs$\mathbf{L}(\alpha)$ reducible to $\overline{\mathbf{STCONN}}$, which is $\mathbf{AC}^0$ reducible to **STCONN**. So *co*-cs$\mathbf{NL}(\alpha) \subseteq$ cs$\mathbf{NL}(\alpha)$. □

Let cs$\mathbf{L}^2(\alpha)$ denote the class of languages computable by a deterministic oracle Turing machine in $\mathcal{O}(\log^2)$ space and constant-height oracle stack.

Corollary 8 (Relativized Savitch's Theorem). *cs*$\mathbf{NL}(\alpha) \subseteq$ *cs*$\mathbf{L}^2(\alpha)$.

Proof. The corollary follows from Lemma 6 and the fact that the composition of a cs$\mathbf{L}(\alpha)$ function and a $(\log^2)$ space function (for **STCONN**) is a cs$\mathbf{L}^2(\alpha)$ function. □

3 Relativized Theories

3.1 Two-Sorted Languages and Complexity Classes

Our theories are based on a two-sorted vocabulary, and it is convenient to reinterpret the complexity classes using this vocabulary [CN06, NC05]. Our two-sorted language has variables $x, y, z, ...$ ranging over $\mathbb{N}$ and variables $X, Y, Z, ...$ ranging over finite subsets of $\mathbb{N}$ (interpreted as bit strings). Our basic two-sorted vocabulary $\mathcal{L}_A^2$ includes the usual symbols $0, 1, +, \cdot, =, \leq$ for arithmetic over $\mathbb{N}$, the length function $|X|$ on strings, the set membership relation $\in$, and string equality $=_2$ (where we usually drop mention of the subscript 2). The function $|X|$ denotes 1 plus the largest element in the set X, or 0 if X is empty (roughly the length of the corresponding string). We will use the notation $X(t)$ for $t \in X$, and we will think of $X(t)$ as the t-th bit in the string X.

Number terms of $\mathcal{L}_A^2$ are built from the constants 0,1, variables $x, y, z, ...$, and length terms $|X|$ using $+$ and $\cdot$. The only *string terms* are string variables $X, Y, Z, ...$ The atomic formulas are $t = u$, $X = Y$, $t \leq u$, $t \in X$ for any number terms t, u and string variables X, Y. Formulas are built from atomic formulas using $\wedge, \vee, \neg$ and both number and string quantifiers $\exists x, \exists X, \forall x, \forall X$.

Bounded number quantifiers are defined as usual, and the bounded string quantifier $\exists X \le t\ \varphi$ stands for $\exists X(|X| \le t \wedge \varphi)$ and $\forall X \le t\ \varphi$ stands for $\forall X(|X| \le t \supset \varphi)$, where X does not occur in the term t.

$\mathbf{\Sigma}_0^B$ is the set of all $\mathcal{L}_A^2$-formulas in which all number quantifiers are bounded and with no string quantifiers. $\mathbf{\Sigma}_1^B$ (corresponding to *strict* $\Sigma_1^{1,b}$ in [Kra95]) formulas begin with zero or more bounded existential string quantifiers, followed by a $\mathbf{\Sigma}_0^B$ formula. These classes are extended to $\mathbf{\Sigma}_i^B$, $i \ge 2$, (and $\mathbf{\Pi}_i^B$, $i \ge 0$) in the usual way.

We use the notation $\mathbf{\Sigma}_0^B(\mathcal{L})$ to denote $\mathbf{\Sigma}_0^B$ formulas which may have two-sorted function and predicate symbols from the vocabulary $\mathcal{L}$ in addition to the basic vocabulary $\mathcal{L}_A^2$.

Two-sorted complexity classes contain relations $R(\vec{x}, \vec{X})$ (and possibly number-valued functions $f(\vec{x}, \vec{X})$ or string-valued functions $F(\vec{x}, \vec{X})$), where the arguments $\vec{x} = x_1, \ldots, x_k$ range over $\mathbb{N}$, and $\vec{X} = X_1, \ldots, X_\ell$ range over finite subsets of $\mathbb{N}$. In defining complexity classes using machines or circuits, the number arguments x_i are presented in unary notation (a string of x_i ones), and the arguments X_i are presented as bit strings. Thus the string arguments are the important inputs, and the number arguments are small auxiliary inputs useful for indexing the bits of strings.

As mentioned before, uniform $\mathbf{AC}^0$ has several equivalent characterizations [Imm99], including **LTH** (the log time hierarchy on alternating Turing machines) and **FO** (describable by a first-order formula using predicates for plus and times). Thus in the two-sorted setting we can define $\mathbf{AC}^0$ to be the class of relations $R(\vec{x}, \vec{X})$ such that some alternating Turing machine accepts R in time $O(\log n)$ with a constant number of alternations, using the input conventions for numbers and strings given above. Then from the **FO** characterization of $\mathbf{AC}^0$ we obtain the following nice connection between $\mathbf{AC}^0$ and our two-sorted $\mathcal{L}_A^2$-formulas.

Theorem 9 ($\mathbf{\Sigma}_0^B$ Representation Theorem). *A relation $R(\vec{x}, \vec{X})$ is in $\mathbf{AC}^0$ iff it is represented by some $\mathbf{\Sigma}_0^B$ formula $\varphi(\vec{x}, \vec{X})$.*

In general, if $\mathbf{C}$ is a class of relations (such as $\mathbf{AC}^0$) then we want to associate a class $\mathbf{FC}$ of functions with $\mathbf{C}$. Here $\mathbf{FC}$ will contain string-valued functions $F(\vec{x}, \vec{X})$ and number-valued functions $f(\vec{x}, \vec{X})$. We require that these functions be p-bounded; i.e. for each F and f there is a polynomial $g(n)$ such that $|F(\vec{x}, \vec{X})| \le g(max(\vec{x}, |\vec{X}|))$ and $f(\vec{x}, \vec{X}) \le g(max(\vec{x}, |\vec{X}|))$.

We define the *bit graph* $B_F(i, \vec{x}, \vec{X})$ by

$$B_F(i, \vec{x}, \vec{X}) \leftrightarrow F(\vec{x}, \vec{X})(i) \tag{4}$$

Definition 10. *If $\mathbf{C}$ is a two-sorted complexity class of relations, then the corresponding function class $\mathbf{FC}$ consists of all p-bounded number functions whose graphs are in $\mathbf{C}$, together with all p-bounded string functions whose bit graphs are in $\mathbf{C}$.*

For example, binary addition $F_+(X, Y) = X + Y$ is in $\mathbf{FAC}^0$, but binary multiplication $F_\times(X, Y) = X \cdot Y$ is not.

Definition 11. *A string function is* $\mathbf{\Sigma}_0^B$-definable *from a collection* $\mathcal{L}$ *of two-sorted functions and relations if it is p-bounded and its bit graph is represented by a* $\mathbf{\Sigma}_0^B(\mathcal{L})$ *formula. Similarly, a number function is* $\mathbf{\Sigma}_0^B$-definable *from* $\mathcal{L}$ *if it is p-bounded and its graph is represented by a* $\mathbf{\Sigma}_0^B(\mathcal{L})$ *formula.*

It is not hard to see that $\mathbf{FAC}^0$ is closed under $\mathbf{\Sigma}_0^B$-definability, meaning that if the bit graph of F is represented by a $\mathbf{\Sigma}_0^B(\mathbf{FAC}^0)$ formula, then F is already in $\mathbf{FAC}^0$.

In order to define complexity classes such as $\mathbf{AC}^0(m)$ and $\mathbf{TC}^0$, as well as relativized classes such as $\mathbf{AC}^0(\alpha)$, we need to iterate $\mathbf{\Sigma}_0^B$-definability to obtain the notion of $\mathbf{AC}^0$ reduction.

Definition 12. *We say that a string function* F *(resp. a number function* f*) is* $\mathbf{AC}^0$*-reducible to* $\mathcal{L}$ *if there is a sequence of string functions* $F_1, \ldots, F_n$ $(n \geq 0)$ *such that*

$$F_i \text{ is } \mathbf{\Sigma}_0^B\text{-definable from } \mathcal{L} \cup \{F_1, \ldots, F_{i-1}\}, \text{ for } i = 1, \ldots, n; \tag{5}$$

and F *(resp.* f*) is* $\mathbf{\Sigma}_0^B$*-definable from* $\mathcal{L} \cup \{F_1, \ldots, F_n\}$*. A relation* R *is* $\mathbf{AC}^0$*-reducible to* $\mathcal{L}$ *if there is a sequence* $F_1, \ldots, F_n$ *as above, and* R *is represented by a* $\mathbf{\Sigma}_0^B(\mathcal{L} \cup \{F_1, \ldots, F_n\})$ *formula.*

In other words, F is $\mathbf{AC}^0$-reducible to $\mathcal{L}$ if there is a uniform constant-depth polysize circuit family that computes F, where the circuits are allowed gates (each of depth one) which compute the functions and predicates in $\mathcal{L}$ (as well as the Boolean connectives).

For each class $\mathbf{C}$ in

$$\{\mathbf{TC}^0, \mathbf{AC}^0(m), \mathbf{NC}^1, \mathbf{L}, \mathbf{NL}\} \tag{6}$$

we consider a natural complete relation $R_{\mathbf{C}}$ as follows:

- $\mathbf{TC}^0$: Numones(k, X) holds iff k is the number of 1 bits in the binary string X.
- $\mathbf{AC}^0(m)$: Mod$_m(X)$ holds iff the number of 1 bits in X is 1 modulo m.
- $\mathbf{NC}^1$: Mfvp(X) holds iff X codes a true balanced monotone Boolean sentence. ("Mfvp" stands for "monotone formula value problem")
- $\mathbf{L}$: Spath(s, t, G) holds iff G codes a directed graph with outdegree at most 1, and s, t are two vertices of G, and there is a path in G from s to t.
- $\mathbf{NL}$: Conn(s, t, G) holds iff G is a directed graph that contains a path from s to t.

The following result follows easily from the definitions of the complexity classes and well-known complete problems:

Theorem 13. *Each class* $\mathbf{C}$ *in* (6) *is the class of relations* $\mathbf{AC}^0$*-reducible to* $R_{\mathbf{C}}$.

Recall the relativized classes given in Definitions 1, 2, and 3. A proof of the next theorem is given in [ACN07].

Theorem 14. *For each class* $\mathbf{C}(\alpha)$ *in*

$$\{\mathbf{TC}^0(\alpha), \mathbf{AC}^0(m,\alpha), \mathbf{NC}^1(\alpha), cs\mathbf{L}(\alpha), cs\mathbf{NL}(\alpha)\}$$

$\mathbf{C}(\alpha)$ *is the class of relations* $\mathbf{AC}^0$*-reducible to* $\{R_{\mathbf{C}}, \alpha\}$.

The following corollary is immediate from the two preceding theorems and the transitivity of $\mathbf{AC}^0$-reducibility. It generalizes results in [Wil89].

Corollary 15. *For any* $\mathbf{C}_1, \mathbf{C}_2$ *in* (6) $\mathbf{C}_1 = \mathbf{C}_2$ *if and only if for all* α, $\mathbf{C}_1(\alpha) = \mathbf{C}_2(\alpha)$, *where* $\mathbf{L}(\alpha)$ *means* $cs\mathbf{L}(\alpha)$ *and* $\mathbf{NL}(\alpha)$ *means* $cs\mathbf{NL}(\alpha)$.

3.2 Nonrelativized Theories

The theory $\mathbf{V}^0$ (essentially $\mathbf{\Sigma}_0^p$-*comp* in [Zam96], and $\mathbf{I\Sigma}_0^{1,b}$ (without #) in [Kra95]) is the theory over $\mathcal{L}_A^2$ that is axiomatized by the axioms listed in Figure 1 together with the axiom scheme $\mathbf{\Sigma}_0^B(\mathcal{L}_A^2)$-**COMP**, i.e. the set of all formulas of the form

$$\exists X \le y \forall z < y (X(z) \leftrightarrow \varphi(z)), \tag{7}$$

where $\varphi(z)$ is any formula in $\mathbf{\Sigma}_0^B(\mathcal{L}_A^2)$, and X does not occur free in $\varphi(z)$.

B1. $x + 1 \ne 0$	**B7.** $(x \le y \wedge y \le x) \supset x = y$
B2. $x + 1 = y + 1 \supset x = y$	**B8.** $x \le x + y$
B3. $x + 0 = x$	**B9.** $0 \le x$
B4. $x + (y + 1) = (x + y) + 1$	**B10.** $x \le y \vee y \le x$
B5. $x \cdot 0 = 0$	**B11.** $x \le y \leftrightarrow x < y + 1$
B6. $x \cdot (y + 1) = (x \cdot y) + x$	**B12.** $x \ne 0 \supset \exists y \le x (y + 1 = x)$
L1. $X(y) \supset y < \|X\|$	**L2.** $y + 1 = \|X\| \supset X(y)$
SE. $[\|X\| = \|Y\| \wedge \forall i < \|X\| (X(i) \leftrightarrow Y(i))] \supset X = Y$	

Fig. 1. 2-**BASIC**

Using the the $\mathbf{\Sigma}_0^B$ Representation Theorem 9, it can be shown that a p-bounded function is in $\mathbf{FAC}^0$ if and only if it is provably total (i.e., $\mathbf{\Sigma}_1^B$ definable) in $\mathbf{V}^0$.

More generally, for various subclasses $\mathbf{C}$ of $\mathbf{P}$, a theory $\mathbf{VC}$ is developed in [CN06, Chapter 9] that characterizes $\mathbf{C}$ in the sense that the functions in $\mathbf{FC}$ are precisely the provably total functions of $\mathbf{VC}$. (The theory for $\mathbf{AC}^0(m)$ is $\mathbf{V}^0(m)$.) The theory $\mathbf{VC}$ is axiomatized by the axioms of $\mathbf{V}^0$ together with an axiom that formalizes a polytime computation of an answer for a complete problem of $\mathbf{C}$, asuming the parameters as given inputs. For a class $\mathbf{C}$ in (6), the complete problem is $R_{\mathbf{C}}$.

To formulate these axioms we introduce the pairing function $\langle y, z\rangle$, which stands for the term $(y + z)(y + z + 1) + 2z$. This allows us to interpret a string X as a two-dimensional bit array, using the notation

$$X(y, z) \equiv X(\langle y, z\rangle) \tag{8}$$

For example, a graph with a vertices can be encoded by a pair (a, E) where $E(u, v)$ holds iff there is an edge from u to v, for $0 \leq u, v < a$. The theory **VNL** is axiomatized by $\mathbf{V}^0$ and $CONN \equiv \forall a \forall E \exists Y\, \delta_{CONN}(a, E, Y)$. The formula $\delta_{CONN}(a, E, Y)$ states that for the graph encoded by (a, E), Y encodes a polytime computation of the nodes that are reachable from nodes 0: $Y(z, x)$ holds iff there is a path from 0 to x of length $\leq z$.

$$\delta_{CONN}(a, E, Y) \equiv Y(0,0) \wedge \forall x < a(x \neq 0 \supset \neg Y(0,x)) \wedge \\ \forall z < a \forall x < a \left[Y(z+1, x) \leftrightarrow (Y(z,x) \vee \exists y < a,\ Y(z,y) \wedge E(y,x))\right].$$

The additional axioms for other theories are listed below. Here $(Z)^x$ is the x-th element of the sequence of numbers encoded by Z:

$$y = (Z)^x \leftrightarrow (y < |Z| \wedge Z(x,y) \wedge \forall z < y \neg Z(x,z)) \vee \\ (\forall z < |Z| \neg Z(x,z) \wedge y = |Z|)$$

Also, $\log a$, or $|a|$, denotes the integral part of $\log_2(a)$. Note that this function is provably total in $\mathbf{V}^0$. The $\mathbf{\Sigma}_0^B$ formulas below contain the functions $(Z)^x$ and $|a|$, but these functions can be eliminated using their $\mathbf{\Sigma}_0^B$ defining axioms.

- **VTC**^{0}: $NUMONES \equiv \forall X \forall x \exists Y\, \delta_{NUM}(x, X, Y)$ where

$$\delta_{NUM}(x, X, Y) \equiv\ (Y)^0 = 0 \wedge \\ \forall z < x \left[(X(z) \supset (Y)^{z+1} = (Y)^z + 1) \wedge (\neg X(z) \supset (Y)^{z+1} = (Y)^z)\right]$$

  (For $z \geq 1$, $(Y)^z$ is the number of 1 bits in $X(0), X(1), \ldots, X(z-1)$.)
- $\mathbf{V}^0(m)$ (the theory for $\mathbf{AC}^0(m)$): $\mathbf{MOD}_m \equiv \forall X \forall x \exists Y\, \delta_{\mathbf{MOD}_m}(x, X, Y)$ where

$$\delta_{\mathbf{MOD}_m}(x, X, Y) \equiv Y(0,0) \wedge \\ \forall z < x \left[(X(z) \supset (Y)^{z+1} = ((Y)^z + 1) \bmod m)) \wedge (\neg X(z) \supset (Y)^{z+1} = (Y)^z)\right].$$

  (For $z \geq 1$, $(Y)^z$ is the number of 1 bits in $X(0), X(1), \ldots, X(z-1)$ modulo m.)
- **VNC**^{1}: $MFVP \equiv \forall a \forall G \forall I \exists Y\, \delta_{MFVP}(a, G, I, Y)$ where

$$\delta_{MFVP}(a, G, I, Y) \equiv\ \forall x < a\,[Y(x+a) \leftrightarrow I(x) \wedge \\ 0 < x \supset [Y(x) \leftrightarrow [(G(x) \wedge Y(2x) \wedge Y(2x+1)) \vee (\neg G(x) \wedge (Y(2x) \vee Y(2x+1)))]]]$$

  (For the formula viewed as a balanced binary tree encoded by (a, G)—node x's children are $2x$ and $2x+1$, and $G(x)$ indicates whether node x is an $\vee$ or $\wedge$ node— $Y(x)$ is the value of node x when the inputs are given by I.)
- **VL**: *SinglePATH* is the axiom

$$[\forall x < a \exists ! y < a E(x,y)] \supset \exists P\ \left[(P)^0 = 0 \wedge \forall v < a E((P)^v, (P)^{v+1})\right]$$

  ($(P)^v$ is the vertex of distance v from 0.)
- **VAC**^{k}: $\forall a \forall E \forall G \forall I \exists Y\, \delta_{MCVP}(a, |a|^k, E, G, I, Y)$ where

$$\delta_{MCVP}(w, d, E, G, I, Y) \equiv \forall x < w \forall z < d\, [(Y(0, x) \leftrightarrow I(x)) \wedge$$
$$(Y(z+1, x) \leftrightarrow [[G(z+1, x) \wedge \forall u < w\, (E(z, u, x) \supset Y(z, u))] \vee$$
$$[\neg G(z+1, x) \wedge \exists u < w\, (E(z, u, x) \wedge Y(z, u))]])]$$

("MCVP" stands for "monotone circuit value problem". Here the formula $\delta_{MCVP}(w, d, E, G, I, Y)$ states that given input I to a circuit encoded by (w, d, E, G)—there are w gates on each of the the d layers, the gate connection is given by E and the gates are specified by G—Y encodes an evaluation of the gates.)

- $\mathbf{VNC}^k$ (for $k \geq 2$):

$$\forall a \forall E \forall G \forall I (\mathit{Fanin2}(a, |a|^k, E) \supset \exists Y\, \delta_{MCVP}(a, |a|^k, E, G, I, Y))$$

Here $\mathit{Fanin2}(w, d, E)$ states that the gates have fanin at most 2:

$$\forall z < d \forall x < w \exists u_1, u_2 < w \forall v < w\, (E(z, v, x) \supset v = u_1 \vee v = u_2)$$

Showing that the functions in **FC** are precisely the provably total functions of **VC** can be done by first developing an universal theory $\overline{\mathbf{VC}}$ whose underlying vocabulary consists of all functions in **FC** with their defining axioms. The provably total functions of $\overline{\mathbf{VC}}$ are precisely the functions in **FC**, so it remains to show that $\overline{\mathbf{VC}}$ is a conservative extension of **VC** [CN06, Corollary 9.33].

Our goal for the remainder of this section is to obtain relativized theories $\mathbf{VC}(\alpha)$ that characterize the relativized classes discussed in Section 2. We will use the results of [CN06, Chapter 9] and the fact that the axioms in $\mathbf{VC}(\alpha)$ encode the polytime computation of corresponding $\mathbf{AC}^0$-complete problems of the classes (cf. Theorem 14).

3.3 Relativized Theories

First note that a sequence of strings can be encoded using the string function *Row*, where

$$\mathit{Row}(x, Z)(i) \leftrightarrow i < |Z| \wedge Z(x, i)$$

($\mathit{Row}(x, Z)$ will be also written as $Z^{[x]}$.)

Notation. For a predicate α, let $\mathbf{\Sigma}_0^B(\alpha)$ denote the class of $\mathbf{\Sigma}_0^B$ formulas in $\mathcal{L}_A^2 \cup \{\mathit{Row}, \alpha\}$.

Definition 16. $\mathbf{V}^0(\alpha) = \mathbf{V}^0 + \mathbf{\Sigma}_0^B(\alpha)$-**COMP**. *For each class* **C** *in* (6), *the theory* $\mathbf{VC}(\alpha)$ *is defined as* **VC** *with* $\mathbf{\Sigma}_0^B$-**COMP** *replaced by* $\mathbf{\Sigma}_0^B(\alpha)$-**COMP**.

Notice that natural relativized versions of the additional axioms of **VC**, such as *CONN*, are already provable in $\mathbf{VC}(\alpha)$. For example, let $\mathit{CONN}(\alpha)$ be the axiom scheme

$$\forall a \exists Y\ [Y(0, 0) \wedge \forall x < a (x \neq 0 \supset \neg Y(0, x)) \wedge$$
$$\forall z < a \forall x < a,\ Y(z+1, x) \leftrightarrow (Y(z, x) \vee \exists y < a,\ Y(z, y) \wedge \varphi(y, x))].$$

where φ is a $\mathbf{\Sigma}_0^B(\alpha)$ formula. Then it is easy to use $\mathbf{\Sigma}_0^B(\alpha)$-**COMP** to show that $\mathbf{VNL}(\alpha) \vdash \mathit{CONN}(\alpha)$.

Theorem 17. *For a class* $\mathbf{C}$ *in* $\{\mathbf{AC}^0, \mathbf{AC}^0(m), \mathbf{TC}^0, \mathbf{NC}^1, \mathbf{L}, \mathbf{NL}\}$, *a function is in* $\mathbf{FC}(\alpha)$ *if and only if it is provably total in* $\mathbf{VC}(\alpha)$.

The theorem can be proved using Theorem 14 by induction on n in Definition 12. Details can be found in [ACN07].

Now we present the theories $\mathbf{VAC}^k(\alpha)$ (for $k \geq 1$) and $\mathbf{VNC}^k(\alpha)$ (for $k \geq 2$). Note that the problem of evaluating uniform $\mathbf{AC}^k(\alpha)$ (or $\mathbf{NC}^k(\alpha)$) circuits is $\mathbf{AC}^0$-complete for the corresponding relativized class. Thus $\mathbf{VAC}^k(\alpha)$ (or $\mathbf{VNC}^k(\alpha)$) will be axiomatized by $\mathbf{V}^0$ together with an additional axiom that formalizes a polytime computation that solves the respective complete problem.

First we formalize a polytime evaluation of an oracle circuit $C = (w, d, E, G)$ given input I. Since the order of inputs to an oracle gate is important, the edge relations of the underlying graph is now encoded by a string variable E, where $E(z, t, u, x)$ indicates that gate u on layer z is the t-th input to gate x on layer $z + 1$. The condition we need for E is

$$Proper(w, d, E) \equiv \forall z < d \forall t, x, u_1, u_2 < w,\ (E(z, t, u_1, x) \wedge E(z, t, u_2, x)) \supset u_1 = u_2$$

In the formula $\delta^{\alpha}_{MCVP}(w, d, E, G, I, Q, Y)$ defined below, $Q^{[z+1,x]}$ encodes the query to the oracle gate x on layer $z + 1$. Here the type of gate x on layer z is specified by $(G)^{\langle z,x\rangle}$.

Definition 18. *The formula* $\delta^{\alpha}_{MCVP}(w, d, E, G, I, Q, Y)$ *is the formula*

$$\begin{aligned}
&\forall z < d \forall x < w\\
&[\forall t < w(Q^{[z+1,x]}(t) \leftrightarrow (\exists u < w,\ E(z, t, u, x) \wedge Y(z, u)))] \ \wedge\ [Y(0, x) \leftrightarrow I(x)] \wedge\\
&\quad [Y(z + 1, x) \leftrightarrow (((G)^{\langle z+1,x\rangle} = \text{“}\wedge\text{”} \wedge \forall t, u < w,\ E(z, t, u, x) \supset Y(z, u)) \vee\\
&\qquad ((G)^{\langle z+1,x\rangle} = \text{“}\vee\text{”} \wedge \exists t, u < w,\ E(z, t, u, x) \wedge Y(z, u)) \vee\\
&\qquad\qquad ((G)^{\langle z+1,x\rangle} = \text{“}\alpha\text{”} \wedge \alpha(Q^{[z+1,x]}))))]
\end{aligned}$$

Definition 19 ($\mathbf{VAC}^k(\alpha)$). *For* $k \geq 1$, $\mathbf{VAC}^k(\alpha)$ *is the theory over the vocabulary* $\mathcal{L}^2_A \cup \{Row, \alpha\}$ *and is axiomatized by the axioms of* $\mathbf{V}^0$ *and the following axiom:*

$$\forall w, E, G, I(Proper(w, d, E) \supset \exists Q, Y \delta^{\alpha}_{MCVP}(w, (\log w)^k, E, G, I, Q, Y))$$

To specify an $\mathbf{NC}^k(\alpha)$ circuit, we need to express the condition that $\wedge$ and $\vee$ gates have fanin 2. Here we use the following formula $Fanin2'(w, d, E, G)$:

$$\forall z < d \forall x < w((G)^{\langle z,x\rangle} \neq \text{“}\alpha\text{”} \supset \exists u_1, u_2 < w \forall t, v < w,\ E(z, t, v, x) \supset v = u_1 \vee v = u_2)$$

Moreover, the nested depth of oracle gates in circuit (w, d, E, G) needs to be bounded. The formula $OHeight(w, d, h, E, G, H)$ below states that this nested depth is bounded by h:

$$\begin{aligned}
&(\forall z \leq d \forall x < w \exists! s \leq h\, H(z, x, s)) \wedge (\forall x < w H(0, x, 0)) \wedge\\
&\forall z < d \forall x < w \exists m[m = max\{\ell : \exists t, u < w E(z, t, u, x) \wedge H(z, u, \ell)\} \wedge\\
&[((G)^{\langle z+1,x\rangle} = \text{“}\alpha\text{”} \supset H(z + 1, x, m + 1)) \wedge ((G)^{\langle z+1,x\rangle} \neq \text{“}\alpha\text{”} \supset H(z + 1, x, m))]]
\end{aligned}$$

Definition 20 ($\mathbf{VNC}^k(\alpha)$). *For $k \geq 2$, $\mathbf{VNC}^k(\alpha)$ is the theory over $\mathcal{L}_A^2 \cup \{Row, \alpha\}$ and is axiomatized by $\mathbf{V}^0$ and the axiom*

$$\forall w \forall E, G, I, H,\ [Proper(w,d,E) \wedge Fanin2'(w,|w|^k,E,G) \wedge$$
$$OHeight(w,d,|w|^{k-1},E,G,H)] \supset \exists Q, Y \delta_{MCVP}^{\alpha}(w,(\log w)^k,E,G,I,Q,Y)$$

The next theorem can be proved in the same way as Theorem 17.

Theorem 21. *For $k \geq 1$, the functions in $\mathbf{FAC}^k(\alpha)$ are precisely the provably total functions of $\mathbf{VAC}^k(\alpha)$. The same holds for $\mathbf{FNC}^k(\alpha)$ and $\mathbf{VNC}^k(\alpha)$, for $k \geq 2$.*

4 Separation Results

One of the obvious benefits of considering relativized complexity classes is that separations are at hand. Even though the unrelativized inclusion $\mathbf{AC}^1 \subseteq \mathbf{PH}$ is strongly conjectured to be strict, no proof is currently known. On the other hand, relative to an oracle the $\mathbf{AC}^k$-hierarchy is strict. Here we reconstruct a technique used by Takeuti [Tak95] to separate theories in weak bounded arithmetic in a circuit-theoretic setting. Using the hierarchy result together with the witnessing theorem we obtain an unconditional separation of our relativized theories.

The idea is that computing the k'th iterate $f^k(0) = f(f(\dots f(0)))$ of a function f is essentially a sequential procedure, whereas shallow circuits represent parallel computation. So a circuit performing well in a sequential task has to be deep. To avoid that the sequential character of the problem can be circumvented by precomputing all possible values, the domain of f is chosen big enough; we will consider functions $f\colon [2^n] \to [2^n]$.

Of course with such a big domain, we cannot represent such functions simply by a value table. That's how oracles come into play: oracles allow us to provide a predicate on strings as input, without the need of having an input bit for every string. In fact, the number of bits potentially accessible by an oracle gate is exponential in the number of its input wires.

Therefore we represent the i'th bit of $f(x)$ for $x \in \{0,1\}^n$ by whether or not the string $x\underline{i}$ belongs to the language of the oracle. Here $\underline{i}$ is some canonical coding of the natural number i using $\log(n)$ bits.

Our argument can be summarized as follows. We assume a circuit of height h be given that supposedly computes the ℓ'th iterate of any function f given by the oracle. Then we construct, step by step, an oracle that fools this circuit, if $\ell > h$. To do so, for each layer of the circuit we decide how to answer the oracle questions, and we do this in a way that is consistent with the previous layers and such that all the circuit at layer i knows about f is at most the value of $f^i(0)$. Of course, to make this step-by-step construction possible we have to consider partial functions during our construction.

If A and B are sets we denote by $f\colon A \rightharpoonup B$ that f is a partial function from A to B. In other words, f is a function, its domain $\mathrm{dom}(f)$ is a subset of A and its range $\mathrm{rng}(f)$ is a subset of B.

Definition 22. A partial function $f\colon [2^n] \rightharpoonup [2^n]$ is called *ℓ-sequential* if for some $k \leq \ell$ it is the case that $0, f(0), f^2(0), \ldots, f^k(0)$ are all defined, but $f^k(0) \notin \mathrm{dom}(f)$.

Note that in Definition 22 it is necessarily the case that $0, f(0), f^2(0), \ldots,$ $f^k(0)$ are distinct. For the easy proof of the next lemma, see [ACN07].

Lemma 23. *Let $n \in \mathbb{N}$ and $f\colon [2^n] \rightharpoonup [2^n]$ be an ℓ-sequential partial function. Moreover, let $M \subset [2^n]$ such that $|\mathrm{dom}(f) \cup M| < 2^n$. Then there is a $(\ell + 1)$-sequential extension $f' \supseteq f$ with $\mathrm{dom}(f') = \mathrm{dom}(f) \cup M$.*

Definition 24. To any natural number n and any partial function $f\colon [2^n] \rightharpoonup [2^n]$ we associate a its *bit graph* $\alpha_{n,f}$ as a partial function $\alpha_{n,f}\colon \{0,1\}^{n+\log n} \rightharpoonup \{0,1\}$ in the obvious way. More precisely, $\alpha_{n,f}(uv)$ is the i'th bit of $f(x)$ if $f(x)$ is defined, and undefined otherwise, where u is a string of length n coding the natural number x and v is a string of length $\log n$ coding the natural number i.

If $f\colon [2^n] \to [2^n]$ is a total function, we define the set $\mathrm{A}_f = \{x \mid \alpha_{n,f}(x) = 1\} \subseteq \{0,1\}^{n+\log n}$.

Immediately from Definition 24 we note that f can be uniquely reconstructed from A_f. If $A \subseteq \{0,1\}^*$ is a set of bitstrings, we denote by $A^{[n]} = \{x \in A \mid |x| = n + \log n\}$ the set of all strings in A of length $n + \log n$.

In what follows, circuits refer to oracle circuits as discussed in Section 2.1. We are mainly interested in circuits with no Boolean inputs, so the output depends only on the oracle.

Theorem 25. *Let C be any circuit of depth h and size strictly less then 2^n. If C on oracle A computes correctly $f^\ell(0)$ for the (uniquely determined) $f\colon [2^n] \to [2^n]$ such that $\mathrm{A}_f = A^{[n]}$, and this is true for all oracles A, then $\ell \leq h$.*

Proof. Assume that such a circuit computes $f^\ell(0)$ correctly for all oracles. We have to find an oracle that witnesses $\ell \leq h$. First fix the oracle arbitrarily on all strings of length different from $n + \log n$. So, in effect we can assume that the circuit only uses oracle gates with $n + \log n$ inputs.

By induction on $k \geq 0$ we define partial functions $f_k\colon [2^n] \rightharpoonup [2^n]$ with the following properties. (Here we number the *levels* of the circuit $0, 1, \ldots, h-1$.)

- $f_0 \subseteq f_1 \subseteq f_2 \subseteq \ldots$
- The size $|\mathrm{dom}(f_k)|$ of the domain of f_k is at most the number of oracle gates in levels strictly smaller than k.
- α_{n,f_k} determines the values of all oracle gates at levels strictly smaller than k.
- f_k is k-sequential.

We can take f_0 to be the totally undefined function, since $f^0(0) = 0$ by definition. As for the induction step let M be the set of all x of length n such that, for some $i < n$, the string $x\underline{i}$ is queried by an oracle gate at level k and let f_{k+1} be a $k+1$-sequential extension of f_k to domain $\mathrm{dom}(f_k) \cup M$ according to Lemma 23.

For $k = h$ we get the desired bound. As α_{n,f_h} already determines the values of all gates, the output of the circuit is already determined, but $f^{h+1}(0)$ is still undefined and we can define it in such a way that it differs from the output of the circuit. □

Inspecting the proof of Theorem 25 we note that it does not at all use what precisely the non-oracle gates compute, as long as the value only depends on the input, not on the oracle. In particular, the proof still holds if we consider subcircuits without oracle gates as a single complicated gate. Thus we have the following corollary of Theorem 5.

Corollary 26. *cs*$\mathbf{NL}(\alpha)$ *can iterate a function given by an oracle only constantly far. In particular, cs*$\mathbf{NL}(\alpha)$ *is a strict subclass of* $\mathbf{AC}^1(\alpha)$.

Having obtained a lower bound on the depth of an individual circuit, it is a routine argument to separate the corresponding circuit classes. In other words, we are now interested in finding one oracle that simultaneously witnesses that the $\mathbf{AC}^k(\alpha)$-hierarchy is strict. For the uniform classes this is possible by a simple diagonalization argument; in fact, the only property of uniformity we need is that there are at most countably many members in each complexity class. So we will use this as the definition of uniformity. It should be noted that this includes all the known uniformity notions.

Definition 27. If $g\colon \mathbb{N} \to \mathbb{N}$ is a function from the natural numbers to the natural numbers, and $A \subseteq \{0,1\}^*$ an oracle, we define the language

$$\mathcal{L}_g^A = \{x \mid \text{the last bit of } f^{g(n)}(0) \text{ is } 1, \\ \text{where } n = |x| \text{ and } f \text{ is such that } A^{[n]} = \mathrm{A}_f\}$$

We note that in Definition 27 the f is uniquely determined by A and the length of x. Also, for logspace-constructible g the language $\mathcal{L}_g^A$ can be computed by logspace-uniform circuits of depth $g(n)$ and size $n \cdot g(n)$.

Recall that a circuit family is a sequence $(C_n)_{n\in\mathbb{N}}$ of circuits, such that C_n has n inputs and one output. The language of a circuit family $(C_n)_{n\in\mathbb{N}}$ is the set of all strings $x \in \{0,1\}^*$ such that the output of $C_{|x|}$ with input x is 1.

Definition 28. A *notion of uniformity* is any countable set $\mathcal{U}$ of circuit families.

Let $\mathcal{U}$ be a notion of uniformity, and $h, s\colon \mathbb{N} \to \mathbb{N}$ functions. The $\mathcal{U}$-uniform h, s-circuits are those circuit families $(C_n)_{n\in\mathbb{N}} \in \mathcal{U}$ of $\mathcal{U}$ such that C_n has depth at most $h(n)$ and size at most $s(n)$.

By a simple diagonalization argument we obtain the following theorem.

Theorem 29. *Let* $\mathcal{U}$ *be a notion of uniformity and* h_c *a family of functions such that for all* $c \in \mathbb{N}$ *the function* h_{c+1} *eventually strictly dominates* h_c. *Moreover, let* s_c *be a family of strictly subexponentially growing functions. Then there is a single oracle* $A \subseteq \{0,1\}^*$ *that simultaneously witnesses that* $\mathcal{L}_{h_{c+1}}^A$ *cannot be computed by* $\mathcal{U}$*-uniform* h_c, s_c*-circuits.*

Corollary 30. *There is a single oracle $A \subseteq \{0,1\}^*$ for which the relativized versions of* AC^k *form a strict hierarchy.*

Corollary 31. *The theories* $\mathbf{VAC}^k(\alpha)$ *form a strict hierarchy.*

References

[ACN07] Aehlig, K., Cook, S., Nguyen, P.:Relativizing Small Complexity Classes and their Theories (2007), `http://www.cs.toronto.edu/~pnguyen/studies.html`

[BIS90] Mix Barrington, D.A., Immerman, N., Straubing, H.: On Uniformity within NC^1. Journal of Computer and System Sciences 41, 274–306 (1990)

[Bus86] Buss, J.: Relativized Alternation. In: Proceedings, Structure in Complexity Theory Conference, Springer, Heidelberg (1986)

[CN06] Cook, S., Nguyen, P.: Foundations of Proof Complexity: Bounded Arithmetic and Propositional Translations. Book in progress (2006)

[Coo85] Cook, S.: A Taxonomy of Problems with Fast Parallel Algorithms. Information and Control 64(1-3), 2–21 (1985)

[Imm99] Immerman, N.: Descriptive Complexity. Springer, Heidelberg (1999)

[Kra95] Krajíček, J.: Bounded Arithmetic, Propositional Logic, and Complexity Theory. Cambridge University Press, Cambridge (1995)

[LL76] Ladner, R., Lynch, N.: Relativization of questions about log space computability. Mathematical Systems Theory 10, 19–32 (1976)

[NC05] Nguyen, P., Cook, S.: Theory for TC^0 and Other Small Complexity Classes. Logical Methods in Computer Science 2, 1 (2005)

[Orp83] Orponen, P.: General Nonrelativizability Results for Parallel Models of Computation. In: Proceedings, Winter School in Theoretical Computer Science, pp. 194–205 (1983)

[RST84] Ruzzo, W., Simon, J., Tompa, M.: Space-Bounded Hierarchies and Probabilistic Computations. Journal of Computer and System Sciences 28(2), 216–230 (1984)

[Sim77] Simon, I.: On some subrecursive reducibilities. PhD thesis, Stanford University (1977)

[Tak95] Takeuti, G.: Separations of Theories in Weak Bounded Arithmetic. Annals of Pure and Applied Logic 71, 47–67 (1995)

[Wil88] Wilson, C.: A Measure of Relativized Space Which Is Faithful with Respect to Depth. Journal of Computer and System Sciences 36, 303–312 (1988)

[Wil89] Wilson, C.: Relativized NC. Mathematical Systems Theory 20, 13–29 (1989)

[Zam96] Zambella, D.: Notes on Polynomially Bounded Arithmetic. Journal of Symbolic Logic 61(3), 942–966 (1996)

Subexponential Time and Fixed-Parameter Tractability: Exploiting the Miniaturization Mapping

Yijia Chen[1] and Jörg Flum[2]

[1] Shanghai Jiaotong University, China
yijia.chen@cs.sjtu.edu.cn
[2] Albert-Ludwigs-Universität Freiburg, Germany
joerg.flum@math.uni-freiburg.de

Abstract. Recently a mapping $\mathscr{M}$, the so-called miniaturization mapping, has been introduced and it has been shown that $\mathscr{M}$ faithfully translates subexponential parameterized complexity into (unbounded) parameterized complexity [2]. We determine the preimages under $\mathscr{M}$ of various (classes of) problems and show that they coincide with natural reparameterizations which take into account the amount of nondeterminism needed to solve them.

1 Introduction

The idea of fixed-parameter tractability is to approach hard algorithmic problems by isolating problem parameters that can be expected to be small in certain applications and then develop algorithms that are polynomial except for an arbitrary dependence on the parameter. More precisely: A parameterized problem is a pair (Q, κ), where Q is a classical problem, say, over the alphabet Σ and $\kappa : \Sigma^* \to \mathbb{N}$ is a polynomial time computable function assigning to every $x \in \Sigma^*$ its *parameter* $\kappa(x)$. The problem (Q, κ) is fixed-parameter tractable if it can be solved by an algorithm the running time of which is bounded by $f(\kappa(x)) \cdot |x|^{O(1)}$, where f is an arbitrary computable function. The class of all fixed-parameter tractable problems is denoted by FPT.

There are natural problems that are fixed-parameter tractable, but require a parameter dependence f of the form $f(k) := 2^{2^k}$ or even worse. However, even for small values of the parameter $\kappa(x)$, a running time such as $2^{2^{2^{\kappa(x)}}} \cdot |x|$ is prohibitive. Hence, besides the *unbounded* fixed-parameter tractability more restrictive notions of tractability have been studied obtained by simply putting upper bounds on the growth of the "parameter dependence" f, the most restrictive one considered so far being $f \in 2^{o(k)}$.[1] The corresponding class of "tractable" problems has been denoted by SUBEPT and the corresponding theory by *subexponential parameterized complexity*.

It is a beautiful aspect of subexponential parameterized complexity theory that it can be faithfully translated into unbounded parameterized complexity theory via the *miniaturization mapping* $\mathscr{M}$: Let (Q, κ) be a parameterized problem over the alphabet Σ. The *miniaturization* $\mathscr{M}(Q, \kappa)$ of (Q, κ) is the parameterized problem:

[1] In the precise definition given in Section 2 we have to require $f \in 2^{o^{\text{eff}}(k)}$, an effective version of $f \in 2^{o(k)}$.

J. Duparc and T.A. Henzinger (Eds.): CSL 2007, LNCS 4646, pp. 389–404, 2007.

> $\mathscr{M}(Q, \kappa)$
> *Instance:* $x \in \Sigma^*$ and $m \in \mathbb{N}$ in unary.
> *Parameter:* $\left\lceil \frac{\kappa(x)}{\log m} \right\rceil$.
> *Question:* Is $x \in Q$?

The mapping $\mathscr{M}$ strongly preserves reducibilities and every parameterized problem in XP has a preimage under the miniaturization map. Here XP denotes the class of parameterized problems (Q, κ) solvable in time $O(|x|^{f(\kappa(x))})$. To state the precise result, proven in [2], we write $(Q, \kappa) \leq^{\text{serf}} (Q', \kappa')$ and $(Q, \kappa) \leq^{\text{fpt}} (Q', \kappa')$ if there is a serf-reduction and an fpt-reduction, respectively, from (Q, κ) to (Q', κ') (compare Section 2 for the definitions of all concepts introduced informally in this introduction).

Theorem 1 (Miniaturization Theorem). *(1) Let (Q, κ) be a parameterized problem. Then*

$$(Q, \kappa) \in \text{SUBEPT} \iff \mathscr{M}(Q, \kappa) \in \text{FPT}.$$

(2) Let (Q, κ) and (Q', κ') be parameterized problems. Then

$$(Q, \kappa) \leq^{\text{serf}} (Q', \kappa') \iff \mathscr{M}(Q, \kappa) \leq^{\text{fpt}} \mathscr{M}(Q', \kappa').$$

(3) Let $(Q, \kappa) \in$ XP. Then there exists a problem (Q', κ') such that

$$\mathscr{M}(Q', \kappa') \equiv^{\text{fpt}} (Q, \kappa).$$

The class XP comprises the most important classes of intractable parameterized problems, namely the classes of the W-hierarchy. We recall a property of these classes. For a set Γ of propositional formulas the *parameterized weighted satisfiability problem* for Γ is defined by

> p-WSAT(Γ)
> *Instance:* A propositional formula $\alpha \in \Gamma$ and $k \in \mathbb{N}$.
> *Parameter:* k.
> *Question:* Does α have a satisfying assignment of weight k?

Then for the classes W[1], W[2], ... of the W-hierarchy the following holds:

– If $t, d \geq 1$ and $t{+}d \geq 3$, then p-WSAT($\Gamma_{t,d}$) is W[t]-complete under fpt-reductions.

Here $\Gamma_{t,d}$ contains the propositional formulas that are big conjunctions of big disjunctions of big conjunctions ... (t alternations) of conjunctions (if t is even) and disjunctions (if t is odd) of d literals (see [5] for the precise definition).

In [2] the authors consider the reparameterization *form*-WSAT(Γ) of p-WSAT(Γ), where

> *form*-WSAT(Γ)
> *Instance:* A formula $\alpha \in \Gamma$ and $r \geq 1$.
> *Parameter:* $r \cdot \lceil \log |\alpha| \rceil$.
> *Question:* Does α have a satisfying assignment of weight r?

They show for the Γ's relevant for the W[t]'s that the problem *form*-WSAT(Γ) is the preimage of p-WSAT(Γ) under the miniaturization mapping, that is:

– If $t, d \geq 1$ and $t + d \geq 3$, then $\mathscr{M}$(*form*-WSAT($\Gamma_{t,d}$)) $\equiv^{\text{fpt}}$ p-WSAT($\Gamma_{t,d}$).

Already an analysis of the proof of part (3) of the Miniaturization Theorem shows that the preimage of a parameterized problem (Q, κ) under the miniaturization mapping essentially is a reparameterization of (Q, κ). How do we get this reparameterization? Is there a natural reparameterization that is a preimage? Can we determine the preimage for a large class of problems? These are the problems we address in this paper. In [2] the authors prove that the reparameterization of problems obtained by multiplying the parameter by the logarithm of the size of the instance gives an inverse for the miniaturization mapping for all problems satisfying certain technical conditions. In particular, *form*-WSAT(Γ) is this reparameterization of p-WSAT(Γ). We generalize this approach.

Let (Q, κ) be a parameterized problem over the alphabet Σ. Often (the classical problem) Q has a canonical representation of the form

$$x \in Q \iff \text{there is } y \in \{0,1\}^{g(x)} \text{ such that } (x, y) \in Q_0 \qquad (1)$$

for some polynomial time computable function $g : \Sigma^* \to \mathbb{N}$ with $g(x) \leq |x|$ and some $Q_0 \in$ PTIME. (We give such a representation for $Q :=$ WSAT(Γ) below.) Consider the nondeterministic algorithm solving $x \in Q$ by guessing $y \in \{0,1\}^{g(x)}$ and then verifying that $(x, y) \in Q_0$ in polynomial time. The deterministic procedure simulating all possible computation paths of this nondeterministic algorithm takes time $2^{g(x)} \cdot |x|^{O(1)}$. The question arises whether we can do better and solve the problem $x \in Q$ in time

$$2^{o(g(x))} \cdot |x|^{O(1)},$$

thus showing $(Q, g) \in$ SUBEPT. We call (Q, g) the *canonical parameterization* of Q (more precisely, one should speak of the canonical parameterization induced by the representation (1) of Q) and sometimes we say that (Q, g) is the *canonical reparameterization* of (Q, κ).

Clearly, the problem (Q, g) is fixed-parameter tractable. Even more, every parameterized problem where the parameter increases monotonically with the size of the input is fixed-parameter tractable (cf. [5, Proposition 1.7]). The proof of this result heavily relies on the fact that we allow arbitrary computable functions in the definition of FPT. Indeed, from the point of view of the unbounded theory the parameters of such parameterizations are not small. However this is not the case in the subexponential theory.

Let Q be the weighted satisfiability problem WSAT(Γ) for formulas in Γ, that is, $(\alpha, k) \in Q$ if and only if $\alpha \in \Gamma$ and α has a satisfying assignment of weight k. A canonical representation of Q of the form (1) is

$$(\alpha, k) \in Q \iff \text{there is a } y \in \{0,1\}^{k \cdot \lceil \log |\text{var}(\alpha)| \rceil} \text{ such that } ((\alpha, k), y) \in Q_0,$$

where var(α) is the set of variables of α and $((\alpha, k), y) \in Q_0$ means that $\alpha \in \Gamma$, that y contains the binary representation of k distinct variables of α and that the assignment setting exactly these variables to TRUE satisfies α. Hence the canonical parameterization is the problem

> *var*-WSAT(Γ)
> *Instance:* A formula $\alpha \in \Gamma$ and $r \geq 1$.
> *Parameter:* $r \cdot \lceil \log |\text{var}(\alpha)| \rceil$.
> *Question:* Does α have a satisfying assignment of weight r?

We show that in addition to *form*-WSAT($\Gamma_{t,d}$) also *var*-WSAT($\Gamma_{t,d}$) is a preimage of p-WSAT($\Gamma_{t,d}$).

The main question we address is: For what other parameterized problems is the canonical reparameterization a preimage under the miniaturization map?

In Section 3 we show that the answer is positive for the parameterized dominating set problem. In Section 4 we introduce a general framework that allows us to carry out the arguments we use for the dominating set problem. We apply this abstract approach in Section 5 to reprove and extend the results obtained in [2] for the weighted satisfiability problem. Perhaps the most far-reaching positive answer to our main question is obtained in Section 6:

> If $t \geq 1$ and the parameterized problem (Q, κ) is W[t]-complete and Fagin-definable by a Π_t-formula, then the preimage of (Q, κ) under the miniaturization mapping is its canonical reparameterization.

To broaden the range of applicability of this result we deal with *relativized* Fagin-definable problems. As an application we get:

> If one can decide whether a hypergraph $\mathcal{H}$ has a hitting set of size r in time $2^{o(r\cdot\log(|V|+|E|))} \cdot \|\mathcal{H}\|^{O(1)}$, then one can even do it in time $2^{o(r\cdot\log|V|)} \cdot \|\mathcal{H}\|^{O(1)}$.

In the last section we discuss the main question for model-checking problems and give an application to the homomorphism problem (see Corollary 31).

Due to space limitations we have to defer most proofs to the full version of the paper.

2 Preliminaries

The set of natural numbers (that is, nonnegative integers) is denoted by $\mathbb{N}$. For a natural number n let $[n] := \{1, \ldots, n\}$. By $\log n$ we mean $\lceil \log n \rceil$ if an integer is expected. For $n = 0$ the term $\log n$ is undefined. We trust the reader's common sense to interpret such terms reasonably.

2.1 Parameterized Complexity

Recall that a *parameterized problem* is a pair (Q, κ), where Q is a classical problem, say, over the alphabet Σ and $\kappa : \Sigma^* \to \mathbb{N}$ is a polynomial time computable function assigning to every $x \in \Sigma^*$ its *parameter* $\kappa(x)$.

The problem (Q, κ) is *fixed-parameter tractable* if it can be solved by an algorithm in time $f(\kappa(x)) \cdot |x|^{O(1)}$, where f is an arbitrary computable function. The class of all fixed-parameter tractable problems is denoted by FPT.

Let $f, g : \mathbb{N} \to \mathbb{N}$ be computable functions. Then $f \in o^{\text{eff}}(g)$ if there is a computable function h such that for all $\ell \geq 1$ and $k \geq h(\ell)$, we have $f(k) \leq g(k)/\ell$. We often write $f(k) \in o^{\text{eff}}(g(k))$ instead of $f \in o^{\text{eff}}(g)$. A parameterized problem (Q, κ) is in SUBEPT if $x \in Q$ is solvable in time $f(\kappa(x)) \cdot |x|^{O(1)}$ for some computable function f with $f(k) \in 2^{o^{\text{eff}}(k)}$.

We assume that the reader is familiar with the notions of (many-one) reductions for the unbounded and for the subexponential parameterized complexity (cf. [5]), which we call fpt-reduction and serf-reduction, respectively. For $* \in \{\text{fpt}, \text{serf}\}$ we write $(Q, \kappa) \leq^* (Q', \kappa')$ if there is a $*$-reduction from (Q, κ) to (Q', κ'), and we write $(Q, \kappa) \equiv^* (Q', \kappa')$ if $(Q, \kappa) \leq^* (Q', \kappa')$ and $(Q', \kappa') \leq^* (Q, \kappa)$.

2.2 Propositional Logic

Formulas of propositional logic are built as usual. Let PROP be the class of all propositional formulas. We use the notations from [5] and in particular its definitions of the classes $\Gamma_{t,d}$, $\Delta_{t,d}$, $\Gamma^+_{t,d}$, $\Delta^+_{t,d}$ and $\Gamma^-_{t,d}$ and $\Delta^-_{t,d}$.

Let V be a set of propositional variables. Often we tacitly identify an assignment $S : V \to \{\text{TRUE}, \text{FALSE}\}$ with the set $\{X \in V \mid S(X) = \text{TRUE}\}$. The *weight* of an assignment S is $|S|$, the number of variables set to TRUE. A propositional formula α is *k-satisfiable* (where $k \in \mathbb{N}$), if there is an assignment for the set var(α) of variables of α of weight k satisfying α. Recall from the Introduction the definition of the *parameterized weighted satisfiability problem* p-WSAT(Γ) for a set Γ of propositional formulas. We shall use the following well-known result.

Theorem 2. *Let $t, d \in \mathbb{N}$ with $t + d \geq 3$.*
- *If t is even, then p-WSAT($\Gamma^+_{t,d}$) is W[t]-complete under fpt-reductions.*
- *If t is odd, then p-WSAT($\Gamma^-_{t,d}$) is W[t]-complete under fpt-reductions.*
- *p-WSAT(PROP) is W[SAT]-complete under fpt-reductions.*

2.3 First-Order Logic

We assume familiarity with the basic notions of first-order logic: (relational) vocabulary, structure, size of a structure, first-order formula, and the satisfaction relation (see [5]).

If φ is a first-order formula, we write $\varphi(x_1, \ldots, x_m)$ to indicate that the free variables in φ are $x_1, \ldots, x_m$. If $\mathcal{A}$ is τ-structure and $\varphi(x_1, \ldots, x_m)$ a formula, we let

$$\varphi(\mathcal{A}) := \{(a_1, \ldots, a_m) \in A^m \mid \mathcal{A} \models \varphi(a_1, \ldots, a_m)\}.$$

For $t \geq 0$ let Π_t denote the class of all first-order formulas of the form

$$\forall x_{11} \ldots \forall x_{1k_1} \exists x_{21} \ldots \exists x_{2k_2} \;\ldots\; Q x_{t1} \ldots Q x_{tk_t}\; \psi,$$

where $Q = \exists$ if t is even and $Q = \forall$ otherwise, and where ψ is quantifier-free.

For $t \geq 0$, $u \geq 1$ a formula is in $\Pi^0_{t,u}$ if it is in Π_t and all quantifier blocks have length bounded by u. For $s \geq 1$, $\Phi[s]$ denotes the class of formulas in Φ whose vocabulary has arity $\leq s$. The following result is well-known.

Theorem 3. *Let $t \geq 0$, $u \geq 1$, and $s \geq 2$. Then p-MC($\Pi^0_{t,u}$) and p-MC($\Pi^0_{t,u}[s]$) are W[$t + 1$]-complete under fpt-reductions, where*

p-MC(Φ) *Instance:* A structure $\mathcal{A}$ and a formula $\varphi \in \Phi$. *Parameter:* $\|\varphi\|$. *Question:* Is $\varphi(\mathcal{A}) \neq \emptyset$?

Here Φ is a set of first-order formulas and $|\varphi|$ denotes the length of the formula φ.

3 An Example

We consider the parameterized dominating set problem

p-DOMINATING-SET *Instance:* A graph $\mathcal{G} = (V, E)$ and $k \in \mathbb{N}$. *Parameter:* k. *Question:* Does $\mathcal{G}$ have a dominating set of size k?

and its canonical reparameterization

uni-DOMINATING-SET *Instance:* A graph $\mathcal{G} = (V, E)$ and $r \in \mathbb{N}$. *Parameter:* $r \cdot \log \|V\|$. *Question:* Does $\mathcal{G}$ have a dominating set of size r?

Theorem 4. $\mathscr{M}$(*uni*-DOMINATING-SET) $\equiv^{\text{fpt}}$ p-DOMINATING-SET.

Proof. p-DOMINATING-SET $\leq^{\text{fpt}}$ $\mathscr{M}$(*uni*-DOMINATING-SET): Let $(\mathcal{G}, k)$ be an instance of p-DOMINATING-SET with $\mathcal{G} = (V, E)$. We set $r := k$ and $m := \|(\mathcal{G}, k)\|$ and consider the instance $(\mathcal{G}, r, m)$ of $\mathscr{M}$(*uni*-DOMINATING-SET). As

$$\left\lceil \frac{r \cdot \log |V|}{\log m} \right\rceil \leq \left\lceil \frac{k \cdot \log |V|}{\log |V|} \right\rceil = k$$

the mapping $(\mathcal{G}, k) \mapsto (\mathcal{G}, r, m)$ is an fpt-reduction.

$\mathscr{M}$(*uni*-DOMINATING-SET) $\leq^{\text{fpt}}$ p-DOMINATING-SET: Let $(\mathcal{G}, r, m)$ be an instance of $\mathscr{M}$(*uni*-DOMINATING-SET) with $\mathcal{G} = (V, E)$. We can assume that

$$2 \leq r \leq |V|. \tag{2}$$

Recall that the parameter of $(\mathcal{G}, r, m)$ is

$$k := \left\lceil \frac{r \cdot \log |V|}{\log m} \right\rceil. \tag{3}$$

Now we distinguish two cases. If $|V| > m$, then $k > r$ by (3). It follows that we can map $(\mathcal{G}, r, m)$ to the equivalent instance $(\mathcal{G}, r)$ of p-DOMINATING-SET (its parameter r is bounded by k). In case $|V| \leq m$ we have

$$k \leq r. \tag{4}$$

Moreover, $r \cdot \log|V| \le k \cdot \log m$ and hence

$$|V|^{r/k} \le m. \tag{5}$$

We want to get from the instance $(\mathcal{G}, r, m)$ of $\mathscr{M}$(*uni*-DOMINATING-SET) in fpt-time a pair $(\mathcal{G}', k')$ such that

$$\mathcal{G} \text{ has a dominating set of size } r \iff \mathcal{G}' \text{ has a dominating set of size } k'$$

and such that $k' \le g(k)$ for some computable g. The next lemma shows that we can even get such a $(\mathcal{G}', k')$ with $k' = k$ in time polynomial in $|V|^{r/k} + \|\mathcal{G}\|$, and hence by (5), in time polynomial in m. Note that by (4) and (2) the assumptions of the lemma are satisfied. □

Lemma 5. *There is an algorithm that assigns to every graph $\mathcal{G} = (V, E)$ and $r, k \in \mathbb{N}$ with*

$$k \le r \le |V| \text{ and } 2 \le |V|$$

a graph $\mathcal{G}'$ such that

$$\mathcal{G} \text{ has a dominating set of size } r \iff \mathcal{G}' \text{ has a dominating set of size } k$$

in time polynomial in $|V|^{r/k} + \|\mathcal{G}\|$.

As p-DOMINATING-SET is W[2]-complete, we get from Theorem 4:

Corollary 6. W[2] = FPT *if and only if there is an algorithm deciding whether a graph $\mathcal{G} = (V, E)$ has a dominating set of size k in time $2^{o^{\text{eff}}(k \cdot \log|V|)} \cdot |V|^{O(1)}$.*

4 The General Framework

We show that the result obtained in the previous section for the dominating set problem can be generalized to any parameterized problem satisfying an analogue of Lemma 5.

We start with a simple observation that often will be useful.

Proposition 7. *Let (Q, κ) be a parameterized problem over the alphabet Σ and $\kappa' : \Sigma^* \to \mathbb{N}$ a further parameterization of Q such that $\kappa'(x) \le O(\kappa(x) \cdot \log|x|)$. Then:*

$$(Q, \kappa) \le^{\text{fpt}} \mathscr{M}(Q, \kappa').$$

In particular, $(Q, \kappa) \le^{\text{fpt}} \mathscr{M}(Q, \kappa)$.

Corollary 8. *Let (Q, κ) be a parameterized problem over the alphabet Σ and κ' a further parameterization such that $\kappa'(x) \le O(\kappa(x) \cdot \log|x|)$. Furthermore let $t \ge 1$. If (Q, κ) is W[t]-complete under fpt-reductions and $\mathscr{M}(Q, \kappa') \in$ W[t], then $\mathscr{M}(Q, \kappa')$ is W[t]-complete under fpt-reductions, too.*

Now we turn to a generalization of the results of the previous section.

Definition 9. Let (Q, κ) be a parameterized problem over the alphabet Σ. Let $h : \Sigma^* \to \mathbb{N}$ be a function computable in polynomial time. We say that (Q, κ) has the *h-condensation property* if there is an algorithm that for every $x \in \Sigma^*$ and $k \in \mathbb{N}$ with $1 \leq k \leq \kappa(x)$ computes an $x' \in \Sigma^*$ in time polynomial in

$$h(x)^{\kappa(x)/k} + |x|$$

such that $\kappa(x') \leq k$ and $(x \in Q \iff x' \in Q)$.

In [2] the authors introduce the notion of scalable parameterized problem and prove Theorem 12 below for scalable problems. The reader familiar with that paper will easily show that a parameterized problem is scalable if and only if it has the h-condensation property for $h(x) := |x|$.

Examples 10. (a) Lemma 5 shows that p-DOMINATING-SET has the h-condensation property for the function h given by $h(\mathcal{G}, r) := |V|$.[2]

(b) p-INDEPENDENT-SET has the h-condensation property for the function h given by $h(\mathcal{G}, r) := |V|$. The verification is similar to the case of the dominating set problem and is implicit in [2] (note that for $r \leq |V|$ we have $|V| \leq \|(\mathcal{G}, r)\| \leq O(|V|^2)$).

(c) Let $t, d \in \mathbb{N}$ with $t + d \geq 3$. The problem p-WSAT(Γ) for $\Gamma := \Gamma^+_{t,d}$ with even t, $\Gamma := \Gamma^-_{t,d}$ with odd t, and $\Gamma :=$ PROP has the h-condensation property for the function h given by $h(\alpha, r) := |\text{var}(\alpha)|$ (= number of variables of α).

Definition 11. Let (Q, κ) be a parameterized problem over the alphabet Σ. Let $h : \Sigma^* \to \mathbb{N}$ be a function computable in polynomial time. The *h-reparameterization* (Q, κ_h) *of* (Q, κ) is then given by

$$\kappa_h(x) := \kappa(x) \cdot \log h(x).$$

Along the lines of the proof of Theorem 4 one can show:

Theorem 12. *Let (Q, κ) be a parameterized problem over Σ. Let $h : \Sigma^* \to \mathbb{N}$ be a function computable in polynomial time such that $h(x) \leq |x|$ for all $x \in \Sigma^*$. Furthermore assume that (Q, κ) has the h-condensation property. Then*

$$\mathscr{M}(Q, \kappa_h) \equiv^{\text{fpt}} (Q, \kappa).$$

Corollary 13. *Let (Q, κ) be a parameterized problem over Σ. Let $h : \Sigma^* \to \mathbb{N}$ be a function computable in polynomial time such that $h(x) \leq |x|$ for all $x \in \Sigma^*$. Furthermore assume that (Q, κ) has the h-condensation property. Then (Q, κ) is fixed-parameter tractable if and only if $x \in Q$ is solvable in time $2^{o^{\text{eff}}(\kappa(x) \cdot \log h(x))} \cdot |x|^{O(1)}$.*

5 Further Applications

Let Γ be a set of propositional formulas. We consider the h-reparameterizations *var*-WSAT(Γ) and *form*-WSAT(Γ) of the parameterized weighted satisfiability problem p-WSAT(Γ) given by setting $h(\alpha, r) := |\text{var}(\alpha)|$ and $h(\alpha, r) := |\alpha|$, respectively; that is,

[2] Lemma 5 only shows that the relevant inequality holds for $k \neq 1$. As usual, the failure for finitely many values is not relevant here and thus we do not mention it in the following.

var-WSAT(Γ) *Instance:* A formula $\alpha \in \Gamma$ and $r \geq 1$. *Parameter:* $r \cdot \log \|\text{var}(\alpha)\|$. *Question:* Does α have a satisfying assignment of weight r?

and

form-WSAT(Γ) *Instance:* A formula $\alpha \in \Gamma$ and $r \geq 1$. *Parameter:* $r \cdot \log \|\alpha\|$. *Question:* Does α have a satisfying assignment of weight r?

As already remarked in [2]:

Lemma 14. *Let $t, d \in \mathbb{N}$ with $t + d \geq 3$. Then for even t*

$$var\text{-WSAT}(\Gamma_{t,d}) \equiv^{\text{serf}} var\text{-WSAT}(\Gamma^+_{t,d})$$

and for odd t

$$var\text{-WSAT}(\Gamma_{t,d}) \equiv^{\text{serf}} var\text{-WSAT}(\Gamma^-_{t,d}).$$

Theorem 15. *Let $t, d \in \mathbb{N}$ with $t + d \geq 3$. If t is even, then the following problems are* W[t]*-complete under fpt-reductions:*

(1) $\mathcal{M}$(*var*-WSAT($\Gamma_{t,d}$)); *(3)* $\mathcal{M}$(*form*-WSAT($\Gamma_{t,d}$));
(2) $\mathcal{M}$(*var*-WSAT($\Gamma^+_{t,d}$)); *(4)* $\mathcal{M}$(*form*-WSAT($\Gamma^+_{t,d}$)).

For odd t one has to replace $\Gamma^+_{t,d}$ by $\Gamma^-_{t,d}$.

Similarly one gets:

Corollary 16. *The problems $\mathcal{M}$(form-*WSAT(PROP)) *and $\mathcal{M}$(var-*WSAT(PROP)) *are* W[SAT]*-complete under fpt-reductions.*

The previous results have the following surprising consequence. We do not know for what other sets Γ of propositional formulas the following claim holds.

Corollary 17. *Let $t, d \in \mathbb{N}$ with $t + d \geq 3$. Let $\Gamma := \Gamma_{t,d}$ or $\Gamma := \Gamma^+_{t,d}$ for even t or $\Gamma := \Gamma^-_{t,d}$ for odd t, or $\Gamma :=$* PROP. *Then*

$$var\text{-WSAT}(\Gamma) \equiv^{\text{serf}} form\text{-WSAT}(\Gamma).$$

Part (2) of the following result is due to Chen et. al. [1]. Recall the following relation between the classes of the M-*hierarchy* and the classes of the W-hierarchy: for every $t \geq 1$, M[t] $\subseteq$ W[t] $\subseteq$ M[$t + 1$], and M[SAT] = W[SAT] (see [4,5]).

Theorem 18. *Let $t, d \in \mathbb{N}$ with $t + d \geq 3$.*

(1) If W[t] = FPT, *then one can decide* WSAT($\Gamma_{t,d}$) *in time* $|\text{var}(\alpha)|^{o^{\text{eff}}(k)} \cdot |\alpha|^{O(1)}$ *for every instance* (α, k).

(2) If M[t] $\neq$ FPT, *then one cannot decide* WSAT($\Gamma_{t,d}$) *in time* $|\text{var}(\alpha)|^{o^{\text{eff}}(k)} \cdot |\alpha|^{O(1)}$ *for every instance* (α, k).

By the same method one gets:

Theorem 19. W[SAT] = FPT *if and only if one can decide* WSAT(PROP) *in time* $|\text{var}(\alpha)|^{o^{\text{eff}}(k)} \cdot |\alpha|^{O(1)}$ *for every instance* (α, k).

6 Fagin-Definable Problems

Let $\varphi(X)$ be any first-order formula with a (second-order) set variable X. Recall that it *Fagin-defines* the parameterized problem

> p-WD$_\varphi$
> *Instance:* A structure $\mathcal{A}$ and $k \in \mathbb{N}$.
> *Parameter:* k.
> *Question:* Is there a subset S of A such that $\mathcal{A} \models \varphi(S)$ and $|S| = k$?

Its canonical reparameterization is the problem (we denote the universe of a structure $\mathcal{A}$ by A)

> *uni*-WD$_\varphi$
> *Instance:* A structure $\mathcal{A}$ and $r \in \mathbb{N}$.
> *Parameter:* $r \cdot \log |A|$.
> *Question:* Is there a subset S of A such that $\mathcal{A} \models \varphi(S)$ and $|S| = r$?

Our main result reads as follows:

Theorem 20. *Let $\varphi(X)$ be a Π_t-formula and assume that p-WD$_\varphi$ is W[t]-complete under fpt-reductions. Then $\mathscr{M}$(uni-WD$_\varphi$) is W[t]-complete under fpt-reductions.*

This result has many applications. For example, the problems p-CLIQUE and p-SET-PACKING are W[1]-complete problems Fagin-definable by Π_1-formulas and p-TOURNAMENT-DOMINATING-SET and p-KERNEL are W[2]-complete problems Fagin-definable by Π_2-formulas. Thus their canonical reparameterizations are W[1]-complete and W[2]-complete, respectively (cf. [5] for the definitions of the problems).

Before we turn to a proof of Theorem 20, we consider an example which suggests to prove a more general result. HITTING-SET is the problem:

> HITTING-SET
> *Instance:* A hypergraph $\mathcal{H} = (V, E)$ and $k \in \mathbb{N}$.
> *Question:* Does there exist a set $S \subseteq V$ of size k such that $S \cap e \neq \emptyset$ for all $e \in E$?

Its canonical parameterization is:

> *vert*-HITTING-SET
> *Instance:* A hypergraph $\mathcal{H} = (V, E)$ and $r \in \mathbb{N}$.
> *Parameter:* $r \cdot \log |V|$.
> *Question:* Does there exist a set $S \subseteq V$ of size r such that $S \cap e \neq \emptyset$ for all $e \in E$?

In order to rewrite the problem as a Fagin-definable one, we represent a hypergraph $\mathcal{H}$ as a τ_{HG}-structure $\mathcal{A}(\mathcal{H})$, where

$$\tau_{\text{HG}} := \{\mathit{VERT}, \mathit{EDGE}, I\}$$

with unary relation symbols *VERT* and *EDGE* and binary relation symbol *I*: We let

$$\begin{array}{ll} A(\mathcal{H}) := V \cup E, & VERT^{\mathcal{A}(\mathcal{H})} := V, \\ EDGE^{\mathcal{A}(\mathcal{H})} := E, & I^{\mathcal{A}(\mathcal{H})} := \{(v,e) \mid v \in V, e \in E, \text{ and } v \in e\}. \end{array}$$

The following formula expresses that X is a hitting set:

$$hit_0(X) := \forall x \exists y \big((Xx \rightarrow VERT\,x) \wedge (EDGE\,x \rightarrow (Xy \wedge Iyx))\big).$$

Furthermore, it is not hard to see that there is a Π_2-sentence *hyp* of vocabulary τ_{HG}, which is satisfied exactly by those τ_{HG}-structures that, up to isomorphism, have the form $\mathcal{A}(\mathcal{H})$ for some hypergraph $\mathcal{H}$. We set

$$hit(X) := hyp \wedge hit_0(X).$$

Then $hit(X)$ is (equivalent to) a Π_2-formula and

$$uni\text{-}\mathrm{WD}_{hit} \leq^{\mathrm{serf}} vert\text{-}\textsc{Hitting-Set},$$

as shown by the mapping $(\mathcal{A}(\mathcal{H}), r) \mapsto (\mathcal{H}, r)$. However uni-WD_{hit} does not coincide with *vert*-HITTING-SET, because the former has the parameter $r \cdot \log\,(|V| + |E|)$ and the latter the parameter $r \cdot \log\,|V|$. Note that $|E|$ can be as large as $2^{|V|}$.

Therefore, we consider a more general reparameterization of the problem Fagin-defined by a formula $\varphi(X)$, namely a *relativized* version, where a subset of the universe is part of the instance; this subset must contain the solution:

> $uni\text{-}rela\text{-}\mathrm{WD}_{\varphi}$
> *Instance:* A structure $\mathcal{A}$, a set $U \subseteq A$, and $r \in \mathbb{N}$.
> *Parameter:* $r \cdot \log\,|U|$.
> *Question:* Does there exist a set $S \subseteq U$ of size r such that $\mathcal{A} \models \varphi(S)$?

As shown by the reduction $(\mathcal{A}, r) \mapsto (\mathcal{A}, A, r)$, we have

$$uni\text{-}\mathrm{WD}_{\varphi} \leq^{\mathrm{serf}} uni\text{-}rela\text{-}\mathrm{WD}_{\varphi};$$

hence, by the Miniaturization Theorem

$$\mathscr{M}(uni\text{-}\mathrm{WD}_{\varphi}) \leq^{\mathrm{fpt}} \mathscr{M}(uni\text{-}rela\text{-}\mathrm{WD}_{\varphi}). \tag{6}$$

Furthermore, the reduction $(\mathcal{H}, r) \mapsto (\mathcal{A}(\mathcal{H}), V, r)$ shows that

$$vert\text{-}\textsc{Hitting-Set} \leq^{\mathrm{serf}} uni\text{-}rela\text{-}\mathrm{WD}_{hit}.$$

The main technical lemma of this section reads as follows:

Lemma 21. *Let $t \geq 1$ and $\varphi(X) \in \Pi_t$. Then $\mathscr{M}(uni\text{-}rela\text{-}\mathrm{WD}_{\varphi}) \in \mathrm{W}[t]$.*

Proof of Theorem 20: Proposition 7 and (6) imply

$$p\text{-}\mathrm{WD}_{\varphi} \leq^{\mathrm{fpt}} \mathscr{M}(uni\text{-}\mathrm{WD}_{\varphi}) \leq^{\mathrm{fpt}} \mathscr{M}(uni\text{-}rela\text{-}\mathrm{WD}_{\varphi})$$

Then, Lemma 21 and the W[t]-hardness of p-WD_{φ} yield $\mathscr{M}(uni\text{-}\mathrm{WD}_{\varphi}) \in \mathrm{W}[t]$. □

As an application we get:

Proposition 22. *The miniaturizations of the problem size-*HITTING-SET *and of the problem vert-*HITTING-SET *are* W[2]*-complete under fpt-reductions, where*

size-HITTING-SET
Instance: A hypergraph $\mathcal{H} = (V, E)$ and $r \geq 1$.
Parameter: $r \cdot \log(|V| + |E|)$.
Question: Does there exist a set $S \subseteq V$ of size r such that $S \cap e \neq \emptyset$ for all $e \in E$?

As an immediate consequence, we get:

Corollary 23. *If there is an algorithm solving* HITTING-SET *in time* $2^{o^{\text{eff}}(r \cdot \log(|V|+|E|))} \cdot \|\mathcal{H}\|^{O(1)}$, *then* HITTING-SET *is solvable in time* $2^{o^{\text{eff}}(r \cdot \log |V|)} \cdot \|\mathcal{H}\|^{O(1)}$.

We turn to the clique problem. As the W[1]-complete problem p-CLIQUE is Fagin-definable by a Π_1-formula, by Theorem 20 we see that $\mathscr{M}$(*uni*-CLIQUE) is W[1]-complete. Therefore:

Corollary 24. W[1] = FPT *if and only if one can decide whether a graph* $\mathcal{G} = (V, E)$ *has a clique of size* k *in time* $2^{o^{\text{eff}}(k \cdot \log |V|)} \cdot |V|^{O(1)}$.

Now similarly as Theorem 18, one can show:

Theorem 25. *(1) If* W[1] = FPT, *then one can decide whether a graph* $\mathcal{G} = (V, E)$ *has a clique of size* k *in time* $|V|^{o^{\text{eff}}(k)}$.

(2) If M[1] $\neq$ FPT, *then one cannot decide whether a graph* $\mathcal{G} = (V, E)$ *has a clique of size* k *in time* $|V|^{o^{\text{eff}}(k)}$.

Part (2) is again proved by Chen et. al. [1].

We close this section with two open problems. From Proposition 7 we know that

$$p\text{-WD}_\varphi \leq^{\text{fpt}} \mathscr{M}(uni\text{-WD}_\varphi) \tag{7}$$

holds for every formula $\varphi(X)$. Is there a natural Fagin-definable problem p-WD$_\varphi$, for which $\leq^{\text{fpt}}$ cannot be replaced by $\equiv^{\text{fpt}}$ in (7) (modulo complexity theoretic assumptions). We believe that this could be the case for p-CLIQUE-OR-INDEPENDENT-SET, where

p-CLIQUE-OR-INDEPENDENT-SET
Instance: A graph $\mathcal{G}$ and $k \in \mathbb{N}$.
Parameter: k.
Question: Does $\mathcal{G}$ have a clique or an independent set of size k?

It is known that the problem is fixed-parameter tractable. However, is it solvable in time $2^{o^{\text{eff}}(k \cdot \log |V|)} \cdot |V|^{O(1)}$?

We come to our second open question. We consider the problem of computing the Vapnik–Chervonenkis dimension of a hypergraph. We let

> p-VC-DIMENSION
> *Instance:* A hypergraph $\mathcal{H} = (V, E)$ and $r \in \mathbb{N}$.
> *Parameter:* r.
> *Question:* Is there is a subset Y of V of cardinality r such for every $Z \subseteq Y$ there is a $e \in E$ such that $Z = Y \cap e$?

It is known that p-VC-DIMENSION is W[1]-complete. From Proposition 7 we know that

$$p\text{-VC-DIMENSION} \leq^{\text{fpt}} \mathscr{M}(\textit{uni}\text{-VC-DIMENSION}).$$

Again we do not know whether we can replace $\leq^{\text{fpt}}$ by $\equiv^{\text{fpt}}$. We cannot apply Theorem 12, as we cannot prove the corresponding condensation property and we cannot apply Theorem 20, as p-VC-DIMENSION is not Fagin-definable by a Π_1-formula.

7 Model-Checking Problems

As shown by Theorem 3, model-checking problems of the form p-MC(Φ) for fragments Φ of first-order logic are complete for the classes of the W-hierarchy. In this section we analyze to what extent this holds for the miniaturization of their canonical reparameterizations *var*-MC(Φ), where

> *var*-MC(Φ)
> *Instance:* A structure $\mathcal{A}$ and a formula $\varphi \in \Phi$.
> *Parameter:* $r \cdot \log |A|$, where r is the number of free variables of φ.
> *Question:* Is $\varphi(\mathcal{A})$ nonempty?

By Theorem 3 one might conjecture that for $t \geq 0$, $u \geq 1$, and $s \geq 2$ the problems $\mathscr{M}(\textit{var}\text{-MC}(\Pi^0_{t,u}))$ and $\mathscr{M}(\textit{var}\text{-MC}(\Pi^0_{t,u}[s]))$ are W[$t+1$]-complete under fpt-reductions.

However this is unlikely as the next result shows that these problems are W[SAT]-complete. Recall that PROP denotes the class of all propositional formulas. By definition, the problem p-SAT(PROP), i.e.,

> p-SAT(PROP)
> *Instance:* $\alpha \in$ PROP.
> *Parameter:* $|\text{var}(\alpha)|$.
> *Question:* Is α satisfiable?

is S[SAT]-complete under serf-reductions and hence $\mathscr{M}$(SAT(PROP)) is W[SAT]-complete under fpt-reductions (see [5, Definition 16.36 and Exercise 16.43]).

Theorem 26. *For $t \geq 0$, $u \geq 1$, and $s \geq 2$*

$$\mathscr{M}(\textit{var}\text{-MC}(\Pi^0_{t,u})) \qquad \textit{and} \qquad \mathscr{M}(\textit{var}\text{-MC}(\Pi^0_{t,u})[s])$$

are W[SAT]*-complete under fpt-reductions.*

To obtain a W$[t+1]$-complete problem we have to consider a subclass of $\Pi^0_{t,u}$. Let $t \geq 0$ and $u \geq 1$. A $\Pi^*_{t,u}$-formula is a $\Pi^0_{t,u}$-formula

$$\varphi = \forall x_{11} \ldots \forall x_{1k_1} \exists x_{21} \ldots \exists x_{2k_2} \ldots Q x_{t1} \ldots Q x_{tk_t} \psi$$

such that

- if t is even (and hence $Q = \exists$), then ψ is in disjunctive normal form, and
- if t is odd (and hence $Q = \forall$), then ψ is in conjunctive normal form.

For $t \geq 0$ and $u, s \geq 1$, we have

$$p\text{-MC}(\Pi^*_{t,u}[s]) \equiv^{\text{fpt}} p\text{-MC}(\Pi^0_{t,u}[s]),$$

as in time allowed by an fpt-reduction, the quantifier-free part of a formula in $\Pi^0_{t,u}[s]$ can be transformed into conjunctive or disjunctive normal form. In particular, the same can be expressed by formulas in $\Pi^*_{t,u}[s]$ as by formulas in $\Pi^0_{t,u}[s]$.

Theorem 27. *Let $t \geq 0$, $u \geq 1$ and $s \geq 2$. Then $\mathscr{M}$(var-*MC$(\Pi^*_{t,u}[s])$) *is* W$[t+1]$-*complete under fpt-reductions.*

We give an application of the preceding result. Let $s \geq 1$. The *parameterized homomorphism problem* p-HOM$[s]$ *for structures of arity* $\leq s$ is the following problem:

p-HOM$[s]$
Instance: Structures $\mathcal{A}$ and $\mathcal{B}$ of arity $\leq s$.
Parameter: $\|\mathcal{A}\|$.
Question: Is there a homomorphism from $\mathcal{A}$ to $\mathcal{B}$?

The following is known (cf. [5]):

Theorem 28. *Let $s \geq 2$. Then* p-HOM$[s]$ *is* W$[1]$-*complete under fpt-reductions.*

The canonical reparameterization of p-HOM$[s]$ is the problem

uni-HOM$[s]$
Instance: Structures $\mathcal{A}$ and $\mathcal{B}$ of arity $\leq s$.
Parameter: $\lvert A\rvert \cdot \log \lvert B\rvert$.
Question: Is there a homomorphism from $\mathcal{A}$ to $\mathcal{B}$?

A further reparameterization of p-HOM$[s]$ is

size-HOM$[s]$
Instance: Structures $\mathcal{A}$ and $\mathcal{B}$ of arity $\leq s$.
Parameter: $\|\mathcal{A}\| \cdot \log \|\mathcal{B}\|$.
Question: Is there a homomorphism from $\mathcal{A}$ to $\mathcal{B}$?

By Proposition 7 we have:

$$p\text{-}\textsc{Hom}[s] \le^{\text{fpt}} \mathcal{M}(\mathit{size}\text{-}\textsc{Hom}[s]) \le^{\text{fpt}} \mathcal{M}(\mathit{uni}\text{-}\textsc{Hom}[s]).$$

Using the previous theorem one can show that the three problems are equivalent:

Theorem 29. *Let $s \ge 2$. Then*

$$p\text{-}\textsc{Hom}[s] \equiv^{\text{fpt}} \mathcal{M}(\mathit{size}\text{-}\textsc{Hom}[s]) \equiv^{\text{fpt}} \mathcal{M}(\mathit{uni}\text{-}\textsc{Hom}[s]).$$

Corollary 30. *Let $s \ge 2$. $\mathcal{M}(\mathit{size}\text{-}\textsc{Hom}[s])$ and $\mathcal{M}(\mathit{uni}\text{-}\textsc{Hom}[s])$ are* W[1]*-complete under fpt-reductions.*

Corollary 31. *Let $s \ge 2$. If there is an algorithm solving the problem* $\textsc{Hom}[s]$ *in time $2^{o^{\text{eff}}(\|\mathcal{A}\| \cdot \log \|\mathcal{B}\|)} \cdot \|\mathcal{B}\|^{O(1)}$, then* $\textsc{Hom}[s]$ *is solvable in time $2^{o^{\text{eff}}(|A| \cdot \log |B|)} \cdot \|\mathcal{B}\|^{O(1)}$.*

7.1 Model-Checking on Trees

We have seen that for many natural parameterized problem (Q, κ) whose instances are of the form (x, y) with $\kappa(x, y) = |y|$ we have

$$\begin{aligned} (Q,\kappa) \text{ is } & \text{ fixed-parameter tractable} \\ & \iff Q \text{ is decidable in time } 2^{o^{\text{eff}}(|y| \cdot \log |x|)} \cdot |x|^{O(1)}. \end{aligned} \tag{8}$$

One can show that the direction from right to left is always true, independent of the choice of (Q, κ). To establish the converse direction, we mostly needed the condensation property for each individual problem. There are problems that (apparently) do not have this property, for example p-WSAT($\Gamma_{t,d}$). Nevertheless, the equivalence (8) still holds for p-WSAT($\Gamma_{t,d}$) by Theorem 15 and the Miniaturization Theorem.

It is not hard to construct an artificial parameterized problem not satisfying (8). We should mention that the parameterized model-checking problem p-MC(TREE, MSO) for monadic second-order logic (MSO) on trees (see [5] for a definition) is a more natural example: in fact, by [3,7] the problem p-MC(TREE, MSO) is fixed-parameter tractable, but one can show that the right side of the equivalence of (8) does not hold for p-MC(TREE, MSO), unless the exponential time hypothesis fails.

8 Conclusions

We introduce the notion of canonical (re)parameterization of a (parameterized) problem and show, among others, that this reparameterization is the preimage under the miniaturization mapping of Fagin-definable problems complete for some level of the W-hierarchy. One of the main open questions is whether this is true for *all* Fagin-definable problems. For every $t \ge 1$ we determine a class of first-order formulas such that the miniaturization of the corresponding parameterized model-checking problem is W[t]-complete under fpt-reductions.

References

1. Chen, J., Chor, B., Fellows, M., Huang, X., Juedes, D., Kanj, I., Xia, G.: Tight lower bounds for certain parameterized NP-hard problems. In: Proceedings of The 19th IEEE Annual Conference on Computational Complexity (CCC'04), pp. 150–160 (2004)
2. Chen, Y., Grohe, M.: An isomorphism between subexponential and parameterized complexity theory. Journal version submitted. A partial conference version appeared In: Proceedings of the 21st IEEE Conference on Computational Complexity (CCC'06), pp. 314–330 (2006), Available at `http://basics.sjtu.edu.cn/~chen/papers/`
3. Doner, J.: Tree acceptors and some of their applications. Journal of Computer and System Sciences 4, 406–451 (1970)
4. Flum, J., Grohe, M.: Parameterized complexity and subexponential time. In: the Complexity Column of the Bulletin of the European Association for Theoretical Computer Science (EATCS), p. 84 (2004)
5. Flum, J., Grohe, M.: Parameterized Complexity Theory. Springer, Heidelberg (2006)
6. Flum, J., Grohe, M., Weyer, M.: Bounded fixed-parameter tractability and log ^2n nondeterministic bits. Journal of Computer and System Sciences 72, 34–71 (2006)
7. Thatcher, J.W., Wright, J.B.: Generalised finite automata theory with an application to a decision problem of second-order logic. Mathematical Systems Theory 2, 57–81 (1968)

From Proofs to Focused Proofs: A Modular Proof of Focalization in Linear Logic

Dale Miller and Alexis Saurin

INRIA & LIX/École Polytechnique, Palaiseau, France
dale.miller@inria.fr
saurin@lix.polytechnique.fr

Abstract. Probably the most significant result concerning cut-free sequent calculus proofs in linear logic is the completeness of focused proofs. This completeness theorem has a number of proof theoretic applications — e.g. in game semantics, Ludics, and proof search — and more computer science applications — e.g. logic programming, call-by-name/value evaluation. Andreoli proved this theorem for first-order linear logic 15 years ago. In the present paper, we give a new proof of the completeness of focused proofs in terms of proof transformation. The proof of this theorem is simple and modular: it is first proved for MALL and then is extended to full linear logic. Given its modular structure, we show how the proof can be extended to larger systems, such as logics with induction. Our analysis of focused proofs will employ a proof transformation method that leads us to study how focusing and cut elimination interact. A key component of our proof is the construction of a *focalization graph* which provides an abstraction over how focusing can be organized within a given cut-free proof. Using this graph abstraction allows us to provide a detailed study of atomic *bias assignment* in a way more refined that is given in Andreoli's original proof. Permitting more flexible assignment of bias will allow this completeness theorem to help establish the completeness of a number of other automated deduction procedures. Focalization graphs can be used to justify the introduction of an inference rule for *multifocus* derivation: a rule that should help us better understand the relations between sequentiality and concurrency in linear logic.

1 Introduction

Linear Logic was introduced 20 years ago by Girard and since then it has led to many developments in proof theory, computational logic, and programming language theory. Much proof theoretic analyses and applications of linear logic have concentrated on the nature and dynamics of cut-elimination via the geometry of interactions, game semantics, interactions, etc. Less has been studied about the structure of cut-free proofs themselves: the main result in that area is probably the completeness of focused proofs due to Andreoli [3,4]. This completeness theorem has a number of applications in computer science: for example, focused proofs have been used to design and formalize logic programming languages [2,20], to formalize proof systems that allow for both forward-chaining and backward-chaining [15,19], and should be behind the dualities between call-by-name and call-by-value evaluation in the λ-calculus [6]. The structure of focused proofs is also a key ingredient in the development of Polarized Logic [17,18] and Ludics [13].

J. Duparc and T.A. Henzinger (Eds.): CSL 2007, LNCS 4646, pp. 405–419, 2007.

Andreoli's result, however, is wrapped up in one theorem about one logic. This seems an unfortunate situation for a number of reasons.

- Various extensions to linear logic are known (based on higher-order quantification [11], induction and co-induction [5], different kinds of exponentials [7,12,16], etc.) and it is likely that one will want to know if focusing can be proved for them.
- When examining the issues behind the assignment of polarity to literals (a necessary *annotation* step needed to define focused proofs), it is clear that there is a lot of flexibility allowed in providing such annotations, certainly more than what is technically allowed in Andreoli's proof system.
- Other logics exhibit focusing behaviors. In particular, there are focused proof systems for classical logic, namely LKQ/LKT [8] and LK_p^η [9], and for intuitionistic logic, namely, LJT [14], LJQ calculus [14,10], and LJF [19].
- In [4], focusing is not seen as a process. There appears to be advantages to consider the process of transforming proofs into focused proofs: mixing this process with the process of doing cut-elimination should also be rather interesting.

These reasons suggest that the notions surrounding the "completeness of focused proofs" is both more general and more flexible than what is captured in the original theorem and its proof. Thus, we take on the task in this paper of attempting to develop an approach to proving focusing results by getting after the essential conditions for "focalization" to hold and by analyzing those conditions more broadly. By analogy, once the importance of cut-elimination was appreciated, Gentzen single cut-elimination theorem was analyzed in ways to uncover the essentially features that now allow researchers to prove cut-elimination for a number of logics.

This paper is organized as follows. In the next Section, we state some basic definitions and results for linear logic, including the original focused proof system (Figure 3). In Section 3, we present the key elements of our methodology, in particular, the *focalization graph* and a flexible *bias assignment* scheme, on the multiplicative and additive subset of linear logic (MALL). Section 4 considers how this methodology can account for additional structure within linear logic, including the exponentials and quantifiers. In Section 5, we briefly consider adding to the sequent calculus proofs the *multifocus* inference rule. Finally, we conclude in Section 6.

2 Linear Logic Preliminaries

The formulas of LL are made from literals which are atoms $(a, b, \dots)$ or negations of atoms $(a^\perp, b^\perp, \dots)$ and multiplicative $(\otimes, \parr, \mathbf{1}, \perp)$, additive $(\oplus, \&, \mathbf{0}, \top)$ and exponential $(!, ?)$ connectives as well as (first-order) quantifiers $(\exists, \forall)$, following the grammar:

$$\begin{aligned} F ::= \ & a \mid F \otimes F \mid F \oplus F \mid \mathbf{1} \mid \mathbf{0} \mid \exists x.F \mid\ !F \\ & a^\perp \mid F \parr F \mid F \,\&\, F \mid \perp \mid \top \mid \forall x.F \mid\ ?F \end{aligned}$$

For notational convenience we will write $A^\perp$ for the negation normal form of A (that is, where negations have only atomic scope) and we will work with one-sided sequents. We give in Figure 1 the inference rules for Linear Logic. The initial rule can be restricted to

$$\frac{}{\vdash A, A^\perp}\ initial \qquad \frac{\vdash \Gamma, A \quad \vdash \Delta, A^\perp}{\vdash \Gamma, \Delta}\ cut$$

$$\frac{}{\vdash \mathbf{1}}\ \mathbf{1} \quad \frac{\vdash \Gamma, A \quad \vdash \Delta, B}{\vdash \Gamma, \Delta, A \otimes B}\ \otimes \quad \frac{\vdash \Gamma, A_1}{\vdash \Gamma, A_1 \oplus A_2}\ \oplus_1 \quad \frac{\vdash \Gamma, A_2}{\vdash \Gamma, A_1 \oplus A_2}\ \oplus_2 \quad \frac{\vdash \Gamma, A[t/x]}{\vdash \Gamma, \exists x.A}\ \exists$$

$$\frac{\vdash \Gamma}{\vdash \Gamma, \bot}\ \bot \quad \frac{\vdash \Gamma, A, B}{\vdash \Gamma, A \parr B}\ \parr \quad \frac{\vdash \Gamma, A \quad \vdash \Gamma, B}{\vdash \Gamma, A \,\&\, B}\ \& \quad \frac{}{\vdash \Gamma, \top}\ \top \quad \frac{\vdash \Gamma, A[c/x]}{\vdash \Gamma, \forall x.A}\ \forall \ \text{ if } c \text{ is new}$$

$$\frac{\vdash ?\Gamma, B}{\vdash ?\Gamma, !B}\ ! \quad \frac{\vdash \Gamma}{\vdash \Gamma, ?B}\ ?w \quad \frac{\vdash \Gamma, ?B, ?B}{\vdash \Gamma, ?B}\ ?c \quad \frac{\vdash \Gamma, B}{\vdash \Gamma, ?B}\ ?d$$

Fig. 1. Inference rules for LL

literals without a loss of completeness. We shall assume this restriction to atomic initial rules in the following. In Figure 2 we give an example of a sequent proof.

The logical connectives of linear logic can be divided into two sets: the *asynchronous* connectives ($\top, \bot, \&, \parr, ?, \forall$) and the *synchronous* connectives ($\mathbf{1}, \mathbf{0}, \otimes, \oplus, !, \exists$) (they are de Morgan duals of the asynchronous connectives). Reading the rules bottom-up, the rules for the asynchronous connectives are invertible (their application is independent from the context) whereas the synchronous have rules for which application depends on the surrounding context. Formulas built with a topmost asynchronous connective are also called negative, the ones built with a synchronous connective are positive.

$$\dfrac{\dfrac{\dfrac{\dfrac{\dfrac{\dfrac{}{\vdash q, q^\perp}\ ini \quad \dfrac{}{\vdash r, r^\perp}\ ini}{\vdash q \otimes r, q^\perp, r^\perp}\ \otimes}{\vdash q \otimes r, q^\perp \parr r^\perp}\ \parr \quad \dfrac{}{\vdash s, s^\perp}\ ini}{\vdash q \otimes r, s \otimes (q^\perp \parr r^\perp), s^\perp}\ \otimes}{\vdash p \oplus (q \otimes r), s \otimes (q^\perp \parr r^\perp), s^\perp}\ \oplus \quad \dfrac{}{\vdash \mathbf{1}}\ \mathbf{1}}{\vdash p \oplus (q \otimes r), s \otimes (q^\perp \parr r^\perp), s^\perp \otimes \mathbf{1}}\ \otimes$$

Fig. 2. Example of a LL proof

The search for a focused proof can utilize this division of inference rules. If we read inference rules from conclusion to premiss, we can apply invertible rules in any order (no need for backtracking) and when only synchronous rules are available we can focus on a certain formula and its positive subformulas. Such a chain of synchronous rules, usually called a *focused phase*, terminates when it reaches an asynchronous formula. Proof search can then alternate between applications of asynchronous rules and chains of synchronous rules.

A second aspect of focused proofs is that the synchronous/asynchronous classification of non-atomic formulas must be extended to atomic formulas. The arbitrary assignment of positive (synchronous) and negative (asynchronous) *bias* to atomic formulas must be made before the notion of focused proof is complete. How this bias is assigned does not affect the existence of a focused proof but does impact the size and shape of the resulting focused proofs. We shall sometimes think of such an assignment of bias to atomic formulas as an *annotation* of the atoms in the formula.

The focusing proof system for linear logic, presented in Figure 3, contains two kinds of sequents. In the sequent $\Psi : \Delta \Uparrow L$, the "zones" Ψ and Δ are multisets and L is a list. This sequent encodes the usual one-sided sequent $\vdash ?\Psi, \Delta, L$ (here, we assume the natural coercion of lists into multisets). This sequent will also satisfy the invariant that

$$\frac{\Psi : \Delta \Uparrow L}{\Psi : \Delta \Uparrow \bot, L}\ \bot \qquad \frac{\Psi : \Delta \Uparrow F, G, L}{\Psi : \Delta \Uparrow F \parr G, L}\ \parr \qquad \frac{\Psi, F : \Delta \Uparrow L}{\Psi : \Delta \Uparrow ?F, L}\ ?$$

$$\frac{}{\Psi : \Delta \Uparrow \top, L}\ \top \qquad \frac{\Psi : \Delta \Uparrow F, L \quad \Psi : \Delta \Uparrow G, L}{\Psi : \Delta \Uparrow F \,\&\, G, L}\ \& \qquad \frac{\Psi : \Delta \Uparrow B[y/x], L}{\Psi : \Delta \Uparrow \forall x.B, L}\ \forall$$

$$\frac{}{\Psi : \cdot \Downarrow \mathbf{1}}\ \mathbf{1} \qquad \frac{\Psi : \Delta_1 \Downarrow F \quad \Psi : \Delta_2 \Downarrow G}{\Psi : \Delta_1, \Delta_2 \Downarrow F \otimes G}\ \otimes \qquad \frac{\Psi : \cdot \Uparrow F}{\Psi : \cdot \Downarrow !F}\ !$$

$$\frac{\Psi : \Delta \Downarrow F_1}{\Psi : \Delta \Downarrow F_1 \oplus F_2}\ \oplus_1 \qquad \frac{\Psi : \Delta \Downarrow F_2}{\Psi : \Delta \Downarrow F_1 \oplus F_2}\ \oplus_2 \qquad \frac{\Psi : \Delta \Downarrow B[t/x]}{\Psi : \Delta \Downarrow \exists x.B}\ \exists$$

$$\frac{\Psi : \Delta, F \Uparrow L}{\Psi : \Delta \Uparrow F, L}\ R\Uparrow \qquad \frac{}{\Psi : K^\bot \Downarrow K}\ I_1 \qquad \frac{\Psi : \Delta \Downarrow F}{\Psi : \Delta, F \Uparrow \cdot}\ D_1$$

$$\frac{\Psi : \Delta \Uparrow F}{\Psi : \Delta \Downarrow F}\ R\Downarrow \qquad \frac{}{\Psi, K^\bot : \cdot \Downarrow K}\ I_2 \qquad \frac{\Psi, F : \Delta \Downarrow F}{\Psi, F : \Delta \Uparrow \cdot}\ D_2$$

Fig. 3. The Σ_3 focused proof system of [4] for linear logic. *The provisos on the rules are the following: In the $\forall$-rule variable y is not free in the conclusion. In $R \Uparrow F$ is not asynchronous while in $R \Downarrow F$ is either asynchronous or a negative literal. In I_1 and I_2, K is a positive literal. In D_1 and D_2, F is not a negative literal.*

requires Δ to contain only literals and synchronous formulas. In the sequent $\Psi : \Delta \Downarrow F$, the zone Ψ is a multiset of formulas and Δ is a multiset of literals and synchronous formulas, and F is a single formula.

The main result about focused proofs is that they are complete for linear logic. The following theorem was proved in [4].

Theorem 1. *Given Ψ a set of formulas, Γ a multiset of non-asynchronous formulas and Δ an arbitrary list of formulas, $\vdash ?\Psi, \Gamma, \Delta$ is provable in LL if and only if the sequent $\Psi : \Gamma \Uparrow \Delta$ is provable in the Σ_3 proof system (given in figure 3).*

3 Focalization in MALL

In this part we will prove Focalization for MALL only in order to deal with a smaller system when introducing our proof technique. We will later extend the result to full LL. In doing so, we are driven by the will for simplicity but also by the particular interest for focalization in MALL for it is the system on which are built the basic objects of Ludics [13], the designs. It is actually the initial motivation of our work: finding a simpler and shorter proof of Focalization for MALL for Ludics purpose.

But still, our main concern is simplicity and that is why we first consider cut-free MALL proofs and we intend to demonstrate that Focalization is actually a fairly simple result, although the size of Σ_3 often makes it difficult to grasp.

3.1 Permutation of Rules in LL

The sequential structure of sequent calculus proofs records the precise ordering of the application of inference rules, even when that ordering is not particularly important or when other orders result in similar proofs. Such sequentialization is responsible for not only an explosion in the space of proofs but also for the possibility of providing a precise analysis of the relationship between proof rules. In other words, what makes it difficult to determine if two sequent proofs are essentially the same or different is what provides us with powerful analysis tools for developing an approach to "causation" *a la* focalization. Systems like proof nets which get rid of the first difficulty have trouble when it comes to checking whether a proof structure is a proof net or whether a link in a proof net depends on another link.

Definition 1 (Permutation of inference rules). *We define two notions of permutability: (i)* α/β**-permutability:** *there is an* α/β*-permutability if, given a sequent* $\mathcal{S}$ *containing two formulas A and B, then for any proof* Π *of* $\mathcal{S}$ *starting with the* α *rule (on formula A) right before the* β *rule (on formula B) is applied, there exists a proof* Π' *of* $\mathcal{S}$ *in which the two rules have been exchanged: the* β *rule comes first, immediately followed by the* α *rule (there is of course a degenerate case for rules with no premiss, like* $\top$*). (ii)* $\alpha\backslash\beta$**-permutability:** *we speak of* $\alpha\backslash\beta$*-permutability when there is both* α/β*-permutability and* β/α*-permutability.*

Given two sets of inference rules $\mathcal{N}$ *and* $\mathcal{P}$*, we say that, with respect to those two sets,* $\mathcal{P}$ *has* **weak permutability** *if given two rules* α,β *of* $\mathcal{P}$ *we have* $\alpha\backslash\beta$*-permutability. We say that* $\mathcal{N}$ *has* **full permutability** *when it has weak permutability and when in addition for any pair of rules* $(\alpha,\beta)\in\mathcal{P}\times\mathcal{N}$*, we have* α/β*-permutability.*

Proposition 1 (Permutabilities of linear logic inference rules). *Let* $\mathcal{N}$ *be the set of inference rules attached to the MALL asynchronous connectives and* $\mathcal{P}$ *be the set of inference rules attached to the MALL synchronous connectives.* $\mathcal{N}$ *has full permutability while* $\mathcal{P}$ *has weak permutability.*

The proof is trivial either by introducing cuts and then reducing them or by doing small step permutations. Notice that the synchronous connectives do not have full permutability: sequent $\vdash a^\perp \parr b^\perp, a\otimes b$ has no cut-free proof that begins with a $\otimes$-rule.

3.2 Focalization Graph

The introduction of the *Focalization Graph* structure brings us to the heart of our result. The acyclicity of the graph will be crucial in establishing focalization.

Definition 2. *A MALL sequent containing at least a negative non-literal formula is* **negative***. It is* **positive** *when it contains no negative non-literal formula and at least one positive non-literal formula. Otherwise it is* **atomic***.*

Definition 3 (Positive Trunks). *Given a MALL proof* Π *of a positive sequent* $\mathcal{S}$ *we define the* **Positive Trunk** Π^+ *as the maximal prefix of the tree* Π *containing only positive rules, that is the tree starting at the root of* Π *and whose leaves are the bottom sequents of the first non-positive rules encountered on every branch of the tree, if such a rule exists. The* **Border** *of a Positive Trunk is the set of its leaves. The border contains only negative or atomic sequents. The* **Active Formulas** *of a Positive Trunk* Π^+ *are the formulas of the base that are decomposed into subformulas within the considered Trunk.*

Remark 1. When addressing the case of the exponentials, we will see that we can add a condition to shorten a branch in the positive trunk, this condition can also be regarded as expressing the fact that the rule for ! is bipolarized, being both positive and negative.

We now define a relation on occurrences of formulas involved in Π: $F \prec G$ iff G is a subformula (or sub-occurrence) of F in the precise sense that occurrence G is obtained from the decomposition of F along a branch of Π.

$$\dfrac{\dfrac{\dfrac{\boxed{\vdash q \otimes r, q^{\perp} \parr r^{\perp}} \quad \boxed{\vdash s, s^{\perp}}}{\vdash q \otimes r, s \otimes (q^{\perp} \parr r^{\perp}), s^{\perp}}\otimes}{\vdash p \oplus (q \otimes r), s \otimes (q^{\perp} \parr r^{\perp}), s^{\perp}}\oplus \qquad \dfrac{}{\vdash \mathbf{1}}\mathbf{1}}{\vdash p \oplus (q \otimes r), s \otimes (q^{\perp} \parr r^{\perp}), s^{\perp} \otimes \mathbf{1}}\otimes$$

Fig. 4. Positive Trunk associated to figure 2

Definition 4 ($\prec$-relation). *The* **suboccurrence relation** *(written $\prec$) on occurrences of formulas appearing in Π is the reflexive and transitive closure of the binary relation $\prec^1$ defined by $F \prec^1 G$ if there exists in Π a rule α with conclusion sequent $\mathcal{S}$ and premisses $(\mathcal{S}_i)_{i\in I}$ such that F is the principal formula of $\mathcal{S}$ and G is a subformula of F produced by the rule α in some of the $\mathcal{S}_i$.*

If $F \prec G$ we will say that G is a **$\prec$-subformula** *of F or a* **descendent** *of F.*

The following lemma will help us proving our main result:

Lemma 1. *Let Π^+ be a Positive Trunk with root $\mathcal{S}$ and border $\mathcal{B}$. For any $\mathcal{S}' \in \mathcal{B}$ the relation $\prec$ defines a one-to-one function from $\mathcal{S}'$ to $\mathcal{S}$.*

Proof. We actually prove a stronger result: the result holds for any sequent appearing in the trunk, not only for sequents in $\mathcal{B}$.

The result is proved by induction on the height of the considered sequent in Π^+:

- The base case is trivial since the considered sequent is $\mathcal{S}$ itself (recall $\prec$ is reflexive).
- Suppose the result is true for a given height $n \leq h(\Pi^+)$ and suppose $n+1 \leq h(\Pi^+)$. Let $\mathcal{S}_{n+1}$ be a sequent of height $n+1$ and let α be the rule of which $\mathcal{S}_{n+1}$ is a premiss and call $\mathcal{S}_n$ its conclusion. By induction hypothesis $\mathcal{S}_n$ satisfies the condition. We can define a one-to-one function ι_n from $\mathcal{S}_{n+1}$ (as set of occurences of formulas) to $\mathcal{S}_n$ as follows: let G be a formula of $\mathcal{S}_{n+1}$. if $F \prec^1 G$ for some $F \in \mathcal{S}_n$ then fix F to be the image of G by ι_n. If no such formula exists, then an occurrence of G is also present in $\mathcal{S}_n$ then associate the two occurrences of G. The function built in this way is one-to-one thanks to the fact that every MALL positive rule produces at most (and actually exactly) one subformula of the principal formula in every premiss of the rule. Composing the function we just defined with the one-to-one function provided by the induction hypothesis we see that $\mathcal{S}_{n+1}$ satisfies the condition.

By induction we get the result we expected. □

Lemma 2. *A formula which is not active in the Positive Trunk appears in exactly one sequent of the border.*

An active formula F to which no branching rule is applied in Π^+ (nor to its $\prec$-subformulas) – speak of a non-branching formula wrt. Π^+ – is such that there exists exactly one formula G in one of the sequents of the border such that $F \prec G$.

To a Positive Trunk we associate a graph as follows:

Definition 5 (Focalization graph). *Given a Positive Trunk Π^+ we define the* **Focalization Graph** $\mathcal{G}$ *to be the graph whose vertices are the active formulas of the Trunk and such that there is an edge from F to G iff there is a sequent $\mathcal{S}'$ in the border containing a negative $\prec$-subformula F' of F and a positive $\prec$-subformula G' of G.*

Example 1. The Focalization graph associated with our example proof is:

$$s^{\perp} \otimes \mathbf{1} \qquad s \otimes (q^{\perp} \parr r^{\perp}) \longrightarrow p \oplus (q \otimes r)$$

This graph is acyclic. In the following we will show that it is true in general and this will be crucial for focalization.

Lemma 3. *If $\mathcal{S}'$ and $\mathcal{S}''$ are sequents occurring in different branches of Π^+, then there is at most one formula in the root of Π^+ which has $\prec$-subformulas in both $\mathcal{S}'$ and $\mathcal{S}''$.*

Proof. If this was not the case, let $\mathcal{S}' \wedge \mathcal{S}''$ be their highest predecessor in the tree. This sequent would necessarily have at least two formulas that would be $\prec$-subformulas of the same formula in the root which is impossible thanks to lemma 1. □

Proposition 2. *The Focalization Graphs are acyclic.*

Proof. We prove the result by *reductio ad absurdum.*
Let $\mathcal{S}$ be a positive sequent with a proof Π. Let Π^+ be the corresponding positive trunk and $\mathcal{G}$ the associated Focalization Graph. Suppose that $\mathcal{G}$ has a cycle and consider such a cycle of minimal length $(F_1 \rightarrow F_2 \rightarrow \cdots \rightarrow F_n \rightarrow F_1)$ in $\mathcal{G}$ and let us consider $\mathcal{S}_1, \ldots, \mathcal{S}_n$ sequents of the border justifying the arrows of the cycle.

Thanks to lemma 3 these sequents are actually uniquely defined. With the same idea we can immediately notice that the cycle is necessarily of length $n \geq 3$: thanks to lemma 1 two $\prec$-subformulas of the same formula can never be in the same sequent in the border of the positive trunk and by lemma 3 there cannot be a cycle of length 2.

Let $\mathcal{S}_0$ be $\bigwedge_{i=1}^{n} \mathcal{S}_i$ be the highest sequent in Π such that all the $\mathcal{S}_i$ are leaves of the tree rooted in $\mathcal{S}_0$. We will obtain the contradiction by studying $\mathcal{S}_0$ and we will reason by case on the rule applied to this sequent $\mathcal{S}_0$:

- the rule cannot be a **1** rule since this rule produces no premiss and thus we would have an empty cycle which is nonsense. Any rule with no premiss would lead to the same contradiction.
- If the rule is one of the $\oplus$-rules, then the premiss $\mathcal{S}_0'$ of the rule would also satisfy the condition required for $\mathcal{S}_0$ (all the $\mathcal{S}_i$ would be part of the proof tree rooted in $\mathcal{S}_0'$) contradicting the maximality of $\mathcal{S}_0$. If the rule is any other non-branching rule, maximality of $\mathcal{S}_0$ would also be contradicted.
- Thus the rule shall be branching: it shall be a $\otimes$-rule. Write $\mathcal{S}_L$ and $\mathcal{S}_R$ for the left and right premisses of $\mathcal{S}_0$. Let $G = G_L \otimes G_R$ be the principal formula in $\mathcal{S}_0$ and let F be the active formula of the Trunk such that $F \prec G$. There are two possibilities:

(i) either $F \in \{F_1, \ldots, F_n\}$ and F is the only formula of the cycle having at the same time $\prec$-subformulas in the left premiss and in the right premiss,
(ii) or $F \notin \{F_1, \ldots, F_n\}$ and no formula of the cycle has $\prec$-subformulas in both premisses.

Let thus I_L (resp. I_R) be the set of indices of the active formulas of the root $\mathcal{S}$ having ($<$-related) subfomulas only in the left (resp. right) premiss. Clearly neither I_L nor I_R is empty since it would contradict the maximality of $\mathcal{S}_0$. Indeed if $I_L = \emptyset$, then $\mathcal{S}_R$ satisfies the condition of being dominated by all the $\mathcal{S}_i$, $1 \leq i \leq n$ and $\mathcal{S}_0$ is not maximal anymore. By definition of the two sets of indices we have of course $I_L \cap I_R = \emptyset$ and the only formula of the cycle possibly not in $I_L \cup I_R$ is F if we are in the case (i): all other formulas in the cycle have their index either in I_L or in I_R.

As a consequence there must be an arrow in the cycle (and thus in the graph) from a formula in I_L to a formula in I_R (or the opposite). Let $i \in I_L$ and $j \in I_R$ be such indices (say for instance $F_i \to F_j$ in $\mathcal{G}$) and let $\mathcal{S}'$ be the sequent of the border responsible for this edge. $\mathcal{S}'$ contains F'_i and F'_j and by definition of the sets I_L and I_R, $\mathcal{S}'$ cannot be in the tree rooted in $\mathcal{S}_0$ which is in contradiction with the way we constructed $\mathcal{S}_0$.

Then there cannot be any cycle in the graph. □

3.3 Pre-focalization Process

What the previous result actually tells us is that the Focalization Graph has a source, a formula that is not pointed to by any other formula in the graph, that is a formula such that whenever a sequent of the border contains one of its $<$-subformulas F, the subformula is not positive or the sequent is positive. To put things in other terms, there is a positive active formula in the root sequent whose positive layer of connective is completely decomposed during the Positive Trunk, independentely of any focusing discipline. This can be regarded as a kind of implicit focusing result. In some sense that tells us there is a formula which is already implicitely focused in the positive trunk.

Thanks to full permutability of the negatives, weak permutability of the positives and the acyclicity of the focalization graphs we know that, given a MALL proof Π of a sequent $\mathcal{S}$, we can transform it to another proof satisfying the following conditions:

Pre-Focalization Process:

1. **Asynchronous phase:** thanks to full permutability of negatives, if $\mathcal{S}$ is negative then we can permute down all the negative rules so that Π is transformed to a proof Π' where the bottom part of the proof tree is made only of negative rules up to the point where the branches of the tree reach positive or atomic sequents;
2. **Synchronous phase:** if $\mathcal{S}$ is positive, the associated Focalization Graph allows us to select a source of the graph, let us say P, as a focus and thanks to weak permutability, we can have the positive rules on $<$-subformulas of P permuted down so that Π is transformed into a tree Π' for which the maximal prefix containing only rules applied to P and its positive $<$-subformulas decomposes P up to its negative or literal subformulas. We are thus left with negative or atomic sequents, or positive sequents where the subformulas of P are literals.
3. **if $\mathcal{S}$ is atomic**, we can only apply an initial rule and thus close the tree.

This process is clearly terminating thanks to easy arguments on the complexity/size of the considered sequents in terms of number of polarity layers, for instance.

3.4 Dealing with Bias Assignments

The method described in the previous section shows a proof transformation technique that results almost in focused proofs but not exactly. Indeed we will now see that Andreoli's system forces more cosntraints on the proofs in that the use of the initial rule is more constrained. We shall now generalize our technique to capture exactly Andreoli's focusing disciplin as well as a more general focusing disciplin with a different management of the atoms. The freedom we get on Bias Assignment can be crucial for several applications in proof search.

In Σ_3, the initial rule has two versions, I_1 and I_2 (see figure 3). The initial rule can be applied only during a focusing phase on *positive literals*. In particular, the sequent $\vdash a^\perp \oplus \mathbf{0}, a \oplus \mathbf{0}$ would have only one focused proof whereas the technique of the Focalization Graph presented previously would have led to two different focused proofs. Andreoli system adds more constraints to the proof search while remaining complete. We now introduce Bias Assignments in order to treat this.

Definition 6. *Given a provable sequent $\mathcal{S}$, we call $\mathcal{P}_\mathcal{S}$ (for available positions for $\mathcal{S}$) the set containing all the branches of all possible proof trees for $\mathcal{S}$. We write $O_\mathcal{S}$ for the set of occurrences of literals occurring in $\mathcal{S}$.*

Definition 7 (Bias assignment $\mathcal{B}_\mathcal{S}$). *A bias assignment for a provable sequent $\mathcal{S}$, written $\mathcal{B}_\mathcal{S}$, is a partial function from $\mathcal{P}_\mathcal{S} \times O_\mathcal{S}$ to $\{-;+\}$*

Example 2. We give here some examples of typical bias assignments:[3pt]

- The bias assignment which is defined ***nowhere*** corresponds to the previous situation.
- ***Andreoli's bias assignment***. $\mathcal{B}^{\Sigma_3}$ is the function defined as: for any atom a, $\mathcal{B}^{\Sigma_3}(\_, a) = +$ and $\mathcal{B}^{\Sigma_3}(\_, a^\perp) = -$. More generally the bias assignments may not be sensitive to their first component and give the same polarity to different occurrences of the same litteral. In that case, we speak of an ***atom-based*** bias assignment.
- We can consider bias assignments which are *sensitive* to the position in the tree where the considered literal is. For such assignments b, $b(p, a)$ may be different from $b(q, a)$. In this case we speak of an ***occurrence-based*** bias assignment. We can consider coherence conditions on the assignments. For instance, moving upwards on a branch, we may want to ensure that the polarity won't change once it is set: if p and q are two branches, p being an extension of q and if $b(q, a) \searrow$ then $b(p, a) \searrow$ and $b(p, a) = b(q, a)$. But on the other hand we may also want to consider totally arbitrary assignments.

Definition 8 ($\mathcal{B}$-Focalization Graphs). *Given a positive sequent $\mathcal{S}$, a proof Π of $\mathcal{S}$ and a bias assignment $\mathcal{B}$ for $\mathcal{S}$, we define the $\mathcal{B}$-Focalization Graph $\mathcal{G}_\mathcal{S}^\mathcal{B}$ as in the previous subsection but considering as negative formulas the literals which are assigned polarity $-$ in a sequent $\mathcal{S}'$ of the border and as positive formulas the literals which are assigned polarity $+$. The literals for which $\mathcal{B}$ is not defined in $\mathcal{S}'$ are treated as before: they do not contribute to the graph.*

The bias assignment results in more arcs in the Focalization Graph. For instance, with $\mathcal{B}^{\Sigma_3}$ our example of figure 2 has the following focalization graph:

$$s^\perp \otimes \mathbf{1} \longrightarrow s \otimes (q^\perp \parr r^\perp) \longrightarrow p \oplus (q \otimes r)$$

This might also produce cycles. The following proposition ensures it does not:

Proposition 3. *Given a positive sequent $\mathcal{S}$ and a proof Π of $\mathcal{S}$, whatever bias assignment $\mathcal{B}$ we choose, the $\mathcal{B}$-Focalization Graph $\mathcal{G}_{\mathcal{S}}^{\mathcal{B}}$ is acyclic.*

It is essentially sufficient to notice that adding these arcs will have no effect on the arguments we used previously since they were only concerned with the splitting structure of the branching rules. We can now state our main results concerning Focalization:

$$\dfrac{\dfrac{\dfrac{\dfrac{\dfrac{\dfrac{}{\vdash q, q^\perp}\,ini \quad \dfrac{}{\vdash r, r^\perp}\,ini}{\vdash q \otimes r, q^\perp, r^\perp}\,\otimes}{\vdash p \oplus (q \otimes r), q^\perp, r^\perp}\,\oplus}{\vdash p \oplus (q \otimes r), q^\perp \parr r^\perp}\,\parr \quad \dfrac{}{\vdash s, s^\perp}\,ini}{\vdash p \oplus (q \otimes r), s \otimes (q^\perp \parr r^\perp), s^\perp}\,\otimes \quad \dfrac{}{\vdash \mathbf{1}}\,\mathbf{1}}{\vdash p \oplus (q \otimes r), s \otimes (q^\perp \parr r^\perp), s^\perp \otimes \mathbf{1}}\,\otimes$$

Fig. 5. focalized proof of figure 2

Theorem 2 **($\mathcal{B}$-Focalization for MALL).** *Let $\mathcal{S}$ be a MALL sequent. To any proof Π of $\mathcal{S}$ and bias assignment $\mathcal{B}$, we can associate a new proof satisfying the following constraints depending on the sequent $\mathcal{S}$:*

(i) if it is a negative sequent starts by decomposing negative formulas;
(ii) when a positive sequent is encountered, a positive formula is chosen as a focus and is hereditarily decomposed until its negative or literal subformulas are found. if the subformula is negative we use the previous item, if the formula is a litteral, the behaviour depends on the bias which is assigned to the literal.

Theorem 3 **(Andreoli's Focalization for MALL).** *If we consider the bias assignment $\mathcal{B}^{\Sigma_3}$, the focalization process produces proofs which are focused in Andreoli's Σ_3 sense.*

4 Focalization for Full LL and Larger Extensions

Our analysis was first restricted to the case of cut-free propositional MALL, mainly for simplicity purposes. We now extend the result to richer fragments of Linear Logic and present how to treat the cut, the exponentials and the quantifiers.

4.1 Quantifiers

The proof in the previous section can be directly adapted to the quantifiers: they are connectives with non-branching rules and with the appropriate permutabilities (full-permutabilities for the $\forall$ which is negative and weak-permutability for the $\exists$ which is positive). The first-order case is thus treated trivially. The higher-order case requires some additional care for Bias Assignments in order to verify that bias assignments are still meaningful in this case but our abstract definition of Bias Assignments allows us to define the needed constraints on bias assignments. The details are beyond the scope of this paper.

4.2 MALL with Cut

Dealing with the cut-rule in an analysis of focusing is not critical when one is driven by completeness purposes only. But since we want to study a dynamic process of focalization, adressing the cut becomes important and even crucial. For instance, we may be interested in studying how the cut-reduction and the focalization process interact.

Our solution is inspired by what Andreoli does [4] but is slightly simplified. The basic idea is to notice that the cut-rule is very similar to a $\otimes$ rule: replacing a cut rule on A in a proof Π of $\vdash \Gamma$ results in an object which is *almost*[1] a proof of $\vdash \Gamma, A \otimes A^\perp$.

In fact we do not even need to use the proof itself. We will simply use this analogy in order to find how to adapt the Focalization graph to proofs with cuts. Our analogy simply suggests to treat the cut rule as a positive and, as a consequence, positive trunks may contain cut rules and the Focalization Graph will have new vertices of the form $Cut(A)$. The relation $\prec$ is extended in a straightforward way ($A \prec Cut(A)$ and $A^\perp \prec Cut(A)$) and the edges are created with the same conditions as we did in the previous section.

As before, we can prove that the Focalization graph is acyclic and then:

Theorem 4. *The Focalization Graph method produces focused proofs from MALL proofs with cuts.*

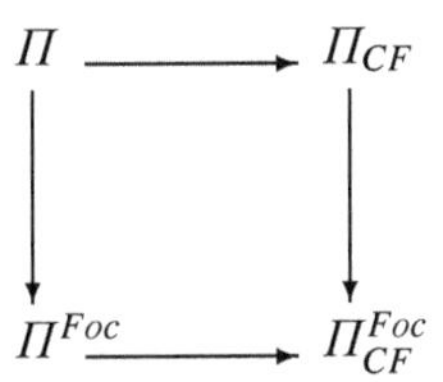

Fig. 6.

We think that the difference between our approach and Andreoli's starts really to make sense at this point: we always stayed in the same proof system, LL, and we worked by proof transformation. In our mind Focalization is really a process for transforming proofs. The interaction between this process and other transformation processes, like cut-reduction for instance shall now be studied.

Pushing this discussion further would be beyond the scope of this paper, but we would like to give an idea of the kind of question we can now try to adress: Given a proof Π in MALL with cuts, two processes are available: focalization and cut-reduction. Do the two processes commutes? Are we in the situation described by figure 6 where vertical arrows correspond to Focalization process while horizontal arrows correpond to the cut-reduction?

4.3 Exponentials

As it comes to exponentials we cannot carry our construction as straightforwardly as we did for the cut since it is not possible to attribute a polarity to the exponentials in a

[1] It is only almost a proof since the &-rule, the ! rule and the $\forall$-rule may cause trouble. Andreoli fixes this by considering the formula $?A \otimes A^\perp$ instead of $A \otimes A^\perp$ which is fine for & and ! but inefficient for the $\forall$ quantifier...

In our setting, we will get a proof of $\vdash \Gamma, A \otimes A^\perp$: we are only interested in the cut rules which are performed within the positive trunk. We can easily check that if Π^+ is a positive trunk for $\vdash \Gamma$ containing a cut rule on A then replacing the cut rule with a tensor rule on $A \otimes A^\perp$ leads straightforwardly to a positive trunk Π'^+ on $\vdash \Gamma, A \otimes A^\perp$.

simple way: they do not have the right permutation rules in order to have full or weak permutability. In order to extend our result we have to adapt the sequent calculus in a way which is pretty similar to what is done by Andreoli with his dyadic sequents [4]. For this change not to seem too *ad hoc* we quickly justify this by considering the !-rule (see figure 1). This rule has a peculiar shape because, contrarily to other inference rules, it depends on the toplevel structure of every formula in the context: one formula has to be banged while all the other shall be question-marked. This indicates a special level of knowledge about the sequent structure which is not the usual one we use in sequent calculus. This is reflected in the way the !-rule is implemented in linear logic programming systems or by the boxing construction in Proof Nets.

We actually see two kinds of operations performed with the !-rule: (i) classifying F as a question-marked formula on the one hand and removing the ! on $!G$ when $!G$ is the only non-question-marked formula in the sequent. This can be reflected by the paradoxical example following: considering $\vdash\ ?\Gamma, F, !G$, can you apply the !-rule to this sequent? There could be two answers: "it depends on F" or "no, at least not yet". Both answers carry the same idea that ! can be applied only if F is $?F'$ but they are different from the operational point of view: the second answer suggests that there is some more work to do in order to apply the !-rule: F should first be recognized as $?F'$. This remark suggests to introduce a separate context that will store those formulas that have been recognized as having a "?": $\vdash \Gamma \mid \Delta$. The two operations discussed earlier and dereliction now become the following rules:

$$\frac{\vdash \Gamma, A \mid \Delta}{\vdash \Gamma \mid\ ?A, \Delta}\ ? \qquad \frac{\vdash \Gamma \mid A}{\vdash \Gamma \mid\ !A}\ ! \qquad \frac{\vdash \Gamma, A \mid A, \Delta}{\vdash \Gamma, A \mid \Delta}\ der$$

We then have to adapt all the usual MALL rules in the obvious way.

?-rule will be considered as negative whereas ! and dereliction will be considered as positive. We can now extend the positive trunks to LL proofs with exponentials:

Definition 9 (Exponential Positive Trunk). *Given a positive sequent $\mathcal{S}$ and a proof Π of $\mathcal{S}$, an exponential positive trunk (or positive trunk for short) for a positive sequent is a maximal subtree of Π containing only positive rules and such that* !*-rules produce leaves of the tree (the branches are cut as soon as a* !*-rule is applied).*

The reader may be surprised by the fact that the branches of the positive trunk are cut as soon as a ! rule is encoutered. This is reminiscent of the bipolar character of the exponentials: the ? is decomposed into two rules (one negative, the other positive) and for its dual connective, the !, the rule is positive but the focusing phase is stopped.

In order to build the Focalization graph, we first notice that each *der*-rule in the positive trunk produces an occurrence of a formula, say A, that might be chosen as a focus. We have to distinguish such occurrences and to do so we will index them as (A, i). The index i will refer to the place in the tree where the dereliction rule has been applied. Notice also the $<$-relation is straightforwardly extended to exponential sequents.

Definition 10. *Let $\vdash \Gamma \mid \Delta$ be a positive sequent[2], Π be a proof of the sequent and Π^+ be the associated (Exponential) Positive Trunk. The* **Exponential Focalization Graph** *extends the definition of standard Focalization graphs as follows:*

[2] Straightforward extension of the one for MALL sequents.

(i) *The vertices of the graph are the active formulas of Δ and the active occurrence of formulas in Γ, ie. of the form (A, i).*
(ii) *The arcs are given by the sequents of the border in the same way as usually (including the bias assignment if any)*[3].

The following allows us to extend our Focalization result to the exponential setting:

Proposition 4. *The Exponential Focalization Graphs are acyclic.*

4.4 Further Extensions

The proof we presented is modular in the sense that it relies on a series of simple results which can be adapted to richer settings. It is what is done in [5] in order to extend Focalization to an extension to LL with Fixpoints. We shall consider in future works other extensions. In particular, non-commutative logics and light logics should be good candidates to test the methodology of this paper.

5 Multi-focalization

The question of Multi-Focalization naturally arises from the structure of Focalization Graphs. Indeed, the only two ingredients needed in our proof are (i) appropriate permutability properties (full and weak permutabilities) and (ii) the acyclicity of the Focalization Graph $\mathcal{G}$ which ensures us of the existence of a source which can be taken as a focus in the proof we are building.

In this last section we consider briefly this question of multi-focalization although most details on an analysis on Multi-Focalization are beyond the scope of this paper and will be postponed to future work. We only intend to introduce this notion and outline what could be the first step to a general theory of multifocalization.

We know that $\mathcal{G}$ has a source, but nothing forbids $\mathcal{G}$ to have multiple sources. In such a case, we would have several formulas (say $F_1, \ldots, F_k$) for which the topmost positive layer of connectives is totally decomposed within the positive trunk. Weak permutability allows to conclude that the proof Π can be transformed to a proof where the bottom part of the tree is made only of positive rules on the F_i's and their subformulas up to a point where all the F_i's are turned to negative formulas (or literals).

This is enough to consider a notion of multifocalization and this leads us to associated sequent rules that we are currently investigating with Kaustuv Chaudhuri and which can be presented in a Σ_3 inspired sequent presentation as

$$\frac{\Psi, F_1, \ldots F_k \colon \Delta \Downarrow F_1^{i_1}, \ldots F_k^{i_k}, F_1', \ldots F_l'}{\Psi, F_1, \ldots F_k \colon \Delta, F_1', \ldots F_l' \Uparrow \cdot}\; MultiFoc$$

with the proviso that during a multifocusing section, only positive rules can be applied: the negative rules that could be present would be frozen until all the positive formulas under focus have been decomposed.

[3] We do not need to take care of the premisses of !-rules since these sequents contain exactly *one* subformula of an active formula of the root: A is the only formula in the linear part of this sequent of the border.

There is much to do in order to understand precisely this notion of Multi-Focalization but we can already draw some comments:

- completeness is not an issue for multi-focalization since it extends focalization;
- more interesting would be to understand how to obtain proofs which have been multifocalized as much as possible. In particular, is there such an interesting notion of maximality in the world of multi-focused proofs?
- clearly, multi-focused proofs have a taste of concurrency: having F and G as foci actually means that we are focussing on the two formulas at the same time, even though we keep the sequent syntax. It would thus be pretty interesting to compare this with works on concurrent or asynchronous games [1];
- This notion of Multifocalization might have interesting consequences for proof search allowing, for instance, to detect failures of the proof search earlier.

6 Conclusion and Future Works

We have presented a new proof of the completeness of focused proofs for linear logic. We first focused on MALL fragment in which rather elementary considerations of the permutability of inference rules allowed us to define a focalization graph. The fact that such a graph is acyclic allows us to build sequent calculus proofs. There are many possibilities for building such proof: a flexible bias assignment mechanism allows edges to be added to the focalization graph, which, in turn, constrains the space of sequent calculus proofs that can be produced. The techniques developed for MALL can be lifted directly to providing focusing results much stronger logics, in particular, full first-order and higher-order linear logic and linear logic with fixed points. Given the centrality of the focalization graph and since such graphs may have more than one source, we have also considered adding to a focused proof system the multifocusing inference rule that can capture such multiplicity of foci.

The structure of Focalization Graph we introduced in this paper and the consideration of Focalization as a process for transforming proofs suggest we study several developments for future works:

- The interaction between Focalization process and cut-reduction shall be made clear;
- We would like to extend our results to richer logics such as non-commutative logics or light logics as a test for our methodology;
- We would be interested in adapting focalization result directly to logics such as LJ;
- The study of Multi-Focalization is a direction that seems to be fruitful and to relate focalization with interesting topics of concurrent view of proofs;
- In a more applied setting, we should pursue the classification of Bias Assignments since it seems to be meaningful for applications in proof search and other settings;

Acknowledgments. We thank David Baelde, Kaustuv Chaudhuri, Pierre-Louis Curien, Claudia Faggian and Olivier Laurent for helpful discussions. We also want to thank the anonymous referees for their comments on a previous version of this paper. This work has been supported in part by INRIA through the "Equipes Associées" Slimmer.

References

1. Abramsky, S., Melliès, P.-A.: Concurrent games and full completeness. In: LICS 1999, pp. 431–442. IEEE Computer Society Press, Los Alamitos (1999)
2. Andreoli, J.-M., Pareschi, R.: Linear objects: Logical processes with built-in inheritance. In: Proceeding of ICLP 1990, Jerusalem (1990)
3. Andreoli, J.-M.: Proposal for a Synthesis of Logic and Object-Oriented Programming Paradigms. PhD thesis, University of Paris VI (1990)
4. Andreoli, J.-M.: Logic programming with focusing proofs in linear logic. J. of Logic and Computation 2(3), 297–347 (1992)
5. Baelde, D., Miller, D.: Least and greatest fixed points in LL (Submitted April 2007)
6. Curien, P.-L., Herbelin, H.: The duality of computation. In: ICFP '00. Proceedings of the fifth ACM SIGPLAN international conference on Functional programming, New York, NY, USA, pp. 233–243. ACM Press, New York (2000)
7. Danos, V., Joinet, J.-B., Schellinx, H.: The structure of exponentials: Uncovering the dynamics of linear logic proofs. In: Mundici, D., Gottlob, G., Leitsch, A. (eds.) KGC 1993. LNCS, vol. 713, pp. 159–171. Springer, Heidelberg (1993)
8. Danos, V., Joinet, J.-B., Schellinx, H.: LKQ and LKT: sequent calculi for second order logic based upon dual linear decompositions of classical implication. In: Girard, Lafont, Regnier (eds.) Workshop on Linear Logic. London Mathematical Society Lecture Notes 222, pp. 211–224. Cambridge University Press, Cambridge (1995)
9. Danos, V., Joinet, J.-B., Schellinx, H.: A new deconstructive logic: Linear logic. Journal of Symbolic Logic 62(3), 755–807 (1997)
10. Dyckhoff, R., Lengrand, S.: LJQ: a strongly focused calculus for intuitionistic logic. In: Beckmann, A., Berger, U., Löwe, B., Tucker, J.V. (eds.) CiE 2006. LNCS, vol. 3988, pp. 173–185. Springer, Heidelberg (2006)
11. Girard, J.-Y.: Linear logic. Theoretical Computer Science 50, 1–102 (1987)
12. Girard, J.-Y.: Light linear logic. Information and Computation 143 (1998)
13. Girard, J.-Y.: Locus solum. MSCS 11(3), 301–506 (2001)
14. Herbelin, H.: Séquents qu'on calcule: de l'interprétation du calcul des séquents comme calcul de lambda-termes et comme calcul de stratégies gagnantes. PhD thesis, Université Paris 7 (1995)
15. Jagadeesan, R., Nadathur, G., Saraswat, V.: Testing concurrent systems: An interpretation of intuitionistic logic. In: Ramanujam, R., Sen, S. (eds.) FSTTCS 2005. LNCS, vol. 3821, Springer, Heidelberg (2005)
16. Lafont, Y.: Soft linear logic and polynomial time. TCS 318(1-2), 163–180 (2004)
17. Laurent, O.: Etude de la polarisation en logique. Thèse de doctorat, Université Aix-Marseille II (March 2002)
18. Laurent, O.: A proof of the focalization property of LL. Unpublished Note (May 2004)
19. Liang, C., Miller, D.: Focusing and polarization in intuitionistic logic. In: Duparc, J., Henzinger, T.A. (eds.) CSL 2007. LNCS, vol. 4646, pp. 451–465. Springer, Heidelberg (2007)
20. Miller, D.: Forum: A multiple-conclusion specification logic. Theoretical Computer Science 165(1), 201–232 (1996)

Linear Realizability

Naohiko Hoshino

RIMS, Kyoto University
naophiko@kurims.kyoto-u.ac.jp

Abstract. We define a notion of relational linear combinatory algebra (rLCA) which is a generalization of a linear combinatory algebra defined by Abramsky, Haghverdi and Scott. We also define a category of assemblies as well as a category of modest sets which are realized by rLCA. This is a linear style of realizability in a way that duplicating and discarding of realizers is allowed in a controlled way. Both categories form linear-non-linear models and their coKleisli categories have a natural number object. We construct some examples of rLCA's which have some relations to well known PCA's.

1 Introduction

A category of realizability with respect to a partial combinatory algebra (PCA) A, for example assemblies $\mathbf{Ass}(A)$ or modest sets $\mathbf{Mod}(A)$, is a category of sets and functions that are implemented by the calculating system A. For computer science, these categories are models of the (second order) lambda calculus and PCF [Jac99],[Lon95]. Moreover, a realizability model provides a strong normalization proof of the second order lambda calculus [HO93]. In this paper we develop a linear variant of realizability by using another algebra in place of PCA, anticipating models of (2nd order) linear lambda calculus.

Properties of these categories, for example being cartesian closed categories (CCC), mainly come from the combinatory completeness of PCA. The combinatory completeness is informally stated as "for any lambda term t, there is an element which works like t", which allows arbitrary copying and discarding of terms. For our purpose of giving a linear variant of realizability, we should consider an algebra in which we can restrict this copying or discarding.

First, such an algebra should at least have elements b,i which satisfy $bxyz = x(yz)$ and $ix = x$ since when we use this algebra as a realizer of a category of assemblies, this algebra must realize identity and composition of two realizers of some functions.

An algebra which is called BCI algebra if it is applied with the above two elements and also an element c satisfying $cxyz = (xz)y$ is called BCI algebra, has combinatory completeness for the untyped pure linear lambda calculus. In the untyped pure linear lambda calculus, no terms are copied or discarded and in fact a category of assemblies of the untyped pure linear lambda calculus forms a symmetric monoidal closed category (SMCC). So a category of assemblies of a BCI algebra is a model of the typed pure linear lambda calculus.

J. Duparc and T.A. Henzinger (Eds.): CSL 2007, LNCS 4646, pp. 420–434, 2007.

The pure linear lambda calculus is too simple, however. We seek for a category of assemblies that is a model of the typed linear lambda calculus which has an exponential comonad !, so we should add some structures corresponding to the exponential comonad to BCI algebras. One algebra that has such a structure is *linear combinatory algebra(LCA)* in [AHS02]. This algebra is a BCI algebra which has an operator ! and some elements which work like natural transformations for the exponential comonad ! of a linear category [Bie95]. LCA's have good relations to SK algebras and there are many important examples of LCA's. For example, Abramsky and Lenisa used an LCA constructed from a set of partial involutions to show that the PER model of this algebra is fully complete w.r.t the fragment of System F consisting of ML-types [AL05].

In [Lon95], Longley introduced applicative morphisms, which are morphisms between applicative structures. Using applicative morphisms, we define relational LCA (rLCA) which is a generalization of LCA. Although the notion of rLCA seems slightly strange as a combinatory algebra, the definition of rLCA seems suitable because there are examples of rLCA's that have an adjoint pair (defined in Def 10) between important PCA's (cf. Sect 5.4, Sect 5.5) and no LCA has an adjoint pair between them (cf. Proposition 11).

The structure of this paper is as follows. In the Sect. 2, we recall the notions of combinatory algebras, applicative morphisms and categories of assemblies and modest sets. In Sect. 3 we show assemblies and modest sets realized by a BCI algebra form SMCC's. In Sect. 4 we show assemblies and modest sets realized by an rLCA form adjoint models. Sect. 5 is for some examples of rLCA's.

2 Background

2.1 Combinatory Algebras

In this section we recall some notions such as partial combinatory algebra, BCI algebra and their combinatory completeness.

Definition 1. *Let* $(A, \cdot)$ *be a pair of a set* A *and a partial binary application* $\cdot : A \times A \rightharpoonup A$. $(A, \cdot)$ *is a* partial combinatory algebra *(PCA) if* A *has elements* s, k *which satisfy:*

$$s \cdot x \downarrow, \quad s \cdot x \cdot y \downarrow, \quad s \cdot x \cdot y \cdot z \simeq (x \cdot z) \cdot (y \cdot z)$$
$$k \cdot x \downarrow, \quad k \cdot x \cdot y \simeq x$$

where $x \cdot y \downarrow$ *means that the value of* $x \cdot y$ *is defined and* $\simeq$ *means that if one side of the equation is defined then the other side is also defined and are equal.*

If the application of a PCA is total then we call this PCA an SK algebra.

Definition 2. *Let* $(A, \cdot)$ *be a pair of a set* A *and a total binary application* $\cdot : A \times A \to A$. $(A, \cdot)$ *is a* BCI algebra *when* A *has elements* b, c, i *satisfying:*

$$b \cdot x \cdot y \cdot z = x \cdot (y \cdot z) \qquad c \cdot x \cdot y \cdot z = (x \cdot z) \cdot y \qquad i \cdot x = x$$

for all $x, y, z \in A$.

In the following we write b, c, i for elements of a BCI algebra as above and s, k for such elements of an SK algebra.

Let A be a PCA or a BCI algebra, a *polynomial* over A is a syntactic expression generated by variables, elements of A and applications. The applicative structure of A induces an evident denotational relation between polynomials and elements of A.

Proposition 1. *Let $(A, \cdot)$ be a PCA and M be a polynomial over A. Then there exists a polynomial $\lambda^* x.M$ whose free variables are just those of M excluding x, which is defined as a polynomial over A and $(\lambda^* x.M)a \simeq M[a/x]$ for all $a \in A$.*

Proposition 2. *Let $(A, \cdot)$ be a BCI algebra and M be a polynomial over A in which x appears exactly once. Then there exists a polynomial $\lambda^* x.M$ whose free variables are just those of M excluding x such that $(\lambda^* x.M)a = M[a/x]$.*

In the following, for elements $p, q, r, \cdots, s$ of a BCI algebra or a PCA, we write $p \cdot q \cdot r \cdots s$ for $(\cdots((p \cdot q) \cdot r) \cdots s)$, and we write $[p, q]$ for $\lambda^* x.xpq$, and

$$\mathtt{let}\ [u, v] = x\ \mathtt{in}\ r(u, v)$$

for $x(\lambda^* u.\lambda^* v.r(u, v))$ where $r(u, v)$ is a polynomial containing u,v as its free variables and having exactly one occurrence of each u and v. From these definitions, we have

$$\mathtt{let}\ [u, v] = [p, q]\ \mathtt{in}\ r(u, v) = r(p, q) \ .$$

Remark 1. Although it is natural to define "partial BCI algebra" which may be defined in a similar way of PCA, we consider only total ones in this paper: we can construct a total BCI algebra from a partial one by adding $\perp$ and define a new application by

$$a \bullet b = \begin{cases} a \cdot b & a \text{ and } b \text{ are not } \perp \text{ and } a \cdot b \text{ is defined} \\ \perp & \text{else} \end{cases}$$

and a category of assemblies, defined in Sect 2.3, realized by the partial BCI algebra is a full subcategory of a category of assemblies realized by the new total BCI algebra.

2.2 Applicative Morphisms

In [Lon95], applicative morphisms and a preorder between them are defined for PCA's. Here, the same definitions of applicative morphisms and preorder are given for BCI algebras and PCA's.

Definition 3. *Let A,B be BCI algebras or PCA's. An* applicative morphism γ *from A to B is a total relation from A to B such that :*

$$\exists r \in B, \forall p, q \in A, pq \textit{ is defined } \Rightarrow \forall s \in \gamma(p), t \in \gamma(q)\ rst \in \gamma(pq) \ .$$

If the domain A is a BCI algebra, "pq is defined " is not necessary since BCI algebra is total. This r is called a realizer for γ and we say γ is realized by r.

We say an applicative morphism γ is *functional* when $\gamma(p)$ is always a singleton and we say an adjoint pair $\delta \dashv \gamma$ is functional when both δ and γ are functional. If γ is functional and $\gamma(p) = \{q\}$, we write $\gamma(p) = q$ and $\gamma(p)$ for q in equations.

Definition 4. *Let A,B be BCI algebras or PCA's and $\gamma, \delta : A \to B$ be applicative morphisms. We write $\gamma \preceq \delta$ if there exists $r \in B$ such that:*

$$\forall p \in A, q \in \gamma(p).\ rq \in \delta(p) \ .$$

We say that two applicative morphisms γ, δ are equivalent if $\gamma \preceq \delta$ and $\delta \preceq \gamma$. It is proved in [Lon95] that PCA's, applicative morphisms and preorder between them form a preorder enriched category. This is also the case for BCI algebras. We have to check that the constructions of realizers in the proof of [Lon95] can also be carried out for BCI algebras. In fact all "lambda term"s in the proof are linear. Notice that an identity on a BCI algebra or a PCA A is $\{(a, a) | a \in A\}$.

Proposition 3. *BCI algebras, PCA's and applicative morphisms and preorder between them form a preorder enriched category.*

2.3 Assemblies, Modest Sets

Categories of *assemblies* and *modest sets* are defined for PCA's in [Lon95]. We can also define these categories for BCI algebras.

Definition 5. *Let A be a BCI algebra. The category of assemblies* **Ass**(A) *consists of*

- *objects : $(X, ||-||_X)$ where X is a set and $||-||_X$ is a map from X to a set of non empty subsets of A.*
- *arrows : $f : (X, ||-||_X) \to (Y, ||-||_Y)$ where f is a realizable map from X to Y .*

Here a realizable map is a map such that there exists $r \in A$ which satisfies for all $a \in ||x||_X$, $ra \in ||fx||_Y$. We say r is a realizer of f or f is realized by r.

Modest sets **Mod**(A) is a full subcategory of **Ass**(A) whose object $(X, ||-||_X)$ satisfies:

$$x \neq y \Rightarrow ||x||_X \bigcap ||y||_X = \phi \ .$$

We sometimes omit $||-||_X$ and write X for an object of assemblies or modest sets. We write $|X|$ for the underlying set of X, and we write $|f|$ for the underlying map of f.

2.4 Categorical Models of Intuitionistic Linear Logic

We recall some models of intuitionistic linear logic.

Definition 6. [Bie95] *A* linear category *is an SMCC with a monoidal comonad $(!, \epsilon, \delta, m_{A,B}, m_I)$ such that*

- *There are two distinguished monoidal natural transformations with components* $e_A :!A \to I$ *and* $d_A :!A \to !A \otimes !A$ *for every free !-coalgebra* $(!A, \delta_A)$ *which form a commutative comonoid and are coalgebra morphisms,*
- *Whenever* $f : (!A, \delta_A) \to (!B, \delta_B)$ *is a coalgebra morphism between free coalgebras, then it is also a comonoidal morphism.*

Definition 7. [Ben94] *A* linear-non-linear model *is a symmetric monoidal adjunction* $F \dashv G : \mathcal{L} \to \mathcal{C}$ *where* $\mathcal{C}$ *is a CCC and* $\mathcal{L}$ *is an SMCC.*

3 Realizability of BCI Algebras

3.1 Assemblies and Modest Sets Realized by BCI Algebras

Proposition 4. *Let* A *be a BCI algebra, then* $\mathbf{Ass}(A)$ *is a symmetric monoidal closed category.*

Proof. (outline) We may take for objects X,Y in $\mathbf{Ass}(A)$,

- $|I| = \{*\}$ and $||*||_I = \{i\}$
- $|X \otimes Y| = |X| \times |Y|$ and $||(x, y)||_{X \otimes Y} = \{[p, q] | p \in ||x||_X, q \in ||y||_Y\}$
- $|X \multimap Y| = \{f : |X| \to |Y| \mid f$ is realizable. $\}$ and $||f||_{X \multimap Y} = \{r | r$ realize $f\}$.

For example, an isomorphism $\rho_X : X \otimes I \to X$ is $(x, *) \mapsto x$ realized by $\lambda^* x.\mathtt{let}\ [p, q] = x\ \mathtt{in}\ qp$ and an evaluation $ev_{X,Y} : (X \multimap Y) \otimes X \to Y$ is $(f, x) \mapsto fx$ realized by $\lambda^* x.\mathtt{let}\ [p, q] = x\ \mathtt{in}\ pq$ □

Proposition 5. *Inclusion functor* $J : \mathbf{Mod}(A) \to \mathbf{Ass}(A)$ *has a left adjoint* Δ. *Hence* $\mathbf{Mod}(A)$ *is a reflective full subcategory of* $\mathbf{Ass}(A)$.

Proof. Let X be an object of $\mathbf{Mod}(A)$. We define $\simeq$ as a transitive closure of $x \sim y$ where $x \sim y$ iff $||x||_X \bigcap ||y||_X \neq \phi$ and $|\Delta X| = |X|/\simeq$, $||[x]||_{\Delta X} = \bigcup_{y \simeq x} ||y||_X$ where $[x]$ is an equivalence class of x. If f is a morphism of $X \to Y$ then $|\Delta f|([x]) = [fx]$ realized by a realizer r of f. If $a \in ||x|| \bigcap ||y||$ then $ra \in ||fx|| \bigcap ||fy||$ hence this is well defined. Let $\eta : 1 \to J\Delta$ be $\eta_X(x) = [x]$ and $\epsilon : \Delta J \to 1$ be the identity; both are realized by i. These natural transformations form unit and counit of the adjunction. □

Lemma 1. *Let* X *be an object in* $\mathbf{Mod}(A)$ *and* Y *be an object in* $\mathbf{Ass}(A)$. *Then* $\eta_{Y \multimap JX} : Y \multimap JX \to J\Delta(Y \multimap JX)$ *is an isomorphism.*

Proof. We show $Y \multimap JX$ is a modest set. Let $f, g \in |Y \multimap JX|$. If $r \in ||f|| \bigcap ||g||$ then for any $a \in ||y||_Y$, $ra \in ||fx||_{JX}$ and $ra \in ||gx||_{JX}$. Since X is a modest set, this implies $fx = gx$. Hence if $f \neq g$ then $||f||_{Y \multimap JX} \bigcap ||g||_{Y \multimap JX} = \phi$. □

In general, if a reflective full subcategory $\Delta \dashv J : \mathcal{C} \to \mathcal{D}$ of an SMCC $\mathcal{D}$ satisfies that for any object X of $\mathcal{C}$ and Y of $\mathcal{D}$, $\eta_{Y \multimap JX} : Y \multimap JX \simeq J\Delta(Y \multimap JX)$ then $\mathcal{C}$ forms an SMCC, whose monoidal product is $\Delta(JX \otimes JY)$, the unit is ΔI, the exponential is $\Delta(JX \multimap JY)$ and $\Delta \dashv J$ is a monoidal adjunction.

Therefore $\mathbf{Mod}(A)$ is an SMCC and J,Δ are monoidal functors. Moreover if $\mathbf{Ass}(A)$ has (co)limits then $\mathbf{Mod}(A)$ also has (co)limits.

3.2 Functors from Applicative Morphisms

In this section, we use A and B for a BCI algebra or a PCA. If γ is an applicative morphism from A to B, we can construct a functor $\gamma_* : \mathbf{Ass}(A) \to \mathbf{Ass}(B)$. This definition is the same as the one in [Lon95]

Definition 8. *Let $\gamma : A \to B$ be an applicative morphism. The functor γ_* sends $(X, ||-||_X)$ in $\mathbf{Ass}(A)$ to $(X, \gamma(||-||_X))$ in $\mathbf{Ass}(B)$ and a morphism $f : X \to Y$ to $f : \gamma_* X \to \gamma_* Y$ whose underlying map is $|f|$.*

If $f : X \to Y$ is realized by $s \in B$ then $\gamma_*(f)$ is realized by $\lambda^* x.rs'x$ where r is a realizer of γ and s' is an element of $\gamma(s)$.

Proposition 6. *Let $\gamma : A \to B$ be an applicative morphism. Then γ_* is a lax monoidal functor from $\mathbf{Ass}(A) \to \mathbf{Ass}(B)$.*

Proof. Underlying maps of two natural transformations $m_{X,Y} : \gamma_*(X) \otimes \gamma_*(Y) \to \gamma_*(X \otimes Y)$ and $m_I : I \to \gamma_*(I)$ are both identity. If we choose an element $a \in \gamma(\lambda^* xy.[x, y])$ and a realizer r of γ, $\lambda^* pq.r(rap)q$ realizes m_X, Y. A realizer of m_I is $\lambda^* x.xi'$ where i' is an element of $\gamma(i)$. It is easy to see that these natural transformations satisfy coherence diagrams. □

If $\gamma \preceq \delta$ are applicative morphisms related by the preorder between them, then there is a monoidal natural transformation from γ_* to δ_*.

Definition 9. *If $\gamma \preceq \delta : A \to B$ are applicative morphisms related by the preorder, $\alpha_* : \gamma_* \to \delta_* : \mathbf{Ass}(A) \to \mathbf{Ass}(B)$ is a natural transformation such that an underlying map of α_{*X} is identity.*

From the definition of preorder, we can see that a realizer of $\gamma \preceq \delta$ realizes $\alpha_{*X} : \gamma_*(X) \to \delta_*(X)$. For any morphism $f : X \to Y$, $|\alpha_{*Y}\gamma_*(f)|$ is $|f|$ and $|\delta_*(f)\alpha_{*X}|$ is also $|f|$. Hence, $\alpha_{*Y}\gamma_*(f) = \delta_*(f)\alpha_{*X}$. It is easy to see that this is a monoidal natural transformation.

This construction is a 2-functor from a 2-category of BCI algebras and PCA's to a 2-category of categories of assemblies realized by BCI algebras and PCA's.

4 Realizability of Relational Combinatory Algebras

4.1 Relational Linear Combinatory Algebras

In Sect. 2.2 we recalled the definitions of applicative morphisms and preorder between them. In this section we define adjoint pair and comonadic applicative morphism. The same definition of adjoint pair for PCA's is given in [Lon95].

Definition 10. *Let A,B be BCI algebras or PCA's. $\delta \dashv \gamma : A \to B$ is an* adjoint pair *if $\delta : B \to A$ and $\gamma : A \to B$ are applicative morphisms satisfying $\delta\gamma \preceq 1_A$ and $1_B \preceq \gamma\delta$.*

Definition 11. *Let A be a BCI algebra or a PCA. $\gamma : A \to A$ is a* comonadic applicative morphism *if γ is an applicative morphism satisfying $\gamma \preceq 1_A$ and $\gamma \preceq \gamma\gamma$. Notice that any comonadic applicative morphism γ is equivalent to $\gamma\gamma$ and γ_* is idempotent.*

Just as a comonad functor can be constructed from an adjunction, we can construct a comonadic applicative morphism from an adjoint pair.

Proposition 7. *Let A,B be BCI algebras or PCA's and $\delta \dashv \gamma : A \to B$ is an adjoint pair. Then $\epsilon = \delta\gamma : A \to A$ is a comonadic applicative morphism.*

Proof. Since $\delta \dashv \gamma$, $\delta\gamma \preceq 1_A$ and $1_B \preceq \gamma\delta$. Hence $\epsilon \preceq 1_A$ and $\epsilon = \delta\gamma \preceq \delta\gamma\delta\gamma = \epsilon\epsilon$. □

We define 'relational LCA', which is an analogue of linear category.

Definition 12. *A* relational linear combinatory algebra *(rLCA) $(A, !)$ consists of a BCI algebra A and a comonadic applicative morphism $! : A \to A$ such that*

$$! \preceq [!, !] \quad ! \preceq k_i$$

where $[!, !]$ is an applicative morphism such that $[!, !](p) = \{[u, v] | u, v \in !(p)\}$ which is realized by

$$\lambda^* pq.\texttt{let } [p_1, p_2] = p \texttt{ in let } [q_1, q_2] = q \texttt{ in } [rp_1q_1, rp_2q_2]$$

using a realizer r of $!$; and k_i is an applicative morphism such that $k_i(a) = \{i\}$ which is realized by i.

In the same way as we can construct a linear category from a linear-non-linear model, we can construct an rLCA from an adjoint pair between a BCI algebra and a PCA.

Proposition 8. *Let A be a BCI algebra, B be a PCA and $\delta \dashv \gamma : A \to B$ be an adjoint pair. Then $(A, !)$ is an rLCA where $! = \delta\gamma : A \to A$.*

Proof. From Proposition 7, $!$ is a comonadic applicative morphism. Let s be a realizer for $\delta\gamma \preceq 1_A$, t be a realizer for $1_B \preceq \gamma\delta$, r_δ a realizer for δ and r_γ a realizer for γ.

Choose $u \in \gamma(\lambda^* xy.[x, y])$ and $v \in \delta(\lambda^* x.r_\gamma(r_\gamma u(tx))(tx))$ and we will show that $! \preceq [!, !]$ is realized by $\lambda^* x.s(r_\delta vx)$; let $p' \in \delta\gamma(p)$. Then $(\lambda^* x.s(r_\delta vx))p' \simeq s(r_\delta vp')$. There exists $p'' \in \gamma(p)$ such that $p' \in \delta(p'')$ and $r_\delta vp'$ is an element of $\delta(r_\gamma(r_\gamma u(tp''))(tp''))$ if $r_\gamma(r_\gamma u(tp''))(tp'')$ is defined. Since $tp'' \in \gamma\delta\gamma(p)$ there exists $q, q' \in \delta\gamma(p)$,

$$r_\gamma(r_\gamma u(tp''))(tp'') \in \gamma((\lambda^* xy.[x, y])qq')$$

if $(\lambda^* xy.[x, y])qq'$ is defined. Since a BCI algebra A is total, this $(\lambda^* xy.[x, y])qq'$ is defined and is equal to $[q, q']$. Hence, $r_\delta vp'$ is an element of $\delta\gamma([q, q'])$ and $t(r_\delta vp') \in \{[q, q']\} \subseteq [!, !](p)$.

$! \preceq k_i$ is realized by $\lambda^* x.t(r_\delta hx)$ where h is an element of $\delta(\lambda^* x.i')$ taking an element $i' \in \gamma(i)$: If $p' \in \delta\gamma(p)$, there exists $q \in \gamma(p)$ and $r_\delta hp' \in \delta((\lambda^* x.i')q)$ if $(\lambda^* x.i')q$ is defined. In fact this is defined and equal to i'. Hence $r_\delta hp'$ is an element of $\delta\gamma(i)$, $t(r_\delta hp')$ is an element of $\{i\}$. □

It seems not the case in general that we can construct a PCA from an rLCA. However if we restrict applicative morphisms to be functional, we can construct a PCA. We recall the notion of *linear combinatory algebra* (LCA) which appears in [AHS02].

Definition 13. *A BCI algebra* $(A, \cdot)$ *is a* linear combinatory algebra *if it has a map* $! : A \to A$ *and* $k, w, d, \delta, f \in A$ *satisfying:*

$$kx!y = x \quad \delta!x = !!x \quad d!x = x \quad wx!y = x!y!y \quad f!x!y = !(xy)$$

If we define γ as $\gamma(p) = \{!p\}$, this is a comonadic applicative morphism. $! \preceq 1$ is realized by d, $! \preceq !!$ is realized by δ and a realizer of γ is f. Hence (A, γ) is an rLCA since $! \preceq [!, !]$ is realized by $\lambda^* z.w(\lambda^* xy.[x, y])z$ and $! \preceq k_i$ is realized by $\lambda^* x.kix$. Hence we can think LCA as a special case of rLCA whose ! is functional.

Proposition 9. *Let* $\delta \dashv \gamma : A \to B$ *be a functional adjoint pair from a BCI algebra A to a PCA B. Then* $(A, !)$ *is an LCA where* $!p = p'$ *iff* $\delta\gamma(p) = \{p'\}$.

Proof. From Proposition 8 and since LCA is a special case of rLCA. □

If we have an LCA then we can construct a functional adjoint pair.

Lemma 2. *Let* $(A, !)$ *be an LCA then an applicative structure* $(A_!, \bullet)$ *is an SK algebra where* $|A_!| = A$ *and* $p \bullet q = p \cdot !q$.

Proof. Let s and k' be elements of $A_!$ such that $s = \lambda^* xyz.w(\lambda^* uv.dxu(fy(\delta v)))z$ and $k' = \lambda^* xy.k(dx)y$. Then s and k' satisfy $s!x!y!z = x(!z)!(y(!z))$ and $k'!x!y = k(d(!x))(!y) = x$. □

Proposition 10. *Let* $(A, !)$ *be an LCA. Then there is a functional adjoint pair between A and* $A_!$.

Proof. Let $\rho : A \to A_!$ and $\sigma : A_! \to A$ be applicative morphisms such that $\rho(p) = p$ and $\sigma(p) = !p$. ρ and σ are realized by $\lambda^* xy.(dx)(dy)$ and $\lambda^* xy.fx(\delta y)$ respectively and $1_{A_!} \preceq \rho\sigma$ is realized by i, $\sigma\rho \preceq 1_A$ is realized by d. □

The next lemma can be proved as Theorem 3.1.8 and Corollary 3.1.9 of [Lon95].

Lemma 3. *There is no SK algebra which has a decidable equality.*

Here a PCA A has a decidable equality when there is an element $d \in A$ such that

$$dxy = \begin{cases} \lambda^* uv.u & \text{if } x = y \\ \lambda^* uv.v & \text{if } x \neq y \end{cases}$$

The generalization of LCA to rLCA enables us to treat PCA which has a decidable equality.

Proposition 11. *If a PCA A has a decidable equality then there is no functional adjoint pair from any BCI algebra to A.*

Proof. We suppose there is a functional adjoint pair $\delta \dashv \gamma : B \to A$ where B is a BCI algebra. By Proposition 10, we have a SK algebra $B_{\delta\gamma}$. We show $B_{\delta\gamma}$ has decidable equality. Then by the above lemma, we obtain contradiction.

Let c be an element of A which decide the equality of A. Then

$$\lambda^* xy.d((((\delta(c) * x) * y) * \delta\gamma(true)) * \delta\gamma(false))$$

decides the equality of $B_{\delta\gamma}$ where $x * y = r_\delta xy$ for a realizer r_δ of δ, d is a realizer of $\delta\gamma \preceq 1$, $true$ and $false$ is $\lambda^* uv.u$ and $\lambda^* uv.v$ of $B_{\delta\gamma}$ and $\lambda^* xy.\cdots$ is a lambda abstraction of B. □

4.2 Assemblies and Modest Sets Realized by rLCA's

Let $(A, !)$ be an rLCA. As we have seen in the previous section, we have a monoidal comonad $!_* : \mathbf{Ass}(A) \to \mathbf{Ass}(A)$ with natural transformations $w_X : !_* X \to I$ which sends x to $*$ and $c_X : !_* X \to !_* X \otimes !_* X$ which sends x to (x, x).

Proposition 12. *Let A be an rLCA then* $\mathbf{Ass}(A)$ *is finitely complete and cocomplete and there are natural isomorphisms $!_* X \otimes !_* Y \simeq !_*(X \times Y)$ and $I \simeq !_* 1$.*

Proof. Definitions of terminal, initial, equalizer and coequalizer are the same in [Lon95].

For products, let X,Y be objects of $\mathbf{Ass}(A)$. Then we define $X \times Y$ as

$$|X \times Y| = |X| \times |Y|$$
$$||(x, y)||_{X \times Y} = \{[a, [p, q]] | \exists r, s. p \in !r, q \in !s, ra \in ||x||_X, sa \in ||y||_Y\}$$

First projection $\pi : X \times Y \to X$ which sends (x, y) to x and is realized by

$$\lambda^* x.\texttt{let } [t, u] = x \texttt{ in let } [v, w] = u \texttt{ in } (kw)(dv)t$$

where k is a realizer of $! \preceq k_i$ and d is a realizer of $! \preceq 1_A$. Second projection π' is similar. Let $f : Z \to X$ and $g : Z \to Y$ be morphisms realized by m, n respectively. Then we have a map $h : Z \to X \times Y$ which sends z to (fz, gz). This morphism is realized by $\lambda^* x.[x, [m', n']]$ where $m' \in !m$ and $n' \in !n$. h satisfies $\pi h = f$, $\pi' h = g$ and we can see uniqueness from underlying maps of these morphisms.

$!_*(\langle \theta(d_X \otimes w_Y), \theta'(w_X \otimes d_Y) \rangle) m_{X,Y}(\delta_X \otimes \delta_Y)$ and $(!_*\pi \otimes !_*\pi') c_{X \times Y}$ form an isomorphism of $!_* X \otimes !_* Y \simeq !_*(X \times Y)$, $!_*(u) m_I$ and w_1 form an isomorphism of $I \simeq !_* 1$ since underlying maps of those morphisms are identity. Here $\theta : X \otimes I \simeq X$ and $\theta' : I \otimes Y \simeq Y$, m is a natural transformation of $!_*$ and $u : I \to 1$ is a unique morphism.

For coproducts, let X,Y be objects of $\mathbf{Ass}(A)$. If k realizes $! \preceq k_i$ and d realizes $! \preceq 1_A$ then

$$|X + Y| = |X| + |Y|$$
$$||(0, x)||_{X+Y} = \{[p, r] | r \in ||x||_X\}$$
$$||(1, y)||_{X+Y} = \{[q, s] | s \in ||y||_Y\}$$

where p is $\lambda^* xy.(ky)(dx)$ and q is $\lambda^* xy.(kx)(dy)$. Inclusion $in_1 : X \to X + Y$ sends x to $(0, x)$. This morphism is realized by $\lambda^* x.[p, x]$. Inclusion $in_2 : Y \to X + Y$ is similar. Let m, n realize $f : X \to Z$ and $g : Y \to Z$ respectively. Then a morphism $h : X + Y \to Z$ which sends $(0, x)$ to fx and $(1, y)$ to gy is realized by $\lambda^* x.\mathtt{let}\ [u, v] = x\ \mathtt{in}\ (um'n')v$ where $m' \in !m$ and $n' \in !n$. For example, if we apply $[p, a] \in ||(0, x)||$ to this realizer then we get $pm'n'a$. From definition of p, this is $(kn')(dm')a = ma \in ||fx||$ since $kn' = i$, $dm' = m$. □

Hence the coKleisli category $\mathbf{Ass}(A)_{!_*}$ is a CCC and $\mathbf{Ass}(A)_{!_*}$ also has finite coproducts since $!_*$ is an idempotent comonad on $\mathbf{Ass}(A)$. Initial object is the same one of $\mathbf{Ass}(A)$ and a coproduct of X and Y is $!_*X + !_*Y$.

Proposition 13. *Let* $(A, !)$ *be an rLCA.* $\mathbf{Ass}(A)$ *and* $\mathbf{Ass}(A)_{!_*}$ *form a linear-non-linear model.*

Proof. By Proposition 12, $\mathbf{Ass}(A)_{!_*}$ is a CCC. We show the left adjoint G is strong monoidal since if the left adjoint is strong monoidal then the adjunction is monoidal. G is a strong monoidal functor since $\alpha : !_*X \otimes !_*Y \to !_*(X \times Y)$ and $\beta : I \to !_*1$ where α and β are isomorphisms given in Proposition 12. Required diagrams commute since underlying maps of these morphisms are identity. □

Since there is a monoidal adjunction $\Delta \dashv J : \mathbf{Mod}(A) \to \mathbf{Ass}(A)$, $\Delta !_* J$ is a monoidal comonad on $\mathbf{Mod}(A)$. If X is a modest set and $a \in ||x||_{!_*X} \cap ||y||_{!_*X}$ then $ra \in ||x||_X \cap ||y||_X$ where r is a realizer of $! \preceq 1_A$, hence $x = y$. Hence if X is a modest set then $!_*JX$ is a modest set and $!_*JX \simeq J\Delta !_* JX$.

From Proposition 5, $\mathbf{Mod}(A)$ is also finite complete and cocomplete. For any object X, Y in $\mathbf{Mod}(A)$, $\Delta !_* J(X \times Y) \simeq \Delta !_*(JX \times JY) \simeq \Delta(!_*JX \otimes !_*JY) \simeq \Delta !_* JX \bar{\otimes} \Delta !_* JY$ and $\Delta !_* J1 \simeq \Delta !_* 1 \simeq \Delta I \simeq \bar{I}$ since Δ is a left adjoint and especially strong monoidal where we write the monoidal product of $\mathbf{Mod}(A)$ as $\bar{\otimes}$ and the unit of $\mathbf{Mod}(A)$ as $\bar{I}$. This means $\mathbf{Mod}(A)$ satisfies Proposition 12 and its coKleisli category $\mathbf{Mod}(A)_{!_*}$ is a CCC. $\mathbf{Mod}(A)_{!_*}$ also has finite coproducts since $\Delta !_* J$ is idempotent and $\mathbf{Mod}(A)$ has finite coproducts.

Let $G' \dashv J'$ be an adjunction of $\mathbf{Mod}(A) \to \mathbf{Mod}(A)_{!_*}$. Then $G' = \Delta G L$ where $L : \mathbf{Mod}(A)_{!_*} \to \mathbf{Ass}(A)_{!_*}$ is a comparison functor of comonad $\Delta !_* J$. L is a strong monoidal functor since LX is JX for any object X of $\mathbf{Mod}(A)_{!_*}$ and a product of X and Y in $\mathbf{Mod}(A)_{!_*}$ is the product of $\mathbf{Mod}(A)$ and J preserves finite products. Δ and G are also strong monoidal functors by Proposition 13 and since Δ is a left adjoint. Hence G' is a strong monoidal functor and $G' \dashv J'$ is a monoidal adjunction.

Proposition 14. *Let* $(A, !)$ *be an rLCA.* $\mathbf{Mod}(A)$ *and its coKleisli category* $\mathbf{Mod}(A)_{!_*}$ *form a linear-non-linear model.*

4.3 Natural Number Object in Ass(A)

Let A be an rLCA. We construct a natural number object of $\mathbf{Ass}(A)_{!_*}$. First we define some notations.

Definition 14. *Let P be a polynomial over A. If P has a variable x which appears in P exactly once. Then $\lambda^* x.P$ is defined as*

$$\begin{aligned}\lambda^* x.x &= i\\ \lambda^* x.PQ &= b\cdot(\lambda^* x.P)\cdot Q \quad (\textit{if } P \textit{ has } x)\\ \lambda^* x.PQ &= c\cdot P\cdot(\lambda^* x.Q) \quad (\textit{if } Q \textit{ has } x)\end{aligned}$$

For any polynomial P over A, $\lambda^* !x.P$ is defined as

$$\begin{aligned}\lambda^* !x.x &= d\\ \lambda^* !x.a &= k\cdot a\\ \lambda^* !x.y &= k\cdot y \quad (\textit{if } y\neq x)\\ \lambda^* !x.PQ &= w\cdot(\lambda^* xy.((\lambda^* !x.P)\cdot x)\cdot((\lambda^* !x.Q)\cdot y))\end{aligned}$$

When free variables of a polynomial P are only x, $\lambda^* !x.P$ has no free variables. Then we treat $\lambda^* !x.P$ as an element of A. By induction of the definition, we can see for any $a' \in !a$, $(\lambda^* !x.P)a' = P[a/x]$.

Let $\bar{n}$ be $\lambda^* !fx.\overbrace{f(\cdots(f}^{n}\,x)\cdots)$. We define $|N|$ as a set of natural numbers and $||n||_N = \{\bar{n}\}$. $0 : 1 \to N$ which sends $*$ to $0 \in |N|$ is realized by $\lambda^* !x.\bar{0}$. $S : N \to N$ is realized by $\lambda^* !n.w(\lambda^* xy.((\lambda^* !f.cf)x)(ny))$ since

$$\overline{n+1} = \lambda^* !f.\overbrace{(cf)(\cdots((cf}^{n+1}\,i)\cdots) = w(\lambda^* xy.((\lambda^* !f.cf)x)(\bar{n}y))$$

In $\mathbf{Ass}(A)_{!_*}$, given $x : 1 \to X$ and $f : X \to X$, a morphism $h : N \to X$ which sends n to $f^n(x)$ is realized by $\lambda^* !n.nra$ where r is an element of $!s$ for a realizer s of f and a is an element of $||x||_X$. By an induction of the definition of $\bar{n}$, we can show h is well defined and uniqueness follows from that $|N|$ is a natural number object of **Set**. Notice that N is a modest set if A has more than two elements: Let a, b are two different elements of A and B be an object such that $|B| = 2$ and $||0||_B = \{a\}$, $||1||_B = \{b\}$. For any $n < m \in N$, since N is a natural number object, $f : N \to B$ which sends i to 0 if $i \leq n$ and 1 if $i > n$ is realizable. Hence $\bar{n} \neq \bar{m}$.

We have $!_*(1) \simeq I$ since $!_* : \mathbf{Ass}(A)_{!_*} \to \mathbf{Ass}(A)$ is a left adjoint. Hence for any $x : I \to X$,$f : X \to X$ in $\mathbf{Ass}(A)$, there exists a unique morphism $h : !_*N \to X$ such that $h!_*(0) = x$ and $h!_*S = fh$.

5 Examples of rLCA

5.1 Linear Lambda Calculus

The untyped linear lambda calculus is defined in [Sim05]. Terms of the untyped linear lambda calculus is defined as

$$t = x|tt|\lambda x.t|\lambda !x.t|!t$$

t of $\lambda x.t$ is required to have exactly one x which is not in any scope of !. A set of closed terms up to reductions given in [Sim05] forms an LCA.

5.2 BCK Algebra with a Structure of ω-cppo

This is inspired by examples in [AHS02]. A *BCK algebra* is a BCI algebra which also has an element k which satisfies $kxy = x$. for all x, y. Let A be a BCK algebra with a structure of ω-complete pointed poset (ω-cppo) and the application of A is continuous with this structure, and $\bot x = \bot$ for all $x \in A$. We define $! : A \to A$ as $!a = \mu x.[a, [x, x]]$ where $\mu x.fx$ is the least fixed point of f. Then $(A, !)$ is an LCA.

It is easy to see $! \preceq 1_A$, $! \preceq k_i$ and $! \preceq [!, !]$ and by the following propositions, $!$ is an applicative morphism and $! \preceq !!$.

If t is constructed from a variable x and elements of A and λ abstractions then t is a continuous function from A to A since the application of A is continuous.

Proposition 15. $!$ *is realized by*

$$f = \mu z.\lambda^* xy.\ \texttt{let}\ [p, [q, r]] = x\ \texttt{in}\ \texttt{let}\ [u, [v, w]] = y\ \texttt{in}\ [pu, [zqv, zrw]]$$

Proof. Let $T_a = [a, [x, x]]$ for $a \in A$. We have $f \bot \bot = \bot$ since $\bot x = \bot$ for all $x \in A$, and for $a, b \in A$, $fT_a^{n+1}(\bot)T_b^{n+1}(\bot)$ is equal to

$$\texttt{let}\ [p, [q, r]] = T_a^{n+1}(\bot)\ \texttt{in}\ \texttt{let}\ [u, [v, w]] = T_b^{n+1}(\bot)\ \texttt{in}\ [pu, [fqv, frw]]$$

which is, by induction, $T_{ab}(fT_a^n(\bot)T_b^n(\bot)) = T_{ab}^{n+1}(\bot)$. Hence by continuity of the application,

$$f!a!b = \bigvee_n fT_a^n(\bot)T_b^n(\bot) = \bigvee_n T_{ab}^n(\bot) = !(ab)$$

□

Proposition 16. $! \preceq !!$ *is realized by*

$$\delta = \mu z.\lambda^* x.\texttt{let}\ [p, [q, r]] = x\ \texttt{in}\ \texttt{let}\ [u, [v, w]] = r\ \texttt{in}\ [w, [zq, zv]]\ .$$

Proof. From properties of the least fixed point,

$$\begin{aligned} !a &= \mu x.\mu y.[a, [x, y]] \quad \cdots (*) \\ &= \mu x.[a, [x, [a, [x, \mu y.[a, [x, y]]]]]] \\ &= \mu x'.\mu x.[a, [x, [a, [x, \mu y.[a, [x', y]]]]]] \\ &= \mu x.[a, [x, [a, [x, \mu y.[a, [!a, y]]]]]] \end{aligned}$$

From $(*)$, $!a = \mu x.[a, [!a, x]]$ and hence $!a = \mu x.[a, [x, [a, [x, !a]]]]$. Let $S_a(x) = [a, [x, [a, [x, !a]]]]$ then $!a = \bigvee_n S_a^n(\bot)$. By induction we can see $\delta S_a^n(\bot) = T_{!a}^n(\bot)$. Hence we have

$$\delta !a = \bigvee_n \delta S_a^n(\bot) = \bigvee_n T_{!a}^n(\bot) = !!a$$

□

For example, let $\mathcal{C}$ be **Rel**(sets and relations) or **Pfn**(sets and partial functions). Then $\mathcal{C}$ has an object $\mathbb{N}$(set of natural numbers) which satisfies $\mathbb{N} \times \mathbb{N} \lhd \mathbb{N}$. Since $\mathcal{C}$ is a traced monoidal category whose monoidal product is coproduct, we can construct a new category by the GoI construction [AHS02], [JSV96]. Let this GoI-category be $\mathcal{G}(\mathcal{C})$. In $\mathcal{G}(\mathcal{C})$, $(\mathbb{N}, \mathbb{N})$ is a reflexive object as

$$(\mathbb{N}, \mathbb{N})^* \multimap (\mathbb{N}, \mathbb{N}) \simeq (\mathbb{N}, \mathbb{N}) \otimes (\mathbb{N}, \mathbb{N}) \lhd (\mathbb{N}, \mathbb{N}) \ .$$

Since $\mathcal{G}(\mathcal{C})$ is an SMCC, $\mathcal{G}(\mathcal{C})(I, (\mathbb{N}, \mathbb{N}))$ forms a BCI algebra. However in fact this is a BCK algebra and $\mathcal{G}(\mathcal{C})(I, (\mathbb{N}, \mathbb{N})) = \mathcal{C}(\mathbb{N}, \mathbb{N})$ forms ω-cppo by inclusion order whose application is continuous and for any $f \in \mathcal{C}(\mathbb{N}, \mathbb{N})$, $\phi \cdot f = \phi$. Hence this is an LCA.

By Lemma 2, we can construct an SK algebra from this algebra. If $\mathcal{C}$ is **Rel** then this algebra is isomorphic to $\mathcal{P}(\omega)$.

These examples are already in [AHS02]. However the LCA's we obtain are a little bit different. Comonoidal applicative morphisms of examples in [AHS02] are different from ours. Although, comonads constructed from these applicative morphisms are equivalent and we can think examples of here and ones in [AHS02] are the same.

5.3 $\mathcal{P}(\omega)_{lin}$

$\mathcal{P}(\omega)_{lin}$ is an LCA defined as

$$\begin{array}{ll}
\mathcal{P}(\omega)_{lin} = \mathcal{P}(\mathbb{N}) & \alpha \cdot \beta = \{n | \langle m, n \rangle \in \alpha, m \in \beta\} \\
!\alpha = \{n | e_n \subseteq \alpha\} & \\
d = \{\langle n, m \rangle | m \in e_n\} & \delta = \{\langle n, m \rangle | \bigcup_{i \in e_m} e_i \subseteq e_n\} \\
w = \{\langle l, \langle m, n \rangle \rangle | l = \langle i, \langle j, n \rangle \rangle \wedge e_i \bigcup e_j \subseteq e_m\} & f = \{\langle l, \langle m, n \rangle \rangle | e_n \subseteq e_l \cdot e_m\}
\end{array}$$

here $\langle -, - \rangle$ is a bijection from $\mathbb{N} \times \mathbb{N}$ to $\mathbb{N}$ and e is a bijection from $\mathbb{N}$ to the set of finite subsets of $\mathbb{N}$. If an object X of a compact closed category $\mathcal{C}$ satisfies $X^* = X$ and $X \otimes X \lhd X$ then $\mathcal{C}(I, X)$ form a BCI algebra, especially $\mathbb{N}$ of **Rel**. Hence this $\mathcal{P}(\omega)_{lin}$ is a BCI algebra and forms an LCA with above ! and d, δ, w, f. It is easy to see the SK algebra obtained from this LCA is $\mathcal{P}(\omega)$. This example can be modified to recursive enumerable subsets of $\mathbb{N}$ and we write this rLCA as $\mathcal{P}(\omega)_{lin,re}$.

5.4 Kleene's First Algebra $\mathbf{K_1}$

In [Lon95] it is proved that there is an adjoint pair $\varphi \dashv \psi : \mathcal{P}(\omega)_{re} \to \mathrm{K}_1$. Let $\delta \dashv \gamma$ be an adjoint pair from $\mathcal{P}(\omega)_{lin,re}$ to $\mathcal{P}(\omega)_{re}$ then $\delta\varphi \dashv \psi\gamma$ is an adjoint pair from $\mathcal{P}(\omega)_{lin,re}$ to K_1 and $(\mathcal{P}(\omega)_{lin,re}, \delta\varphi\psi\gamma)$ is an rLCA. Notice that there is no adjoint pair between any BCI algebra and K_1 by Proposition 11.

5.5 Kleene's Second Algebra $\mathbf{K_2}$

Definition 15. *K_2 is a set of functions from $\mathbb{N}$ to $\mathbb{N}$ whose application is*

$$f \cdot g = \begin{cases} f * g & (\textit{if } f * g \textit{ is a total function}) \\ \textit{undefined} & (\textit{else}) \end{cases}$$

where $f * g(n) = m$ *iff*

$$\exists k.\forall i < k.f([n, g(0), \cdots, g(i)]) = 0 \wedge f([n, g(0), \cdots, g(k)]) = m + 1$$

here $[-, \cdots, -]$ *is a bijection from a set of finite lists of natural numbers to* $\mathbb{N}$. *For details, see* [KV65].

Let $A_{\mathbf{Pfn}}$ be an LCA constructed from **Pfn** in Sec 5.2 and write $(A_{\mathbf{Pfn}})_!$ for the SK algebra constructed as in Lemma 2. Then there is an adjoint pair $\gamma \dashv \delta : (A_{\mathbf{Pfn}})_! \to \mathrm{K}_2$. Since there is an adjoint pair from $A_{\mathbf{Pfn}}$ to $(A_{\mathbf{Pfn}})_!$ we obtain an adjoint pair from $A_{\mathbf{Pfn}}$ to K_2. $\gamma : \mathrm{K}_2 \to (A_{\mathbf{Pfn}})_!$ is defined as $\gamma(f) = \{f\}$ and $\delta : (A_{\mathbf{Pfn}})_! \to \mathrm{K}_2$ is defined as

$$\delta(f) = \{g | \langle g0 \rangle \leq \langle g1 \rangle \leq \cdots \leq f, \bigvee_n \langle gn \rangle = f\}$$

where $\langle - \rangle$ is bijection from $\mathbb{N}$ to a set of partial functions of $\mathbb{N}$ to $\mathbb{N}$ whose domain is finite. The order $\leq$ is an inclusion order of its graph and $\bigvee$ is a union of graphs. Notice that K_2 is another example of PCA that has no adjoint pair between any BCI algebra.

Acknowledgement. I am grateful to Masahito Hasegawa and anonymous reviewers for their valuable advice.

6 Concluding remarks

- In order to model second order linear lambda calculus, we can use a category of partial equivalence relation (PER). By a similar argument of PER realized by PCA, PER realized by an rLCA provides a model of the second order linear lambda calculus.
- We can restrict the modality of ! of rLCA to only $! \preceq [\overbrace{id, id, \cdots, id}^{n}]$ for $n \geq 0$. These modalities are what soft linear logic [Laf04] has. One example is the untyped soft linear lambda calculus whose terms are

  $$t = x | tt | \lambda x.t | \lambda !x.t | !t$$

  where t of $\lambda x.t$ is required to have exactly one appearance of x which is not in any scope of ! and t of $\lambda !x.t$ is required to have exactly one appearance of x which is in a scope of at most one ! or to have no x which is in a scope of !. Then the untyped soft linear lambda calculus strongly normalizes in polynomial steps in "weight" of a term and morphisms of a category of assemblies realized by the untyped soft linear lambda calculus is computable in polynomial time in some sense.

Some further considerations are found in the author's MSc thesis [Hos07].

References

[AHS02] Abramsky, S., Haghverdi, E., Scott, P.J.: Geometry of interaction and linear combinatory algebras. Mathematical Structures in Computer Science 12(5), 625–665 (2002)

[AL05] Abramsky, S., Lenisa, M.: Linear realizability and full completeness for typed lambda-calculi. Ann. Pure Appl. Logic 134(2-3), 122–168 (2005)

[Ben94] Benton, P.N.: A mixed linear and non-linear logic: Proofs, terms and models. In: Pacholski, L., Tiuryn, J. (eds.) CSL 1994. LNCS, vol. 933, pp. 121–135. Springer, Heidelberg (1995) (extended abstract)

[Bie95] Bierman, G.M.: What is a categorical model of intuitionistic linear logic? In: Dezani-Ciancaglini, M., Plotkin, G. (eds.) TLCA 1995. LNCS, vol. 902, pp. 78–93. Springer, Heidelberg (1995)

[HO93] Hyland, J.M.E., Ong, C.-H.L.: Modified realizability toposes and strong normalization proofs. In: Bezem, M., Groote, J.F. (eds.) TLCA 1993. LNCS, vol. 664, pp. 179–194. Springer, Heidelberg (1993)

[Hos07] Hoshino, N.: Linear realizability. Master's thesis, Kyoto University (2007)

[Jac99] Jacobs, B.: Categorical Logic and Type Theory. Studies in Logic and the Foundations of Mathematics, vol. 141. North Holland, Amsterdam (1999)

[JSV96] Joyal, A., Street, R., Verity, D.: Traced monoidal categories. Math. Proc. Cambridge Phil. Soc. 119(3) (1996)

[KV65] Kleene, S.C., Vesley, R.E.: The Foundations of Intuitionistic Mathematics, especially in relation to recursive functions. North-Holland, Amsterdam (1965)

[Laf04] Lafont, Y.: Soft linear logic and polynomial time. Theor. Comput. Sci. 318(1-2), 163–180 (2004)

[Lon95] Longley, J.: Realizability Toposes and Language Semantics. PhD thesis, Edinburgh University (1995)

[Sim05] Simpson, A.K.: Reduction in a linear lambda-calculus with applications to operational semantics. In: Giesl, J. (ed.) RTA 2005. LNCS, vol. 3467, pp. 219–234. Springer, Heidelberg (2005)

Correctness of Multiplicative (and Exponential) Proof Structures is *NL*-Complete

Paulin Jacobé de Naurois* and Virgile Mogbil*

LIPN – UMR7030, CNRS – Université Paris 13,
99 av. J-B Clément, F–93430 Villetaneuse, France
denaurois@lipn.univ-paris13.fr,
virgile.mogbil@lipn.univ-paris13.fr

Abstract. We provide a new correctness criterion for unit-free MLL proof structures and MELL proof structures with units. We prove that deciding the correctness of a MLL and of a MELL proof structure is *NL*-complete. We also prove that deciding the correctness of an intuitionistic multiplicative essential net is *NL*-complete.

Introduction

The *proof nets* [Gir87, DR89] of Linear logic (LL) are a parallel syntax for logical proofs without all the bureaucracy of sequent calculus. They are a non-sequential graph-theoretic representation of proofs, where the order in which some rules are used in a sequent calculus derivation, when irrelevant, is neglected. The unit-free multiplicative proof nets are inductively defined from sequent calculus rules of unit-free Multiplicative Linear Logic (MLL). A *proof structure* is freely built on the same syntax as proof nets, without any reference to a sequent calculus derivation.

In LL we are mainly interested in the following decision problems: Deciding the provability of a given formula, which gives the expressiveness of the logic; deciding if two given proofs reduce to the same normal form, i.e. the cut-elimination problem which corresponds to program equivalence using the Curry-Howard isomorphism; and deciding the correctness of a given proof structure, i.e. whether it comes from a sequent calculus derivation. For this last decision problem, one uses a *correctness criterion* to distinguish proof nets among proof structures. We recall the following main results [Kan92, Mai] and we complete (in bold) the correctness cases:

fragment		decision problem		
	units	provability	cut-elimination	correctness
MLL	no	*NP*-complete	*P*-complete	**NL-complete**
MELL	yes	open	(at most non-elementary)	

* Work supported by project NO-CoST (ANR).

J. Duparc and T.A. Henzinger (Eds.): CSL 2007, LNCS 4646, pp. 435–450, 2007.

Correctness is equivalent to provability for unit only MLL because proof nets are formulae syntactic trees. However it is not so obvious for the propositional case as one can observe following the long story of correctness criteria:

- Long-trip [Gir87] is based on travels and was the first one.
- Acyclic-Connected [DR89] is a condition is based on switchings i.e. the choice of one premise for each $\parr$ connective. The condition is that all the associated graphs are trees. A naive implementation of Acyclic-Connected uses exponential time.
- Contractibility [Dan90] is done in quadratic time by repeating two graph rewriting rules until one obtains a simple node.
- Graph Parsing [Laf95] is a strategy for Contractibility which is implemented in linear time as a sort of unification [Gue99].
- Dominator Tree [MO00, MO06] is a linear time correctness criterion for essential nets, to which proof structures correctness reduces in linear time.
- Ribbon [Mel04] is a topological condition requiring homeomorphism to the disk.

For other fragments of Linear Logic, some of these criteria apply or are extended as for MELL[1] [Dan90, GM01].

A feature of these criteria is that they successively lower the complexity of sequential, deterministic algorithms deciding correctness for MLL. Switching from proof structures to paired graphs, that is undirected graphs with a distinguished set of edges, we give a new correctness criterion for MLL and more generally for MELL. This new correctness criterion gives us a lower bound for the correctness decision problem for both MLL and MELL. This lower bound yields an exact characterization of the complexity of this problem, and induces naturally efficient parallel and randomized algorithms for it.

The paper is organized as follows: we recall preliminary definitions and results in linear logic and complexity theory in Section 1. Section 2 is devoted to the exposition of a new correctness criterion for unit-free MLL and MELL with units. We prove its *NL*-completeness in Section 3, and the *NL*-completeness of the criterion for IMLL in Section 4.

1 Background

1.1 MLL and Proof Structures

Roman capitals A, B stand for MLL formulae, which are given by the following grammar, where $\otimes$ and $\parr$ are duals for the negation $^\perp$, accordingly to De Morgan laws:

$$F ::= A \mid A^\perp \mid F \otimes F \mid F \parr F$$

[1] As usual M, A and E denote respectively for Multiplicative, Additive and Exponential fragments of LL.

Greek capitals Γ, Δ stand for sequents, which are multiset of formulae, so that exchange is implicit. The MLL sequent calculus is given by the following rules:

$$\frac{}{\vdash A, A^{\perp}}\,(ax) \qquad \frac{\vdash \Gamma, C \quad \vdash \Delta, C^{\perp}}{\vdash \Gamma, \Delta}\,(cut) \qquad \frac{\vdash \Gamma, A \quad \vdash \Delta, B}{\vdash \Gamma, \Delta, A \otimes B}\,\otimes \qquad \frac{\vdash \Gamma, A, B}{\vdash \Gamma, A \parr B}\,\parr$$

Definition 1. *A* MLL proof structure *is a finite directed acyclic graph (DAG)*[2] *whose nodes, called* links, *are defined together with an arity and a coarity, i.e. a given number of incident edges called the* premises *of the node and a given number of emergent edges called the* conclusions *of the node. Moreover the proof structure edges are labelled by formulae and every edge is conclusion of exactly one link and premise of at most one link. The links of are the following:*

nodes		*ax*		*cut*		$\otimes$		$\parr$	
arity	*edge labels*	*0*	$\emptyset$	*2*	$A, A^{\perp}$	*2*	A, B	*2*	A, B
coarity	*edge labels*	*2*	$A, A^{\perp}$	*0*	$\emptyset$	*1*	$A \otimes B$	*1*	$A \parr B$

We allow edges with a source but no target (i.e pending or dandling edges), they are called the conclusions *of the proof-structure.*

A MLL proof net *is a* MLL *proof structure inductively defined as follows:*

- *an ax-link is a proof net with conclusions* $A, A^{\perp}$*,*
- *if* P *is a proof net with conclusions* Γ, A, B *then* P *extended with a* $\parr$*-link of premises* A *and* B *is a proof net with conclusions* $\Gamma, A \parr B$*.*
- *if* P_1 *and* P_2 *are disjoint proof nets with respective conclusions* Γ, A *and* Δ, B *then* $P_1 \cup P_2$ *extended with a* $\otimes$*-link of premises* A *and* B *is a proof net with conclusions* $\Gamma, A \otimes B, \Delta$*.*

It follows from the definition that MLL proof structures and proof nets have a non-empty set of conclusions, which corresponds to a MLL sequent. The inductive definition of MLL proof nets corresponds to a graph theoretic abstraction of the derivation rules of MLL; any proof net is sequentializable, i.e. corresponds to a MLL derivation: given a proof net P of conclusion Γ, there exists a sequent calculus proof of $\vdash \Gamma$ which infers P.

Definition 2. *A* paired graph *is an undirected graph* $G = (V, E)$ *with a set of* pairs $C(G) \subseteq E \times E$ *which are pairwise disjoint couples of edges with the same target, called a* pair-node, *and two (possibly distinct) sources called the* premise-nodes.

A switching S *of* G *is the choice of an edge for every pair of* $C(G)$*. With each switching* S *is associated a subgraph* $S(G)$ *of* G*: for every pair of* $C(G)$*, erase the edges which are not selected by* S*. When* S *selects the (abusively speaking) left edge of each pair,* $S(G)$ *is denoted as* $G[\forall \mapsto \backslash^{\cdot}]$*. Also,* $G[\forall \mapsto \cdot\!\!.\!\cdot]$ *stands for* $G \setminus \{e, e' \mid (e, e') \in C(G)\}$*.*

[2] For convenience the edges are oriented up-down, so we do not mention the orientation.

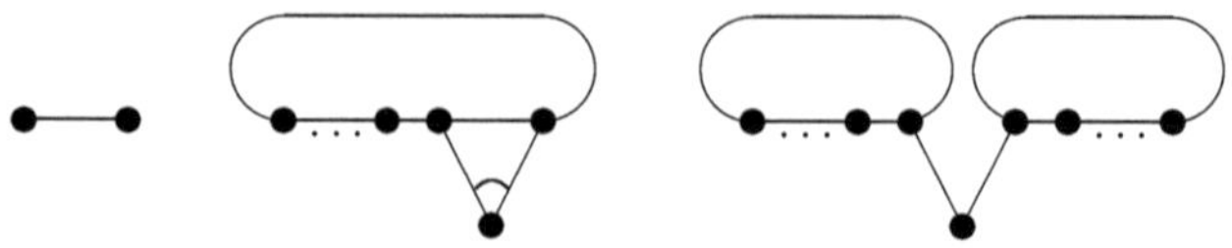

Fig. 1. Paired graph constructors associated to MLL proof nets: *ax*-link, ⅋-link and ⊗-link

Let $R = (V, E)$ be a MLL proof structure. To R, we naturally associate the paired graph $G_R = (V, E')$ where E' is the set of non-pending edges of E and $C(G_R)$ contains the premises of each ⅋-link of R (Figure 1). For a pair of edges $(v, x), (w, x)$, we adopt the representation of Figure 1, where the two edges of the pair are joined by an arc.

We define the following graph rewriting rules $\rightarrow_c$ of Figure 2 on paired graphs where all the nodes are distinct and rule $\rightarrow_{R_2}$ applies only for a non-pair edge:

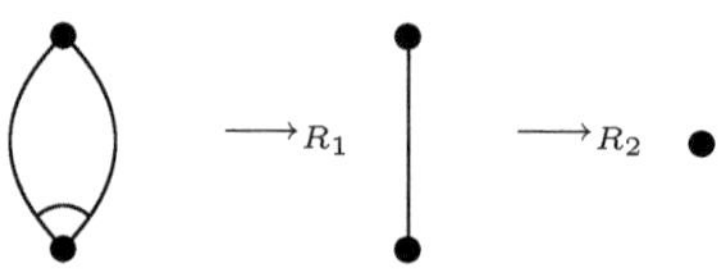

Fig. 2. Contraction rules $\rightarrow_c$

We denote by $G \rightarrow_c \bullet$ the fact that G contracts to a single vertex with no edge.

Definition 3. *A* MLL *proof structure R is* DR-correct *if for all switching S of G_R, the graph $S(G_R)$ is acyclic and connected.*
A MLL *proof structure R is* contractile *if $G_R \rightarrow_c^* \bullet$.*

Theorem 1. *[DR89, Dan90] A* MLL *proof structure R is a* MLL *proof net iff R is DR-correct iff R is contractile*[3].

We define the following decision problem MLL-CORR:
GIVEN: A MLL proof structure R
PROBLEM: Is R a MLL proof net?

1.2 MELL and Proof Structures

The definition of MELL formulae follows that of MLL formulae in Section 1.1, with ! and ? duals for the negation $^\perp$, as well as the neutral elements 1 and $\perp$:

$$\text{MELL:} \qquad F ::= A \mid A^\perp \mid F \otimes F \mid F ⅋ F \mid !F \mid ?F \mid 1 \mid \perp$$

[3] The criteria in [DR89, Dan90] are expressed for switchings and contraction rules for proof structures only. The equivalence with Definition 3 is left to the reader.

The MELL sequent calculus contains the rules of the MLL sequent calculus, as well as the following rules:

$$\frac{\vdash \Gamma}{\vdash \Gamma, \bot}\ \bot \qquad \frac{}{\vdash 1}\ 1 \qquad \frac{\vdash \Gamma}{\vdash \Gamma, ?A}\ ?W \qquad \frac{\vdash \Gamma, ?A, ?A}{\vdash \Gamma, ?A}\ ?C \qquad \frac{\vdash \Gamma, A}{\vdash \Gamma, ?A}\ ?D \qquad \frac{\vdash ?\Gamma, A}{\vdash ?\Gamma, !A}\ !P$$

Definition 4. MELL structures *are defined similarly to* MLL *proof structures (Definition 1), with the additional links, where the* $?W$*-link subsumes both* $?W$ *and* $\bot$*rules:*

nodes		1		$?W$		$?C$		$?D$		$!P$	
arity	*edge labels*	*0*	$\emptyset$	*0*	$\emptyset$	*2*	$?A, ?A$	*1*	A	*1*	A
coarity	*edge labels*	*1*	1	*1*	$\bot$ *or* $?A$	*1*	$?A$	*1*	$?A$	*1*	$!A$

Definition 5. *An* exponential box *is a* MELL *structure whose conclusions are all ?-formulae but one, its* principal door, *which is conclusion of a* $!P$*-link. Similarly, a* weakening box *is a* MELL *structure with a distinguished conclusion, its* principal door, *which is conclusion of a* $?W$*-link. A* box *is either an exponential or a weakening box.*

Definition 6. *A* MELL proof structure (R, B) *is given by a* MELL *structure* R *and a box mapping* B*, which associates to any link* l *of* R *a box* b_l *or* R*. Moreover, boxes may nest but may not partially overlap, and a unique exponential (respectively weakening) box is associated to each* $!P$*-(resp.* $?W$*-)link. By convention, when a link belongs to several boxes, the mapping returns the innermost box to which it belongs, otherwise it returns* R*.*

It follows from the definition that, for any $!P$ (respectively $?W$)-link, the box mapping associates the exponential (resp. weakening) box to which it naturally corresponds. The whole proof structure R is treated as a particular box, and is associated to all links that do not belong to any exponential or weakening box.

Let (R, B) be a MELL proof structure, with boxes $b_1, \ldots, b_n$. Let $b_0 = R$. We define as follows the family $\mathcal{G}_{(R,B)} = \{G^i_{(R,B)}\}_{i=0\ldots n}$ of paired-graphs:

- $G^i_{(R,B)}$ contains a node l for every link l of $R \setminus \{?W\text{-links}\}$ with $B(l) = b_i$, and an edge (l, l') for all links l, l' of $R \setminus \{?W\text{-links}\}$ with $B(l) = B(l') = b_i$. $C(G^i_{(R,B)})$ contains the premises of each $⅋$-link and $?C$-link l of R with $B(l) = b_i$.
- Assume b_j is an outermost box included in b_i. A node $\overline{b_j} \in G^i_{(R,B)}$ is associated to b_j, and an edge $(\overline{b_j}, l) \in G^i_{(R,B)}$ for all link l conclusion of a link in b_j and such that $B(l) = b_i$.

Essentially, $G^i_{(R,B)}$ is the paired graph corresponding to the box b_i, where all inner boxes are considered contracted to a single node. Moreover, for the sake of connectivity, the $?W$-link (if there is any) corresponding to b_i is removed.

Definition 7. *A* MELL *proof structure* (R, B) *with boxes* $b_1, \ldots, b_n$ *is* contractile *if* $\forall i \in \{0, \ldots, n\}$, $G^i_{(R,B)} \to^*_c \bullet$.

As for MLL, one may inductively define particular MELL proof structures, called MELL proof nets, which exactly correspond to derivations in MELL sequent calculus. Theorem 2 allows to distinguish MELL proof nets among proof structures:

Theorem 2. *[GM01] A* MELL *proof structure* (R, B) *is a* MELL *proof net iff* (R, B) *is contractile*[4].

We define the following decision problem MELL-CORR:
GIVEN: A MELL proof structure (R, B)
PROBLEM: Is (R, B) a MELL proof net?

1.3 Intuitionistic Multiplicative Linear Logic and Essential Nets

The intuitionistic fragment of MLL (IMLL) is the ($\otimes$, $\multimap$)-fragment of Linear logic, where linear implication is no more defined by $A \multimap B = A^\perp \parr B$ but is a connective. The sequent calculus corresponds to the MLL sequent calculus but with two-sided sequents (using linear negation) and at most one (distinguished) formula on the right.

For this sub-logic, Lamarche has proposed a version of proof structures called *essential nets.* They are built on the links given in Figure 3 where, to each $\parr^+$-labelled node, one associates a negatively-labelled node (left premise) called the sink of p. They also have a distinguished link called the *root.*

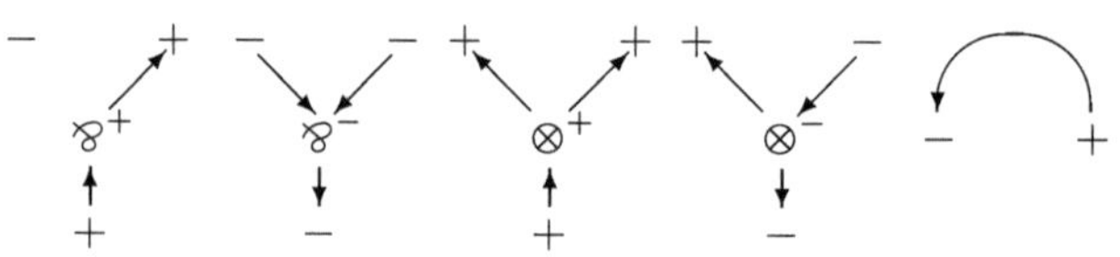

Fig. 3. Essential net links: $\parr^+$, $\parr^-$, $\otimes^+$, $\otimes^-$ and Axiom

Definition 8. *An essential net of a linearly balanced*[5] IMLL *sequent is L-correct if it is acyclic, every node is reachable from the root, and every* $\parr^+$*-node* p *satisfies the L-condition: every path from the root that reaches the sink of* p *passes through* p.

Lamarche has shown that the essential net of an IMLL sequent denotes a IMLL sequent derivation if and only if it is L-correct. For additional information on essential net correction, including translation to proof structures, we refer the reader to [MO06].

We define the following decision problem IMLL-CORR:
GIVEN: A multiplicative essential net N of a linearly balanced IMLL sequent
PROBLEM: Is N correct?

[4] The criterion in [GM01] uses contraction rules for MELL proof structures only. As for Theorem 1, the proof of the equivalence with Definition 7 is left to the reader.

[5] I.e. every atom that occurs in the sequent does so exactly twice, once positively and once negatively

1.4 Complexity Classes and Related Problems

We will mention several major complexity classes below P, some of which having natural complete problems that we will use in this paper. Let us briefly recall some basic definitions and results.

Definition 9. *Complexity classes:*

- AC^0 *(respectively* AC^1*) is the class of problems solvable by a uniform family of circuits of constant (resp. logarithmic) depth and polynomial size, with NOT gates and AND, OR gates of unbounded fan-in.*
- *L is the class of problems solvable by a deterministic Turing machine which only uses a logarithmic working space.*
- *NL (respectively coNL) is the class of problems solvable by a non-deterministic Turing machine which only uses a logarithmic working space, such that:*
  1. *If the answer is "yes," at least one (resp. all) computation path accepts.*
  2. *If the answer is "no," all (resp. at least one) computation paths reject.*

Theorem 3. *[Imm88, Sze87] NL = coNL.*

The following inclusion results are also well known:

$$AC^0 \subset L \subset NL \subset AC^1 \subset P, \tag{1}$$

where it remains unknown whether any of these inclusions is strict.

It is important to note that Theorems 6 and 7 give NL-completeness results under constant-depth (actually AC^0) reductions. From (1) above, it should be clear to the reader that the reductions lie indeed in a class small enough for being relevant. For a good exposition of constant-depth reducibility, see [CSV84].

In the sequel, we will often use the notion of a *path* in a directed -or undirected- graph. A path is a sequence of vertices such that there is an edge between any two consecutive vertices in the path. A path will be called *elementary* when any node occurs at most once in the path.

Let us now list some graph-theoretic problems that will be used in this paper.

IS TREE (IT): Given an undirected graph $G = (V, E)$, is it a tree?
IT is L-complete under constant-depth reductions [JLM97].

SOURCE-TARGET CONNECTIVITY (STCONN): Given a directed graph $G = (V, E)$ and two vertices s and t, is there a path from s to t in G ?
STCONN is NL-complete under constant-depth reductions [JLL76].

UNDIRECTED SOURCE-TARGET CONNECTIVITY (USTCONN):
Given an undirected graph $G = (V, E)$ and two vertices s and t, do s and t belong to the same connected component of G ?
USTCONN is L-complete under constant-depth reductions [Rei05].

UNIVERSAL SOURCE DAG (SDAG):
Given a directed graph $G = (V, E)$, is it acyclic and does there exist a source node s such that there is a path from s to each vertex ?

Proposition 1. SDAG $\in NL$.

Proof. Given $G = (V, E)$ a directed graph, its acyclicity can be expressed as follows:

$$\forall (x, y) \in V^2 : \neg \text{STCONN}(G, x, y) \vee \neg \text{STCONN}(G, y, x).$$

Since $NL = coNL$ (Theorem 3) and STCONN $\in NL$, acyclicity is clearly in NL. Checking whether each vertex can be reached from a vertex s can also be done with STCONN subroutines, therefore SDAG is in NL. □

Proposition 2. SDAG *is coNL-hard under constant-depth reductions.*

Proof. Let $\mathcal{L}$ be any language in $coNL$. $\mathcal{L}$ is then decided by a non-deterministic Turing machine M in space less than $k \log(n)$ on inputs of size n, for some $k \geq 0$.

Let $\mathcal{C}_n$ be the set of configurations of M of size less or equal to $k \log(n)$, and define $T = |\mathcal{C}_n|$. Clearly, $T \leq n^k$ is an upper bound for the computation time of M on inputs of size n. Without loss of generality, we assume that every configuration of M has at least one outgoing transition, possibly towards itself, and that the result of the computation is given by the state reached by M after exactly T computation steps. A configuration is thus either accepting or rejecting.

Let us consider the following directed graph $G_n = (V_n, E_n)$, where:

$V_n = \bigcup_{c \in \mathcal{C}_n, t \in [0,T]} \{(c, t)\} \cup \{c_A\} \cup \{c_R\} \cup \{s\}$.
For $(c, t), (c', t+1) \in V_n$, $((c', t+1) \to (c, t)) \in E_n$ if and only if $c \to c'$ is a transition of M.
For $c \in \mathcal{C}_n$, $(c_A \to (c, T)) \in E_n$ iff c is an accepting configuration of M.
For $c \in \mathcal{C}_n$, $(c_R \to (c, T)) \in E_n$ iff c is a rejecting configuration of M.
$(s \to c_A) \in E_n$, $(s \to c_R) \in E_n$.

A path $(c_1, t_1) \to \cdots \to (c_k, t_k)$ in G_n follows by construction a sequence $t_1, \ldots, t_k$ that is strictly decreasing. Since there is no edge $(c, t) \to c_A$, $(c, t) \to c_R$ nor $(c, t) \to s$, it is then clear that G_n is acyclic.

Moreover, since every configuration of M has at least one outgoing transition, every vertex $(c, t), t < T$ in G_n has at least one parent node $(c', t+1)$. By induction on t, it follows that every vertex in G_n is reachable from s. Therefore G_n satisfies SDAG.

Let x be an input of size n to M. An initial configuration $c_x \in \mathcal{C}_n$ of M is naturally associated to this input x. Consider now the directed graph $H_n^x = G_n \cup \{(c_x, 0) \to c_R\}$.

Then, H_n^x satisfies SDAG if and only if $x \in \mathcal{L}$. Indeed, by Definition 9, $x \in \mathcal{L}$ if and only if there exists no computation path $c_x \to \cdots \to c_r$ of length T in M, where c_r is a rejecting configuration. By construction of G_n, such a path corresponds to a path $(c_r, T) \to \cdots \to (c_x, 0)$ in G_n. Then $x \in \mathcal{L}$ if and only if there exists no path $c_R \to \cdots \to (c_x, 0)$ in G_n, if and only if H_n^x is acyclic. Since G_n satisfies SDAG, it follows that H_n^x satisfies SDAG if and only if $x \in \mathcal{L}$.

Moreover, it is well known that the configuration graph of a Turing machine can be computed with constant-depth circuits. Computing H_n^x from the configuration graph of M requires only purely local rewriting rules, that can all be performed in parallel. Therefore, H_n^x can also be computed with constant-depth circuits. □

Propositions 1 and 2, and Theorem 3 yield the following result:

Theorem 4. SDAG *is NL-complete under constant-depth reductions.*

2 New Correctness Criteria for MLL and MELL

For a given proof net, the following notion of dependency graph provides a partial order among its ⅋-nodes corresponding to some valid contraction sequences accordingly to rule R_1.

Definition 10. *Let G be a paired graph. The* dependency graph $D(G)$ *of G is the directed graph (V_G, E_G) defined as follows:*

- $V_G = \{v \mid v$ *is a pair-node in* $G\} \cup \{s\}$.
- *Let x be a pair-node in G, with premise-nodes x_l and x_r. The edge $(s \to x)$ is in E_G if and only if:*
  1. *There exists an elementary path $p_x = x_l, \ldots, x_r$ in $G[\forall \mapsto \backslash \cdot]$,*
  2. *$x \notin p_x$, and for all pair-node y in G, $y \notin p_x$.*
- *Let x be a pair-node in G, with premise-nodes x_l and x_r, and let $y \neq x$ be another pair-node in G. The edge $(y \to x)$ is in E_G if and only if:*
  1. *There exists an elementary path $p_x = x_l, \ldots, x_r$ in $G[\forall \mapsto \backslash \cdot]$,*
  2. *$x \notin p_x$, and for every elementary path $p_x = x_l, \ldots, x_r$ in $G[\forall \mapsto \backslash \cdot]$ with $x \notin p_x$, $y \in p_x$.*

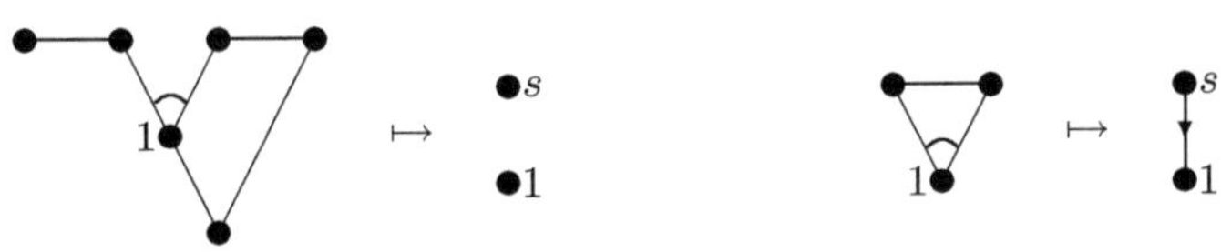

Fig. 4. Dependency graph examples: source is needed!

Remark 1. Other definitions of dependency graph are possible:
- One may choose to consider not only elementary paths, but *any* path, which yields a stronger contraction order. Surprisingly enough, Theorem 6 holds also for this relaxed version of dependency.
- One may also consider the dependency graph only for a paired graph G where $G[\forall \mapsto \backslash \cdot]$ is a tree. In that case, if there exists an elementary path $p_x = x_l, \ldots, x_r$ which does not contain x, this elementary path is unique. The results of Section 3 hold also for this version of dependecy, yet Lemma 4 does not rely on [Rei05], but rather on the fact that checking reachability in a forest is L-complete [CM87].

Lemma 1. *Let G and H be paired graphs, with $G \to_c H$. Then, $G[\forall \mapsto \backslash \cdot] \to_c^* H[\forall \mapsto \backslash \cdot]$, and $G[\forall \mapsto \backslash \cdot]$ is a tree if and only if $H[\forall \mapsto \backslash \cdot]$ is a tree.*

Proof. If $G \to_{R_1} H$ denote by v the redex pair-node in G, with premise w. The reduced pattern in H is the non-pair edge (v, w), therefore $G[\forall \mapsto \backslash \cdot] = H[\forall \mapsto \backslash \cdot]$.

If $G \to_{R_2} H$, it is clear that $G[\forall \mapsto \backslash \cdot] \to_{R_2} H[\forall \mapsto \backslash \cdot]$ with the same redex. It is also clear that rule $\to_{R_2}$ preserves connectivity and acyclicity. □

Lemma 2. *If $G \to_c^* \bullet$ then $D(G)$ satisfies* SDAG.

Proof. Since $\bullet[\forall \mapsto \backslash\cdot]$ is a tree, by Lemma 1 so is $G[\forall \mapsto \backslash\cdot]$. Therefore, for any pair-node x with premise-nodes x_l and x_r in G, there exists a unique elementary path $p_x = x_l - \cdots - x_r$ in $G[\forall \mapsto \backslash\cdot]$. It follows by construction of $D(G)$ that x has at least one parent node in $D(G)$. Moreover, a path $x \to \cdot\cdot \to y$ in $D(G)$ induces by construction an elementary path $x_l - \cdots - y$ in $G[\forall \mapsto \backslash\cdot]$. Therefore a cycle $x \to \cdot\cdot \to y, y \to \cdot\cdot \to x$ in $D(G)$ induces a cycle $x_l - \cdots - y, y_l - \cdots - x$ in $G[\forall \mapsto \backslash\cdot]$. Since $G[\forall \mapsto \backslash\cdot]$ is a tree, $D(G)$ is acyclic. Since every vertex of $D(G)$ but s has at least one parent node and $D(G)$ is acyclic, $D(G)$ satisfies SDAG. □

Lemma 3. *Let G be a paired graph such that $G[\forall \mapsto \backslash\cdot]$ is a tree. If the dependency graph $D(G)$ of G satisfies* SDAG *then $G \to_c^* \bullet$.*

Proof. let $d(v)$, the depth of a pair-node $v \in G$, be the length of the longest path from the source s of $D(G)$ to the vertex $v \in D(G)$. Assume that $D(G)$ satisfies SDAG, and let $X^d = \{x \text{ pair-node in } G | d(x) = d\}$ and $Y^d = \cup_{d' \leqslant d} X^{d'}$.

By induction on the depth we prove that there exists a sequence of contractions $\mathcal{C}_d$ such that $G \to^{\mathcal{C}_d} G^d$ satisfies:

$$\text{Each pair-node } y \in G \text{ s.t. } d(y) \leqslant d \text{ is contracted in } G^d. \tag{2}$$

The proof by induction is as follows:

- For $d = 1$, let $x \in X^1$, with premise-nodes x_l and x_r. By definition of X^1, there exists an elementary path $p_x = x_l - \cdots - x_r$ in $G[\forall \mapsto \backslash\cdot]$ such that $x \notin p_x$ and for any pair-node y in $G[\forall \mapsto \backslash\cdot]$, $y \notin p_x$. The same holds for the path $p_x = x_l - \cdots - x_r$ in G, with respect to any pair-node $y \in G$.

  Let $\mathcal{E}_x^1 = \{e \text{ edge of } p_x \mid x \in X^1\}$. The set of contractions $\mathcal{R}_x^1 = \{e \to_c \bullet \mid e \in \mathcal{E}_x^1\}$ contracts the edges of p_x, and let $\mathcal{R}^1 = \cup_{x \in X^1} \mathcal{R}_x^1$. Clearly, $x_l = x_r \neq x$ in the contracted paired graph obtained from G by $\mathcal{R}_x^1$. Since $x \notin p_y$ for any $y \in X^1$, the same holds for the paired graph obtained from G by $\mathcal{R}^1$.

  Let $\mathcal{C}_1$ be the sequence $\mathcal{R}^1$, followed by the set of contraction rules of the pair-nodes $x \in X^1$. Define G^1 such that $G \to^{\mathcal{C}_1} G^1$. It is clear that G^1 satisfies (2).
- Assume by induction that there exists a sequence of contractions $\mathcal{C}_d$ such that $G \to^{\mathcal{C}_d} G^d$ satisfies (2).
  Let $x \in X^{d+1}$, with premise-nodes x_l and x_r.
  Since $G \to^{\mathcal{C}_d} G^d$ and $G[\forall \mapsto \backslash\cdot]$ is a tree, Lemma 1 applies:

  $$G[\forall \mapsto \backslash\cdot] \to^{\mathcal{C}'_d} G^d[\forall \mapsto \backslash\cdot], \text{ and } G^d[\forall \mapsto \backslash\cdot] \text{ is a tree.} \tag{3}$$

  By definition of X^{d+1}, there exists an elementary path $p_x = x_l - \cdots - x_r$ in $G[\forall \mapsto \backslash\cdot]$ such that $x \notin p_x$ and, for every pair-node $y \in G$ of depth $d(y) > d$, $y \notin p_x$.
  Define p_x^d such that $p_x \to^{\mathcal{C}'_d} p_x^d$. By (3), p_x^d is an elementary path in $G^d[\forall \mapsto \backslash\cdot]$ such that $x \notin p_x^d$ and, for every pair-node $y \in G^d[\forall \mapsto \backslash\cdot]$ of depth $d(y) > d$,

$y \notin p_x^d$. The same holds for p_x^d in G^d, with respect to any pair-node $y \in G^d$, since, by induction, for any pair-node $y \in G^d$, $d(y) > d$.

Let $\mathcal{E}_x^{d+1} = \{e \text{ edge of } p_x \mid x \in X^{d+1}\}$. The set of contractions $\mathcal{R}_x^{d+1} = \{e \rightarrow_c \bullet \mid e \in \mathcal{E}_x^{d+1}\}$ contracts the edges of p_x^d, and let $\mathcal{R}^{d+1} = \cup_{x \in X^{d+1}} \mathcal{R}_x^{d+1}$. Clearly, $x_l = x_r \neq x$ in the contracted paired graph obtained from G by $\mathcal{R}_x^{d+1}$. Since $x \notin p_y$ for any $y \in X^{d+1}$, the same holds for the contracted paired graph obtained from G by $\mathcal{R}^{d+1}$.

Let $\mathcal{C}_{d+1}$ be the sequence $\mathcal{C}_d$, followed by $\mathcal{R}_{d+1}$, and followed by the set of contraction rules of the pair-nodes $x \in X^{d+1}$. Define G^{d+1} such that $G \rightarrow^{\mathcal{C}_{d+1}} G^{d+1}$. G^{d+1} satisfies (2).

Since $D(G)$ satisfies SDAG, the maximal depth $m = max\{d(x) | x \in D(G)\}$ is well-defined and every pair-node x of G belongs to X^m. Therefore, $G \rightarrow^{\mathcal{C}_m} G^m$ and G^m satisfies (2). Since $G[\forall \mapsto \backslash\cdot]$ is a tree, by Lemma 1 so is $G^m[\forall \mapsto \backslash\cdot] = G^m$. It follows that $G \rightarrow_c^* \bullet$. □

Lemmas 2 and 3 and Theorems 1 and 2 imply the Theorem:

Theorem 5 (Correctness Criteria).
A MLL *proof structure R is a* MLL *proof net if and only if:*

1. *$D(G_R)$ satisfies* SDAG, *and*
2. *$G_R[\forall \mapsto \backslash\cdot]$ is a tree.*

A MELL *proof structure (R, B) with boxes $b_1, \ldots, b_n$ is a* MELL *proof net if and only if:*

1. *$\forall i \in \{0, \ldots, n\}$, $D(G^i_{(R,B)})$ satisfies* SDAG, *and*
2. *$\forall i \in \{0, \ldots, n\}$, $G^i_{(R,B)}[\forall \mapsto \backslash\cdot]$ is a tree.*

3 *NL*-Completeness of the Criteria for MLL and MELL

Let DepGRAPH be the function: $G \mapsto D(G)$, which associates its dependency graph to a paired graph G.

Lemma 4. DepGRAPH $\in$ *FL*.

Proof. The following functions can easily be computed in *FL*:

- $G, x \in G \mapsto (G[\forall \mapsto \backslash\cdot]) \setminus \{x\}$
- $G, x \in G \mapsto (G[\forall \mapsto \cdot.\cdot]) \setminus \{x\}$
- $G, x \in G, y \in G \mapsto (G[\forall \mapsto \backslash\cdot]) \setminus \{x, y\}$

Consider now the following algorithm for DepGRAPH:

```
INPUT (G)
FOR ALL x pair-node in G, with premise-nodes x_l and x_r DO
        IF USTCONN((G[∀ ↦·.·]) \ {x}, x_l, x_r) THEN OUTPUT (s → x) ∈ D(G)
FOR ALL ( x pair-node in G, with premise-nodes x_l and x_r, y pair-node in G) DO
        IF ¬USTCONN((G[∀ ↦\·]) \ {x, y}, x_l, x_r)
        AND USTCONN((G[∀ ↦\·]) \ {x}, x_l, x_r) THEN
                OUTPUT (y → {x}) ∈ D(G).
```

- USTCONN$((G[\forall \mapsto \cdot.\cdot]) \setminus \{x\}, x_l, x_r)$ tests whether there exists an elementary path $p_x = x_l - \cdots - x_r$ such that $x \notin p_x$ and, for all pair-node y in G, $y \notin p_x$.
- $\neg$ USTCONN$((G[\forall \mapsto \cdot\backslash\cdot]) \setminus \{x, y\}, x_l, x_r)$ tests whether any elementary path $p_x = x_l - \cdots - x_r$ such that $x \notin p_x$ contains y.
- USTCONN$((G[\forall \mapsto \cdot\backslash\cdot]) \setminus \{x\}, x_l, x_r)$ tests whether there exists a path $p_x = x_l - \cdots - x_r$ in G' such that $x \notin p_x$. From the previous point, if such a path p_x exists, $y \in p_x$.

It follows that this algorithm computes DepGRAPH. Since USTCONN $\in L$, this algorithm belongs to FL^L (the class of functions computable in logarithmic space, with oracles in L). Since $FL^L = FL$, DepGRAPH $\in FL$. □

Proposition 3. *MELL* – corr $\in$ *NL*.

Proof. Let (R, B) be a MELL-proof structure with boxes $b_1, \ldots, b_n$. Each function $(R, B), i \in \{0, \ldots, n\} \mapsto G^i_{(R,B)}$ can be easily be computed in *FL*. Checking that $G^i_{(R,B)}[\forall \mapsto \cdot\backslash\cdot]$ is a tree is doable in L since IT $\in L$. Checking that $D(G^i_{(R,B)})$ satisfies SDAG can be done in *NL*, by composing the function DepGRAPH in *FL* (Lemma 4) with an *NL* algorithm for SDAG (Theorem 4).

Since the number of paired graphs $G^i_{(R,B)}$ is linearly bounded, it suffices to sequentially perform these tasks for $i = 0, \ldots, n$, with a counter i of logarithmic size. □

Note that the previous best algorithms [Laf95, Gue99] are not likely to be implemented in logarithmic space, since they require on-line modification of the structure they manipulate. The purpose of our criterion of Theorem 5 is precisely that it allows a space-efficient implementation.

Proposition 4. *MLL*-corr *is NL-hard under constant-depth reductions.*

Proof. We actually reduce SDAG to MLL-corr. Let G be a directed graph, and consider the proof structure S_G defined as follows (see Figure 5), and let G_{S_G} be its associated paired graph:

1. To any vertex v of G, we associate a $\otimes$-link $\overline{v}$ with parent links $\overline{v}_{in}$ and $\overline{v}_{out}$.
2. If there are $i > 0$ in-going edges to v, $\overline{v}_{in}$ is a $⅋$-link of arity i, with parent links $\overline{v}^1_{in}, \ldots, \overline{v}^i_{in}$. If v has no in-going edge, $\overline{v}_{in}$ is one conclusion of an axiom-link Ax^v_{in}, the other conclusion of Ax^v_{in} being a conclusion of S^x_n.
3. If there are $j > 0$ outgoing edges from v, $\overline{v}_{out}$ is a $\otimes$-link of arity j, with parent links $\overline{v}^1_{out}, \ldots, \overline{v}^j_{out}$. If v has no outgoing edge, $\overline{v}_{out}$ is one conclusion of an axiom-link Ax^v_{out}, the other conclusion of Ax^v_{out} being a conclusion of S^x_n.
4. To an edge $v \to w$ of G, we associate an axiom-link $Ax^{(v \to w)}$ with conclusions $Ax^{(v \to w)}_{in}$ and $Ax^{(v \to w)}_{out}$. Moreover, if $v \to w$ is the kth outgoing edge from v, $Ax^{(v \to w)}_{in}$ is $\overline{v}^k_{out}$. If $v \to w$ is the lth in-going edge to w, $Ax^{(v \to w)}_{out}$ is $\overline{w}^l_{in}$.

It is quite clear that this reduction can be computed by constant-depth circuits. We now claim that S_G is correct if and only if G satisfies SDAG.

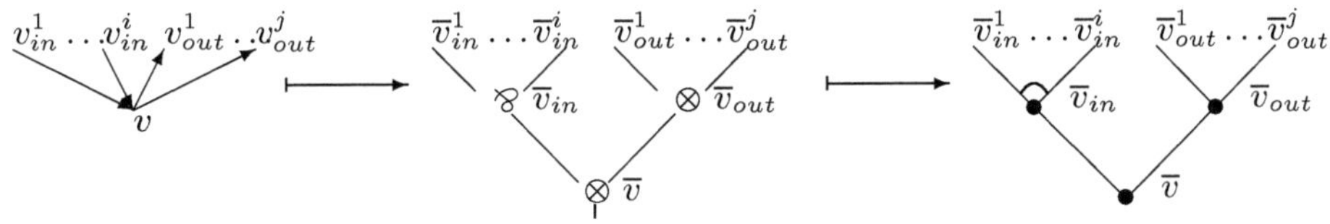

Fig. 5. Construction of S_G and G_{S_G}

Assume G contains a cycle. There exists then an elementary path $p = x_1 \to \cdot\cdot \to x_l$, with $x_l \to x_1 \in G$. Then, for any edge $x_t \to x_{t+1} \in p$, there exists a switching of the pair-node $\overline{x_{t+1}}_{in}$ in G_{S_G}, which connects $\overline{x_t}$ and $\overline{x_{t+1}}$. Similarly for the edge $x_l \to x_1 \in G$. Since p is elementary, these pair-nodes are all different; therefore there exists cyclic switching of G_{S_G} and S_G is not correct.

It is clear that if G is acyclic, it has at least one node of arity 0. Moreover, if G is acyclic and has only one node of arity 0, a proof by induction shows that G satisfies SDAG.

Assume therefore that G is acyclic and has at least two nodes, r and s, of arity 0. Let S' be any switching of G_{S_G}, and assume that there exists an elementary path p from $\overline{r}$ to $\overline{s}$ in S'. Let $p' = \overline{r}, \overline{x_1}, \ldots, \overline{x_k}, \overline{s}$ be the sequence of non pair-nodes of p corresponding to vertices of G. p' follows by construction edges of G, accordingly to their orientation or not. Since r and s have arity 0, there exist three nodes $\overline{x_t}, \overline{x_{t+1}}, \overline{x_{t+2}}$ in p' such that $(x_t \to x_{t+1})$ and $(x_{t+2} \to x_{t+1})$ are edges of G. By construction of G_{S_G}, $\overline{x_t}$ and $\overline{x_{t+2}}$ are then premise-nodes of the same pair-node $\overline{x_{t+1}}_{in}$ in G_{S_G}, which contradicts that p is a path in S'. Therefore, S' is not connected, and S_G is not correct.

Assume now that G satisfies SDAG and let $d(v)$, the depth of a vertex v of G, be the length of the longest path from the source s of G to v. Denote by G^d the subgraph of G consisting only in the vertices of depth less than d, and by $G_{S_G^d}$ the corresponding paired graph. It is easy to see that the rules of Figure 1 can be turned in a n-ary version, and that $G_{S_G^{d+1}}$ can be obtained from $G_{S_G^d}$ by these n-ary rules. By induction on d, it follows that S_G is correct. □

Since MLL is a subsystem of MELL, Propositions 3 and 4 immediately yield the following result:

Theorem 6. *MLL-*CORR *and MELL-*CORR *are NL-complete under constant-depth reductions.*

4 *NL*-Completeness of the Criterion for IMLL

Proposition 5. $IMLL - \text{CORR} \in NL$.

Proof. For a given essential net N, denote by $r(N)$ its root. For a given $⅋^+$-link x in N, denote by $s(x)$ its sink. Consider now the following algorithm for IMLL-CORR:

INPUT (N)
IF $\neg\mathrm{SDAG}(N, r(N))$ THEN REJECT
FOR ALL x $\parr^+$-link in N DO
  IF $\mathrm{STCONN}(N \setminus \{x\}, r(N), s(x))$ THEN REJECT
ELSE ACCEPT.

This algorithm checks that N satisfies SDAG, and that the L-condition applies. Since SDAG $\in$ *NL* and STCONN $\in$ *NL*, and *NL* = *coNL* (Theorem 3), this algorithm belongs clearly to *NL*. □

Proposition 6. *IMLL-CORR is NL-hard under constant-depth reductions.*

Proof. We actually reduce SDAG to IMLL-CORR. Let G be a directed graph, and consider the essential net N_G defined as follows (see Figure 6):

1. To any vertex v of G of arity $i > 0$, we associate a $\otimes^-$-node $\overline{v}$, with parent node (right premise) $\overline{v}_{in}$ of polarity $-$ and child node (left premise) $\overline{v}_{out}$ of polarity $+$. To v of arity 0 we associate a node $\overline{v} = \overline{v}_{out}$ of polarity $+$.
2. If there are $i > 0$ in-going edges to v , $\overline{v}_{in}$ is a $\parr^-$-node of arity i, with parent nodes $\overline{v}^1_{in}, \ldots, \overline{v}^i_{in}$ of polarity $-$.
3. If there are $j > 0$ outgoing edges from v, $\overline{v}_{out}$ is a $\otimes^+$-node of arity j, with child nodes $\overline{v}^1_{out}, \ldots, \overline{v}^j_{out}$ of polarity $+$.
4. If there is no outgoing edge from v, $\overline{v}_{out}$ is a $\parr^+$-node with child node (right premise) $\overline{v}^1_{out}$ of polarity $+$, and sink node (left premise) $\overline{v}^{sink}_{out}$, of polarity $-$. There is moreover an axiom-edge $\overline{v}^1_{out} \to \overline{v}^{sink}_{out}$.
5. Let $v \to w$ be an edge of G. Assume $v \to w$ is the kth outgoing edge from v, and the lth in-going edge to w. To $v \to w$, we associate an axiom-edge from $\overline{v}^k_{out}$ of polarity $+$ to $\overline{w}^l_{in}$ of polarity $-$.

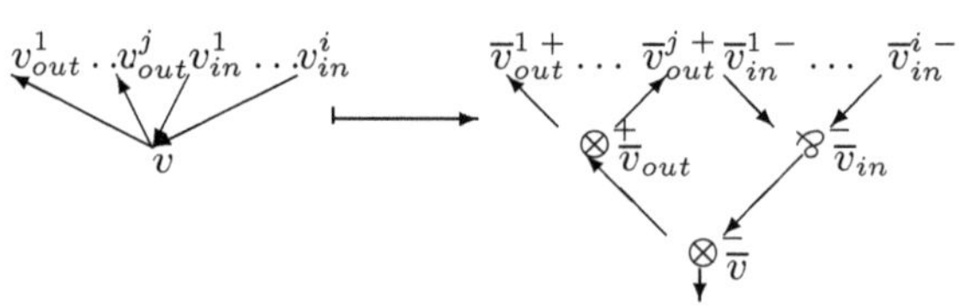

Fig. 6. Construction of N_G

It is clear that the reduction is constant-depth. Since the only $\parr^+$-links of N_G correspond to leaves of G, N_G satisfies the L-condition by construction. Therefore, it is L-correct if and only if G satisfies SDAG. □

Propositions 5 and 6 immediately yield the following result:

Theorem 7. *IMLL-CORR is NL-complete under constant-depth reductions.*

Note that [MO06] provides a linear-time reduction from MLL-CORR to IMLL-CORR, which yields a linear-time algorithm for MLL-CORR. This reduction actually occurs to be linear-space, and cannot be used for directly proving Proposition 6.

5 Conclusion and Acknowledgments

Deciding the correctness of unit-free MLL proof structures, MELL proof structures, and unit-free IMLL essential nets where problems known to be decidable in deterministic, sequential linear time. We have shown their NL-completeness, thus establishing that it would be most unlikely to find better sequential deterministic algorithms. As a byproduct, we obtain efficient parallel algorithms for both problems, namely AC^1 algorithms. Moreover, since $NL = RL = ZPL$, we also naturally obtain Monte-Carlo and Las-Vegas logarithmic space random algorithms, by simply using random walks for our graph reachability procedures. It remains to be checked whether our approach can be extended to MALL.

We are grateful to Harry Mairson for raising the question of the exact complexity of the correctness problems, and to the members of the No-Cost project for useful discussions and comments. We also thank the anonymous referees for their comments.

References

[CM87] Cook, S.A., McKenzie, P.: Problems complete for deterministic logarithmic space. J. Algorithms 8(3), 385–394 (1987)

[CSV84] Chandra, A.K., Stockmeyer, L.J., Vishkin, U.: Constant depth reducibility. SIAM J. Comput. 13(2), 423–439 (1984)

[Dan90] Danos, V.: Une application de la logique linéaire à l'étude des processus de normalisation (principalement de λ-calcul). PhD thesis, Université Denis Diderot, Paris 7 (1990)

[DR89] Danos, V., Regnier, L.: The structure of multiplicatives. Archive for Mathematical Logic 28(3), 181–203 (1989)

[Gir87] Girard, J.-Y.: Linear logic. Theoretical Computer Science 50(1), 1–102 (1987)

[GM01] Guerrini, S., Masini, A.: Parsing MELL proof nets. Theoretical Computer Science 254(1–2), 317–335 (2001)

[Gue99] Guerrini, S.: Correctness of multiplicative proof nets is linear. In: Proc. of the annual Symp. on Logic in Computer Science (LICS'99), pp. 454–463. IEEE Computer Society Press, Los Alamitos (1999)

[Imm88] Immerman, N.: Nondeterministic space is closed under complementation. SIAM J. Comput., 17(5), 935–938 (1988)

[JLL76] Jones, N.D., Lien, Y.E., Laaser, W.T.: New problems complete for nondeterministic logspace. Mathematical Syst. Theory 10, 1–17 (1976)

[JLM97] Jenner, B., Lange, K.-J., McKenzie, P.: Tree isomorphism and some other complete problems for deterministic logspace. DIRO 1059, Univ. de Montréal (1997)

[Kan92] Kanovich, M.I.: Horn programming in linear logic is NP-complete. In: Proc. of the annual Symp. on Logic in Computer Science (LICS'92), pp. 200–210. IEEE Computer Society Press, Los Alamitos (1992)

[Laf95] Lafont, Y.: From proof-nets to interaction nets. In: Girard, J.-Y., Lafont, Y., Regnier, L. (eds.) Advances in Linear Logic, vol. 222, pp. 225–247. Cambridge University Press, Cambridge (1995)

[Mai] Mairson, H.G.: Normalization bounds for multiplicative linear logic are axiom-sensitive. Presentation at GEOCAL'06 Workshop of Implicit Computational Complexity. Slides available at `http://www-lipn.univ-paris13.fr/~baillot/GEOCAL06/SLIDES/Mairson.pdf`

[Mel04] Melliès, P.-A.: A topological correctness criterion for non-commutative logic. London Mathematical Society Lecture Notes Series, vol. 316. Cambridge University Press, Cambridge (2004)

[MO00] Murawski, A., Ong, L.: Dominator trees and fast verification of proof nets. In: Proc. of the annual Symp. on Logic in Computer Science (LICS'00), pp. 181–191. IEEE Computer Society Press, Los Alamitos (2000)

[MO06] Murawski, A., Ong, L.: Fast verification of MLL proof nets via IMLL. ACM Trans. Comput. Logic 7(3), 473–498 (2006)

[Rei05] Reingold, O.: Undirected st-connectivity in log-space. In: Gabow, H.N., Fagin, R. (eds.) STOC, pp. 376–385. ACM, New York (2005)

[Sze87] Szelepcsényi, R.: The method of forcing for nondeterministic automata. Bulletin of the EATCS 33, 96–99 (1987)

Focusing and Polarization in Intuitionistic Logic

Chuck Liang[1] and Dale Miller[2]

[1] Department of Computer Science, Hofstra University, Hempstead, NY 11550
chuck.liang@hofstra.edu
[2] INRIA & LIX/Ecole Polytechnique, Palaiseau, France
dale.miller@inria.fr

Abstract. A focused proof system provides a normal form to cut-free proofs that structures the application of invertible and non-invertible inference rules. The focused proof system of Andreoli for linear logic has been applied to both the *proof search* and the *proof normalization* approaches to computation. Various proof systems in literature exhibit characteristics of focusing to one degree or another. We present a new, focused proof system for intuitionistic logic, called *LJF*, and show how other proof systems can be mapped into the new system by inserting logical connectives that prematurely stop focusing. We also use *LJF* to design a focused proof system for classical logic. Our approach to the design and analysis of these systems is based on the completeness of focusing in linear logic and on the notion of polarity that appears in Girard's LC and LU proof systems.

1 Introduction

Cut-elimination provides an important normal form for sequent calculus proofs. But what normal forms can we uncover about the structure of cut-free proofs? Since cut-free proofs play important roles in the foundations of computation, such normal forms might find a range of applications in the proof normalization foundations for functional programming or in the proof search foundations of logic programming.

1.1 About Focusing

Andreoli's *focusing* proof system for linear logic (the *triadic* proof system of [1]) provides a normal form for cut-free proofs in linear logic. Although we describe this system, here called *LLF*, in more detail in Section 2, we highlight two aspect of focusing proofs here. First, linear logic connectives can be divided into the *asynchronous* connectives, whose right-introduction rules are invertible, and the *synchronous* connectives, whose right introduction rules are not (generally) invertible. The search for a focused proof can capitalize on this classification by applying (reading inference rules from conclusion to premise) all invertible rules in any order (without the need for backtracking) and by applying a chain of non-invertible rules that focus on a given formula and its positive subformulas. Such a chain of applications, usually called a *focus*, terminates when it reaches

J. Duparc and T.A. Henzinger (Eds.): CSL 2007, LNCS 4646, pp. 451–465, 2007.

an asynchronous formula. Proof search can then alternate between applications of asynchronous introduction rules and chains of synchronous introduction rules.

A second aspect of focusing proofs is that the synchronous/asynchronous classification of non-atomic formulas must be extended to atomic formulas. The arbitrary assignment of positive (synchronous) and negative (asynchronous) *bias* to atomic formulas can have a major impact on, not the existence of focused proofs, but the shape of focused proofs. For example, consider the Horn clause specification of the Fibonacci series:

$$\mathit{fib}(0,0) \wedge \mathit{fib}(1,1) \wedge \forall n \forall f \forall f'[\mathit{fib}(n,f) \wedge \mathit{fib}(n+1,f') \supset \mathit{fib}(n+2,f+f')].$$

If all atomic formulas are given negative bias, then the only focused proofs of $\mathit{fib}(n, f_n)$ are those that can be classified as "backward chaining" (the size of the smallest one being exponential in n). On the other hand, if all atomic formulas are given positive bias, then the only focused proofs are those that can be classified as "forward chaining" (the size of the smallest one being linear in n).

1.2 Results

The contributions of this paper are the following. First, we introduce in Section 5 a new focusing proof system *LJF* and show that it is sound and complete for intuitionistic logic. Notable features of *LJF* are that it allows for atoms of different bias and it contains two versions of conjunction: while these conjunctions are logically equivalent, they are affected by focusing differently. Second, in Section 6, we show how several other focusing proof systems can be captured in *LJF*, in the sense of *full completeness* (one-to-one correspondence between proofs in different systems). One should note that while there are many focusing proof systems for intuitionistic logic in the literature, we appear to be the first to provide a single (intuitionistic) framework for capturing many of them. Third, in Section 7, we use *LJF* to derive *LKF*, a focusing system for classical logic.

1.3 Methodology and Related Work

There are a number of sequent calculus proof systems known to be complete for intuitionistic logic that exhibit characteristics of focusing. Some of these proof systems are based on fixing globally on either forward chaining or backward chaining. The early work on *uniform proofs* [17] and the *LJT* proof system [11] are both backward chaining calculi (all atoms have negative bias). The *LJQ* calculus [11,7] similarly selects the global preference to be forward chaining (all atoms have positive bias). Less has been published about systems that allow for mixing bias on atoms. The λRCC proof system of Jagadeesan, Nadathur, and Saraswat [13] allows for both forward chaining and backward chaining in a superset of the hereditary Harrop fragment of intuitionistic logic. Chaudhuri, Pfenning, and Price in [3] observed that focusing proofs with mixed biases on atoms can form a declarative basis for mixing forward and backward chaining within Horn clauses. The PhD theses of Howe [12] and Chaudhuri [2] also explored various focusing proof systems for linear and intuitionistic logic.

We are interested in providing a flexible and unifying framework that can collect together important aspects of many of these proof systems. There are several ways to motivate and validate the design of such a system. One approach stays entirely within intuitionistic logic and works directly with invertibility and permutability of inference rules. Such an approach has been taken in many papers, such as [17,18,7]. Our approach uses linear logic, with its exponential operators ! and ?, as a unifying framework for looking at intuitionistic (and classical) logic. The fact that Andreoli's focused system was defined for full linear logic provides us with a convenient platform for exploring the issues around focusing and polarity. We translate intuitionistic logic into linear logic, then show that proof systems for intuitionistic logic match focused proofs of the translated image (Section 3). A crucial aspect of understanding focusing in intuitionistic logic is provided by identifying the precise relationship between Andreoli's notion of polarity with Girard's notion of polarity found in the LC [9] and LU [10] systems (Section 4).

Another system concerning polarity and focusing is found in the work of Danos, Joinet and Schellinx [5,6]. Many techniques that they developed, such as *inductive decorations,* are used throughout our analysis. Our work diverges from theirs in the adaptation of Andreoli's system (LLF) as our main instrument of construction. The LK_p^η system of [6] describes focused proofs for classical logic. Its connections to polarization and focusing were further explored and extended by Laurent, Quatrini and de Falco [14] using *polarized proof nets.* It may be tempting to speculate that the best way to arrive at a notion of intuitionistic focusing is by simple modifications to these systems, such as restricting them to single-conclusion sequents. Closer examination however, reveal intricate issues concerning this approach. For example, the notion of classical *polarity* appears to be distinct from and *contrary* to intuitionistic polarity, especially at the level of atoms (see Sections 4 and 7). Resolving this issue would be central to finding systems that support combined forward and backward chaining. Although the relationship between LK_p^η and our systems is interesting, we chose for this work to derive intuitionistic focusing from focusing in linear logic as opposed to classical logic.

Much of the research into focusing systems has been motivated by their application. For example, the papers [13,17,12,2] are motivated by foundational issues in logic programming and automated deduction. The papers [11,5,6,14] are motivated by foundational issues in functional programming and the λ-calculus. Also, Levy [15] presents focus-style proof systems for typing in the λ-calculus and Curien and Herbelin [4] (among others) have noted the relationship between forward chaining and call-by-value evaluation and between backward chaining and call-by-name evaluation.

Our work can be extended to second order logic, although this paper is concerned mainly with first-order quantification.

Many details missing from this paper can be found in the report [16].

2 Focusing in Linear Logic

We summarize the key results from [1] on focusing proofs for linear logic.

A *literal* is either an atomic formula or the negation of an atomic formula. A linear logic formula is in *negation normal form* if it does not contain occurrences of $\multimap$ and if all negations have atomic scope. If K is literal, then $K^\perp$ denotes its complement: in particular, if K is $A^\perp$ then $K^\perp$ is A.

Connectives in linear logic are either *asynchronous* or *synchronous*. The asynchronous connectives are $\bot$, $⅋$, $?$, $\top$, $\&$, and $\forall$ while the synchronous connectives are their de Morgan dual, namely, $\mathbf{1}$, $\otimes$, $!$, $\mathbf{0}$, $\oplus$, and $\exists$. Asynchronous connectives are those where the right-introduction rule is always invertible. Formally, a formula in negation normal form is of three kinds: literal, asynchronous (*i.e.*, its top-level connective is asynchronous), and synchronous (*i.e.*, its top-level connective is synchronous).

$$\frac{\Psi : \Delta \Uparrow L}{\Psi : \Delta \Uparrow \bot, L}\ [\bot] \qquad \frac{\Psi : \Delta \Uparrow F, G, L}{\Psi : \Delta \Uparrow F ⅋ G, L}\ [⅋] \qquad \frac{\Psi, F : \Delta \Uparrow L}{\Psi : \Delta \Uparrow ?F, L}\ [?]$$

$$\frac{}{\Psi : \Delta \Uparrow \top, L}\ [\top] \qquad \frac{\Psi : \Delta \Uparrow F, L \quad \Psi : \Delta \Uparrow G, L}{\Psi : \Delta \Uparrow F \,\&\, G, L}\ [\&] \qquad \frac{\Psi : \Delta \Uparrow B[y/x], L}{\Psi : \Delta \Uparrow \forall x.B, L}\ [\forall]$$

$$\frac{\Psi : \Delta, F \Uparrow L}{\Psi : \Delta \Uparrow F, L}\ [R\Uparrow] \quad \text{provided that F is not asynchronous}$$

$$\frac{}{\Psi : \cdot \Downarrow \mathbf{1}}\ [\mathbf{1}] \qquad \frac{\Psi : \Delta_1 \Downarrow F \quad \Psi : \Delta_2 \Downarrow G}{\Psi : \Delta_1, \Delta_2 \Downarrow F \otimes G}\ [\otimes] \qquad \frac{\Psi : \cdot \Uparrow F}{\Psi : \cdot \Downarrow !F}\ [!]$$

$$\frac{\Psi : \Delta \Downarrow F_1}{\Psi : \Delta \Downarrow F_1 \oplus F_2}\ [\oplus_l] \qquad \frac{\Psi : \Delta \Downarrow F_2}{\Psi : \Delta \Downarrow F_1 \oplus F_2}\ [\oplus_r] \qquad \frac{\Psi : \Delta \Downarrow B[t/x]}{\Psi : \Delta \Downarrow \exists x.B}\ [\exists]$$

$$\frac{\Psi : \Delta \Uparrow F}{\Psi : \Delta \Downarrow F}\ [R\Downarrow] \quad \text{provided that } F \text{ is either asynchronous or a negative literal}$$

$$\text{If } K \text{ a positive literal:} \quad \frac{}{\Psi : K^\perp \Downarrow K}\ [I_1] \qquad \frac{}{\Psi, K^\perp : \cdot \Downarrow K}\ [I_2]$$

$$\text{If } F \text{ is not a negative literal:} \quad \frac{\Psi : \Delta \Downarrow F}{\Psi : \Delta, F \Uparrow \cdot}\ [D_1] \qquad \frac{\Psi, F : \Delta \Downarrow F}{\Psi, F : \Delta \Uparrow \cdot}\ [D_2]$$

Fig. 1. The focused proof system LLF for linear logic

As mentioned in Section 1.1, the classification of non-atomic formulas as asynchronous or synchronous is pushed to literals by assigning a fixed but arbitrary *bias* to atoms: an atom given a *negative bias* is linked to asynchronous behavior while an atom given *positive bias* is linked to synchronous behavior. In Andreoli's original presentation of LLF [1] all atoms were classified as "positive" and their negations "negative." Girard made a similar assignment for LC [9]. In a classical setting, such a choice works fine since classical negation simply flips bias. In intuitionistic systems, however, a more natural treatment is to assign an arbitrary bias directly to atoms. This bias of atoms is extended to literals: negating

Table 1. The 0/1 translation used to encode LJ proofs into linear logic

B	B^1	B^0	$(B^0)^\perp$
atom Q	Q	Q	$Q^\perp$
true	1	$\top$	0
false	0	0	$\top$
$P \wedge Q$	$!(P^1 \& Q^1)$	$!P^0 \& !Q^0$	$?(P^0)^\perp \oplus ?(Q^0)^\perp$
$P \vee Q$	$!P^1 \oplus !Q^1$	$!P^0 \oplus !Q^0$	$?(P^0)^\perp \& ?(Q^0)^\perp$
$P \supset Q$	$!(?(P^0)^\perp \parr Q^1)$	$!P^1 \multimap !Q^0$	$!P^1 \otimes ?(Q^0)^\perp$
$\neg P$	$!(0 \parr ?(P^0)^\perp)$	$!P^1 \multimap 0$	$!P^1 \otimes \top$
$\exists x P$	$\exists x\, !P^1$	$\exists x\, !P^0$	$\forall x\, ?(P^0)^\perp$
$\forall x P$	$!\forall x P^1$	$\forall x\, !P^0$	$\exists x\, ?(P^0)^\perp$

a negative atom yields a positive literal and negating a positive atom yields a negative literal.

The focusing proof system LLF for linear logic, presented in Figure 1, contains two kinds of sequents. In the sequent $\Psi\colon \Delta \Uparrow L$, the "zones" Ψ and Δ are multisets and L is a list. This sequent encodes the usual one-sided sequent $\vdash ?\Psi, \Delta, L$ (here, we assume the natural coercion of lists into multisets). This sequent will also satisfy the invariant that requires Δ to contain only literals and synchronous formulas. In the sequent $\Psi\colon \Delta \Downarrow F$, the zone Ψ is a multiset of formulas and Δ is a multiset of literals and synchronous formulas, and F is a single formula. Notice that the bias of literals is explicitly referred to in the $[R \Uparrow]$ and initial rules: in particular, in the initial rules, the literal on the right of the $\Downarrow$ must be positive.

Changes to the bias assigned to atoms does not affect provability of a linear logic formula: instead it affects the structure of focused proofs.

3 Translating Intuitionistic Logic

Table 1 contains a translation of intuitionistic logic into linear logic. This translation induces a bijection between arbitrary LJ proofs and LLF proofs of the translated image in the following sense. First notice that this translation is *asymmetric*: the intuitionistic formula A is translated using A^1 if it occurs on the right-side of an LJ sequent and as A^0 if it occurs on the left-side. Since this translation is used to capture cut-free proofs, such distinctions are not problematic. Since the left-hand side of a sequent in LJ will be negated when translated to a one-sided linear logic sequent, $(B^0)^\perp$ is also shown. The following Proposition essentially says that, via this translation, linear logic focusing can capture arbitrary proofs in LJ. Here, $\vdash_I$ denotes an intuitionistic logic sequent.

Proposition 1. *Let $(\Gamma^0)^\perp$ be the multiset $\{(D^0)^\perp \mid D \in \Gamma\}$. The focused proofs of $\vdash (\Gamma^0)^\perp :\Uparrow R^1$ are in bijective correspondence with the LJ proofs of $\Gamma \vdash_I R$.*

For a detailed proof, see [16]. We simply illustrate one case in constructing the mapping from a collection of LLF rules to an LJ inference rule.

$$\cfrac{\cfrac{\cfrac{\vdash (\Gamma^0)^\perp, (D_i^0)^\perp : R^1 \Uparrow}{\vdash (\Gamma^0)^\perp : R^1 \Uparrow ?(D_i^0)^\perp}\,[?]}{\vdash (\Gamma^0)^\perp : R^1 \Downarrow ?(D_i^0)^\perp}\,[R\Downarrow]}{\vdash (\Gamma^0)^\perp : R^1 \Downarrow ?(D_1^0)^\perp \oplus ?(D_2^0)^\perp}\,[\oplus] \quad \longmapsto \quad \cfrac{\Gamma, D_i \vdash_I R}{\Gamma, D_1 \wedge D_2 \vdash_I R}\,[\wedge L]$$

The liberal use of ! in this translation *throttles* focusing. This translation is reminiscent of the earliest embedding of classical into intuitionistic logic of Kolmogorov, which uses the double negation in a similarly liberal fashion.

The 0/1 translation can be used as a starting point in establishing the completeness of other proof systems. These systems can be seen as *induced* from alternative translations of intuitionistic logic.

Consider, for example, the *LJQ′* proof system presented in [7]. The translation for this system is given here using the "q/j" mapping. The "$\otimes 1$" device is another way to control focusing, or the lack thereof. *All atoms must be given positive bias* for this translation.

F	F^q (right)	F^j (left)
atom C	C	C
false	0	0
$A \wedge B$	$A^q \otimes B^q$	$!\,A^j \otimes !\,B^j$
$A \vee B$	$A^q \oplus B^q$	$!\,A^j \oplus !\,B^j$
$A \supset B$	$(!\,A^j \multimap B^q) \otimes 1$	$A^q \multimap !\,B^j$

With minor changes, Girard's original (non-polarized) translation of intuitionistic logic [8] induces the complement to *LJQ′* called *LJT* [11], which is itself derived from LKT [5] (where this connection was noted implicitly.) *All atoms must be given negative bias* for this translation.

Given a translation such as that of *LJQ′*, one can give a completeness proof for the system using a *"grand tour"* through linear logic as follows:

1. Show that a proof under the 0/1 translation can be converted into a proof under the new translation. This usually follows from cut-elimination.
2. Define a mapping between proofs in the new system (such as LJQ) and LLF proofs of its translation.
3. Show soundness of the new system with respect to LJ. This is usually trivial. The "tour" is now complete, since proofs in LJ map to proofs under the 0/1 translation.

An intuitionistic system that contains atoms of both positive and negative bias is λRCC [13]. Two special cases of the $\supset L$ rule are distinguished involving $E \supset D$ for positive atom E and $G \supset A$ for negative atom A. Each rule requires that the complementary atom (E on the left, A on the right) is present when applied, thus terminating one branch of the proof. One can translate these special cases using forms $E \multimap !\,D'$ and $!\,G' \multimap A$, respectively, in linear logic. The strategy outlined above can then be used to not only prove its completeness but also extend it with more aggressive focusing features.

Our interest here is not the construction of individual systems but the building of a unifying framework for focusing in intuitionistic logic. Such a task requires a closer examination of *polarity* and its connection to focusing.

4 Permeable Formulas and Their Polarity

Focused proofs in linear logic are characterized by two different phases: the invertible (asynchronous) phase and the non-invertible (synchronous) phase. These two phases are characterized by introduction rules for dual sets of formulas. In order to construct a general focusing scheme for intuitionistic logic, the non-linear (exponential) aspects of proofs need special attention, especially in light of the fact that the [!] rule stops a bottom-up construction of focused application of synchronous rules (the arrow $\Downarrow$ in the conclusion flips to $\Uparrow$ in the premise).

For our purposes here, a particularly flexible way to deal with the exponentials in the translations of intuitionistic formulas is via the notion of *permeation* that is used in LU [10]. In particular, there are essentially three grades of *permeation*. The formula B is *left-permeable* if $B \equiv\, !\,B$, is *right-permeable* if $B \equiv\, ?\,B$, and *neutral* otherwise. Within sequent calculus proofs, a formula is left-permeable if it admits structural rules on the left and right-permeable if it admits structural rules on the right. An example of a left-permeable formula is $\exists x\, !A$. All left-permeable formulas are synchronous and all right-permeables asynchronous. In the LU system, both the left and right sides of sequents contain two zones — one that treats formulas linearly and one that permits structural rules. A left-permeable (resp., right-permeable) formula is allowed to move between both zones on the left (right). In addition, LU introduces atoms that are inherently left or right-permeable or neutral. Although they appear to properly extend linear logic, one can simulate LU in "regular" linear logic by translating left-permeable atoms A as $!\,A$ and right-permeable ones as $?\,A$.

To preserve the focusing characteristics of permeable atoms as positively or negatively biased atoms, we use the following LU-inspired asymmetrical translation. The superscript –1 indicates the left-side translation and +1 indicates the right-side translation:

$P^{-1} = \,!\,P$ and $P^{+1} = P$, for left-permeable (positive) atom P.
$N^{-1} = N$ and $N^{+1} = \,?\,N$, for right-permeable (negative) atom N.
$B^{-1} = B^{+1} = B$, for neutral atom B.

The ! rule of LLF causes a loss of focus in all circumstances, and is the main reason why we use an asymmetrical translation. The translation of positive atoms above preserves *permeation on the left* while allowing for *focus on the right.* That is, left-permeable atoms can now be interpreted meaningfully as positively biased atoms in focused proofs, and dually for right-permeable atoms. Furthermore, the permeation of positive atoms is *"one-way only:"* they cannot be selected for focus again once they enter the non-linear context.

Intuitionistic logic uses the left-permeable and neutral formulas and atoms. LU defines a translation for intuitionistic logic so that *all synchronous formulas are left-permeable.* For example, $\vee$ is translated as follows (here, P, Q are positive and N, M are negative): $(P \vee Q)^{-1} = P^{-1} \oplus Q^{-1}$, $(P \vee N)^{-1} = P^{-1} \oplus !N^{-1}$, $(N \vee P)^{-1} = !N^{-1} \oplus Q^{-1}$, and $(N \vee M)^{-1} = !N^{-1} \oplus !M^{-1}$. The final element of intuitionistic polarity is that *neutral atoms should be assigned negative bias in focused proofs.* Neutral atoms that are introduced into the left context (e.g. by a $\supset L$ rule) must immediately end that branch of the proof in an identity rule. Otherwise, the unique *stoup* is lost when multiple non-permeable atoms accumulate in the linear context.

The LU and LLF systems serve as a convenient platform for the unified characterization of polarity and focusing in all three logics. We can now understand the terminology of "positive" and "negative" formulas in each logic as follows:

Linear logic: *Positive* formulas are synchronous formulas and positively biased neutral atoms. *Negative* formulas are asynchronous formulas and negatively biased neutral atoms.

Intuitionistic logic: *Positive* formulas are left-permeable formulas and *negative* formulas are asynchronous neutral formulas and negatively biased neutral atoms.

Classical logic: *Positive* formulas are left-permeable formulas.*Negative* formulas are right-permeable formulas.

5 The *LJF* Sequent Calculus

Since the polarities of intuitionistic logic observe stronger invariances, intuitionistic focused proofs are more well-structured than LLF proofs. The non-linear context of LLF contains both synchronous and asynchronous formulas, whereas in intuitionistic logic sequents can be clearly divided into zones respecting polarity. That is, when translating an intuitionistic sequent into a LLF sequent, synchronous formulas on the left are placed in the linear context.

We also make an adjustment on the LU translation of intuitionistic logic. Instead of using & or $\otimes$ depending on the polarities of the subformulas, we construct two versions of intuitionistic conjunction, which has the following meaning in linear logic (P, Q for positives, N, M for negatives, A, B arbitrary):

$$\begin{array}{ll}
(P \wedge^+ Q)^{-1} = P^{-1} \otimes Q^{-1} & (A \wedge^+ B)^{+1} = A^{+1} \otimes B^{+1} \\
(P \wedge^+ N)^{-1} = P^{-1} \otimes !N^{-1} & \\
(N \wedge^+ P)^{-1} = !N^{-1} \otimes P^{-1} & (A \wedge^- B)^{-1} = A^{-1} \& B^{-1} \\
(N \wedge^+ M)^{-1} = !N^{-1} \otimes !M^{-1} & (A \wedge^- B)^{+1} = A^{+1} \& B^{+1}
\end{array}$$

The connectives $\wedge^-$ and $\wedge^+$ are equivalent in intuitionistic logic in terms of provability but differ in their impact on the structure of focused proofs. The use of two conjunctions means that the top-level structure of formulas completely determines their polarity. *Polarity* in intuitionistic logic is defined as follows.

Definition 1. *Atoms in LJF are arbitrarily positive or negative.* Positive formulas *are among positive atoms, true, false,* $A \wedge^+ B$, $A \vee B$ *and* $\exists x A$. Negative formulas *are among negative atoms,* $A \wedge^- B$, $A \supset B$ *and* $\forall x A$.

$$\frac{[N,\Gamma] \xrightarrow{N} [R]}{[N,\Gamma] \longrightarrow [R]}\ Lf \qquad \frac{[\Gamma] -_{P}\!\!\rightarrow}{[\Gamma] \longrightarrow [P]}\ Rf \qquad \frac{[\Gamma], P \longrightarrow [R]}{[\Gamma] \xrightarrow{P} [R]}\ R_l \qquad \frac{[\Gamma] \longrightarrow N}{[\Gamma] -_{N}\!\!\rightarrow}\ R_r$$

$$\frac{[C,\Gamma],\Theta \longrightarrow \mathcal{R}}{[\Gamma],\Theta, C \longrightarrow \mathcal{R}}\ []_l \qquad \frac{[\Gamma],\Theta \longrightarrow [D]}{[\Gamma],\Theta \longrightarrow D}\ []_r$$

$$\frac{}{[P,\Gamma] -_{P}\!\!\rightarrow}\ I_r,\ \text{atomic } P \qquad \frac{}{[\Gamma] \xrightarrow{N} [N]}\ I_l,\ \text{atomic } N$$

$$\frac{}{[\Gamma],\Theta, \mathit{false} \longrightarrow \mathcal{R}}\ \mathit{falseL} \qquad \frac{[\Gamma],\Theta \longrightarrow \mathcal{R}}{[\Gamma],\Theta, \mathit{true} \longrightarrow \mathcal{R}}\ \mathit{trueL} \qquad \frac{}{[\Gamma] -_{\mathit{true}}\!\!\rightarrow}\ \mathit{trueR}$$

$$\frac{[\Gamma] \xrightarrow{A_i} [R]}{[\Gamma] \xrightarrow{A_1 \wedge^- A_2} [R]}\ \wedge^- L \qquad \frac{[\Gamma],\Theta, A, B \longrightarrow \mathcal{R}}{[\Gamma],\Theta, A \wedge^+ B \longrightarrow \mathcal{R}}\ \wedge^+ L$$

$$\frac{[\Gamma],\Theta \longrightarrow A \quad [\Gamma],\Theta \longrightarrow B}{[\Gamma],\Theta \longrightarrow A \wedge^- B}\ \wedge^- R \qquad \frac{[\Gamma] -_{A}\!\!\rightarrow \quad [\Gamma] -_{B}\!\!\rightarrow}{[\Gamma] -_{A \wedge^+ B}\!\!\rightarrow}\ \wedge^+ R$$

$$\frac{[\Gamma],\Theta, A \longrightarrow \mathcal{R} \quad [\Gamma],\Theta, B \longrightarrow \mathcal{R}}{[\Gamma],\Theta, A \vee B \longrightarrow \mathcal{R}}\ \vee L \qquad \frac{[\Gamma] -_{A_i}\!\!\rightarrow}{[\Gamma] -_{A_1 \vee A_2}\!\!\rightarrow}\ \vee R$$

$$\frac{[\Gamma] -_{A}\!\!\rightarrow \quad [\Gamma] \xrightarrow{B} [R]}{[\Gamma] \xrightarrow{A \supset B} [R]}\ \supset L \qquad \frac{[\Gamma],\Theta, A \longrightarrow B}{[\Gamma],\Theta \longrightarrow A \supset B}\ \supset R$$

$$\frac{[\Gamma],\Theta, A \longrightarrow \mathcal{R}}{[\Gamma],\Theta, \exists y A \longrightarrow \mathcal{R}}\ \exists L \qquad \frac{[\Gamma] -_{A[t/x]}\!\!\rightarrow}{[\Gamma] -_{\exists x A}\!\!\rightarrow}\ \exists R \qquad \frac{[\Gamma] \xrightarrow{A[t/x]} [R]}{[\Gamma] \xrightarrow{\forall x A} [R]}\ \forall L \qquad \frac{[\Gamma],\Theta \longrightarrow A}{[\Gamma],\Theta \longrightarrow \forall y A}\ \forall R$$

Fig. 2. The Intuitionistic Sequent Calculus LJF. Here, P is positive, N is negative, C is a negative formula or positive atom, and D a positive formula or negative atom. Other formulas are arbitrary. Also, y is not free in Γ, Θ, or R.

The above translation induces the sequent calculus *LJF* for intuitionistic logic, shown in Figure 2. Sequents in *LJF* can be interpreted as follows:

1. $[\Gamma],\Theta \longrightarrow \mathcal{R}$ (end sequent): this is an *unfocused sequent.* Γ contains negative formulas and positive atoms. $\mathcal{R}$ represents either a formula R or $[R]$.
2. $[\Gamma] \longrightarrow [R]$: this represents a sequent in which all asynchronous formulas have been decomposed, and is ready for a formula to be selected for focus.
3. $[\Gamma] \xrightarrow{A} [R]$: this is a *left-focusing* sequent, with focus on formula A. The meaning of this sequent remains $\Gamma, A \vdash_I R$.
4. $[\Gamma] -_{A}\!\!\rightarrow$: this is a *right-focusing* sequent on formula A, with the meaning $\Gamma \vdash_I A$.

Theorem 1. *LJF is sound and complete with respect to intuitionistic logic.*

Proof. Using the "grand tour" strategy. See [16, Section 6] for details.

Given the different forms of sequents, the cut rule for *LJF* takes many forms:

$$\frac{[\Gamma],\Theta \longrightarrow P \quad [\Gamma'],\Theta',P \longrightarrow \mathcal{R}}{[\Gamma\Gamma'],\Theta\Theta' \longrightarrow \mathcal{R}}\ Cut^{+} \qquad \frac{[\Gamma],\Theta \longrightarrow C \quad [C,\Gamma'],\Theta' \longrightarrow \mathcal{R}}{[\Gamma\Gamma'],\Theta\Theta' \longrightarrow \mathcal{R}}\ Cut^{-}$$

$$\frac{[\Gamma] \xrightarrow{B} [P] \quad [\Gamma'],P \longrightarrow [R]}{[\Gamma\Gamma'] \xrightarrow{B} [R]}\ Cut_1^{\leftarrow} \qquad \frac{[\Gamma] \longrightarrow N \quad [N,\Gamma'] \xrightarrow{B} [R]}{[\Gamma\Gamma'] \xrightarrow{B} [R]}\ Cut_2^{\leftarrow}$$

$$\frac{[\Gamma] -_{C}\!\!\rightarrow \quad [C,\Gamma'] -_{R}\!\!\rightarrow}{[\Gamma\Gamma'] -_{R}\!\!\rightarrow}\ Cut^{\rightarrow}$$

Notice that the last three cut rules retain focus in the conclusion. These rules extend those of *LJQ'* [7], which were shown to be useful for studying term-reduction systems. See [16] for a proof of the admissibility of these rules.

Like LLF, a key characteristic of *LJF* is the assignment of arbitrary polarity to atoms. To illustrate the effect of these assignments on the structure of focused proofs, consider the sequent $a, a \supset b, b \supset c \vdash c$ where a, b and c are atoms. This sequent can be proved either by *forward chaining* through the clause $a \supset b$, or *backward chaining* through the clause $b \supset c$. Assume that atoms a and b are assigned positive polarity and that c is assigned negative polarity. This assignment effectively adopts the forward chaining strategy, reflected in the following *LJF* proof segment (here, Γ is the set $\{a, a \supset b, b \supset c\}$):

$$\frac{\dfrac{}{[\Gamma] -_{a}\!\!\rightarrow}\ I_r \qquad \dfrac{\dfrac{\dfrac{\dfrac{\dfrac{}{[b,\Gamma] -_{b}\!\!\rightarrow}\ I_r \quad \dfrac{}{[b,\Gamma] \xrightarrow{c} [c]}\ I_l}{[b,\Gamma] \xrightarrow{b\supset c} [c]}\ \supset L}{[b,\Gamma] \longrightarrow [c]}\ Lf}{[\Gamma], b \longrightarrow [c]}\ []_l}{[\Gamma] \xrightarrow{b} [c]}\ R_l}{[\Gamma] \xrightarrow{a\supset b} [c]}\ \supset L$$

The polarities of a and c do not fundamentally affect the structure of the proof in this example. However, assigning negative polarity to atom b would restrict the proof to use the backward chaining strategy:

$$\frac{\dfrac{\dfrac{\dfrac{\dfrac{\dfrac{}{[\Gamma] -_{a}\!\!\rightarrow}\ I_r \quad \dfrac{}{[\Gamma] \xrightarrow{b} [b]}\ I_l}{[\Gamma] \xrightarrow{a\supset b} [b]}\ \supset L}{[\Gamma] \longrightarrow [b]}\ Lf}{[\Gamma] \longrightarrow b}\ []_r}{[\Gamma] -_{b}\!\!\rightarrow}\ R_r \qquad \dfrac{}{[\Gamma] \xrightarrow{c} [c]}\ I_l}{[\Gamma] \xrightarrow{b\supset c} [c]}\ \supset L$$

6 Embedding Intuitionistic Systems in LJF

The *LJF* proof system can be used to "host" other focusing proof system for intuitionistic logic. One obvious restriction to *LJF* is its purely negative fragment, which essentially corresponds to LJT. In the negative fragment one also finds *uniform proofs,* where the right "goal" formula is always fully decomposed before any left rule is applied. Various other proof systems can be embedded into *LJF* by mapping intuitionistic formulas to intuitionistic formulas in such a way that focusing features in *LJF* are stopped by the insertion of *delay* operators. In particular, if we define $\partial^-(B) = true \supset B$ and $\partial^+(B) = true \wedge^+ B$, then B, $\partial^-(B)$, and $\partial^+(B)$ are all logically equivalent but $\partial^-(B)$ is always negative and $\partial^+(B)$ is always positive.

Proofs in the LJQ' system can be embedded into *LJF* by translating all left-side formulas (l) as negatives and all right-side formulas (r) as positives: in particular, for atom B, $B^l = B^r = B$, $false^l = \partial^-(false)$, $false^r = false$, $(A \wedge B)^l = \partial^-(A^l \wedge^+ B^l)$, $(A \wedge B)^r = A^r \wedge^+ B^r$, $(A \vee B)^l = \partial^-(A^l \vee B^l)$, $(A \vee B)^r = A^r \vee B^r$, $(A \supset B)^l = A^r \supset \partial^+(B^l)$, $(A \supset B)^r = \partial^+(A^l \supset B^r)$.

Arbitrary LJ proofs can be embedded within *LJF* by inserting sufficient delaying operators. The table here provides the translation (redefining the superscripts l and r, for convenience). Together with cut-elimination, the embedding also suggests a completeness proof for *LJF* independently of linear logic. The following example embeds the $\wedge R$ rule in *LJF*:

F	F^l (left)	F^r (right)
atom C	C	C
false	$\partial^-(false)$	*false*
true	$\partial^-(true)$	*true*
$A \wedge B$	$\partial^+(A^l) \wedge^- \partial^+(B^l)$	$\partial^+(A^r \wedge^- B^r)$
$A \vee B$	$\partial^-(A^l \vee B^l)$	$\partial^-(A^r) \vee \partial^-(B^r)$
$A \supset B$	$\partial^-(A^r) \supset \partial^+(B^l)$	$\partial^+(A^l \supset B^r)$
$\exists x A$	$\partial^-(\exists x A^l)$	$\exists x \partial^-(A^r)$
$\forall x A$	$\forall x \partial^+(A^l)$	$\partial^+(\forall x A^r)$

$$
\dfrac{\dfrac{\dfrac{\dfrac{[\Gamma] \longrightarrow [A^r]}{[\Gamma] \longrightarrow A^r}\,[]_r \quad \dfrac{[\Gamma] \longrightarrow [B^r]}{[\Gamma] \longrightarrow B^r}\,[]_r}{[\Gamma] \longrightarrow A^r \wedge^- B^r}\,\wedge^- R}{[\Gamma] \xrightarrow{A^r \wedge^- B^r}}\,R_r \quad \dfrac{}{[\Gamma] \xrightarrow{true}}\,trueR}{\dfrac{[\Gamma] \xrightarrow{(A^r \wedge^- B^r) \wedge^+ true}}{[\Gamma] \longrightarrow [\partial^+(A^r \wedge^- B^r)]}\,Rf}\,\wedge^+ R
$$

The system λRCC also presents interesting choices. In particular, it may not always be the best choice to focus maximally. Forward chaining may generate a new formula or "clause" that may need to be used multiple times. In a $\supset L$ rule on formulas $E \supset D$ where E is a positive atom, one may not wish to decompose the formula D immediately. This is accomplished in the linear translation with a !. It can also be accomplished by using formulas $E \supset \partial^+(D)$ in case D is negative, and $E \supset \partial^+(\partial^-(D))$ in case D is positive. Note that unlike the l/r translations for LJQ and LJ above, these simple devices do not hereditarily alter the structure of D.

7 Embedding Classical Logic in LJF

We can use *LJF* to formulate a focused sequent calculus for classical logic that reveals the latter's constructive content in the style of LC. While it is possible to derive such a system again using linear logic, classical logic can also be embedded within intuitionistic logic using the *double-negation* translations of Gödel, Gentzen, and Kolmogorov. These translations do not, however, yield significant focusing features. Girard's *polarized* version of the double negation translation for LC approaches the problem of capturing duality in a more subtle way. Following the style of *LJF*, we wish to define dual versions of each propositional connective, which leads to a more usable calculus. We thus modify the LC translation in a natural way, which is consistent with its original intent. The proof system we derive is called *LKF*.

We must first separate classical from intuitionistic polarity since these are different notions (see the end of Section 4).

Definition 2. *Atoms are arbitrarily classified as either positive or negative. The literal* $\neg A$ *has the opposite polarity of the atom* A. Positive formulas *are among positive literals,* $\mathcal{T}$, $\mathcal{F}$, $A \wedge^+ B$, $A \vee^+ B$, $A \supset^+ B$ *and* $\exists x A$. Negative formulas *are among negative literals,* $\neg\mathcal{T}$, $\neg\mathcal{F}$, $A \wedge^- B$, $A \vee^- B$, $A \supset^- B$ *and* $\forall x A$. *Negation* $\neg A$ *is defined by de Morgan dualities* $\neg A/A$, $\wedge^+/\vee^-$, $\wedge^-/\vee^+$ *and* $\forall/\exists$. *Negative implication* $A \supset^- B$ *is defined as* $\neg A \vee^- B$ *and* $A \supset^+ B$ *is defined as* $\neg A \vee^+ B$. *Formulas are assumed to be in negation normal form (that is, formulas that do not contain implications and negations have atomic scope).*

The constants $\mathcal{T}$, $\mathcal{F}$, $\neg\mathcal{T}$ and $\neg\mathcal{F}$ are best described, respectively, as $1, 0, \bot$ and $\top$ in linear logic. Just as we have dual versions of each connective, we also have dual versions of each identity. But this is not linear logic as the formulas are polarized *in the extreme.* The distinction between the positive and negative versions of each connective affects only the structure of proofs and not provability.

Let $\sim\! A$ represent the intuitionistic formula $A \supset \phi$ where ϕ is *some unspecified positive atom.* The "$\approx$" embedding of classical logic is found in Table 2. Variations are possible on the embedding. Note that the classical $\wedge^-$ is not defined in terms of the intuitionistic $\wedge^-$. The embeddings are selected to enforce the dualities $\wedge^-/\vee^+$ and $\wedge^+/\vee^-$. Alternatives may also work, but will increase the complexity of the derivation. Here, the cases all follow the pattern P or $\sim P$ where P is a positive intuitionistic formula. In particular, negative intuitionistic atoms are not used in the embedding.

The $\approx$ embedding induces the *LKF* sequent calculus in Figure 3 from the image of *LJF* proofs, analogous to how *LJF* was derived from LLF. Here is one sample correspondence between a *LJF* rule and a LKF rule:

$$\frac{[\Delta], \Psi, A, B \longrightarrow [\phi]}{[\Delta], \Psi, A \wedge^+ B \longrightarrow [\phi]} \wedge^+ L \qquad \longmapsto \qquad \frac{\vdash [\Theta], \Gamma, A, B}{\vdash [\Theta], \Gamma, A \vee^- B} \vee^-$$

Sequents of the form $\vdash [\Theta], \Gamma$ are unfocused while those of the form $\mapsto [\Theta], A$ focus on the *stoup* formula A.

Table 2. Polarized embedding of classical logic. The $(\cdot)^{\approx}$ translation on compound formulas is given above (there, A, B represent formulas not preceded by $\sim$). For positive classical atom P, $P^{\approx} = P$; for negative classical atom N, $N^{\approx} = \sim N$; (where both P and N are assigned positive intuitionistic polarity), and for the logical constants $\mathcal{T}^{\approx} = true$, $\mathcal{F}^{\approx} = false$, $(\neg\mathcal{T})^{\approx} = \sim true$, $(\neg\mathcal{F})^{\approx} = \sim false$.

$\mathcal{A}^{\approx}$	$\mathcal{B}^{\approx}$	$(\mathcal{A} \wedge^{+} \mathcal{B})^{\approx}$	$(\mathcal{A} \wedge^{-} \mathcal{B})^{\approx}$	$(\mathcal{A} \vee^{+} \mathcal{B})^{\approx}$	$(\mathcal{A} \vee^{-} \mathcal{B})^{\approx}$	$(\neg\mathcal{A})^{\approx}$
A	B	$A \wedge^{+} B$	$\sim(\sim A \vee \sim B)$	$A \vee B$	$\sim(\sim A \wedge^{+} \sim B)$	$\sim A$
A	$\sim B$	$A \wedge^{+} \sim B$	$\sim(\sim A \vee B)$	$A \vee \sim B$	$\sim(\sim A \wedge^{+} B)$	$\cdot$
$\sim A$	B	$\sim A \wedge^{+} B$	$\sim(A \vee \sim B)$	$\sim A \vee B$	$\sim(A \wedge^{+} \sim B)$	A
$\sim A$	$\sim B$	$\sim A \wedge^{+} \sim B$	$\sim(A \vee B)$	$\sim A \vee \sim B$	$\sim(A \wedge^{+} B)$	$\cdot$

$\mathcal{A}^{\approx}$	$\mathcal{B}^{\approx}$	$(\mathcal{A} \supset^{+} \mathcal{B})^{\approx}$	$(\mathcal{A} \supset^{-} \mathcal{B})^{\approx}$	$(\forall x \mathcal{A})^{\approx}$	$(\exists x \mathcal{A})^{\approx}$
A	B	$\sim A \vee B$	$\sim(A \wedge^{+} \sim B)$	$\sim(\exists x \sim A)$	$\exists x A$
A	$\sim B$	$\sim A \vee \sim B$	$\sim(A \wedge^{+} B)$	$\cdot$	$\cdot$
$\sim A$	B	$A \vee B$	$\sim(\sim A \wedge^{+} \sim B)$	$\sim(\exists x A)$	$\exists x \sim A$
$\sim A$	$\sim B$	$A \vee \sim B$	$\sim(\sim A \wedge^{+} B)$	$\cdot$	$\cdot$

The following correctness theorem for LKF can be proved by relating it to the Gödel-Gentzen translation (see [16, Section 9] for more details).

Theorem 2. *LKF is sound and complete with respect to classical logic.*

We have constructed this embedding of classical logic as a further demonstration of the abilities of *LJF* as a hosting framework. The embedding also revealed interesting relationships between classical and intuitionistic polarity. It is also possible to derive LKF from linear logic: each connective needs to be defined as either wholly positive or negative. For example, the translation of $(A \vee^{-} B)^p$ is $A^p \parr B^p$ if A^p and B^p are both negative; is $A^p \parr\, ?B^p$ if only A^p is negative; is $?A^p \parr B^p$ if only B^p is negative; and is $?A^p \parr ?B^p$, if A^p and B^p are both positive. This translation is called the *"polaro"* translation in [6], where it was used to formulate LK_p^{η}, the first focused proof system for classical logic. Like the $\approx$ translation, the polaro translation is a derivative of the LC/LU analysis of polarity. Except for the treatment of atoms, LKF is derivable from LLF using the polaro translation in the same manner that *LJF* is derived.

LK_p^{η} was extended to $LK_{pol}^{\eta,\rho}$ in [14]. These systems were formulated independently of Andreoli's results. The authors of [6] opted not to present LK_p^{η} as a sequent calculus because they feared that it will have the cumbersome size of LU. Such cumbersomeness can, in fact, be avoided by adopting LLF-style *reaction* rules.

Given our goals, the choice in adopting Andreoli's system is justified in that LKF and LJF have the form of compact sequent calculi ready for implementation. More significantly perhaps, LK_p^{η} and $LK_{pol}^{\eta,\rho}$ define focusing for classical logic. They map to polarized forms of linear logic (LLP and LL_{pol}). LLF is defined for full classical linear logic. LKF is embedded within LLF in the same way that LC is embedded within LU. LLF is well suited for hosting other logics.

$$\frac{\vdash [\Theta, C], \Gamma}{\vdash [\Theta], \Gamma, C} [] \qquad \frac{\mapsto [P, \Theta], P}{\vdash [P, \Theta]} \; Focus \qquad \frac{\vdash [\Theta], N}{\mapsto [\Theta], N} \; Release$$

$$\frac{}{\mapsto [\neg P, \Theta], P} \; ID^+, \text{ atomic } P \qquad \frac{}{\mapsto [N, \Theta], \neg N} \; ID^-, \text{ atomic } N$$

$$\frac{}{\mapsto [\Theta], \mathcal{T}} \; indeed \qquad \frac{}{\vdash [\Theta], \Gamma, \neg \mathcal{F}} \; absurd \qquad \frac{\vdash [\Theta], \Gamma}{\vdash [\Theta], \Gamma, \neg \mathcal{T}} \; trivial$$

$$\frac{\vdash [\Theta], \Gamma, A \quad \vdash [\Theta], \Gamma, B}{\vdash [\Theta], \Gamma, A \wedge^- B} \; \wedge^- \qquad \frac{\vdash [\Theta], \Gamma, A, B}{\vdash [\Theta], \Gamma, A \vee^- B} \; \vee^-$$

$$\frac{\vdash [\Theta], \Gamma, B, \neg A}{\vdash [\Theta], \Gamma, A \supset^- B} \; \supset^- \qquad \frac{\vdash [\Theta], \Gamma, A}{\vdash [\Theta], \Gamma, \forall x A} \; \forall$$

$$\frac{\mapsto [\Theta], A \quad \mapsto [\Theta], B}{\mapsto [\Theta], A \wedge^+ B} \; \wedge^+ \qquad \frac{\mapsto [\Theta], A_i}{\mapsto [\Theta], A_1 \vee^+ A_2} \; \vee^+ \qquad \frac{\mapsto [\Theta], A[t/x]}{\mapsto [\Theta], \exists x A} \; \exists$$

$$\frac{\mapsto [\Theta], \neg A}{\mapsto [\Theta], A \supset^+ B} \; \supset^+ \qquad \frac{\mapsto [\Theta], B}{\mapsto [\Theta], A \supset^+ B} \; \supset^+$$

Fig. 3. The Classical Sequent Calculus LKF. Here, P is positive, N is negative, C is a positive formula or a negative literal, Θ consists of positive formulas and negative literals, and x is not free in Θ, Γ. End-sequents have the form $\vdash [], \Gamma$.

8 Conclusion and Future Work

We have studied focused proof construction in intuitionistic logic. The key to this endeavor is the definition of polarity for intuitionistic logic. The *LJF* proof system captures focusing using this notion of polarity. We illustrate how systems such as LJ, LJT, LJQ, and λRCC can be captured within *LJF* by assigning polarity to atoms and by adding to intuitionistic logic formulas annotations on conjunctions and delaying operators. We also use *LJF* to derive and justify the *LKF* focusing proof system for classical logic.

It remains to examine the impact of these focusing calculi on typed λ-calculi, logic programming, and theorem proving. Given the connections observed between LJQ/LJT and call-by-name/value, the *LJF* system could provide a framework for λ-term evaluations that combine the eager and lazy evaluation strategies. In the area of theorem proving, there are a number of completeness theorems for various restrictions to resolution: it would be interesting to see if any of these are captured by an appropriate mapping into *LKF*.

Acknowledgments. We would like to thank the reviewers of an earlier version of this paper for their comments. The work reported here was carried out while the first author was on sabbatical leave from Hofstra University to LIX. The second author was supported by INRIA through the "Equipes Associées" Slimmer and by the Information Society Technologies program of the European Commission, Future and Emerging Technologies under the IST-2005-015905 MOBIUS project.

References

1. Andreoli, J.-M.: Logic programming with focusing proofs in linear logic. J. of Logic and Computation 2(3), 297–347 (1992)
2. Chaudhuri, K.: The Focused Inverse Method for Linear Logic. PhD thesis, Carnegie Mellon University,Technical report CMU-CS-06-162 (December 2006)
3. Chaudhuri, K., Pfenning, F., Price, G.: A logical characterization of forward and backward chaining in the inverse method. In: Furbach, U., Shankar, N. (eds.) IJCAR 2006. LNCS (LNAI), vol. 4130, pp. 97–111. Springer, Heidelberg (2006)
4. Curien, P.-L., Herbelin, H.: The duality of computation. In: ICFP '00. Proceedings of the fifth ACM SIGPLAN international conference on Functional programming, New York, NY, USA, pp. 233–243. ACM Press, New York (2000)
5. Danos, V., Joinet, J.-B., Schellinx, H.: LKQ and LKT: sequent calculi for second order logic based upon dual linear decompositions of classical implication. In: Girard, Lafont, Regnier (eds.) Workshop on Linear Logic. London Mathematical Society Lecture Notes 222, pp. 211–224. Cambridge University Press, Cambridge (1995)
6. Danos, V., Joinet, J.-B., Schellinx, H.: A new deconstructive logic: Linear logic. Journal of Symbolic Logic 62(3), 755–807 (1997)
7. Dyckhoff, R., Lengrand, S.: LJQ: a strongly focused calculus for intuitionistic logic. In: Beckmann, A., Berger, U., Löwe, B., Tucker, J.V. (eds.) CiE 2006. LNCS, vol. 3988, pp. 173–185. Springer, Heidelberg (2006)
8. Girard, J.-Y.: Linear logic. Theoretical Computer Science 50, 1–102 (1987)
9. Girard, J.-Y.: A new constructive logic: classical logic. Math. Structures in Comp. Science 1, 255–296 (1991)
10. Girard, J.-Y.: On the unity of logic. Annals of Pure and Applied Logic 59, 201–217 (1993)
11. Herbelin, H.: Séquents qu'on calcule: de l'interprétation du calcul des séquents comme calcul de lambda-termes et comme calcul de stratégies gagnantes. PhD thesis, Université Paris 7 (1995)
12. Howe, J.M.: Proof Search Issues in Some Non-Classical Logics. PhD thesis, University of St Andrews, Available as University of St Andrews Research Report CS/99/1 (December 1998)
13. Jagadeesan, R., Nadathur, G., Saraswat, V.: Testing concurrent systems: An interpretation of intuitionistic logic. In: Ramanujam, R., Sen, S. (eds.) FSTTCS 2005. LNCS, vol. 3821, Springer, Heidelberg (2005)
14. Laurent, O., Quatrini, M., de Falco, L.T.: Polarized and focalized linear and classical proofs. Ann. Pure Appl. Logic 134(2-3), 217–264 (2005)
15. Levy, P.B.: Jumbo lambda-calculus. In: Bugliesi, M., Preneel, B., Sassone, V., Wegener, I. (eds.) ICALP 2006. LNCS, vol. 4052, Springer, Heidelberg (2006)
16. Liang, C., Miller, D.: On focusing and polarities in linear logic and intuitionistic logic. Unpublished report (December 2006)
17. Miller, D., Nadathur, G., Pfenning, F., Scedrov, A.: Uniform proofs as a foundation for logic programming. Annals of Pure and Applied Logic 51, 125–157 (1991)
18. Pfenning, F.: Automated theorem proving. Lecture notes (March 2004)

Incorporating Tables into Proofs

Dale Miller and Vivek Nigam

INRIA & LIX/École Polytechnique, Palaiseau, France
nigam@lix.inria.fr, dale.miller@inria.fr

Abstract. We consider the problem of automating and checking the use of previously proved lemmas in the proof of some main theorem. In particular, we call the collection of such previously proved results a *table* and use a partial order on the table's entries to denote the (provability) dependency relationship between tabled items. Tables can be used in automated deduction to store previously proved subgoals and in interactive theorem proving to store a sequence of lemmas introduced by a user to direct the proof system towards some final theorem. Tables of literals can be incorporated into sequent calculus proofs using two ideas. First, cuts are used to incorporate tabled items into a proof: one premise of the cut requires a proof of the lemma and the other branch of the cut inserts the lemma into the set of assumptions. Second, to ensure that lemma is not reproved, we exploit the fact that in *focused proofs*, atoms can have different *polarity*. Using these ideas, simple logic engines that do focused proof search (such as logic programming interpreters) are able to check proofs for correctness with guarantees that previous work is not redone. We also discuss how a table can be seen as a proof object and discuss some possible uses of tables-as-proofs.

1 Introduction

A sequence of well chosen lemmas is often an important part of presenting a proof in, at least, informal mathematics. In some situations, one might feel that the sequence of lemmas itself could constitute an actual proof, particularly if the reader of the proof has significant mathematical means to fill in the gaps between the lemmas. Of course, as lemmas at the beginning of the list are proved, they can be used to help prove lemmas later in the list.

Although generating lemmas is a well known and critical activity in mathematical proof, producing and using such lemmas can be important in, say, logic programming, deductive databases, and model checking. In such settings, the underlying proofs that such systems attempt to build are usually *cut-free* (that is, they lack the use of lemmas). That does not mean, however, that lemmas (and, hence, the cut-inference rule) do not have a role in improving the search for or the presentation of proofs.

Consider attempting to prove the conjunctive query $B \wedge C$ from a logic program Γ. This attempt can be reduced to first attempting to prove B from Γ and then C from Γ. It might well be the case that during the attempt to prove C, many subgoals might need to be proved that were previously established during

J. Duparc and T.A. Henzinger (Eds.): CSL 2007, LNCS 4646, pp. 466–480, 2007.

the attempt to prove B. Of course, if proved subgoals can be remembered from the first conjunct to the second, then it might be possible to build smaller proofs and these might be easier to find and to check for correctness. Some implemented logic programming systems already use tables in this fashion: for example, in XSB [16] and in Twelf [15], it is possible to specify that some predicates should be *tabled*: that is, whenever an atomic formula with such a predicate is successfully proved, that atomic formula is remembered, so that, any other time a proof of that atom is attempted, the proof process can be stopped with a success.

In this paper, we consider a general notion of *table* and attempt to show how proof theory can account for the following two salient aspects of tables.

(*i*) *Entering tabled formulas into the proof context.* Proofs will be sequent calculus proofs, and tables will be partially ordered collections of formulas. In a straightforward fashion, the cut-inference rule is used to state the obligation to prove a tabled lemma as well as insert it into the main proof context.

(*ii*) *Avoiding reproving of tabled formulas.* It is easy to provide algorithmic means for making certain that formulas are not reproved (for example, prior to attempting a proof of a formula, check if that formula is in the table). More challenging is to find a purely proof theoretic solution in which the only proofs that can be built are those in which reproving cannot happen. We achieve this by first restricting tables to be literals (a typical assumption in implementations of tabling). Second, we exploit some recent developments in the understanding of *focused* proofs in intuitionistic logic that allow literals to be given different *polarity*. Polarity can be used to signal that a literal is in or out of the table. Focused proof search can then be organized so that a tabled literal is not reproved.

This paper is structured as follows. Section 2 presents a couple of examples that help to motivate particular connections between tables and proofs. Section 3 illustrates how tables can be inserted into proofs by using the multicut inference rule (a simple generalization of the cut rule). Section 4 presents the main technical background of our approach: namely, the notions of focusing and polarity in intuitionistic logic. In Section 5, we show how focusing and polarity can be exploited to ensure that reproving already proved atoms is avoided, and later, in Section 6, we extend this result to literals. Section 7 discusses the possible merits of considering tables as proof objects themselves.

2 Two Motivating Examples

Consider the graph depicted in Figure 1, and assume that its arcs are represented by atomic facts of the form $(arr\ N_1\ N_2)$, where N_1 and N_2 are adjacent nodes in the graph. Consider also the following two Horn clauses for describing a path in this graph: $\forall x(path\,x\,x)$ and $\forall x \forall y \forall z(arr\ x\ z \wedge path\ z\ y \supset path\ x\ y)$,

Now consider attempting a proof of the conjunctive query $path\ a_1\ a_4 \wedge path\ a_2\ a_4$. The usual goal-directed logic interpreter will attempt to prove the two conjuncts independently. After making suitable backward chaining steps, both independent attempts will give rise to the same subgoal $path\ a_3\ a_4$. The logic interpreter will then proceed to construct two (possibly identical) proofs of

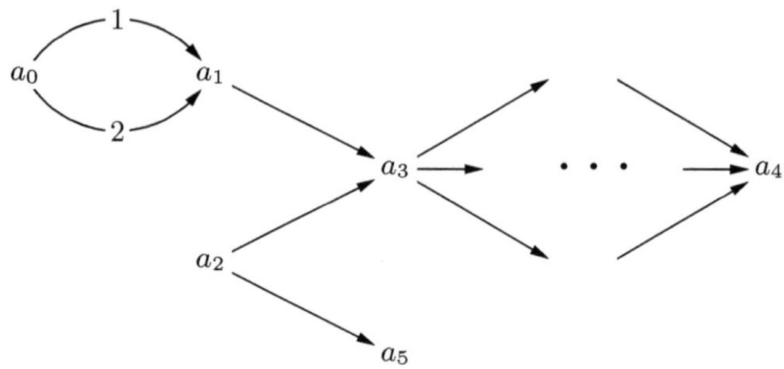

Fig. 1. A directed graph used to illustrate how tables can be included in proofs. The ellipses represent a section of the graph with a large number of paths from a_3 to a_4.

this subgoal. Clearly, a superior approach to proving this conjunctive goal would be to first prove the "lemma" *path* a_3 a_4, and then make that lemma available to the proof of the original conjunctive goal.

A basic problem still persists: how does one ensure that the assumed lemma is not reproved? If there are special algorithmic connections between the logic interpreter and the tabling mechanism, as exist in, say, XSB [16] and Twelf [15], then there are simple solutions to this problem of reproving lemmas. The question we are concerned with here, however, is whether or not there is an implementation independent and proof-theoretic solution to this problem.

For a second example, consider the following possible approach to *memoization* that one could attempt to use in logic programming languages (such as λProlog) that contain implicational goals [10]. Assume that the formula A is atomic and that we wish to prove the conjunction $A \wedge G$, for some general goal formula G. Since the attempt to prove G can reduce to several attempts to prove A, one might be tempted to rewrite the original conjunctive goal as the logically equivalent goal $A \wedge (A \supset G)$. During the attempt to prove G, the assumption A is available to establish any subgoal A immediately. Unfortunately, when moving from $A \wedge G$ to $A \wedge (A \supset G)$, one is making proof search more non-deterministic since for every proof that proves A by matching with the assumed version of A, there is another proof where A is, in fact, reproved. As a result, this naive approach to memoization has never been successfully used in λProlog.

This example also allows us to notice that our concern for not reproving previously proved formulas is different from the concerns of relevance logic [2], a logic in which the nature of implication is changed so that hypotheses are *necessary* for the proof of conclusions. In the example above, if the attempt to prove G succeeds without using the assumption A, the implication is still true even if the assumption A is not "relevant" to the conclusion G. The logic of this paper is intuitionistic.

Both of these examples illustrate a need for not only making proved atoms available for reuse but also enforcing that they are not reproved.

3 Tables as Multicut Derivations

In its most general form, we consider a table as a partially ordered finite set of formulas.

Definition 1. *A* table *is a tuple* $\mathcal{T} = \langle \mathcal{A}, \preceq \rangle$*, where* $\mathcal{A}$ *is some finite set of formulas, and* $\preceq$ *is a partial order relation over the elements of* $\mathcal{A}$*.*

The intended meaning of a table is that it is a structured collection of provable formulas (from some assumed context, say, Γ). The order relationship $B \preceq C$ denotes the fact that the proof of the formula B is available for reuse during a proof attempt of C: that is, if an attempt to prove the formula C results in the subgoal B then proof search can stop immediately since B has a proof.

The following inference rule, called the *multicut* rule, is often used as a technical generalization to the cut rule to help prove cut-elimination theorems (see, for example, [5,19]).

$$\frac{\Delta_1 \longrightarrow B_1 \quad \cdots \quad \Delta_n \longrightarrow B_n \qquad B_1, \ldots, B_n, \Gamma \longrightarrow C}{\Delta_1, \ldots, \Delta_n, \Gamma \longrightarrow C}\ mc \quad (n \geq 0)$$

Notice that if $n = 1$, this rule reduces to the usual cut rule (for a single conclusion calculus), and if $n = 0$, this rule is essentially a simple "repetition." If $n \geq 1$, then this rule can be seen as encoding n separate applications of the cut-rule.

The following definition describes how a table can be translated to a collection of multicut inference rules.

Definition 2. *Let* $\mathcal{T} = \langle \mathcal{A}, \preceq \rangle$ *be a table. The* multicut derivation *for* $\mathcal{T}$ *and the sequent* $\mathcal{S} = \Gamma \longrightarrow G$*, written as* $mcd(\mathcal{T}, \mathcal{S})$*, is defined inductively as follows: if* $\mathcal{A}$ *is empty, then* $mcd(\mathcal{T}, \mathcal{S})$ *is the derivation containing just the sequent* $\Gamma \longrightarrow G$*. Otherwise, if* $\{A_1, \ldots, A_n\}$ *is the collection of* $\preceq$*-minimal elements in* $\mathcal{A}$ *and if* Π *is the multicut derivation for the smaller table* $\langle \mathcal{A} \setminus \{A_1, \ldots, A_n\}, \preceq \rangle$ *and the sequent* $\Gamma, A_1, \ldots, A_n \longrightarrow G$*, then* $mcd(\mathcal{T}, \mathcal{S})$ *is the derivation*

$$\frac{\Gamma \longrightarrow A_1 \quad \cdots \quad \Gamma \longrightarrow A_n \qquad \begin{array}{c}\Pi \\ \Gamma, A_1, \ldots, A_n \longrightarrow G\end{array}}{\Gamma \longrightarrow G}\ mc$$

Multicut derivations are always open *derivations (that is, they contain leafs that are not proved). A* proof of a multicut derivation *is any (closed) proof that extends this open derivation.*

To illustrate this definition, consider the graph example in Section 2: let Γ contain the encoding of the original adjacency information as well as the specification of the *path* predicate, and consider the table that contains just the atomic formula $path\ a_3\ a_4$. The following is the multicut derivation for $\Gamma \longrightarrow path\ a_1\ a_4 \wedge path\ a_2\ a_4$:

$$\frac{\Gamma \longrightarrow path\ a_3\ a_4 \qquad \Gamma, path\ a_3\ a_4 \longrightarrow path\ a_1\ a_4 \wedge path\ a_2\ a_4}{\Gamma \longrightarrow path\ a_1\ a_4 \wedge path\ a_2\ a_4}\ mc$$

By using the cut, it was possible to introduce the lemma *path* a_3 a_4 in the context of the rightmost branch. The left premise requires showing that there is, in fact, a path from a_3 to a_4 while the right branch attempts to show the original conjunctive goal under the assumption that the existence of that path is granted. Unfortunately, there are proofs of this right-most premise where this lemma is not used but is reproved. In the next section, we introduce the notions of *focusing* and *polarity* in order to provide means for enforcing reuse.

4 Focusing and Polarities

In order to present a focused proof system, we first classify the connectives $\wedge$, $\exists$, *true* and $\perp$ as *synchronous* (their right introduction is not necessarily invertible) and the connectives $\supset$, and $\forall$ as *asynchronous* (their right introduction rules are invertible). This dichotomy must also be extended to atomic formulas: some atoms are considered asynchronous and the rest are considered synchronous. Since the terms "asynchronous" and "synchronous" do not apply well to atomic formulas, we shall instead use the slightly more general notions of *polarity* for a formula. In particular, a formula is positive if its main connective is synchronous or it is a *positive atom* and is negative if its main connective is asynchronous or it is a *negative atom*. The polarity of atoms is not necessarily fixed: we shall assign different polarities to atoms to achieve different purposes.

Although the notion of *focused proof* was originally given by Andreoli for linear logic [3], we shall use the recently designed LJF focused proof system for intuitionistic logic [7] displayed in Figure 2. This system has four types of sequents.

1. The sequent $[\Gamma] -_A\rightarrow$ is a *right-focusing* sequent (the focus is A);
2. The sequent $[\Gamma] \xrightarrow{A} [R]$: is a *left-focusing* sequent (with focus on A);
3. The sequent $[\Gamma], \Theta \longrightarrow \mathcal{R}$ is an *unfocused sequent.* Here, Γ contains negative formulas and positive atoms, and $\mathcal{R}$ is either in brackets, written as $[R]$, or without brackets;
4. The sequent $[\Gamma] \longrightarrow [R]$ is an instance of the previous sequent where Θ is empty.

As an inspection of the inference rules of LJF reveals, the search for a *focused* proof is composed of two alternating phases, and these phases are governed by polarities. The *asynchronous phase* applies invertible (asynchronous) rules until exhaustion: no backtracking during this phase of search is needed. The asynchronous phase uses the third type of sequent above (the unfocused sequents): in that case, Θ contains positive or negative formulas. If Θ contains positive formulas, then an introduction rule (either $\wedge_l, \exists_l$, $true_l$, or $false_l$) is used to decompose it; if it is negative, then the formula is moved to the Γ context (by using the $[]_l$ rule). The end of the asynchronous phase is represented by the fourth type of sequent. Such a sequent is then established by using one of the decide rules, D_r or D_l. The application of one of these decide rules then selects a formula for focusing and switches proof search to the *synchronous phase* or *focused phase.*

$$\frac{[N,\Gamma] \xrightarrow{N} [R]}{[N,\Gamma] \longrightarrow [R]}\ D_l \qquad \frac{[\Gamma] -_{P}\!\!\rightarrow}{[\Gamma] \longrightarrow [P]}\ D_r \qquad \frac{[\Gamma], P \longrightarrow [R]}{[\Gamma] \xrightarrow{P} [R]}\ R_l \qquad \frac{[\Gamma] \longrightarrow N}{[\Gamma] -_{N}\!\!\rightarrow}\ R_r$$

$$\frac{[\Gamma, N_a], \Theta \longrightarrow \mathcal{R}}{[\Gamma], \Theta, N_a \longrightarrow \mathcal{R}}\ []_l \qquad \frac{[\Gamma], \Theta \longrightarrow [P_a]}{[\Gamma], \Theta \longrightarrow P_a}\ []_r$$

$$\frac{}{[\Gamma] \xrightarrow{A_n} [A_n]}\ I_l \qquad \frac{}{[\Gamma, A_p] -_{A_p}\!\!\rightarrow}\ I_r$$

$$\frac{}{[\Gamma], \Theta, \bot \longrightarrow \mathcal{R}}\ false_l \qquad \frac{[\Gamma], \Theta \longrightarrow \mathcal{R}}{[\Gamma], \Theta, true \longrightarrow \mathcal{R}}\ true_l \qquad \frac{}{[\Gamma] -_{true}\!\!\rightarrow}\ true_r$$

$$\frac{[\Gamma], \Theta, A, B \longrightarrow \mathcal{R}}{[\Gamma], \Theta, A \wedge B \longrightarrow \mathcal{R}}\ \wedge_l \qquad \frac{[\Gamma] -_{A}\!\!\rightarrow \quad [\Gamma] -_{B}\!\!\rightarrow}{[\Gamma] -_{A \wedge B}\!\!\rightarrow}\ \wedge_r$$

$$\frac{[\Gamma] -_{A}\!\!\rightarrow \quad [\Gamma] \xrightarrow{B} [R]}{[\Gamma] \xrightarrow{A \supset B} [R]}\ \supset_l \qquad \frac{[\Gamma], \Theta, A \longrightarrow B}{[\Gamma], \Theta \longrightarrow A \supset B}\ \supset_r$$

$$\frac{[\Gamma], \Theta, A \longrightarrow \mathcal{R}}{[\Gamma], \Theta, \exists y A \longrightarrow \mathcal{R}}\ \exists_l \qquad \frac{[\Gamma] -_{A[t/x]}\!\!\rightarrow}{[\Gamma] -_{\exists x A}\!\!\rightarrow}\ \exists_r \qquad \frac{[\Gamma] \xrightarrow{A[t/x]} [R]}{[\Gamma] \xrightarrow{\forall x A} [R]}\ \forall_l \qquad \frac{[\Gamma], \Theta \longrightarrow A}{[\Gamma], \Theta \longrightarrow \forall y A}\ \forall_r$$

Fig. 2. The *LJF* system [7] originally has one disjunction and two conjunctions, $\wedge^+, \wedge^-$. In this paper, we only need one conjunction: we will drop $\wedge^-$ and write $\wedge$ for $\wedge^+$. Here A_n denotes a negative atom, A_p a positive atom, P a positive formula, N a negative formula, N_a a negative formula or an atom, and P_a a positive formula or an atom. All other formulas are arbitrary and y is not free in Γ, Θ or R.

This focused phase then proceeds by applying sequences of inference rules on focused formulas: in general, backtracking may be necessary in this phase of search. The focusing phase ends with one of the *release rule* R_l or R_r.

As is pointed out in [7], if all atoms are given negative polarity, the resulting proof system models backward chaining proof search and uniform proofs [11]. If positive atoms are permitted as well, then forward chaining steps can also be accommodated.

We now present the LJF^t proof system that extends LJF by adding a multicut rule and by allowing atoms to have different polarity on the different branches of the multicut rule. In particular, occurrences of atoms in LJF^t proofs are assigned polarities in the following fashion: all atoms are initially given negative polarity: thus proof search with such atoms is the usual goal-directed search. When an atom is inserted into a proof context via a multicut inference rule, that atom's occurrences on the right-most branch will have positive polarity: in principle, a forward chaining discipline is used on that atom on that branch, and it is this discipline that is used to implement the reuse policy on that part of the multicut derivation.

The sequents in LJF^t are the same four kinds of sequents except that we add a polarity declaration, $\mathcal{P}$, to all of them: if an atom appears in the set of atoms

$\mathcal{P}$, then it is considered positive; otherwise it is considered negative. Recall also that literals are either atomic formulas or negated atomic formulas (and that $\neg A$ is encoded as $A \supset \bot$). The multicut rule is the only rule that can change the declaration $\mathcal{P}$. In particular, the *polarized version* of the multicut rule is given as

$$\frac{\mathcal{P}; [\Gamma] \longrightarrow [L_1] \qquad \cdots \qquad \mathcal{P}; [\Gamma] \longrightarrow [L_n] \qquad \mathcal{P} \cup \Delta_P; [\Gamma \cup \Delta_L] \longrightarrow [R]}{\mathcal{P}; [\Gamma] \longrightarrow [R]}\ mc.$$

Here, $\Delta_L = \{L_1, \ldots, L_n\}$ is a set of literals and $\Delta_P = \{A \mid A \in \Delta_L \text{ or } \neg A \in \Delta_L\}$ is the set of all atoms in Δ_L. Notice that the literals in Δ_L are cut-formulas and that the atoms in Δ_P switch their polarity from negative in the conclusion of this rule to positive in the right-most premise. Whenever we use this multicut inference rule, we shall arrange things so that the sets Δ_P and $\mathcal{P}$ are disjoint.

As the notion of polarity of an atom is now declared via $\mathcal{P}$ instead of being globally fixed as in LJF, the inference rules in LJF^t must be adapted accordingly from LJF: for example, the LJF^t rule I_r^t will be derived from the LJF rule I_r as follows:

$$\frac{}{\mathcal{P}; [\Gamma, A_p] -_{A_p}\rightarrow}\ I_r^t, \text{ where } A_p \in \mathcal{P}.$$

In general, if the name of a rule is R in LJF, the corresponding rule in LJF^t is R^t. The following proposition can be proved by a simple induction on the depth of the cut free proofs.

Proposition 1. *LJF^t is sound and complete with respect to LJF.*

5 Tables of Finite Successes

In this section, we restrict our attention in two directions. First, we shall only consider tables containing atomic formulas. Such a restriction is familiar from such implemented tabling systems as [16,15] where the only items placed in a table are atomic formulas. Second, we shall only allow logic specifications to be Horn clauses, which are defined as D-formulas in the following grammar.

$$G := true \mid A \mid G_1 \wedge G_2 \mid \exists x\, G \qquad\qquad D := A \mid G \supset A \mid D_1 \wedge D_2 \mid \forall x\, D$$

As a consequence of these restrictions, we shall only be tabling atoms if they are proved by "finite success": this contrasts with the situation addressed in the next section where tables can contain negated atoms if "finite failure" is successful to prove them. The restriction to Horn clause formulas is critical for the results here since such clauses ensure that the goal-reduction phase can be seen as completely synchronous. Goals with implications and universal quantifiers causes goal-reduction to mix synchronous and asynchronous phases. Therefore, allowing them can cause the focus of proof search to be broken before positive atomic formulas are encountered.

The following proposition states that a multicut derivation of a provable sequent (using the polarized version of the multicut rule) can be extended to a valid focused proof.

Proposition 2. *Let Γ be a collection of Horn clauses, G be a G-formula, and let $\mathcal{T}$ be a table of atoms, all of which are provable from Γ (the partial order is not restricted). The sequent $\Gamma \longrightarrow G$ is intuitionistically provable if and only if the open derivation $mcd(\mathcal{T}, \Gamma \longrightarrow G)$ can be extended to a proof in LJF^t.*

Proof. The proof in the forward direction is by induction on the length of the longest path in the table's partial order. The converse is proved by forgetting the polarity information and using cut-elimination. □

The next proposition shows that polarities can be used to guarantee that any tabled atomic formula that has been proved once (and, hence, has positive polarity) will not be reproved. This proposition is proved by induction on the depth of the proof tree.

Proposition 3. *Let Γ be a set of Horn clauses, $A \in \mathcal{P} \cap \Gamma$, and Ξ be an arbitrary LJF^t proof tree for $\mathcal{P}; [\Gamma] -_G\!\rightarrow$. Then every occurrence of a sequent with right-hand side the atom A is the conclusion of an I_r^t rule.*

Since all the lemmas of a table are included as positive atoms in the right branch of its multicut derivation, all the proofs of any lemma in this branch will be composed of a single rule I_r^t.

Consider again the example in Section 2, where the subgoal *path* a_3 a_4 is tabled. Any proof of the rightmost branch of the multicut derivation obtained, will never reprove the lemma *path* a_3 a_4:

$$\dfrac{\dfrac{\dfrac{\dfrac{}{\mathit{path}\ a_3\ a_4; [\Gamma, \mathit{path}\ a_3\ a_4] \xrightarrow{\mathrm{arr}\, a_1\, a_3} [\mathrm{arr}\, a_1\, a_3]}\, I_l^t}{\mathit{path}\ a_3\ a_4; [\Gamma, \mathit{path}\ a_3\ a_4] \longrightarrow [\mathrm{arr}\, a_1\, a_3]}\, D_l^t}{\mathit{path}\ a_3\ a_4; [\Gamma, \mathit{path}\ a_3\ a_4] -_{\mathrm{arr}\, a_1\, a_3}\!\rightarrow}\, R_r^t \qquad \dfrac{}{\mathit{path}\ a_3\ a_4; [\Gamma, \mathit{path}\ a_3\ a_4] -_{\mathit{path}\ a_3\ a_4}\!\rightarrow}\, I_r^t}{\dfrac{\mathit{path}\ a_3\ a_4; [\Gamma, \mathit{path}\ a_3\ a_4] -_{\mathrm{arr}\, a_1\, a_3 \wedge \mathit{path}\ a_3\ a_4}\!\rightarrow}{\mathit{path}\ a_3\ a_4; [\Gamma, \mathit{path}\ a_3\ a_4] \longrightarrow [\mathit{path}\ a_1\ a_4]}\, D_l^t, \forall_l^t, \forall_l^t, \forall_l^t, \supset_l^t}\, \wedge_r^t$$

The memoization example of Section 2 can be addressed similarly: instead of doing the goal reduction illustrated on the left below, use a multicut as is illustrated on the right:

$$\dfrac{\Gamma \longrightarrow A \qquad \Gamma \longrightarrow G}{\Gamma \longrightarrow A \wedge G} \quad \Longrightarrow \quad \dfrac{\mathcal{P}; [\Gamma] \longrightarrow [A] \qquad \mathcal{P} \cup \{A\}; [\Gamma, A] \longrightarrow [A \wedge G]}{\mathcal{P}; [\Gamma] \longrightarrow [A \wedge G]}\, mc.$$

In this way, all attempts to prove A on the right will be trivial applications of the initial rule.

When the asynchronous phase of proof search ends, that is, when all the invertible rules have been applied, the decide rules, namely D_l^t and D_r^t, chose a formula on which search should focus. Since logic programs generally contain many formulas, the choice made by these decide rules is a form of *don't know* non-determinism, which is a potential source of backtracking. For example, while the sequent $[A_1, A_1 \supset A_0, A_2 \supset A_0] \longrightarrow [A_0]$ has four formulas on which to focus, a valid LJF proof can be built on by focusing on the formula $A_1 \supset A_0$ (here, A_0, A_1, A_2 are atomic formulas).

While we are mainly interested in the use of tables and not with their discovery, we consider briefly one example of how a table can be built. In particular, a cut-free LJF proof Ξ of $\Gamma \longrightarrow G$ can be made into a table as follows. The table consists of all atoms that are on the right-hand side of some sequent in Ξ. The occurrences of proved atoms in Ξ can be ordered using postorder traversal (*i.e.*, process a node's premises before processing the node). The final order used for the table (which is on atomic formulas and not their occurrences) is then obtained from this postorder traversal by retaining only the first occurrence of any repeated atomic formula. The following proposition shows that it is trivial to extend a multicut derivation that is built in this way from a complete proof: the following definition helps to formalize what we mean as trivial here.

Definition 3. *The* decide-depth *of an LJF^t proof Ξ is the maximum number of occurrences of decide rules (*i.e., *D_r and D_l) on any path from the root to a leaf in Ξ.*

Proposition 4. *Let Ξ be a LJF proof of $\Gamma \longrightarrow G$ and let $\mathcal{T}$ be a table obtained from Ξ using the postorder traversal described above. There exists a proof for $mcd(\mathcal{T}, [\cdot]\Gamma \longrightarrow G)$ such that all of its added subproofs have decide-depth of at most one.*

Proof. Proof by induction on the length of the table's longest chain. □

Given that it is simple to check if a table is derived from a cut-free proof, one might consider that the table is, in fact, a legitimate proof object. Within the proof carrying code framework [12], it might be more interesting to send an ordered collection of atoms to represent a proof than to send some more complex representation of a sequent calculus proof tree. We will return to this aspect of tables in Section 7.

6 Tables of Finite Failures

We now generalize LJF^t by including a proof theoretic notion of *fixed points* that is treated technically using a notion of *definitions*. A *definition* is a countable set of *clauses*, written as $\forall\bar{x}[p\,\bar{t} \stackrel{\Delta}{=} B]$: here p is a predicate, every free variable of B (the *body* of the clause) is also free in the atom $p\,\bar{t}$ (the *head* of the clause), and all variables free in $p\,\bar{t}$ are contained in the list $\bar{x}$ of variables. The symbol $\stackrel{\Delta}{=}$ is not a logical connective but is used to indicate a definitional clause. The left and right introduction rules for defined atoms, namely Def_l and Def_r, are shown in Figure 3. Notice that all free variables in a sequent are *eigenvariables* (no *logical* variables appear here). We shall call LJ^{Δ} the result of adding to Gentzen's LJ calculus the unpolarized versions of Def_l and Def_r (this logic is a first-order version of the logic $FO\lambda^{\Delta}$ in [8,9]). The polarized version of this proof system $LJF^{\Delta t}$ results from adding the inference rules in Figure 3 to LJF^t.

As is shown in [18,6], this notion of definition can yield a proof theoretic approach to negation-as-failure. We shall use this aspect of definitions to extend

$$\frac{\{\mathcal{P};[\Gamma\theta],\Theta\theta,B\theta \longrightarrow \mathcal{R}\theta \mid \theta = mgu(H,A) \text{ for some clause } H \triangleq B\}}{\mathcal{P};[\Gamma],\Theta,A \longrightarrow \mathcal{R}}\ \mathit{Def}_l, A \notin \mathcal{P}$$

$$\frac{\mathcal{P};[\Gamma]\text{—}_{B\theta}\rightarrow}{\mathcal{P};[\Gamma]\text{—}_{A_n}\rightarrow}\ \mathit{Def}_r, A_n \notin \mathcal{P},\ \text{where } H \triangleq B, \text{ and } H\theta = A_n$$

$$\frac{\mathcal{P};[\Gamma]\text{—}_{P_a}\rightarrow}{\mathcal{P};[\Gamma] \longrightarrow [P_a]}\ D_r^t$$

Fig. 3. The rules for introducing atomic formulas and for selecting a formula on the right. Remember that A_n denotes a negative atom and that P_a denotes a positive formula or an atom (positive or negative).

the notion of table of finite success in Section 5 to also contain finite failures. As a consequence, tables will now contain both atoms and negated atoms (*i.e.* literals). The literal $\neg A$ is always of negative polarity since it is defined by the asynchronous formula $A \supset \bot$: notice that the atom A can be either of positive or negative polarity.

The proof theoretic characterization of negation-as-failure is obtained by the Def_l rule. When this rule is used to introduce the atom A on the left of a sequent, a premise for each possible way that the definition could entail A is created in one step. Since all possible instances must be considered, this rule is part of the asynchronous phase of proof search. On the other hand, the Def_r rule's behavior is similar to that of the backward chaining rule of a logic interpreter and, therefore, is applied only in the synchronous phase. We extend the idea of the previous section and consider that backward chaining (that is the Def_r rules) is applied only to negative atoms and forward chaining to positive atoms. Hence, we allow focusing on negative atoms, but do not allow Def_r to be applied on positive atoms.

Proposition 5. *$LJF^{\Delta t}$ is sound and complete with respect to LJ^{Δ}.*

Proof. Soundness follows simply by dropping polarity information from sequents and by using cut-elimination. To prove completeness, assume that a sequent $\Gamma \longrightarrow B$ is provable in LJ^{Δ}. All we need to show is that $[\cdot]\Gamma \longrightarrow B$ (an unfocused sequent with no classified formulas) has an $LJF^{\Delta t}$ proof with an empty table (that is, without any occurrence of the multicut inference rule). As a result, completeness is proved by showing that any cut-free proof in LJ^{Δ} can be made into a focused proof by permutations of inference rules following standard argument lines, such as those in [7,9]. □

Assume again here that all definitions are based on Horn clauses: in particular, all definition clauses are of the form $\forall\bar{x}[A \triangleq G]$ where G is a goal formula defined as at the start of Section 5. For example, the specification of the *path* predicate in Section 2 is written as the two-clause definition

$$\forall x \forall y[\mathit{path}\ x\ y \triangleq \exists z(\mathit{arr}\ x\ z \wedge \mathit{path}\ z\ y)] \quad \text{and} \quad \forall x[\mathit{path}\ x\ x \triangleq \mathit{true}].$$

Since definitions are considered to be global, they are not included in sequents: as a consequence, the left-hand side of sequents contains only the formulas inserted by multicuts.

In the previous section, we used decide-depth as a measure of proof complexity (from the point-of-view of discovering the proof). In the logic considered in this section, it seem more sensible to use the following measure instead.

Definition 4. *The* Def_r-depth *of an* LJF^t *proof* Ξ *is the maximum number of occurrences of the* Def_r *rule on any path from the root to a leaf in* Ξ.

The next proposition, which can be proved by induction on proof trees, guarantees that a sequent that does a right-hand focusing on a literal built from a positive polarity atom yields proofs with small Def_r-depth. Notice that a proof with small Def_r-depth is not necessarily small since the Def_l inference rule can be used without bound: uses of the Def_l, however, are always invertible.

Proposition 6. *Let* $\mathcal{D}$ *be a set of definitions,* Γ *be a set of literals built on positive polarity atoms, and* $L \in \Gamma$. *If* Ξ *is an* $LJF^{\Delta t}$ *proof of* $\mathcal{P}; [\Gamma] -_G\!\!\rightarrow$ *then all occurrences of sequents in* Ξ *that have* L *as their right-focus formula are the conclusion of a proof with* Def_r*-depth at most 1.*

In particular, if L is $\neg A$ and Γ' is Γ with L removed, then an attempt to prove $\mathcal{P}; [\neg A, \Gamma'] -_{\neg A}\!\!\rightarrow$ can only yield an "immediate" proof: the proof of this sequent reduces to $\mathcal{P}; [\neg A, \Gamma', A] \longrightarrow \bot$ and this sequent is provable if and only if there is an atomic B such that $B \supset \bot$ and B are in $\Gamma \cup \{A\}$ (given that B has positive polarity).

If we know that a certain atom A is not intuitionistically provable from a set of assumptions Γ (using finite-failure, for example) then it is possible to organize focused proof search to fail immediately when an attempt to prove A is made. The following proposition, which is proved by induction on the depth of the proof tree, provides that conclusion since it states that such attempts on A are not the conclusion of any $LJF^{\Delta t}$ inference rule.

Proposition 7. *Let* A *be an atom such that* $\Gamma \longrightarrow A$ *is not provable in* LJ^{Δ} *and let* $A \in \mathcal{P}$. *Let* Ξ *be an arbitrary* $LJF^{\Delta t}$ *derivation for* $\mathcal{P}; [\Gamma] -_G\!\!\rightarrow$. *Then all sequents in* Ξ *with right-hand side* A *are open leafs.*

To illustrate this proposition, assume that we have proved the lemma $\neg A$ from Γ. On the right premise of the cut-rule used to insert $\neg A$ as an additional assumption, the atom A is given positive polarity. If one attempts to prove A (with left-hand side Γ) then Def_r cannot be applied. Similarly, the only other way to prove such a sequent is the I_r^t rule, but this implies that the positive atom A is in Γ, which is only possible if A was, in fact, proved from Γ, which is explicitly ruled out. Thus, using polarity, it is possible to "immediately" recognize a failure to prove A.

We can transplant the graph example in Section 2 to this section by mapping the Horn clause specifications for the *path*-atoms and *arr*-atoms into the corresponding definitions. Assume that the table contains only the literal $\neg path\, a_1\, a_5$.

The proof of the multicut derivation for the query $\neg path\ a_0\ a_5$ is illustrated below. Here, $\mathcal{P}$ is the set $\{path\ a_1\ a_5\}$ and A is an eigenvariable of the proof. (The $\star$ annotation indicates that two identical premises have been replaced by one.)

$$
\dfrac{\dfrac{\dfrac{\dfrac{\dfrac{\dfrac{\dfrac{}{\mathcal{P}; [\Gamma, path\ a_1\ a_5] \xrightarrow{path\ a_1\ a_5}}\ I_r^t \qquad \dfrac{\dfrac{}{\mathcal{P}; [\Gamma, path\ a_1\ a_5], \bot \longrightarrow [\bot]}\ false_l^t}{\mathcal{P}; [\Gamma, path\ a_1\ a_5] \xrightarrow{\bot} [\bot]}\ R_l^t}{\mathcal{P}; [\Gamma, path\ a_1\ a_5] \xrightarrow{\neg\, path\ a_1\ a_5} [\bot]}\ \supset_l^t}{\mathcal{P}; [\Gamma, \neg\, path\ a_1\ a_5], path\ a_1\ a_5 \longrightarrow \bot}\ []_l^t, D_l^t}{\mathcal{P}; [\Gamma, \neg\, path\ a_1\ a_5], arr\ a_0\ A, path\ A\ a_5 \longrightarrow \bot}\ Def_l \star}{\mathcal{P}; [\Gamma, \neg\, path\ a_1\ a_5], \exists z[arr\ a_0\ z \wedge path\ z\ a_5] \longrightarrow \bot}\ \exists_l, \wedge_l^t}{\mathcal{P}; [\Gamma, \neg\, path\ a_1\ a_5], path\ a_0\ a_5 \longrightarrow \bot}\ Def_l}{\mathcal{P}; [\Gamma, \neg\, path\ a_1\ a_5] \longrightarrow \neg path\ a_0\ a_5}\ \supset_r^t
$$

7 Table as Proof Objects

We have illustrated how tables can be incorporated into proofs. To what extent can we think of tables as proofs themselves? Of course, this question is best addressed when one knows what one will do with a proof.

In the *proof carrying code* setting [12], proof objects are transmitted together with mobile code to assure that some (safety) properties are satisfied by these programs. Before a client executes the transmitted code the client checks that the proof that that code is carrying proves the program's safety. Thus, proof objects must be engineered so that they are not too large (in order to reduce transmission costs) and not too complex to check (in order to reduce resource requirements on client proof checkers).

Tables might well be a good format for proofs in this setting for several reasons. First, tables represent declarative information and not procedural information: in particular, tables only describe what is provable and does not go into detail about how things are proved. Proof checking can then be organized around simple proof search engines that implement, for example, LJF. The trade-offs between proof size and proof checking time are fairly clear: if the producer of a proof tables all successfully proved atoms (as in Proposition 4) then tables can be large but proof checking can be simple (only proofs of decide-depth 1 must be considered in extending a multicut derivation). On the other hand, if some atomic formulas are not tabled, then the client may have to reprove them: clearly, reproving some atomic formulas might be rather straightforward and something that a client might be willing to do to help reduce the size of a transmitted proof.

In [17], Roychoudhury *et.al.* propose using tables to build *justifications* that can be seen as a kind of proof. In their setting, these proof objects serve to explain why a logic program can or cannot prove a given atom. They argue that their justification can be used within model checkers and parsers. It seems likely that our use of tables as proofs can be used in these settings as well.

We now consider two examples where tables relate to more than just proofs: they can also be simulations (Example 1) and winning strategies (Example 2). These examples also illustrate that non-Horn examples can also be used in the framework that was described above.

Example 1. Encode a noetherian abstract labeled transition system as a definition by writing the transition $P \xrightarrow{A} P'$ as the clause $one(p, a, p') \stackrel{\Delta}{=} true$. McDowell *et.al.* showed in [9] that the additional definition clause

$$\forall P, Q[sim(P, Q) \stackrel{\Delta}{=} \forall A, P'.one(P, A, P') \supset \exists Q'.one(Q, A, Q') \wedge sim(P', Q')]$$

can be used to compute the simulation relation. In particular, processes P is simulated by Q if and only if the atomic formula $sim(P, Q)$ is provable. (Bisimulation can be encoded using a slightly more complex definition.) Moreover, if Ξ is a cut-free proof of that atomic formula and if $\mathcal{S}$ is the set of all pairs $\langle t, s\rangle$ such that Ξ contains a subproof of $sim(t, s)$, then $\mathcal{S}$ is a simulation. Furthermore, let $\preceq$ be the postorder relation on $\mathcal{S}$ derived from Ξ as described in Section 5. Notice that it is now a simple matter to check that $\mathcal{S}$ is, in fact, a simulation by treating it as a table and considering extending its induced multicut derivation to a complete proof. In particular, let $\langle p, q\rangle \in \mathcal{S}$ and let $\mathcal{P} = \{sim(t, s) \mid \langle t, s\rangle \in \mathcal{S}$, and $sim(t, s) \prec sim(p, q)\}$. An attempt to extend the sequent $\mathcal{P}; [\mathcal{P}] \longrightarrow sim(p, q)$ yields a proof of the form

$$\dfrac{\dfrac{\cdots \quad \dfrac{\dfrac{\dfrac{\dfrac{\dfrac{}{\mathcal{P}; [\mathcal{P}] -_{true}\!\rightarrow}\ true_r^t}{\mathcal{P}; [\mathcal{P}] -_{one(q,a,q')}\!\rightarrow}\ Def_r \qquad \dfrac{}{\mathcal{P}; [\mathcal{P}] -_{sim(p',q')}\!\rightarrow}\ I_r^t}{\mathcal{P}; [\mathcal{P}] -_{one(q,a,q')\wedge sim(p',q')}\!\rightarrow}\ \wedge_r^t}{\mathcal{P}; [\mathcal{P}] -_{\exists Q'.one(q,a,p')\wedge sim(p',Q')}\!\rightarrow}\ \exists_r^t}{\mathcal{P}; [\mathcal{P}] \longrightarrow [\exists Q'.one(q, a, Q') \wedge sim(p', Q')]}\ D_r \quad \cdots}{\mathcal{P}; [\mathcal{P}], one(p, A, P') \longrightarrow [\exists Q'.one(q, A, Q') \wedge sim(P', Q')]}\ Def_l}{\mathcal{P}; [\mathcal{P}] \longrightarrow \forall A, P'.\ one(p, A, P') \supset \exists Q'.\ one(q, A, Q') \wedge sim(P', Q')}\ \forall_l^t, \supset_l^t, []_r^t$$

The ellipses represents that there are other premises generated by the Def_l rule that introduces the atom $one(p, A, P')$: there is one premise for each pair $\langle a', p'\rangle$ such that $p \xrightarrow{a'} p'$ (if there is none, then the proof is completed at this point). Notice that the only Def_r rule in this proof is on the one-step transition and since these are given via a simple list of clauses, finding a q' such that $q \xrightarrow{a'} q'$ is a simple computation.

Example 2. Consider a game between two players, named 1 and 2, who alternate in playing (consider tic-tac-toe) and that one player wins when the other player cannot move. We assume that the state of the game is encoded as a term in the logic and that the binary predicate $move(P, Q)$ encodes the fact that there is move from position P to Q. Furthermore, assume that there are no infinite plays. Then there is a winning strategy from the position P if and only if the atom $win(P)$ is provable from a definition that includes the clause

$$\forall P[win(P) \stackrel{\Delta}{=} \forall P'.\ move(P, P') \supset \exists Q.\ move(P', Q) \wedge win(Q)]$$

as well as the (Horn clause) definition of $move(P, Q)$. As with the previous example, let Ξ be a proof of the atom $win(p)$, let $\mathcal{W}$ be the set of atoms of the form $win(P)$ that are proved in subproofs of Ξ, and let $\preceq$ be the postorder traversal ordering of $\mathcal{W}$ based on Ξ. It is now a simple matter to verify that $\mathcal{W}$ encodes a winning strategy: simply build the multicut derivation associated to the table $\mathcal{W}$ and extend it to a complete proof. This later step is essentially the same kind of restricted proof search that is presented for the previous example based on simulation.

8 Conclusions and Future Work

This paper is part of a project to use focused proofs as a framework for relating a variety of proof representations. Here we showed a connection between tables and sequent calculus proofs. We expect that similar results will also allow us to relate sequent calculus proofs to other proof objects, e.g., the *oracles* of Necula and Rahul [13] and the fixpoints in the *Abstraction Carrying Code* [1].

Clearly, it should be possible to put more in tables than literals: for example, it seems easy to account for universally quantified literals in table. The Twelf system [14,15] and the Bedwyr system [4] are two examples of implementations of logics in which tables of atoms are used to improve proof search but where goals can be much richer, including specifically universal quantifiers and implications. It would be interesting to find a way to extend our work here to allow such goal formulas to be tabled as well.

In this paper, we investigated the problem of automating and checking the use of previously proved lemmas (or table) in the proof of some main theorem. After motivating the use of focusing and the polarity of atoms, we presented two focused systems; one involving Horn clauses, and another adding negation-as-failure. We also showed that by using a partial ordering relation over the elements of the table, we could define a multicut derivation which could be easily extended to a proof. With these systems, we were also able to give a declarative interpretation to the *memoization* procedure. And finally, we proposed to use tables as a proof objects and illustrated this with some examples.

Acknowledgments. We thank David Baelde and the anonymous reviewers for their comments on an earlier draft of this paper. This work has been supported in part by INRIA through the "Equipes Associées" Slimmer and by the Information Society Technologies programme of the European Commission, Future and Emerging Technologies under the IST-2005-015905 MOBIUS project.

References

1. Albert, E., Puebla, G., Hermenegildo, M.V.: Abstraction-carrying code. In: Baader, F., Voronkov, A. (eds.) LPAR 2004. LNCS (LNAI), vol. 3452, pp. 380–397. Springer, Heidelberg (2005)
2. Anderson, A.R., Belnap, N.D.: Entailment: The Logic of Relevance and Necessity. Princeton University Press, Princeton (1975)
3. Andreoli, J.-M.: Logic programming with focusing proofs in linear logic. J. of Logic and Computation 2(3), 297–347 (1992)
4. Baelde, D., Gacek, A., Miller, D., Nadathur, G., Tiu, A.: The Bedwyr system for model checking over syntactic expressions. In: CADE-21 (to appear, 2007)
5. Gentzen, G.: Investigations into logical deductions. In: Szabo, M.E. (ed.) The Collected Papers of Gerhard Gentzen, pp. 68–131. North-Holland, Amsterdam (1969)
6. Girard, J.-Y.: A fixpoint theorem in linear logic. An email posting to the mailing list linear@cs.stanford.edu (February 1992)
7. Liang, C., Miller, D.: Focusing and polarization in intuitionistic logic. In: Duparc, J., Henzinger, T.A. (eds.) CSL 2007. LNCS, vol. 4646, pp. 451–465. Springer, Heidelberg (2007)
8. McDowell, R., Miller, D.: Cut-elimination for a logic with definitions and induction. Theoretical Computer Science 232, 91–119 (2000)
9. McDowell, R., Miller, D., Palamidessi, C.: Encoding transition systems in sequent calculus. Theoretical Computer Science 294(3), 411–437 (2003)
10. Miller, D.: A logical analysis of modules in logic programming. Journal of Logic Programming 6(1-2), 79–108 (1989)
11. Miller, D., Nadathur, G., Pfenning, F., Scedrov, A.: Uniform proofs as a foundation for logic programming. Annals of Pure and Applied Logic 51, 125–157 (1991)
12. Necula, G.C.: Proof-carrying code. In: Conference Record of the 24th Symposium on Principles of Programming Languages 97, Paris, France, pp. 106–119. ACM Press, New York (1997)
13. Necula, G.C., Rahul, S.P.: Oracle-based checking of untrusted software. In: POPL, pp. 142–154 (2001)
14. Pfenning, F., Schürmann, C.: System description: Twelf — A meta-logical framework for deductive systems. In: Ganzinger, H. (ed.) Automated Deduction - CADE-16. LNCS (LNAI), vol. 1632, pp. 202–206. Springer, Heidelberg (1999)
15. Pientka, B.: Tabling for higher-order logic programming. In: 20th International Conference on Automated Deduction, Talinn, Estonia, pp. 54–69. Springer, Heidelberg (2005)
16. Ramakrishna, Y.S., Ramakrishnan, C.R., Ramakrishnan, I.V., Smolka, S.A., Swift, T., Warren, D.S.: Efficient model checking using tabled resolution. In: Grumberg, O. (ed.) CAV 1997. LNCS, vol. 1254, pp. 143–154. Springer, Heidelberg (1997)
17. Roychoudhury, A., Ramakrishnan, C.R., Ramakrishnan, I.V.: Justifying proofs using memo tables. In: PPDP, pp. 178–189 (2000)
18. Schroeder-Heister, P.: Definitional reflection and the completion. In: Dyckhoff, R. (ed.) ELP 1993. LNCS, vol. 798, pp. 333–347. Springer, Heidelberg (1994)
19. Slaney, J.: Solution to a problem of Ono and Komori. Journal of Philosophic Logic 18, 103–111 (1989)

A Cut-Free and Invariant-Free Sequent Calculus for PLTL*

Joxe Gaintzarain[1], Montserrat Hermo[1], Paqui Lucio[1], Marisa Navarro[1], and Fernando Orejas[2]

[1] Dpto de Lenguajes y Sistemas Informáticos, Universidad del País Vasco, 20080-San Sebastián, Spain
[2] Dpto de Lenguajes y Sistemas Informáticos, Universidad Politécnica de Catalunya, 08034-Barcelona, Spain

Abstract. Sequent calculi usually provide a general deductive setting that uniformly embeds other proof-theoretical approaches, such as tableaux methods, resolution techniques, goal-directed proofs, etc. Unfortunately, in temporal logic, existing sequent calculi make use of a kind of inference rules that prevent the effective mechanization of temporal deduction in the general setting. In particular, temporal sequent calculi either need some form of cut, or they make use of invariants, or they include infinitary rules. This is the case even for the simplest kind of temporal logic, propositional linear temporal logic (PLTL). In this paper, we provide a complete finitary sequent calculus for PLTL, called $\mathcal{FC}$, that not only is cut-free but also invariant-free. In particular, we introduce new rules which provide a new style of temporal deduction. We give a detailed proof of completeness.

1 Introduction

The development of automated deduction systems for temporal logic has followed two main proof-theoretical approaches: tableaux (see [12]) and resolution (see [1]), which are both refutational proof methods. Sequent calculi are usually used to provide a general deductive setting that uniformly embeds refutational methods and other deduction techniques such as goal-directed proofs or natural deduction. In temporal logic, tableaux methods generate graphs instead of the classical trees and resolution methods require more involved normal forms and inference rules than the classical clausal form and the classical resolution rule. This complicates the association of a sequent calculus proof to each tableaux graph or each resolution proof. In addition, existing sequent calculi for temporal logic (cf. [6,8,11]) make use of a kind of inference rules that prevents this correspondence and complicates the implementation of temporal deduction in the general setting. In particular, temporal sequent calculi either need some form of cut (classical cut or invariant-based cut) or they include infinitary rules. Cut rules imply the "invention" of lemmata, called cut formula, for their application. Invariants are particular cut formulas for proving temporal eventualities. This is the case even for the simplest kind

* This work has been partially supported by Spanish Project TIN2004-079250-C03.

J. Duparc and T.A. Henzinger (Eds.): CSL 2007, LNCS 4646, pp. 481–495, 2007.

of temporal logic, propositional linear temporal logic (PLTL). In this sense, the formulation of a cut-free, invariant-free finitary sequent calculus, can be considered a relevant open problem that is solved in this paper.

More precisely, in [6] and [11], two sequent calculi for PLTL with invariant-based rules are presented. In fact, in both approaches, they present a system including also a cut rule and then prove cut elimination. However, invariant-based rules for temporal connectives cannot be avoided. In [8] various sequent calculi are presented for PLTL without the until operator (this means that the logic considered has a limited expressive power). He provides completeness and cut-elimination proofs, together with various interesting reductions among the various calculi. However, every calculus includes either some infinitary rule or some invariant-based rule. Other proof-theoretic approaches for PLTL include its first axiomatization á la Hilbert presented in [2], and the first detailed description of a tableaux method for deciding the satisfiability of any PLTL-formula presented in [12]. The satisfiability problem for PLTL is PSPACE-complete (cf. [10]). See [9] for a good survey about theorem-proving in PLTL and its extensions.

In this paper, we provide a complete finitary sequent calculus for PLTL, called $\mathcal{FC}$, that not only is cut-free but also invariant-free. In particular, we introduce a new rule for the until operator that provides a new style of temporal deduction for eventualities. Moreover, deduction for "always"-formulas is also affected by this new style.

In order to show completeness, we have not followed the standard approach of, first, proving completeness including a cut rule in the calculus and, then, showing a cut elimination result (cf. [3]). Actually, the first part of that approach, proving completeness of $\mathcal{FC}$ plus the cut rule, is quite easy. In particular, just with the rules in $\mathcal{FC}$ it is easy to derive every axiom (except the modus ponens rule) in the system proved complete in [5]. Obviously, with the addition of the cut rule one can easily derive modus ponens. Unfortunately, we have been unable to directly prove cut elimination. Instead, we have directly proved the completeness of $\mathcal{FC}$, which indirectly means that the cut rule is not needed. The proof is partially inspired by the tableaux method proposed in [5]. In particular their notion of maximal strongly connected components has been very useful in our proof. However, unlike [5], we use a filtration technique for constructing models from saturated consistent sets of formulas (as states).

The paper is organized as follows. Section 2 is a basic introduction to PLTL. In sections 3 and 4 we introduce our calculus $\mathcal{FC}$, proving its soundness. More precisely, in section 3 we describe the basic rules for describing the next ($\circ$) and until ($\mathcal{U}$) connectives, while in section 4 we present some useful derived rules describing, in particular, the rest of the temporal connectives. Section 5 presents the completeness proof of $\mathcal{FC}$. Finally, in section 6 we draw some concluding remarks.

2 PLTL: Language and Model Theory

A PLTL-formula is built using the constant proposition F, propositional variables (denoted by lowercase letters $p, q, \ldots$) from a set Prop, the classical connectives $\neg$ and $\vee$, and the temporal connectives $\circ$ and $\mathcal{U}$. A lowercase Greek letter $(\varphi, \psi, \chi, \gamma, \ldots)$ denotes a formula and an uppercase one $(\Phi, \Delta, \Gamma, \Psi, \Omega, \ldots)$ denotes a finite set of PLTL-formulas. PLTL-formulas of the form p and $\neg p$, where $p \in$ Prop, are called *literals* and PLTL-formulas that do not begin with the connective $\neg$ are called *positive*. As usual other connectives can be defined in terms of the previous ones: $\mathrm{T} \equiv \neg\mathrm{F}$, $\varphi \wedge \psi \equiv \neg(\neg\varphi \vee \neg\psi)$, $\diamond\varphi \equiv \mathrm{T}\,\mathcal{U}\,\varphi$, $\Box\varphi \equiv \neg\diamond\neg\varphi$. PLTL-formulas of the form $\varphi\,\mathcal{U}\,\psi$ and $\diamond\varphi$ are called *eventualities*. In the rest of this paper, we simply say *formula* instead of PLTL-formula. The operator next translates any set of formulas into another (possibly empty) set of formulas $\mathsf{next}(\Phi) = \{\varphi \mid \circ\varphi \in \Phi\}$.

It is well known that PLTL is a non-compact logic. As a consequence, strong completeness requires an infinitary proof system, whose deduction rules may require infinitely many premises. Our calculus is finitary, hence, as usual (see, e.g. [6], [2] and [11]), our completeness result is in this sense, weak. Therefore, along this paper, every set of PLTL-formulas is assumed to be finite. Given any (finite) set $\Phi = \{\varphi_1, \ldots, \varphi_n\}$ we will use $\Phi^\neg$ to denote the formula $\neg\varphi_1 \vee \ldots \vee \neg\varphi_n$. In particular, $\Phi^\neg$ is the constant F when Φ is empty.

Definition 1. *A* PLTL-structure $\mathcal{M}$ *is a pair* $(S_\mathcal{M}, V_\mathcal{M})$ *such that* $S_\mathcal{M}$ *is a denumerable sequence of states* $s_0, s_1, s_2, \ldots$ *and* $V_\mathcal{M}$ *is a map* $V_\mathcal{M} : S_\mathcal{M} \to 2^{\mathsf{Prop}}$. ■

Intuitively, $V_\mathcal{M}$ specifies which atomic propositions are (necessarily) true in each state.

Definition 2. *The truth of a formula* φ *in the state* s_j *of a* PLTL*-structure* $\mathcal{M}$*, which is denoted by* $\langle\mathcal{M}, j\rangle \models \varphi$*, is inductively defined as follows:*

- $\langle\mathcal{M}, j\rangle \not\models \mathrm{F}$
- $\langle\mathcal{M}, j\rangle \models p$ *iff* $p \in V_\mathcal{M}(s_j)$ *for* $p \in \mathsf{Prop}$
- $\langle\mathcal{M}, j\rangle \models \neg\varphi$ *iff* $\langle\mathcal{M}, j\rangle \not\models \varphi$
- $\langle\mathcal{M}, j\rangle \models \varphi \vee \psi$ *iff* $(\langle\mathcal{M}, j\rangle \models \varphi$ *or* $\langle\mathcal{M}, j\rangle \models \psi)$
- $\langle\mathcal{M}, j\rangle \models \circ\varphi$ *iff* $\langle\mathcal{M}, j+1\rangle \models \varphi$
- $\langle\mathcal{M}, j\rangle \models \varphi\,\mathcal{U}\,\psi$ *iff* $\langle\mathcal{M}, k\rangle \models \psi$ *for some* $k \geq j$ *and* $\langle\mathcal{M}, i\rangle \models \varphi$ *for every* $j \leq i < k$. ■

This is extended to sets in the usual way: $\langle\mathcal{M}, j\rangle \models \Phi$ iff $\langle\mathcal{M}, j\rangle \models \varphi$ for all $\varphi \in \Phi$. We say that $\mathcal{M}$ is a model of Φ, in symbols $\mathcal{M} \models \Phi$, iff $\langle\mathcal{M}, 0\rangle \models \Phi$. A satisfiable set of PLTL-formulas has at least one model, otherwise it is unsatisfiable. The *logical consequence* relation between a set of formulas Φ and a formula χ, denoted as $\Phi \models \chi$, is defined in the following way:

$$\Phi \models \chi \text{ iff for every PLTL-structure } \mathcal{M} \text{ and every } j \in I\!N : \\ \text{if } \langle\mathcal{M}, j\rangle \models \Phi \text{ then } \langle\mathcal{M}, j\rangle \models \chi$$

3 The Sequent Calculus $\mathcal{FC}$

In this section, we introduce a sound and complete sequent calculus, called $\mathcal{FC}$, that is fully free of cut. That is, in $\mathcal{FC}$ there are neither classical cut rules nor invariant-based rules for temporal connectives. The calculus $\mathcal{FC}$ uses asymmetric sequents, i.e. sequents formed by a set of assumptions and a single conclusion. The former set is called the antecedent of the sequent and the latter formula is called the consequent. We write

Classical connectives rules

$$(\neg L)\ \frac{\Delta \vdash \varphi}{\Delta, \neg\varphi \vdash \chi} \qquad (R\neg)\ \frac{\Delta, \varphi \vdash \mathbf{F}}{\Delta \vdash \neg\varphi} \qquad (\vee L)\frac{\Delta, \varphi \vdash \chi \quad \Delta, \psi \vdash \chi}{\Delta, \varphi \vee \psi \vdash \chi} \qquad (R\vee)\frac{\Delta \vdash \varphi}{\Delta \vdash \varphi \vee \psi} \quad \frac{\Delta \vdash \psi}{\Delta \vdash \varphi \vee \psi}$$

Temporal connectives rules

$$(R\circ L)\ \frac{\mathsf{next}(\Delta) \vdash \varphi}{\Delta \vdash \circ\varphi} \qquad (\neg\circ L)\ \frac{\Delta, \circ\neg\varphi \vdash \chi}{\Delta, \neg\circ\varphi \vdash \chi} \qquad (R\circ\neg)\ \frac{\Delta \vdash \neg\circ\varphi}{\Delta \vdash \circ\neg\varphi}$$

$$(\mathcal{U}\,L)_i \frac{\Delta, \psi \vdash \chi \quad \Delta, \varphi, \neg\psi, \circ(\delta_i \,\mathcal{U}\, \psi) \vdash \chi}{\Delta, \varphi \,\mathcal{U}\, \psi \vdash \chi} : \begin{cases} \delta_1 = \varphi \\ \delta_2 = \varphi \wedge (\Delta^{\neg} \vee \chi) \end{cases} \qquad (R\,\mathcal{U}\,)\ \frac{\Delta, \neg\varphi \vdash \psi \quad \Delta, \varphi, \neg\circ(\varphi \,\mathcal{U}\, \psi) \vdash \psi}{\Delta \vdash \varphi \,\mathcal{U}\, \psi}$$

Structural rules

$$(As)\ \Delta, \varphi \vdash \varphi \qquad (Wk)\ \frac{\Delta \vdash \chi}{\Delta, \Delta' \vdash \chi} \qquad (Cd)\ \frac{\Delta, \neg\varphi \vdash \mathbf{F}}{\Delta \vdash \varphi} \qquad (\circ\mathbf{F})\ \frac{\Delta \vdash \circ\mathbf{F}}{\Delta \vdash \chi}$$

Fig. 1. The sequent calculus $\mathcal{FC}$

$\Delta \vdash \chi$ to represent a sequent whose antecedent is Δ and whose consequent is χ. We have preferred to formulate the calculus by means of asymmetric (or one-conclusion) sequents, instead of symmetric (multiple-conclusioned) sequents, because the former are closer to natural deduction and captures better our intuition in logical reasoning. A multiple-conclusioned system can be easily obtained from $\mathcal{FC}$. For getting rid of some rules and giving a more compact presentation, we could also take the one-sided sequent approach (also known as Tait-style). However, it requires to keep formulas in negation normal form and results a bit more unusual and unnatural at first sigth.

The calculus $\mathcal{FC}$ consists of the primitive rules that are summarized in Fig. 1. We have split these rules into three packages. Two of them consist of rules for classical and temporal connectives, respectively. These rules follow the traditional style of introduction of the connective in the left/right part of the sequent. In addition we need some structural rules which form the third package.

The rules for classical connectives are classical. With respect to the temporal connectives, the three rules for the next operator, $(R\circ L)$, $(\neg\circ L)$ and $(R\circ\neg)$, are well known in the literature of PLTL. Besides, by means of $(\mathcal{U}\,L)_i$ we represent two rules for two different δ_i where $i = 1$ or $i = 2$. The rules $(\mathcal{U}\,L)_1$ and $(R\,\mathcal{U}\,)$ are also well known. Both are included in the existing Gentzen systems where other invariant-based rules for the until operator are given (cf.[6,11]). Instead, we add a rule $(\mathcal{U}\,L)_2$ which does not require invariant generation. This rule $(\mathcal{U}\,L)_2$, which up to our knowledge is completely new, can be considered quite peculiar, since the second premise includes a formula which depends on the whole conclusion of the rule.[1] In addition $(\mathcal{U}\,L)_2$ leads to a new deduction style that is opposite, in some sense, to the invariant-based reasoning.

[1] Remember that Δ is always assumed to be a finite set and that $\Delta^{\neg}$ is $\mathbf{F}$ whenever Δ is empty.

The underlying idea in the rule $(\mathcal{U}\, L)_2$ is that the sequences of states along which the satisfaction of an eventuality is delayed should be ever-changing sequences. In the proof of the soundness theorem, we show in detail that the rule $(\mathcal{U}\, L)_2$ is correct. We believe that this correctness proof reflects the intuition behind the rule.

Regarding structural rules, $(\circ\mathrm{F})$ is the only rule that is not a classical rule. At first sight, the introduction of the weakening rule (Wk) in the structural package could be surprising since very commonly (Wk) is an elementary property and an admissible rule. However, the form of the rule $(\mathcal{U}\, L)_2$ prevents that traditional methods for proving admissibility (cf. [7]) could be applied to the calculus $\mathcal{FC}$. Although experimental work (see Example 6) indicates that (Wk) could be admissible in $\mathcal{FC}$, this is still an interesting open problem. This work is mainly focused in completeness, the minimality of the calculus remains as future work.

An $\mathcal{FC}$-proof is a tree (written right side up, with its root on the bottom) labelled with sequents. The sequent to be proved labels its root, the leaves are labelled with axioms (which are rules without premises), and all the local subtrees must be accepted by some inference rule in $\mathcal{FC}$. In the Examples 4 and 5, we give a sequence of sequents that ends with the root (the proved sequent) and add additional information for describing the structure of the tree.

The expression $\Gamma \vdash_{\mathcal{FC}} \chi$ is used to denote that there exists an $\mathcal{FC}$-proof of the sequent $\Gamma \vdash \chi$. We say that a set of formulas Γ is $\mathcal{FC}$-consistent if and only if $\Gamma \not\vdash_{\mathcal{FC}} \mathrm{F}$. The soundness of $\mathcal{FC}$ means that every $\mathcal{FC}$-provable sequent, namely $\Gamma \vdash \chi$, is correct regarding to logical consequence. In particular, every satisfiable set of formulas is $\mathcal{FC}$-consistent.

Theorem 3. *For any set of formulas $\Gamma \cup \{\chi\}$, if $\Gamma \vdash_{\mathcal{FC}} \chi$ then $\Gamma \models \chi$.*

Proof. By induction on the length of the $\mathcal{FC}$-proof, it suffices to prove that every primitive rule of $\mathcal{FC}$ (see Fig. 1) is correct in the sense of preserving the logical consequence relation between the antecedent and the consequent.

Now, the correctness proof of most rules is just routine. Actually, the only correctness proof that poses some difficulties is the proof of the rule $(\mathcal{U}\, L)_2$. Hence, we only give the details for this rule.

We will show that, if we assume that $\Delta \cup \{\varphi\mathcal{U}\psi, \neg\chi\}$ is satisfiable, then we would build a countermodel for some of the two premises of the rule $(\mathcal{U}\, L)_2$. Let $\langle \mathcal{M}, i\rangle \models \Delta \cup \{\varphi\mathcal{U}\psi, \neg\chi\}$ and s_1 the least $s \geq i$ such that $\langle \mathcal{M}, s\rangle \models \psi$. If $s_1 = i$ then $\langle \mathcal{M}, i\rangle$ serves as countermodel for the first premise. Otherwise, if $s_1 > i$, let s_2 be the greatest s such that $i \leq s < s_1$ and $\langle \mathcal{M}, s\rangle \models \Delta \cup \{\varphi\mathcal{U}\psi, \neg\chi\}$. As a consequence of the choice of s_1 and s_2, it holds $\langle \mathcal{M}, s_2\rangle \models \circ((\varphi \wedge (\Delta^{\neg} \vee \chi))\mathcal{U}\psi)$. Then, $\langle \mathcal{M}, s_2\rangle$ is a countermodel of the second premise. ∎

4 Derived Rules and Proofs

In this section we present some derived rules that can be used as a shortcut for several lines of primitive-rules-only proofs. Actually, some of these rules are used below in the proof of the completeness theorem.

The first group of derived rules, including the contraposition rules $(Cp1)$ and $(Cp2)$, can be derived in a standard way from the classical primitive rules in $\mathcal{FC}$.

$$(Cp1)\ \frac{\Delta, \neg\varphi \vdash \psi}{\Delta, \neg\psi \vdash \varphi} \qquad (Cp2)\ \frac{\Delta, \varphi \vdash \psi}{\Delta, \neg\psi \vdash \neg\varphi} \qquad (\mathbf{F}L)\ \Delta, \mathbf{F} \vdash \chi$$

$$(CdL)\ \Delta, \varphi, \neg\varphi \vdash \chi \qquad (\neg\neg L)\ \frac{\Delta, \varphi \vdash \chi}{\Delta, \neg\neg\varphi \vdash \chi} \qquad (\neg \vee L)\ \frac{\Delta, \neg\varphi, \neg\psi \vdash \chi}{\Delta, \neg(\varphi \vee \psi) \vdash \chi}$$

For the temporal connectives, the following derived rules will be used later:

$$(\circ L)\ \frac{\mathsf{next}(\Delta) \vdash \mathbf{F}}{\Delta \vdash \chi} \qquad (\neg\ \mathcal{U} L)\ \frac{\begin{array}{c}\Delta, \neg\varphi, \neg\psi \vdash \chi \\ \Delta, \varphi, \neg\psi, \neg\circ(\varphi \mathcal{U} \psi) \vdash \chi\end{array}}{\Delta, \neg(\varphi \mathcal{U} \psi) \vdash \chi}$$

It is easy to check that $(\circ L)$ is derived by $(R\circ L)$ and $(\circ\mathbf{F})$ and $(\neg\ \mathcal{U} L)$ by $(Cp1)$ and $(R\mathcal{U})$.

Other derived rules allow us to reason about the rest of the classical or temporal connectives, which have been introduced as a shorthand to abbreviate some formulas. For instance, since $\varphi \wedge \psi$ stands for $\neg(\neg\varphi \vee \neg\psi)$, the classical sequent rules for $\wedge$ can be derived:

$$(\wedge L)\ \frac{\Delta, \varphi, \psi \vdash \chi}{\Delta, \varphi \wedge \psi \vdash \chi} \qquad (R\wedge)\ \frac{\Delta \vdash \varphi \quad \Delta \vdash \psi}{\Delta \vdash \varphi \wedge \psi}$$

Likewise, using the abbreviations $\Diamond\varphi$ and $\Box\varphi$ for $\mathbf{T}\mathcal{U}\,\varphi$ and $\neg\Diamond\neg\varphi$, respectively, we are also able to derive the following useful rules:

$$(\Diamond L)_i\ \frac{\begin{array}{c}\Delta, \varphi \vdash \chi \\ \Delta, \neg\varphi, \circ(\delta_i \mathcal{U} \varphi) \vdash \chi\end{array}}{\Delta, \Diamond\varphi \vdash \chi} : \begin{cases}\delta_1 = \mathbf{T} \\ \delta_2 = \Delta^{\neg} \vee \chi\end{cases} \qquad (R\Diamond)\ \frac{\Delta, \neg\circ\Diamond\varphi \vdash \varphi}{\Delta \vdash \Diamond\varphi}$$

$$(\Box L)\ \frac{\Delta, \varphi, \circ\Box\varphi \vdash \chi}{\Delta, \Box\varphi \vdash \chi} \qquad (R\Box)_i\ \frac{\begin{array}{c}\Delta \vdash \varphi \\ \Delta, \circ(\delta_i \mathcal{U} \neg\varphi) \vdash \neg\varphi\end{array}}{\Delta \vdash \Box\varphi} : \begin{cases}\delta_1 = \mathbf{T} \\ \delta_2 = \Delta^{\neg}\end{cases}$$

Note also that, by $(\Box L)$ and $(\neg\circ L)$, the following contradiction rule is also derivable:

$$(Cd\Box)\ \Delta, \Box\varphi, \neg\circ\Box\varphi \vdash \chi.$$

It is well known that the until operator $\mathcal{U}$ is not expressible in temporal logic with only $\circ$, $\Box$, and $\Diamond$ as temporal operators (cf. [4,2]). As a consequence a complete calculus for the sublogic that uses $\Diamond$ instead of $\mathcal{U}$ cannot be derived (by abbreviation) from $\mathcal{FC}$, since the rule $(\Diamond L)_2$ needs the until operator for expressing its second premise.

Let us now illustrate the $\mathcal{FC}$-style of natural reasoning by means of some examples of $\mathcal{FC}$-proofs. In order to allow easier reading, we have underlined, at each step, the formulas that are related with the applied deduction rule.

Example 4. The following proof shows that $p, \Box(\neg p \vee \circ p) \vdash_{\mathcal{FC}} \Box p$. This is a typical property of *induction on time*. We have used $\Box\varphi$ to abbreviate $\Box(\neg p \vee \circ p)$.

1. $- \underline{p}, \Box\varphi \vdash \underline{p}$ by (As)
2. $- \underline{p}, \Box\varphi, \underline{\neg p}, \neg\neg p, \circ((\neg p \vee \neg\Box\varphi)\,\mathcal{U}\,\neg p) \vdash \mathbf{F}$ by (CdL)
3. $- p, \underline{\Box\varphi, \neg\Box\varphi}, \neg\neg p, \circ((\neg p \vee \neg\Box\varphi)\,\mathcal{U}\,\neg p) \vdash \mathbf{F}$ by (CdL)
4. $- \underline{p}, \Box\varphi, \underline{\neg p} \vdash \mathbf{F}$ by (CdL)
5. $- \underline{p}, \Box\varphi, \underline{\neg p \vee \neg\Box\varphi}, \neg\neg p, \circ((\neg p \vee \neg\Box\varphi)\,\mathcal{U}\,\neg p) \vdash \mathbf{F}$ by 2, 3 and $(\vee L)$
6. $- p, \Box\varphi, \underline{(\neg p \vee \neg\Box\varphi)\,\mathcal{U}\,\neg p} \vdash \mathbf{F}$ by 4, 5 and $(\mathcal{U}\,L)_1$
7. $- p, \underline{\neg p}, \circ\Box\varphi, \circ((\neg p \vee \neg\Box\varphi)\,\mathcal{U}\,\neg p) \vdash \underline{\neg p}$ by (As)
8. $- p, \underline{\circ p, \circ\Box\varphi, \circ((\neg p \vee \neg\Box\varphi)\,\mathcal{U}\,\neg p)} \vdash \neg p$ by 6 and $(\circ L)$
9. $- p, \underline{\neg p \vee \circ p}, \circ\Box\varphi, \circ((\neg p \vee \neg\Box\varphi)\,\mathcal{U}\,\neg p) \vdash \neg p$ by 7, 8 and $(\vee L)$
10. $- p, \underline{\Box\varphi}, \circ((\neg p \vee \neg\Box\varphi)\,\mathcal{U}\,\neg p) \vdash \neg p$ by 9 and $(\Box L)$
11. $- p, \Box\varphi \vdash \underline{\Box p}$ by 1, 10 and $(R\Box)_2$. ■

It is worthy to note that $\{\Box\beta, \circ((\varphi \vee \neg\Box\beta)\,\mathcal{U}\,\psi)\}$ and $\{\Box\beta, \circ(\varphi\,\mathcal{U}\,\psi)\}$ are equivalent sets of formulas. As a consequence, the above proof could be simplified if the sequent to be derived at step 10 were $p, \Box\varphi, \circ(\neg p\,\mathcal{U}\,\neg p) \vdash \neg p$ instead of

$$p, \Box\varphi, \circ((\neg p \vee \neg\Box\varphi)\,\mathcal{U}\,\neg p) \vdash \neg p.$$

A practical implementation of $\mathcal{FC}$ should apply the rules $(\mathcal{U}\,L)_2$ (and also $(\Box L)_2$ and $(\Diamond L)_2$) yielding as subgoal $\circ(\varphi\,\mathcal{U}\,\psi)$ instead of $\circ((\varphi \vee \neg\Box\beta)\,\mathcal{U}\,\psi)$. In general, the rule $(\mathcal{U}\,L)_2$ should take into account the equivalence of the following two sets of formulas:

$$\{\Box\alpha, \neg(\alpha\,\mathcal{U}\,\beta), \circ((\varphi \vee (\alpha\,\mathcal{U}\,\beta))\,\mathcal{U}\,\psi)\} \text{ and } \{\Box\alpha, \neg(\alpha\,\mathcal{U}\,\beta), \circ(\varphi\,\mathcal{U}\,\psi)\}.$$

Note that the former pair of equivalent sets is a particular case of the latter one.

Example 5. The following is an $\mathcal{FC}$-proof of the sequent $p\,\mathcal{U}\,q, \neg q \vdash \circ\Diamond q$:

1. $- \underline{q, \neg q} \vdash \circ\Diamond q$ by (CdL)
2. $- \underline{q}, \neg\circ\Diamond q \vdash \underline{q}$ by (As)
3. $- p, \underline{\circ\Diamond q}, \neg q, \circ((p \wedge (\neg\neg q \vee \circ\Diamond q))\,\mathcal{U}\,q), \underline{\neg\circ\Diamond q} \vdash q$ by (CdL)
4. $- p, \underline{\neg\neg q, \neg q}, \circ((p \wedge (\neg\neg q \vee \circ\Diamond q))\,\mathcal{U}\,q), \neg\circ\Diamond q \vdash q$ by (CdL)
5. $- p, \underline{\neg\neg q \vee \circ\Diamond q}, \neg q, \circ((p \wedge (\neg\neg q \vee \circ\Diamond q))\,\mathcal{U}\,q), \neg\circ\Diamond q \vdash q$ by 3, 4 and $(\vee L)$
6. $- \underline{p \wedge (\neg\neg q \vee \circ\Diamond q)}, \neg q, \circ((p \wedge (\neg\neg q \vee \circ\Diamond q))\,\mathcal{U}\,q), \neg\circ\Diamond q \vdash q$ by 5 and $(\wedge L)$
7. $- \underline{(p \wedge (\neg\neg q \vee \circ\Diamond q))\,\mathcal{U}\,q}, \neg\circ\Diamond q \vdash q$ by 2, 6 and $(\mathcal{U}\,L)_1$
8. $- (p \wedge (\neg\neg q \vee \circ\Diamond q))\,\mathcal{U}\,q \vdash \underline{\Diamond q}$ by 7 and $(R\Diamond)$
9. $- p, \neg q, \underline{\circ((p \wedge (\neg\neg q \vee \circ\Diamond q))\,\mathcal{U}\,q)} \vdash \underline{\circ\Diamond q}$ by 8 and $(R\circ L)$
10. $- \underline{p\,\mathcal{U}\,q}, \neg q \vdash \circ\Diamond q$ by 1, 9 and $(\mathcal{U}\,L)_2$

It is easy to check that using only the rule $(\mathcal{U}\,L)_1$ we cannot prove the sequent. ∎

Example 6. Consider the sequent $q, p\mathcal{U}\,\mathbf{F} \vdash \mathbf{F}$. It is easy to give an $\mathcal{FC}$-proof of $p\mathcal{U}\,\mathbf{F} \vdash \mathbf{F}$ since by $(\mathcal{U}\,L)_2$ it should be proved $\mathbf{F} \vdash \mathbf{F}$ and $p, \neg\mathbf{F}, \circ(p \wedge (\mathbf{F} \vee \mathbf{F}))\mathcal{U}\,\mathbf{F} \vdash \mathbf{F}$. The latter is easily proved by $(R\circ L)$ and $(\mathcal{U}\,L)_1$. Finally, by (Wk), $q, p\mathcal{U}\,\mathbf{F} \vdash \mathbf{F}$ is derived from $p\mathcal{U}\,\mathbf{F} \vdash \mathbf{F}$.

It could be believed that (Wk) is essential for proving this kind of sequents, where some part of the antecedent is unnecessary for entailing the consequent. However, the following is a scketch of an $\mathcal{FC}$-proof of the sequent $q, p\mathcal{U}\,\mathbf{F} \vdash \mathbf{F}$ that does not use the rule (Wk):
The first two main goals are: $q, \mathbf{F} \vdash \mathbf{F}$ and $q, p, \neg\mathbf{F}, \circ((p \wedge (\neg q \vee \mathbf{F}))\mathcal{U}\,\mathbf{F}) \vdash \mathbf{F}$. The former is an instance of (As), while the latter reduces to

$$(p \wedge (\neg q \vee \mathbf{F}))\mathcal{U}\,\mathbf{F} \vdash \mathbf{F}$$

by $(\circ\mathbf{F})$ and $(R\circ L)$. From this, by $(\mathcal{U}\,L)_2$ and $(\wedge L)$, we obtain two new goals. The first is $\mathbf{F} \vdash \mathbf{F}$, which is an (As). The second goal is

$$p, \neg q \vee \mathbf{F}, \circ((p \wedge (\neg q \vee \mathbf{F}) \wedge (\mathbf{F} \vee \mathbf{F}))\mathcal{U}\,\mathbf{F}) \vdash \mathbf{F}$$

Then,

$$(p \wedge (\neg q \vee \mathbf{F}) \wedge (\mathbf{F} \vee \mathbf{F}))\mathcal{U}\,\mathbf{F} \vdash \mathbf{F}$$

is obtained by $(\circ\mathbf{F})$ and $(R\circ L)$. Finally, $(\mathcal{U}\,L)_1$, $(\wedge L)$ and (As) suffice. ∎

This (Wk)-free deduction style can be easily generalized to any sequent of the form $\Delta, \varphi\mathcal{U}\,\mathbf{F} \vdash \mathbf{F}$, since the maximum number of nested next operators in Δ, φ is finite. In fact, we conjecture that (Wk) is admissible in $\mathcal{FC}$.

5 The Completeness of $\mathcal{FC}$

In this section, we prove that $\mathcal{FC}$ is a complete calculus using the technique of filtration. In particular, we define a notion of saturated set of formulas that enables the construction of a model for any set of formulas Φ such that $\Phi \nvdash_{\mathcal{FC}} \mathbf{F}$. To this end, we first build a nondeterministic structure in which this model is embedded. The idea of using maximal strongly connected components, inspired by [5], is crucial in handling eventualities in this nondeterministic structure.

In the first subsection, we introduce a notion of saturation for sets of formulas which preserves $\mathcal{FC}$-consistency. In the second subsection, we show how to associate a nondeterministic structure to any $\mathcal{FC}$-consistent set of formulas. Finally, we prove the completeness of the calculus $\mathcal{FC}$.

5.1 Saturated Sets of Formulas

The closure of a set of formulas Φ consists of all formulas that we may use for constructing a model of Φ.

Definition 7. *Let Φ be a set of formulas. Let* $\mathsf{subform}(\Phi)$ *be the set of all the subformulas of the formulas in Φ. Let* $\mathsf{basic}(\Phi) = \mathsf{subform}(\Phi) \cup \{\neg\varphi \mid \varphi \in \mathsf{subform}(\Phi)\}$. *The* closure set of Φ, *denoted* $\mathsf{clo}(\Phi)$, *is the extension of* $\mathsf{basic}(\Phi)$ *with the following two sets of formulas:*

$$\{\circ(\varphi\,\mathcal{U}\,\psi), \neg\circ(\varphi\,\mathcal{U}\,\psi), \circ\neg(\varphi\,\mathcal{U}\,\psi) \mid \varphi\,\mathcal{U}\,\psi \in \mathsf{basic}(\Phi)\}$$
$$\{\circ\neg\varphi \mid \neg\circ\varphi \in \mathsf{basic}(\Phi)\}.$$ ■

For example, if Φ is the singleton $\{p \wedge (p\,\mathcal{U}\,\neg\circ q)\}$ then $\mathsf{clo}(\Phi)$ consists of the union of the following four sets:

$$\{p \wedge (p\,\mathcal{U}\,\neg\circ q), p, p\,\mathcal{U}\,\neg\circ q, \neg\circ q, \circ q, q\}$$
$$\{\neg(p \wedge (p\,\mathcal{U}\,\neg\circ q)), \neg p, \neg(p\,\mathcal{U}\,\neg\circ q), \neg\neg\circ q, \neg q\}$$
$$\{\circ(p\,\mathcal{U}\,\neg\circ q), \neg\circ(p\,\mathcal{U}\,\neg\circ q), \circ\neg(p\,\mathcal{U}\,\neg\circ q)\}$$
$$\{\circ\neg q\}$$

where the first set is $\mathsf{subform}(\Phi)$, whose joint with the second set constitutes $\mathsf{basic}(\Phi)$. The last two sets respectively correspond with the two final extensions in the above definition.

Now, we define a successor relation on sets of formulas.

Definition 8. *Let Ω_1 and Ω_2 be two subsets of* $\mathsf{clo}(\Phi)$ *for some set Φ. We say that Ω_2 is a Φ-*successor *of Ω_1 iff $\varphi \in \Omega_2$ for all $\circ\varphi \in \Omega_1$. The set of Φ-successors of a given set of formulas Ω is*

$$\mathsf{succ}_\Phi(\Omega) = \{\Omega' \subseteq \mathsf{clo}(\Phi) \mid \Omega' \textit{ is a } \Phi\textit{-successor of } \Omega\}.$$ ■

Definition 9. *We say that a set Ω of formulas is saturated iff it satisfies the following conditions:*

1. *If $\varphi \vee \psi \in \Omega$ then $\varphi \in \Omega$ or $\psi \in \Omega$*
2. *If $\neg(\varphi \vee \psi) \in \Omega$ then $\neg\varphi \in \Omega$ and $\neg\psi \in \Omega$*
3. *If $\varphi\,\mathcal{U}\,\psi \in \Omega$ then $\psi \in \Omega$ or $\{\varphi, \neg\psi, \circ(\varphi\,\mathcal{U}\,\psi)\} \subseteq \Omega$*
4. *If $\neg(\varphi\,\mathcal{U}\,\psi) \in \Omega$ then $\{\neg\psi, \neg\varphi\} \subseteq \Omega$ or $\{\varphi, \neg\psi, \neg\circ(\varphi\,\mathcal{U}\,\psi)\} \subseteq \Omega$*
5. *If $\neg\neg\varphi \in \Omega$ then $\varphi \in \Omega$.*
6. *If $\neg\circ\varphi \in \Omega$ then $\circ\neg\varphi \in \Omega$.*

Given a set Φ, we denote by $\mathsf{satur}(\Phi)$ *the set of all saturated subsets of* $\mathsf{clo}(\Phi)$. *For any $\Gamma \subseteq \mathsf{clo}(\Phi)$, we denote by* $\mathsf{satur}^\Gamma(\Phi)$ *the subset of* $\mathsf{satur}(\Phi)$ *that includes all the supersets of Γ. In particular,* $\mathsf{satur}(\Phi) = \mathsf{satur}^\emptyset(\Phi)$ *where $\emptyset$ denotes the empty set.* ■

For the additionally defined connectives, the saturation conditions are easily deduced from Definition 9.

Proposition 10. *The saturation conditions for $\wedge$, $\diamond$ and $\square$ are:*

- *If $\varphi \wedge \psi \in \Omega$ then $\varphi \in \Omega$ and $\psi \in \Omega$*
- *If $\neg(\varphi \wedge \psi) \in \Omega$ then $\neg\varphi \in \Omega$ or $\neg\psi \in \Omega$*

- *If* $\diamond\varphi \in \Omega$ *then* $\varphi \in \Omega$ *or* $\{\neg\varphi, \circ\diamond\varphi\} \subseteq \Omega$
- *If* $\neg\diamond\varphi \in \Omega$ *then* $\{\neg\varphi, \neg\circ\diamond\varphi\} \subseteq \Omega$
- *If* $\square\varphi \in \Omega$ *then* $\{\varphi, \circ\square\varphi\} \subseteq \Omega$
- *If* $\neg\square\varphi \in \Omega$ *then* $\{\varphi, \neg\circ\square\varphi\} \subseteq \Omega$ *or* $\neg\varphi \in \Omega$. ■

Note that if Φ is finite so is $\mathsf{clo}(\Phi)$. As a consequence, every $\Omega \in \mathsf{satur}(\Phi)$ is also finite.

The following lemma states that any subset of a $\mathcal{FC}$-consistent set can be extended to a saturated set while preserving the consistency of the whole set.

Lemma 11. *For all sets of formulas* Φ, Ψ, Γ *such that* $\Gamma \subseteq \mathsf{clo}(\Phi)$ *and* $\Gamma, \Psi \not\vdash_{\mathcal{FC}} \mathbf{F}$, *there exists at least one* $\widehat{\Gamma} \in \mathsf{satur}^{\Gamma}(\Phi)$ *such that* $\widehat{\Gamma}, \Psi \not\vdash_{\mathcal{FC}} \mathbf{F}$.

Proof. Suppose that $\widehat{\Gamma}, \Psi \vdash_{\mathcal{FC}} \mathbf{F}$ for all $\widehat{\Gamma} \in \mathsf{satur}^{\Gamma}(\Phi)$. Then, a $\mathcal{FC}$-proof of $\Gamma, \Psi \vdash \mathbf{F}$ can be easily built using these sequents as leaves and the rules $(\vee L)$, $(\neg \vee L)$, $(\mathcal{U}\, L)_1$, $(\neg\, \mathcal{U}\, L)$, $(\neg\neg L)$ and $(\neg\circ L)$. ■

Note that Ψ (in the above lemma) is not required to be a subset of the closure of Φ. It could be seen as the context of Γ and, in particular, it could be empty.

Corollary 12. *If* $\Phi \not\vdash_{\mathcal{FC}} \mathbf{F}$ *then there exists* $\Omega \in \mathsf{satur}^{\Phi}(\Phi)$ *such that* $\Omega \not\vdash_{\mathcal{FC}} \mathbf{F}$. ■

5.2 Nondeterministic Models of $\mathcal{FC}$-Consistent Sets

We are going to build a model whose states are $\mathcal{FC}$-consistent saturated sets. We use the following notion of nondeterministic PLTL-structure for representing collections of PLTL-structures. In fact, each infinite path in a nondeterministic PLTL-structure is a PLTL-structure.

Definition 13. A nondeterministic PLTL-structure *(nd-*PLTL*-structure, for short)* $\mathcal{G}$ *is a triple* $(S_{\mathcal{G}}, R_{\mathcal{G}}, V_{\mathcal{G}})$ *such that:*

- $S_{\mathcal{G}}$ *is a finite non-empty set of states*
- $R_{\mathcal{G}} \subseteq S_{\mathcal{G}} \times S_{\mathcal{G}}$ *is called* reachability relation
- $V_{\mathcal{G}}$ *is a map* $V_{\mathcal{G}} : S_{\mathcal{G}} \longrightarrow 2^{\mathsf{Prop}}$.

A path π *in a nd-*PLTL*-structure* $\mathcal{G}$ *is a non-empty sequence of states* $s_0, s_1, \ldots \in S_{\mathcal{G}}$ *and* $s_i \in R_{\mathcal{G}}(s_{i-1})$ *for all* $i \geq 1$. ■

We denote by $R_{\mathcal{G}}^{+}$ and $R_{\mathcal{G}}^{*}$ the transitive closure and the reflexive-transitive closure of the reachability relation $R_{\mathcal{G}}$, respectively.

Definition 14. *The truth of a formula* φ *in a state* s *of a nd-*PLTL*-structure* $\mathcal{G}$, *denoted by* $\langle \mathcal{G}, s\rangle \models \varphi$, *is defined as in the Definition 2, except for the temporal operators:*

- $\langle \mathcal{G}, s\rangle \models \circ\varphi$ *iff for all* $s' \in R_{\mathcal{G}}(s)$ $\langle \mathcal{G}, s'\rangle \models \varphi$
- $\langle \mathcal{G}, s\rangle \models \varphi\,\mathcal{U}\,\psi$ *iff there exists a finite path* $s_0, s_1, \ldots, s_n$ *in* $S_{\mathcal{G}}$ *such that* $s = s_0$, $\langle \mathcal{G}, s_n\rangle \models \psi$ *and* $\langle \mathcal{G}, s_i\rangle \models \varphi$ *for every* $0 \leq i \leq n-1$. ■

Note that, the above satisfaction definition of $\mathcal{U}$ only requires the existence of a path because nd-PLTL-structures could contain infinite paths that repeat infinitely many times a subsequence of states and do not reach some other finitely reachable states.

Now, we associate a nondeterministic structure to any consistent set.

Definition 15. *For any given $\mathcal{FC}$-consistent set of formulas Φ, $\mathcal{G}_\Phi = (S_{\mathcal{G}_\Phi}, R_{\mathcal{G}_\Phi}, V_{\mathcal{G}_\Phi})$ is the nd-PLTL-structure where*

- $S_{\mathcal{G}_\Phi} = \{\Omega \mid \Omega \in \mathsf{satur}(\Phi)$ *and* $\Omega \not\vdash_{\mathcal{FC}} \mathbf{F}\}$
- $\Omega' \in R_{\mathcal{G}_\Phi}(\Omega)$ *iff* $\Omega' \in \mathsf{succ}_\Phi(\Omega)$ *for all* $\Omega, \Omega' \in S_{\mathcal{G}_\Phi}$
- $V_{\mathcal{G}_\Phi}(\Omega) = \{p \mid p \in \Omega$ *and* $p \in \mathsf{Prop}\}$. ■

Note that, according to Corollary 12, $S_{\mathcal{G}_\Phi}$ cannot be empty. In the rest of this section we will assume that Φ is always an $\mathcal{FC}$-consistent set of formulas and $\mathcal{G}_\Phi$ is its associated nd-PLTL-structure. Now, we will show how the notion of maximal strongly connected components [5] yields a partition in $S_{\mathcal{G}_\Phi}$.

Definition 16. *A* strongly connected component *(scc, for short) is a subset $\mathcal{S}$ of $S_{\mathcal{G}_\Phi}$ such that every pair formed by two different states $\Omega_1, \Omega_2 \in \mathcal{S}$ satisfies that $\Omega_2 \in R^+_{\mathcal{G}_\Phi}(\Omega_1)$ and $\Omega_1 \in R^+_{\mathcal{G}_\Phi}(\Omega_2)$.*

A maximal *scc* (mscc, *for short) is an scc $\mathcal{S}$ such that there is no scc $\mathcal{S}' \subseteq S_{\mathcal{G}_\Phi}$ that satisfies $\mathcal{S} \subsetneq \mathcal{S}'$.*

We will denote by $[\Omega]$ the mscc where Ω is included and $\mapsto$ is the binary relation induced by $R_{\mathcal{G}_\Phi}$ as follows:

$[\Omega_1] \mapsto [\Omega_2]$ iff there exist $\Omega'_1 \in [\Omega_1]$, $\Omega'_2 \in [\Omega_2]$ such that $\Omega'_2 \in R_{\mathcal{G}_\Phi}(\Omega'_1)$. ■

Note that an mscc $[\Omega]$ could consist just of the state Ω. In such case $[\Omega]$ can represent (on its own) a model only when $\Omega \in R_{\mathcal{G}_\Phi}(\Omega)$. An mscc that consists of exactly one state Ω such that $\Omega \notin R_{\mathcal{G}_\Phi}(\Omega)$ is called *trivial*. Otherwise, we say that it is a *nontrivial* mscc (*nt-mscc*, for short).

Definition 17. *A path $\pi = \Omega_0, \Omega_1, \ldots$ in $S_{\mathcal{G}_\Phi}$ is* fulfilling *if for every $\Omega_i \in \pi$ and every $\varphi \mathcal{U} \psi \in \Omega_i$ there exists some $j \geq i$ such that $\psi \in \Omega_j$ and for every $i \leq k \leq j-1$, $\varphi \in \Omega_k$.*

An scc $\mathcal{S}$ in $S_{\mathcal{G}_\Phi}$ is self-fulfilling *if for every $\Omega \in \mathcal{S}$ and every formula $\varphi \mathcal{U} \psi \in \Omega$, there exists a finite path $\Omega_0, \Omega_1, \ldots, \Omega_n$ in $\mathcal{S}$ such that $\Omega_0 = \Omega$, $\psi \in \Omega_n$ and $\varphi \in \Omega_i$ for every $0 \leq i \leq n-1$.* ■

Lemma 18. *For every $\Omega \in S_{\mathcal{G}_\Phi}$ the set $R_{\mathcal{G}_\Phi}(\Omega)$ is non-empty.*

Proof. If $\Omega \in S_{\mathcal{G}_\Phi}$ then $\Omega \not\vdash_{\mathcal{FC}} \mathbf{F}$. Hence $\mathsf{next}(\Omega) \not\vdash_{\mathcal{FC}} \mathbf{F}$ holds by rules $(R{\circ}L)$ and $(\circ\mathbf{F})$. From Lemma 11 there exists at least one $\Omega' \in S_{\mathcal{G}_\Phi}$ such that $\Omega' \in \mathsf{succ}_\Phi(\Omega)$. ■

Corollary 19. *For every $\Omega \in S_{\mathcal{G}_\Phi}$ there is at least one infinite path $\Omega_0, \Omega_1, \ldots$ such that $\Omega = \Omega_0$.* ■

Now, we will show that $\mathcal{G}_\Phi$ satisfies, by construction, the adequate properties for handling eventualities. In particular, in the next proposition we show that non-satisfied eventualities are kept in paths at least until they are fulfilled.

Proposition 20. *Let $\Omega \in S_{\mathcal{G}_\Phi}$ such that $\varphi \mathcal{U} \psi \in \Omega$. For every finite path $\Omega_0, \Omega_1, \ldots, \Omega_n$ in $S_{\mathcal{G}_\Phi}$ such that $\Omega_0 = \Omega$ and every $1 \leq i \leq n$: if $\varphi \mathcal{U} \psi \notin \Omega_i$ then $\psi \in \Omega_k$ for some $0 \leq k < i$ and $\varphi \in \Omega_j$ for all $0 \leq j \leq k-1$.*

Proof. By induction on n. The case $n = 0$ trivially holds. For $n \geq 1$, we distinguish the following cases. First, if either $i = n$ and there exists $j \leq n-1$ such that $\varphi \mathcal{U} \psi \notin \Omega_j$ or $1 \leq i < n$, then the property holds by the induction hypothesis. Second, if $i = n$ and $\varphi \mathcal{U} \psi \in \Omega_j$ for all $0 \leq j \leq n-1$, then $\psi \in \Omega_j$ or $\{\varphi, \neg\psi, \circ(\varphi \mathcal{U} \psi)\} \subseteq \Omega_j$, since each Ω_j is saturated. This implies that $\psi \in \Omega_{n-1}$ because otherwise $\circ(\varphi \mathcal{U} \psi) \in \Omega_{n-1}$ which would mean $\varphi \mathcal{U} \psi \in \Omega_n$. ∎

The next proposition shows how negated eventualities propagate in $\mathcal{G}_\Phi$.

Proposition 21. *Let $\Omega \in S_{\mathcal{G}_\Phi}$ such that $\neg(\varphi \mathcal{U} \psi) \in \Omega$. Then, every finite path $\pi = \Omega_0, \Omega_1, \ldots, \Omega_n$ in $S_{\mathcal{G}_\Phi}$ such that $\Omega_0 = \Omega$ satisfies one of the two following properties:*

(a) $\{\varphi, \neg\psi, \neg(\varphi \mathcal{U} \psi)\} \subseteq \Omega_i$ *for any* $i \in \{0..n\}$
(b) *There exists* $0 \leq j \leq n$ *such that* $\{\neg\varphi, \neg\psi\} \subseteq \Omega_j$ *and* $\{\varphi, \neg\psi, \neg(\varphi \mathcal{U} \psi)\} \subseteq \Omega_i$ *for any* $i \in \{0..j-1\}$.

Proof. By induction on n. Since Ω is saturated, the case $n = 0$ is trivial. For $n \geq 1$, the induction hypothesis guarantees that the path $\pi' = \Omega_0, \Omega_1, \ldots, \Omega_{n-1}$ satisfies one of the properties (a) or (b). If π' satisfies (b), so does π. If π' satisfies (a) then by definition of $S_{\mathcal{G}_\Phi}$ we have $\{\varphi, \neg\psi, \neg(\varphi \mathcal{U} \psi)\} \subseteq \Omega_n$ or $\{\neg\varphi, \neg\psi\} \subseteq \Omega_n$. Hence, π verifies (a) or (b) respectively. ∎

Now we will prove that for any $\Omega \in S_{\mathcal{G}_\Phi}$, either the mscc $[\Omega]$ is a self-fulfilling nt-mscc or there exists a self-fulfilling nt-mscc that is reachable from $[\Omega]$.

Lemma 22. *For any non-self-fulfilling mscc $[\Omega]$ in $S_{\mathcal{G}_\Phi}$, there exists (at least) one $\Omega' \in S_{\mathcal{G}_\Phi}$ such that $\Omega' \notin [\Omega]$ and $[\Omega] \mapsto [\Omega']$.*

Proof. For a trivial mscc, this is an easy consequence of Lemma 18. Hence, we assume $[\Omega]$ to be a nt-mscc which is not self-fulfilling. That is, there is some $\Omega_0 \in [\Omega]$ and some formula $\varphi \mathcal{U} \psi \in \Omega_0$ such that there does not exist a finite path $\Omega_0, \Omega_1, \ldots, \Omega_n$ in $[\Omega]$ such that $\psi \in \Omega_n$ and $\varphi \in \Omega_i$ for every $0 \leq i < n$. Then, for all $\Delta \in [\Omega]$:

$$\{\varphi, \neg\psi, \varphi \mathcal{U} \psi, \circ(\varphi \mathcal{U} \psi)\} \subseteq \Delta$$

Let us consider the subset of $S_{\mathcal{G}_\Phi}$ formed by all the states that are successors of some state in $[\Omega]$:

$$\mathcal{S}([\Omega]) = \bigcup_{\Delta \in [\Omega]} R_{\mathcal{G}_\Phi}(\Delta)$$

Since $[\Omega]$ is a nt-mscc it must verify $[\Omega] \subseteq \mathcal{S}([\Omega])$. If $[\Omega] \subsetneq \mathcal{S}([\Omega])$ the lemma holds trivially. On the contrary, if $[\Omega] = \mathcal{S}([\Omega])$ we show that there is a contradiction as follows. Consider any state $\Delta \in [\Omega] \subseteq S_{\mathcal{G}_\Phi}$. Since Δ is $\mathcal{FC}$-consistent, then $\Delta \not\vdash_{\mathcal{FC}} \mathbf{F}$. Hence, by rules $(\mathcal{U}\, L)_2$ and $(\neg L)$, we have that

$$\Delta, \circ((\varphi \wedge \Delta^{\neg}) \mathcal{U} \psi) \not\vdash_{\mathcal{FC}} \mathbf{F}$$

Hence, by $(R\circ L)$, the set $\mathsf{next}(\Delta) \cup \{(\varphi \wedge \Delta^{\neg})\mathcal{U}\,\psi\}$ is also $\mathcal{FC}$-consistent. Then, by Lemma 11, there exists at least one set $\Delta' \in \mathsf{satur}^{\mathsf{next}(\Delta)}(\Phi)$ such that $\Delta', (\varphi \wedge \Delta^{\neg})\mathcal{U}\,\psi \not\vdash_{\mathcal{FC}} \mathbf{F}$. By (Wk), Δ' is also $\mathcal{FC}$-consistent. Hence, $\Delta' \in S_{\mathcal{G}_\Phi}$ and, by construction $\Delta' \in R_{\mathcal{G}_\Phi}(\Delta) \subseteq \mathcal{S}([\Omega])$. Therefore, $\Delta' \in [\Omega]$, since we are supposing that $[\Omega] = \mathcal{S}([\Omega])$. It is worthy to note that $R_{\mathcal{G}_\Phi}(\Delta)$ should be non-empty by Lemma 18. Besides, since $\neg\psi \in \Delta'$ (by construction) and $\Delta', (\varphi \wedge \Delta^{\neg})\mathcal{U}\,\psi \not\vdash_{\mathcal{FC}} \mathbf{F}$, the rules $(\mathcal{U}\,L)_2$ and (CdL) allow us to conclude that

$$\Delta', \varphi \wedge \Delta^{\neg}, \circ((\varphi \wedge \Delta^{\neg} \wedge (\Delta')^{\neg})\mathcal{U}\,\psi) \not\vdash_{\mathcal{FC}} \mathbf{F}$$

Hence, by (Wk), we have obtained from Δ an $\mathcal{FC}$-consistent set Δ' such that $\Delta' \cup \{\circ((\varphi \wedge \Delta^{\neg} \wedge (\Delta')^{\neg})\mathcal{U}\,\psi)\}$ is also $\mathcal{FC}$-consistent. Starting with any $\Delta_0 \in [\Omega]$ and repeating the above procedure we can construct a path $\pi = \Delta_0, \Delta_1, \ldots$ of states in $[\Omega]$ such that for every $i \geq 1$

$$\Delta_i, (\varphi \wedge \Delta_0^{\neg} \wedge \Delta_1^{\neg} \wedge \ldots \wedge \Delta_{i-1}^{\neg})\mathcal{U}\,\psi \not\vdash_{\mathcal{FC}} \mathbf{F}$$

By finiteness of $[\Omega]$, there must exist $n \geq 1$ such that $\Delta_n = \Delta_i$ for some $0 \leq i \leq n-1$. In particular, for such n we have that

$$\Delta_n, (\varphi \wedge \Delta_0^{\neg} \wedge \Delta_1^{\neg} \wedge \ldots \wedge \Delta_{n-1}^{\neg})\mathcal{U}\,\psi \not\vdash_{\mathcal{FC}} \mathbf{F}$$

But this is a contradiction, by $(\mathcal{U}\,L)_1$, $(\wedge L)$ and (Wk), because $\Delta_n, \Delta_n^{\neg} \vdash_{\mathcal{FC}} \mathbf{F}$ can be easily derived using $(\vee L)$ and (CdL). ■

Corollary 23. *For any $\Omega \in S_{\mathcal{G}_\Phi}$, either the mscc $[\Omega]$ is a self-fulfilling nt-mscc or there exists $\Omega' \in S_{\mathcal{G}_\Phi}$ such that $\Omega' \notin [\Omega]$, $\Omega' \in R^+_{\mathcal{G}_\Phi}(\Omega)$ and $[\Omega']$ is a self-fulfilling nt-mscc.*

Proof. By finiteness of $S_{\mathcal{G}_\Phi}$, if $[\Omega]$ is not a self-fulfilling non-trivial mscc, then Lemma 22 guarantees the existence of $[\Omega']$. In the case of a trivial mscc, also Lemma 18 should be used. ■

Lemma 24. (Nondeterministic Model Existence) *For every $\Omega \in S_{\mathcal{G}_\Phi}$ it holds that if $\varphi \in \Omega$ then $\langle \mathcal{G}_\Phi, \Omega \rangle \models \varphi$.*

Proof. By structural induction on φ. The case of literals is trivial by definition of $\mathcal{G}_\Phi$.

For formulas of the form $\neg\neg\varphi$, $\varphi \vee \psi$, $\neg(\varphi \vee \psi)$, $\circ\varphi$ and $\neg\circ\varphi$ it holds by definition of $\mathcal{G}_\Phi$ and the induction hypothesis on $\{\varphi\}$, $\{\varphi, \psi\}$, $\{\neg\varphi, \neg\psi\}$, $\{\varphi\}$ and $\{\neg\varphi\}$, respectively.

For $\varphi\mathcal{U}\,\psi$, by the above Proposition 20 and Corollary 23 there exists a finite path $\Omega_0, \Omega_1 \ldots \Omega_n$ in $S_{\mathcal{G}_\Phi}$ such that $\Omega_0 = \Omega, \psi \in \Omega_n$ and $\varphi \in \Omega_i$ for every $0 \leq i \leq n-1$. By the induction hypothesis, $\langle \mathcal{G}_\Phi, \Omega_n \rangle \models \psi$ and $\langle \mathcal{G}_\Phi, \Omega_i \rangle \models \varphi$ for every $0 \leq i \leq n-1$ and consequently $\langle \mathcal{G}_\Phi, \Omega \rangle \models \varphi\mathcal{U}\,\psi$.

For $\neg(\varphi\mathcal{U}\,\psi)$ formulas, by the above Proposition 21 and the induction hypothesis there does not exist any finite path $\Omega_0, \Omega_1 \ldots \Omega_n$ in $S_{\mathcal{G}_\Phi}$ such that $\Omega_0 = \Omega, \langle \mathcal{G}_\Phi, \Omega_n \rangle \models \psi$ and $\langle \mathcal{G}_\Phi, \Omega_i \rangle \models \varphi$ for every $0 \leq i \leq n-1$. Consequently $\langle \mathcal{G}_\Phi, \Omega \rangle \not\models \varphi\mathcal{U}\,\psi$ and hence $\langle \mathcal{G}_\Phi, \Omega \rangle \models \neg(\varphi\mathcal{U}\,\psi)$. ■

5.3 Model Existence and Completeness

Using the nondeterministic structure $\mathcal{G}_\Phi$ (which was defined in the previous subsection), we are now able to build a model of any $\mathcal{FC}$-consistent set.

Lemma 25. (Path Existence) *For every $\Omega \in S_{\mathcal{G}_\Phi}$ there exists at least one infinite fulfilling path $\pi = \Omega_0, \Omega_1, \ldots$ where $\Omega_0 = \Omega$.*

Proof. Let us show how to build the path π depending on the mscc to which Ω belongs. If $[\Omega]$ is a self-fulfilling mscc, then choose π' to be any finite path that covers all the states in $[\Omega]$. Then, the infinite path $\pi = \pi', \pi', \pi', \ldots$ is fulfilling. Otherwise, if $[\Omega]$ is not a self-fulfilling mscc, by Corollary 23, there exists $\Omega' \in S_{\mathcal{G}_\Phi}$ such that $\Omega' \notin [\Omega]$, $\Omega' \in R^+_{\mathcal{G}_\Phi}(\Omega)$ and $[\Omega']$ is a self-fulfilling mscc. Let π_1 be any finite path from Ω to Ω' and let π_2 be the infinite path in $[\Omega']$ constructed as in the previous case. Then, $\pi = \pi_1, \pi_2$ is an infinite fulfilling path. ■

Lemma 26. (Model Existence) *Let $\pi = \Omega_0, \Omega_1 \ldots$ an infinite fulfilling path in $S_{\mathcal{G}_\Phi}$. Then, the* PLTL*-structure $\mathcal{M}_\pi$ defined by*

- $S_{\mathcal{M}_\pi} = \Omega_0, \Omega_1, \ldots$
- $V_{\mathcal{M}_\pi}(\Omega_i) = \{p \mid p \in \Omega_i\}$

satisfies that $\langle \mathcal{M}_\pi, i\rangle \models \varphi$ for every $i \in \mathbb{N}$ and every $\varphi \in \Omega_i$.

Proof. Immediate consequence of Lemma 24. ■

Finally, we are able to prove the completeness of $\mathcal{FC}$.

Theorem 27. (Completeness of $\mathcal{FC}$) *For any set of formulas $\Gamma \cup \{\chi\}$, if $\Gamma \models \chi$ then $\Gamma \vdash_{\mathcal{FC}} \chi$.*

Proof. Suppose that $\Gamma \nvdash_{\mathcal{FC}} \chi$. Then, by rule (Cd), $\Gamma, \neg\chi \nvdash_{\mathcal{FC}} \mathbf{F}$. Hence, by Corollary 12, Lemma 25 and Lemma 26 there exists a model of $\Gamma \cup \{\neg\chi\}$. Therefore, $\Gamma \not\models \chi$. ■

6 Concluding Remarks

We have introduced a sound and complete (finitary) sequent calculus $\mathcal{FC}$ for the logic PLTL. The calculus $\mathcal{FC}$ is cut-free and invariant-free and it leads to a new deduction style in temporal logic. We are working on the mechanization of the calculus $\mathcal{FC}$ in the generic proof-assistant Isabelle (cf. `http://isabelle.in.tum.de`) in order to allow the interactive formalization of $\mathcal{FC}$-proofs for temporal properties. Tableaux and resolution methods are better suited for completely automatic theorem proving. In this regard, the rules $(\mathcal{U}\,L)_2$ and $(\Diamond L)_2$ give rise to new ideas for improving the existing methods of temporal tableaux and temporal resolution. Following these ideas, we are also working on avoiding the construction of the whole states-graph in the tableaux framework and the construction of invariants in the resolution setting. These methods should manage formulas of the form $(\Delta^\neg \wedge \varphi)\,\mathcal{U}\,\psi$ such that Δ is also part of the set of formulas to be processing. Hence, from the point of view of efficiency, shared formulas would be very useful for practical implementation. Additional future work includes the extension of this ideas to the branching case, the first-order case (in spite of its incompleteness) or its complete fragments.

References

1. Fisher, M.: A resolution method for temporal logic. In: IJCAI, pp. 99–104 (1991)
2. Gabbay, D., Pnueli, A., Shelah, S., Stavi, J.: On the temporal analysis of fairness. In: POPL '80. Proceedings of the 7th ACM SIGPLAN-SIGACT symposium on Principles of programming languages, New York, NY, USA, pp. 163–173. ACM Press, New York (1980)
3. Gentzen, G.: Untersuchungen uber das logische schliessen. In: Szabo, M.E. (ed.) Mathematische Zeitschrift, vol. 39, pp. 176–210, 405–431, 1934–1935. North-Holland, Amsterdam (1969) (English translation in The collected papers of Gerhard Gentzen, pp. 68–131)
4. Kamp, J.A.W.: Tense Logic and the Theory of Linear Order. PhD. Thesis, University of California, Los Angeles (1968)
5. Lichtenstein, O., Pnueli, A.: Propositional temporal logics: Decidability and completeness. Logic Journal of the IGPL 8(1) (2000)
6. Paech, B.: Gentzen-systems for propositional temporal logics. In: Börger, E., Kleine Büning, H., Richter, M.M. (eds.) CSL 1988. LNCS, vol. 385, pp. 240–253. Springer, Heidelberg (1989)
7. Pfenning, F.: Structural cut elimination: I. intuitionistic and classical logic. Inf. Comput. 157(1-2), 84–141 (2000)
8. Pliuskevicius, R.: Investigation of finitary calculus for a discrete linear time logic by means of infinitary calculus. In: Baltic Computer Science, pp. 504–528 (1991)
9. Reynolds, M., Dixon, C.: Theorem-proving for discrete temporal logic. In: Handbook of Temporal Reasoning in Artificial Intelligence, pp. 279–314. Elsevier, Amsterdam (2005)
10. Sistla, A.P., Clarke, E.M.: The complexity of propositional linear temporal logics. J. ACM 32(3), 733–749 (1985)
11. Szalas, A.: Temporal logic of programs: A standard approach. In: Time and Logic. A Computational Approach, pp. 1–50. UCL Press Ltd., (1995)
12. Wolper, P.: Temporal logic can be more expressive. Information and Control 56(1–2), 72–99 (1983)

Unbounded Proof-Length Speed-Up in Deduction Modulo

Guillaume Burel

Université Henri Poincaré & LORIA*
Campus scientifique BP 239 — 54506 Vandœuvre-lès-Nancy Cedex France
guillaume.burel@ens-lyon.org

Abstract. In 1973, Parikh proved a speed-up theorem conjectured by Gödel 37 years before: there exist arithmetical formulæ that are provable in first order arithmetic, but whose shorter proof in second order arithmetic is arbitrarily smaller than any proof in first order. On the other hand, resolution for higher order logic can be simulated step by step in a first order narrowing and resolution method based on deduction modulo, whose paradigm is to separate deduction and computation to make proofs clearer and shorter.

We prove that $i+1$-th order arithmetic can be linearly simulated into i-th order arithmetic modulo some confluent and terminating rewrite system. We also show that there exists a speed-up between i-th order arithmetic modulo this system and i-th order arithmetic without modulo. All this allows us to prove that the speed-up conjectured by Gödel does not come from the deductive part of the proofs, but can be expressed as simple computation, therefore justifying the use of deduction modulo as an efficient first order setting simulating higher order.

Keywords: proof theory, rewriting, higher order logic, arithmetic.

1 Introduction

Even if two logical systems are shown to be expressively equivalent, i.e. they can prove exactly the same formulæ, they can lead to very different proofs, in particular in terms of length. For instance, it is shown that Frege systems have an exponential speed-up over resolution for propositional logic [5]. However in mechanized theorem proving, the length of proofs has an importance: First, computers have limited capacities, and this can lead to a difference between the practical expressiveness of theoretically equivalent systems. Even if computing power is always increasing, so that one is no longer afraid to use SAT-solvers within verification tools (mainly because worst cases do not often occur in practice), it is not conceivable to build an automated theorem prover that produces proofs of non-elementary length. Second, the length of a proof is one (among others) criterion for defining the quality of a proof. Indeed, a smaller proof is often more readable and, in the case for instance of software certification and proof

* UMR 7503 CNRS-INPL-INRIA-Nancy2-UHP

J. Duparc and T.A. Henzinger (Eds.): CSL 2007, LNCS 4646, pp. 496–511, 2007.

engineering, more communicable and often also more maintainable. In [10,2], this notion of "good proofs" is translated through a proof ordering, which of course may correspond to the comparison of proof lengths.

Obtaining a speed-up can also have a theoretical interest, because, as remarked by Parikh in the introductory paragraph of [16], "the celebrated P=NP? question can itself be thought of as a speed-up question." (See [7].) All this explains the research for new formalisms whose deductive systems provide smaller proofs, such as for instance the calculus of structures w.r.t. the sequent calculus [17].

In this paper, the length of a proof will correspond to its number of steps (sometimes called lines), whatever the actual size of the formulæ appearing in them is. Considering only the minimal length of proofs, the definition of a speed-up is the following: given some function h over natural numbers, a system has a speed-up for h over another one, if there exists an infinite set of formulæ provable in both of them, such that, if the length of the proofs in the first system is l and the length in the second system is k, then $k > h(l)$.

In 1936, Gödel [16] conjectured that there exists such a speed-up for all recursive functions between i-th order and $i+1$-th order arithmetic, no matter the formal system actually used. In other words, he stated that for all recursive functions h, it is possible to find an infinite set of formulæ such that, for each of them, denoted by P, if k is the minimal number of steps in the proofs of P in the i-th order arithmetic (k is assumed to exist, so that P is provable in it), and l is the minimal number of steps in the proofs of P in the $i+1$-th order arithmetic, then $k > h(l)$.

This result was proved for first-order arithmetic by Parikh [21], who actually proved a stronger theorem: this proof-length speed-up exists in fact also for non-recursive functions. This was generalized to all orders by Krajíček , and was proved for the true language of arithmetic by Buss [6] (the former results used an axiomatization of arithmetic using ternary predicates to represent addition and multiplication). The theorem proved by Buss is stated as follow:

Theorem 1 ([6, Theorem 3]). *Let $i \geq 0$. Then there is an infinite family $\mathcal{F}$ of $\prod_1^0$-formulæ such that*

1. *for all $P \in \mathcal{F}$, $Z_i \vdash P$*
2. *there is a fixed $k \in \mathbb{N}$ such that for all $P \in \mathcal{F}$, $Z_{i+1} \vdash_{\overline{k\ steps}} P$*
3. *there is no fixed $k \in \mathbb{N}$ such that for all $P \in \mathcal{F}$, $Z_i \vdash_{\overline{k\ steps}} P$.*

Z_i corresponds to the $i+1$-th order arithmetic (so Z_0 is in fact first order arithmetic), and $Z_i \vdash_{\overline{k \text{ steps}}} P$ means that P can be proved in at most k steps within a schematic system —i.e. a Hilbert-type (or Frege) system with a finite number of axiom schemata and inference rules— for $i+1$-th order arithmetic. (In fact, Buss proved this theorem also for weakly schematic systems, i.e. schematic systems in which every tautology can be used as an axiom, as well as generalizations of axioms, but we will not use this fact here.)

Because this theorem is concerned in arithmetic, an intuitive notion of computation take place in the proofs. Indeed, as remarked by Poincaré, establishing that $2+2=4$ using the definition of the addition is just a verification, and not a demonstration, so that in a proof occur in fact not only pure deduction but also computation. Therefore, the question arises whether this speed-up comes from the deductive or the computational part of the proofs, or both of them. Of course, the difference between computation and deduction cannot be clearly determined. Because of the Curry-Howard correspondence, the whole content of the proofs could be considered as computation. Here, this difference must be thought of as the distinction between what is straightforward (at least decidable), and what must be reasoned out.

Deduction modulo [12] is a presentation of a given logic —and the formalisms associated with it— identifying what corresponds to computation. The computational part of a proof is put in a congruence between formulæ modulo whom the application of the deduction rules takes place. This leads to the sequent calculus modulo and the natural deduction modulo. The congruence is better represented as a set of rewrite rules that can rewrite terms but also *atomic propositions*: indeed, one wants for instance to consider the definition of the addition or multiplication using rewrite rules over terms as part of the computation, but also the following rewrite rule:

$$x \times y = 0 \ \rightarrow \ x = 0 \vee y = 0$$

which rewrites an atomic proposition to a formula, so that the following simple proof of $t \times t = 0$ can be deduced from a proof π of $t = 0$:

$$\vee\text{-i}\ \frac{\begin{array}{c}\pi \\ t = 0\end{array}}{t \times t = 0}\ {\scriptstyle t \times t = 0 \ \longrightarrow \ t = 0 \vee t = 0}$$

Deduction modulo is logically equivalent to the considered logic [12, Proposition 1.8], but proofs are often considered as simpler, because the computation is hidden, letting the deduction clearly appear. Proofs are also claimed to be smaller for the same reason. Nevertheless, this fact was never quantified. This paper answers this issue. Of course, if there are no restriction on the rewrite rules that are used (for instance if it is allowed to use a rewrite system semi-deciding the validity of formulæ), it is not surprising that the length of the proofs can be unboundedly reduced. Notwithstanding, we will consider in this paper only very simple rewrite systems: they will be finite, terminating, confluent (i.e. deterministic) and linear (variables in the left-hand side only appear once).

Besides, it is possible, in deduction modulo, to build proofs of Higher Order Logic using a first order system [11]. Using this, a step of higher order resolution is completely simulated by a step of ENAR, the resolution and narrowing method based on deduction modulo. Therefore, it seems reasonable to think that deduction modulo is able to give the same proof-length speed-ups as the ones occurring between $i+1$-th and i-th order arithmetic. This paper therefore investigates how to relate proof-length speed-ups in arithmetic with the computational content of the proofs.

To prove that the speed-up theorem of Buss comes from the computational part of the proofs, we first define a linear translation between proofs in $i+1$-th order arithmetic and i-th order arithmetic modulo some rewrite system $\mathcal{R}_i$. Second, using this translation and Buss' theorem, we prove that there is no proof-length speed-up between $i+1$-th order arithmetic and i-th order arithmetic modulo, whereas there exists such a speed-up between i-th order arithmetic modulo and i-th order arithmetic without modulo. Therefore, we conclude that the speed-up between $i+1$-th order arithmetic and i-th order arithmetic lies in the modulo, i.e. the computational part of the proofs.

In the next section, we will recall the definition of a schematic system, and we will present such a system for i-th order arithmetic. The section 3 will define formally what deduction modulo, and in particular natural deduction modulo consists of. In Section 4 we will give the exact translations between a proof in the schematic system for i-th order arithmetic and a proof in natural deduction, modulo or not, as well as the simulation in natural deduction of $i+1$-th order arithmetic in i-th order arithmetic modulo. An upper bound of the increase in the length of the proofs due to these translations will be given. Finally, in Section 5 we will use these translations to determine the origin of the speed-up in arithmetic, and we will conclude about the interest of working within a first-order system modulo to simulate higher order. All the details can be found in the full version of this paper [4].

2 A Schematic System for i-th Order Arithmetic

2.1 Schematic Systems

We recall here, using Buss' terminology [6], what a schematic system consists in. It is essentially an Hilbert-type (or Frege) proof system, i.e. valid formulæ are derived from a finite number of axiom schemata using a finite number of inference rules. Theorem 1 is true on condition that proofs are performed using a schematic system.

First, we recall how to build many-sorted first order formulæ, mainly to introduce the notations we will use. A (first order) many-sorted signature consists in a set of function symbols and a set of predicates, all of them with their arity (and co-arity for function symbols). We denote by $\mathcal{T}(\Sigma, V)$ the set of *terms* built from a signature Σ and a set of variables V. An *atomic proposition* is given by a predicate symbol A of arity $[i_1, \ldots, i_n]$ and by n terms $t_1, \ldots, t_n \in \mathcal{T}(\Sigma, V)$ with matching sorts. It is denoted $A(t_1, \ldots, t_n)$. *Formulæ* can be built using the following grammar[1]:

$$\mathcal{P} \quad \stackrel{!}{=} \quad \bot \mid A \mid \mathcal{P} \wedge \mathcal{P} \mid \mathcal{P} \vee \mathcal{P} \mid \mathcal{P} \Rightarrow \mathcal{P} \mid \forall x.\, \mathcal{P} \mid \exists x.\, \mathcal{P}$$

where A ranges over atomic propositions and x over variables. $P \Leftrightarrow Q$ will be used as a syntactic sugar for $(P \Rightarrow Q) \wedge (Q \Rightarrow P)$. Positions in a term or a formula, free variables and substitutions are defined as usual (see [1]). The replacement of a variable x by a term t in a formula P is denoted by $\{t/x\}P$, the

[1] $\stackrel{!}{=}$ is used for definitions.

restriction of a term or proposition t at the position $\mathfrak{p}$ by $t_{|\mathfrak{p}}$, and its replacement in t by a term or proposition s by $t[s]_{\mathfrak{p}}$.

Then, given a many-sorted signature of first order logic, we can consider infinite sets of *metavariables* α^i for each sort i (which will be substituted by variables), of *term variables* τ^i for each sort i (which will be substituted by terms) and *proposition variables* $A(x_1, \ldots, x_n)$ for each arity $[i_1, \ldots, i_n]$ (which will be substituted by formulæ).

Metaterms are built like terms, except that they can contain metavariables and term variables. Metaformulæ are built like formulæ, except that they can contain proposition variables (which play the same role as predicates) and metaterms.

A *schematic system* is a finite set of inference rules, where an inference rule is a triple of a finite set of metaformulæ (the *premises*), a metaformulæ (the *conclusion*), and a set of side conditions of the forms α^j *is not free in* Φ or s *is freely substitutable for* α^j *in* Φ where Φ is a metaformula and s a metaterm of sort j. It is denoted by

$$\frac{\Phi_1 \qquad \ldots \qquad \Phi_n}{\Psi}\ (R)$$

An inference with an empty set of premises will be called an *axiom schema*.

2.2 *i*-th Order Arithmetic

i-th order arithmetic (Z_{i-1}) is a many-sorted theory with the sorts $0, \ldots, i-1$ and the signature

$$\begin{array}{rclrclrcl} 0 & : & 0 & + & : & [0;0] \to 0 & = & : & [0;0] \\ s & : & [0] \to 0 & \times & : & [0;0] \to 0 & \in^j & : & [j;j+1] \end{array}.$$

The schematic system we use can be found in its totality in the full version [4]. The most representative inference rules are given here as examples:

$14 + 2 \times$ i axiom schemata of classical logic. We take the one used by Gentzen [15, Chapter 5] to prove the equivalence of his formalisms with an Hilbert-type proof system:

$$(A \Rightarrow A \Rightarrow B) \Rightarrow A \Rightarrow B \tag{1}$$

$$(A \Rightarrow B) \Rightarrow (B \Rightarrow C) \Rightarrow A \Rightarrow C \tag{2}$$

$$(A \wedge B) \Rightarrow A \tag{3}$$

$$A(\tau^j) \Rightarrow \exists \alpha^j.\ A(\alpha^j) \quad (\tau^j \text{ is freely substitutable for } \alpha^j \text{ in } A(\alpha^j)) \tag{4}$$

$$A \vee (A \Rightarrow \bot) \tag{5}$$

$1 + 2 \times$ i inference rules of classical logic. Again, we consider the one of [15]:

$$\frac{A \qquad A \Rightarrow B}{B} \tag{6}$$

$$\frac{B(\beta^j) \Rightarrow A}{(\exists \alpha^j.\ B(\alpha^j)) \Rightarrow A}\ (\beta^j \text{ is not free in } (\exists \alpha^j.\ B(\alpha^j)) \Rightarrow A) \tag{7}$$

7 identity axiom schemata. They define the particular relation =:

$$\forall \alpha^0.\ \alpha^0 = \alpha^0 \tag{8}$$

$$\forall \alpha^0 \beta^0 \gamma^0.\ \alpha^0 = \beta^0 \Rightarrow \alpha^0 + \gamma^0 = \beta^0 + \gamma^0 \tag{9}$$

$$\forall \alpha^0 \beta^0.\ \alpha^0 = \beta^0 \Rightarrow A(\alpha^0) \Rightarrow A(\beta^0) \tag{10}$$

7 Robinson's axioms. They are the axioms defining the function symbols of arithmetic [19]:

$$\forall \alpha^0 \beta^0.\ s(\alpha^0) = s(\beta^0) \Rightarrow \alpha^0 = \beta^0 \tag{11}$$

$$\forall \alpha^0.\ \alpha^0 \times 0 = 0 \tag{12}$$

$$\forall \alpha^0 \beta^0.\ \alpha^0 \times s(\beta^0) = \alpha^0 \times \beta^0 + \alpha^0 \tag{13}$$

$i + 1$ induction and comprehension axiom schemata.

$$A(0) \Rightarrow \left(\forall \beta^0.\ A(\beta^0) \Rightarrow A(s(\beta^0))\right) \Rightarrow \forall \alpha^0.\ A(\alpha^0) \tag{14}$$

For all $0 \leq j < i - 1$,

$$\exists \alpha^{j+1}.\ \forall \beta^j.\ \beta^j \in^j \alpha^{j+1} \Leftrightarrow A(\beta^j) \quad (\alpha^{j+1} \text{ is not free in } A) \tag{15}$$

From this point on, we will denote by $Z_{i-1} \vdash^{\mathsf{S}}_{k} P$ the fact that there exists a proof of P of length at most k in this schematic system, i.e. P can be derived using at most k instances of these inference rules.

3 Deduction Modulo

3.1 Rewriting Formulæ

In this section, we recall the definition of deduction modulo, as can be found in [12,13]. In deduction modulo, formulæ are considered modulo some congruence defined by some rules that rewrite not only terms but also formulæ. We use standard definitions, as can be found in [1], and extend them to proposition rewriting [12].

A *term rewrite rule* is the pair of terms l, r such that all free variables of r appear in l. It is denoted $l \rightarrow r$. A *term rewrite system* is a set of term rewrite rules. A term s can be rewritten to a term t by a term rewrite rule $l \rightarrow r$ if there exists some substitution σ and some position $\mathfrak{p}$ in s such that $\sigma l = s_{|\mathfrak{p}}$ and $t = s[\sigma r]_{\mathfrak{p}}$. An atomic proposition $A(s_1, \ldots, s_i, \ldots, s_n)$ can be rewritten to the atomic proposition $A(s_1, \ldots, t_i, \ldots, s_n)$ by a term rewrite rule $l \rightarrow r$ if s_i can be rewritten to t_i by $l \rightarrow r$. This relation is extended by congruence to all formulæ.

$$\Rightarrow\text{-i}\ \frac{\begin{array}{c}[A]\\ B\end{array}}{C}\ \text{if } C \xleftrightarrow[\mathcal{R}]{*} A \Rightarrow B \qquad \Rightarrow\text{-e}\ \frac{A \qquad C}{B}\ \text{if } C \xleftrightarrow[\mathcal{R}]{*} A \Rightarrow B$$

$$\wedge\text{-i}\ \frac{A \qquad B}{C}\ \text{if } C \xleftrightarrow[\mathcal{R}]{*} A \wedge B \qquad \wedge\text{-e}\ \frac{C}{A}\ \text{if } C \xleftrightarrow[\mathcal{R}]{*} A \wedge B \text{ or } C \xleftrightarrow[\mathcal{R}]{*} B \wedge A$$

$$\vee\text{-i}\ \frac{A}{C}\ \text{if } C \xleftrightarrow[\mathcal{R}]{*} A \vee B \text{ or } C \xleftrightarrow[\mathcal{R}]{*} B \vee A \qquad \vee\text{-e}\ \frac{C \qquad \begin{array}{c}[A]\\ D\end{array} \qquad \begin{array}{c}[B]\\ D\end{array}}{D}\ \text{if } C \xleftrightarrow[\mathcal{R}]{*} A \vee B$$

$$\forall\text{-i}\ \frac{\{y/x\}A}{B}\ \begin{array}{l}\text{if } B \xleftrightarrow[\mathcal{R}]{*} \forall x.\ A \text{ and } y \text{ is not}\\ \text{free in } A \text{ nor in the assump-}\\ \text{tions of the proof above}\end{array} \qquad \forall\text{-e}\ \frac{A}{B}\ \text{if } A \xleftrightarrow[\mathcal{R}]{*} \forall x.\ C \text{ and } B \xleftrightarrow[\mathcal{R}]{*} \{t/x\}C$$

$$\exists\text{-i}\ \frac{B}{A}\ \text{if } A \xleftrightarrow[\mathcal{R}]{*} \exists x.\ C \text{ and } B \xleftrightarrow[\mathcal{R}]{*} \{t/x\}C \qquad \exists\text{-e}\ \frac{B \qquad \begin{array}{c}[\{y/x\}A]\\ C\end{array}}{C}\ \begin{array}{l}\text{if } B \xleftrightarrow[\mathcal{R}]{*} \exists x.\ A \text{ and } y \text{ is}\\ \text{not free in } C \text{ nor in the}\\ \text{assumption of the proof}\\ \text{above except } \{y/x\}A\end{array}$$

$$\mathit{classical}\ \frac{}{B}\ \text{if } A \xleftrightarrow[\mathcal{R}]{*} B \vee (B \Rightarrow \bot) \qquad \bot\text{-e}\ \frac{A}{B}\ \text{if } A \xleftrightarrow[\mathcal{R}]{*} \bot$$

Fig. 1. Inference Rules of Natural Deduction Modulo

A *proposition rewrite rule* is the pair of an atomic proposition A and a formula P, such that all free variables of P appear in A. It is denoted $A \rightarrow P$. A *proposition rewrite system* is a set of proposition rewrite rules. A formula Q can be rewritten to a formula R by a proposition rewrite rule $A \rightarrow P$ if there exists some substitution σ and some position $\mathfrak{p}$ in Q such that $\sigma A = Q_{|\mathfrak{p}}$ and $R = Q[\sigma P]_{\mathfrak{p}}$. Semantically, this proposition rewrite relation must be seen as a logical equivalence between formulæ.

A *rewrite system* is the union of a term rewrite system and a proposition rewrite system. The fact that P can be rewritten to Q either by a term or by a proposition rewrite rule of a rewrite system $\mathcal{R}$ will be denoted by $A \xrightarrow[\mathcal{R}]{} P$. The transitive (resp. reflexive transitive) closure of this relation will be denoted by $\xrightarrow[\mathcal{R}]{*}$ (resp. $\xleftrightarrow[\mathcal{R}]{*}$).

3.2 Natural Deduction Modulo

Using some equivalence $\xleftrightarrow[\mathcal{R}]{*}$ defined by a rewrite system $\mathcal{R}$, we can define natural deduction modulo as in [13]. Its inference rules are represented in Fig. 1. They are the same as the one introduced by Gentzen [15], except that we work modulo the rewrite relation. Leaves of a proof that are not introduced by some inference rules (contrary to A in $\Rightarrow$-i for instance) are the assumptions of the proof. Note that if we do not work modulo, $\Rightarrow$-e is exactly the same as (6).

The length of a proof is the number of inferences used in it. We will denote by $\mathcal{T} \vdash^{\mathsf{N}}_{k\,\mathcal{R}} P$ the fact that there exists a proof of P of length at most k using a finite subset of $\mathcal{T}$ ($\mathcal{T}$ can be infinite) as assumptions. In the case where $\mathcal{R} = \emptyset$, we are back to pure natural deduction, and we will use $\mathcal{T} \vdash^{\mathsf{N}}_{k} P$. Abusing notations, we will write $Z_i \vdash^{\mathsf{N}}_{k\,\mathcal{R}} P$ to say that there is a proof of P of length at most k using as assumptions a finite subset of instances of the axiom schemata (8) to (15).

4 Translations

4.1 From $Z_i \vdash^{\mathsf{S}}$ to $Z_i \vdash^{\mathsf{N}}$

We want to translate a proof in the schematic system of Z_i into a proof in pure natural deduction using as assumptions instances of the axiom schemata (8) to (15).

For the axiom schemata and inference rules of classical logic, we use the same translation as Gentzen, for instance the axiom schema (4) is translated into the natural deduction proof

$$\Rightarrow\text{-i}\ \dfrac{\exists\text{-i}\ \dfrac{A(\tau^j)\ \text{(i)}}{\exists \alpha^j.\ A(\alpha^j)}}{A(\tau^j) \Rightarrow \exists \alpha^j.\ A(\alpha^j)}\ \text{(i)}$$

and the inference rule (7) into

$$\exists\text{-e}\ \dfrac{\exists \alpha^j.\ B(\alpha^j)\ \text{(i)} \qquad \Rightarrow\text{-e}\ \dfrac{B(\beta^j)\ \text{(ii)} \qquad B(\beta^j) \Rightarrow A}{A}}{\Rightarrow\text{-i}\ \dfrac{A}{\exists \alpha^j.\ B(\alpha^j) \Rightarrow A}\ \text{(i)}}\ \text{(ii)}$$

(note that the side condition ensures that it is possible to consider that what will be substituted for β^j is free in A and the assumptions of the proof above $B(\beta^j) \Rightarrow A$). All these inference rules have a translation whose length does not depend on the formulæ finally substituted in the proof.

In a schematic system proof, there is also a finite number of instances of the axioms schemata for identity, Robinson's axioms and induction and comprehension schemata. We keep these instances as assumptions in natural deduction, so that we obtain a proof in natural deduction using as assumptions a finite subset of instances of the axiom schemata (8) to (15), and whose length is linear compared to the schematic system proof:

Proposition 1. *It is possible to translate a proof of length n in the schematic system for Z_i into a proof of length $O(n)$ in (pure) natural deduction using assumptions in Z_i.*

$$Z_i \vdash^{\mathsf{S}}_{k} P \quad \rightsquigarrow \quad Z_i \vdash^{\mathsf{N}}_{O(k)} P$$

4.2 From $Z_i \vdash^{\mathsf{N}}$ to $Z_i \vdash^{\mathsf{S}}$

In this section, we consider a proof of P in natural deduction, using as assumption finite instances of (8) to (15) in the language of Z_i. We translate it into a proof in the schematic system for Z_i.

This is essentially a generalization of the translation from the λ-calculus to combinatory logic (see [9]). We define mutually recursively two functions by induction on the inference rules: T transforms a proof of P in natural deduction using assumptions Γ into a proof of P in the schematic system (1) to (7) plus Γ. T_A transform a proof of P in natural deduction using assumptions Γ, A into a proof of $A \Rightarrow P$ in the schematic system (1) to (7) plus Γ.

Due to lack of space, the definition of T and T_A is given here only for the existential quantifier, but can be entirely found in the full version of this paper [4].

$$\mathrm{T}\left(\exists\text{-i}\,\dfrac{\begin{array}{c}\pi\\ \{t/x\}A\end{array}}{\exists x.\ A}\right) \stackrel{!}{=} (6)\,\dfrac{\begin{array}{c}\mathrm{T}\,(\pi)\\ \{t/x\}A\end{array} \qquad \{t/x\}A \Rightarrow \exists x.\ A \;(4)}{\exists x.\ A}$$

$$\mathrm{T}\left(\exists\text{-e}\,\dfrac{\begin{array}{c}\pi_1\\ \exists x.\ A\end{array} \qquad \pi_2\left\{\begin{array}{c}[\{y/x\}A]\\ B\end{array}\right.}{B}\right) \stackrel{!}{=} (6)\,\dfrac{\begin{array}{c}\mathrm{T}\,(\pi_1)\\ \exists x.\ A\end{array} \qquad (7)\,\dfrac{\begin{array}{c}\mathrm{T}_A\,(\pi_2)\\ \{y/x\}A \Rightarrow B\end{array}}{(\exists x.\ A) \Rightarrow B}}{B}$$

$$\mathrm{T}_A\left(\exists\text{-i}\,\dfrac{\pi\left\{\begin{array}{c}[A]\\ \{t/x\}B\end{array}\right.}{\exists x.\ B}\right) \stackrel{!}{=}$$

$$(6)\,\dfrac{\{t/x\}B \Rightarrow \exists x.\ B \;(4) \qquad (6)\,\dfrac{\begin{array}{c}\mathrm{T}_A\,(\pi)\\ A \Rightarrow \{t/x\}B\end{array} \qquad \cdots\ (2)}{(\{t/x\}B \Rightarrow \exists x.\ B) \Rightarrow A \Rightarrow \exists x.\ B}}{A \Rightarrow \exists x.\ B}$$

$$\mathrm{T}_A\left(\exists\text{-e}\,\dfrac{\pi_1\left\{\begin{array}{c}[A]\\ \exists x.\ B\end{array}\right. \qquad \pi_2\left\{\begin{array}{c}[A, \{y/x\}B]\\ C\end{array}\right.}{C}\right) \stackrel{!}{=}$$

$$(6)\,\dfrac{(6)\,\dfrac{(7)\,\dfrac{\begin{array}{c}\mathrm{T}_{\{y/x\}B}\begin{pmatrix}\mathrm{T}_A\,(\pi_2)\\ A \Rightarrow C\end{pmatrix}\\ \{y/x\}B \Rightarrow A \Rightarrow C\end{array}}{\exists x.\ B \Rightarrow A \Rightarrow C} \qquad (6)\,\dfrac{\begin{array}{c}\mathrm{T}_A\,(\pi_1)\\ A \Rightarrow \exists x.\ B\end{array} \qquad \cdots\ (2)}{(\exists x.\ B \Rightarrow A \Rightarrow C) \Rightarrow A \Rightarrow A \Rightarrow C}}{A \Rightarrow A \Rightarrow C} \qquad \cdots\ (1)}{A \Rightarrow C}$$

It can be verified that this definition transforms a proof of size n into a proof of size $O(3^n)$. Due to [7, Corollary 3.4], we could have found, at least for the propositional part, a polynomial translation. Nevertheless all we need in this paper is the fact that the increase of the proof length in the translation is bounded.

Proposition 2. *It is possible to translate a proof of length n in the (pure) natural deduction using assumptions in Z_i into a proof of length $O(3^n)$ in the schematic system for Z_i.*

$$Z_i \vdash^{\mathsf{N}}_{k} P \quad \rightsquigarrow \quad Z_i \vdash^{\mathsf{S}}_{O(3^k)} P$$

4.3 From $Z_{i+1} \vdash^{\mathsf{S}}$ and $Z_{i+1} \vdash^{\mathsf{N}}$ to $Z_i \vdash^{\mathsf{N}}_{\mathcal{R}_i}$

This time, we translate a proof in the schematic system for Z_{i+1} into a proof in natural deduction modulo using as assumption instances of the axiom schemata (8) to (15), but in the language of Z_i. The point is that, using modulo, it is possible to downshift an order.

We follow the translation of Section 4.1, except for the axiom schemata (10), (14) and (15) that are instantiated by formulæ that are in the language of Z_{i+1} but not in the language of Z_i. To translate these schemata, we will use the work of F. Kirchner [18] which permits to express first-order theories using a finite number of axioms. The idea is to transform some metaformula $A(t_1, \ldots, t_n)$ used in an axiom schema into a formula of the form $\langle t_1, \ldots, t_n \rangle \in \gamma$ where γ will be some term representing what formula will be actually substituted for A.

Following F. Kirchner's method, we add new sorts ℓ for lists and c for classes, as well as new function symbols and predicate

$$\begin{array}{lll} 1^j &:& j \\ S^j &:& [j] \rightarrow j \\ \cdot[\cdot]^j &:& [j;\ell] \rightarrow j \end{array} \quad \begin{array}{lll} nil &:& \ell \\ ::^j &:& [j;\ell] \rightarrow \ell \\ \dot{=} &:& [0;0] \rightarrow c \\ \dot{\in}^j &:& [j;j+1] \rightarrow c \end{array} \quad \begin{array}{lll} \cup &:& [c;c] \rightarrow c \\ \cap &:& [c;c] \rightarrow c \\ \supset &:& [c;c] \rightarrow c \end{array} \quad \begin{array}{lll} \emptyset &:& c \\ \mathcal{P}^j &:& [c] \rightarrow c \\ \mathcal{C}^j &:& [c] \rightarrow c \\ \epsilon &:& [\ell;c] \end{array} .$$

$\langle \alpha_1, \ldots, \alpha_n \rangle$ will be syntactic sugar for $\alpha_1 ::^{j_1} \cdots :: \alpha_n ::^{j_n} nil$ for the appropriate j_m. We change the axiom schemata (10), (14) and (15) into the following *axioms*:

$$\forall \gamma^c.\ \forall \alpha^0 \beta^0.\ \alpha^0 = \beta^0 \Rightarrow \langle \alpha^0 \rangle \in \gamma^c \Rightarrow \langle \beta^0 \rangle \in \gamma^c \tag{16}$$

$$\forall \gamma^c. \langle 0 \rangle \in \gamma^c \Rightarrow \left(\forall \beta^0.\ \langle \beta^0 \rangle \in \gamma^c \Rightarrow \langle s(\beta^0) \rangle \in \gamma^c \right) \Rightarrow \forall \alpha^0.\ \langle \alpha^0 \rangle \in \gamma^c \tag{17}$$

For all $0 \leq j < i$,

$$\forall \gamma^c.\ \exists \alpha^{j+1}.\ \forall \beta^j.\ \beta^j \in^j \alpha^{j+1} \Leftrightarrow \langle \beta^j \rangle \in \gamma^c \tag{18}$$

The rewrite system $\mathcal{R}_i$ is then the following:

$$\begin{array}{rcl} t[nil]^j &\rightarrow& t \\ 1^j[t ::^j l]^j &\rightarrow& t \\ S^j(n)[t ::^j l]^j &\rightarrow& n[l]^j \\ s(n)[l]^0 &\rightarrow& s(n[l]^0) \\ (t_1 + t_2)[l]^0 &\rightarrow& t_1[l]^0 + t_2[l]^0 \\ (t_1 \times t_2)[l]^0 &\rightarrow& t_1[l]^0 \times t_2[l]^0 \\ l \in\ \dot{=}(t_1, t_2) &\rightarrow& t_1[l]^0 = t_2[l]^0 \end{array} \qquad \begin{array}{rcl} l \in \dot{\in}^j(t_1, t_2) &\rightarrow& t_1[l]^j \in^j t_2[l]^{j+1} \\ l \in A \cup B &\rightarrow& l \in A \vee l \in B \\ l \in A \cap B &\rightarrow& l \in A \wedge l \in B \\ l \in A \supset B &\rightarrow& l \in A \Rightarrow l \in B \\ l \in \emptyset &\rightarrow& \bot \\ l \in \mathcal{P}^j(A) &\rightarrow& \exists x.\ x ::^j l \in A \\ l \in \mathcal{C}^j(A) &\rightarrow& \forall x.\ x ::^j l \in A \end{array}$$

Note that this system is finite, terminating (either the size of a list decreases, or else a $\cdot[\cdot]$ or an ϵ goes more inside or disappears), confluent (the only critical pairs, of the form: $f(t_1, \ldots, t_n) \underset{\mathcal{R}_i}{\longleftarrow} f(t_1, \ldots, t_n)[nil] \underset{\mathcal{R}_i}{\longrightarrow} f(t_1[nil], \ldots, t_n[nil])$, are easily joinable), and linear (variables appears only once in the left hand side of the rewrite rules).

Proposition 2 of [18] says that it is possible, for any formula P of the language of i-th order arithmetic, to prove $\exists E.\ \forall x_1 \cdots x_n.\ \langle x_1, \ldots, x_n \rangle \in E \Leftrightarrow P$.

$$\forall\text{-e}\ \frac{\forall\gamma^c.\ \forall\alpha^0\beta^0.\ \alpha^0 = \beta^0 \Rightarrow \langle\alpha^0\rangle \in \gamma^c \Rightarrow \langle\beta^0\rangle \in \gamma^c\ (16)}{\forall\alpha^0\beta^0.\ \alpha^0 = \beta^0 \Rightarrow A(\alpha^0) \Rightarrow A(\beta^0)}\ \langle\alpha^0\rangle \in E^x_A \Rightarrow \langle\beta^0\rangle \in E^x_A \xrightarrow[\mathcal{R}_i]{*} A(\alpha^0) \Rightarrow A(\beta^0)$$

$$\forall\text{-e}\ \frac{\forall\gamma^c.\langle 0\rangle \in \gamma^c \Rightarrow (\forall\beta^0.\ \langle\beta^0\rangle \in \gamma^c \Rightarrow \langle s(\beta^0)\rangle \in \gamma^c) \Rightarrow \forall\alpha^0.\ \langle\alpha^0\rangle \in \gamma^c\ (17)}{A(0) \Rightarrow (\forall\beta^0.\ A(\beta^0) \Rightarrow A(s(\beta^0))) \Rightarrow \forall\alpha^0.\ A(\alpha^0)}\ \text{for all } t,\ \langle t\rangle \in E^x_A \xrightarrow[\mathcal{R}_i]{*} A(t)$$

$$\forall\text{-e}\ \frac{\forall\gamma^c.\ \exists\alpha^{j+1}.\ \forall\beta^j.\ \beta^j \in^j \alpha^{j+1} \Leftrightarrow \langle\beta^j\rangle \in \gamma^c\ (18)}{\exists\alpha^{j+1}.\ \forall\beta^j.\ \beta^j \in^j \alpha^{j+1} \Leftrightarrow A(\beta^j)}\ \langle\beta^j\rangle \in E^x_A \xrightarrow[\mathcal{R}_i]{*} A(\beta^j)$$

Fig. 2. Translations of the axiom schemata (10), (14) and (15)

Moreover, the proof of this proposition shows us how to construct the witness E. We will denote it by $E_P^{x_1,\dots,x_n}$. Then, one can prove that $\langle t_1,\dots,t_n\rangle \in E_P^{x_1,\dots,x_n} \xrightarrow[\mathcal{R}_i]{*} \{t_1/x_1,\dots,t_n/x_n\}P$. For instance, consider the formula $x = 0 \vee \exists y.\ x \in^0 y$, which will be denoted by P. Then E^x_P equals $\dot{=}(1, S(0)) \cup \mathcal{P}^1\left(\dot{\in}^0(S(1),1)\right)$ and $\langle t\rangle \in E^x_P$ can be rewritten to $t = 0 \vee \exists x.\ t \in^0 x$.

Consequently, the axiom schemata (10), (14) and (15) for formulæ of the language of Z_{i+1} but not in the language of Z_i are replaced by the proofs in Fig. 2. In these translations, we need to instantiate γ^c with some E^x_A. It is well-known that the instantiations are the most problematic rules in deductive systems, at least for automated provers (e.g. they are what leads to nondeterminism and/or nontermination of tableaux methods for first order logic), because the instantiated term must be somehow guessed. Nevertheless, the instantiation here is entirely and automatically determined by the formula used in the schema, so that no harm is done.

Using this, a proof in the schematic system for Z_{i+1} can be translated into a proof of P in natural deduction modulo $\mathcal{R}_i$ using as assumptions the axioms (10), (14) and (15) as well as a finite subset of instances the axiom schemata (8) to (15) for i-th order arithmetic, and whose length is linear compared to the schematic system proof:

Proposition 3. *It is possible to translate a proof of length n in the schematic system for Z_{i+1} into a proof of length $O(n)$ in the natural deduction modulo $\mathcal{R}_i$ using assumptions in Z_i, (16), (17) and (18).*

$$Z_{i+1} \vdash^{\mathsf{S}}_{k} P \quad \rightsquigarrow \quad Z_i, (16), (17), (18) \vdash^{\mathsf{N}}_{O(k)}{}_{\mathcal{R}_i} P$$

This result can also be stated entirely in natural deduction

Theorem 2. *For all $i \geq 0$, there exists a (finite) rewrite system $\mathcal{R}_i$ and a finite set of axioms Γ such that for all formulæ P, if $Z_{i+1} \vdash^{\mathsf{N}}_{k} P$ then $Z_i, \Gamma \vdash^{\mathsf{N}}_{O(k)}{}_{\mathcal{R}_i} P$.*

Proof. Let Γ be $\{(16),(17),(18)\}$. We replace the instance of the axiom schemata (10), (14) and (15) by the axioms (16), (17) and(18) as indicated in Fig. 2. □

Note that, contrarily to HOL-$\lambda\sigma$ which permits to simulate Higher Order Logic, the rewrite system purposed here is finite and terminating.

The fact to add the finite set of axioms Γ could be seen as some deceit, because we do not work really in Z_i, but in a theory strictly stronger. By the way, due to Theorem 2, it is possible to prove the consistency of Z_i in Z_i, Γ modulo $\mathcal{R}_i$. Nevertheless, the point here is that it is possible, by working modulo $\mathcal{R}_i$, to simulate Z_{i+1} using a finite set of axioms, and not axiom schemata, without exploding the length of the proofs. If we were not working modulo this rewrite system, but using a finite theory compatible with it (i.e. proving exactly the same formulæ), then it would not be possible to give a bound to the translation:

Proposition 4. *For all $i \geq 0$, for all finite theories T_i compatible with $\mathcal{R}_i$, there is an infinite family $\mathcal{F}$ such that*

1. *for all $P \in \mathcal{F}$, $Z_i, \Gamma, T_i \vdash^{\mathsf{S}} P$*
2. *there is a fixed $k \in \mathbb{N}$ such that for all $P \in \mathcal{F}$, $Z_{i+1} \vdash^{\mathsf{S}}_{k\ steps} P$*
3. *there is no fixed $k \in \mathbb{N}$ such that for all $P \in \mathcal{F}$, $Z_i, \Gamma, T_i \vdash^{\mathsf{S}}_{k\ steps} P$.*

It could also have been possible to translate the formulæ that one wants to prove, as is done in [14], where a formula of first order arithmetic is transformed by adding the information that some variable n is an integer using some predicate $N(n)$ which can be rewritten into an axiom corresponding to the induction schema for first order arithmetic. Here, P could be translated into (16) $\Rightarrow$ (17) $\Rightarrow$ (18) $\Rightarrow P$.

5 Application to Speed-Ups in Arithmetic

5.1 Bypassing Buss' Speed-Up Using Modulo

The goal of this section is to prove that one can work in Z_i modulo some rewrite system $\mathcal{R}_i$ to be able to build proof as small as the one of Z_{i+1}. Indeed, Theorem 2 permits to show that Gödel's theorem does not extend if one works modulo $\mathcal{R}_i$ (what is formulated here in a positive way):

Corollary 1 (of Theorem 2). *For all $i \geq 0$, there exists a (finite) rewrite system $\mathcal{R}_i$ and a finite set of axioms Γ such that for all infinite family $\mathcal{F}$ of $\prod_1^0$-formulæ, if*

- *for all $P \in \mathcal{F}$, $Z_i \vdash^{\mathsf{N}} P$*
- *there is a fixed $k \in \mathbb{N}$ such that for all $P \in \mathcal{F}$, $Z_{i+1} \vdash^{\mathsf{N}}_{k\ steps} P$*

then there is a fixed $k' \in \mathbb{N}$ such that for all $P \in \mathcal{F}$, $Z_i, \Gamma \vdash^{\mathsf{N}}_{k'\ steps\ \mathcal{R}_i} P$.

5.2 Speed-Up Due to Computation

On the contrary, we want to show that it is possible to achieve the same speed-up as the one between i-th order and $i+1$-th order arithmetic just by working modulo some rewrite system in i-th order arithmetic:

Theorem 3. *For all $i \geq 0$, there is a rewrite system $\mathcal{R}_i$ such that there is an infinite family $\mathcal{F}$ such that*

1. *for all $P \in \mathcal{F}$, $Z_i \vdash^{\mathsf{N}} P$*
2. *there is a fixed $k \in \mathbb{N}$ such that for all $P \in \mathcal{F}$, $Z_i \vdash^{\mathsf{N}}_{k\ steps\ \mathcal{R}_i} P$*
3. *there is no fixed $k \in \mathbb{N}$ such that for all $P \in \mathcal{F}$, $Z_i \vdash^{\mathsf{N}}_{k\ steps} P$.*

Proof. The rewrite system $\mathcal{R}_i$ is the one defined in Section 4.3. Let $\mathcal{F}$ be the family of formulæ obtained by Theorem 1. Let $\mathcal{F}' \stackrel{!}{=} \{(16) \Rightarrow (17) \Rightarrow (18) \Rightarrow P : P \in \mathcal{F}\}$. Then:

1. For all $P' \in \mathcal{F}'$, $Z_i \vdash^{\mathsf{N}} P'$: we know that $Z_i \vdash^{\mathsf{S}} P$, therefore using Proposition 1, $Z_i \vdash^{\mathsf{N}} P$ and, adding to this proof $2+i$ times $\Rightarrow$-i, $Z_i \vdash^{\mathsf{N}} P'$.
2. There is a k such that for all $P' \in \mathcal{F}'$, $Z_i \vdash^{\mathsf{N}}_{k\ \mathcal{R}_i} P'$: there exists some k such that for all $P \in \mathcal{F}$, $Z_{i+1} \vdash^{\mathsf{S}}_{k} P$. Using Proposition 3, there exists some K such that for all $P \in \mathcal{F}$, we have $Z_i, (16), (17), (18) \vdash^{\mathsf{S}}_{K\ \mathcal{R}_i} P$ and one can add $2+i$ times $\Rightarrow$-i to obtain a proof of P'.
3. There is no k such that for all $P' \in \mathcal{F}'$, $Z_i \vdash^{\mathsf{N}}_{k} P'$: Suppose by contradiction that there is a k such that for all $P' \in \mathcal{F}'$, $Z_i \vdash^{\mathsf{N}}_{k} P'$, then using $2+i$ times $\Rightarrow$-e, we have $Z_i, (16), (17), (18) \vdash^{\mathsf{N}}_{k+2+i} P$. But (16), (17) and (18) use function symbols not appearing in P nor Z_i (for instance ϵ). Therefore they cannot be used in a proof of P in Z_i, so that in fact $Z_i \vdash^{\mathsf{N}}_{k+2+i} P$. Then, using Proposition 2, $Z_i \vdash^{\mathsf{S}}_{O(3^k)} P$, and that will be in contradiction with the fact that there is no K such that for all P, $Z_i \vdash^{\mathsf{S}}_{K} P$.

Schematically,

$$\begin{array}{lcl} Z_{i+1} \vdash^{\mathsf{S}}_{k} P & \overset{\text{Prop. 3}}{\rightsquigarrow} & Z_i, (16), (17), (18) \vdash^{\mathsf{N}}_{K\ \mathcal{R}_i} P \rightsquigarrow Z_i \vdash^{\mathsf{N}}_{K+2+i\ \mathcal{R}_i} P' \\ \text{Theo. 1} \updownarrow & & \\ Z_i \vdash^{\mathsf{S}}_{\not{3}^{\not{k}}} P & \underset{\text{Prop. 2}}{\overset{\text{Prop. 1}}{\rightsquigarrow \atop \leftsquigarrow}} & Z_i, (16), (17), (18) \vdash^{\mathsf{N}}_{\not{k}} P \rightsquigarrow Z_i \vdash^{\mathsf{N}}_{\not{k}} P' \end{array}$$

$\square$

Note that it is possible to get speed-ups in deduction modulo w.r.t pure natural deduction with systems much more simpler than for higher order arithmetic. (Take for instance the rule $s(x)+y \rightarrow x+s(y)$ and consider the formulæ $\underline{n}+\underline{n} = \underline{n+n}$ where $\underline{n}$ denotes the usual representation of the natural number n using 0 and s, for all natural numbers n.) Our last result however, combined with Corollary 1, permits to conclude that proof-length speed-ups *in arithmetic* result from the computational part of the proofs, which is expressed by the rewrite systems $\mathcal{R}_i$.

6 Conclusion and Perspectives

We have first proved that it is possible to use some rewrite system to simulate the difference between i-th and $i+1$-th order arithmetic at the condition to add three extra axioms which replace the missing axiom schemata. This simulation is linear in terms of proof length, which permits to prove that there is no proof-length speed-up between $i+1$-th order arithmetic and its simulation, on the contrary to without modulo as it is expressed in Buss' theorem. Furthermore, this simulation allows to get the same proof speed-up for deduction modulo over non modulo systems than the one of Buss' theorem. Together with the first result, this proves that the gap between i-th and $i+1$-th order arithmetic is in fact due to the computational part of the proofs. In this particular case, we also clearly identify the computation occurring in the proofs with a finite, terminating and confluent (so, in a sense, deterministic) rewrite system. This is not surprising, because, if one looks carefully, the proof of Theorem 1 given by Buss in [6] deeply relies on the fact that it is possible to define some truth predicate for the formulæ of the preceding order. Therefore, in a sense, it is possible, in $i+1$-th order arithmetic, to compute the validity of a formula in i-th order arithmetic.

Speed-ups in deduction modulo must not be considered as cheating, by hiding part of the proofs in the congruence. This must be thought of as a way to separate what is deduced and what is computed. To find a proof, both parts need to be built. To check the proof however, only the deductive part is necessary, because the rest can be effectively computed during the verification (hence the need to have a decidable congruence, even better if it is determined by simple deterministic algorithm). This can be applied to automated and interactive theorem proving, as well as in representation of proofs in natural language (where all computational details are often implicitly left the reader).

These results are encouraging indicators that it is as good to work directly in higher order logics, as is done in the current interactive theorem provers, such as Coq [22] or Isabelle/HOL [20], or using a first order implementation of these logics, as could be done in a proof assistant based on deduction modulo (or on its sequel named superdeduction, see [3]). This paper gives clues to answer positively this question, although we were interested in the step between i-th order and $i+1$-th order arithmetic, and not between first order and higher order logic. The fact that higher order resolution can be simulated step by step by ENAR [11] is not a solution, because there may exist some other higher order proof system that produce proofs that cannot be conveniently translated in a first order system modulo. So, our next challenge will be, starting from the current results, to investigate how exactly higher order logic prevails or not over first order logic, by studying more closely the simulation of higher order logic.

A first direction to do so will be to prove that it is possible to apply transitivity between the simulation of Z_{i+1} in Z_i and the one of Z_{i+2} in Z_{i+1}, in order to get a simulation of Z_{i+2} in Z_i, for instance by combining $\mathcal{R}_i$ and $\mathcal{R}_{i+1}$. In addition to the expression of first order arithmetic as a theory modulo [14], this would lead to the linear simulation of higher order arithmetic entirely as a theory modulo. It should however be noted that one of the main advantage of our rewrite systems w.r.t. HOL-$\lambda\sigma$, i.e. its finiteness, will be lost because of the need for a rule to decompose $\dot{\in}^i$ for all orders i.

Another direction would be to look directly at the difference of the lengths of proofs in the expression of HOL in the sequent calculus modulo [11], or of every PTS in $\lambda\Pi$ modulo [8].

Acknowledgments. The author wishes to thank G. Dowek, T. Hardin, C. Kirchner and the anonymous referees for many discussions and comments about this paper.

References

1. Baader, F., Nipkow, T.: Term Rewriting and all That. CUP (1998)
2. Bonacina, M.P., Dershowitz, N.: Abstract canonical inference. ACM Trans. Comput. Logic 8 (2007)
3. Brauner, P., Houtmann, C., Kirchner, C.: Principles of superdeduction. In: LICS, IEEE Computer Society, Los Alamitos (to appear, 2007)
4. Burel, G.: Unbounded proof-length speed-up in deduction modulo. Research report (2007), Available at `http://hal.inria.fr/inria-00138195`
5. Buss, S.R.: Polynomial size proofs of the propositional pigeonhole principle. The Journal of Symbolic Logic 52, 916–927 (1987)
6. Buss, S.R.: On Gödel's theorems on lengths of proofs I: Number of lines and speedup for arithmetics. The Journal of Symbolic Logic 59, 737–756 (1994)
7. Cook, S.A., Reckhow, R.A.: The relative efficiency of propositional proof systems. The Journal of Symbolic Logic 44, 36–50 (1979)
8. Cousineau, D., Dowek, G.: Embedding pure type systems in the lambda-pi-calculus modulo. In: TLCA (to appear, 2007)
9. Curry, H.B., Feys, R., Craig, W.: Combinatory Logic, vol. 1. Elsevier Science Publishers, B. V., North-Holland, Amsterdam (1958)
10. Dershowitz, N., Kirchner, C.: Abstract Canonical Presentations. Theoretical Computer Science 357, 53–69 (2006)
11. Dowek, G., Hardin, T., Kirchner, C.: HOL-$\lambda\sigma$ an intentional first-order expression of higher-order logic. Math. Structures Comput. Sci. 11, 1–25 (2001)
12. Dowek, G., Hardin, T., Kirchner, C.: Theorem proving modulo. Journal of Automated Reasoning 31, 33–72 (2003)
13. Dowek, G., Werner, B.: Proof normalization modulo. The Journal of Symbolic Logic 68, 1289–1316 (2003)
14. Dowek, G., Werner, B.: Arithmetic as a theory modulo. In: Giesl, J. (ed.) RTA 2005. LNCS, vol. 3467, pp. 423–437. Springer, Heidelberg (2005)
15. Gentzen, G.: Untersuchungen über das logische Schliessen. Translated In: Szabo, M.E. (ed.) Mathematische Zeitschrift, vol. 39, pp. 176–210, 405–431 (1934) (The Collected Papers of Gerhard Gentzen as Investigations into Logical Deduction)
16. Gödel, K.: On the length of proofs. In: Feferman, S., et al. (eds.) Kurt Gödel: Collected Works, vol. 1, pp. 396–399. Oxford University Press, Oxford (1986)
17. Guglielmi, A.: Polynomial size deep-inference proofs instead of exponential size shallow-inference proofs (2004), Available at `http://cs.bath.ac.uk/ag/p/AG12.pdf`
18. Kirchner, F.: A finite first-order theory of classes (2006), Available at `http://www.lix.polytechnique.fr/Labo/Florent.Kirchner/doc/fotc2006.pdf`

19. Mostowski, A., Robinson, R.M., Tarski, A.: Undecidable Theories. In: Studies in Logic and the Foundations of Mathematics, North-Holland, Amsterdam (1953)
20. Nipkow, T., Paulson, L.C., Wenzel, M.: Isabelle/HOL — A Proof Assistant for Higher-Order Logic. In: Nipkow, T., Paulson, L.C., Wenzel, M. (eds.) Isabelle/HOL. LNCS, vol. 2283, Springer, Heidelberg (2002)
21. Parikh, R.J.: Some results on the length of proofs. Transactions of the ACM 177, 29–36 (1973)
22. The Coq Development Team: The Coq Proof Assistant Reference Manual. INRIA. Version 8.0 (2006), available at `http://coq.inria.fr/doc/main.html`

Propositional Logic for Circuit Classes

Klaus Aehlig[1,⋆] and Arnold Beckmann[2]

[1] Department of Computer Science
University of Toronto
10 King's College Road, Toronto, ON M5S 3G4, Canada
klausa@cs.toronto.edu

[2] Department of Computer Science
University of Wales Swansea
Singleton Park, Swansea, SA2 8PP, United Kingdom

Abstract. By introducing a parallel extension rule that is aware of independence of the introduced extension variables, a calculus for quantified propositional logic is obtained where heights of derivations correspond to heights of appropriate circuits. Adding an uninterpreted predicate on bit-strings (analog to an oracle in relativised complexity classes) this statement can be made precise in the sense that the height of the most shallow proof that a circuit can be evaluated is, up to an additive constant, the height of that circuit.

The main tool for showing lower bounds on proof heights is a variant of an iteration principle studied by Takeuti. This reformulation might be of independent interest, as it allows for polynomial size formulae in the relativised language that require proofs of exponential height.

1 Introduction and Related Work

In systems like "extended Frege" there is a rule that allows one to introduce a new variable by a defining clause $p \leftrightarrow A$. If several variables are to be introduced, several instances of this rule have to be used. This holds regardless of the presence or absence of dependencies between these variables.

However, such dependencies are known to make a big difference in the world of computation. Both, uniform AC^0 and polynomial time can be described by families of polynomial size circuits. Nevertheless, AC^0 has much smaller computational power. The reason is that the nodes in AC^0 circuits are constrained to be arranged in a finite number of layers.

Even though various propositional calculi are known for small complexity classes, none reflects correctly the height of circuits. We suggest a calculus that has this property and can serve as a framework for investigating the "circuit strength" of various propositional calculi and small complexity classes; the latter come in via propositional translations of appropriate theories of Bounded Arithmetic [4,6].

⋆ Supported by DFG grant Ae 102/1-1. Part of the work was carried out while affiliated with University of Wales Swansea and supported by EPSRC grant EP/D03809X/1.
Corresponding author.

J. Duparc and T.A. Henzinger (Eds.): CSL 2007, LNCS 4646, pp. 512–526, 2007.

We consider relativised circuit classes [15]. That is, our circuits will not only contain logical gates but also gates that query an oracle. There are several motivations for doing so. Hardly any separations of unrelativised small complexity classes are known, but separating relativised circuit classes is straightforward. So, in order to precisely state that the calculus adequately reflects the differences between different circuit classes, we need to consider the relativised forms; an absolute separation of the levels of the AC^k hierarchy seems out of reach at the moment. Moreover, this calculus is intended as a target for propositional translations of theories of Bounded Arithmetic. Following standard proof theoretical practise [11], a better classification of theories can be obtained for the variants relativised to an uninterpreted predicate.

Quantified propositional logic in relation to complexity classes and bounded arithmetic has been studied by Krajíček and Pudlák [9]. They introduced various dag-like (G_1, G_2,..., G) and tree-like systems (G_1^*, G_2^*,..., G^*). Cook and Morioka (in a slightly modified setting) identified [5] G_0 and G_0^* which relate to NC^1. One motivation for the study of restricted propositional proof systems is the relation to (weak) theories of bounded arithmetic [4,6]. For various complexity classes, corresponding proof systems [10,12] have been identified. However, a unifying framework for the propositional systems still seems to be missing. We suggest a calculus which is flexible enough to allow for embedding of various theories, but is still strict enough that the *height* of proofs is a meaningful measure.

Studying the height of proofs is a standard approach in ordinal informative proof theory [11] and has been adopted to the Bounded Arithmetic setting by Beckmann [1]. It was also implicitly used by Krajíček [8].

Our research presented here investigates a particular form of the iteration principle. An important source for this has been Takeuti's investigations [14] where he obtained separations of some versions of bounded arithmetic theories [3] related to circuit complexity classes. A different form of the iteration principle has been introduced and studied by Buss and Krajíček [2] to obtain separations between bounded arithmetic theories related to (relativised) polynomial time and polynomial local search (PLS).

This article is organized as follows. In Section 2 we define our calculus AC^0-Tait. In Section 3 we consider the formula expressing that a circuit of height h can be evaluated. We note that this formula can be proven by a proof of height $h+\mathcal{O}(1)$. For the other direction we need a few preparations, that are interesting results in their own right. First we study in Section 4 a formula expressing that a function can be iterated ℓ times; we show that a proof of this formula requires height at least ℓ. As the iteration formula is a polynomial size $\Sigma_1^q(\alpha)$-formula and can express exponentially long iterations, this establishes an exponential lower bound for the calculus with cuts on arbitrary quantifier-free formulae. In Section 5 we study a version of the calculus, extended with cuts, and prove cut-elimination. The cut-lemma will allow us to transform a proof that a particular circuit can be evaluated into a proof of the iteration principle without increasing the height by more than a constant. Putting things together in Section 6 shows that there are circuits of height h where a proof that they can be evaluated requires height at least $h - \mathcal{O}(1)$.

2 Quantified Propositional Logic and Definition of the Calculus

In this section we will introduce our calculus. It will be in the style of Tait [13], that is, roughly, one-sided sequent calculus. Following standard simplifications, a sequent is a set of formulae, and negation is an operation on formulae, not a logical symbol.

Definition 1. The *atoms of propositional logic* are variables $p, q, r, \ldots$, their negations $\bar{p}, \bar{q}, \bar{r}, \ldots$, as well as the constants T and F for truth and falsity.

The set of all propositional atoms is denoted by $\mathcal{A}$ and we use $\wp$ to range over elements of $\mathcal{A}$.

The set of *quantified propositional formulae* $A, B, C, \ldots$ is built up from the atoms of propositional logic and parameter $\alpha_k \wp_1 \ldots \wp_k$ and negated parameter $\bar{\alpha}_k \wp_1 \ldots \wp_k$ where $\wp_1, \ldots, \wp_k$ are propositional atoms, by conjunctions $\bigwedge_k A_1 \ldots A_k$ and disjunctions $\bigvee_k A_1 \ldots A_k$, and universal $\forall_k p_1 \ldots p_k A$ and existential $\exists_k p_1 \ldots p_k A$ quantification.

Here $k \geq 1$ is a natural number on the meta level. The variables $p_1, \ldots, p_k$ and their negations $\bar{p}_1, \ldots, \bar{p}_k$ are bound in $\forall_k p_1 \ldots p_k A$ and $\exists_k p_1 \ldots p_k A$.

Syntactical equality is denoted by $\equiv$. A quantified propositional formula without any quantifications is called a *propositional formula.* We use the expression *purely propositional formula* for a propositional formula, if we want to emphasise that it is not quantified.

We write $\wedge$ and $\vee$ for $\bigwedge_2$ and $\bigvee_2$, respectively. We use $A \wedge B$ and $A \vee B$ as abbreviations for $\wedge AB$ and $\vee AB$, respectively, if there is no danger of confusion. Also, parentheses may be used to facilitate reading or to disambiguate these abbreviations.

By induction on A a formula $\neg A$ is defined according to the de Morgan rules in the obvious way, e.g., $\neg p \equiv \bar{p}$, $\neg \bar{p} \equiv p$, $\neg(\alpha_k \wp_1 \ldots \wp_k) \equiv \bar{\alpha}_k \wp_1 \ldots \wp_k$, $\neg(\bigwedge_k A_1 \ldots A_k) \equiv \bigvee_k (\neg A_1) \ldots (\neg A_k)$, and so on. A simple induction on A shows that $\neg\neg A \equiv A$.

If A is a quantified propositional formula, $\vec{p}$ are pairwise disjoint propositional variables, and $\vec{B}$ are quantified propositional formulae, then by $A[\vec{B}/\vec{p}]$ we denote the simultaneous capture-free substitution of all p_i by B_i and of all $\bar{p}_i$ by $\neg B_i$.

We use the notation $A(\vec{p})$ to distinguish certain variables of A, in order to be able to use $A(\vec{B})$ as a shorthand for the substitution $A[\vec{B}/\vec{p}]$. This notation does not imply that these variables actually do occur free and the list $\vec{p}$ does not necessarily exhaust all the free variables of A.

We use $\Gamma, \Delta, \ldots$ to denote finite sets of formulae.

Definition 2. The *propositional rules* are the following rules.

$$\frac{}{\Gamma, p, \bar{p}} \qquad \frac{}{\Gamma, \mathrm{T}} \qquad \frac{}{\Gamma, \alpha_k(\wp_1, \ldots, \wp_k), \bar{\alpha}_k(\wp_1, \ldots, \wp_k)}$$

$$\frac{\Gamma, A_i}{\Gamma, \bigvee_k A_1 \ldots A_k} \qquad \frac{\ldots \Gamma, A_i \ldots \quad (1 \leq i \leq k)}{\Gamma, \bigwedge_k A_1 \ldots A_k}$$

Definition 3. The *rules of parameter extensionality* are the following rules.

$$\frac{\Gamma, \alpha_k(\wp_1, \ldots, \wp_k) \qquad \ldots \Gamma, \wp_i \leftrightarrow \wp_i' \ldots \qquad (1 \leq i \leq k)}{\Gamma, \alpha_k(\wp_1', \ldots, \wp_k')}$$

$$\frac{\Gamma, \bar{\alpha}_k(\wp_1, \ldots, \wp_k) \qquad \ldots \Gamma, \wp_i \leftrightarrow \wp_i' \ldots \qquad (1 \leq i \leq k)}{\Gamma, \bar{\alpha}_k(\wp_1', \ldots, \wp_k')}$$

Definition 4. The rules of quantification are the following rules.

$$\frac{\Gamma, A(\vec{a})}{\Gamma, \forall_k \vec{p} A(\vec{p})} \qquad \frac{\Gamma, A(\vec{\wp})}{\Gamma, \exists_k \vec{p} A(\vec{p})}$$

Here $\vec{a}$ have to be pairwise distinct eigenvariables. The $\vec{\wp}$ may be arbitrary propositional atoms.

Definition 5. The *cut rule* is the following rule.

$$\frac{\Gamma, A \qquad \Gamma, \neg A}{\Gamma}$$

The formula A in the cut rule is called the "cut formula".

One of the problems that can be solved in AC^0 is the following:

> Given truth values $p_1, \ldots, p_n$ and $q_1, \ldots, q_n$, output q_i if i is the smallest index such that p_i is true.

A similar task in standard calculi of propositional logic would require a sequence of cuts, thus artificially increasing the height. As our investigations are essentially based on differences like constant versus logarithmic height, we cannot afford this increase. We therefore introduce a new rule allowing multiple cuts at once.

The presence of this rule will be essential in Corollary 44 where it is used to obtain a proof of *constant* height.

Definition 6. The multi-cut rule is the rule

$$\frac{\ldots \qquad \Gamma, \Delta_i \qquad \ldots}{\Gamma}$$

where the Δ_i are sets of purely propositional formulae such that from the collection of the Δ_i the empty sequent can be derived by cuts only. The weight of the multi-cut rule is $\sum_i |\Delta_i|$, where $|\Delta_i|$ is the cardinality of the set Δ_i.

In other words, if from an arbitrary number of sequents, a sequent Γ can be derived by cuts on only purely propositional formulae, then this derivation of Γ counts as a single application of the multi-cut rule. For the calculus obtained to be a proof system in the sense of Cook and Reckhow [7] we require that the sequence of cuts be annotated in notations for proofs. However, as we are only interested in the number of rules applied we will never deal with notations for proofs.

Remark 7. Using the multi-cut rule it is possible to prove purely propositional induction in constant depth. In fact, from proofs of $\Gamma, \neg A_i, A_{i+1}$ for all i, we can conclude by a single inference $\Gamma, \neg A_0, A_k$.

Next we will define the comprehension rule. It is motivated by the extension rule of extended Frege calculus. There, a new propositional variable may be introduced by the axiom $p \leftrightarrow \varphi$, if p is new, that is, does not occur anywhere earlier in the derivation. The extension rule says that if Γ can be derived from the assumption $\exists p(p \leftrightarrow \varphi)$, then it can also be derived without. Note that $\neg(\exists p(p \leftrightarrow \varphi)) \equiv \forall p \neg(p \leftrightarrow \varphi)$. As usual, the universal quantifier is expressed by the eigenvariable condition. As discussed in the introduction, we allow the introduction of several extension variables at the same time.

Definition 8. The *$\mathcal{F}$-comprehension rule of width k* is the rule

$$\frac{\Gamma, \neg(p_1 \leftrightarrow \varphi_1), \ldots, \neg(p_k \leftrightarrow \varphi_k)}{\Gamma}$$

where $\varphi_1, \ldots, \varphi_k \in \mathcal{F}$ and $p_1, \ldots, p_k$ are pairwise distinct eigenvariables, that is, variables that do not occur (free) in Γ or any of the φ_i's.

The variables p_i are also called "extension variables" and the φ_i "extension formulae".

The name "$\mathcal{F}$-Comprehension Rule" is justified by the fact, that it allows simple proofs of (propositional translations of) the comprehension axiom for formulae in $\mathcal{F}$. Consider the following derivation (where we omit some side formulae; note that weakening is admissible).

$$\dfrac{\dfrac{\dfrac{\ldots \qquad \overline{\overline{(p_i \leftrightarrow \varphi_i), \neg(p_i \leftrightarrow \varphi_i)}} \qquad \ldots}{\bigwedge_k (p_i \leftrightarrow \varphi_i), \neg(p_1 \leftrightarrow \varphi_1), \ldots, \neg(p_k \leftrightarrow \varphi_k)} \bigwedge_k}{\exists_k \vec{p} \bigwedge_k (p_i \leftrightarrow \varphi_i), \neg(p_1 \leftrightarrow \varphi_1), \ldots, \neg(p_k \leftrightarrow \varphi_k)} \exists_k}{\exists_k \vec{p} \bigwedge_k (p_i \leftrightarrow \varphi_i)} \mathcal{F}\text{-comprehension}$$

It should be noted that the height of this derivation only depends on the φ_i and is independent of k. Proposition 13 will provide the needed proofs of the first sequents and will actually show that the heights depend only on the depths of φ_i's.

Note that in all the rules we may always assume without loss of generality that the conclusion is already contained in the premise (i.e., is an element of the context Γ already). For example, a typical instance of the or-rule would in fact be

$$\frac{\Gamma, A_0 \vee A_1, A_i}{\Gamma, A_0 \vee A_1}$$

Definition 9. The AC^0-Tait calculus is given by the rules considered so far, that is, it is given by the propositional rules, the parameter extensionality rule, the rules of quantification, the cut rule with cut-formulae restricted to purely propositional formulae, the multi-cut rule, and the comprehension rule for purely propositional formulae.

We assume all our proofs to be tree-like. This is not a restriction, as we only look at the height (not the size) of proofs.

Immediately by inspection of the rules, we note that weakening is admissible. This will be used tacitly in the sequel.

Definition 10. An AC^0-Tait proof is called *w, c-slim*, if all formulae occurring in the proof have size at most w, each multi-cut rule has weight at most c, and each comprehension rule has at most c extension variables.

We write $\vdash^h_{w,c} \Gamma$ to denote that Γ has an AC^0-Tait proof of height h that is w, c-slim.

The calculus AC^0-Tait is our analogue to what in usual proof theoretic investigations corresponds to cut-free proofs. So we also consider a variant with proper cuts. In Section 5 we will show how they can be eliminated.

Definition 11. If $\mathcal{C}$ is a set of formulae that contains all the purely propositional formulae and is closed under substitution of propositional atoms for propositional atoms we define the calculus "AC^0-Tait with $\mathcal{C}$-cuts" to be AC^0-Tait, but with the cut rule liberalised to formulae in $\mathcal{C}$.

We write $d \vdash^h_{\mathcal{C};w,c} \Gamma$ to denote that d is an AC^0-Tait with $\mathcal{C}$-cuts proof of Γ of height h that is w, c-slim.

Definition 12. The size $\mathsf{sz}(A)$ and depth $\mathsf{dp}(A)$ of a formula A are defined to be the number of occurrences of atoms and connectives in A, and the length of a longest path in the syntax tree of A, respectively. In particular, $\mathsf{dp}(\mathrm{T}) = \mathsf{dp}(p) = \mathsf{dp}(\alpha_k \vec{\wp}) = 1$, $\mathsf{dp}(\bigvee_k \vec{A}) = 1 + \max\{\mathsf{dp}(A_i) \mid 1 \le i \le k\}$, $\mathsf{sz}(\mathrm{T}) = \mathsf{sz}(p) = \mathsf{sz}(\alpha_k \vec{\wp}) = 1$, $\mathsf{sz}(\bigvee_k \vec{A}) = 1 + \sum_{1 \le i \le k} \mathsf{sz}(A_i)$.

By a simple induction on A one shows

Proposition 13. $\vdash^{\mathcal{O}(\mathsf{dp}(A))}_{\mathsf{sz}(A),0} A, \neg A$

A reader familiar with theories [4] like V^0 will note that proofs in V^0 translate into families of AC^0-Tait proofs of *constant* height. In fact, Δ^B_0-comprehension in V^0 can be translated using the comprehension rule in AC^0-Tait as discussed after Definition 8. The induction implicit in the $|\cdot|$-function can be handled by the multi-cut rule, compare Remark 7. "Wide" conjunction and disjunction and quantifying blocks of propositional variables with their corresponding rules are in one-to-one correspondence with first and second order quantifiers in two-sorted Bounded Arithmetic.

3 On Evaluating Circuits

We consider the problem of proving that a circuit, possibly with oracle gates, can be evaluated.

Definition 14. Let C be a circuit with nodes $n_1, \dots, n_k$. Then we define the *evaluation formula associated with* C as the formula $\Psi_C(\vec{p})$ where $p_1, \dots, p_k$ are propositional variables associated with nodes $n_1, \dots, n_k$, respectively. Ψ_C is the conjunction of the conditions for each node. If the node i is an $\wedge$-gate, then the associated condition is

$$p_i \leftrightarrow \bigwedge_{\ell} p_{i_1} \cdots p_{i_\ell}$$

where $n_{i_1}, \dots, n_{i_\ell}$ are the inputs for node i; the condition for an $\vee$-gate is similar. In the special cases of an $\wedge$ or $\vee$-gate without inputs, we use the constants T and F, respectively.

For an oracle gate, the condition is

$$p_i \leftrightarrow \alpha_\ell(p_{i_1}, \dots, p_{i_\ell})$$

where, again, $n_{i_1}, \dots, n_{i_\ell}$ are, in that order, the inputs to node i. Similarly for a negated oracle gate.

It should be noted that Ψ_C is a formula of *constant* depth, irrespectively of the shape of the circuit. However, as we shall see, the height of the proof needed to prove that this circuit can be evaluated depends on the actual structure of the circuit.

Lemma 15. *If C is a circuit of height h, then there is a proof of height $h + \mathcal{O}(1)$ for $\exists_k \vec{p}\,\Psi_C(\vec{p})$.*

Proof. For $0 \leq \ell < h$ let $p_{i_1}^{(\ell)}, \dots, p_{i_{k_\ell}}^{(\ell)}$ be the variables associated with the nodes of level ℓ. So a variable $p_i^{(\ell)}$ depends only on variables $p_j^{(\ell')}$ for some $\ell' < \ell$. We write $C_i^{(\ell)}$ for the condition associated with $p_i^{(\ell)}$. Then the derivation

$$\dfrac{\dfrac{\dfrac{\dfrac{\dfrac{\dfrac{\cdots \quad \dfrac{}{p_i^{(\ell)} \leftrightarrow C_i^{(\ell)}, \neg(p_{i_j}^{(\ell)} \leftrightarrow C_{i_j}^{(\ell)})} \quad \cdots \quad \begin{matrix}(1 \leq \ell < h)\\ (0 \leq j < k_\ell)\end{matrix}}{\Psi_C, \neg(\vec{p}^{\,(h)} \leftrightarrow \vec{C}^{(h)}), \dots, \neg(\vec{p}^{\,(2)} \leftrightarrow \vec{C}^{(2)}), \neg(\vec{p}^{\,(1)} \leftrightarrow \vec{C}^{(1)})}\ \bigwedge}{\exists_k \vec{p}\,\Psi_C, \neg(\vec{p}^{\,(h)} \leftrightarrow \vec{C}^{(h)}), \dots, \neg(\vec{p}^{\,(2)} \leftrightarrow \vec{C}^{(2)}), \neg(\vec{p}^{\,(1)} \leftrightarrow \vec{C}^{(1)})}\ (\exists_k)}{\exists_k \vec{p}\,\Psi_C, \neg(\vec{p}^{\,(h)} \leftrightarrow \vec{C}^{(h)}), \dots, \neg(\vec{p}^{\,(2)} \leftrightarrow \vec{C}^{(2)})}\ (\text{comp})}{\cdots}\ (\text{comp})}{\exists_k \vec{p}\,\Psi_C, \neg(\vec{p}^{\,(h)} \leftrightarrow \vec{C}^{(h)})}\ (\text{comp})}{\exists_k \vec{p}\,\Psi_C}\ (\text{comp})$$

is as desired.

4 Sequential Iteration in Quantified Propositional Logic

For n a natural number we write $[n]$ for the set $\{0, 1, \ldots, n-1\}$. That is, in set theoretic terms, $[n] = n$. If A and B are sets we denote by $f\colon A \rightharpoonup B$ that f is a *partial function from A to B*. In other words, f is a function, its domain $\mathsf{dom}(f)$ is a subset of A and its range $\mathsf{rng}(f)$ is a subset of B.

By abuse of notation we identify a list $\langle \wp_0, \ldots, \wp_{n-1} \rangle \in \{\mathrm{T}, \mathrm{F}\}^n$ of n boolean values with an element of $[2^n]$ in the following way, assuming the n is understood from the context. $\langle \wp_0, \ldots, \wp_{n-1} \rangle = \sum_{i=0}^{n-1} \chi_{\wp_i} \cdot 2^i$, where we set $\chi_\mathrm{T} = 1$ and $\chi_\mathrm{F} = 0$.

For the rest of this section we assume that n is big enough, so that $n + \log(n)$ and $2n$ are different. Note that this is the case if $n \geq 1$. The intended meaning of $\alpha_{n+\log n}$ and α_{2n} is that they fix the values of a function $f\colon [2^n] \to [2^n]$ in the following way: $\alpha_{n+\log n}(i, x)$ is true iff the i-th bit of $f(x)$ is 1, and $\alpha_{2n}(i, x)$ is true iff $f^i(0) = x$, where $f^i(0)$ is the result of computing the ith iterative of f on 0. Storing f by its bitgraph $\alpha_{n+\log n}$ automatically guarantees that a total function of $[n]$ is described, a property which would otherwise require adding more complex quantification to our principle.

Definition 16. We write "$f(p_1, \ldots, p_n) = q_1, \ldots, q_n$" for $\bigwedge_{i<n}(q_i \leftrightarrow \alpha_{\tilde{n}}(i, \vec{p}))$ where $\tilde{n} = n + \log(n)$. We write "$\vec{p} = \vec{q}$" for $\bigwedge_{i<n}(p_i \leftrightarrow q_i)$

Definition 17. We write "$f^{p_1,\ldots,p_n}(0) = q_1, \ldots, q_n$" for $\alpha_{2n}(\vec{p}, \vec{q})$.

It should be noted that "$f(0) = \vec{q}$" and "$f^1(0) = \vec{q}$" are not only different formulae, but are not even logically equivalent.

Definition 18. We write "$p_0, \ldots, p_{n-1} = q_0, \ldots, q_{n-1} + 1$" for the obvious AC^0-formulation of the successor relation, that is, for

$$\bigvee_i (\bigwedge_{j<i} p_j \wedge \neg p_i \wedge \bigwedge_{j<i} \neg q_j \wedge q_i \wedge \bigwedge_{j>i} (p_j \leftrightarrow q_j)) \ .$$

Fix $\ell \leq n$. Our iteration principle will express that α_{2n} stores the graph of $i \mapsto f^i(0)$ for $i = 0, \ldots, \ell$. Using the common idea that $\exists x. f^i(0) = x$ expresses that $f^i(0)$ can be computed, we can argue as follows. If $f^0(0)$ can be computed but $f^\ell(0)$ cannot, then there must be some i such that $f^i(0)$ can be computed but $f^{i+1}(0)$ cannot. The crux is now that this can be expressed using existential quantifiers only, which makes use of the trick that we are storing f by it's bitgraph. If $f^0(0) = 0$ and no m exists with $f^\ell(0) = m$, then there are m, m', i, i' with $i' = i + 1$ and $f^i(0) = m$ and $f(m) = m'$ and not $f^{i'}(0) = m'$. Prenexing this description and identifying the two independent occurrences of m gives us the following iteration formula and principle.

Definition 19. *The n, ℓ-iteration formula $\Phi_{n,\ell}$* is the following purely propositional formula

$$\begin{aligned}
&\Phi_{n,\ell}(\vec{p}, \vec{p}\,', \vec{q}, \vec{q}\,') \ \equiv \\
&\quad \text{“}f^\ell(0) = \vec{p}\text{”} \vee \neg\text{“}f^0(0) = 0\text{”} \\
&\vee (\text{“}\vec{q}\,' = \vec{q} + 1\text{”} \ \wedge \ \text{“}f^{\vec{q}}(0) = \vec{p}\text{”} \ \wedge \ \text{“}f(\vec{p}) = \vec{p}\,'\text{”} \ \wedge \ \neg\text{“}f^{\vec{q}\,'}(0) = \vec{p}\,'\text{”})
\end{aligned}$$

The n, ℓ*-iteration principle* is the formula

$$\exists_{4n} \vec{p}\vec{p}'\vec{q}\vec{q}'. \, \Phi_{n,\ell}(\vec{p},\vec{p}',\vec{q},\vec{q}') \, .$$

Definition 20. A *partial propositional assignment* is a finite partial mapping from the propositional variables to $\{\mathrm{T},\mathrm{F}\}$.

A *partial parameter assignment* is any partial mapping (not necessarily finite) from atomic parameters $\alpha_k(\vec{\wp})$, with $\wp_i \in \{\mathrm{T},\mathrm{F}\}$, to $\{\mathrm{T},\mathrm{F}\}$.

In the context of propositional logic, we use "valuation" as another word for partial (propositional or parameter) assignment. We use η to range over valuations. In accordance with set theoretic notions we write the empty valuation as $\emptyset$.

Definition 21. A quantified propositional formula is α*-free*, if it does not contain any propositional parameter α_n, for any n. It is called *closed*, if it does not contain any free propositional variables.

Note that any closed, α-free quantified propositional formula has a standard truth value T of F in the obvious way.

Definition 22. If A is a quantified propositional formula and η a partial propositional assignment, we define $A\eta$ by induction on A. For p a propositional variable with $p \in \mathsf{dom}(\eta)$ we set $p\eta \equiv \eta(p)$ and $\bar{p}\eta \equiv \neg\eta(p)$. For $p \notin \mathsf{dom}(\eta)$ we set $p\eta \equiv p$ and $\bar{p}\eta \equiv \bar{p}$. The remaining cases are defined homomorphically, e.g., $(\bigwedge_k \vec{A})\eta \equiv \bigwedge_k \vec{A}\eta$. In particular $\alpha_k(\wp_1,\ldots,\wp_k)\eta \equiv \alpha_k(\wp_1\eta,\ldots,\wp_k\eta)$.

If A is a closed purely propositional formula and η a partial parameter assignment, we define $A\eta$ by induction on A. For $\alpha_k(\vec{\wp})$ with $\alpha_k(\vec{\wp}) \in \mathsf{dom}(\eta)$ we set $(\alpha_k\vec{\wp})\eta \equiv \eta(\alpha_k(\vec{\wp}))$ and $(\bar{\alpha}_k\vec{\wp})\eta \equiv \neg\eta(\alpha_k(\vec{\wp}))$. Otherwise we set $(\alpha_k\vec{\wp})\eta \equiv \alpha_k(\vec{\wp})$ and $(\bar{\alpha}_k\vec{\wp})\eta \equiv \bar{\alpha}_k(\vec{\wp})$. The remaining cases are defined homomorphically.

If $\Gamma = \{A_1,\ldots,A_k\}$ is a set of formulae we write $\Gamma\eta$ for $\{A_1\eta,\ldots,A_k\eta\}$.

Lemma 23. *If $\eta \subset \eta'$ are partial propositional assignments and A is a quantified propositional formula such that $A\eta$ is closed, then $A\eta \equiv A\eta'$.*

If $\eta \subset \eta'$ are partial parameter assignments and A is a closed purely propositional formula such that $A\eta$ is α-free, then $A\eta \equiv A\eta'$.

Definitions 24 and 26 encode the crucial idea of our proof of the boundedness theorem (Theorem 32). Eventually we will be working upwards through a single path of a given proof, and partially define a function $f\colon [2^n] \rightharpoonup [2^n]$ in order to falsify all quantifier free formulae on this path. We want to do this in such a way, that, at level h, only $0, f(0), \ldots, f^{h-1}(0)$ are defined. But, to assign a truth value to a quantifier free formula, we not only have to set the parameter bits that encode the relation "$f(x) = y$", but also those that encode the iterations of f of the form "$f^k(0) = y$".

The idea is to assign them values consistent with what we have so far and also consistent with our strategy on how we plan to extend f. As we want to keep $f^h(0)$ undefined, all the values in $\mathsf{dom}(f)$ are "forbidden" anyway for the next extension of f. Note that, if $f^i(0)$ is defined and $f^i(0) = f^j(0)$ for some $i < j$, then all the values $f^k(0)$ are already defined.

Definition 24. A partial function $f\colon [2^n] \rightharpoonup [2^n]$ is called *ℓ-sequential* if for some $k \le \ell$ it is the case that $0, f(0), f^2(0), \ldots, f^k(0)$ are all defined, but $f^k(0) \notin \mathsf{dom}(f)$.

Example 25. The empty function is ℓ-sequential for any $\ell \in \mathbb{N}$. If f is a partial function with $f(0) = 0$ then f is *not* ℓ-sequential for any ℓ.

Definition 26. If $n \in \mathbb{N}$ is a natural number and $f\colon [2^n] \rightharpoonup [2^n]$ a partial function, we associate to f, or actually to the pair n, f, a partial parameter assignment η_f as follows.

For $j \in [n]$, $x \in [2^n]$ with $f(x)$ defined, say $f(x) = \langle \vec{r} \rangle \in [2^n]$, we set $\eta_f(\alpha_{n+\log(n)}(j,x)) = r_j$. Otherwise $\eta_f(\alpha_{n+\log(n)}(j,x))$ is undefined.

For $x, \ell \in [2^n]$ we set $\alpha_{2n}(\ell, x) = \mathrm{T}$ if $f^\ell(0)$ is defined and equal to x; otherwise we set $\alpha_{2n}(\ell, x) = \mathrm{F}$ if $x \in \mathsf{dom}(f)$; otherwise $\alpha_{2n}(\ell, x)$ is undefined.

For $k \notin \{2n, n + \log(n)\}$ we set $\eta_f(\alpha_k(\vec{\wp}))$ arbitrarily, say F. Also, if $\vec{p} \in \{\mathrm{T},\mathrm{F}\}^{\log n} \setminus [n]$, we set $\alpha_{n+\log n}(\vec{p}, \vec{q})$ arbitrarily, say F.

"Good extensions" of partial functions are those that comply with the above idea, that is, those that do not assign new values that are already in the domain.

Definition 27. If $f, f'\colon [2^n] \rightharpoonup [2^n]$ are partial functions, and $f \subset f'$ then f' is called a *good extension* of f, if $\forall x \in \mathsf{dom}(f')(x \in \mathsf{dom}(f) \vee f'(x) \notin \mathsf{dom}(f))$.

Remark 28. If $f \subset f'$ and $f' \subset f''$ are good extensions, then so is $f \subset f''$.

Proposition 29. *If $f \subset f'$ is a good extension, then $\eta_f \subset \eta_{f'}$.*

Lemma 30. *Let $n \in \mathbb{N}$ and $f\colon [2^n] \rightharpoonup [2^n]$ be an ℓ-sequential partial function. Moreover, let $M \subset [2^n]$ such that $|\mathsf{dom}(f) \cup M| < 2^n$. Then there is an $(\ell+1)$-sequential good extension f' of f with $\mathsf{dom}(f') = \mathsf{dom}(f) \cup M$.*

Proof. Let $a \in [2^n] \setminus (M \cup \mathsf{dom}(f))$. Such an a exists by our assumption on the cardinality of $M \cup \mathsf{dom}(f)$. Let f' be f extended by setting $f'(x) = a$ for all $x \in M \setminus \mathsf{dom}(f)$. This f' is as desired.

Indeed, assume that $0, f'(0), \ldots, f'^{\ell+1}(0), f'^{\ell+2}(0)$ are all defined. Then, since $a \notin \mathsf{dom}(f')$, all the $0, f'(0), \ldots, f'^{\ell+1}(0)$ have to be different from a. Hence these values have already been defined in f. But this contradicts the assumption that f was ℓ-sequential. □

Lemma 31. *For every closed, purely propositional, formula A of size ℓ there is a set $M \subset [2^n]$ such that $|M| \le \ell$ and for every function f with $M \subset \mathsf{dom}(f)$ it holds that $A\eta_f$ is α-free.*

Proof. Let M be the set of all $x \in [2^n]$ such that an atom of the form $\alpha_{n+\log(n)}(j,x)$ or $\alpha_{2n}(k,x)$ occurs in A.

Note that $x \in \mathsf{dom}(f)$ forces $\eta_f(\alpha_{2n}(k,x))$ to have a definite value (F unless $f^k(0) = x$, in which case it would be T). □

Theorem 32. *Let k, n, w, c be natural numbers with $c \cdot w \geq 2$. Assume $\vdash^h_{w,c} \Gamma$ with $\Gamma = \Delta, \exists_{4n} \vec{r} \Phi_{n,\ell}(\vec{r})$, where $\Phi_{n,\ell}$ is the n, ℓ-iteration formula. Let η be a partial propositional assignment and $f\colon [2^n] \rightharpoonup [2^n]$ be k-sequential. Assume $|\mathsf{dom}(f)| + cwh < 2^n$. If $\Delta\eta\eta_f$ is purely propositional, closed, α-free, and false then $\ell \leq k + h$.*

The special case $\Delta = \emptyset$, $\eta = \emptyset$, $f = \emptyset$ and $k = 0$ yields

Corollary 33. *If $\vdash^h_{w,c} \exists_{4n} \vec{r} \Phi_{n,\ell}(\vec{r})$ and $cwh < 2^n$ for some c, w with $cw \geq 2$ then $h \geq \ell$.*

Proof (of the theorem). We argue by induction on h with case distinction according to the last rule of the proof.

The last rule cannot be a propositional axiom, as axioms cannot have $\exists_{4n} \vec{r} \Phi_{n,\ell}(\vec{r})$ as a main formula; however, all the formulae in $\Delta\eta\eta_f$ are false so Δ cannot be a tautology, as it would have to be, as the calculus is sound. In the case of an $\bigvee_k$-inference apply the induction hypothesis, in the case of an $\bigwedge_k$-inference, the induction hypothesis is applicable to at least one of the subderivations. The last rule cannot be an $\forall_j$-rule as this would require a quantified formula in Δ.

If the last rule is a multi-cut rule

$$\frac{\ldots \Gamma, \Delta_i \ldots}{\Gamma}$$

we know, since the proof is w, c-slim, that $\bigcup_i \Delta_i$ contains at most c formulae of size at most w. Let $\eta' \supset \eta$ such that all $\Delta_i \eta'$ are closed. Let M be the union of the sets asserted by Lemma 31 for the formulae in $\bigcup_i \Delta_i \eta'$. Then $|M| \leq c \cdot w$. We extend f in a good way to some $(k+1)$-sequential f' with $\mathsf{dom}(f') = \mathsf{dom}(f) \cup M$. Noting that all the $\Delta_i \eta' \eta_{f'}$ are sets of α-free, closed, purely propositional formulae we can assign them truth values. Since, by cuts we can derive the empty sequent from the sets Δ_i, and hence also from the sets $\Delta_i \eta' \eta_{f'}$, one of them has to contain only false formulae. Apply the induction hypothesis to this subderivation.

The case of a cut rule is similar, but easier.

Assume that the last rule was a parameter extensionality rule as follows.

$$\frac{\Gamma, \alpha_k(\wp_1, \ldots, \wp_j) \qquad \ldots \Gamma, \wp_i \leftrightarrow \wp'_i \ldots \qquad (1 \leq i \leq j)}{\Gamma, (\alpha_k(\wp'_1, \ldots, \wp'_k))}$$

Extend η to some η' assigning values to all the $\vec{\wp}$. If for some $1 \leq i \leq j$ we have $\wp\eta' \neq \wp'\eta'$ we can apply the induction hypothesis to the corresponding subderivation. Otherwise $(\alpha_k(\vec{\wp}))\eta'\eta_f \equiv (\alpha_k(\vec{\wp}'))\eta'\eta_f$ and we can apply the induction hypothesis to the first subderivation.

Assume that the last inference rule was an $\exists_j$-rule.

$$\frac{\Gamma, \Phi_{n,\ell}(\vec{\wp}, \vec{\wp}', \vec{\wp}'', \vec{\wp}''')}{\Gamma} \exists_{4n}$$

We can extend η to η' such that there are natural numbers m, m', i, i' such that $\vec{\wp}\eta' = m$, $\vec{\wp}'\eta' = m'$, $\vec{\wp}''\eta' = i$ and $\vec{\wp}'''\eta' = i'$. If $\ell \leq k$ there is nothing to show. Otherwise, we will argue as follows that $\Phi_{n,\ell}(m, m', i, i')\eta_{f'}$ can be falsified by choosing an appropriate $(k{+}1)$-sequential good extension f' of f. Since $\ell > k$, for every good $(k{+}1)$-sequential extension f' of f we have $f'^{(\ell+1)}(0)$ undefined. Hence for any such f' with $m \in \mathsf{dom}(f')$ we know that $f'^{(\ell)}$ is either undefined or different from m (for otherwise $f'^{(\ell+1)}(0)$ would be defined). In either case $\eta_{f'}(\alpha_{2n}(\ell, m)) = \mathrm{F}$. Recall that adding a value m to the domain of f' ensures that $\eta_{f'}(\alpha_{2n}(\ell, m))$ has a definite value. The second disjunct $\neg$"$f^0(0) = 0$" is falsified by $\eta_{f'}$ for any f'. For the last disjunct "$i' = i + 1$" $\wedge$ "$f^i(0) = m$" $\wedge$ "$f(m) = m'$" $\wedge\, \neg$"$f^{i'}(0) = m'$", we may assume that $i' = i + 1$, for otherwise it is falsified anyway. For any f' with $m, m' \in \mathsf{dom}(f')$ we know that $\eta_{f'}$ assigns definite truth values to "$f^i(0) = m$", "$f(m) = m'$", and "$f^{i'}(0) = m'$". If the first two conjuncts are assigned T, than this can only be if $f'^i(0) = m$ and $f'(m) = m'$. But in this case $f'^{i+1}(0) = m'$, so $\neg$"$f^{i+1}(0) = m'$" is assigned F. Altogether we can take any $(k{+}1)$-sequential good extension f' of f with $\mathsf{dom}(f') = \mathsf{dom}(f) \cup \{m, m'\}$. Then $\Phi_{n,\ell}(\vec{\wp}, \ldots)\eta'\eta_{f'}$ is α-free, closed, purely propositional and false and we can apply the induction hypothesis (recalling that we assumed $wc \geq 2$).

The last remaining case is that the last rule was a comprehension rule

$$\frac{\Gamma, \neg(p_1 \leftrightarrow \varphi_1), \ldots, \neg(p_j \leftrightarrow \varphi_j)}{\Gamma}$$

where the φ_i are purely propositional, the $\vec{p}$ are eigenvariables, and, since the proof is w, c-slim, $j \leq c$. Let $\eta'' \supset \eta$ be such that all $\varphi_i\eta''$ are closed. Let M_i be the set asserted by Lemma 31 for $\varphi_i\eta''$. Extend in a good way f to a $(k{+}1)$-sequential f' with $\mathsf{dom}(f') = \mathsf{dom}(f) \cup \bigcup_i M_i$. Due to the eigenvariable condition we can assume without loss of generality that $\vec{p} \notin \mathsf{dom}(\eta'')$. Extend η'' to η' by setting p_i to the truth value of $\varphi_i\eta''\eta_{f'}$. We then can apply the induction hypothesis.

This finishes the proof. □

As a proof complexity consequence of the above theorem we can make the following observation.

Corollary 34. *There is a family of polynomial size $\Sigma_1^q(\alpha)$-formulae, i.e., formulae of the shape of existentially quantified purely propositional formulae, such that every AC^0-Tait proof with polynomially branching rules and polynomial size formulae requires exponential height.*

Any proof of this family requires exponential size.

Proof. As Corollary 33 shows, the family $(\exists_{4n}\vec{r}\Phi_{n,2^n-1}(\vec{r}))_{n\in\mathbb{N}}$ is as desired. It should be noted that these formuale indeed only grow polynomially, as, of course, the number $2^n - 1$ can be represented by n bits. □

5 Cut-Elimination

We now show how cuts on more complicated formulae can be reduced to quantifier-free cuts. We will obtain the typical increase in height occurring in proof-theoretic cut-elimination.

Definition 35. A substitution σ is called an *atomic substitution*, if for every propositional variable p we have $\sigma(p) \in \mathcal{A}$. In other words, a substitution is atomic, if the range only contains propositional atoms.

Lemma 36. *If* $\vdash^h_{\mathcal{C};w,c} \Gamma$ *then* $\vdash^h_{\mathcal{C};w,c} \Gamma\sigma$ *for every atomic substitution* σ.

Lemma 37. *If* $\vdash^h_{\mathcal{C};w,c} \Gamma, \forall_k \vec{p} A(\vec{p})$, *then* $\vdash^h_{\mathcal{C};w,c} \Gamma, A(\vec{\wp})$ *for arbitrary propositional atoms* $\vec{\wp}$.

Proof. Induction on the derivation. We can identically reproduce any rule that does not have $\forall_k \vec{p} A(\vec{p})$ as main formula.

In case $\Delta, \forall_k \vec{p} A(\vec{p})$ was concluded from $\Delta, \forall_k \vec{p} A(\vec{p}), A(\vec{a})$ with pairwise distinct eigenvariables $\vec{a}$, we may, by Lemma 36, assume without loss of generality that the $\vec{a}$ are disjoint from $\vec{\wp}$. First apply the induction hypothesis to the premise, obtaining $\Delta, A(\vec{\wp}), A(\vec{a})$ and then apply Lemma 36 to obtain $\Delta, A(\vec{\wp})$. Note that the eigenvariable property ensures that Δ is not affected by this substitution. □

Lemma 38. *Assume* $A \in \mathcal{C}$. *If* $\vdash^h_{\mathcal{C};w,c} \Gamma, \forall_k \vec{p} A(\vec{p})$ *and* $\vdash^{h'}_{\mathcal{C};w,c} \Gamma, \exists_k \vec{p} \neg A(\vec{p})$ *then* $\vdash^{h+h'}_{\mathcal{C};w,c} \Gamma$.

Proof. Let $\vec{q}$ be new and pairwise distinct variables. By Lemma 37 we get $\vdash^h_{\mathcal{C};w,c} \Gamma, A(\vec{q})$. Now argue by Induction on the second derivation, or, equivalently, by induction on h'. Every rule of the second derivation can be reproduced identically, except for an $\exists_k$-introduction with conclusion $\exists_k \vec{p} \neg A(\vec{p})$.

So assume that $\vdash^{h''+1}_{\mathcal{C};w,c} \Gamma, \exists_k \vec{p} \neg A(\vec{p})$ was concluded from $\vdash^{h''}_{\mathcal{C};w,c} \Gamma, \exists_k \vec{p} \neg A(\vec{p}), \neg A(\vec{\wp})$ with the $\vec{\wp}$ necessarily propositional atoms, by the restriction of the quantification rules (Definition 4). First apply the induction hypothesis to $\vdash^{h''}_{\mathcal{C};w,c} \Gamma, \exists_k \vec{p} \neg A(\vec{p}), \neg A(\vec{\wp})$ and obtain $\vdash^{h''+h}_{\mathcal{C};w,c} \Gamma, \neg A(\vec{\wp})$. Also, apply Lemma 36 to $\vdash^h_{\mathcal{C};w,c} \Gamma, A(\vec{q})$ and obtain $\vdash^h_{\mathcal{C};w,c} \Gamma, A(\vec{\wp})$ using that the $\vec{q}$ are fresh, i.e., in particular not free in Γ. Then conclude $\vdash^{h''+h+1}_{\mathcal{C};w,c} \Gamma$ by a cut on $A(\vec{\wp})$ which is allowed as $A \in \mathcal{C}$ and $\mathcal{C}$ is closed under atomic substitutions. □

Corollary 39. *If* $\vdash^h_{\exists\mathcal{C};w,c} \Gamma$ *then* $\vdash^{2^h}_{\mathcal{C};w,c} \Gamma$ *where* $\exists\mathcal{C} = \mathcal{C} \cup \{\exists_k \vec{p} A(\vec{p}) \mid A(\vec{p}) \in \mathcal{C}\}$.

6 Circuit Evaluation and Iteration

We now have all the preparations needed to show the following lower bound. Consider a proof that a circuit can be evaluated. If the circuit has height h, then the proof has to have height at least $h - \mathcal{O}(1)$.

Obviously, a circuit of height h can compute the h'th iterate of the function given by α. From the fact that this circuit can be evaluated, we can conclude that the h-iteration principle holds.

In the following let C_h be the circuit canonically iterating the function given by the oracle. We also assume the size parameter n to be understood; we set $\tilde{n} = n + \log n$. Immediately from the definition we get

Proposition 40. *$\Psi_{C_h}(\vec{w})$ is the conjunction of the clauses $w_j^{(0)} \leftrightarrow \mathrm{F}$ and the conjuncts "$f(\vec{w}^{(\ell)}) = \vec{w}^{(\ell+1)}$". The latter is built of the formulae $w_j^{(\ell+1)} \leftrightarrow \alpha_{\tilde{n}}(j, \vec{w}^{(\ell)})$ of $0 \leq \ell < h-1$ and $0 \leq j < n$.*

Proposition 41. *$\mathsf{sz}(\Psi_{C_h}) \in \mathcal{O}(n \cdot h)$ and $\mathsf{dp}(\Psi_{C_h}) \in \mathcal{O}(1)$.*

For $1 \leq \ell < n$ we set $\Delta_\ell \equiv \neg$"$f^{\ell-1}(0) = \vec{w}^{(\ell-1)}$", "$f^{\ell}(0) = \vec{w}^{(\ell)}$". Recall that "$f^{\vec{p}}(0) = \vec{q}$" is a shorthand for $\alpha_{2n}(\vec{p}, \vec{q})$. So, by resolution the Δ_ℓ imply $\neg$"$f^0(0) = \vec{w}^{(0)}$", "$f^h(0) = \vec{w}^{(h)}$".

Proposition 42. *If $i \in [n-1]$ and $\vec{p} = i$ and $\vec{q} = i+1$ then $\vdash^{\mathcal{O}(1)}_{\mathcal{O}(\log n),1}$ "$\vec{q} = \vec{p} + 1$".*

Lemma 43. *$\vdash^{\mathcal{O}(1)}_{\mathcal{O}(nh),1} \Delta_\ell, \exists_{4n}\vec{u}\Phi_{n,h}(\vec{u}), \forall_{h\cdot n}\vec{w}\neg\Psi_{C_h}(\vec{w})$ with $\Phi_{n,h}(\vec{u})$ the n,h-iteration Formula, as defined in Definition 19.*

Proof. First note, that there are constant height proofs of the following sequents.

- "$f(\vec{w}^{(\ell-1)}) = \vec{w}^{(\ell)}$", $\neg\Psi_{C_h}(\vec{w})$
- "$f^{\ell-1}(0) = \vec{w}^{(\ell-1)}$", $\neg$"$f^{\ell-1}(0) = \vec{w}^{(\ell-1)}$"
- "$f^{\ell}(0) = \vec{w}^{(\ell)}$", $\neg$"$f^{\ell}(0) = \vec{w}^{(\ell)}$"
- "$(\ell+1) = \ell + 1$"

Therefore applications of an $\bigwedge_4$-rule followed by an $\vee$-rule and an $\exists_{4n}$-rule gives us $\exists_{4n}\vec{u}\Phi_{n,h}(\vec{u}), \neg\Psi_{C_h}(\vec{w}), \Delta_\ell$ from where we get the desired derivation by an $\forall_{nh}$-rule.

Corollary 44. $\vdash^{\mathcal{O}(1)}_{\mathcal{O}(nh),\mathcal{O}(h)} \exists_{4n}\vec{u}\Phi_{n,h}(\vec{u}), \forall_{h\cdot n}\vec{w}\neg\Psi_{C_h}$

Proof. Apply a mutli-cut rule to the derivations of Lemma 43 to obtain $\neg$"$f^0(0) = \vec{w}^{(0)}$", "$f^h(0) = \vec{w}^{(h)}$", $\exists_{4n}\vec{u}\Phi_{n,h}(\vec{u}), \forall_{h\cdot n}\vec{w}\neg\Psi_{C_h}(\vec{w})$. Two $\vee$-rules and an $\exists_{4n}$-rule finish the proof.

Theorem 45. *There are natural numbers c, C such that forall sufficiently large n, h whenever $c^2 \cdot h^2 n < 2^n$ and $\vdash^{h'}_{c\cdot nh,ch} \exists_{nh}\vec{w}\Psi_{C_h}(\vec{w})$ then $h' \geq h - C$.*

Proof. Assume $\vdash^{h'}_{\tilde{w},\tilde{c}} \exists_{nh}\vec{w}\Psi_{C_h}$. By Corollary 44 we have (for sufficiently large n) a derivation $\vdash^{c_2}_{c_1 nh, c_1 h} \exists_{4n}\vec{u}\Phi_{n,h}(\vec{u}), \forall_{hn}\vec{w}\neg\Psi_{C_h}$. Therefore, by Lemma 38, we get $\vdash^{h'+c_2}_{\max\{w,c_1 nh\},\max\{\tilde{c},c_1 h\}} \exists_{4n}\vec{u}\Phi_{n,h}(\vec{u})$. So, by Corollary 33, we get $h' + c_2 \geq h$, provided $\max\{\tilde{w}, c_1 nh\} \cdot \max\{\tilde{c}, c_1 h\} < 2^n$.

An immediate consequence of Theorem 45 is that a proof of Ψ_{C_h} requires height $h - \mathcal{O}(1)$, for h growing sub-exponentially with n.

References

1. Beckmann, A.: Dynamic ordinal analysis. Archive for Mathematical Logic 42, 303–334 (2003)
2. Buss, S.R., Krajíček, J.: An application of boolean complexity to separation problems in bounded arithmetic. Proceedings of the London Mathematical Society 69(3), 1–27 (1994)
3. Clote, P., Takeuti, G.: First order bounded arithmetic and small Boolean circuit complexity classes. In: Clote, P., Remmel, J. (eds.) Feasible Mathematics II, Birkhäuser, Boston, MA. Progr. Comput. Sci. Appl. Logic, vol. 13, pp. 154–218 (1995)
4. Cook, S.A.: Theories for complexity classes and their propositional translations. In Krajíček, J. (ed.) Complexity of computations and proofs, Quaderni die Matematica, pp. 175–227. Dipartimento di Matematica, Seconda Universitá degli Studi di Napoli (2003)
5. Cook, S.A., Morioka, T.: Quantified propositional calculus and a second-order theory for NC^1. Arch. Math. Logic 44(6), 711–749 (2005)
6. Cook, S.A., Nguyen, P.: Foundations of proof complexity: Bounded arithmetic and propositional translations. draft of a book, available at `http://www.cs.toronto.edu/~sacook/csc2429h/book/`
7. Cook, S.A., Reckhow, R.A.: The relative efficiency of propositional proof systems. The Journal of Symbolic Logic 44(1) (1979)
8. Krajíček, J.: Lower bounds to the size of constant-depth propositional proofs. The Journal of Symbolic Logic 59(1), 73–86 (1994)
9. Krajíček, J., Pudlák, P.: Quantified propositional calculi and fragments of bounded arithmetic. Zeitschrift für mathematische Logik und Grundlagen der Mathematik 36, 29–46 (1990)
10. Perron, S.: A propositional proof system for log space. In: Ong, C.-H.L. (ed.) CSL 2005. LNCS, vol. 3634, pp. 509–524. Springer, Heidelberg (2005)
11. Pohlers, W.: Proof theory. An introduction, Lecture Notes in Mathematics, vol. 1407, Springer, Heidelberg (1989)
12. Skelley, A.: Propositional PSPACE reasoning with Boolean programs versus quantified Boolean formulas. In: Díaz, J., Karhumäki, J., Lepistö, A., Sannella, D. (eds.) ICALP 2004. LNCS, vol. 3142, pp. 1163–1175. Springer, Heidelberg (2004)
13. Tait, W.W.: Normal derivability in classical logic. In: Barwise, J. (ed.) The Syntax and Semantics of Infinitatry Languages. Lecture Notes in Mathematics, vol. 72, pp. 204–236. Springer, Heidelberg (1968)
14. Takeuti, G.: Separations of theories in weak bounded arithmetic. Annals of Pure and Applied Logic 71, 47–67 (1995)
15. Wilson, C.B.: A measure of relativized space which is faithful with respect to depth. Journal of Computer and System Sciences 36(3), 303–312 (1988)

Game Characterizations and the PSPACE-Completeness of Tree Resolution Space

Alexander Hertel and Alasdair Urquhart*

Department of Computer Science,
University of Toronto, Toronto ON M5S 3G4, Canada
{ahertel, urquhart}@cs.toronto.edu

Abstract. The Prover/Delayer game is a combinatorial game that can be used to prove upper and lower bounds on the size of Tree Resolution proofs, and also perfectly characterizes the space needed to compute them. As a proof system, Tree Resolution forms the underpinnings of all DPLL-based SAT solvers, so it is of interest not only to proof complexity researchers, but also to those in the area of propositional reasoning. In this paper, we prove the PSPACE-Completeness of the Prover/Delayer game as well as the problem of predicting Tree Resolution space requirements, where space is the number of clauses that must be kept in memory simultaneously during the computation of a refutation. Since in practice memory is often a limiting resource, researchers developing SAT solvers may wish to know ahead of time how much memory will be required for solving a certain formula, but the present result shows that predicting this is at least as hard as it would be to simply find a refutation.

1 Introduction and Motivation

The Tree Resolution (T-RES) proof system has been studied extensively, and is understood quite well. Its algorithmic incarnation, DPLL, forms the basis of many SAT-solvers including clause learning algorithms. Any lower bounds proved for T-RES immediately imply lower bounds for real-world DPLL algorithms. However, compared with Resolution proof size, the amount of space needed to compute a proof has not been studied nearly as extensively. The usual definition of a Resolution proof is that it is a simple sequence of clauses, but we can redefine this so that only a small number of clauses are stored in a working memory space, typically much smaller than the size of the proof itself.

The problem of determining the space required for Resolution proofs seems to have been suggested for the first time by Armin Haken in 1998, and was explored in a paper by Alekhnovich, Ben-Sasson, Razborov and Wigderson [1] and independently by Esteban and Torán [7]; Ben-Sasson proved an important tradeoff between Resolution size and space in [3]. In addition, a number of other results relating Resolution space and width are known [2,11].

* The authors gratefully acknowledge the support of NSERC and the University of Toronto Department of Computer Science.

J. Duparc and T.A. Henzinger (Eds.): CSL 2007, LNCS 4646, pp. 527–541, 2007.

From a practical point of view, DPLL and clause learning algorithms have been highly successful at solving SAT and SAT-related problems. The main limiting factor on these algorithms is space, namely the size of the cache used for memoization. This has inspired much research into methods for pruning space in a Resolution search. Thus there are both practical and theoretical motivations for understanding Resolution space as a resource.

In this paper we prove that for T-RES, determining clause space requirements is $\mathcal{PSPACE}$-Complete, which unfortunately implies that computing the T-RES space requirements for a formula is at least as hard as actually refuting it, and shows that it is therefore probably not feasible for researchers working with DPLL-based SAT solvers to predict a formula's memory needs. In addition, there are many formulas on which these SAT solvers fail, and it would be very useful to be able to tell ahead of time if this will be the case. If there was any hope of using T-RES space bounds to predict this, then the present result casts some serious doubt on that plan.

The key to our main theorem is the intimate connection between Resolution space and games, and we combine three previous results and ideas in order to prove that both the Prover/Delayer game and T-RES clause space problem are $\mathcal{PSPACE}$-Complete. The high-level idea behind this paper is as follows: From [7], Prover/Delayer number is known to equal T-RES clause space, and from [10], pebbling Lingas circuits is known to be $\mathcal{PSPACE}$-Complete. But the former deals with formulas, and the latter uses circuits, so we use pebbling contradiction formulas to bridge this gap. Specifically, we give tight bounds relating the black pebbling number of any Lingas circuit C to the Prover/Delayer number of the formula $Peb^2(C)$, thereby proving the $\mathcal{PSPACE}$-Completeness of the Prover/Delayer game (and T-RES clause space by the equivalence from [7]).

2 Definitions

We assume that the reader is familiar with basic proof complexity, circuit complexity, general complexity theory, and use [5] as our standard reference. By a *DAG* we mean a directed acyclic graph in which all nodes except source nodes have fan-in two, but unbounded fan-out, and there is a unique node (the *sink*, *target* or *output* node) with fan-out zero. A *monotone circuit* is defined as a DAG in which all nodes except the sources are labelled with AND or OR.

2.1 Resolution Space

The definition of T-RES clause space requires a non-standard definition of T-RES proof that depends on the notion of configuration:

Definition 1 (Configuration-Style T-RES Proof). *A configuration $\mathbb{C}$ is a set of clauses. The sequence of configurations $\pi = \mathbb{C}_0, \mathbb{C}_1, ..., \mathbb{C}_k$ is a T-RES proof of C from the formula F if $\mathbb{C}_0 = \emptyset$, $C \in \mathbb{C}_k$, and for each $i < k$, $\mathbb{C}_{i+1}$ is obtained from $\mathbb{C}_i$ by deleting one or more of its clauses, adding the resolvent of two clauses of $\mathbb{C}_i$* ***and deleting both parent clauses****, or adding one or more of the clauses of F (initial clauses).*

This leads us to our definition of space. Intuitively, space is the amount of memory required in order to compute π:

Definition 2 (Clause Space). *Let F be a set of clauses and π be a configuration-style* T-RES *proof of clause C from F. The tree clause space of π, denoted $TCS(\pi)$ is the maximum number of clauses in any configuration of π. The tree clause space of resolving C from F, denoted $TCS(F \vdash_{\text{T-RES}} C)$, is the minimum $TCS(\pi)$ over all* T-RES *proofs π of C from F.*

2.2 Pebbling Circuits and Games

The investigation of Resolution space is closely associated with the well-known pebbling game and pebbling number of a DAG, originally explored in [6] as a means of investigating bounds on storage requirements.

We define the *generalized black pebbling game* as a single-player game in which the goal is to 'pebble' the target node of a monotone circuit C. The game has the following rules: It starts with no pebbles on the circuit. At any point, the player may place a pebble onto any source node, or remove a pebble from any node. For any AND gate v, if both of v's immediate predecessors have pebbles on them, then the player may place a pebble on v, or slide a pebble from a predecessor to v. Similarly, for any OR gate v, if at least one of v's immediate predecessors has a pebble on it, then the player may place a pebble on v, or slide a pebble from a predecessor to v. The game ends once the target node has a pebble on it.

Most of the black pebble games found in the literature can be viewed as restricted versions of this generalization. In particular, if all gates in the circuit are AND gates, then this game is equivalent to the original pebbling game on DAGs defined by Cook and Sethi [6].

Definition 3 (Black Pebbling Number of a Monotone Circuit). *The 'black pebbling number' of a monotone circuit C, denoted B-Peb(C), is the minimum number of pebbles needed in order to pebble C's target node using the above rules.*

Each version of the pebbling game can either be defined with or without sliding; for the purposes of this paper, we shall always allow sliding.

2.3 Pebbling Contradictions

The formulas we call 'pebbling contradictions' are based on the various forms of the pebbling game. Ben-Sasson gives a brief history of these formulas in [3]. The particular contradictory formulas we employ here are a generalization of those used by Ben-Sasson, Impagliazzo and Wigderson to give a near-optimal separation of tree-like and general resolution [4]. They can be interpreted as making the following contradictory claim: "The input nodes of a monotone circuit C are all set to true, but the output node is set to false.", where we represent "node v is set to true" by the disjunction $v_0 \vee v_1$.

Definition 4 (Pebbling Contradictions Based on Circuits). *Let C be a monotone circuit. The set of clauses $Peb^2(C)$ contains the following clauses:*

1. *For each source node s, the clause $\{s_0, s_1\}$;*
2. *For each AND node c with immediate predecessors a and b, four propagation clauses $\{\neg a_0, \neg b_0, c_0, c_1\}$, $\{\neg a_0, \neg b_1, c_0, c_1\}$, $\{\neg a_1, \neg b_0, c_0, c_1\}$, and $\{\neg a_1, \neg b_1, c_0, c_1\}$;*
3. *For each OR node c with immediate predecessors a and b, four propagation clauses $\{\neg a_0, c_0, c_1\}$, $\{\neg b_0, c_0, c_1\}$, $\{\neg a_1, c_0, c_1\}$, and $\{\neg b_1, c_0, c_1\}$;*
4. *For the target node t, two singleton clauses $\{\neg t_0\}$ and $\{\neg t_1\}$.*

2.4 The Prover/Delayer Game

The Prover/Delayer game (P/D game), described in [4,12], is a combinatorial game between two players, the 'Prover', and the 'Delayer,' and is played on an unsatisfiable CNF formula F. The goal of the Prover is to falsify some initial clause of F. Since the formula is unsatisfiable, this is inevitable.

The game proceeds in rounds. Each round starts with the Prover querying the value of a variable. The Delayer can give one of three answers: 'True', 'False', or 'You Choose'. If the Delayer says 'You Choose', then the Prover gets to decide the value of that variable, and the Delayer wins one point. This is the only way in which points can be scored. The game finishes when any clause has been falsified. The Delayer's aim is to win as many points as possible, while the Prover aims to minimize this quantity.

Definition 5. *Let F be an unsatisfiable CNF formula. The Prover/Delayer number of F, denoted $PD(F)$, is the greatest number of points the Delayer can score on F with the Prover playing optimally.*

3 Results Related to Resolution Space

3.1 Space and Games

The pebbling game is closely related to Resolution clause space. Let F be any arbitrary unsatisfiable formula, π a configuration-style RES refutation of F, and G the DAG underlying the structure of π. The clause space used in computing π is exactly equal to $B\text{-}Peb(G)$. More specifically, $CS(F \vdash_{\mathsf{RES}} \emptyset)$ is equal to the pebbling number of the DAG with the smallest pebbling number of all DAGs underlying valid RES refutations of F. Of course, the analogous idea holds for T-RES and tree clause space.

Esteban & Torán proved that the tree clause space of any unsatisfiable formula is essentially the same as its Prover/Delayer number:

Theorem 1 ([8]). *For any unsatisfiable CNF formula F, $TCS(F \vdash_{\mathsf{T\text{-}RES}} \emptyset) = PD(F) + 1$.*

In addition to doing some of the pioneering research in the area of Resolution space, Esteban & Torán also showed that an upper bound on tree clause space yields an upper bound on the size of T-RES proofs [7]. Combining this with Theorem 1 and a little extra work shows that an upper bound on the number of points scored by the Prover immediately gives an upper bound on T-RES size:

Corollary 1. *If F is a contradictory formula with $PD(F) \leq k$, then F has a* T-RES *proof of size* $\leq \binom{n}{k+1}$.

The P/D game can also be used to give a lower bound on T-RES size; in [4], Ben-Sasson, Impagliazzo and Wigderson prove that lower bounds on $PD(F)$ can be used to prove lower bounds on the size of T-RES proofs:

Theorem 2 ([4]). *If the Delayer has a strategy guaranteed to win $> k$ points on F, then every* T-RES *refutation of F has size $> 2^k$.*

In other words, upper and lower bounds on $PD(F)$ not only immediately imply tightly corresponding upper and lower bounds on $TCS(F \vdash_{\text{T-RES}} \emptyset)$, but they also imply upper and lower bounds for T-RES proof size, showing not only that the P/D game is intimately connected to the T-RES proof system, but also that T-RES size and space are closely related.

3.2 Complexity of Pebbling

A number of complexity results involving pebbling are known. In 1978, Andrzej Lingas proved the following:

Theorem 3 ([10]). *Given a monotone circuit C and an integer k, the problem of determining if C can be black-pebbled with k pebbles is $\mathcal{PSPACE}$-Complete.*

Gilbert, Lengauer and Tarjan extended this result to DAGs in 1980:

Theorem 4 ([9]). *Given a DAG G and an integer k, the problem of determining if G can be black-pebbled with k pebbles is $\mathcal{PSPACE}$-Complete.*

These results are particularly interesting because the pebbling game only has one player, whereas most $\mathcal{PSPACE}$-Complete games have two.

4 Main Results

4.1 An Easy Case of the Pebbling Game

Although pebbling circuits is $\mathcal{PSPACE}$-Complete in general, in this section we define an interesting set of binary DAGs whose pebbling numbers can be computed in polynomial time. To this end, we first need to define the concept of the pebbling number of a vertex in a graph:

Definition 6 (Pebbling Number of a Vertex). *In a DAG, the pebbling number of a vertex x, $Peb(x)$, is the pebbling number of the subgraph rooted at x, with x set to be the target node.*

A source node has a pebbling number of 1, while if c has predecessors a and b, then $\text{Peb}(c) \leq \max(\text{Peb}(a), \text{Peb}(b)) + 1$. This is because we can always pebble c with the following strategy: Assuming that $\text{Peb}(a) \leq \text{Peb}(b)$, first pebble b using $\text{Peb}(b)$ pebbles, then remove all the pebbles except the one on b, next pebble a using $\text{Peb}(a)$ extra pebbles, then slide the pebble on a to c.

Definition 7 (Increasing DAGs). $\mathcal{G_I} = \{G \mid G$ *is a DAG such that there is no* $c \in V(G)$ *with predecessors* a *and* b *with* a, b, *and* c *all having the same pebbling number*$\}$

Although the pebbling game on binary DAGs is $\mathcal{PSPACE}$-Complete in general, when we restrict ourselves to inputs from $\mathcal{G_I}$, the problem becomes much easier:

Theorem 5. *Given a binary DAG* $G \in \mathcal{G_I}$ *and integer* k, *the problem of determining if* G *can be pebbled using at most* k *pebbles is in* $\mathcal{P}$.

Proof: Since the pebbling number of each node in G is uniquely determined by those of its predecessors, simply start at the source nodes and set their pebbling numbers to 1. Then find a vertex x for which the pebbling numbers of both predecessors have been determined (since G is a DAG, such a node always exists), and set x's pebbling number. Repeat until the target node's number has been determined. Clearly this can be done in polynomial time. □

This gives some insight into why the pebbling game is so difficult. An inspection of the constructions from both [9] and [10] shows that the graphs resulting from the reductions contain a large number of vertices that have the same pebbling numbers as their predecessors, so the obvious algorithm above fails.

4.2 Prover Strategy for the $\mathcal{G_I}$ DAGs

In this section we describe an efficient Prover strategy for the $\mathcal{G_I}$ graphs:

Lemma 1. *For any binary DAG* $G \in \mathcal{G_I}$ *with pebbling number* k, *the Prover has a strategy limiting the Delayer to at most* $k = B\text{-}Peb(G)$ *points playing on the formula* $Peb^2(G)$.

Proof: In the interests of concision, we omit a detailed proof, and instead provide the following proof sketch: The Prover first labels each node in G with its pebbling number, and first queries the variables associated with the target. The Delayer must set them both to False, or else the game is over. The Prover then follows a path towards the source nodes, forcing the Delayer to set both variables c_0 and c_1 associated with each node c on the path to False.

Given any node c on the path, let us suppose that it has both variables c_0 and c_1 set to False, and that the predecessors of c are a and b. In all cases, it is possible for the Prover to query the variables associated with a and b in such a way that sets both a_0 and a_1, or b_0 and b_1 to False while giving up at most one point every time the pebbling number decreases. Eventually the path will reach a node with two source nodes as its children, in which case it is easy to see that the Delayer can win at most 2 more points. Since the Delayer wins at most 1 point every time the pebbling number decreases, and since only 2 points can be won in the base case, the Prover limits the Delayer to at most $B\text{-}Peb(G)$ points, as required. □

4.3 Prover Strategy for the Lingas Circuits

Lingas' 1978 paper [10] shows that the black pebbling game on monotone circuits is $\mathcal{PSPACE}$-Complete by reducing from the 3-QBF problem with alternating quantifiers. This is done by taking a formula F and outputting a binary circuit C and integer k such that F is true if and only if C has a pebbling number of at most k.

The reduction builds a number of 'widgets' based on F. The original example from [10] can be seen below in Figure 1. The construction includes one widget for each quantifier, literal, and clause, as well as one pyramid graph that acts as a conjunction between the clauses. Each clause widget is incident on the widgets corresponding to the literals contained in the clause. The literal widgets are shown using shorthand notation; they are the pyramid graphs from [6], and an example is given in the upper corner of the diagram.

The pebbling number output by the reduction is $k = 2U + E + M$, where U, E, and M are the respective number of universal quantifiers, existential quantifiers, and clauses in F. Intuitively, the reduction works by forcing any pebbling to simulate a brute force QBF proof system. For each universal quantifier, the pebbling must travel up both sides, requiring that the entire circuit below that point be re-pebbled, thereby ensuring that F is true.

We refer to the circuits corresponding to true QBF formulas as the 'Lingas circuits', defined formally as follows:

Definition 8 (Lingas Circuits). $\mathcal{C_L} = \{C \mid \exists$ *a true* 3-QBF *formula* F *with alternating quantifiers such that applying Lingas's reduction to* F *yields the circuit* $C\}$

We shall make use of the $\mathcal{PSPACE}$-Completeness of the Lingas circuits to prove the $\mathcal{PSPACE}$-Completeness of the P/D game as well as the T-RES clause space problem. To this end we must first prove that for any Lingas circuit C, when playing on $Peb^2(C)$, the Prover has a strategy limiting the Delayer to at most $B\text{-}Peb(C)$ points. The strategy is quite simple: the Prover first forces the Delayer to admit that both variables associated with the target node must be False. As with the $\mathcal{G_I}$ DAGs, the Prover then proceeds to propagate this 'Double False' setting downwards.

We show that it is possible to traverse each existential widget while giving up at most one point, each universal widget while giving up at most two, and finally that it is possible to traverse the conjunctive pyramid and derive the ultimate contradictions while giving up at most M points, where M is the number of clauses in the formula underlying the circuit.

Each of these steps shall be proven by giving a decision tree showing the order in which the Prover queries variables. Each branch in this tree shall terminate

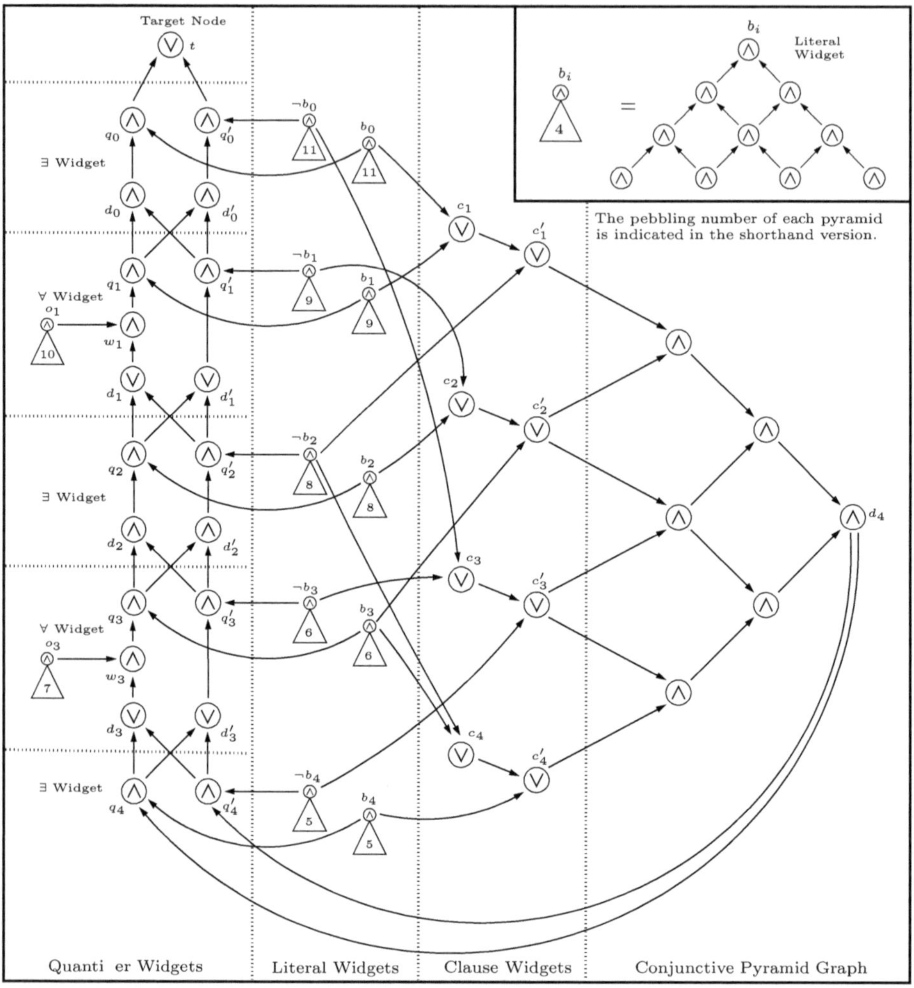

Fig. 1. The monotone circuit resulting from applying Lingas's reduction to the true formula $F = \exists x_0 \forall x_1 \exists x_2 \forall x_3 \exists x_4\ (x_0 \vee x_1 \vee \neg x_2) \wedge (\neg x_1 \vee x_2 \vee x_3) \wedge (\neg x_0 \vee \neg x_3 \vee \neg x_4) \wedge (\neg x_2 \vee x_3 \vee x_4)$. An example of a pyramid graph is shown in the upper corner. In this case $k = 2 \times 2 + 3 + 4 = 11$, but if F were false, then the circuit would require $k + 1$ pebbles and could not be pebbled using only k.

with the Delayer scoring no more than $B\text{-}Peb(C)$ points. A contradiction can be obtained (thereby ending the game) by falsifying an initial clause of $Peb^2(C)$.

The Prover's strategy is interesting in that whenever the Delayer responds 'You Choose' to a query, the Prover shall always respond with 'True'. Note that the Delayer has the choice of three different responses after each variable is queried, but our decision tree only needs to be binary, because if the Prover is willing to give the Delayer one point by responding to the Delayer's 'You Choose' answer, then there is no sense in exploring the path in which the Delayer gives the same answer without winning a point.

Theorem 6. *For any binary circuit $C \in \mathcal{C}_{\mathcal{L}}$ with pebbling number k, when playing on the formula $Peb^2(C)$, the Prover has a strategy limiting the Delayer to at most $k = B\text{-}Peb(C)$ points.*

Proof. The Prover's strategy proceeds as follows: first query t_0 and then t_1. The Delayer must set these variables to False, or else the game is over. The Delayer has therefore scored no points, and we enter the first quantifier widget with k points remaining. Next the Prover inductively traverses the quantifier widgets in order, propagating the 'Double False' setting downwards towards the conjunctive pyramid. The Prover has a strategy that gives up at most one point while traversing an existential widget, and at most two points when traversing a universal widget.

The existential widget and the Prover's strategy for traversing it are shown in Figure 2. We assume that we have j points remaining before the Delayer reaches k points, and that both variables associated with either the node d_{i-1} or d'_{i-1} have been set to False. Decision tree edges are labelled with the different possibilities at each step during the P/D game; 'YC,T' represents the case when the Delayer says 'You Choose', and the Prover replies 'True', and 'F' represents the case when the Delayer says 'False'. Leaves are labelled in the P/D game outcomes which they lead to; those labelled with numbers represent situations in which the game is over since an initial clause has been falsified, and the number indicates how many points were scored. Although we said that the Prover may give up at most 1 point while traversing this widget, some leaves end with the Delayer scoring 2 points. This is not a problem because even if this is the final quantifier widget, the formula has at least one clause, so we have at least one extra point's leeway.

The leaf corresponding to both of b_i's variables being set to False corresponds to the game entering the pyramid graph associated with variable b_i that has pebbling number j. Since each pyramid graph belongs to the $\mathcal{G}_{\mathcal{I}}$ family, the Delayer can score at most j points on it by Lemma 1. The remaining leaf is the one in which the widget has been traversed with only 1 point scored. Note that this decision tree corresponds to the strategy in which the Prover traversed the widget while setting one of the variables associated with b_i to True and both of the variables associated with d_{i+1} to False, thereby leading to the next widget. The case in which the Prover traverses the widget while setting one of the variables associated with $\neg b_i$ to True and both of the variables associated with d'_{i+1} to False is completely symmetrical. In other words, the Prover has complete control over which of the literal widgets is set to True. This is important because it allows the Prover to set the literal widgets in such a way so as to make the formula underlying the circuit true.

The universal widget and the Prover's strategy for traversing it are shown in Figure 2. We have j points remaining before the Delayer reaches k points, and both variables associated with either the node d_i or d'_i have been set to False. Once again, some paths lead to the Delayer scoring 3 points, but this is all right

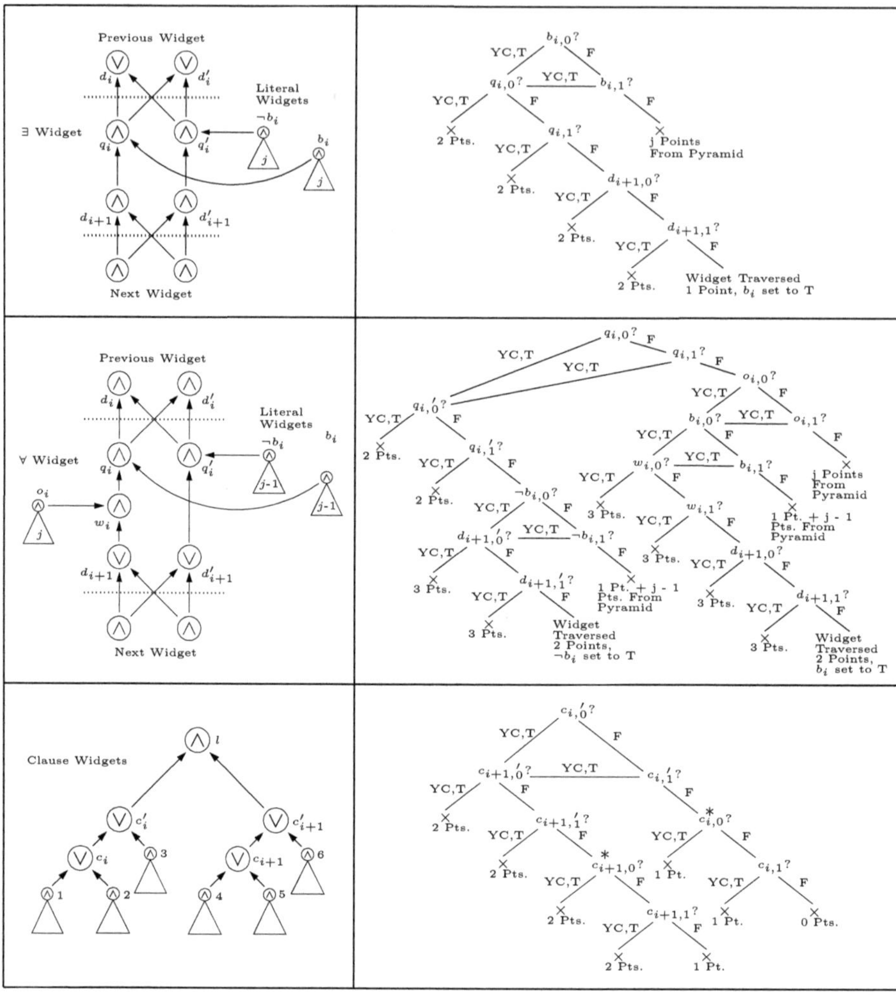

Fig. 2. The different widgets from Lingas's construction together with the decision trees showing the prover's strategies for traversing them. Duplicate subtrees are indicated by having multiple parents.

even if this is the final quantifier widget, since we have at least one point's leeway. Also as before, the pyramid graphs corresponding to both the literal widgets and the o_i widget belong to the $\mathcal{G}_{\mathcal{I}}$ family, so by Lemma 1, the Delayer can score at most $j-1$, $j-1$, and j points on them, respectively. Note the two leaves in which the widget has been traversed with only 2 points scored. In one case, one of the variables associated with b_i is set to True and both variables associated with d_{i+1} are set to False, and in the other case one of the variables associated with $\neg b_i$ is set to True, and both of the variables associated with d'_{i+1} are set to False, thereby leading to the next widget.

We now show how the Prover traverses the conjunctive pyramid graph. The apex of the conjunctive pyramid is part of the final quantifier widget, so both of its variables have been set to False. Since the formula underlying the Lingas circuit has M clauses and we have successfully traversed all of the clause widgets while giving up only the minimum number of points, we still have M points remaining. The conjunctive pyramid required to join M clauses has $M-1$ AND gates on its base, and therefore has pebbling number $M-1$. Let us assume for a moment that the leaves of the conjunctive pyramid have no children. Since the pyramid belongs to the $\mathcal{G}_{\mathcal{I}}$ family, the Prover has a strategy limiting the Delayer to at most $M-1$ points. However, since each of the conjunctive pyramid's leaves does have two children, we must explore the decision tree rooted at the possibility that the two variables associated with a leaf l are both set to False. Since the Delayer did not say 'You Choose' on either of l's variables, the number of points scored in setting them both to False is at most $M-2$. This is the worst-case scenario branching down the conjunctive pyramid, so the number of points scored on all other paths to leaves is strictly less than $M-2$.

We therefore have at least 2 points left with which to derive a contradiction within the clause widgets, and shall show how to force a contradiction given a conjunction pyramid leaf l that has had both of its associated variables set to False. This strategy is shown in Figure 2.

A key observation to make is that it is possible for the Prover to traverse the quantifier widgets such that the literal widgets that are attached to the existential widgets are set to True in any arbitrary way. Since the formula which the overall circuit is based on is a true QBF formula F, it is therefore possible to set them so that each clause widget is incident on at least one literal widget that has been set to True. This ensures that the Prover's strategy given by the decision tree in Figure 2 will always work. Note that the subtrees in the Prover's strategy labelled with $*$ may or may not be necessary, depending on which literal widgets were set to True.

Since each existential and universal widgets can be traversed while only giving up 1 and 2 points respectively, the conjunctive pyramid can be traversed while only giving up $M-2$ points, and a final contradiction can always be derived within the clause widgets while giving up at most 2 points, it follows that all of the above decision trees can be combined to show that the Prover has a strategy limiting the Delayer to at most $B\text{-}Peb(C)$ points, as required. □

4.4 Delayer Strategy for Monotone Circuits

We now show that for any binary monotone circuit C, when playing the P/D game on $Peb^2(C)$, the Delayer has a strategy that will always win at least $B\text{-}Peb(C)$ points. In [4] the authors give a Delayer strategy that wins at least $B\text{-}Peb(G)-3$ points, where $B\text{-}Peb(G)$ is defined without sliding. We improve this argument to show that the Delayer has a strategy that is guaranteed to win at least $B\text{-}Peb(G)$ points, defined with sliding.

In [4], the Delayer maintains a DAG G, with certain nodes marked as source nodes, and other nodes marked as target nodes. In the improved strategy, which applies to the more general case of monotone circuits, the Delayer also maintains such a marked monotone circuit C, but in addition, certain source nodes have pebbles on them. Thus at each stage of the game, the Delayer maintains a structure of the form $\langle C, S, T, P\rangle$, where C is a monotone circuit, S and T are disjoint subsets of C, and $P \subseteq S$. By $\mathrm{Peb}(C, S, T, P)$, we mean the pebbling number of the marked, pebbled circuit $\langle C, S, T, P\rangle$, where the pebbles on the nodes in P can be used as 'free pebbles'.

The Delayer's strategy proceeds as follows: at the start of the game, P is empty, S is the set of source nodes of the circuit C, and T contains the unique target node of C. At the start of each round in the game, the Prover queries a variable v_i, associated with a vertex v in the circuit C. The Delayer's strategy consists of two stages. In the first stage, the Delayer updates the sets S and T; in the second stage, the Delayer answers the Prover's query, and updates P.

In the first stage, assume that the marked, pebbled circuit at the start of the round is $\langle C, S, T, P\rangle$; the Delayer updates S and T to S' and T' as follows:

Case 1a: If $v \in S \cup T$, then set $S' := S$, and $T' := T$.

Case 1b: If $v \notin S \cup T$, and $\mathrm{Peb}(C, S, T, P) = \mathrm{Peb}(C, S, T \cup \{v\}, P)$, then set $S' := S$ and $T' := T \cup \{v\}$.

Case 1c: If $v \notin S \cup T$, and $\mathrm{Peb}(C, S, T, P) > \mathrm{Peb}(C, S, T \cup \{v\}, P)$, then set $S' := S \cup \{v\}$ and $T' := T$.

In the second stage of the Delayer's strategy, the Delayer updates P to P', and responds to the Prover's query.

Case 2a: If $v \in S'$, and v has no pebble on it, then respond 'You Choose', and place a pebble on v (i.e. $P' := P \cup \{v\}$). If v has a pebble on it, then set the queried variable the value True.

Case 2b: If $v \in T'$, then set the queried variable to False, and set $P' := P$.

Lemma 2. *When the game terminates,* $\mathit{Peb}(C, S, T, P) = 0$.

Proof. When the game terminates, an initial clause is falsified. Because of the strategy followed by the Delayer in the second stage of each round, this clause cannot be a clause associated with one of the initial source nodes of C, nor can it be the clause associated with the target node of C. Consequently, it must be a propagation clause associated with a vertex v and its immediate predecessors u_1, u_2 if v is an $\wedge$ gate, or just one of its immediate predecessors if v is an $\vee$ gate. Let us assume that v is an $\wedge$ gate. Then both variables associated with v must be set to False, from which it follows that $v \in T$, by Case 2a. If u_i is one of the immediate predecessors of v, one of the two variables associated with u_i must be set to True. By Case 2b, u_i cannot be in T, so $u_i \in S$. It follows from this that there must be a pebble on u_i. Hence, we can pebble $\langle C, S, T, P\rangle$ by sliding the pebble on u_i to v, showing that $\mathrm{Peb}(C, S, T, P) = 0$. The second case, where v is an $\vee$ gate, proceeds by essentially the same argument. □

Lemma 3. *If $\langle C, S, T, P\rangle$ is a marked, pebbled circuit, and v a vertex in C, then* $Peb(C, S, T, P) \leq \max\{Peb(C, S, T \cup \{v\}, P), Peb(C, S \cup \{v\}, T, P \cup \{v\}) + 1\}$.

Proof. We employ the following strategy to pebble T from S, using only the free pebbles in P. First, pebble $T \cup \{v\}$ from S, using $\mathrm{Peb}(G, S, T \cup \{v\}, P)$ pebbles. If the result is a pebble in T, then we are through, otherwise the final configuration has a pebble on v. Simply remove the other pebbles, then pebble T from $S \cup \{v\}$; this final step uses a total of $\mathrm{Peb}(C, S \cup \{v\}, T, P \cup \{v\}) + 1$ pebbles. □

Theorem 7. *If at the beginning of a round in the game, the marked, pebbled circuit is $\langle C, S, T, P\rangle$, then the Delayer can score at least $Peb(C, S, T, P)$ more points in the game.*

Proof: We argue by induction on the depth of the game tree. Lemma 2 settles the base case. Assume that the theorem holds for a subtree in which the marked, pebbled circuit is $\langle C, S', T', P'\rangle$; we wish to show that it holds also for its immediate supertree, associated with the marked, pebbled circuit $\langle C, S, T, P\rangle$. If $\mathrm{Peb}(C, S, T, P) = \mathrm{Peb}(C, S', T', P')$, then there is nothing to prove, so we can assume that $\mathrm{Peb}(C, S, T, P) > \mathrm{Peb}(C, S', T', P')$. By Lemma 3, $\mathrm{Peb}(C, S, T, P) = \mathrm{Peb}(C, S', T', P') + 1$. Now consider the round of the game in which the Delayer's initial circuit is $\langle C, S, T, P\rangle$, and the final circuit is $\langle C, S', T', P'\rangle$, and the variable queried was v_i. If v were in T', or if v had a pebble on it at the start of the round, then $\mathrm{Peb}(C, S, T, P) = \mathrm{Peb}(C, S', T', P')$. It follows from this that v is in S', but does not have a pebble on it at the start of the round, so the Delayer scores a point during this round. This proves that the condition of the Theorem also holds for the supertree. □

Corollary 2. *For any monotone circuit C, when playing on the formula $Peb^2(C)$, the Delayer has a strategy that wins at least B-$Peb(C)$ points.*

4.5 $\mathcal{PSPACE}$-Completeness of TCSP and P/D Game

This section contains our main result, namely the $\mathcal{PSPACE}$-Completeness of the T-RES clause space problem as well as the P/D game. The T-RES clause space problem ($TCSP$) asks: Given a formula F and integer k, does F have a T-RES refutation with clause space at most k? The Prover Delayer game Problem ($PDGAME$) asks: Given a formula F and integer k, is $PD(F)$ at most k?

The results from the previous sections allow us to prove the $\mathcal{PSPACE}$-Completeness of $TCSP$ and $PDGAME$ via a straightforward reduction from the $\mathcal{PSPACE}$-Complete problem of pebbling Lingas circuits. However, before proving this result we must first give a $\mathcal{PSPACE}$ algorithm for them these problems which uses following Lemma:

Lemma 4. *Every unsatisfiable CNF formula F has a* T-RES *refutation with clause space at most $n+1$, where n is the number of distinct variables in F.*

Proof. Let F be any arbitrary unsatisfiable CNF formula. Simply take F and choose some ordering for its n variables. Build a complete DPLL tree for F, branching on this ordering. This tree can be viewed as a T-RES proof that can be pebbled with at most $n+1$ pebbles, since any binary tree can be pebbled with $h+1$ pebbles, where h is the height of the tree. These pebbles correspond to which clauses need to be kept in memory in order to verify the proof, showing that for any unsatisfiable F, $TCS(F \vdash_{\text{T-RES}} \emptyset) \leq n+1$, as required. □

Lemma 5. *$TCSP \in \mathcal{PSPACE}$ and $PDGAME \in \mathcal{PSPACE}$*

Proof: Given an input (F, k) we first determine if F is satisfiable. Since $SAT \in \mathcal{NP}$ and $\mathcal{NP} \subseteq \mathcal{PSPACE}$, this is not a problem. If F is satisfiable, then we reject. If it is unsatisfiable, then we look at k. If $k \geq n+1$, where n is the number of distinct variables in F, then by Lemma 4 we simply accept. Otherwise F is unsatisfiable and $k \leq n$, so if F has a (configuration style) T-RES refutation π with $TCS(F \vdash_{\text{T-RES}} \emptyset) \leq k$, then each configuration contains at most k clauses. Since $k \leq n$ and each clause contains at most n variables, each configuration in π requires only polynomial space. Although we don't have access to π, we verify it as follows: Using a non-deterministic algorithm, start with a configuration $\mathbb{C}_0 = \emptyset$. Guess configuration $\mathbb{C}_1$, check to ensure that it follows from $\mathbb{C}_0$ by a legal tree resolution step, and erase configuration $\mathbb{C}_0$. Next, guess configuration $\mathbb{C}_2$, check to make sure that it follows from $\mathbb{C}_1$, and erase configuration $\mathbb{C}_1$. Continue this way until $\mathbb{C}_k$ has been derived. At any time, there are only two configurations in memory, but since each configuration uses at most quadratic space, our computation is in $\mathcal{NPSPACE}$. Finally, we appeal to Savitch's theorem to show that the problem of determining whether or not F has T-RES refutation π with $TCS(F \vdash_{\text{T-RES}} \emptyset) \leq k$ is in $\mathcal{PSPACE}$, thereby completing our $\mathcal{PSPACE}$ algorithm. By Theorem 1, this immediately implies that $PDGAME \in \mathcal{PSPACE}$ as well. □

We are now ready to prove our main result:

Theorem 8. *$PDGAME$ and $TCSP$ are both $\mathcal{PSPACE}$-Complete.*

Proof: By Lemma 5, we know that $PDGAME \in \mathcal{PSPACE}$. Next we show that $PDGAME$ is $\mathcal{PSPACE}$-Hard by reducing from black pebbling the Lingas circuits $\mathcal{C}_{\mathcal{L}}$, which proceeds by simply taking the input (C, k), and outputting $(Peb^2(C), k)$. Clearly this is a polytime reduction, and it is easy to see that it is correct by showing that $(C, k) \in \mathcal{C}_{\mathcal{L}}$ if and only if $(Peb^2(C), k) \in PDGAME$: If $(C, k) \in \mathcal{C}_{\mathcal{L}}$, then $B\text{-}Peb(C) \leq k$, so by Theorem 6, $PD(Peb^2(C)) \leq k$, and $(Peb^2(C), k) \in PDGAME$. On the other hand, if $(C, k) \notin \mathcal{C}_{\mathcal{L}}$, then $B\text{-}Peb(C) > k$, so by Corollary 2, $PD(Peb^2(C)) > k$, and $(Peb^2(C), k) \notin PDGAME$.

Therefore $PDGAME$ is $\mathcal{PSPACE}$-Complete, and by Theorem 1 it is easy to design a $\mathcal{PSPACE}$-Hardness reduction from $PDGAME$ to $TCSP$ as well. □

References

1. Alekhnovich, M., Ben-Sasson, E., Razborov, A.A., Wigderson, A.: Space Complexity in Propositional Calculus. SIAM J. of Comp. 31(4), 1184–1211 (2001)
2. Atserias, A., Dalmau, V.: A Combinatorial Characterization of Resolution Width. In: Proc. of the 18th IEEE Conference on Computational Complexity (2003)
3. Ben-Sasson, E.: Size Space Tradeoffs For Resolution. In: Proceedings of the 34th ACM Symposium on the Theory of Computing, pp. 457–464 (2002)
4. Ben-Sasson, E., Impagliazzo, R., Wigderson, A.: Near Optimal Separation of Tree-like and General Resolution. Combinatorica 24(4), 585–604 (2004)
5. Clote, P., Kranakis, E.: Boolean Functions and Computation Models. Springer, Heidelberg (2001)
6. Cook, S., Sethi, R.: Storage Requirements for Deterministic Polynomial Time Recognizable Languages. J. of Computer & System Sciences, 25–37 (1976)
7. Esteban, J., Torán, J.: Space Bounds for Resolution. Information and Computation 171, 84–97 (2001)
8. Esteban, J., Torán, J.: A Combinatorial Characterization of Treelike Resolution Space. Information Processing Letters 87, 295–300 (2003)
9. Gilbert, J.R., Lengauer, T., Tarjan, R.E.: The Pebbling Problem is Complete in Polynomial Space. SIAM Journal of Computing 9(3), 513–524 (1980)
10. Lingas, A.: A PSPACE-Complete Problem Related to a Pebble Game. In: Proceedings of the Fifth Colloquium on Automata, Languages and Programming, London, UK, pp. 300–321. Springer, Heidelberg (1978)
11. Nordström, J.: Narrow Proofs May Be Spacious: Separating Space and Width in Resolution. In: Proc.of the 38th ACM Symposium on the Theory of Computing (2006)
12. Pudlák, P., Impagliazzo, R.: Lower Bounds for DLL Algorithms for k-SAT. In: Proceedings of SODA 2000 (2000)

Continuous Previsions*

Jean Goubault-Larrecq

LSV, ENS Cachan, CNRS, INRIA Futurs
61, avenue du président-Wilson, 94235 Cachan, France
goubault@lsv.ens-cachan.fr

Abstract. We define strong monads of *continuous (lower, upper) previsions*, and of *forks*, modeling both probabilistic and non-deterministic choice. This is an elegant alternative to recent proposals by Mislove, Tix, Keimel, and Plotkin. We show that our monads are sound and complete, in the sense that they model exactly the interaction between probabilistic and (demonic, angelic, chaotic) choice.

1 Introduction

Moggi's computational λ-calculus [17] has proved useful to define various notions of computations on top of the lambda-calculus: side-effects, input-output, continuations, non-determinism [27], probabilistic computation [20] in particular. But mixing monads is hard, and finding the "right" monad that would combine both non-determinism and probabilistic choice has taken quite some effort. (We review recent progress below.)

The purpose of this paper is to introduce simple monads that do the job well. These are monads of *continuous previsions*, which can be seen as continuation-style monads. The idea of considering previsions comes from economics and statistics [4,12].

Outline. After stating some required preliminaries in Section 2, we recall the notion of *game* introduced in [5], arguing why these are natural extensions of notions of continuous valuations ($\sim$ measures) that also accommodate demonic and angelic non-deterministic choice. These notions induce functors on $\boldsymbol{Top}$, $\boldsymbol{Cpo}$, $\boldsymbol{Pcpo}$ (*pointed* cpos), but fail to yield monads. We analyze this failure in Section 4 by moving, through a Riesz-like representation theorem, to the new notions of collinear previsions, and previsions. We then show that indeed previsions yield strong monads, giving a simple semantics to a rich λ-calculus [17] with both probabilistic and non-deterministic choice. Finally, we show in Section 5 that our monad model is not only sound but complete.

This work is a summary of most of Chapters 10-12 of [6].

Related Work. Finding a monad combining both probabilistic and non-deterministic choice can be done by using general monad combination principles. The right way to combine monads in general is open to discussion. Lüth [11] proposes to combine monads by taking their coproduct in the category of monads. This coproduct exists under relatively mild assumptions [10]. However, in general the coproduct of two monads is

* Partially supported by the INRIA ARC ProNoBis. Part of Section 4 was done while the author was invited at U. Laval, Québec City, Québec, July 2004. We acknowledge their support.

J. Duparc and T.A. Henzinger (Eds.): CSL 2007, LNCS 4646, pp. 542–557, 2007.

an inscrutable object. A simpler, explicit description can be found in specific cases. For example, when taking coproducts of two *ideal* monads [2]. In particular, combining *non-blocking* non-determinism and probabilistic choice falls into this case. The resulting monad is relatively unenlightening, though: it is the monad of all sequences of choices, both probabilistic and non-deterministic [2, exemple 4.3].

Varacca [25,26] also proposed a monad combining non-determinism with probabilistic choice. Ghani and Uustalu [2] note that the above coproduct monad is close to Varacca's synchronization trees. The works closer to ours in computer science are those of Mislove [16] and Tix [23,24]. While this won't be entirely obvious from our definitions, we will establish a formal connection between their models and ours (Section 5). Outside computer science, previsions have their roots in economics and statistics [28]. However, we consider previsions on topological spaces, not just on sets.

This paper can be seen also be seen as a followup to [5], inasmuch as previsions are strongly tied to notions of convex and concave games. Our previsions can also be seen as predicate transformers; as such, and modulo minor details, they were discovered independently by Keimel and Plotkin [9] (K. Keimel, personal communication).

2 Preliminaries

We assume the reader to be familiar with (point-set) topology, in particular topology of T_0 but not necessarily Hausdorff spaces, as encountered in domain theory. See [3,1,15] for background. Let $int(A)$ denote the interior of A, $cl(A)$ its closure. The *Scott topology* on a poset X, with ordering $\leq$, has as opens the upward-closed subsets U (i.e., $x \in U$ and $x \leq y$ imply $y \in U$) such that for every directed family $(x_i)_{i \in I}$ having a least upper bound $\sup_{i \in I} x_i$ inside U, some x_i is already in U. The *way-below* relation $\ll$ is defined by $x \ll y$ iff for any directed family $(z_i)_{i \in I}$ with a least upper bound z such that $y \leq z$, then $x \leq z_i$ for some $i \in I$. A poset is *continuous* iff $\Downarrow y = \{x \in X | x \ll y\}$ is directed, and has y as least upper bound. Then every open U can be written $\bigcup_{x \in U} \Uparrow x$, where $\Uparrow x = \{y \in X | x \ll y\}$, and each of these sets is open.

Every topological space X has a specialization quasi-ordering $\leq$, defined by: $x \leq y$ iff every open that contains x contains y. X is T_0 iff $\leq$ is a (partial) ordering. The specialization ordering of the Scott topology of a quasi-ordering $\leq$ is $\leq$ itself. Every open is upward-closed, and a subset $A \subseteq X$ is *saturated* if and only if A is the intersection of all opens that contain it; equivalently, iff A is upward-closed in $\leq$. Let $\uparrow A$ denote the upward-closure of A under a quasi-ordering $\leq$, $\downarrow A$ its downward-closure. A T_0 space is *sober* iff every irreducible closed subset is the closure $cl\{x\} = \downarrow x$ of a (unique) point x. The Hofmann-Mislove Theorem implies that every sober space is *well-filtered* [8]: given any filtered family of saturated compacts $(Q_i)_{i \in I}$ in X, and any open U, $\bigcap_{i \in I} Q_i \subseteq U$ iff $Q_i \subseteq U$ for some $i \in I$. In particular, $\bigcap_{i \in I} Q_i$ is saturated compact. X is *locally compact* iff whenever $x \in U$ (U open) there is a saturated compact Q such that $x \in int(Q) \subseteq Q \subseteq U$. Every continuous cpo is sober and locally compact in its Scott topology. X is *coherent* iff the intersection of any two saturated compacts is compact. (Some authors take X to be coherent iff finite intersections of saturated compacts are compact. This would imply compactness, which we don't assume.) A coherent, well-filtered locally compact space is called *stably locally compact*. *Stably compact* spaces are those that are additionally compact, and have a wonderful theory (see, e.g., [8]). We consider the space $\mathbb{R}$ of all reals with the Scott topology of its natural ordering $\leq$. Its opens are $\emptyset$, $\mathbb{R}$, and the intervals $(t, +\infty)$, $t \in \mathbb{R}$. Any closed interval of $\mathbb{R}$, e.g., $[0, 1]$, is a stably locally compact, continuous cpo with the Scott topology. Because we equip $\mathbb{R}$ with the Scott topology, our *continuous* functions $f : X \to \mathbb{R}$ are those usually called *lower semi-continuous* in the mathematical literature.

By a *capacity* on X, we mean any function ν from $\mathcal{O}(X)$, the set of all opens of X, to $\mathbb{R}^+$, such that $\nu(\emptyset) = 0$ (a.k.a., a *set function*). A *game* ν is a *monotonic capacity*, i.e., $U \subseteq V$ implies $\nu(U) \leq \nu(V)$[1]. A *valuation* is a *modular* game ν, i.e., one such that $\nu(U \cup V) + \nu(U \cap V) = \nu(U) + \nu(V)$ for every opens U, V. A game is *continuous* iff $\nu(\bigcup_{i \in I} U_i) = \sup_{i \in I} \nu(U_i)$ for every directed family $(U_i)_{i \in I}$ of opens, and *normalized* iff $\nu(X) = 1$. Continuous valuations have a nice theory that fits topology well [7,8].

The *Dirac valuation* δ_x at $x \in X$ is the continuous valuation mapping each open U to 1 if $x \in U$, to 0 otherwise. Continuous valuations are canonically ordered by $\nu \leq \nu'$ iff $\nu(U) \leq \nu'(U)$ for every open U of X.

A *monad* on a category $\mathbb{C}$ may be presented in several different ways. One is based on triples $(\boldsymbol{T}, \boldsymbol{\eta}, \boldsymbol{\mu})$ of an endofunctor on $\mathbb{C}$, a unit, and a multiplication natural transformation. A presentation that is easier to grasp is in terms of Kleisli triples [14]. A *Kleisli triple* is a triple $(\boldsymbol{T}, \boldsymbol{\eta}, \_^\dagger)$, where $\boldsymbol{T}$ maps objects X of $\mathbb{C}$ to objects $\boldsymbol{T}X$ of $\mathbb{C}$, $\boldsymbol{\eta}_X$ is a morphism from X to $\boldsymbol{T}X$ for each X, and $f^\dagger$ (the *extension* of f) is a morphism from $\boldsymbol{T}X$ to $\boldsymbol{T}Y$ for each morphism $f : X \to \boldsymbol{T}Y$, satisfying: (1) ${\boldsymbol{\eta}_X}^\dagger = \mathrm{id}_{\boldsymbol{T}X}$; (2) for every $f : X \to \boldsymbol{T}Y$, $f^\dagger \circ \boldsymbol{\eta}_X = f$; (3) for every $g : X \to \boldsymbol{T}Y$, $f : Y \to \boldsymbol{T}Z$, then $f^\dagger \circ g^\dagger = (f^\dagger \circ g)^\dagger$. Kleisli triples and monads are equivalent.

3 Continuous Games, Convexity, Concavity

We follow [5]. A game ν on X is *convex* iff $\nu(U \cup V) + \nu(U \cap V) \geq \nu(U) + \nu(V)$ for every opens U, V. It is *concave* if the opposite inequality holds. Convex games are a cornerstone of economic theory [4,18].

One fundamental example of a game that is not a valuation is the *unanimity game* $\mathfrak{u}_A$ ($A \neq \emptyset$), defined by $\mathfrak{u}_A(U) = 1$ if $A \subseteq U$, $\mathfrak{u}_A(U) = 0$ otherwise. As we argue in [5], $\mathfrak{u}_A$ is a natural "probability-like" description of demonic non-deterministic choice, in the sense that drawing "at random" according to $\mathfrak{u}_A$ means that some malicious adversary C will choose an element of A for you. This is perhaps best conveyed by a thought experiment. You, the honest player P, would like to draw some element x from X with distribution ν (a game). Imagine you would like to know your chances of getting one from some (open) subset U of X. If ν is a probability distribution, then your chances will be equal to $\nu(U)$. This is standard. For general ν, continue to define your chances as $\nu(U)$. If $\nu = \mathfrak{u}_A$, and U does not contain A, then $\nu(U) = 0$, and your chances are zero: intuitively, C will pick an element in A, but outside U—on purpose. The only case where C is forced to pick an element in A which will suit P (i.e., be in U, too), is when $A \subseteq U$—and then P will be pleased with probability one.

[1] The name "game" is unfortunate, as there is no obvious relationship between this and games as they are usually handled in computer science, in particular with stochastic games. The notion stems from (cooperative) games in economics, where X is the set of players, not of states.

It is clear that $\mathfrak{u}_A$ is convex. It is in fact more. Call a game ν *totally convex* iff:

$$\nu\left(\bigcup_{i=1}^{n} U_i\right) \geq \sum_{I\subseteq\{1,\ldots,n\},I\neq\emptyset} (-1)^{|I|+1}\nu\left(\bigcap_{i\in I} U_i\right) \tag{1}$$

for every finite family $(U_i)_{i=1}^{n}$ of opens ($n \geq 1$), where $|I|$ denotes the cardinality of I. A *belief function* is a totally convex game. The dual notion of *total concavity* is obtained by replacing $\bigcup$ by $\bigcap$ and conversely in (1), and turning $\geq$ into $\leq$. A *plausibility* is a totally concave game. If $\geq$ is replaced by $=$ in (1), then we retrieve the familiar inclusion-exclusion principle from statistics. In particular any (continuous) valuation is a (continuous) belief function. Any belief function is a convex game, but not conversely [4,5].

Every game of the form $\sum_{i=1}^{n} a_i \mathfrak{u}_{Q_i}$, with $a_i \in \mathbb{R}^+$, and Q_i compact saturated and non-empty, is a continuous belief function, which we call *simple* belief function in [5]. When $\sum_{i=1}^{n} a_i = 1$, drawing an element from X "at random" (in the sense illustrated above) according to the simple belief function $\nu = \sum_{i=1}^{n} a_i \mathfrak{u}_{Q_i}$ intuitively corresponds to drawing one compact Q_i at random with probability a_i, then to let the malicious adversary C draw some element, demonically, from Q_i [5].

Let us turn to integration. Let ν be a game on X, and f be continuous from X to $\mathbb{R}^+$ (i.e., lower semi-continuous: $\mathbb{R}^+$ comes with the Scott topology). Assume f bounded, too, i.e., $\sup_{x\in X} f(x) < +\infty$. The *Choquet integral* of f along ν is:

$$\oint_{x\in X} f(x)d\nu = \int_0^{+\infty} \nu(f^{-1}(t,+\infty))dt \tag{2}$$

where the right hand side is an improper Riemann integral. This is well-defined, since $f^{-1}(t,+\infty)$ is open for every $t \in \mathbb{R}^+$ by assumption, and ν measures opens. Also, since f is bounded, the improper integrals above really are ordinary Riemann integrals over some closed intervals. The function $t \mapsto \nu(f^{-1}(t,+\infty))$ is decreasing, and every decreasing (even non-continuous, in the usual sense) function is Riemann-integrable, therefore the definition makes sense.

Alternatively, any *step function* $\sum_{i=0}^{n} a_i \chi_{U_i}$, where $a_0 \in \mathbb{R}^+$, $a_1, \ldots, a_n \in \mathbb{R}^+$, $X = U_0 \supseteq U_1 \supseteq \ldots \supseteq U_n$ is a decreasing sequence of opens, and χ_U denotes the indicator function of U ($\chi_U(x) = 1$ if $x \in X$, $\chi_U(x) = 0$ otherwise) is continuous: its integral along ν then equals $\sum_{i=0}^{n} a_i \nu(U_i)$—for *any* game ν. It is well-known that every bounded continuous function f can be written as the least upper bound of a sequence of step functions $f_K = a + \frac{1}{2^K}\sum_{k=1}^{\lfloor (b-a)2^K \rfloor} \chi_{f^{-1}(a+\frac{k}{2^K},+\infty)}(x)$, $K \in \mathbb{N}$, where $a = \inf_{x\in X} f(x)$, $b = \sup_{x\in X} f(x)$. Then the integral of f along ν is the least upper bound of the increasing sequence of the integrals of f_K along ν.

The main properties of Choquet integration are as follows. First, the integral is increasing in its function argument: if $f \leq g$ then the integral of f along ν is less than or equal to that of g along ν. If ν is continuous, then integration is also Scott-continuous in its function argument. The integral is also monotonic and Scott-continuous in the game ν. Integration is linear in the game, too, so integrating along $\sum_{i=1}^{n} a_i \nu_i$ is the same as taking the integrals along each ν_i, and computing the obvious linear combination.

However, Choquet integration is *not* linear in the function integrated, unless the game ν is a valuation. Still, it is *positively homogeneous*: integrating αf for $\alpha \in \mathbb{R}^+$ yields α times the integral of f. And it is additive on *comonotonic* functions $f, g : X \to \mathbb{R}$ (i.e., there is no pair $x, x' \in X$ such that $f(x) < f(x')$ and $g(x) > g(x')$).

Returning to the example of a simple belief function $\nu = \sum_{i=1}^{n} a_i \mathfrak{u}_{Q_i}$, the properties above imply that the integral of f along ν is $\sum_{i=1}^{n} a_i \min_{x \in Q_i} f(x)$ [5, Proposition 1]. (Note that $f(x)$ indeed attains its minimum over Q_i, which is compact.) Another way to read this is as follows. Imagine P publishes how much money, $f(x)$, she would earn if you picked x. When $\sum_{i=1}^{n} a_i = 1$, it is legitimate to say that the integral of f along ν should be some form of expected income. The formula above states that, when ν is a simple belief function, your expected income is exactly what you would obtain on average by drawing Q_i at random with probability a_i, then letting the malicious adversary C pick some element of Q_i for you—minimizing your earnings $f(x)$. In other words, integrating along a simple belief function computes *average min-payoffs*.

This can be generalized to all continuous, not just simple, belief functions [5, Theorem 4]. More precisely, the space $\mathbf{Cd}_{\leq 1}(X)$ of all continuous belief functions ν on X such that $\nu(X) \leq 1$ is isomorphic to the space $\mathbf{V}_{\leq 1}(\mathcal{Q}(X))$ of continuous valuations ν^* (of total mass at most 1) over the *Smyth powerdomain* $\mathcal{Q}(X)$ of X, provided X is well-filtered and locally compact. $\mathcal{Q}(X)$ is the cpo of non-empty compact saturated subsets of X, ordered by reverse inclusion $\supseteq$, and is a model of demonic non-determinism. (A similar result holds for *normalized* games and valuations ν, i.e., such that $\nu(X) = 1$: $\nu \mapsto \nu^*$ is again an isomorphism from $\mathbf{Cd}_1(X)$ to $\mathbf{V}_1(\mathcal{Q}(X))$.) The construction of ν^* from ν is relatively difficult, however it is noteworthy that when $\nu = \sum_{i=1}^{n} a_i \mathfrak{u}_{Q_i}$ is simple, then ν^* is exactly the simple valuation $\sum_{i=1}^{n} a_i \delta_{Q_i}$, which describes the choice of an element Q_i at random with probability a_i, as intuition would have it.

Similarly, the space $\mathbf{Pb}_{\leq 1}(X)$ of all *continuous plausibilities* ν with $\nu(X) \leq 1$ is isomorphic to $\mathbf{V}_{\leq 1}(\mathcal{H}_u(X))$ when X is stably locally compact, and where $\mathcal{H}_u(X)$ is the topological Hoare powerspace, defined as the set of non-empty closed sets of X, with topology generated by the sub-basic open sets $\Diamond U = \{F \text{ closed} | F \cap U \neq \emptyset\}$, U open in X. (This is the upper topology of $\subseteq$, which in general does not coincide with the Scott topology, unless e.g. X is a continuous cpo.) $\mathcal{H}_u(X)$ is used to model angelic non-determinism. We do not develop this here (see [6]). However, we mention that the corresponding *simple* plausibilities are of the form $\sum_{i=1}^{n} a_i \mathfrak{e}_{F_i}$, where F_i is a non-empty closed subset of X (an element of $\mathcal{H}_u(X)$), and the *example game* $\mathfrak{e}_F$ is defined so that $\mathfrak{e}_F(U) = 1$ if F meets U, $\mathfrak{e}_F(U) = 0$ otherwise: in this case C tries to help you, by finding some element in U that would also be in F, if possible.

Recall that every belief function is convex. One may show that Choquet integration along ν is *super-additive* (the integral of $f + g$ is at least that of f plus that of g) when ν is convex, and *sub-additive* (the integral of $f + g$ is at most that of f plus that of g) when ν is concave. See [4] for the finite case, [6, chapitre 4] for the topological case.

In the sequel, let $\mathbf{J}(X)$, $\nabla \mathbf{J}(X)$, $\Delta \mathbf{J}(X)$ be the spaces of plain, convex and concave continuous games respectively ("plain" meaning with no added property).

4 Continuous Previsions

For any space X, let $\langle X \to \mathbb{R}^+\rangle$ be the space of all bounded continuous functions from X to $\mathbb{R}^+$, with the Scott topology. Each continuous game ν on X gives rise to a functional $\alpha_{\mathcal{C}}(\nu)$ from $\langle X \to \mathbb{R}^+\rangle$ to $\mathbb{R}^+$, mapping f to its Choquet integral along ν.

Think of $f(x)$ again as defining how much money you would earn when x is chosen from X, by some computation process. We intentionally leave the notion of computation process undefined. This may be the process of drawing "at random" along a game ν, as in Section 3. In the sequel, we shall explore the view that x is the output of an arbitrary program, defined in some non-deterministic and probabilitic functional language. I.e., any program returns a value x ($\perp$ on non-termination, say), and if so P earns $f(x)$. For purely probabilistic programs (no non-deterministic choice), a prevision F is essentially a function mapping earning functions f to their average value $F(f)$, over all possible executions. Slightly more generally, for any belief function ν, there is a prevision $\alpha_{\mathcal{C}}(\nu)$ that maps each $f \in \langle X \to \mathbb{R}^+\rangle$ to the average min-payoff we get when our final earnings are given by f. Milking out the properties of $\alpha_{\mathcal{C}}(\nu)$, we arrive at:

Definition 1 (Prevision). *A* prevision *is a functional F from $\langle X \to \mathbb{R}^+\rangle$ to $\mathbb{R}^+$ such that F is* positively homogeneous *(for every $\alpha \geq 0$, $F(\alpha f) = \alpha F(f)$), and* monotonic *(if $f \leq g$ [pointwise], then $F(f) \leq F(g)$).*

F is a lower prevision *if moreover F is* super-additive*, i.e., $F(f+g) \geq F(f)+F(g)$. F is an* upper prevision *iff F is* sub-additive*: $F(f+g) \leq F(f)+F(g)$. F is* collinear *iff F is additive on comonotonic pairs, i.e., if whenever f and g are comonotonic, then $F(f+g) = F(f)+F(g)$. A prevision F is* linear *iff $F(f+g) = F(f)+F(g)$ for every $f, g \in \langle X \to \mathbb{R}^+\rangle$.*

Finally, F is continuous *iff it is Scott-continuous: for every directed family $(f_i)_{i\in I}$ of bounded continuous functions with least upper bound f, $F(\sup_{i\in I} f_i) = \sup_{i\in I} F(f_i)$.*

We write $\mathbf{P}(X)$, $\nabla\,\mathbf{P}(X)$, $\Delta\,\mathbf{P}(X)$ respectively the spaces of all continuous previsions, of continuous lower previsions, of continuous upper previsions equipped with the Scott topology of the pointwise ordering $\leq$. The spaces $\mathbf{P}^{\text{\ding{86}}}(X)$, $\nabla\,\mathbf{P}^{\text{\ding{86}}}(X)$, $\Delta\,\mathbf{P}^{\text{\ding{86}}}(X)$ will be the subspaces of those that are collinear.

We do not quite follow standard naming conventions. Standardly [28], a lower prevision is just a real-valued functional. *Coherent* lower previsions (taking a more readable definition from [13]) are those F such that $F(f) \geq \sum_{i=1}^n \lambda_i F(f_i) + \lambda_0$ whenever $f \geq \sum_{i=1}^n \lambda_i f_i + \lambda_0$, $\lambda_i > 0$, $\lambda_0 \in \mathbb{R}$. In our case, we reserve the "lower" adjective, so as to have a dual notion of *upper* prevision.

It is clear that any continuous game ν defines a continuous collinear prevision $\alpha_{\mathcal{C}}(\nu)$. Moreover, if ν is convex, then $\alpha_{\mathcal{C}}(\nu)$ is lower, and if ν is concave, then $\alpha_{\mathcal{C}}(\nu)$ is upper. The following isomorphism result, akin to Riesz' Representation Theorem, is known as Schmeidler's Theorem for convex games on discrete topologies. Let $\gamma_{\mathcal{C}}(F)$, for any prevision F, be the capacity ν such that $\nu(U) = F(\chi_U)$ for every open U of X. Order previsions pointwise, then:

Theorem 1. *$\alpha_{\mathcal{C}} \dashv \gamma_{\mathcal{C}}$ is a Galois connection from (plain, convex, concave) games into (plain, lower, upper) collinear previsions, with $\alpha_{\mathcal{C}}$ injective. That is, $\alpha_{\mathcal{C}}$ and $\gamma_{\mathcal{C}}$ are monotonic, $\alpha_{\mathcal{C}}(\gamma_{\mathcal{C}}(F)) \leq F$ for every collinear prevision F, and $\gamma_{\mathcal{C}}(\alpha_{\mathcal{C}}(\nu)) = \nu$ for every game ν.*

Moreover, when restricted to continuous previsions and games, $\alpha_{\mathfrak{C}}$ and $\gamma_{\mathfrak{C}}$ define an isomorphism between $\mathbf{J}(X)$ and $\mathbf{P}^{\divideontimes}(X)$, between $\nabla\,\mathbf{J}(X)$ and $\nabla\,\mathbf{P}^{\divideontimes}(X)$, between $\Delta\,\mathbf{J}(X)$ and $\Delta\,\mathbf{P}^{\divideontimes}(X)$.

Proof. That $\gamma_{\mathfrak{C}}(F)$ is a game for any prevision is easy. When F is lower, note that $\chi_{U \cup V}$ and $\chi_{U \cap V}$ are comonotonic, and $\chi_{U \cup V} + \chi_{U \cap V} = \chi_U + \chi_V$. So $\gamma_{\mathfrak{C}}(F)(U \cup V) + \gamma_{\mathfrak{C}}(F)(U \cap V) = F(\chi_{U \cup V} + \chi_{U \cap V})$ (since F is collinear) $= F(\chi_U + \chi_V) \geq F(\chi_U) + F(\chi_V)$ (since F is super-additive) $= \gamma_{\mathfrak{C}}(F)(U) + \gamma_{\mathfrak{C}}(F)(V)$. Similarly, $\gamma_{\mathfrak{C}}(F)$ is concave if F is upper.

For the converse, we first show that: (A) for any collinear prevision F on X, for any step function f, written $a + \sum_{i=1}^{m} a_i \chi_{U_i}$ with $U_1 \supseteq \ldots \supseteq U_m$, $a \in \mathbb{R}$, $a_1, \ldots, a_m \in \mathbb{R}^+$, then the Choquet integral of f along $\gamma_{\mathfrak{C}}(F)$ equals $F(f)$. This is an easy exercise as soon as one realizes that $\sum_{i=0}^{k-1} a_i \chi_{U_i}$ and $a_k \chi_{U_k}$ are comonotonic for every k, $1 \leq k \leq m$. The equality $\gamma_{\mathfrak{C}}(\alpha_{\mathfrak{C}}(\nu))(U) = \nu(U)$ is obvious, $\alpha_{\mathfrak{C}}$ and $\gamma_{\mathfrak{C}}$ are clearly monotonic. To show that $\alpha_{\mathfrak{C}}(\gamma_{\mathfrak{C}}(F)) \leq F$, we must show that the Choquet integral of f along $\gamma_{\mathfrak{C}}(F)$ is less than or equal to $F(f)$. Using the step functions f_K, $K \in \mathbb{N}$, by (A) the Choquet integral of f_K is less than or equal to $F(f_K)$. The least upper bound of the Choquet integrals of f_K, $K \in \mathbb{N}$ is that of f, and the least upper bound of $F(f_K)$ is at most $F(f)$. So $\alpha_{\mathfrak{C}}(\gamma_{\mathfrak{C}}(F))(f) \leq F(f)$. When F is continuous, the least upper bound of $F(f_K)$ is exactly $F(f)$, whence $\alpha_{\mathfrak{C}}(\gamma_{\mathfrak{C}}(F)) = F$. □

One easy, well-known consequence of this is that $\alpha_{\mathfrak{C}}$ and $\gamma_{\mathfrak{C}}$ define an order isomorphism between the space $\mathbf{V}(X)$ of continuous valuations and that $\mathbf{P}^{\triangle}(X)$ of continuous linear previsions ([7, Theorem 6.2], [22, Satz 4.16]). Intuitively, any continuous game ν gives rise to a continuous collinear prevision $\alpha_{\mathfrak{C}}(\nu)$ that computes a generalized form of expectation along ν, and every continuous collinear prevision arises this way.

It is easy to check that $\mathbf{J}$, $\nabla\,\mathbf{J}$, $\Delta\,\mathbf{J}$, $\mathbf{V}$, $\mathbf{P}^{\divideontimes}$, $\nabla\,\mathbf{P}^{\divideontimes}$, $\Delta\,\mathbf{P}^{\divideontimes}$, $\mathbf{P}^{\triangle}$ define functors $\boldsymbol{T}$ from $\boldsymbol{Top}$ to $\boldsymbol{Top}$, where $\boldsymbol{Top}$ is the category of topological spaces.

To define a monad structure on $\boldsymbol{T}$, we need a *unit* $\boldsymbol{\eta}_X : X \to \boldsymbol{T}X$, natural in X. This is defined by $\boldsymbol{\eta}_X(x) = \delta_x$. However, there is in general no extension $f^\dagger$ of $f : X \to \boldsymbol{T}Y$. The natural candidate is:

$$f^\dagger(\nu)(V) = \oint_{x \in X} f(x)(V) d\nu$$

when $\boldsymbol{T}$ is a game functor ($\mathbf{J}$, $\nabla\,\mathbf{J}$, $\Delta\,\mathbf{J}$, $\mathbf{V}$), or $f^\dagger(F)(h) = F(\lambda x \in X \cdot f(x)(h))$ when $\boldsymbol{T}$ is a prevision functor ($\mathbf{P}^{\divideontimes}$, $\nabla\,\mathbf{P}^{\divideontimes}$, $\Delta\,\mathbf{P}^{\divideontimes}$, $\mathbf{P}^{\triangle}$). While this indeed works when $\boldsymbol{T} = \mathbf{V}$ [7, Section 4.2], or when $\boldsymbol{T} = \mathbf{P}^{\triangle}$ using the isomorphism between $\mathbf{V}$ and the latter, it fails for the other functors. To understand why, take $\boldsymbol{T} = \nabla\,\mathbf{P}^{\divideontimes}$, and consider $X = \{1, 2\}$, $Y = \{*_{11}, *_{12}, *_{21}, *_{22}\}$ (with their discrete topologies), $F = \alpha_{\mathfrak{C}}(\mathfrak{u}_{\{1,2\}})$, i.e., $F(h) = \min(h(1), h(2))$ for every $h : Y \to \mathbb{R}^+$, $f : X \to \boldsymbol{T}Y$ defined by $f(1) = \alpha_{\mathfrak{C}}(3/4\delta_{*_{11}} + 1/4\delta_{*_{12}})$ and $f(2) = \alpha_{\mathfrak{C}}(1/3\delta_{*_{21}} + 2/3\delta_{*_{22}})$, so that $f(1)(h) = 3/4h(*_{11}) + 1/4h(*_{12})$ and $f(2)(h) = 1/3h(*_{21}) + 2/3h(*_{22})$ for every $h : Y \to \mathbb{R}^+$. Let h and h' be defined by: $h(*_{11}) = 0.3$, $h(*_{12}) = h(*_{22}) = 0.1$, $h(*_{21}) = 0.7$, $h'(*_{11}) = 0.5$, $h'(*_{12}) = h'(*_{22}) = 0$, $h'(*_{21}) = 0.7$, then $f^\dagger(F)(h) = 0.25$, $f^\dagger(F)(h') = 0.233\ldots$, $f^\dagger(F)(h+h') = 0.533\ldots$, but $f^\dagger(F)(h) + f^\dagger(F)(h') = 0.4833\ldots \neq f^\dagger(F)(h + h')$, although h and h' *are* comonotonic. In other words, $\_^\dagger$ does not preserve collinearity.

In everyday terms, collinear previsions, or more specifically belief functions represent a process where P draws at random first, then C chooses non-deterministically [5]. The example above is a non-deterministic choice (among $\{1, 2\}$) followed by probabilistic choices. In other words, the non-deterministic player C plays first, followed by the probabilistic player P. But it is well-known that you cannot permute non-deterministic and probabilistic choices, and the example above only serves to restate this.

Our cure is simple: drop the collinearity condition. We shall therefore consider monads of continuous (plain, lower, upper) previsions. Let ***Posc*** be the category of posets with Scott-continuous maps, ***Cpo*** its full subcategory of cpos. We consider posets equipped with their Scott topology, whence these two categories are full subcategories of ***Top***. Note that $\mathbf{P}(X)$, $\bigtriangledown \mathbf{P}(X)$, $\bigtriangleup \mathbf{P}(X)$ are only posets, not cpos.

Theorem 2. *Define $\boldsymbol{T}X$ as $\mathbf{P}(X)$, resp. $\bigtriangledown \mathbf{P}(X)$, resp. $\bigtriangleup \mathbf{P}(X)$. Let $\boldsymbol{\eta}_X(x) = \lambda h \in \langle X \to \mathbb{R}^+ \rangle \cdot h(x)$, and $f^\dagger(F)(h) = F(\lambda x \in X \cdot f(x)(h))$ for every $f : X \to \boldsymbol{T}Y$. Then $\boldsymbol{T}$ is a monad on* ***Top****, i.e., $(\boldsymbol{T}, \boldsymbol{\eta}, \_^\dagger)$ is a Kleisli triple. On* ***Posc****, $\boldsymbol{T}$ is a strong monad: $\boldsymbol{t}_{X,Y} : X \times \boldsymbol{T}Y \to \boldsymbol{T}(X \times Y)$ defined as $\boldsymbol{t}_{X,Y}(x, F)(h) = F(\lambda y \in Y \cdot h(x, y))$ is a tensorial strength.*

Proof. We must first show that, for every $f : X \to \boldsymbol{T}Y$, $f^\dagger$ is indeed a continuous map from $\boldsymbol{T}X$ to $\boldsymbol{T}Y$. Foremost, we must make sure that for every continuous (plain, lower, upper) prevision F on X, $f^\dagger(F)$ is a continuous (plain, lower, upper) prevision on Y. This is easy, but relatively tedious verification. Now note that the formulae defining $\boldsymbol{\eta}$, $\_^\dagger$, $\boldsymbol{t}$ are exactly the formulae defining the *continuation monad* [17]. It follows that the Kleisli triple axioms also hold in our case.

Contrarily to what might be expected, $\boldsymbol{t}_{X,Y}$ is not defined on all of ***Top***—it may fail to be continuous. On ***Posc***, this is repaired by the fact that a function of two arguments is continuous iff it is continuous in each argument separately (a fact that fails in ***Top***). The tensorial strength equations [17] are checked as for the continuation monad. □

That the formulae for unit, extension, and tensorial strength are the same as for the continuation monad is no accident. Imagine $F \in \boldsymbol{T}X$ is the semantics of a (probabilistic and non-deterministic) program expected to return a result x of type X. As we have already argued, when $F = \alpha_{\mathcal{C}}(P)$, with P a continuous valuation, then $F(h)$ is the *average* payoff, defined as the (Choquet) integral of $h(x)$ along P. When $F = \alpha_{\mathcal{C}}(\nu)$ with ν a continuous belief function, then $F(h)$ is the average min-payoff, where minima are taken over (demonically) non-deterministic choices. When F is not collinear, then more complicated "averaging" processes are involved. In particular, we allow taking means of mins of means of mins... representing plays where P, C, P, C, ... take turns. That arbitrarily many turns can be chained in a (not necessarily collinear) prevision will be a consequence of the fact that prevision functors define monads, and in particular have a well-defined multiplication. This is standard in the monadic approach to side-effects [17]: multiplication is the key to defining sequential composition—here, of plays.

More explicitly, take n continuous functions $f_1 : X_0 \to \boldsymbol{T}X_1$, $f_2 : X_1 \to \boldsymbol{T}X_2, \ldots,$ $f_n : X_{n-1} \to \boldsymbol{T}X_n$. Then, when $\boldsymbol{T}$ is a monad, $f_n^\dagger \circ f_{n-1}^\dagger \circ \ldots \circ f_2^\dagger \circ f_1 : X_0 \to \boldsymbol{T}X_n$ is the sequential composition $f_1; f_2; \ldots; f_n$ of $f_1, f_2, \ldots, f_{n-1}, f_n$ in this order: given $x_0 \in X_0$, the process $f_1(x_0)$ computes some element $x_1 \in X_1$ (in our case, by drawing it "at random", say; deterministic computations are of course allowed, too), then $f_2(x_1)$

computes some $x_2 \in X_2$, etc. The monad laws then say that composing with the idle process $\boldsymbol{\eta}_X : X \to \boldsymbol{T}X$ does nothing, and that sequential composition is associative.

While Theorem 2 then establishes a form of soundness (which we shall make more precise below), the goal of the next sections will be to show that the prevision axioms are complete, in the sense that there is no junk: every continuous (lower, upper) prevision is a mix of (demonic, angelic) non-deterministic and probabilistic choices.

One may wonder what the equivalent of *normalized* games ($\nu(X) = 1$) and *sub-normalized* games ($\nu(X) \leq 1$) would be through the correspondence of Theorem 1. Requiring $F(\chi_X)$ to be equal (resp. less than or equal) to 1 is the obvious choice. However, this is not preserved by $\_^\dagger$ when F is not collinear. So we define:

Definition 2. *A prevision F on X is* normalized, *resp.* sub-normalized, *iff for every $f \in \langle X \to \mathbb{R}^+\rangle$, for every $a \in \mathbb{R}^+$, $F(a+f) = a+F(f)$ (resp. $F(a+f) \leq a+F(f)$).*

We let $\mathbf{J}_1(X)$, $\bigtriangledown\mathbf{P}_1^{\circledast}(X)$, $\bigtriangledown\mathbf{P}_1(X)$, ..., be the subspaces of normalized games/ previsions, and $\mathbf{J}_{\leq 1}(X)$, $\bigtriangledown\mathbf{P}_{\leq 1}^{\circledast}(X)$, $\bigtriangledown\mathbf{P}_{\leq 1}(X)$, ..., those of sub-normalized games/ previsions.

Proposition 1. *Theorem 1 again holds for normalized (continuous) games and previsions, and for sub-normalized (continuous) games and previsions.*

Now the spaces of sub-normalized and normalized continuous previsions are cpos. The spaces of sub-normalized continuous previsions are *pointed*, i.e., they have a least element $\perp$, the constant 0 function. If X is itself pointed, then the spaces of normalized continuous previsions are pointed, too, with least element $\alpha_{\mathfrak{C}}(\delta_\perp)$ (a continuous linear prevision). The latter maps $h \in \langle X \to \mathbb{R}^+\rangle$ to $h(\perp)$. Let ***Cpo*** the category of cpos, ***Pcpo*** that of pointed cpos. It follows:

Proposition 2. *Let $\boldsymbol{T}X$ be $\mathbf{P}_{\leq 1}(X)$, $\bigtriangledown\mathbf{P}_{\leq 1}(X)$, $\bigtriangleup\mathbf{P}_{\leq 1}(X)$, $\mathbf{P}_1(X)$, $\bigtriangledown\mathbf{P}_1(X)$, or $\bigtriangleup\mathbf{P}_1(X)$. $(\boldsymbol{T}, \boldsymbol{\eta}, \boldsymbol{\mu}, \boldsymbol{t})$ is a strong monad on **Cpo** and on **Pcpo**.*

Theorem 2 allows us to give a semantics to a λ-calculus with both probabilistic and non-deterministic choices. Consider the syntax of terms and types:

$M, N, P ::=$	x	variable
$\mid$	c	constant
$\mid$	MN	application
$\mid$	$\lambda x \cdot M$	abstraction
$\mid$	$()$	empty tuple
$\mid$	(M, N)	pair
$\mid$	fst M	first projection
$\mid$	snd M	second projection
$\mid$	$\texttt{val}\, M$	trivial computation
$\mid$	$\texttt{let val}\, x = M \,\texttt{in}\, N$	let-expression

$\tau ::=$	α	base types
$\mid$	$\mathtt{u}$	type of ()
$\mid$	$\tau \times \tau$	product
$\mid$	$\tau \to \tau$	function types
$\mid$	$T\tau$	computation types

The typing rules, as well as the categorical semantics in a let-CCC, are standard [17]. Note that ***Cpo*** and ***Pcpo*** are Cartesian-closed. Together with the strong monads of Proposition 2, they form let-CCCs. The typing rules for computation types are: if $\Gamma \vdash M : \tau$ then $\Gamma \vdash \texttt{val}\, M : T\tau$; and if $\Gamma \vdash M : T\tau_1$ and $\Gamma, x : \tau_1 \vdash N : T\tau_2$ then $\Gamma \vdash \texttt{let val}\, x = M \,\texttt{in}\, N : T\tau_2$.

As should be expected, the semantics has a strong continuation flavor. For each term M of type τ in context $\Gamma = x_1 : \tau_1, \ldots, x_n : \tau_n$, $[\![M]\!]$ is a morphism (a continuous map) from $[\![\Gamma]\!] = [\![\tau_1]\!] \times \ldots \times [\![\tau_n]\!]$ to $[\![\tau]\!]$. The cases for val and let are given by: $[\![\mathtt{val}\, M]\!]\,(v_1, \ldots, v_n) = \lambda h \in \langle [\![\tau]\!] \to \mathbb{R}^+ \rangle \cdot h([\![M]\!]\,(v_1, \ldots, v_n))$, and $[\![\mathtt{let\,val}\, x = M\ \mathtt{in}\, N]\!]\,(v_1, \ldots, v_n) = \lambda h \in \langle [\![\tau_2]\!] \to \mathbb{R}^+ \rangle \cdot [\![M]\!]\,(v_1, \ldots, v_n)(\lambda v \in [\![\tau_1]\!] \cdot [\![N]\!]\,(v_1, \ldots, v_n, v)(h))$. Let bool be a base type, with $[\![\mathtt{bool}]\!] = \mathbb{S}$, where $\mathbb{S} = \{0, 1\}$ is Sierpiński space ($0 < 1$). Constants c may include a least fixpoint operator in ***Pcpo***, the Boolean constants false, true, a case construct $\mathtt{case} : \mathtt{bool} \times \tau \times \tau \to \tau$ with $[\![\mathtt{case}]\!]\,(0, v_0, v_1) = v_0$ and $[\![\mathtt{case}]\!]\,(1, v_0, v_1) = v_1$. The interpretation of T as a monad of previsions allows us, additionally, to give meaning to a coin-flipping operator $\mathtt{flip} : T\mathtt{bool}$, with $[\![\mathtt{flip}]\!] = \alpha_{\mathcal{C}}(1/2\delta_0 + 1/2\delta_1) = \lambda h \in \langle \mathbb{S} \to \mathbb{R}^+ \rangle \cdot 1/2(h(0) + h(1))$, and a non-deterministic choice operator $\mathtt{amb} : T\mathtt{bool}$. When $\boldsymbol{T}$ is $\bigtriangledown \mathbf{P}_1$, amb is the demonic choice (of a Boolean): $[\![\mathtt{amb}]\!] = \alpha_{\mathcal{C}}(\mathfrak{u}_{\{0,1\}}) = \lambda h \in \langle \mathbb{S} \to \mathbb{R}^+ \rangle \cdot \min(h(0), h(1))$ (the chosen Boolean x is the one that minimizes payoff $h(x)$). When $\boldsymbol{T}$ is $\triangle \mathbf{P}_1$, we get angelic choice: $[\![\mathtt{amb}]\!] = \alpha_{\mathcal{C}}(\mathfrak{e}_{\{0,1\}}) = \lambda h \in \langle \mathbb{S} \to \mathbb{R}^+ \rangle \cdot \max(h(0), h(1))$ (maximize payoff).

One might think that letting $\boldsymbol{T}$ be $\mathbf{P}_1$ would lead to chaotic choice. This certainly accommodates both demonic (min) and angelic choice (max). However, $\mathbf{P}_1$ is a very large space, and seems to contain objects that do not correspond to any mixture of probabilistic and non-deterministic choice. The right notion is suggested by [6, section 7.5].

Definition 3 (Fork). *A* fork *on X is any pair $F = (F^-, F^+)$ where F^- is a lower prevision, F^+ is an upper prevision, and for any $h, h' \in \langle X \to \mathbb{R}^+ \rangle$,*

$$F^-(h + h') \le F^-(h) + F^+(h') \le F^+(h + h') \qquad (3)$$

F is continuous, *resp.* normalized, sub-normalized, collinear, *whenever both F^- and F^+ are.*

While the above definition was found from purely mathematical arguments, Walley [28, Section 2] defines essentially the same notion in finance. However, we allow any pair (F^-, F^+) satisfying these conditions to be a fork. Walley only observes that whenever F^- is a coherent prevision (in his sense), on a discrete space, then letting $F^+(h) = -F^-(-h)$ yields a fork (F^-, F^+).

One may think of F^- as the pessimistic part of F, which will give us the least expected payoff, while F^+ is the optimistic part, yielding the greatest expected payoff. Taking $h' = 0$ in (3) shows indeed that $F^-(h) \le F^+(h)$ for each h. Let $\mathbf{F}(X)$ be the space of continuous forks on X, ordered by $\le \times \le$. The subspaces $\mathbf{F}_1(X)$ and $\mathbf{F}_{\le 1}(X)$ of normalized and sub-normalized forks are cpos. The latter is pointed (with least element $(0, 0)$) and the former is as soon as X is (with least element $(\alpha_{\mathcal{C}}(\delta_\bot), \alpha_{\mathcal{C}}(\delta_\bot))$). The semantics is essentially the pairing of two continuation semantics, e.g., $[\![\mathtt{val}\, M]\!]\,(v_1, \ldots, v_n) = (F^-, F^+)$, where $F^- = F^+ = \lambda h \in \langle [\![\tau]\!] \to \mathbb{R}^+ \rangle \cdot h([\![M]\!]\,(v_1, \ldots, v_n))$ (a *linear* prevision); $[\![\mathtt{let\,val}\, x = M\ \mathtt{in}\, N]\!]\,(v_1, \ldots, v_n) = (\lambda h \in \langle [\![\tau_2]\!] \to \mathbb{R}^+ \rangle \cdot F^-(\lambda v \in [\![\tau_1]\!] \cdot F_v^-(h)), \lambda h \in \langle [\![\tau_2]\!] \to \mathbb{R}^+ \rangle \cdot F^+(\lambda v \in [\![\tau_1]\!] \cdot F_v^+(h)))$, where $(F^-, F^+) = [\![M]\!]\,(v_1, \ldots, v_n)$ and $(F_v^-, F_v^+) = [\![N]\!]\,(v_1, \ldots, v_n, v)$. The constants with the most interesting semantics are amb, where $[\![\mathtt{amb}]\!] = (\lambda h \in \langle \mathbb{S} \to \mathbb{R}^+ \rangle \cdot \min(h(0), h(1)), \lambda h \in$

$\langle \mathbb{S} \to \mathbb{R}^+\rangle \cdot \max(h(0), h(1)))$ (i.e., it computes both pessimistic and optimistic outcomes), and `flip`, where $[\![\texttt{flip}]\!] = (F^-, F^+)$ with $F^- = F^+ = \lambda h \in \langle \mathbb{S} \to \mathbb{R}^+\rangle \cdot 1/2(h(0) + h(1))$. For the rest of the language, we rely on [17] and:

Proposition 3. *Let TX be defined as $\mathbf{F}(X)$, $\mathbf{F}_{\leq 1}(X)$, or $\mathbf{F}_1(X)$. Let $\boldsymbol{\eta}_X(x) = (F^-, F^+)$ with $F^- = F^+ = \lambda h \in \langle X \to \mathbb{R}^+\rangle \cdot h(x)$, and for every $f : X \to \boldsymbol{T}Y$, let $f^\dagger(F^-, F^+) = (\lambda h \in \langle Y \to \mathbb{R}^+\rangle \cdot F^-(\lambda x \in X \cdot f^-(x)(h)), \lambda h \in \langle Y \to \mathbb{R}^+\rangle \cdot F^+(\lambda x \in X \cdot f^+(x)(h)))$, where by convention $f(x) = (f^-(x), f^+(x))$. Then $(\boldsymbol{T}, \boldsymbol{\eta}, \boldsymbol{\mu})$ is a monad on **Top**. Together with $\boldsymbol{t}_{X,Y} : X \times \boldsymbol{T}Y \to \boldsymbol{T}(X \times Y)$ defined by $\boldsymbol{t}_{X,Y}(x, (F^-, F^+)) = (\lambda h \in \langle Y \to \mathbb{R}^+\rangle \cdot F^-(\lambda y \in Y \cdot h(x, y)), \lambda h \in \langle Y \to \mathbb{R}^+\rangle \cdot F^+(\lambda y \in Y \cdot h(x, y)))$, it forms a strong monad on **Cpo** and **Pcpo**.*

Proof. That the strong monad laws are satisfied is obvious. The core of the proof is in showing that unit, extension, and tensorial strength are well-defined. We deal with extension. Recall that $f^\dagger(F^-, F^+) = (F'^-, F'^+)$, where $F'^- = \lambda h \in \langle Y \to \mathbb{R}^+\rangle \cdot F^-(\lambda x \in X \cdot f^-(x)(h))$ and $F'^+ = \lambda h \in \langle Y \to \mathbb{R}^+\rangle \cdot F^+(\lambda x \in X \cdot f^+(x)(h)))$. Then $F'^-(h + h') = \lambda h \in \langle Y \to \mathbb{R}^+\rangle \cdot F^-(\lambda x \in X \cdot f^-(x)(h + h')) \leq \lambda h \in \langle Y \to \mathbb{R}^+\rangle \cdot F^-(\lambda x \in X \cdot f^-(x)(h) + f^+(x)(h'))$ (since $f(x) = (f^-(x), f^+(x)) \in \boldsymbol{T}Y$ and F^- is monotonic) $\leq \lambda h \in \langle Y \to \mathbb{R}^+\rangle \cdot F^-(\lambda x \in X \cdot f^-(x)(h)) + F^+(\lambda x \in X \cdot f^+(x)(h'))$ (since $(F^-, F^+) \in \boldsymbol{T}X) = F'^-(h) + F'^+(h')$. Similarly, $F'^-(h) + F'^+(h') \leq F'^+(h + h')$. □

5 Hearts and Skins: Completeness

One of the fundamental theorems of the theory of cooperative games is Shapley's Theorem, which states that every convex game ν has a non-empty core (on finite discrete X)—the core $Core(\nu)$ being the set of measures p such that $\nu \leq p$ and $\nu(X) = p(X)$. A refinement of this is Rosenmuller's Theorem, which states that a game ν is convex iff its core is non-empty and for every function $f : X \to \mathbb{R}^+$, the integral of f along ν is the minimum of all integrals of f along p, $p \in Core(\nu)$. In particular, there is a measure p such that $\nu \leq p$, $\nu(X) = p(X)$, and integrating f along p gives the same result as integrating it along ν [4].

We show that the same results hold in the continuous case in [6, chapitre 10]. Remember that games correspond to collinear previsions. Our purpose here is to show that similar theorems hold on previsions that need not be collinear (see [6, chapitre 11] for a more complete development). The analogue of measures will be linear previsions. We drop the analogue of the $\nu(X) = p(X)$ condition, however we concentrate on normalized games and previsions, because the technical treatment is slightly easier. We call the analogue of cores hearts, and the dual notion skin.

Definition 4 (Heart, Skin). *For any function F from $\langle X \to \mathbb{R}^+\rangle$ to $\mathbb{R}^+$, its* heart *$Coeur(F)$ is the set of linear functionals G such that $F \leq G$. Its* continuous heart *$CCoeur(F)$ is the subset of those Gs that are continuous. Its* skin *$Peau(F)$ is the set of linear functionals G such that $G \leq F$. Its* continuous skin *$CPeau(F)$ is the subset of those functionals G that are continuous.*

Again, we let $Coeur_1(F)$, $CCoeur_1(F)$, ..., be the subsets of the corresponding spaces consisting of normalized previsions only, and similarly $Coeur_{\leq 1}(F)$, ..., for those consisting of sub-normalized previsions.

Most of the developments below rest on Roth's Sandwich Theorem ([21], [24, Theorem 3.1]), which states that on every ordered cone C, for every positively homogeneous super-additive function $q : C \to \overline{\mathbb{R}}^+$ and every positively homogeneous sub-additive function $p : C \to \overline{\mathbb{R}}^+$ such that $a \leq b$ implies $q(a) \leq p(b)$ (e.g., when $q \leq p$ and either q or p is monotonic), then there is a monotonic linear function $f : C \to \overline{\mathbb{R}}^+$ such that $q \leq f \leq p$. Here $\overline{\mathbb{R}}^+$ is $\mathbb{R}^+$ plus an extra point at infinity $+\infty$. A *cone* is a set C, together with a binary operation $+$ turning it into a commutative monoid, and a scalar multiplication $\cdot$ from $\mathbb{R}^+ \times C$ to C, such that $1 \cdot a = a$, $0 \cdot a = 0$, $(rs) \cdot a = r \cdot (s \cdot a)$, $r \cdot (a + b) = r \cdot a + r \cdot b$, and $(r + s) \cdot a = r \cdot a + s \cdot a$. An *ordered cone* is equipped in addition with a partial ordering $\leq$ making $+$ and $\cdot$ monotonic. We only use Roth's Theorem on ordered cones of the form $\langle X \to \mathbb{R}^+ \rangle$. Our key result is:

Theorem 3. *Let X be a stably locally compact space, F a continuous lower prevision, and f a bounded continuous function from X to $\mathbb{R}^+$. Then there is a continuous linear functional G from $\langle X \to \mathbb{R}^+ \rangle$ to $\overline{\mathbb{R}}^+$ such that $F \leq G$ and $F(f) = G(f)$.*

Proof. Let F be a lower prevision on X, and $f \in \langle X \to \mathbb{R}^+ \rangle$. Define $\breve{F}_f$ by $\breve{F}_f(g) = \inf_{\substack{\lambda \in \mathbb{R}^+ \\ \lambda f \geq g}} \left[F(\lambda f) - \sup_{\substack{h \in \langle X \to \mathbb{R}^+ \rangle \\ g+h \leq \lambda f}} F(h) \right]$, taking this to be $+\infty$ is there is no $\lambda \in \mathbb{R}^+$ such that $\lambda f \geq g$. One checks that $\breve{F}_f$ is monotonic, positively homogeneous, sub-additive, above F ($\breve{F}_f(g) \geq F(g)$ for all g), touches F at f ($\breve{F}_f(f) = F(f)$). Roth's Sandwich Theorem then gives us a monotonic linear functional G_0 such that $F \leq G_0$ and $F(f) = G_0(f)$. However, G_0 may fail to be continuous. One now observes that $\langle X \to \mathbb{R}^+ \rangle$ is a continuous poset, with a basis B consisting of step functions. By Scott's Formula, the functional G defined by $G(f) = \sup_{g \in B, g \ll f} G_0(g)$ is continuous; in fact, the largest continuous functional below G_0. It follows that $F \leq G$ and $F(f) = G(f)$. The most difficult part of the proof is showing that G is linear. This rests on the fact that $\ll$ is multiplicative i.e., for any $a > 0$, $f \ll g$ iff $a \cdot f \ll a \cdot g$, and additive, i.e., if $h, f, g \in \langle X \to \mathbb{R}^+ \rangle$ are such that $h \ll f + g$, then $h \leq f' + g'$ for some $f', g' \in B$ with $f' \ll f$, $g' \ll g$; and conversely, $f' \ll f$ and $g' \ll g$ imply $f' + g' \ll f + g$. □

Note that G may take the value $+\infty$. We can refine this in the case of normalized previsions (for sub-normalized previsions, see [6, section 11.4]):

Theorem 4. *Let X be a stably locally compact space, F a normalized continuous lower prevision on X, and f a bounded continuous function from X to $\mathbb{R}^+$. Then there is a normalized continuous linear prevision G such that $F \leq G$ and $F(f) = G(f)$.*

Proof. Similar to Theorem 3. However, it may be that $\breve{F}_f$ reaches $+\infty$. Refine this by letting $\breve{\breve{F}}_f(g) = \inf_{\epsilon \in \mathbb{R}^+} \breve{F}_{f+\epsilon}(g)$, and using $\breve{\breve{F}}_f$ instead of $\breve{F}_f$. One checks that, since F is normalized, $\breve{F}_{f+\epsilon}$ is antitone in ϵ. Then $\breve{\breve{F}}_f$ is again monotonic, positively homogeneous, sub-additive (using antitony in ϵ), above F, and touches F at f. Moreover, it is easy to see that $\breve{F}_f(\chi_X) = 1$. We build G_0, then G from $\breve{F}_f$, as in Theorem 3. Additionally, we need X to be compact so as to establish that $G(\chi_X) = 1$. Since G is linear, it follows that G is normalized. □

One can deal with upper previsions instead, see [6, section 11.5], using a notion we call convex-concave duality to reduce to the above. We then obtain [6, théorème 11.5.22] that, when X is stably compact, F is a normalized continuous upper prevision on X, there is a normalized continuous linear prevision G on X such that $G \leq F$. Moreover, for every $f \in \langle X \to \mathbb{R}^+ \rangle$, $F(f) = \sup_{G \in CPeau_1(F)} G(f)$.

Theorem 4 allows us to state a form of Rosenmuller's Theorem:

Theorem 5. *Let X be stably locally compact, F a continuous normalized prevision on X. Then F is lower iff $CCoeur_1(F) \neq \emptyset$ and for every $f \in \langle X \to \mathbb{R}^+ \rangle$, $F(f) = \inf_{G \in CCoeur_1(F)} G(f)$. In this case, the inf is attained: $F(f) = \min_{G \in CCoeur_1(F)} G(f)$.*

There is, of course, a dual theorem on upper previsions and their skins [6, théorème 11.7.5]; infs are replaced by sups, which need not be attained.

Our completeness results to come are based on another topology on spaces of previsions: the *weak topology* is the coarsest that makes the function $F \mapsto F(f)$ continuous, for each $f \in \langle X \to \mathbb{R}^+ \rangle$. The Scott topology is in general finer. Let $\nabla \mathbf{P}_{1\ wk}(X)$ be $\nabla \mathbf{P}_1(X)$ with the weak topology, and similarly for other spaces. Then:

Proposition 4. *Let X be stably compact, F a normalized continuous lower prevision, then $CCoeur_1(F)$ is a non-empty saturated compact convex subset of $\mathbf{P}^{\triangle}_{1\ wk}(X)$.*

Compactness follows from Plotkin's version of the Banach-Alaoglu Theorem [19], or using a technique by Heckmann and Jung [8, Section 3.2], while convexity (i.e., $\alpha F + (1-\alpha)F'$ is in $CCoeur_1(F)$ as soon as F and F' are, $\alpha \in [0,1]$) is clear. That skins are closed is much easier, while non-emptiness is by the dual Rosenmuller Theorem:

Proposition 5. *Let X be a topological space, F a normalized continuous upper prevision, then $CPeau_1(F)$ is a closed convex subset of $\mathbf{P}^{\triangle}_{1\ wk}(X)$. It is non-empty as soon as X is stably compact.*

Finally, call a *lens* of a space X any non-empty intersection $L = Q \cap F$ of a saturated compact Q and a closed subset F. Then:

Proposition 6. *Let X be a stably compact space. The continuous normalized* body *$CCorps_1(F) = CCoeur_1(F^-) \cap CPeau_1(F^+)$ of a continuous normalized fork $F = (F^-, F^+)$ on X is a lens. Moreover, $CCoeur_1(F^-) = \uparrow CCorps_1(F)$ and $CPeau_1(F^+) = \downarrow CCorps_1(F)$.*

Proof. We show that: $(*)$ for each $G \in CCoeur_1(F^-)$, there is a $G' \in CCoeur_1(F^-) \cap CPeau_1(F^+)$ such that $G' \leq G$. Let $F'(h) = \inf_{f,g \in \langle X \to \mathbb{R}^+ \rangle / f+g \geq h} (F^+(f) + G(g))$. One checks that $F^- \leq F' \leq G$, that F' is an upper prevision, so by Roth's Sandwich Theorem, there is a linear monotonic functional G_0 such that $F^- \leq G_0 \leq F'$. Since $G_0 \leq F'$, G_0 does not take the value $+\infty$. Build G from G_0 using Scott's Formula, as before. It is easy to see that G is a continuous, normalized, linear prevision. Since $F^- \leq G'$, $G' \in CCoeur_1(F^-)$. Since $G' \leq F' \leq F^+$, $G' \in CPeau_1(F^+)$. Since $F' \leq F' \leq G$, $G' \leq G$ follows.

By $(*)$, $CCoeur_1(F^-) \cap CPeau_1(F^+)$ is non-empty. That $CCoeur_1(F^-) = \uparrow (CCoeur_1(F^-) \cap CPeau_1(F^+))$ is another easy consequence of $(*)$. Next we show $CPeau_1(F^+) = \downarrow (CCoeur_1(F^-) \cap CPeau_1(F^+))$ in a similar way, by defining $F''(h) = \sup_{\substack{f,g \in \langle X \to \mathbb{R}^+ \rangle \\ f+g \leq h}} (F^-(f) + G(g))$, where $G \in CPeau_1(F^-)$, and using F'' to show that there is some $G' \in CCoeur_1(F^-) \cap CPeau_1(F^+)$ such that $G \leq G'$. □

These three propositions state that any normalized continuous lower prevision, resp. upper prevision, resp. fork F gives rise to an element $CCoeur_1(F)$, resp. $CPeau_1(F)$, resp. $CCorps_1(F)$ of the Smyth powerdomain $\mathcal{Q}(\mathbf{P}^{\triangle}_{1\,wk}(X))$ (demonic non-deterministic choice of a probability law—remember that $\mathbf{P}^{\triangle}_1(X) \cong \mathbf{V}_1(X)$), resp. the Hoare powerdomain $\mathcal{H}(\mathbf{P}^{\triangle}_{1\,wk}(X))$ over $\mathbf{P}^{\triangle}_{1\,wk}(X)$ (angelic), resp. the Plotkin powerdomain over $\mathbf{P}^{\triangle}_{1\,wk}(X)$ (chaotic). This is a form of *completeness*: spaces of previsions and of forks contain no junk, and really are no more than mixes of non-deterministic and probabilistic choice. More explicitly, in the demonic case, let F be any normalized continuous lower prevision on stably compact X: then F is the sequential composition $f_0; f_1 = f_1^\dagger \circ f_0$ (applied to some dummy $*$), where $f_0(*) = \alpha_{\mathcal{C}}(\mathfrak{u}_{CCoeur_1(F)})$ (non-deterministic choice of some G from the heart of F), and $f_1(G) = G$ (drawing at random along $\gamma_{\mathcal{C}}(G)$). We need Proposition 4 for this to be well-defined. $F = (f_0; f_1)(*)$ follows from the definition of the heart and general properties of the Choquet integral. In the angelic case, any normalized continuous upper prevision F is the sequential composition $f_0; f_1$, where $f_0(*) = \alpha_{\mathcal{C}}(\mathfrak{e}_{\mathcal{C}Peau_1(F)})$, and again $f_1(G) = G$.

In the converse direction, still assuming X stably compact, there is a map $\bigsqcap$: $\mathcal{Q}(\mathbf{P}^{\triangle}_{1\,wk}(X)) \to \nabla\mathbf{P}_1(X)$ defined by $\bigsqcap K(f) = \min_{G \in K} G(f)$, and $CCoeur_1 \dashv \bigsqcap$ is a Galois connection consisting of continuous maps [6, théorème 11.7.11], while there is a continuous map $\bigsqcup$: $\mathcal{H}(\mathbf{P}^{\triangle}_{1\,wk}(X)) \to \nabla\mathbf{P}_1(X)$ defined by $\bigsqcup C(f) = \sup_{G \in C} G(f)$, so that $\bigsqcup \dashv CPeau_1$ is a Galois insertion.

We conclude by noticing that, when X is a continuous cpo with a least element, $\mathbf{P}^{\triangle}_{1\,wk}(X)$ is homeomorphic to $\mathbf{V}_1(X)$ with the weak topology, and the latter coincides then with the Scott topology [8]. Apart from spurious details (e.g., we bound our valuations by 1 instead of $+\infty$), there is therefore a strong connection with the models of Mislove [16] and Tix [23,24]. This is explored, under a different light, in [9].

Acknowledgment. Thanks to Nadia Tawbi for inviting me in July 2004 in Québec City. Thanks to Vincent Danos, Josée Desharnais and François Laviolette, through whom I came to study related problems, namely [5]. Thanks to Prakash Panangaden for his warm support, to all people who showed interest in this work [6], to Michael Mislove, and to Klaus Keimel.

References

1. Abramsky, S., Jung, A.: Domain theory. In: Abramsky, S., Gabbay, D.M., Maibaum, T.S.E. (eds.) Handbook of Logic in Computer Science, vol. 3, pp. 1–168. Oxford University Press, Oxford (1994)
2. Ghani, N., Uustalu, T.: Coproducts of ideal monads. Theoretical Informatics and Applications 38(4), 321–342 (2004)

3. Gierz, G., Hofmann, K.H., Keimel, K., Lawson, J.D., Mislove, M., Scott, D.S.: Continuous lattices and domains. In: Encyclopedia of Mathematics and its Applications, vol. 93, Cambridge University Press, Cambridge (2003)
4. Gilboa, I., Schmeidler, D.: Additive representation of non-additive measures and the Choquet integral. Discussion Papers 985, Northwestern University, Center for Mathematical Studies in Economics and Management Science (1992)
5. Goubault-Larrecq, J.: Continuous capacities on continuous state spaces. In: Arge, L., Cachin, C., Jurdzinski, T., Tarlecki, A. (eds.) ICALP 2007. LNCS, vol. 4596, pp. 764–776. Springer, Heidelberg (2007)
6. Goubault-Larrecq, J.: Une introduction aux capacités, aux jeux et aux prévisions. Version 1.7 (June 2007), `http://www.lsv.ens-cachan.fr/~goubault/ProNobis/pp.pdf`
7. Jones, C.: Probabilistic Non-Determinism. PhD thesis, University of Edinburgh, Technical Report ECS-LFCS-90-105 (1990)
8. Jung, A.: Stably compact spaces and the probabilistic powerspace construction. In: Desharnais, J., Panangaden, P. (eds.) Domain-theoretic Methods in Probabilistic Processes. Electronic Lecture Notes in Computer Science, vol. 87, p. 15. Elsevier, Amsterdam (2004)
9. Keimel, K., Plotkin, G.: Predicate transformers for convex powerdomains. Math. Struct. Comp. Sci., p. 42 (Submitted 2007)
10. Max Kelly, G.: A unified treatment of transfinite constructions for free algebras, free monoids, colimits, associated sheaves and so on. Bull. Austr. Math. Soc., 22, 1–83 (1980)
11. Lüth, C.: Categorical Term Rewriting: Monads and Modularity. PhD thesis, University of Edinburgh (1997)
12. Maaß, S.: Coherent lower previsions as exact functionals and their (sigma-)core. In: de Cooman, G., Fine, T., Seidenfeld, T. (eds.) Proc. 2nd Intl. Symp. Imprecise Probabilities and their Applications (ISIPTA'01), pp. 230–236 (2001)
13. Maass, S.: Continuous linear representation of coherent lower previsions. In: Bernard, J.-M., Seidenfeld, T., Zaffalon, M. (eds.) Proc. 3rd Intl. Symp. on Imprecise Probabilities and Their Applications (ISIPTA'03), Carleton Sci. Proc. in Informatics, vol. 18, pp. 371–381 (2003)
14. Manes, E.G.: Algebraic Theories. Graduate Texts in Mathematics, vol. 26. Springer, Heidelberg (1976)
15. Mislove, M.: Topology, domain theory and theoretical computer science. Topology and Its Applications 89, 3–59 (1998)
16. Mislove, M.: Nondeterminism and probabilistic choice: Obeying the law. In: Palamidessi, C. (ed.) CONCUR 2000. LNCS, vol. 1877, pp. 350–364. Springer, Heidelberg (2000)
17. Moggi, E.: Notions of computation and monads. Inf. & Comp., 93, 55–92 (1991)
18. Osborne, M.J., Rubinstein, A.: A Course in Game Theory. MIT Press, Cambridge (1994)
19. Plotkin, G.: A domain-theoretic Banach-Alaoglu theorem. Math. Struct. Comp. Sci., 16, 299–311 (2006)
20. Ramsey, N., Pfeffer, A.: Stochastic lambda calculus and monads of probability distributions. In: Proc. 29th Ann. ACM SIGPLAN-SIGACT Symp. Principles of Programming Languages (POPL'02), pp. 154–165 (2002)
21. Roth, W.: Hahn-Banach type theorems for locally convex cones. Journal of the Australian Mathematical Society 68(1), 104–125 (2000)
22. Tix, R.: Stetige Bewertungen auf topologischen Räumen. Diplomarbeit, TH Darmstadt (June 1995)
23. Tix, R.: Continuous D-Cones: Convexity and Powerdomain Constructions. PhD thesis, Technische Universität Darmstadt (1999)
24. Tix, R., Keimel, K., Plotkin, G.: Semantic domains for combining probability and non-determinism. Electronic Notes in Theoretical Computer Science 129, 1–104 (2005)

25. Varacca, D.: The powerdomain of indexed valuations. In: Proc. 17th Ann. Symp. Logic in Computer Science (LICS'02), pp. 299–308. IEEE Computer Society Press, Los Alamitos (2002)
26. Varacca, D., Winskel, G.: Distributing probability over nondeterminism. Math. Struct. Comp. Sci., 26 (Accepted 2005)
27. Wadler, P.: Comprehending monads. Math. Struct. Comp. Sci. 2, 461–493 (1992)
28. Walley, P.: Statistical Reasoning with Imprecise Probabilities. Chapman and Hall, London (1991)

Bad Variables Under Control

Andrzej S. Murawski*

Oxford University Computing Laboratory,
Wolfson Building, Parks Road, Oxford OX1 3QD, UK

Abstract. We give a fully abstract game model for Idealized Algol with non-local control flow. In contrast to most previous papers on game semantics, we do not need to include the bad-variable constructor **mkvar** to obtain full abstraction. Using the model we show that, unlike in the "control-free" case, the presence of **mkvar** does affect observational equivalence. We conclude by discussing the effect of **mkvar** on nondeterministic and probabilistic variants of Idealized Algol.

1 Introduction

In the computer science classic "The essence of Algol" [1], Reynolds has laid out a series of principles that, in his opinion, should underlie the contemporary evolution of programming languages. He also defined a prototype language, called Idealized Algol, whose design was to be their embodiment. Based on a simple imperative language extended with the procedural mechanism of the call-by-name lambda calculus, Idealized Algol has come to be regarded as a canonical proposal for synthesizing functional and imperative programming. Its elegant definition has lent itself to a systematic analysis leading to significant progress in the field of programming language semantics [2].

One of Reynolds's insights concerned the semantics and type-theoretic treatment of assignable variables. He viewed them as dual in nature: producing values on the one hand (like expressions) and capable of accepting them on the other. This idea is reminiscent of the distinction between l- and r-values, but Reynolds took it much further: he advocated that the variable type be actually *identified* with the product of an "acceptor type" and the expression type. This decomposition enabled him to define the meaning of variables without any commitment to the structure of state, suggesting new abstract approaches to modelling state. Reynolds himself pursued one, based on functor categories, but alternatives have also emerged.

A particularly fruitful way of modelling computation turned out to be based on the idea that programs should be viewed as processes interacting with one another. Reddy calls this the object-based approach to semantics [3]. From this point of view, Reynolds's analysis of variables simply suggests viewing a variable as an object equipped with a reading- and a writing-method which it uses to communicate with the environment. Similar philosophy can be found in encodings of imperative programs into process algebras. In denotational semantics this

* Funded by an Advanced Research Fellowship from the UK EPSRC (EP/C539753/1).

J. Duparc and T.A. Henzinger (Eds.): CSL 2007, LNCS 4646, pp. 558–572, 2007.

approach was adopted in Abramsky and McCusker's game model of Idealized Algol [4] and Reddy's work on coherence spaces [3].

Game semantics has led to many full abstraction results ever since. These amount to defining a model such that equality in the model corresponds to program equivalence, or, in the more general inequational variant, such that observational approximation coincides with a distinguished preorder on the model. A general way of proving such results leads through a definability argument, which demonstrates that all compact elements of the model are definable, i.e. are denotations of some programs. However, when the variable type is a product type, definability fails unless an explicit pairing construct is added to the language. In the case of [4] this is the so-called "bad-variable constructor" **mkvar**, which makes a writing method M and a reading method N into a variable:

$$\frac{\Gamma \vdash M : \mathsf{exp} \to \mathsf{com} \quad \Gamma \vdash N : \mathsf{exp}}{\Gamma \vdash \mathbf{mkvar}(M, N) : \mathsf{var}}.$$

The reduction rules for $\mathbf{mkvar}(M, N)$ simply evaluate N for reading and evaluate Mv when the value v is to be written. Because, in absence of **mkvar**, definability fails, it seems likely that a game model that is (equationally or inequationally) fully abstract for a language with **mkvar**, will not be fully abstract in its absence. Thus it is natural to ask how one can repair existing game models to analyze observational approximation and equivalence in **mkvar**-free settings and whether this is really needed in all cases. This direction is also motivated by the fact that **mkvar** is not really a conventional programming construct, though languages incorporating bad variables as a feature do exist in the literature[1].

This paper uses a version of Idealized Algol that allows for side-effects in expressions and variables, to which we shall henceforth refer as IA. A fully abstract model for IA with **mkvar** was given in [4], where the preorder for inequational full abstraction was simply inclusion of *complete* plays (those in which all questions have been answered). Subsequently, McCusker [7] has introduced a different preorder on complete plays which captures observational approximation in IA. His was actually the only paper so far that has analyzed the game semantics of an IA-like language without **mkvar**. The aim of the present paper is to achieve a full abstraction result for IA enriched with non-local control flow (also without **mkvar**), i.e. in a setting where the transfer of control can violate the usual discipline of block entry and exit. Syntactically, the requisite extension of IA can be achieved by adding a **catch**-construct, in the spirit of Reynolds's escape [1] and Cartwright and Felleisen's control operators [8]. It is well known that violations of the stack discipline can be modelled in game semantics by relaxing the bracketing condition [9]. However, full abstraction for IA + **catch** is not merely a matter of applying McCusker's preorder in the unbracketed setting and we show

[1] In the late 1960s POP-2 [5] was an attempt to define a broad-spectrum language for both numerical and symbolic computation combining the strengths of FORTRAN and LISP. The language was based on *doublets*, which are conceptually the same as **mkvar**-objects. Also, Reynolds's GEDANKEN [6] had support for *implicit references* in the spirit of **mkvar**.

what further refinements are necessary to accomplish it. Our model can then be used to demonstrate that, in contrast to McCusker's result on conservativity of $\mathsf{IA} + \mathbf{mkvar}$ (with respect to observational equivalence), $\mathsf{IA} + \mathbf{catch} + \mathbf{mkvar}$ does *not* extend $\mathsf{IA} + \mathbf{catch}$ conservatively. Next we go on to investigate the impact of **mkvar** on other important extensions of IA, namely, with nondeterminism and probability. It turns out that **mkvar** does affect observational equivalence in $\mathsf{IA} + \mathbf{or}$, but it does not for $\mathsf{IA} + \mathbf{coin}$.

In addition to the already-cited paper by McCusker, techniques based on nominal sets have recently been proposed as a general approach to handling the absence of **mkvar** [10,11]. Because they bring an additional layer of combinatorial complexity to game semantics (names inside moves, name invariance), it seems a valuable goal to understand how the same results can still be achieved by "anonymous" game semantics. This paper explores this direction in the call-by-name setting of Algol-like languages, leaving call-by-value as a challenge for future work.

2 IA, $\mathsf{IA}_{\mathbf{catch}}$

We consider Reynolds's Idealized Algol [1] in which expressions and variables can produce side-effects. Its syntax is given in Figure 1. The types T are formed from the base types

$$B ::= \mathsf{com} \mid \mathsf{exp} \mid \mathsf{var}$$

using the $\rightarrow$ constructor: $T ::= B \mid T \rightarrow T$. The base types represent commands, natural-number-valued expressions and variables respectively. The (call-by-name) operational semantics of the language can be found in [12]. We assume that initially all variables have value 0 and write Ω_T for the divergent term $\mathbf{Y}_T(\lambda x^T.x) : T$. In IA the flow of control can be influenced locally through sequential composition and conditionals. Non-local flow control can be added, for example, via the construct:

$$\frac{\Gamma, x : \mathsf{com} \vdash M : \mathsf{com}}{\Gamma \vdash \mathbf{catch}\, x \,\mathbf{in}\, M : \mathsf{com}}.$$

Operationally, $\mathbf{catch}\, x\, \mathbf{in}\, M$ amounts to encapsulating M in a block so that any occurrence of x in M will trigger a forward jump out of the block. Some typical control constructs found in programming languages, e.g. `break` and `continue` of C, can easily be expressed in IA augmented with **catch**. Using side-effects one can also detect whether an early exit has taken place. For instance, it is possible to simulate Cartwright and Felleisen's control operator $\mathbf{catch}_{\mathsf{exp}}$ [8]:

$$\frac{\Gamma, x : \mathsf{exp} \vdash M : \mathsf{exp}}{\Gamma \vdash \mathbf{catch}_{\mathsf{exp}}\, x \,\mathbf{in}\, M : \mathsf{exp}},$$

which returns 0 for early exit and $n+1$ if M evaluates to n, by

$$\mathbf{new}\, X, Y \,\mathbf{in}\, \mathbf{catch}\, z \,\mathbf{in}\, X := M[(Y := 1; z; 0)/x]; \mathbf{if}\, !Y \,\mathbf{then}\, 0 \,\mathbf{else}\, \mathbf{succ}(!X).$$

$$\frac{}{\Gamma \vdash \mathbf{skip} : \mathsf{com}} \qquad \frac{i \in \mathbb{N}}{\Gamma \vdash i : \mathsf{exp}} \qquad \frac{}{\Gamma, x : T \vdash x : T}$$

$$\frac{\Gamma \vdash M : \mathsf{exp}}{\Gamma \vdash \mathbf{succ}(M) : \mathsf{exp}} \qquad \frac{\Gamma \vdash M : \mathsf{exp}}{\Gamma \vdash \mathbf{pred}(M) : \mathsf{exp}}$$

$$\frac{\Gamma \vdash M : \mathsf{exp} \quad \Gamma \vdash M_0, M_1 : B}{\Gamma \vdash \mathbf{if}\, M\, \mathbf{then}\, M_1\, \mathbf{else}\, M_0 : B} \qquad \frac{\Gamma \vdash M : \mathsf{com} \quad \Gamma \vdash N : B}{\Gamma \vdash M; N : B}$$

$$\frac{\Gamma, x : T \vdash M : T'}{\Gamma \vdash \lambda x^T.M : T \to T'} \qquad \frac{\Gamma \vdash M : T \to T' \quad \Gamma \vdash N : T}{\Gamma \vdash MN : T'}$$

$$\frac{\Gamma \vdash M : \mathsf{var}}{\Gamma \vdash !M : \mathsf{exp}} \qquad \frac{\Gamma \vdash M : \mathsf{var} \quad \Gamma \vdash N : \mathsf{exp}}{\Gamma \vdash M := N : \mathsf{com}}$$

$$\frac{\Gamma, x : \mathsf{var} \vdash M : B}{\Gamma \vdash \mathbf{new}\, x\, \mathbf{in}\, M : B} \qquad \frac{}{\Gamma \vdash \mathbf{Y}_T : (T \to T) \to T}$$

Fig. 1. IA syntax

catch can be regarded as an atom of non-local control: a minimalistic, yet expressive, mechanism capable of modelling the effect of more complicated control constructs such as labelled jumps (`goto`) and call-with-current-continuation (`callcc`) [9].

In the paper we shall consider extensions $\mathcal{L} = \mathsf{IA}+\Box$ of IA, written $\mathsf{IA}_\Box$, where $\Box \subseteq \{\mathbf{catch}, \mathbf{mkvar}, \mathbf{or}\}$. The operational semantics of each of these languages induces a notion of termination $M \Downarrow_\mathcal{L}$ for closed terms $\vdash M : \mathsf{com}$. Then we can define observational approximation and equivalence as follows.

Definition 1. *Suppose $\Gamma \vdash M_1, M_2 : T$ are terms of $\mathcal{L}$. $\Gamma \vdash M_1 : T$ approximates $\Gamma \vdash M_2 : T$ (written $\Gamma \vdash M_1 \sqsubseteq_\mathcal{L} M_2$) iff, for all $\mathcal{L}$-contexts $C[-]$ such that $\vdash C[M_1], C[M_2] : \mathsf{com}$ holds, $C[M] \Downarrow_\mathcal{L}$ implies $C[N] \Downarrow_\mathcal{L}$. Two $\mathcal{L}$-terms are equivalent (written $\Gamma \vdash M_1 \cong_\mathcal{L} M_2$) if they $\sqsubseteq_\mathcal{L}$-approximate each other.*

3 Games

The focus of this section is a full abstraction result for $\mathsf{IA}_{\mathbf{catch}+\mathbf{mkvar}}$, which can readily be synthesized from [12] and [4]. The first to study control operators in game semantics was Laird [9], who discovered that their presence can be modelled by relaxing the *bracketing* condition.

Definition 2. *An* arena *is a triple $A = \langle M_A, \lambda_A, \vdash_A \rangle$, where*

- *M_A is a set of* moves*;*
- *$\lambda_A : M_A \to \{O, P\} \times \{Q, A\}$ is a function indicating to which player (O or P) a move belongs and of what kind it is (question or answer);*
- *$\vdash_A \subseteq (M_A + \{\star\}) \times M_A$ is the so-called* enabling *relation, which must satisfy the following conditions.*

- *If* $\star$ *enables a move then it is an O-question without any other enabler. A move like this is called* initial *and we shall write* I_A *for the set containing all initial moves of A.*
- *If one move enables another then the former must be a question and the two moves must belong to different players.*

Product and arrow arenas can be constructed as follows:

$$\begin{array}{ll} M_{A\times B} = M_A + M_B & M_{A\Rightarrow B} = M_A + M_B \\ \lambda_{A\times B} = [\lambda_A, \lambda_B] & \lambda_{A\Rightarrow B} = [\overline{\lambda}_A, \lambda_B] \\ \vdash_{A\times B} = \vdash_A + \vdash_B & \vdash_{A\Rightarrow B} = \vdash_B + (I_B \times I_A) + (\vdash_A \cap (M_A \times M_A)) \end{array}$$

$\overline{\lambda}_A$ reverses the ownership of moves in A while preserving their kind. Here are the arenas used interpret the base types of IA (the moves at the bottom are answer-moves).

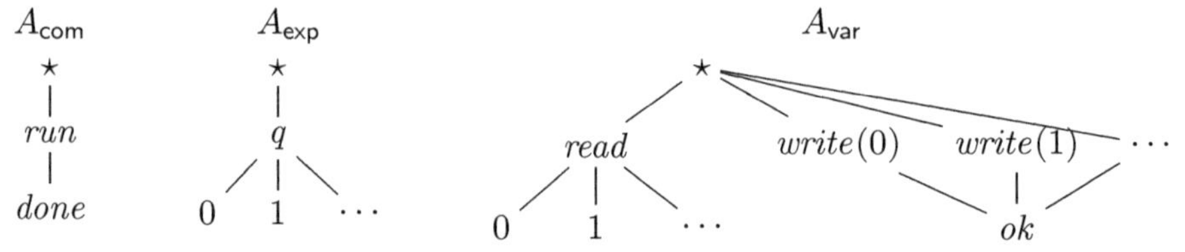

Given an IA-type T, we shall write $[\![T]\!]$ for the corresponding arena obtained compositionally from A_{com}, A_{exp} and A_{var} using the $\Rightarrow$ construction.

A *justified sequence* s in an arena A is a sequence of moves in which every move $m \notin I_A$ must have a pointer to an earlier move n in s such that $n \vdash_A m$. n is then said to be the *justifier* of m. It follows that every justified sequence must begin with an O-question. The view $\ulcorner s \urcorner$ of a justified sequence s is defined by

$$\begin{array}{rll} \ulcorner \epsilon \urcorner & = \epsilon & \\ \ulcorner s m \urcorner & = m & \text{if } m \text{ is initial in } A \\ \ulcorner s_1\, \overset{\frown}{m\, s_2\, n} \urcorner & = \ulcorner s \urcorner\, \overset{\frown}{m\, n} & \end{array}$$

A justified sequence s satisfies the *visibility condition* iff in any prefix $s'm$ of s such that m is not initial, the justifier of m lies in $\ulcorner s' \urcorner$. A justified sequence satisfies the *bracketing* condition if any answer-move is justified by the latest unanswered question that precedes it.

Definition 3. *A justified sequence is a* play *iff O- and P-moves alternate and the visibility condition is satisfied. We write* P_A *for the set of plays in A.*

Note that plays do not satisfy the bracketing condition. This notion of play suffices to define a game model of $\mathsf{IA}_{\mathbf{catch+mkvar}}$, in which terms are interpreted as strategies.

Definition 4. *A* strategy *in an arena A, written* $\sigma : A$*, is a non-empty set of even-length plays in A which is closed under taking even prefixes and satisfies the* determinacy *condition:* $smn_1, smn_2 \in \sigma$ *entails* $n_1 = n_2$ *(targets of pointers from* n_1 *and* n_2 *are also required to be the same).*

For any arena A, the strategy on $A \Rightarrow A$ that copies moves between the two instances of A is called the *identity* strategy. Arenas and strategies form a category in which a morphism between A and B is a strategy on $A \Rightarrow B$. In order to compose two strategies $\sigma : A \Rightarrow B$, $\tau : B \Rightarrow C$, one first defines *interaction sequences* on A, B, C, which are sequences of moves from arenas A, B and C together with justification pointers from all moves except those initial in C. The set of all such sequences will be denoted by $\mathsf{int}(A, B, C)$. Given an interaction sequence u, we write $u \upharpoonright A, B$ for its subsequence consisting of all A- and B-moves as well as pointers between them (pointers from/to moves of C in u are erased, though). $u \upharpoonright B, C$ is defined analogously. $u \upharpoonright A, C$ is defined similarly except that, whenever a pointer from an A-move m_A points at a B-move m_B which in turn has a pointer to a C-move m_C, we add a pointer from m_A to m_C. Then one takes $\sigma; \tau : A \Rightarrow C$ to be

$$\{u \upharpoonright A, C \,|\, u \in \mathsf{int}(A, B, C),\ u \upharpoonright A, B \in \sigma,\ u \upharpoonright B, C \in \tau\}.$$

Arenas and strategies form a category in which identity strategies are indeed the identity maps.

A play is called *single-threaded* if it contains just one occurrence of an initial move. In general, a play may consist of several interleaved single-threaded plays. Strategies determined completely by their single-threaded plays will be called *single-threaded*: they consist of all plays that are interleavings of the single-threaded plays belonging to the strategy.

Arenas and single-threaded strategies turn out to form a cartesian closed category, which provides a canonical interpretation of λ-abstraction and application. The inclusion relation on strategies enriches the category with the structure of a complete partial order needed to interpret recursion. Other features of $\mathsf{IA}_{\mathbf{catch+mkvar}}$ can be interpreted by composition with special designated strategies, which we list in Figure 2 along with their single-threaded complete plays. For illustration, we give the two maximal single-threaded plays of the strategy $\mathsf{catch} : [\![(\mathsf{com}_2 \to \mathsf{com}_1) \to \mathsf{com}_0]\!]$ used to interpret **catch**:

$$run_0\ run_1\ done_1\ done_0 \qquad\qquad run_0\ run_1\ run_2\ done_0.$$

We have used subscripts to indicate the copies of com from which the moves originate. Then $[\![\Gamma \vdash \mathbf{catch}\, x\, \mathbf{in}\, M]\!] = [\![\Gamma \vdash \lambda x^{\mathsf{com}}.M]\!]; \mathsf{catch}$.

Theorem 1. *Arenas and single-threaded strategies ordered by inclusion are an inequationally fully abstract model of* $\mathsf{IA}_{\mathbf{catch+mkvar}}$*: for any* $\mathsf{IA}_{\mathbf{catch+mkvar}}$*-terms* $\Gamma \vdash M_1, M_2 : T$*:*

$$\Gamma \vdash M_1 \sqsubsetapprox_{\mathsf{IA}_{\mathbf{catch+mkvar}}} M_2 \iff [\![\Gamma \vdash M_1]\!] \subseteq [\![\Gamma \vdash M_2]\!].$$

To our knowledge, this theorem has not appeared in the literature so far, though it seems to be known within the community. It can be proved in a similar way to the characterization of program approximation via complete plays for $\mathsf{IA}_{\mathbf{mkvar}}$ [4]. The argument relies on the fact that any finite strategy is definable

$$
\begin{array}{ll}
\mathsf{skip} : [\![\mathsf{com}]\!] & \mathit{run}\ \mathit{done} \\
\mathsf{i} : [\![\mathsf{exp}]\!] & q\, i \\
\mathsf{succ} : [\![\mathsf{exp}]\!]_1 \Rightarrow [\![\mathsf{exp}]\!]_0 & q_0\, q_1\, \sum_{i\in\mathbb{N}} i_1\, (i+1)_0 \\
\mathsf{pred} : [\![\mathsf{exp}]\!]_1 \Rightarrow [\![\mathsf{exp}]\!]_0 & q_0\, q_1\, \sum_{i\in\mathbb{N}^+} i_1\, (i-1)_0 \\
\mathsf{if}_B : [\![\mathsf{exp}]\!]_3 \Rightarrow [\![B]\!]_2 \Rightarrow [\![B]\!]_1 \Rightarrow [\![B]\!]_0 & \\
\qquad \sum_{m\vdash_{[\![B]\!]} n} (m_0\, q_3\, 0_3\, m_1\, n_1\, n_0 + m_0\, q_3\, (\sum_{i\in\mathbb{N}^+} i_3)\, m_2\, n_2\, n_0) & \\
\mathsf{seq}_B : [\![\mathsf{com}]\!]_2 \Rightarrow [\![B]\!]_1 \Rightarrow [\![B]\!]_0 & \sum_{m\vdash_{[\![B]\!]} n} m_0\, \mathit{run}_2\, \mathit{done}_2\, m_1\, n_1\, n_0 \\
\mathsf{deref} : [\![\mathsf{var}]\!]_1 \Rightarrow [\![\mathsf{exp}]\!]_0 & q_0\, \mathit{read}_1\, \sum_{i\in\mathbb{N}} i_1\, i_0 \\
\mathsf{assign} : [\![\mathsf{var}]\!]_2 \Rightarrow [\![\mathsf{exp}]\!]_1 \Rightarrow [\![\mathsf{com}]\!]_0 & \mathit{run}_0\, q_1\, \sum_{i\in\mathbb{N}} i_1\, \mathit{write}(i)_2\, \mathit{ok}_2\, \mathit{done}_0 \\
\mathsf{cell}_B : ([\![\mathsf{var}]\!]_2 \Rightarrow [\![B]\!]_1) \Rightarrow [\![B]\!]_0 & \\
\qquad \sum_{m\vdash_{[\![B]\!]} n} m_0 m_1 (\mathit{read}_2\, 0_2)^* (\sum_{i\in\mathbb{N}} \mathit{write}(i)_2\, \mathit{ok}_2 (\mathit{read}_2\, i_2)^*)^* n_1 n_0 & \\
\mathsf{mkvar} : [\![\mathsf{exp} \to \mathsf{com}]\!]_2 \Rightarrow [\![\mathsf{exp}]\!]_1 \Rightarrow [\![\mathsf{var}]\!]_0 & \\
\qquad \mathit{read}_0\, q_1\, (\sum_{i\in\mathbb{N}} i_1\, i_0) + \sum_{i\in\mathbb{N}} \mathit{write}(i)_0\, \mathit{run}_2\, (q_2\, i_2)^*\, \mathit{done}_2\, \mathit{ok}_0 &
\end{array}
$$

Fig. 2. Special strategies

by a term of $\mathsf{IA}_{\mathbf{catch+mkvar}}$, which can be shown using the techniques of [12]. Soundness and Adequacy ($[\![\vdash M : \mathsf{com}]\!] = [\![\vdash \mathbf{skip}]\!]$ if and only if $M \Downarrow$) can also be proved in the standard way. The goal of this paper is to show an analogous theorem for $\mathsf{IA}_{\mathbf{catch}}$, with the inclusion ordering replaced by a different preorder.

4 The Essence of mkvar

Game semantics interprets local variable allocation in $\Gamma \vdash \mathbf{new}\, x\, \mathbf{in}\, M : \mathsf{com}$ through composition of $[\![\Gamma \vdash \lambda x^{\mathsf{var}}.M]\!]$ with the strategy $\mathsf{cell}_{\mathsf{com}} : [\![(\mathsf{var}_2 \to \mathsf{com}_1) \to \mathsf{com}_0]\!]$. $\mathsf{cell}_{\mathsf{com}}$ itself denotes the term

$$\vdash \lambda f^{\mathsf{var}\to\mathsf{com}}.\mathbf{new}\, x\, \mathbf{in}\, f x : (\mathsf{var} \to \mathsf{com}) \to \mathsf{com}.$$

Single-threaded plays of $\mathsf{cell}_{\mathsf{com}}$ are prefixes of plays of the following shape:

$$\mathit{run}_0\, \mathit{run}_1\, (\mathit{read}_2\, 0_2)^* (\sum_{i\in\mathbb{N}} \mathit{write}(i)_2\, \mathit{ok}_2\, (\mathit{read}_2\, i_2)^*)^*\, \mathit{done}_1 \mathit{done}_0$$

i.e. *read*'s trigger responses consistent with preceding *write*'s. Using **mkvar** we can easily violate the discipline, because unrelated methods can be employed for reading and writing. However, the same effect can also be achieved without it. Indeed, under call-by-name evaluation, whenever a *read* follows a *write*, we cannot really be sure that they refer to the same variable. Here is a term illustrating this behaviour

$$\vdash \lambda f^{\mathsf{var}\to\mathsf{com}}.\mathbf{new}\, X, Y\, \mathbf{in}\, f(\mathbf{if}\, !X\, \mathbf{then}\, Y\, \mathbf{else}\, X),$$

which will produce the play $run_0\, run_1\, write(1)_2\, ok_2\, read_2\, 0_2\, done_1\, done_0$. Thus the essence of **mkvar** does not boil down to breaking the logical link between *read*'s and *write*'s. We argue that it lies in uniformity instead: in absence of **mkvar** each variable operation, whether a read or a write, produces the same side-effects while it is being completed. To formalize this intuition it is useful to observe that, without **mkvar**, subterms of type var (in β-normal form) always have "tails" of the shape $f M_1 \cdots M_k : \mathsf{var}$, which may be combined using conditionals, pre-composed with commands and bound with **new** (if $k = 0$). In game semantics, when O plays a move q_O in order to explore such a subterm, the resultant plays will initially correspond to the associated side-effects (if any). These side-effects will be independent of q_O, which could well be *read*, *write*(0) or *write*(13). Eventually, when the "tail" is reached and f is not bound by **new**, P will play a copy q_P of q_O. Below we introduce new relations on plays and strategies to express an aspect of the uniformity that will turn out useful in subsequent technical arguments.

Definition 5. *Given an arena A corresponding to an* IA*-type and $q, q' \in \{read\} \cup \{write(i) \mid i \in \mathbb{N}\}$, the relation $\diamond_{\mathsf{O}}^{q,q'} \subseteq P_A \times P_A$ is defined as follows: $t \diamond_{\mathsf{O}}^{q,q'} t'$ iff $t = s_1 q s_2$, $t' = s_1 q' s_2$ and q, q' are O-moves from the same copy of A_{var} that have not been answered in t, t' respectively. $\diamond_{\mathsf{P}}^{q,q'}$ is defined in an analogous way, by replacing "O-moves" with "P-moves". We write $\diamond_{\mathsf{O}}$ for the (symmetric) relation $\bigcup_{q,q'} \diamond_{\mathsf{O}}^{q,q'}$. $\diamond_{\mathsf{P}}$ is defined similarly.*

Definition 6. *A strategy $\sigma : A$ is $\diamond$-closed iff, for any $s \in \sigma, t \in P_A$, if there exist q, q' such that $s \diamond_{\mathsf{O}}^{q,q'} t$, then $t \in \sigma$ or there exists $s' \in \sigma$ such that $t \diamond_{\mathsf{P}}^{q,q'} s'$.*

Next we turn to questions that have been answered. If the q_P from the above scenario is eventually answered, say, with a_O, P will immediately answer q_O with a copy a_P of a_O. Afterwards, the play will actually follow independently of the value of q and a. This is in contrast to the case of $k = 0$ and f being bound by **new**. Here after a series of possible side-effects (independent of q_O) P will answer q_O with a_P. Because this case corresponds to examining a genuine storage cell, what happens next could depend on the current value of the variable: if $s_1\, write(0) s_1\, ok s_2 \in \sigma$, it does not have to be the case that $s_1\, write(2) s_2\, ok s_2 \in \sigma$. Similarly, if $s_1\, read s_2 0 s_3 \in \sigma$, we do not know whether $s_1\, write(2) s_2\, ok s_2 \in \sigma$. However, if $s_1\, read s_2 0 s_3 \in \sigma$, we can be sure that $s_1\, write(0) s_2\, ok s_3 \in \sigma$, because overwriting a variable with its current value does not change the state. Below we define another closure property of strategies, which unifies the observations just made about answer-moves. This is essentially a more precise variant of the $\propto$-closure used in [7], adapted to general plays.

Definition 7. *Given an arena A corresponding to an* IA*-type we define $\lhd_{\mathsf{O}} \subseteq P_A \times P_A$ as follows: $t \lhd_{\mathsf{O}} t'$ iff $t = s_1\ read\ \widehat{s_2}\ i\ s_3$ and $t' = s_1\ write(i)\ \widehat{s_2}\ ok\ s_3$ for some $i \in \mathbb{N}$, where read and write(i) are O-moves from the same copy of A_{var}. $\lhd_{\mathsf{P}}$ is defined in an analogous way.*

Definition 8. *A strategy $\sigma : A$ is $\lhd$-closed iff, for any $s \in \sigma, t \in P_A$, if $s \lhd_{\mathsf{O}} t$ then $t \in \sigma$ or there exists $s' \in \sigma$ satisfying $t \lhd_{\mathsf{P}} s'$.*

Lemma 1. *Strategies denoting* $\mathsf{IA}_{\mathbf{catch}}$*-terms are* $\diamond$*- and* $\lhd$*-closed.*

Proof. First one shows that $\diamond$- and $\lhd$-closure are preserved by composition. Then it suffices to show that the basic special strategies used to construct the model satisfy the Lemma.

Note that the strategy corresponding to **mkvar** satisfies neither $\diamond$- nor $\lhd$-closure.

5 What Difference Does mkvar Make?

Because the addition of new syntactic features makes the discriminating power of contexts stronger, it is natural to expect that some approximations in $\mathsf{IA}_{\mathbf{catch}}$ will no longer hold in $\mathsf{IA}_{\mathbf{catch+mkvar}}$. For IA and $\mathsf{IA}_{\mathbf{mkvar}}$, it was shown in [7] that essentially all examples of IA approximations that fail in $\mathsf{IA}_{\mathbf{mkvar}}$ are based on approximating reads with matching writes, as in

$$x : \mathsf{var} \vdash \mathbf{if}\,!x\,\mathbf{then}\,\Omega\,\mathbf{else}\,\mathbf{skip} \precsim_{\mathsf{IA}} x{:=}0.$$

The same idea can also be used to demonstrate the difference between $\mathsf{IA}_{\mathbf{catch}}$ and $\mathsf{IA}_{\mathbf{catch+mkvar}}$, but we will also have another class of examples, relying on variable operations immediately followed by divergence:

$$x : \mathsf{var} \vdash \mathbf{if}\,!x\,\mathbf{then}\,\Omega\,\mathbf{else}\,\Omega \quad \cong_{\mathsf{IA}_{\mathbf{catch}}} \quad x{:=}0; \Omega \quad \cong_{\mathsf{IA}_{\mathbf{catch}}} \quad x{:=}1; \Omega.$$

Because the terms generate different plays, these equivalences do not hold in $\mathsf{IA}_{\mathbf{catch+mkvar}}$ so, in contrast to $\mathsf{IA}_{\mathbf{mkvar}}$, $\mathsf{IA}_{\mathbf{catch+mkvar}}$ turns out not to extend $\mathsf{IA}_{\mathbf{catch}}$ conservatively even for observational equivalence. In the remainder of the paper we show how to capture approximation in $\mathsf{IA}_{\mathbf{catch}}$ with a preorder based on $\diamond_{\mathsf{P}}$ and $\lhd_{\mathsf{P}}$, which will make it clear that equivalences above do hold.

Definition 9. *Suppose* $\sigma, \tau : A$ *are single-threaded. We define* $\sigma \sqsubseteq \tau$ *to hold iff for any* $s \in \sigma$ *there exists* $t \in \tau$ *such that* $s\,(\diamond_{\mathsf{P}} \cup \lhd_{\mathsf{P}})^*\,t$*.*

Because σ and τ are single-threaded, the quantification over $s \in \sigma$ could well range over single-threaded plays only. In the next section we aim to prove:

Theorem 2 (Full abstraction). *For any* $\mathsf{IA}_{\mathbf{catch}}$*-terms* $\Gamma \vdash M_1, M_2 : T$ *we have* $\Gamma \vdash M_1 \precsim_{\mathsf{IA}_{\mathbf{catch}}} M_2$ *if and only if* $[\![\Gamma \vdash M_1]\!] \sqsubseteq [\![\Gamma \vdash M_2]\!]$*.*

6 Proof of Full Abstraction

The left-to-right direction hinges on the fact that $[\![\Gamma \vdash M_1]\!] \sqsubseteq [\![\Gamma \vdash M_2]\!]$ implies $[\![\vdash C[M_1]]\!] \sqsubseteq [\![\vdash C[M_2]]\!]$. Indeed, more generally, one can show that composition of $\diamond$- and $\lhd$-closed strategies is monotone with respect to $\sqsubseteq$, which implies the above.

Lemma 2. *For any* $\lhd$*- and* $\diamond$*-closed strategies* $\sigma_1, \sigma_2 : A \Rightarrow B$ *and* $\tau_1, \tau_2 : B \Rightarrow C$*: if* $\sigma_1 \sqsubseteq \sigma_2$ *and* $\tau_1 \sqsubseteq \tau_2$ *then* $\sigma_1; \tau_1 \sqsubseteq \sigma_2; \tau_2$*.*

Proof. By repeated alternate applications of closure rules for σ_i and τ_i.

Lemma 3. *For any* $\mathsf{IA}_{\mathbf{catch}}$*-terms* $\Gamma \vdash M_1, M_2 : T$ *if* $[\![\Gamma \vdash M_1]\!] \sqsubseteq [\![\Gamma \vdash M_2]\!]$ *then* $\Gamma \vdash M_1 \sqsubsim_{\mathsf{IA}_{\mathbf{catch}}} M_2$.

Proof. Suppose $\vdash \mathcal{C}[M_1] : \mathsf{com} \Downarrow$. Then, by Soundness (of the game model of $\mathsf{IA}_{\mathbf{catch+mkvar}}$), $[\![\vdash \mathcal{C}[M_1] : \mathsf{com}]\!] = [\![\vdash \mathbf{skip}]\!]$. Because $[\![\Gamma \vdash M_1]\!] \sqsubseteq [\![\Gamma \vdash M_2]\!]$, we have $[\![\vdash \mathbf{skip}]\!] = [\![\vdash \mathcal{C}[M_1]]\!] \sqsubseteq [\![\vdash \mathcal{C}[M_2]]\!]$. Hence, $[\![\vdash C[M_2]]\!] = [\![\vdash \mathbf{skip}]\!]$ and, by Adequacy, $\mathcal{C}[M_2] \Downarrow$.

To establish the converse we need a new definability argument. Because the strategies involved are $\diamond$- and $\lhd$-closed, it is no longer impossible to prove definability for single positions. We shall restrict ourselves to plays of the shape $run \cdots done$, as these suffice for the reconstruction of contexts used in the definition of $\sqsubsim_{\mathsf{IA}_{\mathbf{catch}}}$.

Proposition 1. *Let* $s = run \cdots done$ *be a single-threaded play in* $[\![T]\!]$. *Then there exists an* $\mathsf{IA}_{\mathbf{catch}}$*-term* $\vdash M_s : T$ *such that the set of single-threaded complete plays of* $[\![\vdash M_s]\!]$ *equals* $\{t \mid s\ (\diamond_O \cup \lhd_O)^*\ t\}$.

We will say that an answer-move occurring in a play is *well-bracketed*, if the question that justifies it is the most recent unanswered question in the view calculated right before the answer has been played. A strategy is called well-bracketed if in each of its plays any P-answer is well-bracketed. Thanks to the factorization techniques developed in [9,12], which factor out non-well-bracketed P-answers using catch, it suffices to prove the above Proposition for plays s in which P-answers are *well-bracketed.* To that end we first identify a family of *innocent* strategies which are definable in IA without **new**. Innocence, first defined in [13], guarantees that P's responses depend only on the current view of the play rather than the whole play.

6.1 Innocent Strategies Without mkvar

Let $A_\Box = \langle \{\Box_q, \Box_a\}, \{(\Box_q, (Q,O)), (\Box_a, (A,P))\}, \{(\star, \Box_q), (\Box_q, \Box_a)\} \rangle$. Given an IA type T, let $[\![T]\!]_{\mathsf{sym}}$ be the arena obtained compositionally from T using the $\times$ and $\Rightarrow$ constructions in the same way as $[\![T]\!]$ except that occurrences of var are interpreted differently: positive ones by $A_\Box$ and negative ones by $A_{\mathsf{var}} \times A_\Box$. Moves of $A_\Box$ will be called *generic*, while those from A_{var} will be referred to as *concrete*. Thus, only P will have concrete moves at his disposal in $[\![T]\!]_{\mathsf{sym}}$.

Given plays $s_1 \in P_{[\![T]\!]_{\mathsf{sym}}}$ and $s_2 \in P_{[\![T]\!]}$, we shall say that s_2 *matches* s_1 iff s_2 can be obtained from s_1 by replacing each occurrence of $\Box_q$ (respectively $\Box_a$) with a concrete question (respectively answer) coming from the same copy of var as the generic move it replaces. Thus, plays in $[\![T]\!]_{\mathsf{sym}}$ can be viewed as specifications of sets of plays in $[\![T]\!]$. Suppose $\sigma : [\![T]\!]_{\mathsf{sym}}$. Because σ is deterministic and $[\![T]\!]_{\mathsf{sym}}$ does not allow concrete O-moves, a play from $[\![T]\!]$ can match at most one play from σ. This matching can be used to define strategies $\widehat{\sigma} : [\![T]\!]$ that "match" σ but, in general, such extensions will not be unique. In what follows, we introduce a special class of strategies on $[\![T]\!]_{\mathsf{sym}}$ that can be extended to strategies on $[\![T]\!]$ in a canonical way.

Definition 10. *A well-bracketed strategy $\sigma : [\![T]\!]_{\mathsf{sym}}$ is a tail strategy iff it satisfies the following conditions.*

(i) If $s\Box_a \in \sigma$ then the last move of s is a $\Box_a$-move.
(ii) If $s\Box_q \in \sigma$ then for any $s\,\Box_q \widehat{s_1}\, \Box_a \in P_{[\![T]\!]_{\mathsf{sym}}}$ such that $s\Box_q s_1 \in \sigma$, we have

$$s\,\Box_q\, \widehat{s_1}\, \Box_a\, \Box_a \in \sigma.$$

Remark 1. Because tail strategies are well-bracketed, the target of the last justification in clause (*ii*) is uniquely determined by $s\Box_q$. We shall call it the *mate* of $\Box_q$. Of course, the mate of $\Box_q$ must also be a $\Box_q$-move. Similarly, we define the mate of a P-answer $\Box_a$ in $s\Box_a \in \sigma$ to be the last move of s, which by clause (*i*) must be an O-answer of the shape $\Box_a$.

Definition 11. *Let $\sigma : [\![T]\!]_{\mathsf{sym}}$ be a tail strategy. We define the* copy-cat extension *$\widehat{\sigma} : [\![T]\!]$ of σ in the following way.*

- $\epsilon \in \widehat{\sigma}$.
- *If $s \in \widehat{\sigma}$, $sm_1 \in P_{[\![\sigma]\!]}$, $tn_1n_2 \in \sigma$ are such that tn_1 matches sm_1 then:*
  - *if n_2 is not generic then $sm_1n_1 \in \widehat{\sigma}$;*
  - *if n_2 is generic, then $s' = sm_1m_2 \in \widehat{\sigma}$, where s' is the unique play matching tn_1n_2 and such that n_2 is instantiated with the same concrete move as its mate.*

Note that when σ is innocent, so is $\widehat{\sigma}$. An innocent strategy is compactly innocent if it depends only on a finite number of views.

Lemma 4. *Suppose $\sigma : [\![T]\!]_{\mathsf{sym}}$ is a compactly innocent tail strategy. Then there exists a* **new***-free term $\vdash M : T$ of IA such that $[\![\vdash M : T]\!] = \widehat{\sigma}$.*

Proof. Follows the standard definability argument for PCF [13].

6.2 Knowingness Without mkvar

Now we continue with definability for certain knowing, i.e. not necessarily innocent, strategies.

Lemma 5. *Suppose $T = \mathsf{var}_k \to \cdots \to \mathsf{var}_1 \to T'$ and let s be a single-threaded play in $[\![T]\!]_{\mathsf{sym}}$ such that any generic P-question $\Box_q$ in s comes from var_i ($i = 1, \cdots, k$) and no two such questions come from the same var_i. Let σ be the least single-threaded strategy on $[\![T]\!]_{\mathsf{sym}}$ containing s. If σ is a tail strategy, then there exists a compactly-innocent tail strategy $\tau : [\![\mathsf{var} \to T]\!]_{\mathsf{sym}}$ such that $\tau; \mathsf{cell}_T^{\mathsf{sym}} = \sigma$[2]. Consequently, $\widehat{\tau}; \mathsf{cell}_T = \widehat{\sigma}$.*

[2] $\mathsf{cell}_T : [\![(\mathsf{var} \to T) \to T]\!]$ works in the same way as $\mathsf{cell}_{\mathsf{com}}$: moves are being copied from one copy of $[\![T]\!]$ to another and, when O makes a move in the var component, P's responses reflect the behaviour of a storage cell. $\mathsf{cell}_T^{\mathsf{sym}}$ works analogously except that it is a strategy in the game $(A_{\mathsf{var}} \Rightarrow [\![T]\!]_{\mathsf{sym}}) \Rightarrow [\![T]\!]_{\mathsf{sym}}$.

Proof. We modify the factorization argument from [4]. Its key idea is that τ follows σ except that it uses the additional var component for recording the single-threaded history of play. Thus, when an O-move from $[\![T]\!]_{\mathsf{sym}}$ is made following some play $t \in \tau$, τ will always play *read* to find out what the current single-threaded play looks like. The subsequent O-move is then regarded as the code of the play. τ is then able to mimic σ in an innocent way, because the code of the whole single-threaded play will be present in the relevant view. However, before τ imitates σ, it always writes the code of the resultant single-threaded play to the var component. Plays of τ thus have the shape $\cdots\, m_T^O\ \mathit{read}\ s\ \mathit{write}(sab)\ \mathit{ok}\ m_T^P$, i.e. the procedure introduces additional moves between any O-move and the P-move that follows.

Note that, in our case, we cannot afford to adopt the above-described procedure after O-moves of the form $\Box_a$, because we want τ to be a tail strategy. However, because σ is a tail strategy, we know that, after an O-move $\Box_a$, P will also respond with $\Box_a$. Moreover, because generic P-questions can only come from some var_i, all O-answers $\Box_a$ must also come from there. Consequently, thanks to the special shape of the arena, the predecessor of any O-answer $\Box_a$ in s must be the P-question $\Box_q$ that enables it. This opens up the way to modifying the previous factorization: before P plays $\Box_q$ in τ, τ can already write the code of $s\Box_q\Box_a\Box_a$ to the var component, because $\Box_a\Box_a$ will follow anyway. When $\Box_a$ is indeed played by O afterwards, τ will not read or write from var component, but will immediately reply with the same $\Box_a$ as σ would. Note that this behaviour is innocent, because no two generic O-answers come from the same var_i.

Proposition 2. *Suppose $s = \mathit{run} \cdots \mathit{done}$ is a single-threaded play of $[\![T]\!]$ in which P-moves are well-bracketed. Then there exists an IA-term $\vdash M_s : T$ such that the set of single-threaded complete plays of $[\![\vdash M_s]\!]$ is $\{t \mid s\ (\lhd_{\mathsf{O}} \cup \diamond_{\mathsf{O}})^*\ t\}$.*

Proof. Note that the shape of the play means that $T = T_l \to \cdots \to T_1 \to \mathsf{com}$. Let k be the number of concrete P-answers in s. We will first replace s with $s' \in P_{[\![T']\!]_{\mathsf{sym}}}$ such that $T' = \mathsf{var}_k \to \cdots \to \mathsf{var}_1 \to T$ and s' satisfies the assumptions of Lemma 5. s' is obtained from s in the following way.

- Any concrete O-question is replaced by $\Box_q$ from the same copy var as the question.
- Any concrete P-answer is replaced with $\Box_q^j, \Box_a^j\,\Box_a$, $\Box_q^j, \Box_a^j$ come from var_j, $\Box_a$ comes from the same copy of var as the answer and the answer in question is the jth concrete P-answer in s.

By Lemma 5 $\widehat{\sigma} = \widehat{\tau}; \mathsf{cell}_{T'}$, where τ is a compactly-innocent tail strategy on $[\![\mathsf{var} \to T']\!]_{\mathsf{sym}}$. By Lemma 4 there exists a IA-term $\vdash M : \mathsf{var} \to T'$ such that $[\![\vdash M]\!] = \widehat{\tau}$. Thus, putting $M' \equiv \lambda x_k \cdots x_1 y_l \cdots y_1.\mathbf{new}\,X\,\mathbf{in}\,M X x_k \cdots x_1 y_l \cdots y_1$, we get $[\![\vdash M' : T']\!] = \widehat{\sigma}$. To obtain $\vdash M'' : T$ satisfying the current Proposition it now suffices to take

$$\lambda y_l \cdots y_1.\mathbf{new}\,x_1, \cdots, x_k\,\mathbf{in}\,\mathit{INIT}; M' x_k \cdots x_1 y_l \cdots y_1; \mathit{FINIT},$$

where $\mathit{INIT} \equiv \mathit{INIT}_1; \cdots; \mathit{INIT}_k$, $\mathit{FINIT} \equiv \mathit{FINIT}_1; \cdots; \mathit{FINIT}_k$,

$$INIT_j \equiv \begin{cases} x_j{:=}i+1 & j\text{th concrete } P\text{-answer in } s \text{ is } ok \text{ justified by } write(i) \\ x_j{:=}i & j\text{th concrete } P\text{-answer in } s \text{ is } i \text{ (justified by } read) \end{cases}$$

and $FINIT_j \equiv \textbf{if}\,(!x_j = i)\,\textbf{then}\,\textbf{skip}\,\textbf{else}\,\Omega$, where the jth concrete P-answer is i (justified by *read*) or or *ok* justified by $write(i)$.

By previous remarks Proposition 2 implies Proposition 1.

Lemma 6. *For any* $\mathsf{IA}_{\mathbf{catch}}$-*terms* $\Gamma \vdash M_1, M_2 : T$, *if* $\Gamma \vdash M_1 \sqsubseteq_{\mathsf{IA}_{\mathbf{catch}}} M_2$ *then* $[\![\Gamma \vdash M_1]\!] \sqsubseteq [\![\Gamma \vdash M_2]\!]$.

Proof. W.l.o.g. assume that Γ is empty (other cases can be reduced to this case by λ-abstraction). Suppose $s \in [\![\vdash M_1 : T]\!]$. Let $s' = run\, s\, done$. By Proposition 1 there exists a term $\vdash M_{s'} : T \to \mathsf{com}$ whose set of single-threaded and complete plays is $\{t \mid s'\,(\diamond_{\mathsf{O}} \cup \lhd_{\mathsf{O}})^*\,t\}$. Let $\mathcal{C}[-] = M_{s'}([-])$. Then $[\![\vdash \mathcal{C}[M_1]]\!] = [\![\vdash \mathbf{skip}]\!]$, so $\mathcal{C}[M_1] \Downarrow$. Because $\Gamma \vdash M_1 \sqsubseteq_{\mathsf{IA}_{\mathbf{catch}}} M_2$, we also have $\mathcal{C}[M_2] \Downarrow$. Hence, $[\![\vdash \mathcal{C}[M_2]]\!] = [\![\vdash \mathbf{skip}]\!]$. But this implies, by definition of composition of strategies, that there must exists $t \in [\![\vdash M_2]\!]$ such that $s(\diamond_{\mathsf{P}} \cup \lhd_{\mathsf{P}})^* t$.

Putting together Lemmas 3 and 6 we obtain Theorem 2.

7 On Conservativity

Harmer and McCusker have investigated the game semantics of nondeterminism and showed how $\mathsf{IA}_{\mathbf{mkvar+or}}$ can be modelled by nondeterministic strategies [14]. Analogously to $\mathsf{IA}_{\mathbf{catch+mkvar}}$ and $\mathsf{IA}_{\mathbf{mkvar}}$, observational approximation (based on may-convergence) in $\mathsf{IA}_{\mathbf{catch+mkvar+or}}$ and $\mathsf{IA}_{\mathbf{mkvar+or}}$ can be shown to correspond to containment of induced plays and complete plays respectively. So, in these two cases, extensions by **or** are conservative both with respective to observational approximation and equivalence. The same turns out to apply to $\mathsf{IA}_{\mathbf{catch}}$ and IA. Indeed, our argument for $\mathsf{IA}_{\mathbf{catch}}$ is immediately applicable to $\mathsf{IA}_{\mathbf{catch+or}}$: nondeterministic strategies ordered by $\sqsubseteq$ form a fully abstract model of $\mathsf{IA}_{\mathbf{catch+or}}$. Similarly, McCusker's preorder for IA [7] also gives full abstraction for $\mathsf{IA}_{\mathbf{or}}$. Consequently, $\mathsf{IA}_{\mathbf{or}}$ and $\mathsf{IA}_{\mathbf{catch+or}}$ are conservative extensions of IA and $\mathsf{IA}_{\mathbf{catch}}$ respectively.

In contrast, extensions of IA, $\mathsf{IA}_{\mathbf{or}}$, $\mathsf{IA}_{\mathbf{or+mkvar}}$ by **catch** are not conservative, even for observational equivalence. In game semantics this manifests itself in the reliance of full abstraction results for **catch**-free languages on complete plays only.

The inclusion of **mkvar** also turns out to affect observational approximation in any of IA, $\mathsf{IA}_{\mathbf{catch}}$, $\mathsf{IA}_{\mathbf{or}}$, $\mathsf{IA}_{\mathbf{catch+or}}$. For IA the effect of **mkvar** was captured by McCusker [7], $\mathsf{IA}_{\mathbf{catch}}$ was examined in this paper and, as we already mentioned, the two results apply to $\mathsf{IA}_{\mathbf{or}}$ and $\mathsf{IA}_{\mathbf{catch+or}}$. As for observational equivalence, **mkvar** is "conservative" for IA, as shown in [7], but not $\mathsf{IA}_{\mathbf{catch}}$, as demonstrated in this paper. It is interesting to note that, for program equivalence, $\mathsf{IA}_{\mathbf{or+mkvar}}$ does *not* extend $\mathsf{IA}_{\mathbf{or}}$ conservatively either. Let M_1, M_2 be the terms from Section 5 such that $M_1 \sqsubseteq_{\mathsf{IA}} M_2$ and $\neg(M_1 \sqsubseteq_{\mathsf{IA}_{\mathbf{mkvar}}} M_2)$. Then M_1 **or** $M_2 \cong_{\mathsf{IA}_{\mathbf{or}}} M_2$, but the terms are not equivalent in $\mathsf{IA}_{\mathbf{or+mkvar}}$.

Finally, let us consider probabilistic game semantics. Probabilistic strategies were introduced by Danos and Harmer [15] as functions $\sigma : P_A^{\mathrm{ev}} \to [0,1]$ where P_A^{ev} stands for the set of even-length plays on A. For probabilistic programs, instead of presence or lack of termination, one talks about the probability of termination, denoted by $\Downarrow_p$. Observational approximation can then be defined as follows.

Definition 12. *Suppose $\Gamma \vdash M_1, M_2 : T$ are terms of $\mathcal{L} = \mathsf{IA}_{\mathbf{coin}}, \mathsf{IA}_{\mathbf{coin+mkvar}}$. $\Gamma \vdash M_1 : T$ approximates $\Gamma \vdash M_2 : T$ iff, for all $\mathcal{L}$-contexts $C[-]$ such that $\vdash C[M_1], C[M_2] : \mathsf{com}$ holds there exist p, q such that $p \leq q$, $\mathcal{C}[M] \Downarrow_p$ and $\mathcal{C}[N] \Downarrow_q$.*

For $\mathsf{IA}_{\mathbf{mkvar}}$ the above notion can be characterized via complete plays [16]: $\Gamma \vdash M_1 \sqsubseteq_{\mathsf{IA}_{\mathbf{coin+mkvar}}} M_2$ iff for all single-threaded complete plays s we have $[\![\Gamma \vdash M_1]\!](s) \leq [\![\Gamma \vdash M_2]\!](s)$. The left-to-right implication depends on the fact that for any single-threaded complete play s one can construct an $\mathsf{IA}_{\mathbf{mkvar}}$-term $\Gamma \vdash M$ such that s is the only single-threaded complete play in $[\![\Gamma \vdash M]\!]$. Using McCusker's definability result for IA [7], which says that there exists an IA-term generating exactly the single-threaded complete plays from $\{t \mid s \lhd_{\mathsf{O}}^{*} t\}$[3], one can show the following result.

Lemma 7. *For any $\mathsf{IA}_{\mathbf{coin}}$-terms $\Gamma \vdash M_1, M_2 : T$, if $\Gamma \vdash M_1 \sqsubseteq_{\mathsf{IA}_{\mathbf{coin}}} M_2$ then for any single-threaded complete play s we have*

$$\sum_{\{t \mid s \lhd_{\mathsf{P}}^{*} t\}} [\![\Gamma \vdash M_1]\!](t) \leq \sum_{\{t \mid s \lhd_{\mathsf{P}}^{*} t\}} [\![\Gamma \vdash M_2]\!](t).$$

The converse of the Lemma is not true. The conceptual reason for the failure is that a single complete play can be extended to a $\diamond$- and $\lhd$-closed strategy in a number of ways, while the definability result in [7] (and in this paper) only explores the simplest one. This approach is fruitful for languages in which termination requires only a single terminating run, but turns out insufficient in the probabilistic setting, when termination is quantitative and all evaluation paths have to be taken into account when comparing programs. We leave the generalization of the definability results for future research. In any case, although the above lemma could not be extended to a full abstraction result, it has an important application in the proof of our last result.

Proposition 3. *$\mathsf{IA}_{\mathbf{coin+mkvar}}$ is a conservative extension of $\mathsf{IA}_{\mathbf{coin}}$ for observational equivalence.*

Proof. Suppose $\Gamma \vdash M_1 \cong_{\mathsf{IA}_{\mathbf{coin}}} M_2$. By Lemma 7, for any single-threaded complete s, we have $\sum_{\{t \mid s \lhd_{\mathsf{P}}^{*} t\}} [\![\Gamma \vdash M_1]\!](t) = \sum_{\{t \mid s \lhd_{\mathsf{P}}^{*} t\}} [\![\Gamma \vdash M_2]\!](t)$. Note that the set $\{t \mid s \lhd_{\mathsf{P}}^{*} t\}$ is finite and partially-ordered by $\lhd_{\mathsf{P}}^{*}$. By induction with respect to the reverse order, one can then prove that for all complete s we have $[\![\Gamma \vdash M_1]\!](s) = [\![\Gamma \vdash M_2]\!](s)$, from which $\Gamma \vdash M_1 \cong_{\mathsf{IA}_{\mathbf{coin+mkvar}}} M_2$ follows.

[3] Proposition 1 specializes to it when both O- and P-answers are well-bracketed.

Thus, for program equivalence, **mkvar** is conservative for IA and $\mathsf{IA}_{\mathbf{coin}}$, but not for $\mathsf{IA}_{\mathbf{catch}}$ or $\mathsf{IA}_{\mathbf{or}}$.

References

1. Reynolds, J.C.: The essence of Algol. In: de Bakker, J.W., van Vliet, J. (eds.) Algorithmic Languages, pp. 345–372. North Holland, Amsterdam (1978)
2. O'Hearn, P.W., Tennent, R.D. (eds.).: Algol-like Languages. Progress in Theoretical Computer Science. Birkhäuser, Boston (1997) (Two volumes)
3. Reddy, U.S.: Global state considered unnecessary: An introduction to object-based semantics. Lisp and Symbolic Computation 9, 7–76 (1996)
4. Abramsky, S., McCusker, G.: Linearity, sharing and state: a fully abstract game semantics for Idealized Algol with active expressions. In: O'Hearn, P.W., Tennent, R.D. (eds.) Algol-like languages, Birkhaüser, pp. 297–329 (1997)
5. Burstall, R.M., Popplestone, R.J.: Pop-2 reference manual. Machine Intelligence 2, 205–246 (1968)
6. Reynolds, J.C.: GEDANKEN-A simple typeless language based on principle of completeness and reference concept. CACM 13, 308–319 (1970)
7. McCusker, G.: On the semantics of Idealized Algol without the bad-variable constructor. In: Proceedings of MFPS'03, ENTCS 83,
8. Cartwright, R., Felleisen, M.: Observable sequentiality and full abstraction (preliminary version). In: Proceedings of POPL'92
9. Laird, J.: Full abstraction for functional languages with control. In: Proceedings of LICS'97
10. Abramsky, S., Ghica, D.R., Murawski, A.S., Ong, C.H.L., Stark, I.D.B.: Nominal games and full abstraction for the nu-calculus. In: Proceedings of LICS'04
11. Laird, J.: A game semantics of local names and good variables. In: Walukiewicz, I. (ed.) FOSSACS 2004. LNCS, vol. 2987, Springer, Heidelberg (2004)
12. Abramsky, S., McCusker, G.: Game semantics. In: Schwichtenberg, H., Berger, U. (eds.) Logic and Computation. Proceedings of the 1997 Marktoberdorf Summer School, Springer, Heidelberg (1998)
13. Hyland, J.M.E., Ong, C.H.L.: On Full Abstraction for PCF. Information and Computation 163(2), 285–408 (2000)
14. Harmer, R., McCusker, G.: A fully abstract game semantics for finite nondeterminism. In: Proceedings of LICS'99
15. Danos, V., Harmer, R.: Probabilistic game semantics. In: Proceedings of LICS'00
16. Murawski, A.S., Ouaknine, J.: On probabilistic program equivalence and refinement. In: Abadi, M., de Alfaro, L. (eds.) CONCUR 2005. LNCS, vol. 3653, Springer, Heidelberg (2005)

A Games Model of Bunched Implications

Guy McCusker[1] and David Pym[2]

[1] Department of Computer Science, University of Bath
Bath BA2 7AY
United Kingdom
G.A.McCusker@bath.ac.uk
[2] Hewlett-Packard Laboratories
Bristol BS34 8QZ
United Kingdom
david.pym@hp.com

Abstract. A game semantics of the $(-\!\!*, \rightarrow)$-fragment of the logic of bunched implications, **BI**, is presented. To date, categorical models of **BI** have been restricted to two kinds: functor category models; and the category **Cat** itself. The game model is not of this kind. Rather, it is based on Hyland-Ong-Nickau-style games and embodies a careful analysis of the notions of resource sharing and separation inherent in **BI**. The key to distinguishing between the additive and multiplicative connectives of **BI** is a semantic notion of separation. The main result of the paper is that the model is fully complete: every finite, total strategy in the model is the denotation of a term of the $\alpha\lambda$-calculus, the term language for the fragment of **BI** under consideration.

1 Introduction

The logic of bunched implications, **BI** [10, 9, 11], is a substructural logic which treats multiplicative and additive versions of its connectives on an equal footing, and in doing so gives an account of the notions of sharing and separation of resources. As a result, **BI** has clear applications to computer science: using **BI** as a type system gives rise to elegant approaches to interference-control in imperative programming [10] improving on Reynolds's *Syntactic Control of Interference* [13]; a particular model of **BI** has become known as *Separation Logic* [14] and is now a widely studied approach to local reasoning about programs which manipulate memory in challenging ways; and **BI** has also been developed into a Hennessy-Milner style logic for specification and reasoning about resource-sensitive properties of processes and systems [12].

From the outset, **BI** has enjoyed a rich semantic theory based on the notion of cartesian doubly-closed category: models of **BI** are categories possessing two distinct monoidal-closed structures, one of which is cartesian. To date the only known instances of this structure are functor category models and the category of categories, **Cat**.

Game semantics is an approach to modelling logics and programming languages which has seen considerable success in the last fifteen years. The first

J. Duparc and T.A. Henzinger (Eds.): CSL 2007, LNCS 4646, pp. 573–588, 2007.

major results were the fully-abstract games models of PCF [1,4,7], and subsequently a wide variety of programming languages and logics have been modelled, very often with full abstraction or full completeness properties.

In this paper, we use game semantics to give a new model of a fragment of **BI**, in the form of the associated term-language, the $\alpha\lambda$-calculus [10,11,9,8]. Our model is a refinement of the Hyland-Ong-Nickau style model of the λ-calculus. We demonstrate that the model is fully complete, that is, that every finite, total element of the model is the denotation of a term of the language, i.e., a proof in **BI**.

There are many further directions for this work. As well as going on to model larger fragments of **BI**, we would like to combine the approach to resource-sensitivity introduced here with the games models of imperative programming [2], aiming to arrive at a semantic account of interference control extending that of [15]. The model given here incorporates a relational, rather than a syntactic, notion of separation, similar to the language λ_{sep} [3] and we conjecture that our model is also a model for that calculus.

2 Bunched Implications and the $\alpha\lambda$-Calculus

We will not present **BI** directly but instead move straight to the associated term-language, the $\alpha\lambda$-calculus. The types of the $\alpha\lambda$-calculus are as follows:

$$A ::= \gamma \mid A \multimap\!\!\ast\, A \mid A \rightarrow A,$$

where γ ranges over a collection of ground types. Its terms are given by the grammar

$$M ::= x \mid \lambda x.M \mid MM \mid \alpha x.M \mid M @ M.$$

The standard β- and η- reduction apply to both kinds of abstraction and application, e.g. $(\alpha x.M) @ N \rightarrow M[N/x]$ and $(\lambda x.M)N \rightarrow M[N/x]$, with the usual notions of free and bound identifiers, capture-free substitution etc.

The typing rules are based on judgements of the form $\Gamma \vdash M : A$, where M is a term, A is a type, and Γ is a *bunch*: bunches are described by the grammar

$$\Gamma ::= I \mid x : A \mid \Gamma, \Gamma \mid \Gamma; \Gamma.$$

I is the empty bunch; $x : A$ is a singleton bunch; and there are two bunch-forming operations, comma and semicolon. The idea is that in a bunch Γ, Δ there is no resource sharing between Γ and Δ, while in $\Gamma; \Delta$ there may be. Formally, this is achieved by allowing contraction across semicolons but not across commas.

We write $\Gamma(\Delta)$ to indicate a bunch in which Δ appears as a subtree, and then $\Gamma(\Delta')$ indicates the similar tree where Δ is replaced by Δ'. Bunches are identified up to ***coherent equivalence***: $\equiv$ is the smallest equivalence relation on bunches including commutative monoid equations for I and ; , commutative monoid equations for I and , and congruence: if $\Delta \equiv \Delta'$ then $\Gamma(\Delta) \equiv \Gamma(\Delta')$. Note that in our presentation we do not have separate units for the two bunch-forming operations. This is because our version of the $\alpha\lambda$-calculus and our model will incorporate weakening for both connectives. That is, we treat the *affine* version of **BI**, so the units are identified.

The typing rules are as follows:

$$\frac{\Gamma \vdash M : A}{\Delta \vdash M : A}\ \Gamma \equiv \Delta \text{ (coherence)} \qquad \frac{}{x : A \vdash x : A}\text{ (id)}$$

$$\frac{\Gamma(\Delta) \vdash M : A}{\Gamma(\Delta, \Delta') \vdash M : A}\text{ (weakening for ,)} \qquad \frac{\Gamma(\Delta) \vdash M : A}{\Gamma(\Delta ; \Delta') \vdash M : A}\text{ (weakening for ;)}$$

$$\frac{\Gamma, x : A \vdash M : B}{\Gamma \vdash \lambda x.M : A \mathrel{-\!\!*} B}\text{ (λ-abstraction)} \qquad \frac{\Gamma \vdash M : A \mathrel{-\!\!*} B \quad \Delta \vdash N : A}{\Gamma, \Delta \vdash MN : B}\left(\begin{array}{c}\text{multiplicative}\\ \text{application}\end{array}\right)$$

$$\frac{\Gamma ; x : A \vdash M : B}{\Gamma \vdash \alpha x.M : A \to B}\text{ (α-abstraction)} \qquad \frac{\Gamma \vdash M : A \to B \quad \Delta \vdash N : A}{\Gamma ; \Delta \vdash M \mathbin{@} N : B}\left(\begin{array}{c}\text{additive}\\ \text{application}\end{array}\right)$$

$$\frac{\Gamma(\Delta ; \Delta') \vdash M : B}{\Gamma(\Delta) \vdash M[\mathsf{idents}(\Delta)/\mathsf{idents}(\Delta')] : B}\ \Delta \cong \Delta' \text{ (contraction)}$$

In the final rule, $\Delta \cong \Delta'$ means that the bunches Δ and Δ' are the same up to renaming of identifiers, and the substitution $M[\mathsf{idents}(\Delta)/\mathsf{idents}(\Delta')]$ simultaneously replaces each identifier of Δ' appearing in M with the corresponding one from Δ.

The simply-typed λ-calculus is of course the fragment of the $\alpha\lambda$-calculus where only ; is used in bunches, and only α and @ are used as term-formers. Even though our syntax for this fragment uses α as the binder, we will continue to call it the λ-calculus! (Note that α and λ are mnemonics for the αdditive and λinear abstractions respectively.)

We first establish some simple facts about the type system.

If Γ is a bunch containing identifiers x and y, then there is a unique sub-bunch of the form Γ_1, Γ_2 or $\Gamma_1; \Gamma_2$ such that x appears in Γ_1 and y in Γ_2. We say that x and y are ***separated*** in Γ if this sub-bunch has the form Γ_1, Γ_2. That is, x and y are in sub-bunches combined by the comma rather than the semicolon.

Lemma 1. *A term $\Gamma \vdash MN : B$ is typeable if and only if there are typeable terms $\Gamma \vdash M : A \mathrel{-\!\!*} B$ and $\Gamma \vdash N : A$, and all the free identifiers of M are separated from all the free identifiers of N in Γ.*

Proof. The only-if direction is a straightforward structural induction on derivations. For the other direction, we show that Γ can be split as $\Gamma = \Gamma_1, \ldots, \Gamma_n$ where each Γ_i contains identifiers from at most one of M and N, and the result follows easily. □

The standard additive-style application rule for @ is admissible:

Lemma 2 (Additive application). *If $\Gamma \vdash M : A \to B$ and $\Gamma \vdash N : A$ then $\Gamma \vdash M \mathbin{@} N : B$.*

Proof. Rename the variables in N to obtain $\Gamma' \vdash N' : A$ with $\Gamma \cong \Gamma'$, then use the application and contraction rules. □

3 A Game Semantics for the $\alpha\lambda$-Calculus

In this section, we briefly recall the basic definitions and results in Hyland-Ong-Nickau style game semantics for the λ-calculus, including the full completeness argument, before going on to refine the model to handle the subtleties of the **BI**. The definitions we use are essentially those of [6], with one or two small presentational changes to facilitate our later refinements.

3.1 Arenas

A game has two participants: Player (P) and Opponent (O). A *play* of the game consists of a sequence of moves, alternately by O and P. In addition, each move is explicitly *justified* by an earlier move of the play, unless it is a special kind of move, called *initial*, which needs no justification. In the games we consider, O always moves first.

Before embarking on a formal definition, let us fix notation for sequences and operations on them. We use s, t, ... to range over sequences and a, b, ..., m, n, ... over elements of sequences. If s and t are sequences, then st or $s \cdot t$ is their concatenation; ε is the empty sequence. A move a will often be identified with the singleton sequence consisting just of a. The sequence $s_{<a}$ is the prefix of s up to, but not including, an element a, while $s_{\leq a}$ does include a.

An ***arena*** is specified by a triple $A = \langle M_A, \lambda_A, \vdash_A \rangle$ where:

- M_A is a set of ***moves***;
- $\lambda_A : M_A \to \{\mathsf{O}, \mathsf{P}\}$ is a ***labelling*** function which indicates whether a move is by Opponent (O) or Player (P). The function $\overline{\lambda_A}$ is λ_A with the O and P reversed. If $\lambda(a) = \mathsf{O}$, we call a an O-move; otherwise, a is a P-move;
- $\vdash_A$ is a relation between $M_A + \{\star\}$ and M_A, called ***enabling***, which satisfies
  - $\star \vdash_A a \Rightarrow [b \vdash_A a \iff b = \star]$, and
  - $a \vdash_A b \wedge a \neq \star \Rightarrow \lambda_A(a) \neq \lambda_A(b)$.

The enabling relation tells us either that a move a is ***initial*** and needs no justification ($\star \vdash_A a$), or that it can be justified by another move b, if b has been played ($b \vdash_A a$). An arena is called ***negative*** if all its initial moves belong to O.

A ***justified sequence*** s of moves in an arena A is a sequence of moves together with ***justification pointers***: for each move a in s which is not initial, there is a pointer to an earlier move b of s such that $b \vdash_A a$. We say the move b ***justifies*** a, and extend this terminology to say that a move b ***hereditarily justifies*** a if the chain of pointers back from a passes through b.

Given a justified sequence s, the ***view*** $\mathsf{view}(s)$ of s is defined as follows:

$$\mathsf{view}(\varepsilon) = \varepsilon \qquad \begin{array}{r}\mathsf{view}(s \cdot a) = a, \text{if } a \text{ is an initial move} \\ \mathsf{view}(s \cdot a \cdot t \cdot b) = \mathsf{view}(s) \cdot a \cdot b.\end{array}$$

If s is a justified sequence containing a move a, we say that a is ***visible at*** s if a appears in $\mathsf{view}(s)$.

A justified sequence s is a ***legal position*** iff:

- O moves first: if $s = as'$ then $\lambda(a) = \mathsf{O}$;
- s is ***alternating***: if $s = s_1 abs_2$ then $\lambda(a) \neq \lambda(b)$;
- The ***visibility condition*** holds: if $s = s_1 a s_2$, and a is not initial, then the justifier of a is visible at s_1.

The set of all legal positions of an arena A is written L_A.

3.2 Strategies

A strategy for an arena A is a rule telling Player what move to make in a given position. Formally, this can be represented as a set σ of legal positions in which P has just moved, i.e., a nonempty set of even-length positions, such that

$$sab \in \sigma \Rightarrow s \in \sigma \qquad sab, sac \in \sigma \Rightarrow sab = sac.$$

For any arena A, the smallest possible strategy is $\{\varepsilon\}$, which never makes any response. It is called the ***empty strategy*** and denoted $\bot$.

A strategy is ***innocent*** if for all $sab, t \in \sigma$, if $ta \in L_A$ and $\mathsf{view}(sa) = \mathsf{view}(ta)$, then also $tab \in \sigma$, with b justified by the same element of $\mathsf{view}(ta) = \mathsf{view}(sa)$ as in sab.

3.3 Constructions on Arenas

Given negative arenas A and B, the arenas $A \times B$ and $A \vdash B$ are defined as follows:

$$\begin{gathered} M_{A\times B} = M_{A\vdash B} = M_A + M_B \text{ (disjoint union)} \\ \lambda_{A\times B} = [\lambda_A, \lambda_B] \qquad \lambda_{A\vdash B} = [\overline{\lambda}_A, \lambda_B] \\ \star \vdash_{A\times B} m \iff \star \vdash_{A\vdash B} m \iff \star \vdash_A m \vee \star \vdash_B m \\ m \vdash_{A\times B} n \iff m \vdash_{A\vdash B} n \iff m \vdash_A n \vee m \vdash_B n. \end{gathered}$$

The idea here is that the games A and B are played in interleaved parallel fashion. In $A \times B$, labelling and enabling are inherited directly from A and B. In $A \vdash B$, the O/P roles in A are reversed; one impact this has is that a legal position in $A \vdash B$ always begins with a move from B. The unit for $\times$ is the empty arena $I = \langle \emptyset, \emptyset, \emptyset \rangle$. Note that $A \times B$ is negative if A and B are, while $A \vdash B$ is not. In fact the only arenas we will consider which are not negative are those of the form $A \vdash B$, where A and B are negative.

3.4 Composition of Strategies

Let A, B and C be negative arenas. An ***interaction sequence*** on A, B, C is a justified sequence u of moves from $M_A + M_B + M_C$ such that

- $u \restriction A, B \in L_{A\vdash B}$,
- $u \restriction B, C \in L_{B\vdash C}$,
- $u \restriction A, C$ is an alternating sequence of moves in $A \vdash C$, and
- there is at least one move between any A-move and any C-move in u.

We write $\mathsf{int}(A, B, C)$ for the set of all such sequences.

Given strategies σ on $A \vdash B$ and τ on $B \vdash C$ we define the composite strategy $\sigma ; \tau$ on $A \vdash C$ by "parallel composition plus hiding".

$$\sigma ; \tau = \{s \mid \exists u \in \mathsf{int}(A, B, C).u \upharpoonright A, B \in \sigma \wedge u \upharpoonright B, C \in \tau \wedge u \upharpoonright A, C = s\}.$$

Lemma 3. *Composition of strategies is well-defined (that is, $\sigma ; \tau$ as defined above is indeed a strategy on $A \vdash C$). Moreover, it is associative and has identity given by the* **copycat strategy**

$$\mathsf{id}_A = \{s \in L_{A \vdash A'} \mid \forall t \sqsubseteq^{even} s.t \upharpoonright A = t \upharpoonright A'\}.$$

(Here A' is the same arena as A; we use the prime only to distinguish the two occurrences. The copycat strategy simply copies O*'s moves back and forth from one occurrence to the other.) Further, the composite of two innocent strategies is innocent and the identity is itself innocent.*

We can now define two categories of games: $\mathcal{G}$ has negative arenas as objects and strategies on $A \vdash B$ as maps from A to B, with composition and identities as above; $\mathcal{G}_{\text{inn}}$ has the same objects but has only innocent strategies as morphisms.

The $\times$ operation on arenas gives a categorical product in $\mathcal{G}_{\text{inn}}$; the projections $A_1 \times A_2 \to A_i$ are given by copycat strategies.

The category $\mathcal{G}_{\text{inn}}$ is in fact cartesian closed: exponentials are given by the arena $A \Rightarrow B$ defined by

$$\begin{aligned} M_{A \Rightarrow B} &= M_A + M_B & \qquad \star \vdash_{A \Rightarrow B} a &\iff \star \vdash_B a \\ \lambda_{A \Rightarrow B} &= [\overline{\lambda}_A, \lambda_B] & \qquad m \vdash_{A \Rightarrow B} n &\iff m \vdash_A n \vee m \vdash_B n \vee [\star \vdash_B m \wedge \star \vdash_A n] \end{aligned}$$

Compare and contrast with $A \vdash B$: in creating $A \Rightarrow B$ we convert initial moves of A to non-initial moves, justified by initial B-moves. Note that this means $A \Rightarrow B$ is always negative if A and B are. Standard presentations use only negative arenas and use this definition for $A \vdash B$, but when we come to refine the category to model **BI** this will no longer suffice: there is an important distinction between morphisms and elements of the exponential type(s).

Now that we have a cartesian closed category, we can interpret the simply-typed λ-calculus in a standard fashion [5]:

- each type A is interpreted as an object $[\![A]\!]$, that is, as an arena, with $[\![A \to B]\!] = [\![A]\!] \Rightarrow [\![B]\!]$
- a context $\Gamma = x_1 : A_1; \ldots ; x_n : A_n$ is interpreted as the product $[\![A_1]\!] \times \cdots \times [\![A_n]\!]$
- each term $\Gamma \vdash M : A$ is interpreted as a map $[\![\Gamma \vdash M]\!]$ (often abbreviated to $[\![M]\!]$) with domain $[\![\Gamma]\!]$ and codomain $[\![A]\!]$.

The interpretation is fixed as soon as we determine an object $[\![\gamma]\!]$ for each ground type: this is done by taking $[\![\gamma]\!]$ to be a one-move arena, consisting of a single initial opponent move γ.

3.5 Full Completeness for the λ-Calculus

We now sketch the proof of the definability result for λ-calculus which forms the basis of, for example, Hyland, Ong and Nickau's fully abstract models of the programming language PCF [4, 7]. The key to the result is that an innocent strategy σ is determined by the set $\mathsf{branches}(\sigma)$ of legal positions $s \in \sigma$ such that every O-move in s is justified by the immediately preceding P-move, and that such sequences correspond to branches of Böhm-trees. Hence innocent strategies correspond to (partial, potentially infinite) Böhm-trees. Thus innocent strategies which are *total*, meaning they have a response to every move O can make, and *finite*, meaning the set $\mathsf{branches}(\sigma)$ is finite, correspond to ordinary Böhm-trees.

We consider a type A built from a single ground type γ using the $\to$ constructor, and a total, finite innocent strategy σ on the arena $[\![A]\!]$. Our task is to find a closed Böhm-tree M such that $[\![M]\!] = \sigma$. Using the uncurrying isomorphisms, we may consider an arena of the form $A_1 \times \cdots \times A_n \vdash \gamma$ and search for a term of the form $x_1 : A_1, \ldots, x_n : A_n \vdash M : \gamma$ to denote our strategy.

Let $A_i = A_{i,1} \to \cdots \to A_{i,k_i} \to \gamma_i$, where we use the subscript on γ simply to distinguish it from other occurrences of the ground type. Since σ is total, it has a response to the initial move γ, so there is some sequence $\gamma\gamma_i \in \mathsf{branches}(\sigma)$. The term denoting σ will then take the form $x_i M_1 \ldots M_{k_i}$ for some terms M_j.

Every longer sequence in $\mathsf{branches}(\sigma)$ has the form $\gamma\gamma_i\gamma_{i,j}t$ where $\gamma_{i,j}$ is the initial move of $[\![A_{i,j}]\!]$ for some j. The set $\{\gamma_{i,j}t \mid \gamma\gamma_i\gamma_{i,j}t \in S\}$ then determines (the branches of) a finite, total innocent strategy τ_j on the arena $A_1 \times \cdots \times A_n \vdash A_{i,j}$. This requires relabelling the moves a little: moves of the sequences t which are hereditarily justified by $\gamma_{i,j}$ become moves in the right-hand arena $A_{i,j}$, while all others remain in the left-hand side.

The set $\mathsf{branches}(\tau_j)$ is strictly smaller than $\mathsf{branches}(\sigma)$, so we can use an inductive argument to find a term M_j such that $[\![M_j]\!]$ is this strategy. We can then verify that $\sigma = [\![x_i M_1 \ldots M_{k_i}]\!]$ thus completing the full completeness argument.

The correspondence between branches of Böhm-trees and plays in $\mathsf{branches}(\sigma)$ motivates our forthcoming definitions of the model of the $\alpha\lambda$-calculus, so we expand on the accompanying intuition:

- An O-move corresponds to the selection of a subterm to investigate: the initial move γ commences investigation of the term, and after P responds with γ_i, O can choose any of the $\gamma_{i,j}$, which begins investigation of the subterm M_j;
- A P-move corresponds to the choice of head-variable for the subterm under investigation;
- The justifier of a P move is the O-move corresponding to the subterm where the chosen head-variable was introduced by abstraction. Initial P-moves correspond to free variables;
- Suppose we have an O move m and a later P-move n whose justifier appears before m. The subterm M corresponding to m contains the use of a variable x corresponding to n, and this variable appears free in M, since it was abstracted (justified) outside M.

To illustrate, consider the λ-term

$$\lambda f.\lambda x.f(\lambda g.gx) : (((\gamma_1 \to \gamma_2) \to \gamma_3) \to \gamma_4) \to \gamma_5 \to \gamma_6.$$

A typical play in the strategy interpreting this term is

$$\gamma_6 \cdot \gamma_4 \cdot \gamma_3 \cdot \gamma_2 \cdot \gamma_1 \cdot \gamma_5.$$

The P-moves correspond to head-variables: γ_4 picks f; γ_2 picks g; and γ_5 picks x. Note also that move γ_5's justifier is before those of γ_1 and γ_3, since x is free in the subterms being investigated, which are x and $\lambda g.gx$ respectively.

3.6 Refining the Model

Armed with intuition about the correspondence between strategies and Böhm-trees in the λ-calculus, let us investigate how these ideas might be refined to reflect the $\alpha\lambda$-calculus.

Let $\Gamma \vdash M$ be a Böhm-tree in the λ-calculus. If we replace some of the ; constructors in Γ with , so that Γ becomes a nontrivial bunch, and replace some of the $\to$ type constructors with $-\!\!*$, what constraints do the typing rules of the $\alpha\lambda$-calculus place upon M, and how might these be reflected in the games model?

For a term $fM_1 \ldots M_n$ to remain typeable after this refinement of the typing, we require at least that

- if f appears free in one of the M_i, then M_i is an argument to a $\to$-function, and
- if M_i and M_{i+j} share a free variable x, then M_{i+j} is an argument to a $\to$-function.

For instance, if f has type $A -\!\!* B \to C$, then in a term $(fM)@N$, f may appear in N but not M; and M and N may share free variables. However, if f has type $A \to B -\!\!* C$ then in a term $(f @ M)N$, f may appear free in M but not N, and M and N's free variables must be distinct.

The constraints run deeper than this: in a subterm $fM_1 \ldots M_n$ of a closed Böhm-tree, if M_{i+j} is an argument to a $-\!\!*$-function, then the free variables of M_i and M_{i+j} must be separated (Lemma 1). What this means for the term is that, if M_i contains x and M_{i+j} contains y:

- whichever of x and y was introduced by abstraction deeper in the term, must have been introduced by λ-abstraction rather than α-abstraction.

To capture all of this in the games model, we will need:

- a semantic account of the notion of "variable appearing free in a subterm";
- a semantic account of the distinction between arguments to $\to$-functions and $-\!\!*$-functions;

- a semantic account of the idea that "the deeper abstraction must be a λ-abstraction".

We shall now formalize these ideas in the games setting.

We shall now extend our games model with the appropriate structure to allow us to incorporate the above ideas. In the next few paragraphs, remarks appearing in brackets [...] indicate the intuition corresponding to our definitions.

Given moves a and b in a legal position s, with a occurring before b and $\lambda(a) \neq \lambda(b)$, we say b is an ***external move*** to a, writing $b\,\mathbf{ext}\,a$, if the justifier of b occurs before a (or b is initial), and a is visible at $s_{<b}$. [This gives us a semantic account of variables appearing free in subterms.]

Given an arena A, a ***separation relation*** on A is an irreflexive, symmetric binary relation $\#_A$ on the moves of A, subject to the condition that if $a \#_A b$ then either $a \vdash_A b$, $b \vdash_A a$ or there is some c such that $c \vdash a, b$. That is, separation exists only between moves which enable one another, or which share a justifier. [Separation will allow us to distinguish between $-\!\!*$ and $\rightarrow$: the arenas $A -\!\!* B$ and $A \rightarrow B$ will differ only in that initial moves from A are separated from those of B in $A -\!\!* B$.]

We refer to an arena A together with a separation relation $\#_A$ as a ***sep-arena***, and often refer to it just as A rather than the pair $(A, \#_A)$.

If s is a legal position in a sep-arena containing two moves a and b, we write $a \#_s b$ if a and b have the same justifier in s (or both are initial) and also $a \#_A b$. [In the case where these are O-moves, $a \#_s b$ tells us that the two subterms being investigated may not share resources.]

We write $a *_s b$ if any of the following conditions holds:

- $a \#_s b$;
- a is justified by a', $b\,\mathbf{ext}\,a'$ and $a \#_A a'$; or
- b is justified by b', $a\,\mathbf{ext}\,b'$ and $b \#_A b'$.

[In the case where these are P-moves, $a *_s b$ tells us that the variables being chosen do not share resources: in a closed term, the more deeply abstracted of the two variables was λ-abstracted.]

Let A be a sep-arena. A legal position s is ***separation safe*** if

- for any O-moves $a, b \in s$ with $a \#_s b$, if $a_1\,\mathbf{ext}\,a$ and $b_1\,\mathbf{ext}\,b$ then $a_1 *_s b_1$,
- for any P-move a such that $a\,\mathbf{ext}\,b$ in s, if b is justified by b' and $b \#_A b'$ then $a *_s b'$.

Example. Supposing we give **BI**-types to the example term at the end of Section 3.5. In the example play, we have $\gamma_5\,\mathbf{ext}\,\gamma_1$, and if g is a $-\!\!*$-function, we will have $\gamma_1 \# \gamma_2$, so separation safety will require $\gamma_5 * \gamma_2$. This will hold if $\gamma_2 \# \gamma_3$, which will be the case if f is a $-\!\!*$-function, so that g is λ-abstracted rather than α-abstracted.

Given negative sep-arenas A and B, we can define a separation relation on $A \vdash B$ as the disjoint union of $\#_A$ and $\#_B$. A strategy σ on $A \vdash B$ is called separation safe if every $s \in \sigma$ is separation safe. [Separation safe strategies are those which adhere to the typing constraints of the $\alpha\lambda$-calculus.]

The main technical effort of the paper consists in showing that separation-safety is preserved by composition of strategies. This involves a detailed analysis of interaction sequences.

A move m in an interaction sequence u necessarily belongs to at least one of $u \upharpoonright A, B$ and $u \upharpoonright B, C$. In the case where m is in B, we say that $u \upharpoonright A, B$ is the P-***component*** of m if m is a P-move there; otherwise it is the O-component. Similarly for $u \upharpoonright B, C$.

The view $\mathsf{view}(u)$ of an interaction sequence is defined exactly as the view of an ordinary legal position. We note that $\mathsf{view}(u)$ is the same as the view of the O-component of the last move of u. This is because the justifier of a B-move which is an O-move in A, B must be a P-move in A, B, and the move immediately preceding it in u must again be an O-move coming from A, B.

We say that moves m and n in $\mathsf{view}(u)$ have opposite polarity if there is an even number of moves strictly between them in $\mathsf{view}(u)$.

Given an interaction sequence u containing moves m and n, with m occurring before n, we say n is an external move to m, $n\, \mathbf{ext}_u\, m$, if m is visible at $u_{<n}$, m and n have opposite polarity and n is initial or has its justifier before m. This is equivalent to saying that $n\, \mathbf{ext}\, m$ in the P-component of n.

The following lemma relates the **ext** relation in $u \upharpoonright A, C$ to those in the A, B and B, C components.

Lemma 4. *Let $u \in \mathsf{int}(A, B, C)$. Let n be a* P*-move in $u \upharpoonright A, C$ and m an* O*-move in $u \upharpoonright A, C$ such that $n\, \mathbf{ext}\, m$ in $u \upharpoonright A, C$. Then there exist moves $n_1, \ldots, n_k$ in u, all coming from B, such that $n\, \mathbf{ext}_u\, n_1\, \mathbf{ext}_u \cdots \mathbf{ext}_u\, n_k\, \mathbf{ext}\, m$.*

We omit the proof, which involves a careful analysis of views in interaction sequences, similar to the proof that innocence is preserved by composition of strategies given in [6].

Having lifted the definition of **ext** to interaction sequences, the definitions of # and $*$ on legal positions lift directly to interaction sequences, and we have $m * n$ in u iff $m * n$ in the P-component of m.

The following lemma is key to our proof that separation-safety is preserved by composition.

Lemma 5. *Let A, B and C be negative sep-arenas. Let $u \in \mathsf{int}(A, B, C)$ be such that $u \upharpoonright A, B$ and $u \upharpoonright B, C$ are separation-safe. Let m_1 and m_2 be moves in u such that either m_1 and m_2 are* O*-moves in $u \upharpoonright A, C$ with $m_1 \# m_2$, or they are* P*-moves in some component with $m_1 * m_2$. Suppose there are moves $n_{1,1}, \ldots, n_{1,k_1}, n_{2,1}, \ldots, n_{2,k_2}$ such that*

- *$n_{i,1}\, \mathbf{ext}\, n_{i,2}\, \mathbf{ext} \cdots \mathbf{ext}\, n_{i,k_i}\, \mathbf{ext}\, m_i$, for $i = 1, 2$,*
- *$n_{1,1}$ and $n_{2,1}$ are* P*-moves in $u \upharpoonright A, C$,*
- *the intervening moves $n_{i,2}, \ldots, n_{i,k_i}$ are in B,*
- *the justifier of $n_{1,1}$ is visible at $u_{<n_{2,1}} \upharpoonright A, C$ and vice versa.*

Then $n_{1,1} * n_{2,1}$ in $u \upharpoonright A, C$.

Proof. We require an auxiliary definition: the ***pivot*** of separation of $m_1 * m_2$. According to the definition of $*$, there are four possibilities:

– m_1 and m_2 are both initial and $m_1 \# m_2$: then there is no pivot;
– m_1 and m_2 have the same justifier j and $m_1 \# m_2$: then j is the pivot;
– m_1 is justified by some j, $m_1 \# j$ and $m_2\, \mathbf{ext}\, j$: then j is the pivot;
– as above with m_1 and m_2 exchanged: again j is the pivot.

We approach our proof by induction on the length of $u_{\leq j}$, where j is the pivot of $m_1 * m_2$; by convention this is zero if there is no pivot.

The base case, therefore, is the case in which the m_i are initial and $m_1 \# m_2$. If they are initial C-moves, the moves n_{i,k_i} must be initial B-moves and by separation safety of $u \upharpoonright B, C$, $n_{1,k_1} * n_{2,k_2}$, which is to say $n_{1,k_1} \# n_{2,k_2}$. The only external moves of initial B-moves are initial A-moves, so we have that n_{1,k_1-1} and n_{2,k_2-1} are initial A-moves and $n_{1,k_1-1} * n_{2,k_2-1}$ by separation safety in $u \upharpoonright A, B$. It must be the case that $n_{1,1} = n_{1,k_1-1}$ and $n_{2,1} = n_{2,k_2-1}$ completing the argument in this case.

If the m_i are initial B-moves, the second half of the above argument applies.

The inductive step considers the cases where a pivot j exists.

Suppose first that j is the shared justifier of m_1 and m_2, and $m_1 \# m_2$. We first deal with the degenerate case in which one of the k_i (wlog k_1) is zero, i.e., $n_{1,1} = m_1$. In this case the m_i are both P-moves in A or C; suppose wlog they are in A. But then it must also be that $k_2 = 0$ since no B-move can be an external of m_2 by definition. Thus we need only show that $m_1 * m_2$ which is true by hypothesis.

If both k_1 and k_2 are non-zero, we have $n_{i,k_i}\, \mathbf{ext}\, m_i$ for $i = 1, 2$ and $m_1 \# m_2$ in $u \upharpoonright A, B$ or $u \upharpoonright B, C$. Hence by separation safety in this component, we have $n_{1,k_1} * n_{2,k_2}$, with pivot appearing earlier in u than j. We can therefore apply the inductive hypothesis to conclude.

Suppose finally that the pivot j is the justifier of m_1, where $m_1 \# j$ and $m_2\, \mathbf{ext}\, j$. (The case with the m_i exchanged is handled symmetrically.) If $k_1 = 0$, then j is in the view at $u_{<n_{2,1}} \upharpoonright A, C$ and $n_{2,1}$'s justifier appears before that of m_2 and hence before j, so $n_{2,1}\, \mathbf{ext}\, j$. Then by definition, $n_{2,1} * m_1 = n_{1,1}$. Otherwise, $k_1 \neq 0$ so we have n_{1,k_1} **ext** m_1, and separation safety in the P-component of n_{1,k_1} ensures that $n_{1,k_1} * j$. The pivot for this separation must appear before j, so we can apply the inductive hypothesis to this pair to complete the proof. □

Given negative sep-arenas A and B, the sep-arenas $A \times B$ and $A * B$ are defined as follows. Both have the same underlying arena as $A \times B$ defined for ordinary arenas; the difference is in the separation relations:

$$m \#_{A\times B}\, n \iff m \#_A\, n \vee m \#_B\, n$$
$$m \#_{A*B}\, n \iff m \#_A\, n \vee m \#_B\, n \vee [\star \vdash_A m \wedge \star \vdash_B n]$$

plus a clause symmetric in m and n. That is, in $A * B$, the initial moves of A are separated from those of B, which is not the case in $A \times B$.

We can also define $A \to B$ and $A \mathrel{-\!\!*} B$ on negative sep-arenas. The underlying arenas are as for $A \Rightarrow B$. The separation relations are defined as follows:

$$m \#_{A\to B}\, n \iff m \#_A\, n \vee m \#_B\, n \vee [\star \vdash_A m \wedge \star \vdash_B b \vdash_B n \wedge n \#_B\, b]$$

$$m \mathbin{\#_{A \multimap\!\!* B}} n \iff m \mathbin{\#_A} n \vee m \mathbin{\#_B} n \vee [\star \vdash_A m \wedge \star \vdash_B n]$$
$$\vee \quad \star \vdash_A m \wedge \star \vdash_B b \vdash_B n \wedge n \mathbin{\#_B} b$$

plus clauses symmetric in m and n. That is, in both $A \to B$ and $A \multimap\!\!* B$, initial A moves are separated from moves of B which are enabled by and separated from initial moves. Furthermore in $A \multimap\!\!* B$, initial A-moves are separated from the initial B-moves that enable them. Thus for example in $A \to (B \multimap\!\!* C)$, initial A-moves are separated from initial B-moves.

To illustrate the impact of separation safety, consider the sep-arenas $A \times B$ and $A * B$. The identity strategy on the underlying arena is separation-safe on $A * B \vdash A \times B$, but not on $A \times B \vdash A * B$. In the latter case, the initial moves on the right are related by $\#$ but are copied to the initial moves on the left, which are not. Since the initial moves on the left are external to those on the right, this violates separation safety.

Similarly, the identity strategy on the underlying arena gives us a valid strategy on $A \to B \vdash A \multimap\!\!* B$ but not the other way around: to see why consider a typical play

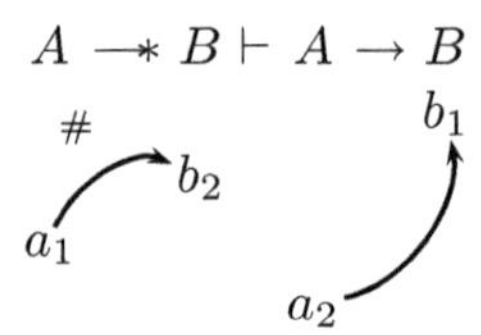

Here we have $a_2 \,\mathbf{ext}\, a_1$, and a_1 is justified by b_2 with $a_1 \# b_2$, so separation safety would require $a_2 * b_2$ which is not the case here. On the other hand if we had $a_2 \# b_1$, which we would if the $\to$ were $\multimap\!\!*$, then indeed $a_2 * b_2$.

Proposition 1. *The identity strategy on a sep-arena is separation safe.*

Proof. Any P-move played by the identity strategy is external only to the immediately preceding O-move, of which it is a copy. Thus if a_1 **ext** b_1 and a_2 **ext** b_2 with $b_1 \# b_2$, we immediately have a_1 **ext** a_2. Similarly if a_1 **ext** b_1 and b_1 is justified by some b_2 with $b_1 \# b_2$, then a_1 is a copy of b_1 and hence is justified by an a_2 with $a_1 \# a_2$. It must also be the case that b_2 is the copy of a_2, so b_2 **ext** a_2 and hence $b_2 * a_1$ as required. □

Proposition 2. *The composition of separation-safe innocent strategies is itself a separation-safe strategy.*

Proof. If $\sigma : A \vdash B$ and $\tau : B \vdash C$ are separation-safe strategies and $s \in \sigma;\tau$ then there exists some $u \in \mathsf{int}(A, B, C)$ such that $u \restriction A, B \in \sigma$, $u \restriction B, C \in \tau$ and $u \restriction A, C = s$.

Suppose $m_1 \#_s m_2$ and $n_i \,\mathbf{ext}_s\, m_i$ for $i = 1, 2$. By Lemma 4, there are moves $n_{i,1}, \ldots, n_{i,k_i}$ in u coming from B such that $n_i \,\mathbf{ext}\, n_{i,1} \,\mathbf{ext} \cdots \mathbf{ext}\, n_{i,k_i} \,\mathbf{ext}\, m_i$. Note also that, since m_i is visible at n_i and m_1 and m_2 share a justifier j, the justifier of each n_i appears in the view $\mathsf{view}(s_{<j})$. Thus n_1's justifier is visible at $s_{<n_2}$ and vice versa. Thus by Lemma 5, $n_1 * n_2$ as required.

Finally suppose m is justified by j with $m \# j$, and some $n \,\mathbf{ext}_s\, m$. Since m is in $\mathsf{view}(s_{<n})$, by the visibility condition the justifier of j is in $\mathsf{view}(s_{<n})$ and the

justifier of n is in $\mathsf{view}(s_{<j})$. Again by Lemma 4 we have a sequence of B-moves in u such that $n\,\mathbf{ext}\,n_1\,\mathbf{ext}\cdots\mathbf{ext}\,n_k\,\mathbf{ext}\,m$. By separation safety of σ and τ we obtain $n_k * j$, and then Lemma 5 tells us that $n * j$ as required. □

We can now define a category $\mathcal{G}_{\mathrm{sep}}$ with negative sep-arenas as objects and separation safe strategies on $A \vdash B$ as maps from A to B.

3.7 Interpreting the $\alpha\lambda$-Calculus

In order to give a semantics of the $\alpha\lambda$-calculus, we must show that $\mathcal{G}_{\mathrm{sep}}$ has enough structure to interpret both ; and , in contexts, and both $\rightarrow$ and $-\!\!*$ in types.

Proposition 3. *The category $\mathcal{G}_{\mathrm{sep}}$ is cartesian closed, with product given by $\times$ and exponential by $\rightarrow$. A second symmetric monoidal structure is provided by $*$ and, if B is a sep-arena with only one initial move and A any sep-arena, then $A -\!\!* B$ is an exponential with respect to $*$.*

Proof. It is not difficult to check that $\times$ is categorical product, just as the corresponding construct is on $\mathcal{G}_{\mathrm{inn}}$. Similarly it is easy to check that $*$ is a monoidal structure: all the structural isomorphisms are given by copycat strategies as usual, and one just has to verify that they are separation safe. The exponentials are where the novelty lies.

In $\mathcal{G}_{\mathrm{inn}}$, the fact that $\Rightarrow$ is the exponential with respect to $\times$ follows from the fact that there is a clear 1-1 correspondence between the legal positions of $A \times B \vdash C$ and $A \vdash B \Rightarrow C$: both consist of moves from A, B and C, and to turn a play in the former arena into one in the latter, one simply alters the justification structure so that initial B-moves are justified by the unique visible initial C-move. This lifts to a natural isomorphism of strategies and is standard in game semantics. We shall argue that the same isomporphism serves for the two exponentials in $\mathcal{G}_{\mathrm{sep}}$.

Our only additional obligation is to show that separation safety is preserved when moving across this isomorphism. Let a_0, b_0, c_0 range over initial moves of A, B and C respectively, and c_1 over moves of C such that some $c_0 \vdash_C c_1$ with $c_0 \#_C c_1$. The only differences between the separation relations on $A \times B \vdash C$ and $A \vdash B \rightarrow C$ is that in the latter, $b_0 \# c_1$ which is not the case in the former. But if s is a position of $A \vdash B \rightarrow C$ containing b_0 and c_1 with the same justifier c_0, so that $b_0 \#_s c_1$, the corresponding position of $A \times B \vdash C$ has $b_o\,\mathbf{ext}\,c_0$ and hence $b_0 * c_1$. A similar argument shows that all instances of $*$ in positions of $A \vdash B \Rightarrow C$ are retained when moving to $A \times B \vdash C$, so separation safety is preserved by the isomorphism of positions.

For the correspondence between $A * B \vdash C$ and $A \vdash B -\!\!* C$ there are two differences in $\#$: first, as above, $b_0 \# c_1$ in the latter but not the former, but this is handled just as above; second, $a_o \# b_0$ in the former but not the latter. For the second, observe that in any position of $A * B \vdash C$ in which $a_0 \# b_0$ is required for separation safety, the initial move c_0 in $\mathsf{view}(s_{<a_0})$ is the same as that in $\mathsf{view}(s_{<b_0})$; this is because C has only one initial move by hypothesis. Therefore

when moving across the isomorphism we have a_0 **ext** c_0 and b_0 justified by c_0 with $b_0 \# c_0$, hence $a_0 * b_0$. A similar analysis shows that moving in the other direction across the isomorphism also preserves separation safety. □

The category $\mathcal{G}_{\mathrm{sep}}$ is therefore *almost* a cartesian doubly-closed category, the notion of categorical model for the $\alpha\lambda$-calculus defined in [9,11,10]. $\mathcal{G}_{\mathrm{sep}}$ lacks certain exponentials, such as those of the form $A \multimap\!\!* (B * C)$, whose right-hand side arena has multiple initial moves. However, those are not necessary for the interpretation of the $\alpha\lambda$-calculus, so we can define the semantics of the language exactly as in [10] using just the structure we have. We briefly sketch the interpretation below; there are no surprises.

Types are interpreted as sep-arenas:

- $[\![\gamma]\!]$ is the arena with a single, initial opponent move γ and empty separation relation;
- $[\![A \multimap\!\!* B]\!] = [\![A]\!] \multimap\!\!* [\![B]\!]$ and $[\![A \to B]\!] = [\![A]\!] \to [\![B]\!]$.

Bunches are also interpreted as sep-arenas:

$$[\![I]\!] = I \qquad [\![\Gamma;\Delta]\!] = [\![\Gamma]\!] \times [\![\Delta]\!] \qquad [\![\Gamma,\Delta]\!] = [\![\Gamma]\!] * [\![\Delta]\!].$$

The coherence of symmetric monoidal categories ensures that if $\Gamma \equiv \Delta$ there is a canonical isomorphism between $[\![\Gamma]\!]$ and $[\![\Delta]\!]$. This allows us to interpret the coherence rule. If $\Gamma \cong \Delta$ then $[\![\Gamma]\!] = [\![\Delta]\!]$, so the diagonal map $\Gamma \to \Gamma \times \Gamma$ can be used to interpret contraction. Since I is terminal and is the unit for both $\times$ and $*$, there is a canonical projection from any $[\![\Gamma(\Delta;\Delta')]\!]$ or $[\![\Gamma(\Delta,\Delta')]\!]$ to $[\![\Gamma(\Delta)]\!]$ which can be used to interpret weakening. Finally, the two abstraction and application rules are interpreted using the currying isomorphisms and the evaluation maps of the two exponentials.

4 Definability

Theorem 1. *Let A be a type of the $\alpha\lambda$-calculus and Γ a bunch, both built over a single ground type γ. Let σ be a finite, total separation-safe innocent strategy on $[\![\Gamma]\!] \vdash [\![A]\!]$. Then there is a term $\Gamma \vdash M : A$ such that $[\![M]\!] = \sigma$.*

Proof. Forgetting the separation relations, we can treat σ as a strategy on the underlying arena and using t he argument for full completeness in the λ-calculus given in Section 3.5, find a term M such that $[\![M]\!] = \sigma$ in the model of ordinary λ-calculus. We must show that this term is typeable in the $\alpha\lambda$-calculus.

It is straightforward to establish that, if we start with a separation-safe strategy σ, the substrategies which give rise to argument terms M_i are themselves separation safe. Thus the inductive hypothesis gives us terms $\Gamma \vdash M_i : A_i$ and a candidate application

$$\Gamma \vdash x_i \bullet M_1 \bullet \cdots \bullet M_k$$

where each $\bullet$ may be a linear or additive application, according to the type of x_i. We must show that this application is typeable in the $\alpha\lambda$-calculus. We proceed by induction on the number of applications. The base case, that of a variable by itself, is trivial. For the inductive step, by inductive hypothesis we have $\Gamma \vdash M_k$ and

$$\Gamma \vdash (x_i \bullet M_1 \bullet \cdots \bullet M_{k-1}).$$

If the final application is additive, the result follows from Lemma 2. In the multiplicative case, by Lemma 1 it suffices to show that the free identifiers of M_k are separated from those of x_i, $M_1, \ldots, M_{k-1}$. In the terminology of Section 3.5, each free identifier of M_k corresponds to a move m in σ such that $m\,\mathbf{ext}\,\gamma_{i_k}$. The x_i corresponds to the move γ_i, while the free identifiers of each M_j corresponds to a move n such that $n\,\mathbf{ext}\,\gamma_{i,j}$. By definition, each $\gamma_{i,j} \# \gamma_{i,k}$ and $\gamma_{i,k} \# \gamma_i$, so by separation safety, $m * n$ and $m * \gamma_i$. All these moves are initial moves in $[\![\Gamma]\!]$, so $*$ can only arise from $\#$. Hence the free identifiers of M_k are separated from those of the M_j and x_i as required. □

References

1. Abramsky, S., Jagadeesan, R., Malacaria, P.: Full abstraction for PCF. Information and Computation 162(2), 409–470 (2000)
2. Abramsky, S., McCusker, G.: Linearity, sharing and state: a fully abstract game semantics for Idealized Algol with active expressions. In: Proceedings of 1996 Workshop on Linear Logic. Electronic notes in Theoretical Computer Science, vol. 3, Elsevier, Amsterdam (1996)
3. Atkey, R.: A λ-calculus for resource separation. In: Díaz, J., Karhumäki, J., Lepistö, A., Sannella, D. (eds.) ICALP 2004. LNCS, vol. 3142, pp. 158–170. Springer, Heidelberg (2004)
4. Hyland, J.M.E., Ong, C.-H.L.: On full abstraction for PCF: I, II and III. Information and Computation 162(2), 285–408 (2000)
5. Lambek, J., Scott, P.J.: Introduction to Higher Order Categorical Logic. Cambridge University Press, Cambridge (1986)
6. McCusker, G.: Games and Full Abstraction for a Functional Metalanguage with Recursive Types. Distinguished Dissertations in Computer Science. Springer, Heidelberg (1998)
7. Nickau, H.: Hereditarily sequential functionals. In: Proceedings of the Symposium on Logical Foundations of Computer Science, Logic at St. Petersburg. Lecture notes in Computer Science, Springer, Heidelberg (1994)
8. O'Hearn, P.W.: Resource interpretations, bunched implications and the $\alpha - \lambda$-calculus. In: Girard, J.-Y. (ed.) TLCA 1999. LNCS, vol. 1581, pp. 258–279. Springer, Heidelberg (1999)
9. O'Hearn, P.W., Pym, D.J.: The logic of bunched implications. Bulletin of Symbolic Logic 5(2), 215–244 (1999)
10. O'Hearn, P.W.: On bunched typing. Journal of Functional Programming 13(4), 747–796 (2003)
11. Pym, D.J.: The Semantics and Proof Theory of the Logic of Bunched Implications. Kluwer Academic, Dordrecht (2002), Errata available at `http://www.cs.bath.ac.uk/~pym/BI-monograph-errata.pdf`

12. Pym, D.J., Tofts, C.: Systems modelling via resources and processes: Philosophy, calculus, semantics, and logic. Electronic Notes in Theoretical Computer Science, vol. 172, pp. 545–587 (2007), Errata available at `http://www.cs.bath.ac.uk/~pym/pym-tofts-fac-errata.pdf`
13. Reynolds, J.C.: Syntactic control of interference. In: Conf. Record 5th ACM Symposium on Principles of Programming Languages, pp. 39–46 (1978)
14. Reynolds, J.C.: Separation logic: a logic for shared mutable data structures. In: Seventeenth Annual IEEE Symposium on Logic in Computer Science (LICS 2002), IEEE Computer Society Press, Los Alamitos (2002) (Invited paper)
15. Wall, M.: Games for Syntactic Control of Interference. PhD thesis, University of Sussex (2005)

The Ackermann Award 2007

M. Grohe, M. Hyland, J.A. Makowsky, and D. Niwinski

Members of EACSL Jury for the Ackermann Award*

The third **Ackermann Award** is presented at this CSL'07. This is the first year in which the EACSL Ackermann Award is generously sponsored. Our sponsor for the next three years is the worlds leading provider of personal peripherals, Logitech S.A., situated in Romanel, Switzerland[1].

Eligible for the 2007 **Ackermann Award** were PhD dissertations in topics specified by the EACSL and LICS conferences, which were formally accepted as PhD theses at a university or equivalent institution between 1.1. 2005 and 31.12. 2006. The Jury received 7 nominations for the **Ackermann Award 2007**. The candidates came from 7 different nationalities from Europe, the Middle East and Asia and received their PhDs in 8 different[2] countries in Europe, North America and Australia.

The topics covered the full range of Logic and Computer Science as represented by the LICS and CSL Conferences. All the submissions were of very high standard and contained outstanding results in their particular domain. The Jury decided finally, to give for the year 2007 three awards, one for work in *game logics*, one for work in *proof theory*, one for work in *automated theorem proving.* The 2007 **Ackermann Award** winners are, in alphabetical order,

- Dietmar Berwanger from Germany, for his thesis
  Games and logical expressiveness,
  issued by the Rheinisch-Westphälische Technische Hochschule Aachen, Germany, in 2005, supervised by Erich Grädel.
- Stephane Lengrand from France, for his thesis
  Normalisation and Equivalence in Proof Theory and Type Theory,
  issued by the University of St Andrews, Scotland and the University Paris VII, France, 2006, jointly supervised by Roy Dyckhoff and Delia Kesner.
- Ting Zhang from China, for his thesis
  Arithmetic Integration of Decision Procedures,
  issued by Stanford University, USA, 2006, jointly supervised by Zohar Manna and Henny Sipma.

---

* We would like to thank H. Barendregt, J. van Benthem R. Dyckhoff, Z. Manna, L. Ong and I. Walukiewicz for their help in preparing the citations.

[1] We would like to thank Daniel Borel, Co-founder and Chairman of the Board of Logitech S.A, for his generous support of the Ackermann Award for the years 2007-2009. For a history of the company, founded in 1981 in Switzerland, consult `http://www.logitech.com`.

[2] Some of the candidates got their degree from two European institutions.

J. Duparc and T.A. Henzinger (Eds.): CSL 2007, LNCS 4646, pp. 589–597, 2007.

The Jury wishes to congratulate the recipients of the Ackermann Award for their outstanding work and wishes them a successful continuation of their career.

The Jury wishes also to encourage all the remaining candidates to continue their excellent work and hopes to see more of their work in the future.

Dietmar Berwanger

Citation. Dietmar Berwanger receives the *2007 Ackermann Award* of the European Association of Computer Science Logic (EACSL) for his thesis

Games and Logical Expressiveness

in which he substantially advanced our understanding of the connections between infinite games and logical definability and solved a long standing open problem by separating Parikh's dynamic game logic from the modal μ-calculus.

Background of the thesis. Close connections between mathematical logic and infinite strategic games have been established long ago, but they are still a topic of active research and bear some deep open problems. The determinacy of infinite two-person games, that is, the question of whether there is always a player who has a winning strategy, has been intensely studied in descriptive set theory. In computer science, a combination of infinite games, logic, and automata forms the theoretical basis for the synthesis and verification of reactive systems. The strategic interaction in two-person games can be used as a simple and elegant model of the interaction between a system and its potentially adverse environment, or the choices of a designer in building a system that reacts to its environment as specified.

The *modal μ-calculus* is a logic that plays a central role in the theory of synthesis and verification. Many other logics that are used as specification languages for reactive systems can be translated into the μ-calculus. A special class of infinite two-person games that is closely related to the μ-calculus is the class of *parity games*. More precisely, the evaluation of a μ-calculus formula on a transition system can be reduced to a parity game and, conversely, the winning regions in parity games are definable in the μ-calculus. It is an important, and still open, question whether these two problems can be solved in polynomial time. While much is understood about this elegant system, many questions remain open, especially concerning the fine-structure that its language provides for analyzing particular types of recursion.

Dynamic game logic was proposed by Parikh as an extension of propositional dynamic logic to the game setting. Dynamic logic is a modal logic where modalities are not just single actions but regular expressions over actions standing for complex games formed using operations of choice, sequential composition, and iteration. In game logic, modalities describe players' powers over sets of outcomes that can be achieved by following different strategies available to them.

In particular, there is a special dualisation operator, which can be used to describe alternation between players in complex games. This duality operator is fundamental for a complete and symmetric treatment of game theoretic statements. Dynamic game logic is intimately related to game-theoretic notions of strategy and equilibrium, which have a mathematical fixed-point character, and hence a comparison with the *modal μ-calculus* is a natural and timely topic of research.

Berwanger's thesis. The core results of the thesis *Games and Logical Expressiveness* are concerned with the expressive power of fragments of the modal μ-calculus and game logic. These results are proved in a very innovative and sometimes surprising way. A notable by-product of the proof of the main theorem is the introduction of a new directed-graph invariant called entanglement, which turned out to be of independent interest. The thesis is written in a concise and elegant style.

The technical part of the thesis begins with an investigation of the expressive power of game logic in the framework of the μ-calculus. Berwanger observes that, as many other modal and temporal logics studied in this context, for example CTL* or PDL, game logic can be translated into the two-variable fragment of the μ-calculus. Surprisingly, he then proves that game logic is nevertheless expressive enough to define the winning regions in parity games, which makes it much more expressive than the other modal and temporal logics mentioned above. This also implies that the model checking problem for game logic is algorithmically as hard as that for the full μ-calculus. Thus, understanding essential aspects of multi-player interaction seems as hard as understanding unlimited recursion.

The main result of the thesis states that the variable hierarchy of the modal μ-calculus is strict, that is, more variables induce more expressive fragments of the logic. As game logic can be embedded into the two-variable fragment, this implies that the μ-calculus is more expressive than game logic, answering an open question asked by Parikh in 1985. But also as a stand-alone result, Berwanger's theorem provides key new insights into the fine-structure of general fixed-point logics, of which the following is an example.

An important and innovative step in the proof of the hierarchy theorem is to "measure" the combinatorial essence of the expressive power of the bounded variable fragments of the μ-calculus by a new directed-graph invariant called *entanglement.* Every finite graph is bisimilar to a tree with back edges. The entanglement of such a tree measures the number of "open" back edges that a node in such a tree can have. The entanglement of a structure is the minimal entanglement of a bisimilar tree with back edges. The notion has a natural characterisation in terms of a search game on directed graphs; similar games are known to characterise other measures for the "tree-likeness" of graphs and directed graphs. Besides applying it in the proof of the main result, Berwanger exploited the connection between entanglement and the μ-calculus in a different way by proving that the model checking problem for the μ-calculus is in polynomial time on structures of bounded entanglement.

The thesis contains a number of further nice results, in particular about the positional determinacy and complexity of *path games*, another family of infinite two-person games related to the classical Banach-Mazur games.

The results on game logic and the variable hierarchy are lasting results in the theory of fixed point logics. The notion of entanglement, introduced as a technical tool in the proof of the hierarchy result, has already turned out to be of independent interest. The thesis introduces a wealth of new ideas and is a pleasure to read.

Biographic Sketch. Dietmar Berwanger was born Lugoj, Romania in 1972. He studied computer science at the RWTH Aachen and Universitá di Roma "La Sapienza" and received his "Diplom" in Aachen in 2000. He wrote his PhD thesis under the supervision of Erich Grädel in Aachen and received his Ph.D. in May 2005. Currently, he is a postdoctoral fellow at the EPFL Lausanne.

Stéphane Lengrand

Citation. Stéphane Lengrand receives the *2007 Ackermann Award* of the European Association of Computer Science Logic (EACSL) for his thesis

Normalisation and Equivalence in Proof Theory and Type Theory

in which he profoundly advanced our understanding of the logical under-pinnings of proof search and programming language semantics.

Background of the thesis. Interactions between logic and computer science provide formal methods for the development of proof assistant software, automated reasoning, high-level programming languages and certified software. Research in this area is concerned with the following key concepts:

- mathematical objects that can formalise the notion of proof,
- computational features of these objects, with notions of *normalisation*,
- equational theories about these objects, i.e. notions of *equivalence*, related to their computational features.

Intellectual direction comes from the *Curry-Howard correspondence*, relating proofs to programs and propositions to types. For a logical system, one attempts first to design an accurate proof-term calculus, and then uses techniques from rewriting theory to establish the desired proof-theoretic properties. Lengrand's wide ranging thesis successfully deploys these ideas in new directions: to versions of the *sequent calculus* appealing for *proof-search*; to powerful type theories; and to the case of classical reasoning.

Lengrand's thesis. The thesis lies on the boundary between proof theory, type theory, λ-calculus and term rewriting. It provides a range of new results relating to normalization, confluence, consistency and equivalence. Several major contributions of the thesis stand out:

- The thesis answers a long-standing challenge arising from the well-known Melliès' counterexample by presenting a calculus with *explicit substitutions*, *erasure* and *duplication* constructors, which refines λ-calculus with an explicit handling of resources. Improving on results of David and Guillaume (in particular with the simulation of β-reduction), such a calculus allows the combination of a full notion of *composition* of substitutions with the property of *Preservation of Strong Normalisation*. In its typed version, the calculus establishes a Curry-Howard correspondence for *multiplicative natural deduction* (with explicit rules of *weakening* and *contraction*).
- A sequent calculus framework, giving a clear and natural theoretical basis to proof-search in type theory, is given. This comes from a formalism, *Pure Type Sequent Calculi* (PTCS) yielding for each type theory expressed in Barendregt's natural deduction style *Pure Type Systems*, a corresponding sequent calculus. Optimised versions of PTSC express, in a natural fashion, proof-search mechanisms found in proof assistants like Coq, in *logic programming*, and in algorithms that enumerate the programs of a given type (i.e. satisfying a given specification).
- The Vorob'ev–Hudelmaier–Dyckhoff sequent calculus G4ip (in which the proofs of every sequent are bounded in depth) is another long-standing issue addressed in the thesis. Lengrand gives the first complete analysis of the computational content of this calculus. This uses rewriting techniques: an internal cut-elimination process, of which the (strong) termination is shown, is expressed as a calculus of proof-terms and rewriting rules. Its semantics is subtle, and capable of eliminating redundancies in proofs to produce proofs of small depth (below the bound).
- The thesis contains a very careful study of the intuitionistic sequent calculus **LJQ** (distinguished by a syntactic restriction on the left rule for implication) and its relation to call-by-value semantics. The main result is an equational correspondence with a slight but delicate modification of Moggi's call-by-value λ-calculus
- A classical version of the system F_ω is shown to be strongly normalising and consistent; this is achieved in an elegant fashion by allowing a purely intuitionistic upper layer of types but a lower, classical, layer of terms.

In general terms, the work on PTSC initiates a particularly promising line of research, in which concepts related to proof-search and logic programming can fruitfully interact with the expressivity of type theories.

The thesis contains many other ideas: a constructive theory of (weak and strong) normalisation; extensions of the simulation technique for proving strong normalisation; notions of proof equality in classical logic relating to computational representations within the framework of *Deep Inference*. It is clearly written, with its many diverse contributions expressed within a general, uniform and abstract framework. Being unusual both in range and in depth, it shows creativity and maturity, and gives a high-level view of fields in logic and computer science.

Biographical Sketch. Stéphane Lengrand was born in 1980 in Paris. He entered the École Normale Supérieure de Lyon in 2000. He took an M.Sc. in Mathematics and the Foundations of Computer Science at Oxford University in 2002. He took his D.E.A. in 2003 and in the same year completed a Licence d'Anglais in Medieval English at the Sorbonne.

Stéphane Lengrand studied for his PhD under cotutelle arrangements at the Université Paris VII and the University of St. Andrews, being supervised jointly by Delia Kesner (Paris VII) and Roy Dyckhoff (St. Andrews). He received his doctorate in December 2006. This year he has been teaching at St. Andrews where he is a Visiting Scholar, and he has been attending the Royal Scottish Academy of Music and Drama in Glasgow from which he hopes to graduate with the Postgraduate Diploma in Cello (Performance).

Ting Zhang

Citation. Ting Zhang receives the *2007 Ackermann Award* of the European Association of Computer Science Logic (EACSL) for his thesis

Arithmetic Integration of Decision Procedures.

His thesis constitutes a *tour de force* in decidability of structures with limited arithmetic capability, and in particular establishes decidability of the first-order theory of Knuth-Bendix order, thus solving a long-standing open problem.

Background of the thesis. Algorithmic decidability of formalized mathematical theories was first studied by Tarski and his students, who discovered that, in spite of general undecidability of mathematics, several basic structures enjoy decidable theories, as, e.g., reals, or integers with just addition (Presburger arithmetics). Whenever it exists, a decision procedure gives us a deeper insight into the structure.

The search for decision procedures remains an active topic in computer science, especially in logic-based verification of programs, where decidable theories underline the fully automatic methods. Computer science has its own menagerie of abstract structures, like words, trees (terms), graphs..., and much effort has been put into understanding decidability issues of these structures. But number-theoretic domains are also needed whenever quantitative questions arrive.

Integration of various discrete structures with arithmetics is a non-trivial task, appearing in many contexts, as, e.g., verification of memory safety properties. The thesis of Ting Zhang meets this challenge by developing a series of decision procedures for first-order theories of algebraic structures integrated with Presburger arithmetics via some functions of measurement. By an ingenious refinement of his method, the author also establishes decidability of the first-order theory of the Knuth-Bendix ordering of terms, thus solving a long-standing open problem in term-rewriting.

Deciding combined theories was previously addressed by Nelson and Oppen (in 1979). These authors proposed a modular combination method under restriction to quantifier-free theories with disjoint signatures. In spite of subsequent attempts by several researchers, these assumptions are hard to remove in general, so the method fails for theories involving measurement functions. In his thesis, rather than searching for general combination scheme, Zhang focuses on specific problems and exploits the algebraic properties of the combined domain.

The idea of a well-ordering of first-order terms originated in Don Knuth's work on "completion procedure" for algebraic theories in early 1970s. Knuth's ideas regarding term inference were subsequently extended by Dershowitz and others, and applied in design and implementation of automated theorem provers. Here the order underlines a rewriting strategy, and the inference rules include side conditions that involve quantified inequalities between terms. The two orderings mainly used in this context are recursive path ordering and Knuth-Bendix ordering. Solving quantified constraints based on the former is however generally undecidable.

The analogous question for the Knuth-Bendix ordering has been believed a difficult open problem. First positive results were given by Korovin and Voronkov, who established NP-completeness of the quantifier-free fragment over arbitrary signature, and decidability of the first-order fragment restricted to unary functions. The proof of Korovin–Voronkov used a reduction to WS2S, and was hard to extend to general case. Let us recall that Konstantin Korovin was among the winners of the Ackermann Award in 2005.

The Knuth-Bendix ordering is of hybrid nature, as it combines a syntactic precedence of function symbols with a linear weight function in a recursive way. It has turned out that the methods developed by Ting Zhang for deciding theories of hybrid structures were powerful enough to establish decidability of the full first-order theory of Knuth-Bendix ordering. As the author remarks: *It is interesting that the combination of term algebra with integer arithmetic can help an open problem in another quite different field.*

Zhang's thesis. The starting point is integration of term algebra and Presburger arithmetic into a one combined structure, additionally equipped with the length-of-term function (which can be replaced by another weight function). The fundamental construction extracts complete integer constraints from term constraints, thus transferring the decision task into arithmetics. This construction is subsequently refined throughout the thesis, yielding the more and more powerful decidability results.

The main contributions are as follows.

- NP-completeness of the quantifier-free fragment and decidability of the full theory of the aforementioned combination of term algebra and Presburger arithmetic. The quantifier elimination procedure involves a block-wise reduction of term quantifiers to integer quantifiers, and eliminates a block of quantifiers of the same kind in one step. This makes the algorithm k-fold exponential for formulas with k quantifier alternations, regardless of the number of quantifiers, which also improves elimination procedure for the pure theory of term algebras.

- Application of the above method to decidability of a theory of term algebra with two integer functions which capture precisely the properties of red-black trees.
- Decision procedure for the first-order theory of queues combined with Presburger arithmetic, and NP-completeness of the quantifier-free fragment extended with prefix predicate.
- Decidability of the first-order theory of Knuth-Bendix order. This is most brilliant achievement of the thesis, based on an extremely sophisticated argument (80 pages), involving many conceptual innovations. The key step is introduction of boundary functions going in the opposite direction: from integers to terms, which helps one to bound the terms being quantified over.

The thesis is based on a number of articles presented to IJCAR'04 (*Best Paper Award*), TPHOLs'04, FSTTCS'05, CADE'05, LFCS'07, and an article in Information & Computation (joint with H.B.Sipma and Z.Manna).

Biographic Sketch. Ting Zhang received his B.Sc. degree in computer science from Peking University, Beijing, China, in 1996, and M.Sc. degree in computer science from the Stanford University, CA, in 2001. He wrote his Ph.D. thesis under the supervision of professor Zohar Manna (the co-advisor being professor Henny Sipma) and obtained the Ph.D. degree in computer science from the Stanford University, in 2006. He is currently a researcher at Microsoft Research Asia in Beijing, China. He also teaches at Tsinghua University.

The Ackermann Award

The EACSL Board decided in November 2004 to launch the EACSL Outstanding Dissertation Award for Logic in Computer Science, the **Ackermann Award**, The award[3]. is named after the eminent logician Wilhelm Ackermann (1896-1962), mostly known for the Ackermann function, a landmark contribution in early complexity theory and the study of the rate of growth of recursive functions, and for his coauthorship with D. Hilbert of the classic *Grundzüge der Theoretischen Logik*, first published in 1928. Translated early into several languages, this monograph was the most influential book in the formative years of mathematical logic. In fact, Gödel's completeness theorem proves the completeness of the system presented and proved sound by Hilbert and Ackermann. As one of the pioneers of logic, W. Ackermann left his mark in shaping logic and the theory of computation.

The **Ackermann Award** is presented to the recipients at the annual conference of the EACSL. The Jury is entitled to give more than one award per year. The award consists of a diploma, an invitation to present the thesis at the CSL conference, the publication of the abstract of the thesis and the citation in the CSL proceedings, and travel support to attend the conference.

[3] Details concerning the Ackermann Award and a biographic sketch of W. Ackermann was published in the CSL'05 proceedings and can also be found at `http://www.dimi.uniud.it/eacsl/award.html`.

The Jury for the **Ackermann Award** consists of eight members, three of them ex officio, namely the president and the vice-president of EACSL, and one member of the LICS organizing committee. The current jury consists of S. Abramsky (Oxford, LICS Organizing Committee), J. van Benthem (Amsterdam), B. Courcelle (Bordeaux), M. Grohe (Berlin), M. Hyland (Cambridge), J.A. Makowsky (Haifa, President of EACSL), D. Niwinski (Warsaw, Vice President of EACSL), and A. Razborov (Moscow and Princeton).

Previous winners of the Ackermann Award were

2005, Oxford:
Mikołaj Bojańczyk from Poland,
Konstantin Korovin from Russia, and
Nathan Segerlind from the USA.

2006, Szeged:
Balder ten Cate from The Netherlands, and
Stefan Milius from Germany

A detailed report on their work appeared in the CSL'05 and CSL'06 proceedings, and is also available via the EACSL homepage.

Author Index

Lecture Notes in Computer Science

Sublibrary 1: Theoretical Computer Science and General Issues

For information about Vols. 1–4424
please contact your bookseller or Springer

Vol. 4743: P. Thulasiraman, X. He, T.L. Xu, M.K. Denko, R.K. Thulasiram, L.T. Yang (Eds.), Frontiers of High Performance Computing and Networking ISPA 2007 Workshops. XXIX, 536 pages. 2007.

Vol. 4742: I. Stojmenovic, R.K. Thulasiram, L.T. Yang, W. Jia, M. Guo, R.F. de Mello (Eds.), Parallel and Distributed Processing and Applications. XX, 995 pages. 2007.

Vol. 4736: S. Winter, M. Duckham, L. Kulik, B. Kuipers (Eds.), Spatial Information Theory. XV, 455 pages. 2007.

Vol. 4732: K. Schneider, J. Brandt (Eds.), Theorem Proving in Higher Order Logics. IX, 401 pages. 2007.

Vol. 4708: L. Kučera, A. Kučera (Eds.), Mathematical Foundations of Computer Science 2007. XVIII, 764 pages. 2007.

Vol. 4707: O. Gervasi, M.L. Gavrilova (Eds.), Computational Science and Its Applications – ICCSA 2007, Part III. XXIV, 1205 pages. 2007.

Vol. 4706: O. Gervasi, M.L. Gavrilova (Eds.), Computational Science and Its Applications – ICCSA 2007, Part II. XXIII, 1129 pages. 2007.

Vol. 4705: O. Gervasi, M.L. Gavrilova (Eds.), Computational Science and Its Applications – ICCSA 2007, Part I. XLIV, 1169 pages. 2007.

Vol. 4703: L. Caires, V.T. Vasconcelos (Eds.), CONCUR 2007 – Concurrency Theory. XIII, 507 pages. 2007.

Vol. 4697: L. Choi, Y. Paek, S. Cho (Eds.), Advances in Computer Systems Architecture. XIII, 400 pages. 2007.

Vol. 4688: K. Li, M. Fei, G.W. Irwin, S. Ma (Eds.), Bio-Inspired Computational Intelligence and Applications. XIX, 805 pages. 2007.

Vol. 4684: L. Kang, Y. Liu, S. Zeng (Eds.), Evolvable Systems: From Biology to Hardware. XIV, 446 pages. 2007.

Vol. 4683: L. Kang, Y. Liu, S. Zeng (Eds.), Intelligence Computation and Applications. XVII, 663 pages. 2007.

Vol. 4681: D.-S. Huang, L. Heutte, M. Loog (Eds.), Advanced Intelligent Computing Theories and Applications. XXVI, 1379 pages. 2007.

Vol. 4671: V. Malyshkin (Ed.), Parallel Computing Technologies. XIV, 635 pages. 2007.

Vol. 4669: J.M. de Sá, L.A. Alexandre, W. Duch, D. Mandic (Eds.), Artificial Neural Networks – ICANN 2007, Part II. XXXI, 990 pages. 2007.

Vol. 4668: J.M. de Sá, L.A. Alexandre, W. Duch, D. Mandic (Eds.), Artificial Neural Networks – ICANN 2007, Part I. XXXI, 978 pages. 2007.

Vol. 4666: M.E. Davies, C.J. James, S.A. Abdallah, M.D Plumbley (Eds.), Independent Component Analysis and Blind Signal Separation. XIX, 847 pages. 2007.

Vol. 4665: J. Hromkovič, R. Kralovič, M. Nunkesser, P. Widmayer (Eds.), Stochastic Algorithms: Foundations and Applications. X, 167 pages. 2007.

Vol. 4664: J. Durand-Lose, M. Margenstern (Eds.), Machines, Computations, and Universality. X, 325 pages. 2007.

Vol. 4662: C. Baranauskas, P. Palanque, J. Abascal, S.D.J. Barbosa (Eds.), Human-Computer Interaction – INTERACT 2007, Part I. XXXIII, 637 pages. 2007.

Vol. 4649: V. Diekert, M.V. Volkov, A. Voronkov (Eds.), Computer Science – Theory and Applications. XIII, 420 pages. 2007.

Vol. 4647: R. Martin, M. Sabin, J. Winkler (Eds.), Mathematics of Surfaces XII. IX, 509 pages. 2007.

Vol. 4646: J. Duparc, T.A. Henzinger (Eds.), Computer Science Logic. XIV, 600 pages. 2007.

Vol. 4644: N. Azemard, L. Svensson (Eds.), Integrated Circuit and System Design. XIV, 583 pages. 2007.

Vol. 4641: A.-M. Kermarrec, L. Bougé, T. Priol (Eds.), Euro-Par 2007 Parallel Processing. XXVII, 974 pages. 2007.

Vol. 4639: E. Csuhaj-Varjú, Z. Ésik (Eds.), Fundamentals of Computation Theory. XIV, 508 pages. 2007.

Vol. 4638: T. Stützle, M. Birattari, H.H. Hoos (Eds.), Engineering Stochastic Local Search Algorithms. X, 223 pages. 2007.

Vol. 4628: L.N. de Castro, F.J. Von Zuben, H. Knidel (Eds.), Artificial Immune Systems. XII, 438 pages. 2007.

Vol. 4627: M. Charikar, K. Jansen, O. Reingold, J.D.P. Rolim (Eds.), Approximation, Randomization, and Combinatorial Optimization. XII, 626 pages. 2007.

Vol. 4624: T. Mossakowski, U. Montanari, M. Haveraaen (Eds.), Algebra and Coalgebra in Computer Science. XI, 463 pages. 2007.

Vol. 4619: F. Dehne, J.-R. Sack, N. Zeh (Eds.), Algorithms and Data Structures. XVI, 662 pages. 2007.

Vol. 4618: S.G. Akl, C.S. Calude, M.J. Dinneen, G. Rozenberg, H.T. Wareham (Eds.), Unconventional Computation. X, 243 pages. 2007.

Vol. 4616: A. Dress, Y. Xu, B. Zhu (Eds.), Combinatorial Optimization and Applications. XI, 390 pages. 2007.

Vol. 4613: F.P. Preparata, Q. Fang (Eds.), Frontiers in Algorithmics. XI, 348 pages. 2007.

Vol. 4600: H. Comon-Lundh, C. Kirchner, H. Kirchner (Eds.), Rewriting, Computation and Proof. XVI, 273 pages. 2007.

Vol. 4599: S. Vassiliadis, M. Berekovic, T.D. Hämäläinen (Eds.), Embedded Computer Systems: Architectures, Modeling, and Simulation. XVIII, 466 pages. 2007.

Vol. 4598: G. Lin (Ed.), Computing and Combinatorics. XII, 570 pages. 2007.

Vol. 4596: L. Arge, C. Cachin, T. Jurdziński, A. Tarlecki (Eds.), Automata, Languages and Programming. XVII, 953 pages. 2007.

Vol. 4595: D. Bošnački, S. Edelkamp (Eds.), Model Checking Software. X, 285 pages. 2007.

Vol. 4590: W. Damm, H. Hermanns (Eds.), Computer Aided Verification. XV, 562 pages. 2007.

Vol. 4588: T. Harju, J. Karhumäki, A. Lepistö (Eds.), Developments in Language Theory. XI, 423 pages. 2007.

Vol. 4583: S.R. Della Rocca (Ed.), Typed Lambda Calculi and Applications. X, 397 pages. 2007.

Vol. 4580: B. Ma, K. Zhang (Eds.), Combinatorial Pattern Matching. XII, 366 pages. 2007.

Vol. 4576: D. Leivant, R. de Queiroz (Eds.), Logic, Language, Information and Computation. X, 363 pages. 2007.

Vol. 4547: C. Carlet, B. Sunar (Eds.), Arithmetic of Finite Fields. XI, 355 pages. 2007.

Vol. 4546: J. Kleijn, A. Yakovlev (Eds.), Petri Nets and Other Models of Concurrency – ICATPN 2007. XI, 515 pages. 2007.

Vol. 4545: H. Anai, K. Horimoto, T. Kutsia (Eds.), Algebraic Biology. XIII, 379 pages. 2007.

Vol. 4533: F. Baader (Ed.), Term Rewriting and Applications. XII, 419 pages. 2007.

Vol. 4528: J. Mira, J.R. Álvarez (Eds.), Nature Inspired Problem-Solving Methods in Knowledge Engineering, Part II. XXII, 650 pages. 2007.

Vol. 4527: J. Mira, J.R. Álvarez (Eds.), Bio-inspired Modeling of Cognitive Tasks, Part I. XXII, 630 pages. 2007.

Vol. 4525: C. Demetrescu (Ed.), Experimental Algorithms. XIII, 448 pages. 2007.

Vol. 4514: S.N. Artemov, A. Nerode (Eds.), Logical Foundations of Computer Science. XI, 513 pages. 2007.

Vol. 4513: M. Fischetti, D.P. Williamson (Eds.), Integer Programming and Combinatorial Optimization. IX, 500 pages. 2007.

Vol. 4510: P. Van Hentenryck, L.A. Wolsey (Eds.), Integration of AI and OR Techniques in Constraint Programming for Combinatorial Optimization Problems. X, 391 pages. 2007.

Vol. 4507: F. Sandoval, A. Prieto, J. Cabestany, M. Graña (Eds.), Computational and Ambient Intelligence. XXVI, 1167 pages. 2007.

Vol. 4501: J. Marques-Silva, K.A. Sakallah (Eds.), Theory and Applications of Satisfiability Testing – SAT 2007. XI, 384 pages. 2007.

Vol. 4497: S.B. Cooper, B. Löwe, A. Sorbi (Eds.), Computation and Logic in the Real World. XVIII, 826 pages. 2007.

Vol. 4494: H. Jin, O.F. Rana, Y. Pan, V.K. Prasanna (Eds.), Algorithms and Architectures for Parallel Processing. XIV, 508 pages. 2007.

Vol. 4493: D. Liu, S. Fei, Z. Hou, H. Zhang, C. Sun (Eds.), Advances in Neural Networks – ISNN 2007, Part III. XXVI, 1215 pages. 2007.

Vol. 4492: D. Liu, S. Fei, Z. Hou, H. Zhang, C. Sun (Eds.), Advances in Neural Networks – ISNN 2007, Part II. XXVII, 1321 pages. 2007.

Vol. 4491: D. Liu, S. Fei, Z.-G. Hou, H. Zhang, C. Sun (Eds.), Advances in Neural Networks – ISNN 2007, Part I. LIV, 1365 pages. 2007.

Vol. 4490: Y. Shi, G.D. van Albada, J. Dongarra, P.M.A. Sloot (Eds.), Computational Science – ICCS 2007, Part IV. XXXVII, 1211 pages. 2007.

Vol. 4489: Y. Shi, G.D. van Albada, J. Dongarra, P.M.A. Sloot (Eds.), Computational Science – ICCS 2007, Part III. XXXVII, 1257 pages. 2007.

Vol. 4488: Y. Shi, G.D. van Albada, J. Dongarra, P.M.A. Sloot (Eds.), Computational Science – ICCS 2007, Part II. XXXV, 1251 pages. 2007.

Vol. 4487: Y. Shi, G.D. van Albada, J. Dongarra, P.M.A. Sloot (Eds.), Computational Science – ICCS 2007, Part I. LXXXI, 1275 pages. 2007.

Vol. 4484: J.-Y. Cai, S.B. Cooper, H. Zhu (Eds.), Theory and Applications of Models of Computation. XIII, 772 pages. 2007.

Vol. 4475: P. Crescenzi, G. Prencipe, G. Pucci (Eds.), Fun with Algorithms. X, 273 pages. 2007.

Vol. 4474: G. Prencipe, S. Zaks (Eds.), Structural Information and Communication Complexity. XI, 342 pages. 2007.

Vol. 4459: C. Cérin, K.-C. Li (Eds.), Advances in Grid and Pervasive Computing. XVI, 759 pages. 2007.

Vol. 4449: Z. Horváth, V. Zsók, A. Butterfield (Eds.), Implementation and Application of Functional Languages. X, 271 pages. 2007.

Vol. 4448: M. Giacobini (Ed.), Applications of Evolutionary Computing. XXIII, 755 pages. 2007.

Vol. 4447: E. Marchiori, J.H. Moore, J.C. Rajapakse (Eds.), Evolutionary Computation, Machine Learning and Data Mining in Bioinformatics. XI, 302 pages. 2007.

Vol. 4446: C. Cotta, J. van Hemert (Eds.), Evolutionary Computation in Combinatorial Optimization. XII, 241 pages. 2007.

Vol. 4445: M. Ebner, M. O'Neill, A. Ekárt, L. Vanneschi, A.I. Esparcia-Alcázar (Eds.), Genetic Programming. XI, 382 pages. 2007.

Vol. 4436: C.R. Stephens, M. Toussaint, D. Whitley, P.F. Stadler (Eds.), Foundations of Genetic Algorithms. IX, 213 pages. 2007.

Vol. 4433: E. Şahin, W.M. Spears, A.F.T. Winfield (Eds.), Swarm Robotics. XII, 221 pages. 2007.

Vol. 4432: B. Beliczynski, A. Dzielinski, M. Iwanowski, B. Ribeiro (Eds.), Adaptive and Natural Computing Algorithms, Part II. XXVI, 761 pages. 2007.

Vol. 4431: B. Beliczynski, A. Dzielinski, M. Iwanowski, B. Ribeiro (Eds.), Adaptive and Natural Computing Algorithms, Part I. XXV, 851 pages. 2007.

MIX
Papier aus verantwortungsvollen Quellen
Paper from responsible sources
FSC® C105338

If you have any concerns about our products,
you can contact us on
ProductSafety@springernature.com

In case Publisher is established outside the EU,
the EU authorized representative is:
Springer Nature Customer Service Center GmbH
Europaplatz 3, 69115 Heidelberg, Germany

Printed by Libri Plureos GmbH
in Hamburg, Germany